W9-BEC-252

GEOGRAPHY

REALMS, REGIONS, AND CONCEPTS

SEVENTH EDITION

H. J. DE BLIJ
PETER O. MULLER

University of Miami

Chapter Opener Maps which appear in this text are from
GOODE'S WORLD ATLAS, copyright © by Rand McNally, R.L.
93-S-115, and are used with permission.

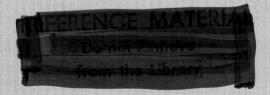

JOHN WILEY & SONS, INC.
New York   Chichester   Brisbane   Toronto   Singapore

Cover Photographs:
(top) Hong Kong/Edward W. Bower/The Image Bank
(bottom) Lombok, Indonesia/Mike Yamashita/Woodfin Camp

ACQUISITIONS EDITOR   Barry Harmon
DEVELOPMENTAL EDITOR   Barbara Heaney
MARKETING MANAGER   Catherine Faduska
PRODUCTION MANAGER   Linda Muriello
SENIOR PRODUCTION EDITOR   Micheline Frederick
DESIGN DIRECTOR   Madelyn Lesure
COVER/TEXT DESIGNER   Laura Ierardi
MANUFACTURING MANAGER   Andrea Price
PHOTO EDITOR   Joan Meisel
PHOTO ASSISTANT   Alissa Mend
ILLUSTRATION COORDINATOR   Edward Starr
Chapter Opener Maps which appear in this text are from
GOODE'S WORLD ATLAS, copyright © by Rand McNally,
R.L. 93-S-115, and are used with permission.

This book was set in 10/12 Roman by V & M Graphics, Inc.
and printed and bound by Von Hoffman Press, Inc. The cover
was printed by Phoenix Color Corp.

Library of Congress Cataloging-in-Publication Data
de Blij, Harm J.
      Geography : realms, regions, and concepts / H.J. de Blij, Peter O.
Muller. — 7th ed.
          p.   cm.
      Rev. ed. of: Geography: regions and concepts. 1991.
      Includes bibliographical references.
      ISBN 0-471-58089-9
      1. Geography.   I. Muller, Peter O.   II. de Blij, Harm J.
Geography: regions and concepts.   III. Title.
      G128.D42   1994
      910—dc20   93-21347
                                                            CIP

Printed in the United States of America
10   9   8   7   6   5   4

TO SPENCER J. FRAZIER
who widened geography's realms
and gave geographers broader horizons

To the Student: A Study Guide for the textbook is available through your college bookstore under the title *Study Guide to Accompany Geography: Realms, Regions, and Concepts*, 7th edition, by H. J. de Blij and Peter O. Muller. The Study Guide can help you with course material by acting as a tutorial, review, and study aid. If the Study Guide is not in stock, ask the bookstore manager to order a copy for you.

PREFACE

Over the more than two decades of its life, this book has been consistently aimed at two objectives. First, it explains the modern world's great geographic realms and their human and physical contents, their assets and liabilities, links and barriers, potentials and prospects. Second, it introduces geography itself, the discipline that links human societies and natural environments through a fascinating perspective that is first detailed in the introductory chapter and then is developed as the book progresses.

As such, the book before you constitutes an antidote to the geographic illiteracy that has lately been so much in the news. Various polls, public surveys, and classroom tests made it clear that our collective geographic knowledge is inadequate in this fast-changing, challenging, competitive world. During the 1980s, the National Geographic Society, the American Express Company, Citibank, and a consortium of scholarly organizations mobilized to combat an educational dilemma that resulted substantially from a neglect of the very topics this book is about. Their efforts and resources have begun to turn the tide, but the real remedy lies in the classroom—with geography teachers, their students, and the fundamentals of the field.

Before we can usefully discuss such commonplace topics as our "shrinking world," our "global village," and our "distant linkages," we should know what the parts are, the elements that are shrinking and linking. This is not merely an academic exercise. Knowledge of the world beyond our borders is a crucial asset in consumer decision-making, international business initiatives, and the global competition that faces a nation with worldwide interests. We can gain this knowledge by studying the layout of our world—i.e., its geography—not by memorization, but by learning where people and things are located and why, how they interact, what impels them to move or migrate, and how they prefer to shape and fashion their abodes.

These are geographic themes, and every time we learn something more about a city, a region, a prevailing climate, or a cluster of resources, we also learn something more about geography itself.

THE APPROACH

This book is organized into twelve chapters: an introduction discusses global distributions, patterns, and trends, and eleven regional chapters follow. The regional discussions are divided into two parts, covering respectively the developed and the developing world.

The format is one that has worked well for more than two decades, linking geography's basic concepts with this overview of the realms and major regions of the world. We have placed more than 150 ideas and concepts in their regional perspective. Most of these concepts are primarily geographical; others are ideas about which we believe students of geography should have some knowledge. Of course, we have not listed on the chapter-opening page every idea and concept used in that chapter (although each term listed is indicated in boldface when it first appears in text). Most teachers, we suspect, will want to make their own region-concept associations, and as readers will readily perceive, the book's organization is quite flexible. In fact, instructors have a great range of opportunities to shape a course to their liking, to transfer concepts, and to focus presentations mainly on conceptual matters or on regionally oriented materials. Moreover, concepts are sometimes raised but not pursued in depth, so that the lecturer may choose to penetrate them in greater detail and/or in a comparative regional context.

THE SEVENTH EDITION

During the 23 years of its life, this book has seen numerous modifications and innovations. But its title never has changed—until now. The framework of geographic realms, introduced with the First Edition in 1971, has (with appropriate revisions) become the standard spatial structure in world regional geography. We strengthen this notion by renaming the book *Geography: Realms, Regions, and Concepts*.

This geographic framework is not static. Realms and their constituent regions do change, although there remains remarkable similarity between the geographic realms laid out in the First Edition and those of the Seventh. One such change is acknowledged through an entirely new chapter, Chapter 4 (The Pacific Rim of Austrasia). From Japan to Australia, the western Pacific margin is in transformation, and Chapter 4 chronicles the emergence of what may, some day, be a new and potent geographical realm on the world stage. At present, the pieces of this nascent realm still are disparate, but a new map is clearly in the making.

The disintegration of the Soviet Union also has transformed the regional map. This momentous development has required considerable adjustment to Chapters 1 (Europe), 2 (Russia is now the focus of this chapter) and, perhaps most incisively, 7 (North Africa/Southwest Asia). The breakdown of the U.S.S.R. came at a time when the notion of a supranational Europe was on the rise, and "republics" long under Soviet sway are reorienting themselves Europeward. Future editions of this book will undoubtedly report on further installments of this significant development. Russia, bereft now of its peripheral empire, is itself an internally fragmented country in the process of restructuring. Chapter 2 focuses on Russia's centrifugal (disunifying) forces, its complicated ethnic and political map, and its worrisome prospects. And another completely rewritten chapter (North Africa/Southwest Asia, Chapter 7), reveals regional consequences of the Soviet collapse. Here we revive an old geographic name for a region long under Moscow's control: *Turkestan*, where Islam is reviving and where an important geopolitical contest is underway.

If the Soviet collapse dominates Eurasia's west, China's ascent is preponderant in the east. In a totally revised and rewritten Chapter 10, the stage is set for a theme that is likely to prevail in future editions of this book: the growing regional and global power of the world's largest nation. China's geography, therefore, is approached from two perspectives: in Chapter 4 as a participant in the economic growth of the Pacific Rim, and in Chapter 10 as the core of an East Asian sphere of influence that extends from Mongolia to Tibet and from eastern Turkestan to the South China Sea.

The South Asia chapter (9) also has been exhaustively reorganized and rewritten. For reasons made clear in the text, this chapter now opens with a discussion of Pakistan, vanguard of Islam on South Asia's western flank. The growing stresses on India's democracy from religious fundamentalists on several sides of the spectrum are discussed in a revised view of that vast country's human geography. The Southeast Asia chapter (11) was revised following the senior author's 1993 field trips to Vietnam, Malaysia, and Thailand.

We have continued the practice of focusing on particular regions in brief, to-the-point vignettes. In addition to three vignettes revised and carried over from the previous edition (Brazil, South Africa, Pacific Realm) we have introduced one to deal with the problem area of Transcaucasia.

In response to recommendations from perceptive reviewers, the North America chapter (3) was not only rewritten but also reorganized to bring Canada into clearer geographic focus. For similar reasons we have continued to highlight the regional geographies of individual countries within world regions, sustaining a trend begun in recent editions of this book.

Still another new feature can be found at the conclusion of each regional chapter: a box headed "The Realm in Transition." Almost everywhere in the world, change is taking place at a growing speed, and we try, in these brief summaries, to gauge the direction of change and its effect on the realm just discussed. Many of these "futures" are thought-provoking and should elicit considerable discussion.

As with every new edition of *Geography: Realms, Regions, and Concepts*, all quantitative information was updated to the year of publication and checked rigorously. In addition to the major revisions described above, hundreds of other modifications were made. For example, the Introduction was substantially condensed and the world regional system was given stronger justification. The sections on the former Czechoslovakia and Yugoslavia (Chapter 1) were brought as up to date as possible. The section on Mexico was rewritten. No paragraph escaped our attention.

ILLUSTRATIONS

Readers who are familiar with earlier editions will note that we have gone to a two-column format. We have taken the opportunity to revise and redraw all cartography, and numerous new thematic maps have been introduced. A major innovation is the introduction of atlas-style opening maps for each regional chapter, all but one drawn from the most recent edition of *Goode's World Atlas* published by Rand McNally. These atlas-format maps enable readers to refer to each realm as a whole; since most of our own maps are thematic in nature, they cannot be as comprehensive. Readers of our book should note that the spelling of some names on these atlas maps is not consistent with our own maps. This is not unusual: various cartographers follows different guidelines when it comes to the spelling of place names. In general, we have followed the standards set by the United States Board of Geographic Names.

PRONUNCIATION GUIDES

Pronunciation Guides are located at the end of each chapter and vignette. In choosing words for inclusion (largely place names), we decided not to list words that were pronounced the way they were written unless we thought mispronunciation was likely. Although we strive for authenticity throughout, we aimed for Americanized rather than native-language pronunciations. For many place names, our initial guide was the current edition of *Webster's New Geographical Dictionary*. In choosing the phonetic presentation method, we kept things as simple as possible by avoiding a formalized symbol system that would have required constant decoding. Accordingly, we employed a syllabic phonetic-spelling system with stress syllables capitalized. (For example, we pronounce our surnames duh-BLAY and MULL-uh.) The most frequently used vowel sounds would translate as follows: *ah* as in f*a*ther, *oh* as in t*o*ne, *au* as in *ou*t, and *uh* as in b*a*n*a*n*a*.

CONTINUING SPECIAL FEATURES OF THIS BOOK

The continuing special features of this textbook are:

◆ **Ten Major Geographic Qualities boxes.** Near the beginning of each regional chapter and vignette, we list, in boxed format, the 10 major *geographic qualities* that best summarize that portion of the earth's surface.

◆ **Focus on a Systematic Field essays.** Also located near the opening of each of the eleven regional chapters is a *Focus on a Systematic Field* essay that covers a major topical subfield of human or physical geography; each of these overviews was carefully selected so that its contents tie in to regional-geographic material subsequently developed in the chapter. (This entire program is depicted in Figure I–15.)

◆ **Appendices and References.** At the end of the book, we have included three appendices: (A) population data on the world's countries, (B) a guide to map reading and interpretation, and (C), an overview of career opportunities in geography. In addition, lengthy lists of *References and Further Readings* are provided for each chapter and vignette. Those lists are followed by the Gazetteer, our *Geographical Index* of map names, and by the main index.

DATA SOURCES

The population figures used in the text are our projections for 1994 (unless otherwise indicated) and are consistent with the national demographic data displayed in Appendix A. The chief source that we used as a basis for developing our projections was the 1992 *World Population Data Sheet* published by the Population Reference Bureau, Inc. The urban population figures—which entail a far greater problem in reliability and comparability—are mainly drawn from the most recent (1993) database published by the United Nations' Population Division. For cities of less than 750,000, we developed our own estimates from a variety of other sources. At any rate, the urban population figures used here are estimates for 1994 and they represent *metropolitan-area totals* unless otherwise specified.

ANCILLARIES

The following ancillaries were prepared to accompany this edition of the book. They may be obtained by contacting John Wiley & Sons.

◆ **Student Study Guide.** Prepared by Peter O. Muller and Elizabeth Muller Hames, University of Miami. This valuable supplement contains objectives, glossary, self-test questions, map exercises, practice exams, term paper pointers, and redrawn maps to provide additional support to the student.

◆ **Instructor's Manual.** Prepared by Laurie Molina, Florida State University. This effective teaching resource provides an outline, brief description, and a list of key terms for each chapter in the text.

◆ **Testbank.** Prepared by Ira Sheskin, University of Miami. The expanded testbank contains over 1,500 test items, including multiple-choice, fill-in-the-blank, matching, and essay questions.

◆ **Computerized Testbank.** The testbank is available in a computerized format for either the IBM or the Macintosh.

◆ **Overhead Transparencies.** Over 75 full-color maps from the text are available as overhead transparencies.

◆ **Slide Set.** The maps available as transparencies are also available as full-color 35 mm slides.

◆ **Supplementary Slides.** This additional slide package contains all photos and more additional maps from the text.

◆ *Good Morning America* **Videotapes.** These videotapes are collections of H. J. de Blij's appearances (as *GMA* Geography Editor) on the popular morning news show. Each segment discusses a geographic issue or other geographically significant world event.

ENVOI

Finally, to the student reader about to embark on the exploration of world geography, we leave you with the following exhortation offered by the renowned author, James Michener, in his 1970 article in *Social Education* (pp. 764–766):

The more I work in the social-studies field the more convinced I become that geography is the foundation of all. . . . When I begin work on a new area—something I have been called upon to do rather frequently in my adult life—I invariably start with the best geography I can find. This takes precedence over everything else, even history, because I need to ground myself in the fundamentals which have governed and in a sense limited human devel-

opment. . . . If I were a young man with any talent for expressing myself, and if I wanted to make myself indispensable to society, I would devote eight or ten years to the real mastery of one of the earth's major regions. I would learn languages, the religions, the customs, the value systems, the history, the nationalisms, and above all the geography [emphasis added], *and when that was completed I would be in a position to write about that region, and I would be invaluable to my nation, for I would be the*

bridge of understanding to the alien culture. We have seen how crucial such bridges can be.

August, 1993   **H. J. de Blij**
                    Georgetown, Washington, D.C.

                    Peter O. Muller
                    Coral Gables, Florida

ACKNOWLEDGMENTS

In the course of this latest revision, we were fortunate to receive advice and assistance from many people.

One of the rewards associated with the publication of a book of this kind is the steady stream of correspondence and other feedback that it generates. Over the years, we have heard from colleagues, students, and lay readers. Geographers, economists, political scientists, education specialists, and others have written us, often with fascinating enclosures. We make it a point to respond personally to every such letter, and our editors have communicated with many of our correspondents as well. We have, moreover, considered every suggestion made—and many who wrote or transmitted their reactions through other channels will see their recommendations in print in the current edition. The list that follows is merely representative of a group of colleagues across North America to whom we are grateful for taking the time to share their thoughts and opinions with us.

Mel Aamodt, California State University-Stanislaus

R. Gabrys Alexson, University of Wisconsin-Superior

Nigel Allan, University of California-Davis

James P. Allen, California State University

John L. Allen, University of Connecticut

Jerry R. Aschermann, Missouri Western State College

Joseph M. Ashley, Montana State University

Theodore P. Aufdemberge, Concordia College (Michigan)

Edward Babin, University of South Carolina-Spartanburg

Marvin W. Baker, University of Oklahoma

Thomas F. Baucom, Jacksonville State Universirty (Alabama)

Gouri Banerjee, Boston University (Massachusetts)

J. Henry Barton, Thiel College (Pennsylvania)

Steven Bass, Paradise Valley Community College (Arizona)

Klaus J. Bayr, University of New Hampshire-Manchester

James Bell, Linn Benton Community College (Oregon)

William H. Berentsen, University of Connecticut

Riva Berleant-Schiller, University of Connecticut

Thomas Bitner, University of Wisconsin

Warren Bland, California State University-Northridge

Davis Blevins, Huntington College (Alabama)

S. Bo Jung, Bellevue College (Nebraska)

Martha Bonte, Clinton Community College (Idaho)

George R. Botjer, University of Tampa (Florida)

Kathleen Braden , Seattle Pacific University (Washington)

R. Lynn Bradley, Belleville Area College (Illinois)

Ken Brehob, Elmhurst (Illinois)

James A. Brey, University of Wisconsin-Fox Valley

Robert Brinson, Santa Fe Community College (Florida)

Reuben H. Brooks, Tennessee State University

Larry Brown, Ohio State University

Lawrence A. Brown, Troy State-Dothan (Alabama)

Robert N. Brown, Delta State University (Mississippi)

Randall L. Buchman, Defiance College (Ohio)

DiAnn Casteel, Tusculum College (Tennessee)

John E. Coffman, University of Houston (Texas)

Dwayne Cole, Grand Rapids Baptist College (Michigan)

Barbara Connelly, Westchester Community College (New York)

Willis M. Conover, University of Scranton (Pennsylvania)

Omar Conrad, Maple Woods Community College (Missouri)

Barbara Cragg, Aquinas College (Michigan)

Georges G. Cravins, University of Wisconsin

John A. Cross, University of Wisconsin-Oshkosh

William Curran, South Suburban (Illinois)

Armando Da Silva, Towson State University (Maryland)

David D. Daniels, Central Missouri State University

Rudolph L. Daniels, Morningside College (Iowa)

Satish K. Davgun, Bemidji State University (Minnesota)

James Davis, Illinois College

Keith Debbage, University of North Carolina-Greensboro

Dennis K. Dedrick, Georgetown College (Kentucky)

Stanford Demars, Rhode Island College

Thomas Dimicelli, William Paterson College (New Jersey)

D.F. Doeppers, University of Wisconsin

Ann Doolen, Lincoln College (Illinois)

Steven Driever, University of Missouri-Kansas City

Keith A. Ducote, Cabrillo Community College (California)

Walter N. Duffet, University of Arizona

Christina Dunphy, Champlain College (Vermont)

Anthony Dzik, Shawnee State University

Dennis Edgell, Firelands BGSU (Ohio)

James H. Edmonson, Union University (Tennessee)

M.H. Edney, State University of New York-Binghamton

Harold M. Elliott, Weber State University (Utah)

James Elsnes, Western State College

Dino Fiabane, Community College of Philadelphia (Pennsylvania)

G.A. Finchum, Milligan College (Tennessee)

Ira Fogel, Foothill College (California)

Robert G. Foote, Wayne State College (Nebraska)

G.S. Freedom, McNeese State University (Louisiana)

Owen Furuseth, University of North Carolina-Charlotte

Richard Fusch, Ohio Wesleyan

Gary Gaile, University of Colorado-Boulder

Evelyn Gallegos, Eastern Michigan University & Schoolcraft College

Jerry Gerlach, Winona State University (Minnesota)

Lorne E. Glaim, Pacific Union College (California)

Sharleen Gonzalez, Baker College (Michigan)

Daniel B. Good, Georgia Southern University

Gary C. Goodwin, Suffolk Community College (New York)

S. Gopal, Boston University (Massachusetts)

Robert Gould, Morehead State University (Kentucky)

Gordon Grant, Texas A&M University

Donald Green, Baylor University (Texas)

Gary M. Green, University of North Alabama

Mark Greer, Laramie County Community College (Wyoming)

Stanley C. Green, Laredo State University (Texas)

W. Gregory Hager, Northwestern Connecticut Community College

Ruth F. Hale, University of Wisconsin-River Falls

John W. Hall, Louisiana State University-Shreveport

Peter L. Halvorson, University of Connecticut

Mervin Hanson, Willmar Community College (Minnesota)

Robert J. Hartig, Fort Valley State College (Georgia)

James G. Heidt, University of Wisconsin Center-Sheboygan

Catherine Helgeland, University of Wisconsin-Manitowoc

Norma Hendrix, East Arkansas Community College

James Hertzler, Goshen College (Indiana)

John Hickey, Inver Hills Community College (Minnesota)

Thomas Higgins, San Jacinto College (Texas)

Eugene Hill, Westminster College (Missouri)

Louise Hill, University of South Carolina-Spartanburg

Miriam Helen Hill, Indiana University Southeast

Suzy Hill, University of South Carolina-Spartanburg

Robert Hilt, Pittsburg State University (Kansas)

Priscilla Holland, University of North Alabama

Robert K. Holz, University of Texas-Austin

R. Hostetler, Fresno City College (California)

Lloyd E. Hudman, Brigham Young University (Utah)

Janis W. Humble, University of Kentucky

William Imperatore, Appalachian State University (North Carolina)

Richard Jackson, Brigham Young University (Utah)

Mary Jacob, Mount Holyoke College (Massachusetts)

Gregory Jeane, Samford University (Alabama)

Scott Jeffrey, Catonville Community College (Maryland)

Jerzy Jemiolo, Ball State University (Indiana)

Sharon Johnson, Marymount College (New York)

David Johnson, University of Southwestern Louisiana

Jeffrey Jones, University of Kentucky

Marcus E. Jones, Claflin College (South Carolina)

Matti E. Kaups, University of Minnesota-Duluth

Colleen Keen, Gustavus Adolphus College (Minnesota)

Gordon F. Kells, Mott Community College

Susanne Kibler-Hacker, Unity College (Maine)

J.W. King, University of Utah

John C. Kinworthy, Concordia College (Nebraska)

Albert Kitchen, Paine College

Ted Klimasewski, Jacksonville State University (Alabama)

Robert D. Klingensmith, Ohio State University-Newark

Lawrence M. Knopp, Jr., University of Minnesota-Duluth

Terrill J. Kramer, University of Nevada

Arthur Krim, Salve Regina College (Rhode Island)

Elroy Lang, El Camino Community College (California)

Christopher Lant, Southern Illinois University-Carbondale

A.J. Larson, University of Illinois-Chicago

Larry League, Dickinson State University (North Dakota)

David R. Lee, Florida Atlantic University

Joe Leeper, Humboldt State University (California)

Yechiel M. Lehavy, Atlantic Community College (New Jersey)

John C. Lewis, Northeast Louisiana University

Caedmon S. Liburd, University of Alaska-Anchorage

T. Ligibel, Eastern Michigan University

Z.L. Lipchinsky, Berea College (Kentucky)

Allan L. Lippert, Manatee Community College (Florida)

John H. Litcher, Wake Forest Universtiy (North Carolina)

Li Liu, Stephen F. Austin State University (Texas)

William R. Livingston, Baker College (Michigan)

Cynthia Longstreet, Ohio State University

Tom Love, Linfield College (Oregon)

K.J. Lowrey, Miami University (Ohio)

Robin R. Lyons, University of Hawaii-Leeward Community College

Susan M. Macey, Southwest Texas State University

Christiane Mainzer, Oxnard College (California)

Harley I. Manner, University of Guam

James T. Markley, Lord Fairfax Community College (Virginia)

Sister Mary Lenore Martin, Saint Mary College (Kansas)

Kent Mathewson, Louisiana State University

Dick Mayer, Maui Community College (Hawaii)

Dean R. Mayhew, Maine Maritime Academy

J.P. McFadden, Orange Coast College (California)

Bernard McGonigle, Community College of Philadelphia (Pennsylvania)

Paul D. Meartz, Mayville State University (North Dakota)

Inez Miyares, Arizona State University

Bob Monahan, Western Washington University

Keith Montgomery, University of Wisconsin-Marathon

John Morton, Benedict College (South Carolina)

Anne Mosher, Louisiana State University

Robert R. Myers, West Georgia College

Yaser M. Najjar, Framingham State College (Massachusetts)

Jeffrey W. Neff, Western Carolina University

David Nemeth, University of Toledo (Ohio)

Raymond O'Brien, Bucks County Community College (Pennsylvania)

John Odland, Indiana University

Patrick O'Sullivan, Florida State University

Bimal K. Paul, Kansas State University

James Penn, Southeastern Louisiana University

Paul Phillips, Fort Hays State University (Kansas)

Michael Phoenix, William Paterson College (New Jersey)

Jerry Pitzl, Macalester College (Minnesota)

Billie E. Pool, Holmes Community College (Mississippi)

Vinton M. Prince, Wilmington College (North Carolina)

Rhonda Reagan, Blinn College (Texas)

Danny I. Reams, Southeast Community College (Nebraska)

Jim Reck, Golden West College (California)

Roger Reede, Southwest State University (Minnesota)

John Ressler, Central Washington University

John B. Richards, Southern Oregon State College

David C. Richardson, Evangel College (Missouri)

Susan Roberts, University of Vermont

Wolf Roder, University of Cincinnati (Ohio)

James Rogers, University of Central Oklahoma

James C. Rose, Tompkins/Cortland Community College (New York)

Thomas E. Ross, Pembroke State University (North Carolina)

Thomas A. Rumney, State University of New York-Plattsburgh

George H. Russell, University of Connecticut

Rajagopal Ryali, Auburn University at Montgomery (Alabama)

Perry Ryan, Mott Community College

Adena Schutzberg, Middlesex Community College (Massachusetts)

Sidney R. Sherter, Long Island University (New York)

Nanda Shrestha, University of Wisconsin-Whitewater

William R. Siddall, Kansas State University

David Silva, Bee County College (Texas)

Richard G. Silvernail, University of South Carolina

Morris Simon, Stillman College (Alabama)

Kenn E. Sinclair, Holyoke Community College (Massachusetts)

Robert Sinclair, Wayne State University (Michigan)

Everett G. Smith, Jr., University of Oregon

Richard V. Smith, Miami University (Ohio)

Carolyn D. Spatta, California State University-Hayward

M.R. Sponberg, Laredo Junior College (Texas)

Donald L. Stahl, Towson State University (Maryland)

Elaine Steinberg, Central Florida Community College

D.J. Stephenson, Ohio University Eastern

Herschel Stern, Mira Costa College (California)

Reed F. Stewart, Bridgewater State College (Massachusetts)

Noel L. Stirrat, College of Lake County (Illinois)

George Stoops, Mankato State University (Minnesota)

Joseph P. Stoltman, Western Michigan University

P. Suckling, University of Northern Iowa

T.L. Tarlos, Orange Coast College (California)

Michael Thede, North Iowa Area Community College

Derrick J. Thom, Utah State University

Curtis Thomson, University of Idaho

S. Toops, Miami University (Ohio)

Roger T. Trindell, Mansfield University of Pennsylvania

Dan Turbeville, East Oregon State College

Norman Tyler, Eastern Michigan University

George Van Otten, Northern Arizona University

C.S. Verma, Weber State College (Utah)

Graham T. Walker, Metropolitan State College of Denver (Colorado)

Deborah Wallin, Skagit Valley College (Washington)

Mike Walters, Henderson Community College (Kentucky)

J.L. Watkins, Midwestern State University (Texas)

P. Gary White, Western Carolina University (North Carolina)

W.R. White, Western Oregon University

Gary Whitton, Fairbanks, Alaska

Gene C. Wilken, Colorado State University

Stephen A. Williams, Methodist College

P. Williams, Baldwin-Wallace College

Morton D. Winsberg, Florida State University

Roger Winsor, Appalachian State University (North Carolina)

William A. Withington, University of Kentucky

A. Wolf, Appalachian State University

Joseph Wood, George Mason University (Virginia)

Richard Wood, Seminole Junior College (Florida)

Sophia Hinshalwood, Montclair State College (New Jersey)

George I. Woodall, Winthrop College (North Carolina)

Stephen E. Wright, James Madison University (Virginia)

Leon Yacher, Southern Connecticut State University

Donald J. Zeigler, Old Dominion University (Virginia)

We also wish to single out a number of people for special mention and thank them publicly. Our University of Miami departmental colleague, Ira M. Sheskin, assisted us in a number of important tasks related to the mechanical preparation of the final manuscript, particularly the tabular display of demographic data in Appendix A. The work of

Stephen S. Birdsall (now Dean of Arts and Sciences as well as Professor of Geography at the University of North Carolina), who in earlier editions contributed much of Chapters 3 and 8, continues to enhance our presentation. E. Willard and Ruby M. Miller of Pennsylvania State University very generously supplied *handwritten* copies of the titles of their latest bibliographic guides to the world's major regions, all of which are listed in the References and Further Readings section. A number of geographers sent us detailed commentaries that showed us where the written text could be made more precise. We are extremely grateful to these colleagues in our discipline's widening national community, and they include: James P. Allen (California State University, Northridge), John M. Crowley (University of Montana), Doug Eyre (University of North Carolina), Fang Yongming (Shanghai, China), Jonathan Lu (University of Northern Iowa), Robert Peplies (East Tennessee State University), Woody Pitts (Sonoma State University [California]), Victor Prescott (University of Melbourne [Australia]), L.A. Reddick (Spoon River College [Illinois]), Harry Schaleman (University of South Florida), and Canute Vander Meer (University of Vermont).

We consulted with a number of individuals on various matters. Carl Haub of the Population Reference Bureau in Washington, D.C. clarified certain demographic information and trends for us, as did Larry Heligman and Nancy Chen of the Population Division of the United Nations in New York City. Our former Miami departmental colleague, John D. Stephens (now Managing Director at the H.M. Gousha Map Company in Comfort, Texas), shared his knowledge of regional development projects in southern South America. Gil Latz of Portland State University advised us on changing regional appellations in Japan. Ann Brittain of the Department of Anthropology at the University of Miami brought us up to date on the terminology for describing the indigenous peoples of the Americas. The junior author's former colleague at Temple University, Henry N. Michael, gave us the benefit of his lengthy experience in studying Russia's cultural geography, helping us to interpret that country's more complicated ethnic spatial patterns. Elizabeth Muller Hames again prepared the Geographical Index, and also found the time (during her first semester as a graduate student in geography) to become co-author of the *Study Guide* that accompanies this book. This seventh edition also owes many debts to those who helped us in the past. The pronunciation guides were originally prepared with the vital input of Melinda S. Meade, (University of North Carolina), Clifton W. Pannell (University of Georgia), and Gerald G. Curtis (University of Miami, Foreign Languages). Ray Henkel (Arizona State University) made available a particularly useful manuscript entitled "Regional Analysis of the Latin American Cocaine Industry." We gratefully acknowledge, too, the comments of several reviewers, all of which were considered in the revision process. (The errors that remain are, of course, ours alone.)

Those reviewers were:
Howard Adkins, Marshall University (West Virginia)
James P. Allen, California State University, Northridge
John Alwin, Central Washington University
Gouri Banerjee, Boston University
David Daniels, Central Missouri State University
Vernon Domingo, Bridgewater State College (Massachusetts)
John D. Eyre, University of North Carolina
Donald Floyd, California Polytechnic State University
Donald Freeman, York University (Ontario)
Mack Gillenwater, Marshall University (West Virginia)
Carmen Harper, St. Cloud State University (Minnesota)
Ray Henkel, Arizona State University
Mary Jacob, Mount Holyoke College (Massachusetts)
Scott Jeffrey, Catonsville Community College (Maryland)
Robert Klingensmith, Ohio State University-Newark
Richard E. Lonsdale, University of Nebraska
Edward J. Malecki, University of Florida
Gary Manson, Michigan State University
Nancy Obermeyer, Indiana State University
Richard Pillsbury, Georgia State University
Mushtaqur Rahman, Iowa State University
Bheru Lal Sukhwal, University of Wisconsin
Canute Vander Meer, University of Vermont

Closer to home, we are again indebted to colleagues who work with us on the Florida Geographic Alliance, especially co-coordinators Ed Fernald and Laurie Molina (who also prepared the *Instructor's Manual* for this edition) of Florida State University in Tallahassee. At the University of Miami's Department of Geography, we are most grateful for the support we continue to receive from everyone. Our faculty colleagues—Tom Boswell, Don Capone, Jan Nijman, Ira Sheskin, and newly-arrived Bin Li—all regularly teach GEG 105, our introductory World Regional Geography course; in such a lively professional environment, we constantly benefit from their feedback, candid advice, and expertise in areas of regional and systematic geography that complement our own specializations. Our enthusiastic office staff tirelessly performs an array of critical supporting tasks, all supervised by our superb senior secretary and office manager, Hilde Al-Mashat.

At Georgetown University's School of Foreign Service, where the senior author serves as the Landegger Distinguished Professor, both of us thank Marta Marschalko for her decade of support of our joint projects, backed by her formidable organizational skills in directing communications that often are required to flow across thousands of miles.

At John Wiley & Sons, the people we work with to transform our rough drafts into an attractive finished book are among the best in the college textbook publishing business. Once again, we benefited from Barbara Heaney's proficiency as Developmental Editor; she probably knows this book as well as we do, and she gave her time and energies unsparingly at every stage of the revision process. Micheline Frederick, Senior Production Supervisor, did an outstanding job in orchestrating the production process while mastering new electronic methodologies to merge the written text with its suite of high-quality, accompanying visual materials; we also wish to take this opportunity to thank her publicly for her truly heroic efforts during early 1992 in masterminding the production of the Revised Sixth Edition, which enabled us to publish a completely updated version of Chapters 1 and 2 less than 100 days after the collapse of the Soviet Union on that memorable Christmas night of 1991. We were most fortunate to have Linda Muriello as our Production Manager, not only because of her insistence on the highest professional standards, but also because of her commitment to this book that stems from her fine work as production supervisor for the Fifth and Sixth Editions.

We owe a particular debt of gratitude to Edward Starr of the Illustration Department, who so smoothly and good-naturedly coordinated our map program. Ed faced the daunting challenge of converting our entire illustration program into computerized format, and he succeeded brilliantly. (On this occasion, we want to thank him, too, for his gracious help in performing the same duties during the accelerated preparation of the Revised Sixth Edition.)

We also recognize the splendid support we received from Ed's colleague, Ishaya Monokoff; our cartographers at Mapping Specialists, Inc. came through with flying colors (no pun intended!), and were most capably supervised by project manager Don Larson.

The photography program was diligently directed by Joan Meisel in association with Photo Research Manager (and our long-time friend) Stella Kupferberg, and her colleague Alissa Mend; we particularly enjoyed working intensively with all three during a most pleasant half-week in Manhattan. Copyediting duties were expeditiously handled by Martha Cooley, and we also thank Elizabeth Swain for her sharp-eyed proofreading of the galleys. We were pleased to have Madelyn Lesure again as our art director (working with Laura Ierardi as our designer), and she deserves most of the credit for the handsome appearance of the finished product.

Beyond the production team, we also received the strong backing of many other Wiley departments. We have worked closely with Geography Editor Barry Harmon since 1989, and we salute him for helping to organize this revision project and coordinating the reviews. Marketing Manager Catherine Faduska spared no effort in supporting this venture, and we thank her and her co-workers for the opportunities they created for us. Supplements Editor Joan Kalkut went to incredible lengths to assure the high quality of our ancillaries, and worked tirelessly to get the total package assembled on time. We are grateful, too, for the administrative support and/or assistance of Kaye Pace, Cynthia Michelsen, Ann Berlin, Brent Peich, Eileen Nava, and Bethany Brooks.

Finally, we thank our wives, Bonnie and Nancy, for their constant encouragement of all our professional activities.

H.J. de Blij
Peter O. Muller

BRIEF CONTENTS

CONTENTS

CHAPTER 2
RUSSIA'S FRACTURING FEDERATION

VIGNETTE
TRANSCAUCASIA: CAULDRON OF CONFLICT

CHAPTER 3
NORTH AMERICA: THE POSTINDUSTRIAL TRANSFORMATION

VIGNETTE
SOUTH AFRICA: CROSSING THE RUBICON   452

CHAPTER 9
SOUTH ASIA: RESURGENT REGIONALISM   461

CHAPTER 10
CHINA: THE LAST EMPIRE?   509

C H A P T E R   1 1

SOUTHEAST ASIA: BETWEEN THE GIANTS

V I G N E T T E

THE PACIFIC REALM

WORLD REGIONAL GEOGRAPHY: PHYSICAL AND HUMAN FOUNDATIONS

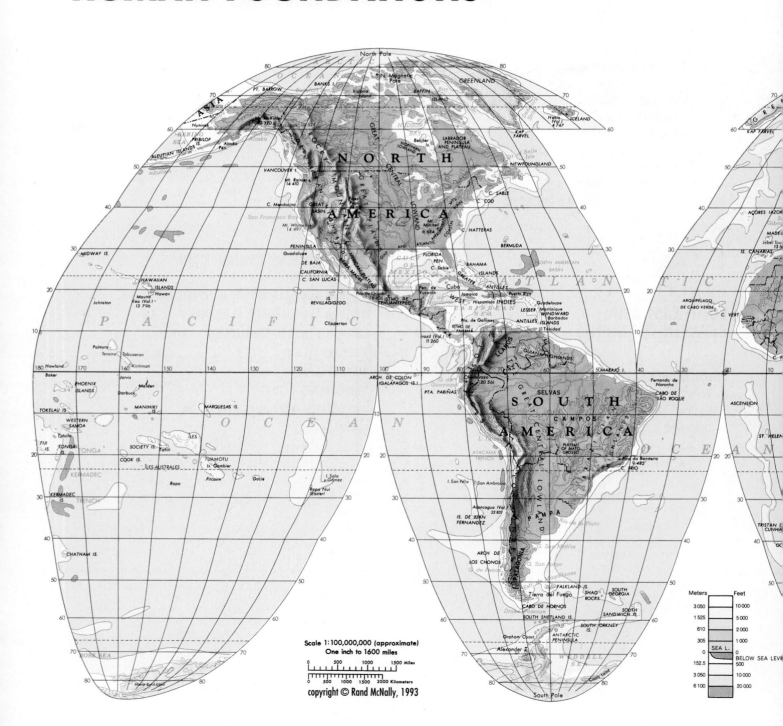

Scale 1:100,000,000 (approximate)
One inch to 1600 miles

0   500   1000   1500 Miles

0   500   1000   1500   2000 Kilometers

copyright © Rand McNally, 1993

IDEAS & CONCEPTS

Geographic realm	Formal region	Population density
Spatial perspective	Spatial system	Urbanization
Taxonomy	Hinterland	State
Transition zone	Functional region	Development
Regional concept	Scale	Iconography
Area	Physiography	Regional geography
Boundaries	Culture	Systematic fields
Location	Cultural landscape	

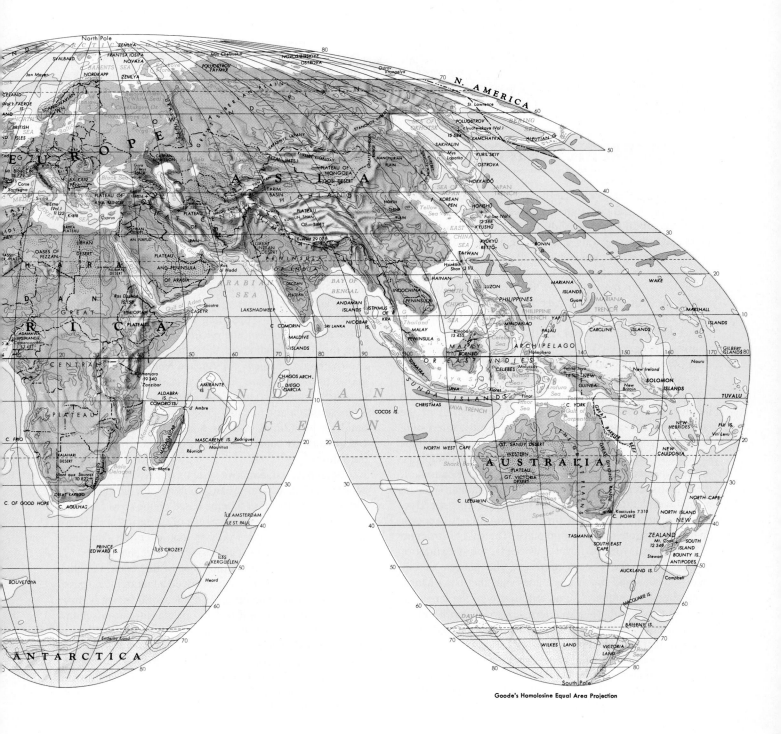

Goode's Homolosine Equal Area Projection

What a time this is to be studying geography! The world is undergoing an historic transformation, one of those momentous political, economic, and social upheavals about which we have read in history books. It is happening today, and as in the past, new alliances are forming, old unions are fracturing, novel ideas are spreading, older notions are fading. We hear from our political leaders about the emergence of a "New World Order," but such pronouncements are premature. A future world order is indeed in the offing, but it is too early to tell what it will be like. Of this we can be sure: the new world map will look quite different from the old. Our task is to understand the ongoing changes, to make sense of the new directions our world is taking. As you will find, geography is your most powerful ally in this mission.

◆ GEOGRAPHIC PERSPECTIVES

In this book we take a penetrating look at the geographic framework of the contemporary world, the grand design that is the product of thousands of years of human achievement and failure, movement and stagnation, revolution and stability, interaction and isolation. Ours is an interconnected world of travel and trade, tourism and television, a global village—but the village still has neighborhoods. Their names are Europe, South America, Southeast Asia, and others familiar to all of us. We call such global neighborhoods **geographic realms**, and when we subject these realms to geographic scrutiny, we find that they each have their own identity and distinctiveness.

Geographers study the locations and distributions of features on the earth's surface. These features may be the landmarks of human occupation or the properties of the natural environment, or both: one of the most interesting themes in geography has to do with the interrelationships of natural environments and human societies. Geographers, in their research, investigate the reasons for, or causes of, these distributions. Their approach to the human and the natural world, therefore, is guided by a **spatial perspective.** Just as historians focus on time and chronology, geographers concentrate on space and place. The spatial structure of cities, the layout of farms and fields, the networks of transportation, the system of rivers, the pattern of climate—all this and much more goes into the examination of a geographic realm. As you will discover, the language of geography is full of spatial terms: area, distance, direction, clustering, proximity, accessibility, isolation, and many others reflect this viewpoint.

In this book we use the geographic perspective and geography's spatial terminology to investigate the world's great geographic realms. We will find that each of these realms possesses a special combination of cultural, organizational, and environmental properties. These characteristic qualities are imprinted on the landscape, giving each realm its own traditional attributes and social settings. As we come to understand the human and natural makeup of those geographic realms, we learn not only *where* they are located (always a crucial question in geography, and often the answer is not as simple as it may appear), but also *why they are located where they are*, how they are constituted, and what their future is likely to be in our changing world.

It would not, however, be enough to study the world from a geographic viewpoint without also learning something about the discipline of geography itself. Beginning with this introductory chapter, we introduce several of the ideas and concepts that make modern geography what it is. Not only will this enhance your awareness of the many dimensions of our complex, multicultural, interconnected world, but many of those geographic ideas will remain useful to you long after you have closed this book. Welcome to geography . . . realms, regions, *and* concepts!

◆ REALMS AND REGIONS

Geographers, like other scholars, seek to establish order from the countless pieces of information (data) with which they are confronted. Biologists have established a system of classification, or **taxonomy**, to categorize millions of plants and animals into a hierarchical system consisting of seven ranks. In descending order, we humans belong to the animal *kingdom*, the *phylum* (division) named chordata, the *class* of mammals, the *order* of primates, the *family* of hominids, the *genus* designated *Homo*, and the *species* known as *Homo sapiens*. Geologists classify the earth's rocks into three major (and many subsidiary) categories and then fit these into a complicated geologic time scale that spans hundreds of millions of years. Historians define eras, ages, and periods to conceptualize the sequence of the events they study.

Geography, too, employs systems of classification. When geographers deal with urban problems, for instance, they employ a classification scheme based on the sizes and functions of the places involved. Some of the terms in this classification are familiar words in our everyday language: megalopolis, metropolis, city, town, village, hamlet.

In regional geography, the focus of this book, the challenge is rather different. We, too, need a hierarchical framework to accommodate the areas of the world we study, from the largest to the smallest. But our classification scheme is horizontal, not vertical. It is *spatial*. Our equivalent of the biologists' overarching kingdoms (of plants, animals) is the earth's natural partitioning into landmasses and oceans. The next level is the division of the inhabited landmasses into geographic realms based, as noted earlier, on human as well as physical (natural) properties.

REALMS AND THEIR CRITERIA

In any classification system, criteria are the key. Not all animals are mammals, and fewer still are primates: the criteria for inclusion in that biological order are more specific and restrictive. A dolphin may look and appear to behave like a fish, but dolphins anatomically and functionally qualify as belonging to the order of mammals.

Geographic realms are based on sets of spatial criteria. First, they are the largest units into which the inhabited world can be divided; smaller units should have a disproportionate influence on the wider world. At present, only one geographic realm fits the latter category—Japan, still so distinct from mainland East Asia that it merits separate status. A century ago, Britain was in a similar position; it lay at the center of a world empire and its industrial strength far exceeded that of adjacent mainland Europe. Today, however, Britain has neither its empire nor its industrial advantage, and we must map it as part of the European geographic realm.

Second, geographic realms are the result of the interaction of human societies and natural environments, a *functional* interaction that is revealed by farms, mines, fishing ports, transport routes, dams, bridges, villages, and countless other features on the landscape. Under this criterion, Antarctica is a continent but not a geographic realm.

Third, geographic realms must represent the most comprehensive and encompassing definition of the great clusters of humankind in the world today. China lies at the heart of such a cluster, as does India. Africa constitutes a geographic realm from the southern margin of the Sahara (an Arabic word for "desert") to the Cape of Good Hope, and from its Atlantic to its Indian Ocean shores.

Figure I-1 displays the 13 world geographic realms based on these criteria. As we will show in more detail later, not only waters but also deserts and mountains still mark the borders of these realms; we will also discuss the position of these boundaries as each realm is examined. For the moment, keep in mind the following:

• *Where geographic realms meet, **transition zones**, not sharp boundaries, mark their contacts.*

We need only remind ourselves of the border zone between the geographic realm in which most of us live, North America, and the adjacent realm of Middle America. The line on Fig. I-1 coincides with the boundary between Mexico and the United States, crosses the Gulf of Mexico, and then separates Florida from Cuba and the Bahamas. But Hispanic influences are strong in North America far to the north of this boundary, and U.S. economic influence is strong to the south of it. The line, therefore, represents an ever-changing zone of regional interaction here. Again, there are many ties between South Florida and the Bahamas, but the Bahamas more strongly resemble a Caribbean society than a North American one.

Cultural transformation in the South Florida borderland. Forty years ago, every one of these signs was in English; now, Spanish is the norm in Miami's "Little Havana." The Hispanic population of metropolitan Miami rose from 4 percent in 1950 to more than 47 percent in 1990.

In Africa, the transition zone from Subsaharan to North Africa is so wide and well defined that we have put it on the world map; elsewhere, transition zones tend to be narrower and less easily represented. In the mid-1990s, such countries as Belarus (between Europe and Russia) and Kazakhstan (between Russia and Muslim Southwest Asia) lie in inter-realm transition zones. Again, over much (though not all) of their length, borders between realms are zones of regional change.

• *Geographic realms change over time.*

Had we drawn Fig. I-1 before Columbus's time, the map would have looked quite different: Amerindian states and peoples would have determined the boundaries in the Americas; Australia and New Guinea—not New Zealand—would have constituted one realm (New Zealand would have been part of the Pacific Realm); Japan, neither an economic nor a technological giant then, would have been part of an East Asian realm. The colonization, Europeanization, and Westernization of the world changed that map dramatically. During the four decades following the end of World War II there was relatively little change, but the

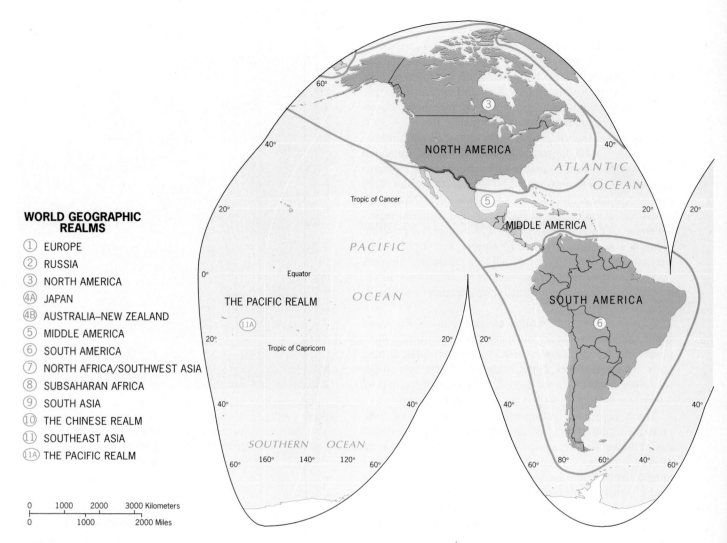

WORLD GEOGRAPHIC REALMS

1. EUROPE
2. RUSSIA
3. NORTH AMERICA
4A. JAPAN
4B. AUSTRALIA–NEW ZEALAND
5. MIDDLE AMERICA
6. SOUTH AMERICA
7. NORTH AFRICA/SOUTHWEST ASIA
8. SUBSAHARAN AFRICA
9. SOUTH ASIA
10. THE CHINESE REALM
11. SOUTHEAST ASIA
11A. THE PACIFIC REALM

FIGURE I-1

years since 1985 have witnessed far-reaching realignments that are still going on. Over six editions of this book, between 1971 and 1991, we needed to make only minor adjustment to a realms framework that was essentially stable. Now, in the mid-1990s, we must make major revisions, underscoring the transformation of our world as the twenty-first century approaches.

As we try to envisage what a New World Order will look like on the map, note that the 13 geographic realms can be divided into two groups: (1) those dominated by one major political entity (North America/United States, Middle America/Mexico, South America/Brazil, South Asia/India, the Chinese Realm/China, and Russia, Japan, and Australia), and (2) those in which there are many countries but no dominant state (Europe, North Africa/ Southwest Asia, Subsaharan Africa, Southeast Asia, and the Pacific Realm). For several decades we have known a world dominated by two major powers, the United States and the former Soviet Union. Will a multipolar world arise from the rubble of

that bipolar world? That is a question to be addressed in the pages that follow.

REGIONS AND THEIR CRITERIA

The spatial division of the world into geographic realms establishes a broad global framework, but for our purposes a more refined level of spatial classification is needed. This brings us to an important organizing concept in geography: the **regional concept**. To continue the analogy with biological taxonomy, we now go from phylum to order. To establish regions within geographic realms, we need more specific criteria.

Let us use the North American realm to demonstrate the regional idea. When we refer to a part of the United States or Canada (e.g., "the South," "the Midwest," or "the Prairie Provinces"), we employ a regional concept—not scientifically, but as a matter of everyday communication. We re-

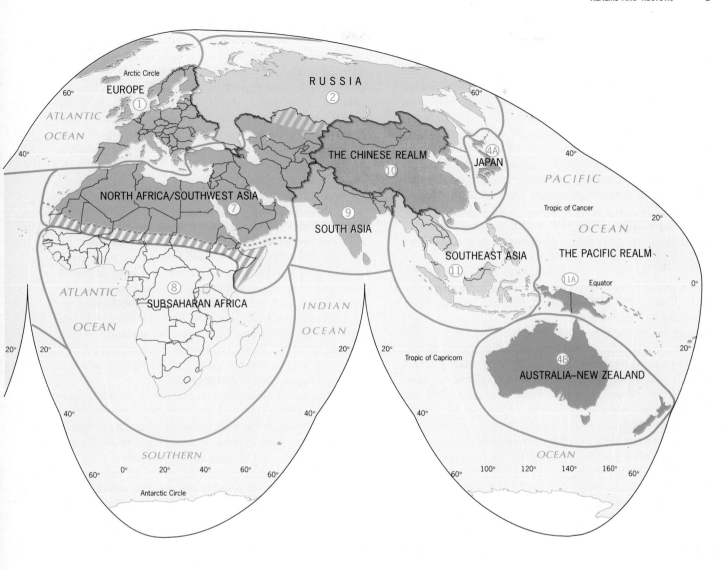

veal our *perception* of local or distant space as well as our mental image of the region to which we make reference.

But what exactly is the Midwest? If you were asked to draw this region on the North American map, how would you do it? Regions are easy to imagine and describe, but they can be very difficult to outline on a map. One way to go about defining the Midwest is to use the borders of states: certain states are part of this region, others not. Another way would be to use agriculture as the chief criterion: where corn and soybeans occupy a certain percentage of the farmland, there lies the Midwest. This obviously would result in a different delimitation; a Midwest based on states is different from a Midwest based on farm production. And therein lies an important principle: regions are scientific devices that allow us to make spatial generalizations, and they are based on artificial criteria we establish for the purpose of constructing them. If you were doing research on the geography behind politics, then a Midwest region based on state boundaries would make sense. If you were study-

ing agricultural distributions, then you would need a different definition of the Midwest.

Given these different dimensions of the same region, are there properties all regions have in common? Let us identify several. To begin with, all regions have **area**. This would seem to be so obvious as not to need elaboration, but there is more to this idea than first meets the eye. Regions may be intellectual constructs, but they are not abstractions: they exist in the real world, and they occupy space on the earth's surface.

It follows that regions have **boundaries**. Occasionally, nature itself draws sharp dividing lines, for example, along the crest of a mountain range or the margin of a forest. More often, regional boundaries are not self-evident and must be determined on the basis of criteria established for that purpose. Take, for instance, the notion of a Corn Belt in the agricultural heartland of the United States. When you travel into that region, you see more and more fields of corn until this crop dominates the farmlands. But where exactly is the limit of this region we call the Corn Belt? That depends on specific criteria. First we would establish a unit area, say, one-quarter square mile. Next we would create a grid consisting of quarter-square-mile units and superimpose it on an area larger than the corn-growing zone. Then we would decide that in order to be part of the Corn Belt, more than 50 percent of the farmland in any unit area must be devoted to growing corn. Additionally, we might decide that a unit area is part of the Corn Belt even if it contains less than 50 percent corn as long as it is completely surrounded by spatial units that do qualify. Result: a Corn Belt region boundary that is both visible and quantitatively defined. Of course, not all regional boundaries can be so specifically delimited, and neighboring regions, like adjacent realms, sometimes display transitional borderlands.

All regions also have **location**. Often the name of a region contains a locational clue, as in Amazon Basin or Indochina (a region of Southeast Asia lying between India and China). Geographers refer to the *absolute location* of a place or region by reference to the earth's grid system; this provides the latitudinal and longitudinal extent of the region with respect to grid coordinates. A far more practical measure is a region's *relative location*, that is, its location with reference to other regions. Again, the names of some regions reveal aspects of their relative locations, as in *Eastern* Europe and *Equatorial* Africa.

Many regions are marked by a certain *homogeneity*, or sameness. This homogeneity may lie in a region's human (cultural) properties or its physical (natural) characteristics, or both. Siberia, a vast region of northern Russia, is marked by a sparse human population residing in widely scattered, small settlements of similar form; frigid climates; extensive areas of permafrost (permanently frozen subsoil); and cold-adapted vegetation. This dominant uniformity makes

it one of Russia's natural and cultural regions, extending from the Ural Mountains in the west to the shores of the Pacific Ocean in the east. When regions display a measurable and often visible internal homogeneity, they are referred to as **formal regions**. But not all formal regions are visibly uniform. For example, a region may be delimited within which one particular language is spoken by, say, 90 percent of the people or more. This cannot be seen in the landscape, but the region is a reality and its boundaries can be drawn quite accurately based on this criterion. It, too, is a formal region.

Other regions are marked not by their internal sameness, but by their functional integration, so that they are **spatial systems**. This type of region is formed by a set of places and their interrelated activities. A large urban area, for example, is likely to have a surrounding zone to which it supplies goods and services, from which it buys farm products, and with which it interacts in numerous ways. The farther away from the city in this region, the weaker are the links to that central *core*—and here in the *periphery* we approach the boundary of the city's **hinterland**, as such a region is called. This hinterland is a region forged by a structured, urban-centered system of interaction, and it is a classic example of a **functional region**.

To a greater or lesser degree, all human-geographic regions are linked to other regions through *interconnections* of various kinds. We noted earlier the transitional nature of geographic-realm borderlands; adjoining regions, too, display trans-border spatial influences. Moreover, there are countless trans-regional interactions involving trade, migration, education, television and radio, and much more. These are links in the rapidly growing interdependence among the world's peoples, and they help to reduce the differences that continue to divide us. Understanding these differences is an important step toward lessening them.

Interconnections in the spatial system: Garden City, Iowa. Roads and grain elevators link the town to its surroundings. The spatial homogeneity of the region is quite evident from the farmscape beyond.

◆ REGIONS AT SCALE

In this book we examine the geography of the world at the level of detail of the region, using regional concepts as we proceed. But all regions are parts of realms, which is why this book is called *Geography: Realms, Regions, and Concepts*.

As we have just noted, regions come in many sizes. Some are huge, for example the Russian formal region of Siberia. Others are comparatively small, as in the case of the functional region constituted by a city and its hinterland. Some geographers have tried to establish a terminology to distinguish between larger and smaller regions; occasionally, you will see a reference to a *subregion* within a region. But no generally accepted nomenclature has ever emerged, so we use the term *region* for bounded spaces large and small. At the level of detail in our maps, we must focus on the larger regions, referring occasionally to small-

er regions "nested" within the larger ones. And this brings us to the geographic concept of scale.

The map is the geographer's strongest ally. It does for geography what taxonomic (and other) classification systems do for the other sciences. Maps display enormous quantities of information; they suggest interrelationships; they answer questions; they lay out spatial problems for researchers to investigate (see Appendix B, *Map Reading and Interpretation*). Many maps, moreover, are simply fascinating. No library is complete without a good atlas.

Maps represent the surface of the earth (and other features of the planet, present and past) at various levels of generalization. These levels of generalization can be gauged by looking at the map's **scale**, that is, the ratio of the distance between two places on a map and the actual distance between those two places on the earth's surface.

Consider the four maps in Fig. I-2. On the first map (upper left), most of the North American realm is shown, but very little spatial information can be provided, al-

EFFECT OF SCALE

0 500 1000 1500 2000 2500 Kilometers

0 500 1000 1500 Miles

1:103,000,000

0 250 500 750 1000 1250 1500 Kilometers

0 250 500 750 Miles

1:53,200,000

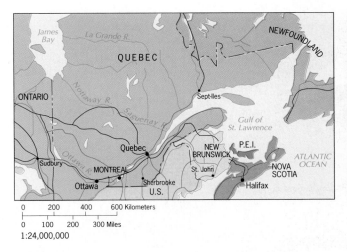

0 200 400 600 Kilometers

0 100 200 300 Miles

1:24,000,000

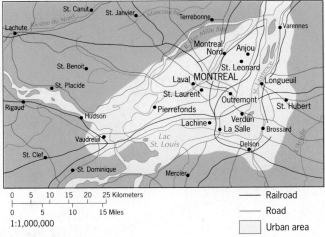

0 5 10 15 20 25 Kilometers

0 5 10 15 Miles

1:1,000,000

—— Railroad

—— Road

Urban area

FIGURE I-2

though the political boundary between Canada and the United States is evident. On the second map (upper right), eastern and central Canada are depicted in sufficient detail to permit display of the provinces, several cities, and some physical features (Manitoba's major lakes) not shown on the first map. The third map (lower left) shows the main surface communications of the province of Quebec and immediate surroundings, the relative location of Montreal, and the St. Lawrence and Hudson/James Bay drainage systems. The fourth map (lower right) reveals the metropolitan layout of Montreal and its adjacent hinterland in considerable detail.

Each of the four maps has a scale designation, which can be shown as a bar graph (in miles and kilometers in this case) and as a fraction—1:103,000,000 on the first map. The fraction is a ratio indicating that one unit of distance on the map (one inch or one centimeter) represents 103 million of the same units on the ground. The smaller the fraction (in other words, the larger the number in the denominator), the smaller the scale of the map. Clearly, this representative fraction (RF) on the first map (1:103,000,000) is the smallest, and that of the fourth map (1:1,000,000) is the largest. Comparing maps number 1 and number 3, we find that on the linear scale, number 3 has a representative fraction that is more than four times larger than that of number 1. When it comes to areal representation, however, 1:24,000,000 is more than 16 times larger than

1:103,000,000 because the linear difference prevails in both dimensions (the length and the breadth of the map).

In a book that surveys the major regions of each world realm, it is obviously necessary to operate at relatively small scales. When studying subregions in greater detail, our ability to specify criteria and to "filter" the factors we employ increases as we work at larger scales; that method will often come into play when urban areas are the topic of concern (as suggested by the map of metropolitan Montreal in Fig. I-2). But most of the time our view will be more macroscopic and general: a small-scale perspective on the world's geographic realms and regions.

◆ THE NATURAL STAGE

Although the world-scale regions we will define and explore in this book are the products of human activity, we will have frequent occasion to refer to the physical setting against which the human drama is being played out. For all our industrial capacity and our technological progress, the natural environment still plays a prominent role in human fortunes.

Significantly, we have learned that environments have changed in the past, sometimes quite rapidly. Archeologists have excavated cities that long ago rose to power in

In Niger, West Africa, the desert encroaches on the pasturelands. Cattle and goats contribute to the process: their grazing, and their hooves, damage and weaken the vulnerable desert margin. Desertification is a cyclic, global problem.

splendor amid plenty of water, farm produce, and raw materials such as wood and fiber. Today their ruins lie buried under desert sand. In Roman times, farmlands near North Africa's Mediterranean coast yielded large amounts of produce, amply watered by rainfall and irrigation systems. Today, the Roman aqueducts lie in ruins, and fields are abandoned. And now we worry about the future: about *desertification*, the drying up of areas in the path of expanding deserts; about global warming, the prospect of warmer temperatures than have prevailed over past decades; and about associated climatic shifts. Will polar icesheets melt and cause sea levels to rise? Will croplands stop yielding their produce? Which regions of the world would be most severely affected by environmental change?

These are important questions. We must not study geographic realms and regions without reference to the physical stage they occupy. Physical geography—the spatial study of the earth's natural phenomena and their systems, processes, and structures—provides us with the necessary information. It was a physical geographer, the climatologist Alfred Wegener, who nearly a century ago used spatial analysis to propose a then-revolutionary hypothesis: the hypothesis of *continental drift*. Wegener studied the outlines of the continents and, marshalling a vast array of evidence, suggested that the landmasses had at one time in the earth's history been united. Since then, they had drifted apart, he argued, but the opposite sides of the Atlantic Ocean still fit together like pieces of a jigsaw puzzle, especially Africa and South America. What was revolutionary about this idea was that the seemingly stable, immovable continents were in fact mobile.

This, Wegener argued in a book published in 1915, explained the great variety of natural landscapes on the continents. For example, as South America drifted westward, away from Africa and into the Pacific, its leading edge crumpled up like the folds of an accordion, forming the Andes Mountains. This, too, is why North America's greatest mountains lie in the west, while those of Australia (which moved eastward into the Pacific) lie in the eastern margin of that landmass.

Wegener did not live to see his hypothesis accepted; many prominent geologists quickly derided it as nonsense. But as is so often the case, a geographer's spatial theory pointed the way for the research of others. Eventually, geologists did discover that the thin crust of the earth consists of a set of *tectonic plates*, great slabs of solid material that not only make up ocean floors but also carry continents. And (to no geographer's surprise), those plates (Fig. I-3) are indeed mobile. They move slowly, a few centimeters (an inch or two) per year, propelled by heat-driven convection cells in the molten rock deep below the crust. A couple of inches per year is not much, but the earth's history is measured in millions of years. So the continental landmasses travel thousands of miles, opening and closing oceans and changing the world map of land and water bodies. As great tectonic plates collide, the one that consists

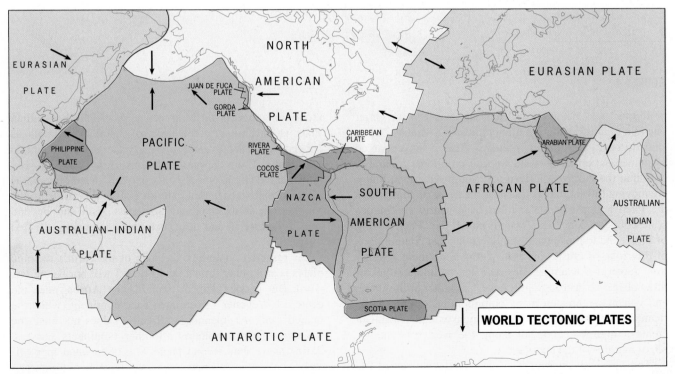

FIGURE I-3

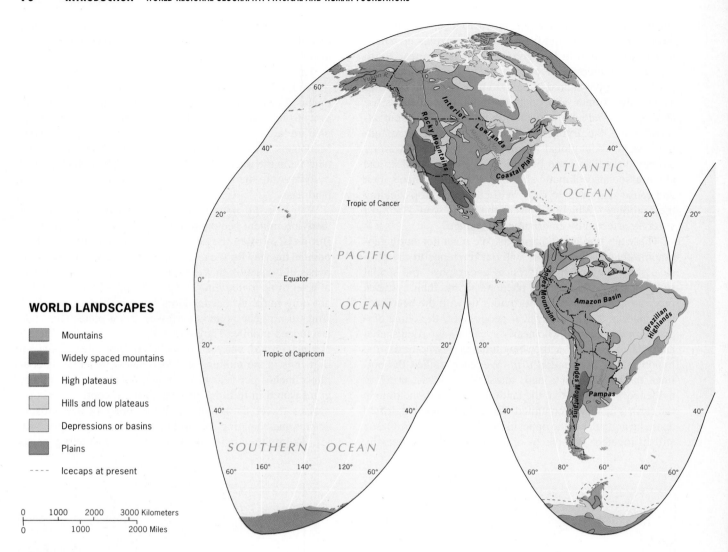

WORLD LANDSCAPES

- Mountains
- Widely spaced mountains
- High plateaus
- Hills and low plateaus
- Depressions or basins
- Plains
- ---- Icecaps at present

```
0    1000    2000    3000 Kilometers
0         1000         2000 Miles
```

FIGURE I-4

of lighter (less heavy) rock rides up over the weightier one in a gigantic collision that creates just what Wegener said it does: great mountain chains. This is accompanied by numerous earthquakes and volcanic eruptions. In Fig. I-3, note that the North American and South American Plates are moving toward the Pacific, as is the Eurasian Plate. Thus the Pacific Ocean is surrounded by a nearly continuous plate-collision zone. This zone is called the "Pacific Ring of Fire," and to those realms and regions that contain parts of this zone of crustal instability, this is no mere abstraction. Japan, for example, lies near the colliding edges of three plates. Disastrous earthquakes are a fact of life there, and volcanoes threaten many communities. On the American side of the Pacific, too, earthquakes and volcanism create a high-risk zone all along the western (leading-edge) margins of the continents.

The physical landscapes of the continents reveal the effects of millions of years of movement. Today, a giant mountain range is forming where the Australian-Indian Plate is pushing into the heart of Asia; the Himalayas capped by Mount Everest (presently the highest mountain on the planet) are the manifestations of this collision. But note that the Himalayas are only one link in a mountainous plate-collision chain that extends southeastward from the Mediterranean to mainland Southeast Asia and beyond (Fig. I-4). Life in and near such zones is always affected by risk.

As research continues, the map of the earth's tectonic plates is still being drawn, and Fig. I-3 will not be the last word. But it does help to explain why Africa, once the center of Wegener's postulated supercontinent (which he named *Pangaea*, meaning "all-earth") does not have the lengthy, linear mountains all other continents possess. Africa, as we shall see in Chapter 8, is often—and appropriately—called the "plateau continent." It is one of the earth's oldest pieces of real estate, geologically speaking.

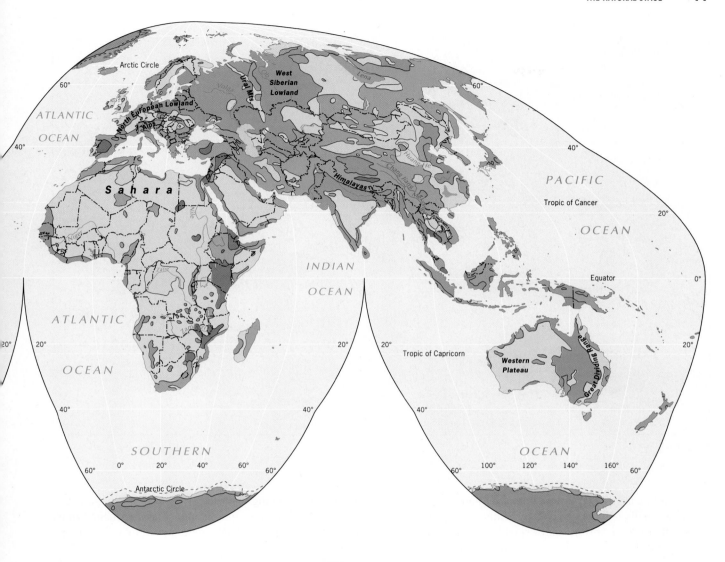

As we investigate each geographic realm, we will take note of the landscapes that form its physical base (Fig. I-4). All the continents contain old geologic cores known as *shields* that often create vast expanses of plains (low-lying flatlands) or plateaus (higher-elevation flatlands). Between these low-relief expanses and the high mountain ranges (Andes, Rockies, Alps, etc.) lie hills of many different sizes. There is no generalizing about the earth's landscapes, and we will have to examine them one realm and region at a time.

GLACIATIONS

To complicate matters still further, the earth—even as the continental landmasses move—periodically undergoes *glaciation*, a time during which global temperatures drop and environments change. A glaciation actually is not just one cooling phase but many. Over millions of years, the temperature swings back and forth, from warmer to colder and back to warmer again; and each time it gets colder, drastic changes occur. Continental-scale, glacial icesheets expand. Sea levels drop. Climates change. Plant life shifts equatorward. Species of animals and plants die out, but others, better adapted, survive and succeed.

Over the past 20 million years, the earth has been cooling, and during the most recent 5 to 6 million years we have been in the grip of a glaciation—the Late Cenozoic glaciation, as geologists have named it. (They used to call it the *Pleistocene* glaciation until they recently discovered that it started well before the Pleistocene epoch, our epoch of emergence, began.) We humans are the products of that glaciation. Our distant ancestors were among species that managed to adapt when, in their East African homeland, the climate cooled, trees died out, drought prevailed, and animals long hunted disappeared. In the several million years that followed, the climate warmed and cooled again many

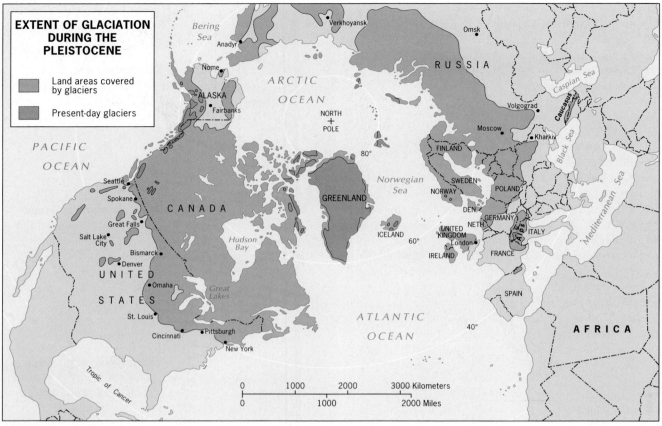

FIGURE I-5

times—as many as two dozen times, perhaps more. We seem repeatedly to have survived by adaptation during cold glacial times, then thrived and expanded during the next *interglaciation* (the global warm-up that occurs between two glaciations).

About 15,000 years ago, the icesheets were back for the last (or rather, most recent) time, and the map of the Northern Hemisphere looked as in Fig. I-5. Since then, the glaciers have again withdrawn, but this has been no ordinary interglaciation. In the geologic eye-blink of the present interglaciation, the entire modern history of human civilization has played itself out—from rock shelters to megacities, from stone tools to space shuttles, from human millions to billions. Geologists call ours the Holocene epoch, as though the Pleistocene has come to an end. There is no evidence for this: compared to earlier global glaciations, the present one has just begun.

These days we worry about global warming, and we are right in doing so if only to reduce the pollutants we are spewing into the atmosphere. But over the long term, the greater threat may come from a return of the glaciers, preceded (as we know it has in the past) by a time of worldwide weather extremes. Some geographers see in recent weather perturbations a pattern similar to those that pre-

saged earlier cooling and glacial expansion. Imagine it: a world with 6 billion people facing a major reduction in its habitable living space. That would make global warming a mere inconvenience by comparison.

WATER—ESSENCE OF LIFE

Water is the key to life on earth. Without water, our planet would be as barren as the moon or Mars. And it would seem we should have no shortage of this vital commodity: more than 70 percent of the earth's surface is covered by the water of the oceans, seas, lakes, and streams. Yet, in fact, water shortages occur in many areas of our planet, even in normally well-watered places. Steadily increasing human populations are putting unprecedented demands on water supplies. Water is carried through modern aqueducts and tunnels from distant rivers and dams to mushrooming metropolises. Farmers complain that the cities' demand is depriving them of needed irrigation water, as is happening in California. In Florida, the megacity centered on Miami depends on an underground reservoir of fresh water tapped by hundreds of wells—but when this *aquifer* is not replenished by rain water, it is invaded by salt water from the adjacent ocean. So Miami, like many other cities, occasionally restricts its inhabitants' water use.

Such problems seem minor compared to what is happening elsewhere. Southern Africa, in the early 1990s, was in the grip of the worst drought in its history; in Zimbabwe, a plateau country in this region's interior, almost all the livestock perished, crops failed, and starvation threatened. Death-dealing droughts have struck West Africa, Northeast Brazil, Ethiopia, Central Asia, and other areas during the past 20 years alone. No place in the world is immune to water shortages. Conflicts loom over actual and proposed dam construction. Wars have been fought over water sources. As human numbers continue to grow and climatic fluctuations increase, the water crisis (for such it already is, for hundreds of millions of us) will become far more significant than recent oil shortages.

Ocean water as such is of little use to humanity: it is made useful through a mechanism that brings moisture from the oceans to the landmasses. This mechanism is the *hydrologic cycle*, which functions as a circulation system. Water evaporates into the air from the salt-water ocean surface, leaving the salt behind; the moisture-laden air mass then drifts over land, where (by various atmospheric processes) condensation occurs and fresh-water precipitation falls. This precipitation replenishes underground reservoirs, moistens soils, sustains plants and animals, and fills streams. Much of it returns to the oceans as runoff via rivers, and the endless cycle repeats itself.

Precipitation Distribution

The hydrologic cycle is neither constant nor infallible. It can and does fail, imperiling life where and when this happens. It also can be enhanced under certain unusual circumstances—for example, during 1992, when a perturbation known as the El Niño–Southern Oscillation (ENSO), a tropical Pacific climate anomaly, sent a series of moisture-laden air masses toward California and Texas. In Southern California, then in the grip of a five-year drought, torrential rains brought mudslides as well as relief. In Texas, where the rains lasted longer, they caused devastating floods that drowned thousands of livestock.

Remember this when you examine the global precipitation map (Fig. I-6): it shows the *average* amount of precipitation received by all areas of the world during the comparatively brief time period in which measurements have been taken. Compare it to Fig. I-1, and you will gain an impression of the overall water availability in each of the world's major geographic realms. But note an important caveat: the amount of precipitation alone is not enough to assess just how plentiful moisture is (see the box entitled "Evapotranspiration").

Figure I-6 reveals an equatorial zone of heavy rainfall, where annual totals often exceed 80 inches (200 cm), extending from mainland Middle America through the Ama-

EVAPOTRANSPIRATION

The map of world precipitation distribution (Fig. I-6) should be viewed in the context of temperature distribution. From the map one might conclude that all the areas that receive over 40 inches (100 cm) of rainfall are equally well supplied with moisture. But there are equatorial areas where temperatures average over 77°F (25°C) that receive 40 inches, and much cooler places—for example, parts of New Zealand and Western Europe—where 40 inches are also recorded. Obviously, evaporation from the ground goes on much more rapidly in the tropics than in the mid-latitude zones.

Similarly, evaporation from vegetation also speeds up in equatorial regions. This evaporation from leaf surfaces is actually a three-stage process. Roots of plants absorb water from the soil. This water is then transmitted through the organism and reaches the leafy parts, which transpire in warm weather much as we humans perspire. From the surface of the leaves, the moisture evaporates. Thus a plant acts like a pump, and the process of evaporation from vegetation is actually a process of transpiration plus evaporation—or *evapotranspiration*.

Thus 40 inches of rainfall in a tropical area may very well be inadequate, because if the amount lost by evaporation and evapotranspiration is calculated, it could exceed 40 inches—which means that the plant would use even more moisture if it were available. Some of those "moist" tropical areas, even those with over 60 inches (150 cm) of rainfall annually, can be shown to be moisture-deficient. In other areas the seasonality of precipitation is so pronounced that there is a deficiency during part of the year. In contrast, in cooler parts of the world, just 30 inches (75 cm) of rainfall may be enough to keep the soil moist and the vegetation adequately supplied. A map such as Fig. I-6 is of necessity a generalization, and it is important to know what it fails to reveal.

zon Basin, across smaller areas of West and Equatorial Africa, and into South and Southeast Asia. This low-latitude zone of high precipitation gives way to dry conditions in both poleward directions. In equator-straddling Africa, for example, the arid Sahara lies to the north of the tropical humid zone and the Kalahari Desert lies to the south. Interior Asia and central Australia also are very dry, as is southwestern North America.

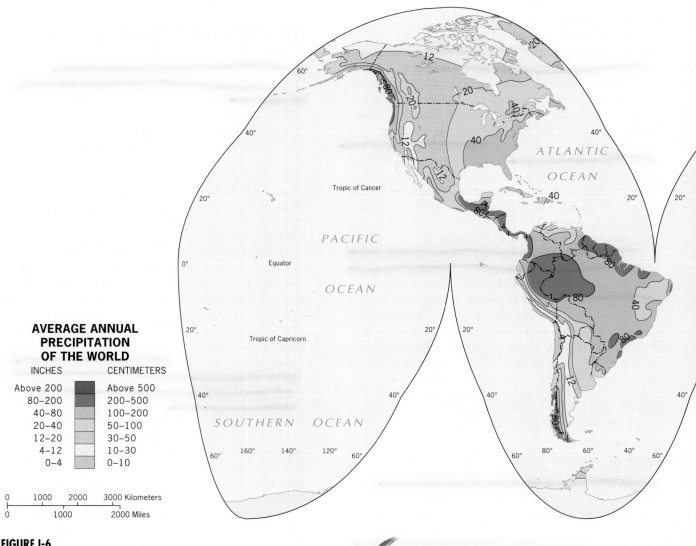

**AVERAGE ANNUAL
PRECIPITATION
OF THE WORLD**

INCHES	CENTIMETERS
Above 200	Above 500
80–200	200–500
40–80	100–200
20–40	50–100
12–20	30–50
4–12	10–30
0–4	0–10

0   1000   2000   3000 Kilometers
0   1000   2000 Miles

FIGURE I-6

The general pattern of equatorial wetness and adjacent dryness is broken along the coasts of all the continents, and it is possible to discern a certain consistency in this spatial distribution of precipitation. Eastern coasts of continents and islands in tropical as well as mid-latitude locations receive comparatively heavy rainfall, as in the southeastern United States, eastern Brazil, eastern Australia, and southeastern China. Furthermore, a narrow zone of higher precipitation exists at higher latitudes on the western margins of the continents, including the Pacific Northwest coast of the United States and Canada, the southern coast of Chile, the southern tip of Africa, the southwestern corner of Australia, and most important, the western exposure of the great Eurasian landmass—Europe.

The distribution of world precipitation, as reflected in Fig. I-6, results from an intricate interaction of global systems of atmospheric and oceanic circulation as well as heat and moisture transfer. Although the analysis of these systems is a part of the subject of physical geography, we should remind ourselves that even a slight change in one of them can have a major effect on a region's habitability. Again, it is important to remember that Fig. I-6 displays the average annual precipitation on the continents, but no place on earth is guaranteed to receive, in any given year, precisely its average rainfall. In general, the variability of precipitation increases as the recorded average total decreases. In other words, rainfall is least dependable just where reliability is needed most—in the drier portions of the inhabited world.

Precipitation and Human Habitation

Even heavy year-round precipitation is no guarantee that an area can sustain large, dense populations. In the equatorial

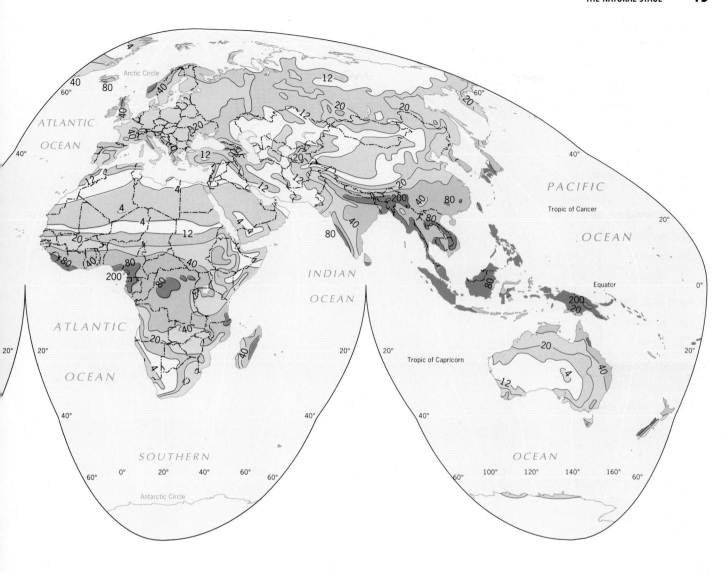

latitudes, heavy precipitation combined with high temperatures leads to the faster destruction of fallen leaves and branches by bacteria and fungi. This severely inhibits the formation of humus, the dark-colored upper soil layer consisting of nutrient-rich decaying organic matter, which is vital to soil fertility. The drenched soil is also subjected to the removal of its best nutrients through leaching—the dissolving and downward transport of nutrients by percolating water—so that only oxides of iron and aluminum remain at the surface to give the tropical soil its characteristic reddish color. Such tropical soils (called oxisols) nurture the dense rainforest, but they cannot support crops without massive fertilization. The rainforest thrives on its own decaying vegetative matter, but when the land is cleared, the leached soil proves to be quite infertile. Not surprisingly, the Amazon and Zaïre (Congo) basins are not among the world's most populous regions.

CLIMATIC REGIONS

It is not difficult to discern the significance of precipitation distribution on the map of world climates (Fig. I-7). Determining climatic regions, however, has always presented problems for geographers. In the first place, climatic records are still scarce, short-term, or otherwise inadequate in many parts of the world. Second, weather and climate tend to change gradually from place to place, but the transitions must be presented as authoritative-looking lines on the map. In addition, there is always room for argument concerning the criteria to be used and how these criteria should be weighed. Vegetation is a response to prevailing climatic conditions. Should boundary lines between climate regions, therefore, be based on vegetative changes observed in the landscape no matter what precipitation and temperature records show? The debate still goes on. World vegeta-

WORLD CLIMATES
After Köppen-Geiger

A HUMID EQUATORIAL CLIMATE

Af no dry season

Am short dry season

Aw dry winter

B DRY CLIMATE

BS semiarid } h=hot
BW arid   } k=cold

C HUMID TEMPERATE CLIMATE

Cf no dry season

Cw dry winter

Cs dry summer

a=hot summer
b=cool summer
c=short cool summer
d=very cold winter

D HUMID COLD CLIMATE

Df no dry season

Dw dry winter

E COLD POLAR CLIMATE

E tundra and ice

H unclassified highlands

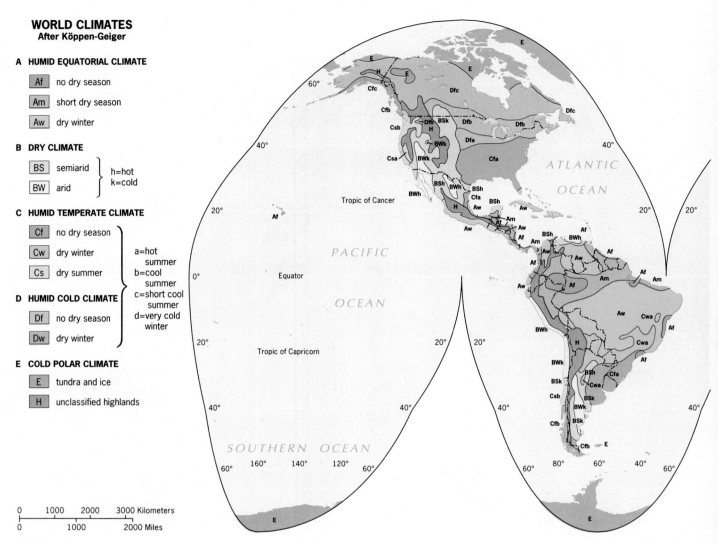

FIGURE I-7

---

tion is mapped in Fig. I-8: compare it to Fig. I-7 and draw your own conclusions!

Figure I-7 displays a regionalization system devised by Wladimir Köppen and modified by Rudolf Geiger. This scheme, which has the advantage of comparative simplicity, is represented by a set of letter symbols. The first (capital) letter is the critical one: the *A* climates are humid and tropical; the *B* climates are dominated by dryness; the *C* climates are humid and comparatively mild; the *D* climates reflect increasing coldness; and the *E* climates mark the frigid polar and near-polar areas.

Humid Equatorial (A) Climates

The humid equatorial, or tropical, climates are marked by high temperatures all year and by heavy precipitation. In the *Af* subtype, the rainfall arrives in substantial amounts

every month; but in the *Am* areas, there is a sudden enormous increase due to the arrival of the annual wet *monsoon* (the Arabic word for "season" [see p. 468]). The *Af* subtype is named after the vegetation association that develops there—the tropical rainforest. The *Am* subtype, prevailing in part of peninsular India, in a coastal area of West Africa, and in sections of Southeast Asia, is appropriately referred to as the monsoon climate. A third tropical climate, the savanna (*Aw*), has a wider daily and annual temperature range and a more strongly seasonal distribution of rainfall. As Fig. I-6 indicates, savanna rainfall totals tend to be lower than those in the rainforest zone, and the associated seasonality is often expressed in a "double maximum." This means that each year produces two periods of increased rainfall separated by pronounced dry spells. In many savanna zones, inhabitants refer to the "long rains" and the "short rains" to identify those seasons; a persistent

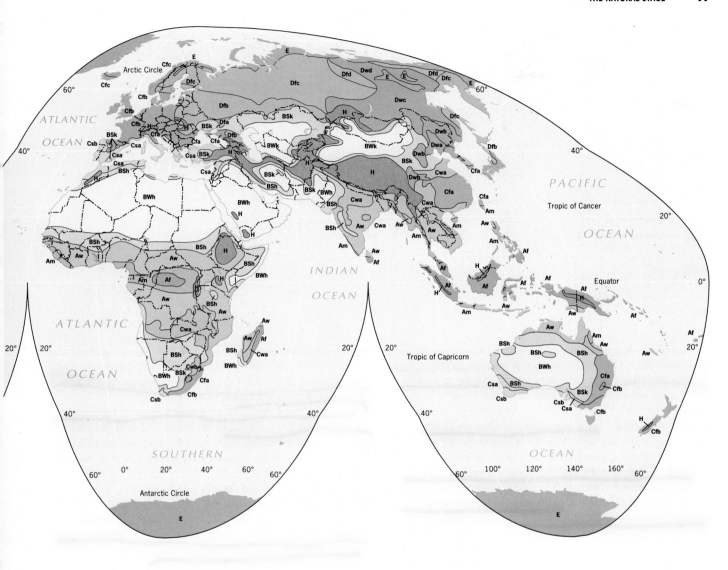

problem in these regions is the unpredictability of the rain's arrival. Savanna soils are not among the most fertile, and when the rains fail the specter of hunger arises on these humid grasslands. Savanna regions are far more densely peopled than rainforest areas, and millions of residents of the savanna subsist on what they manage to cultivate. Rainfall variability under the savanna regime is their principal environmental problem.

Dry (B) Climates

Dry climates occur in lower as well as higher latitudes. The difference between the *BW* (true desert) and the moister *BS* (semiarid steppe) varies but may be taken to lie at about 10 inches (25 cm) of annual precipitation. Parts of the central Sahara in North Africa receive less than 4 inches (10 cm) of rainfall. A pervasive characteristic of

the world's arid areas is an enormous daily temperature range, especially in subtropical deserts. In the Sahara, there are recorded instances of a maximum daytime shade temperature of over 120°F (49°C) followed by a nighttime low of 48°F (9°C). Soils in these arid areas tend to be thin and poorly developed; soil scientists have an appropriate name for them—aridisols.

Humid Temperate (C) Climates

As the map shows, these mid-latitude climate areas almost all lie just beyond the Tropics of Cancer and Capricorn (23½° North and South latitude, respectively.) This is the prevailing climate in the southeastern United States from Kentucky to central Florida, on North America's west coast, in most of Europe and the Mediterranean, in southern Brazil and northern Argentina, in coastal South Africa and Aus-

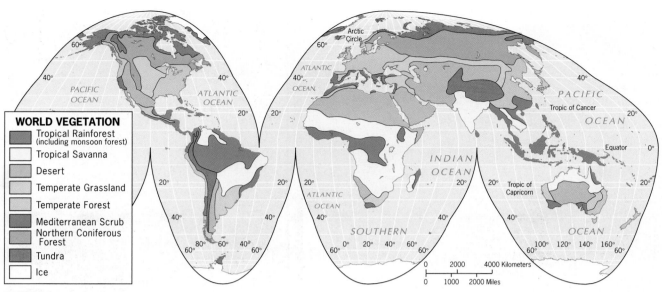

WORLD VEGETATION
- Tropical Rainforest (including monsoon forest)
- Tropical Savanna
- Desert
- Temperate Grassland
- Temperate Forest
- Mediterranean Scrub
- Northern Coniferous Forest
- Tundra
- Ice

FIGURE I-8

tralia, and in eastern China and southern Japan. None of these areas suffers climatic extremes or severity, but the winters can be fairly cold, especially away from temperature-moderating water bodies. These areas lie about midway between the winterless equatorial climates and the summerless polar zones. Some fertile and productive soils have developed under this regime, as we will note in our discussion of the North American and European realms.

The humid temperate climates range from quite moist, as along the densely forested coasts of Oregon, Washington, and British Columbia, to relatively dry, as in the so-called Mediterranean (dry-summer) areas that include not only coastal southern Europe and northwestern Africa but also the southwestern tips of Australia and Africa, central Chile, and Southern California. In these Mediterranean environments, the scrubby, moisture-preserving vegetation creates a natural landscape very different from that of richly green Western Europe.

Humid Cold (*D*) Climates

The humid cold (or "snow") climates may be called the continental climates, for they seem to develop in the interior of large landmasses, as in the heart of Eurasia or North America. No equivalent land areas at similar latitudes exist in the Southern Hemisphere; consequently, no *D* climates occur there at all.

Great annual temperature ranges mark these humid continental climates, and very cold winters and relatively cool summers are the rule. In a *Dfa* climate, for instance, the warmest summer month (July) may average as high as 70°F (21°C), but the coldest month (January) might average only 12°F (-11°C). Total precipitation, a substantial part of which comes as snow, is not very high, ranging from over 30 inches (75 cm) to a steppe-like 10 inches (25 cm). Compensating for this paucity of precipitation are cool temperatures that inhibit the loss of moisture via evaporation and evapotranspiration.

Some of the world's most productive soils lie in areas under humid cold climates, including the U.S. Midwest, parts of southern Russia and Ukraine, and Northeast China. The period of winter dormancy (when all water is frozen) and the accumulation of plant debris during the fall combine to balance the soil-forming and enriching processes. The soil differentiates into well-defined nutrient-rich layers, and a substantial store of organic humus accumulates. Even where the annual precipitation is light, this environment sustains extensive coniferous forests.

Cold Polar (*E*) and Highland (*H*) Climates

Cold polar (*E*) climates are differentiated into true icecap conditions, where permanent ice and snow keep vegetation from gaining a foothold, and the tundra, where up to four months of the year may have average temperatures above freezing. Like rainforest, savanna, and steppe, the term tundra is vegetative as well as climatic, and the boundary between the *D* and *E* climates in Fig. I-7 corresponds quite closely to that between the northern coniferous forests and the tundra (Fig. I-8). Finally, the *H* climates—unclassified highlands mapped in gray—resemble the *E* climates in a number of ways. High elevations and the complex topography of major mountain systems often produce near-Arctic climates above the tree line, even in the lowest latitudes (such as the Andes in western equatorial South America).

Later, when we investigate the reasons behind the location of the world's great clusters of human population, we should remind ourselves of the importance of water to

Nowhere is human settlement tied more closely to water supplies than in Egypt's arid Nile valley. Note how sharply the desert (foreground) is demarcated from the irrigated strip of land that parallels the Nile.

human life—throughout history. The valley and delta of the lower Nile River in northeastern Africa, the basin of the Ganges River in India and Bangladesh, and the plain of the Huang He (Yellow River) in China all contain recent alluvial sols, and through their nearly legendary fertility they sustain many millions of people. Today, more than 95 percent of Egypt's 58 million people live within 12 miles (20 km) of the Nile waterway (see photo above). Hundreds of millions of Indians and Chinese depend directly on alluvial soils in the river basins of the Ganges and Huang He, where crops are grown that range from corn to cotton, wheat to jute, rice to soybeans. These are only the most prominent examples of such alluvium-based agglomerations; in Pakistan, Bangladesh, Vietnam, and many other countries, the alluvial soils of river valleys provide in abundance what the generally infertile upland soils do not.

These riverine population clusters also provide a link with humanity's past. In the fertile valleys of Southwest Asia's rivers, the art and science of irrigation may first have been learned. And two of the world's oldest continuous cultures—Egypt and China—still retain heartlands near their ancient geographic hearths of thousands of years ago.

◆ REGIONS AND CULTURES

Whenever we explore a geographic realm or region, it is important to assess the physical stage that forms its base—its total physical geography or **physiography**. Still, the realms and regions to be discussed in the chapters that follow are defined by human-geographic criteria, and one of those criteria, culture, is especially significant. Therefore, we should carefully consider the culture concept in a regional context.

Anthropologists, when they define the term **culture**, tend to concentrate on abstractions: on learning, knowledge and its transmission, and behavior. Ralph Linton, more than a half century ago, defined culture as "the sum total of the knowledge, attitudes, and habitual behavior patterns shared and transmitted by the members of a society." Marvin Harris, in 1971, wrote that culture is "the learned patterns of thought and behavior characteristic of a population or society." Literally hundreds of definitions of the culture concept exist; as with the regional concept in geography, we are reminded that such definitions are arbitrary and designed for particular purposes.

Geographers are most interested in the imprints of culture and its associated patterns of behavior on the landscape. Thus, we will examine how members of a society perceive and exploit their available resources, the way they maximize the opportunities and adapt to the limitations of their natural environment, and the way they organize the portion of the earth that is theirs. Human works remain etched on the earth's surface for a very long time: Egyptian pyramids, Chinese walls, and Roman roads and bridges still mark countrysides millennia later. Over time, regions take on certain dominant qualities that collectively constitute a regional character, a personality, a distinct atmosphere. This is one of the crucial criteria in our division of the human world into major geographic realms and regions.

CULTURAL LANDSCAPES

As we noted, geographers are particularly concerned with the impress of culture upon the earth's physical surface. Culture is expressed in many ways as it gives visible character to a region. Often a single scene in a photograph or picture can reveal to us, in general terms, in which part of the world it was made. The architecture, forms of transportation, goods being carried, and clothing of the people (all this is part of culture too) reveal enough to permit a guess as to which region of the world is represented. This is so because the people of any culture are active agents of change; when they occupy their portion of the earth's available space, they transform the land by building structures on it, creating lines of transport and communication, parceling out the fields, and tilling the soil (among countless other activities).

This composite of human imprints on the earth's surface is called the cultural landscape, a term that came into general use in geography during the 1920s. Carl Ortwin Sauer (for several decades a professor of geography at the University of California, Berkeley) developed a school of cultural geography that was focused around the concept of **cultural landscape**. In a paper written in 1927 and entitled "Recent Developments in Cultural Geography," Sauer proposed his most straightforward definition of the cultural landscape: *the forms superimposed on the physical landscape by the activities of man*. He stressed that such forms result from the operation of cultural processes—causal forces that shape cultural patterns—that unfold over a long period of time and involve the cumulative influences of successive occupants.

Sometimes these successive groups are not of the same culture. Settlements built by European colonizers over a century ago are now occupied by Africans; minarets of Islam still rise above the buildings of certain Eastern European cities, recalling an earlier period of dominance by the

Cultural landscape in East Africa: a Kikuyu village in Kenya's highlands. In the fields you can see sisal, bananas, corn, and vegetables.

Muslim Ottoman Empire. In 1929, Derwent Whittlesey introduced the term *sequent occupance* to categorize these successive stages in the evolution of a region's cultural landscape (this concept is explored further on page 442).

The durability of the concept of cultural landscape is underscored by its redefinition in 1984 by the eminent scholar John Brinckerhoff Jackson. His statement closely parallels Sauer's: *a composition of man-made or man-modified spaces to serve as infrastructure or background for our collective existence*. Thus the cultural landscape consists of buildings and roads and fields—and more. But it also possesses an intangible quality, an atmosphere or flavor, a sense of place that is often easy to perceive yet difficult to define. The smells and sights and sounds of a traditional African market are unmistakable, but try recording those qualities on maps or in some other objective way for comparative study!

More concrete properties are easier to observe and record. Take, for instance, the urban "townscape"—a prominent element of the overall cultural landscape—and compare a major U.S. city with, say, one in Japan. Visual representations of these two metropolitan scenes would reveal the differences quickly, of course, but so would maps of the two urban places. The American city with its rectangular layout of the *central business district* (*CBD*) and its widely dispersed, now-heavily-urbanized suburbs contrasts sharply with the clustered, space-conserving Japanese metropolis. Using a rural example, the spatially lavish subdivision and ownership patterns of American farmland look unmistakably different from the traditional African countryside, with its irregular, often tiny patches of land surrounding a village (see photo at left). Still, the totality of a cultural landscape can never be captured on a photograph or map because the personality of a region involves far more than its prevailing spatial organization; one must also include its visual appearance, its noises and odors, the shared experiences of its inhabitants, and even their pace of life.

CULTURE AND ETHNICITY

Language, religion, and other cultural traditions often are durable and persistent. Culture is not necessarily based in ethnicity, however, and as we study the world's human-geographic regions, we should be aware that peoples of different ethnic stocks can achieve a common cultural landscape, while people of the same ethnic background can be divided along cultural lines.

The recent events in the former Eastern European country of Yugoslavia are a good case in point. As the old Yugoslavia broke up in the early 1990s, its component parts were first affected and then engulfed by what was often described as "ethnic" conflict. When the crisis reached Bosnia-Herzegovina in the heart of Yugoslavia, three groups fought a bitter civil war. These groups were identified as Bosnians, Serbs, and Muslims—but in fact they were all ethnic Slavs (Yugoslavia means "Land of the South Slavs".) What distanced them from one another, and what kept the conflict going, was cultural tradition, not ethnicity. The Bosnians and the Serbs had developed different communities in different parts of the former Yugoslavia. The Muslims were descendants of Slavs who, a century ago or more, had been converted by the Turks (who once ruled here) to Islam. All three of these groups feared domination by the others, and so South Slav turned against South Slav in what was, in truth, a culture-based conflict.

Even though the post-Cold War world is but a few years old, it has already witnessed numerous intraregional conflicts; in later chapters, we will explain some of these in geographic context. Not all of them have ethnicity at their roots. Culture is a great unifier; it can also be a powerful divider.

◆ REALMS OF POPULATION

Earlier we noted that population numbers by themselves do not define geographic realms or regions. Population distributions, and the functioning society that gives them common ground, are more significant criteria. That is why it is possible to identify one geographic realm with barely more than 20 million people (Australia-New Zealand) and another with 1.3 billion inhabitants (the Chinese Realm). Neither population numbers nor territorial size alone can delimit a geographic realm.

Nevertheless, the map of world population distribution (Fig. I-9) suggests the relative location of several of the world's geographic realms, based on the strong clustering in certain areas. Before examining these clusters in some detail, let us remember that the earth's human population is fast approaching 6 billion—6 thousand million people confined to the landmasses that constitute less than 30 percent of our planet's surface, much of which is arid desert, rugged mountain terrain, or frigid tundra. (Remember that Fig. I-9 is an early-1990s still picture of an ever-changing scene; the explosive growth of humankind continues.) After thousands of years of relatively slow growth, world population during the past two centuries has been expanding at an increasing rate. It took about 17 centuries following the time of the birth of Christ for the world to add 250 million people; now that same number is being added about every *two and a half years*!

Demographic (population-related) issues will arise repeatedly as we survey the world's most crowded realms in later chapters. For the moment we will confine ourselves to the overall global situation, and for that purpose it is interesting to compare the map of world population distribution to the demographic data arrayed in Appendix A (pp. A-1–A-5). This table provides information about the total populations in each of the geographic realms as well as their individual countries. Also, if you compare Fig. I-9 to the maps of terrain (Fig. I-4), precipitation (Fig. I-6), and climate (Fig. I-7) we have already encountered, you will note that the world's largest population concentrations lie in the fertile basins of major rivers (China's Huang He and Chang Jiang, India's Ganges). We live in a modern world, but old ways of life still prescribe the location of tens of millions on this earth.

Figure I-9 reveals the locations of three major population agglomerations in Eurasia, a smaller cluster in eastern North America, and still smaller concentrations in Middle and South America and Africa. The populations of individual countries in these realms are represented on a specially transformed map or *cartogram* (Fig. I-10). This cartogram is not based on traditional representations of scale or area; instead, countries are drawn in proportion to their population so that those containing large numbers are "blown up" in population-space while those containing smaller num-

FIGURE I-9

bers are "shrunk" in size. This method reveals the bulk of China's and India's populations compared to diminutive Australia. It allows us at a glance to gauge relative population sizes rather than countries' territorial dimensions.

THE LEADING POPULATION CLUSTERS

The world's greatest population cluster, *East Asia*, lies centered on China and includes the Pacific-facing Asian coastal zone from the Korean Peninsula to Vietnam. The map indicates that the number of people per unit area—the **population density**—tends to decline from the coastal zone toward the interior. Note, however, the ribbon-like extensions marked *A* and *B* (Fig. I-9). These, as the map of world landscapes (Fig. I-4) confirms, are populations concentrated in the valleys of China's major rivers, the Huang He and the Chang Jiang (the Yellow and Yangzi [or Long], respectively). Here in East Asia, the great majority of the people are farmers, not city dwellers. True, there are great

cities in China (such as Beijing and Shanghai), but the total population of these urban centers is far outnumbered by the farmers—those who live and work on the land and whose crops of rice and wheat feed not only themselves but also the people in those cities.

The *South Asia* population cluster lies centered on India and includes the populous neighboring countries of Bangladesh and Pakistan. This huge agglomeration of humanity focuses on the broad plain of the lower Ganges River (*C* in Fig. I-9). As the cartogram (Fig. I-10) shows, this cluster is nearly as large as that of East Asia and at present growth rates will overtake East Asia early in the next century. As in East Asia, the overwhelming majority of the people are farmers, but here in South Asia the pressure on the land is even greater, and farming is not as efficient. As we note in Chapter 9, the population issue looms very large in this realm.

The third-ranking population cluster, *Europe*, also lies on the world's biggest landmass but at the opposite end from China. The European cluster, including western Rus-

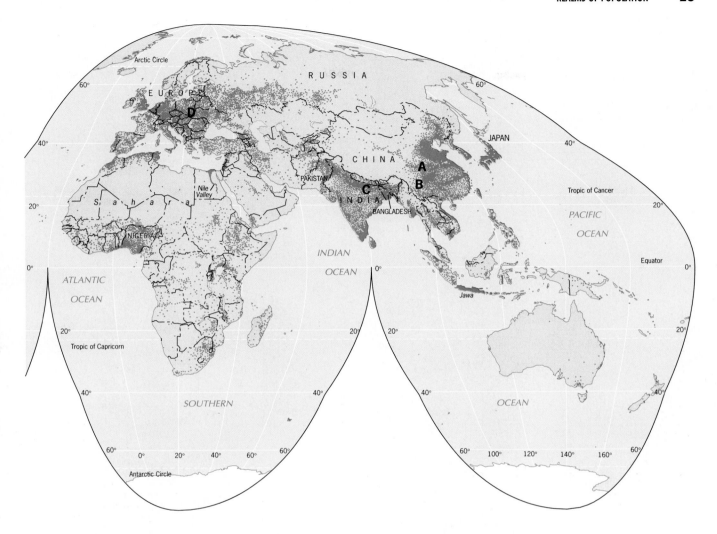

sia, counts over 700 million inhabitants, which puts it in a class with the two larger Eurasian concentrations—but there the similarity ends. Here, the key to the linear, east-west orientation of the axis of population (*D* in Fig. I-9) is not a fertile river basin but a zone of raw materials for industry. Europe is among the world's most highly urbanized and industrialized realms, its human agglomeration sustained by forges and factories rather than paddies and pastures.

The three world population concentrations just discussed (East Asia, South Asia, and Europe) account for about 3 billion of the world's more than 5.6 billion people. Nowhere else on the globe is there any population cluster with a total of even half of any of these. Look at the dimensions of the landmasses on Fig. I-9 and consider that the populations of South America, Subsaharan Africa, and Australia together total less than that of India alone. In fact, the next-ranking cluster is *Eastern North America*, comprising the east-central United States and southeastern Canada; however, it is only about one-quarter the size of the small-

est of the Eurasian concentrations. As Fig. I-9 clearly shows, this region does not possess the large, contiguous, high-density zones of Europe or East and South Asia. Like the European cluster, much of the population of this region is concentrated in several major metropolitan centers; the rural areas remain relatively sparsely settled. The heart of the North American cluster lies in the urban complex that lines the U.S. northeastern seaboard from Boston to Washington, D.C., and includes New York, Philadelphia, and Baltimore—the great multi-metropolitan agglomeration urban geographers refer to as *Megalopolis*.

LESSER POPULATION CONCENTRATIONS

Smaller clusters of population have developed in other parts of the world. In Southeast Asia, the island of Jawa (Java), with an estimated 110 million inhabitants, constitutes a significant human agglomeration. In Africa south of the Sahara, the major cluster lies in the continent's largest

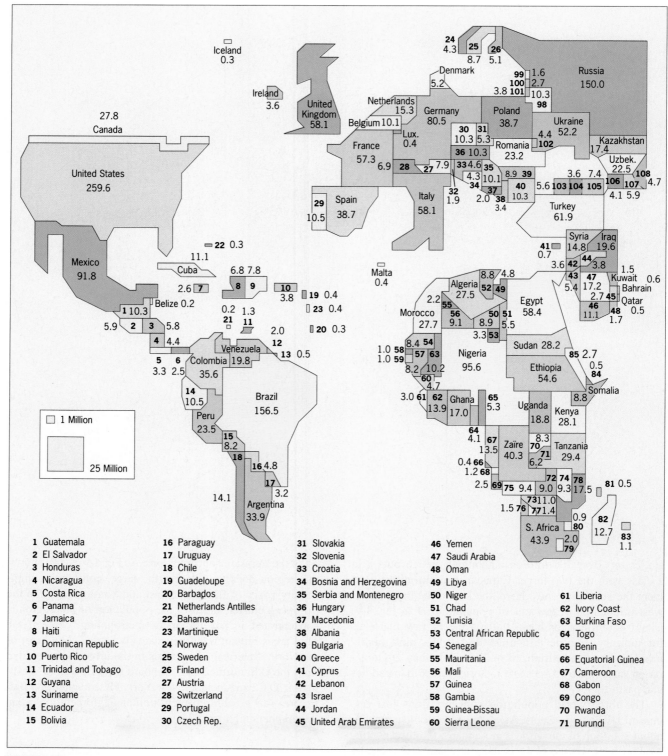

FIGURE I-10

1 Guatemala	16 Paraguay	31 Slovakia	46 Yemen
2 El Salvador	17 Uruguay	32 Slovenia	47 Saudi Arabia
3 Honduras	18 Chile	33 Croatia	48 Oman
4 Nicaragua	19 Guadeloupe	34 Bosnia and Herzegovina	49 Libya
5 Costa Rica	20 Barbados	35 Serbia and Montenegro	50 Niger
6 Panama	21 Netherlands Antilles	36 Hungary	51 Chad
7 Jamaica	22 Bahamas	37 Macedonia	52 Tunisia
8 Haiti	23 Martinique	38 Albania	53 Central African Republic
9 Dominican Republic	24 Norway	39 Bulgaria	54 Senegal
10 Puerto Rico	25 Sweden	40 Greece	55 Mauritania
11 Trinidad and Tobago	26 Finland	41 Cyprus	56 Mali
12 Guyana	27 Austria	42 Lebanon	57 Guinea
13 Suriname	28 Switzerland	43 Israel	58 Gambia
14 Ecuador	29 Portugal	44 Jordan	59 Guinea-Bissau
15 Bolivia	30 Czech Rep.	45 United Arab Emirates	60 Sierra Leone

61 Liberia	
62 Ivory Coast	
63 Burkina Faso	
64 Togo	
65 Benin	
66 Equatorial Guinea	
67 Cameroon	
68 Gabon	
69 Congo	
70 Rwanda	
71 Burundi	

country in terms of population, Nigeria; and another major concentration can be seen to encircle Lake Victoria in East Africa. North of the Sahara, Egypt's Nile valley and delta,

with 58 million inhabitants, resemble in form (if not in size) the river-focused pattern of East and South Asia. In South America, no comparable agglomerations exist. The

CARTOGRAM OF THE WORLD'S NATIONAL POPULATIONS, 1994

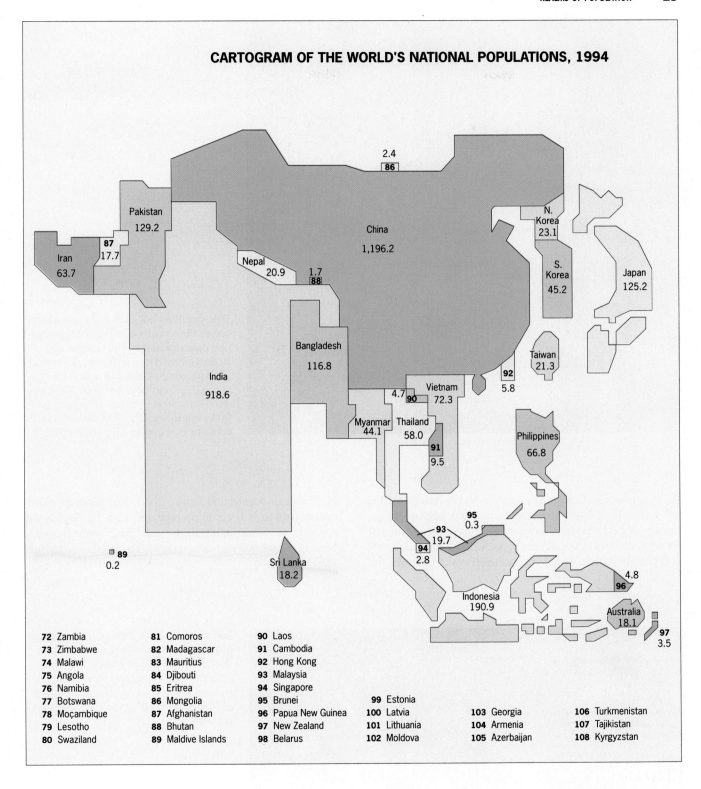

72 Zambia	81 Comoros	90 Laos	
73 Zimbabwe	82 Madagascar	91 Cambodia	
74 Malawi	83 Mauritius	92 Hong Kong	
75 Angola	84 Djibouti	93 Malaysia	
76 Namibia	85 Eritrea	94 Singapore	
77 Botswana	86 Mongolia	95 Brunei	99 Estonia
78 Moçambique	87 Afghanistan	96 Papua New Guinea	100 Latvia
79 Lesotho	88 Bhutan	97 New Zealand	101 Lithuania
80 Swaziland	89 Maldive Islands	98 Belarus	102 Moldova

103 Georgia	106 Turkmenistan
104 Armenia	107 Tajikistan
105 Azerbaijan	108 Kyrgyzstan

sizeable cluster in Middle America is Mexico City, one of the two largest cities in the world but not the core of a major regional human concentration.

Regional geographers are interested not only in the distribution of population but also in its rate of growth. Of the four largest clusters on the world population map,

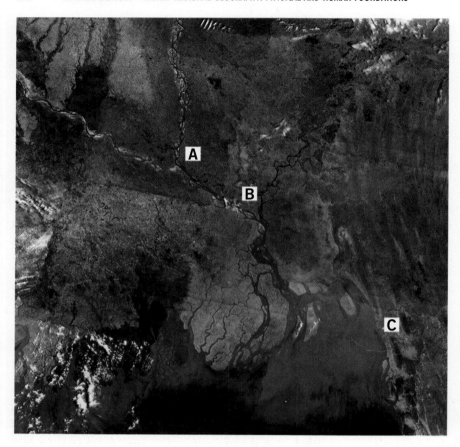

This satellite view of Bangladesh covers one of the world's densest rural population concentrations. The Ganges and Brahmaputra rivers join at *A*; the capital of Bangladesh, Dhaka, is at *B*; Chittagong, the port city, lies at *C*. Silt fills the river waters of this double delta and flows out to sea. The area shown here contains as many as 150 million people.

the two smallest (Europe and Eastern North America) also are growing slowly compared to the two largest, where growth is faster. Size discrepancies among these concentrations, therefore, are increasing as well. In the mid-1990s it appeared that the South Asia cluster would overtake the East Asia concentration before 2010. The issue of population growth will come up repeatedly as we discuss the fortunes of countries in slow- and fast-growing regions, because economic and demographic conditions are closely intertwined.

Nighttime satellite view of the conterminous United States and its immediate neighbors: the use of energy is a good indicator of the concentration of people and activities. Note the sharp definition of Megalopolis along the U.S. northeastern seaboard.

URBANIZATION

As our overview of the leading population clusters suggests, urbanization increasingly marks the current stage of humanity's expansion throughout the world—although not everywhere with the same intensity. Since 1950, the population concentrated in cities worldwide has nearly tripled to a total exceeding 2.5 billion, thereby making urbanites of more than 4 out of every 10 people on earth. The significance of this latest increase in global urbanization can be seen in Fig. I-11.

Throughout this century, urban population has been increasing at an accelerating rate. Table I-1 breaks down this recent urbanization by different parts of the world for each decade since 1950. Observe that urban population growth has been universal, with rapid recent advances occurring in the less developed parts of the world. Keep in mind that although the overall *percentage* of urban population is lower in less developed regions, the much greater *absolute* number of people makes even a 1 percent cityward shift a major movement.

Urbanization as a force of change is now so pervasive in contemporary world regional geography that we shall be considering its various manifestations throughout this book. As a prelude, we offer here some basic ideas about the nature of cities—the spatial focal points containing sizeable concentrations of people and activities that are produced by the operation of world urbanization processes.

Truman Hartshorn defines cities as "centers of power and prestige—economic, political, and social . . . they are where the action is in terms of innovations and control." He further points out that cities are agglomerations of people "with a distinctive way of life, in terms of employment patterns and organization" and that they contain "a high degree of specialized and segregated land uses and a wide variety of social, economic, and political institutions that coordinate the use of [urban] facilities and resources." We are also reminded that cities today have expanded into vast *metropolitan* complexes that are composed of the older core or central city and a surrounding outer suburban city (at least in the developed world) that is rapidly enlarging in size and urban function.

Although this definition by an urban geographer captures a wide range of functions and properties, we should also be aware of additional uses of the term on the world scene. Basically, the term *city* is a political designation and refers to a municipal entity that is governed by some kind of administrative organization. Seen more broadly, the largest cities (especially when they serve as capitals) are nothing less than the foci—indeed complete microcosms—of their national cultures. In succeeding chapters we will introduce additional urban terminology and concepts, including a capsule survey of urban geography itself at the beginning of Chapter 3.

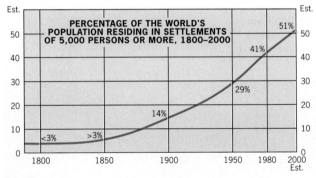

FIGURE I-11

TABLE I-1
Urban Population as a Percentage of Total Population in Different Areas of the World, 1950–2000

Area	2000 (est.)	1990	1980	1970	1960	1950
North America	77	75	74	74	70	64
Europe	77	73	70	67	61	57
Soviet Union*	68	66	63	57	49	39
Latin America	76	72	65	57	49	42
Asia	43	34	26	23	22	16
Africa	41	34	28	23	18	15
WORLD TOTAL	51	45	40	37	34	29

*Former Soviet Union for 2000 estimate.
Source: *World Urbanization Prospects 1990: Estimates and Projections of Urban and Rural Populations and of Urban Agglomerations* (New York: United Nations, Department of International Economic and Social Affairs, ST/ESA/SER.A/121, 1991).

◆ REALMS, REGIONS, AND STATES

Whenever geographers are confronted by the need to define (and justify!) boundaries, they look for precedents—for example, an existing grid that can be put to use or data from a previous effort that may prove helpful in the upcoming challenge. A case in point is the Township-and-Range system of land division in the United States west of the Appalachian Mountains. When you fly over this area, you will see a persistent rectangular pattern on the ground, the product of the Ordinance of 1785, when it was decided to survey and delimit this then-remote area before it would be opened to settlement. The country beyond the Appalachians was divided into squares of 6 by 6 miles, so that each square (or "township") contained 36 square miles. Next, each of these square miles (640 acres) was subdivided into four 160-acre squares, which was to be the smallest parcel of land a settler could purchase (later, smaller parcels were also allowed). Today, the Township-and-Range

system is etched into the entire cultural landscape, having controlled the location of towns, roads, farms, and fields (see photo below). For geographers, the system is a great help: when it comes to delimiting a Corn Belt, a Wheat Belt, or some other economic region, this ready-made grid is invaluable.

On a global scale we have at our disposal only one existing framework, and it is not a regular grid: the international boundary system that marks the territorial limits of the world's 180-plus countries (Fig. I-12). As the map shows, countries range in size from the largest, Russia (6.6 million square miles [17.1 million sq km]), to entities so tiny they cannot be shown at the smallest square used for Fig. I-10. Irregular as this framework may be, it nonetheless helps us in our effort to define regions within the world's great geographic realms. Although we will often refer to the political entities shown in Fig. I-12 as countries, the appropriate geographic term for them is states.

The **state** has been developing for thousands of years, ever since agricultural surpluses made possible the growth

The spatial imprint of the Township-and-Range system on the U.S. landscape, seen here around Lompoc in California's coastal zone. One problem: this land division system is based on the mile and other British measures, creating a legal nightmare when the United States goes metric. It's a geographic reason why the globally-used metric system's adoption here has been slow.

Boundary-making is a global human imperative. The ancient Romans built Hadrian's Wall across Britain to demarcate their domain from that of the Scots to the north. Today this is a relict boundary, but the modern border between England and Scotland lies not far away.

of large and powerful cities that could command hinterlands and control peoples far beyond their walls. But the modern state is a relatively recent phenomenon. Just a little over a century ago, there still were open, unclaimed, unbounded areas that sometimes served as buffer zones between rival states. The boundary framework we see on the world political map today substantially came about during the nineteenth century, and the independence of dozens of former colonies (which made them states as well) occurred during the twentieth century. Today, the *European state model*—a clearly and legally defined territory inhabited by a population governed from a capital city by a representative government—is gaining ground in the aftermath of collapsed colonial and communist empires.

As Figs. I-1 and I-12 suggest, geographic realms mostly are assemblages of states, and the borders between realms frequently coincide with the boundaries between countries—for example, between North America and Middle America along the U.S.-Mexico boundary. But a realm boundary can also cut *across* a state, as in the case of the one between Subsaharan Africa and the Muslim-dominated realm of North Africa/Southwest Asia. Here the boundary takes on the properties of a wide transition zone, but it still divides states such as Chad and Sudan. The transformation

of the margins of the former Soviet Union, too, is creating similar cross-country transitions. Newly independent states such as Belarus (between Eastern Europe and Russia) and Kazakhstan (between Russia and Muslim Southwest Asia) lie in zones of regional change.

Most often, though, geographic realms consist of groups of states whose boundaries also mark the limits of the realms. Look at the case of Southeast Asia, for instance. The northern border of this geographic realm coincides with the political boundary that separates China (a realm practically unto itself) from Vietnam, Laos, and Myanmar (Burma). Its western border is defined by the boundary between Myanmar and Bangladesh (which is part of the South Asian realm). Here, the state boundary framework helps delimit geographic realms.

The world boundary framework is even more useful in the delimitation of regions within geographic realms. We shall discuss regional divisions every time we introduce a geographic realm, but an example is appropriate here. In the Middle America realm, we recognize four regions (see Fig. I-14). Two of these lie on the mainland: Mexico, the giant of the realm, and Central America, the seven comparatively small countries located between Mexico and the Panama-Colombia border (which is also the boundary with

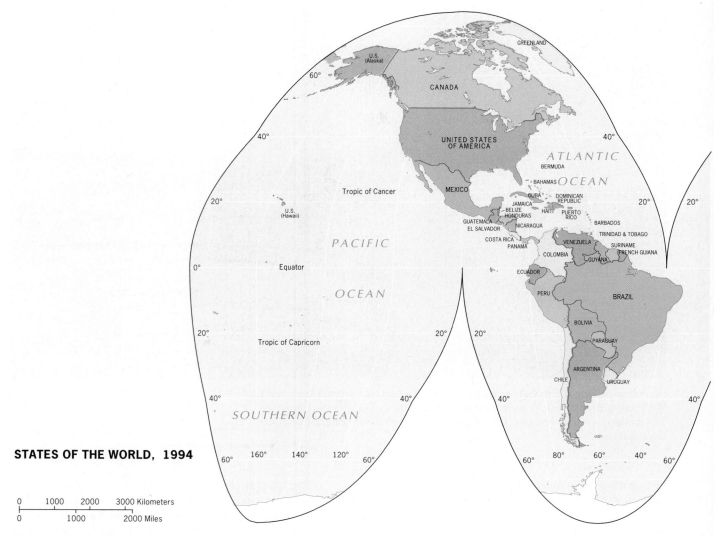

STATES OF THE WORLD, 1994

0	1000	2000	3000 Kilometers
0		1000	2000 Miles

FIGURE I-12

the South American realm). Central America often is mis-defined in news reports; the correct regional definition is based on the politico-geographical framework.

To our earlier criteria of physiography, population distribution, and cultural geography, therefore, we now add political geography as a determinant of world-scale geographic regions. In doing so we should remain aware that the global boundary framework continues to change and that boundaries are created (as in Cyprus between Greeks and Turks) as well as eliminated (between West and East Germany, for example). But the overall system, much of it resulting from colonial and imperial expansionism, has turned out to be quite durable, despite the predictions of some geographers that the "boundaries of imperialism" would be replaced by newly negotiated ones in the post-colonial period.

Toward the end of this book, we will take note of a recent, ominous development in boundary-making: the extension of boundaries onto and into the oceans and seas. It is a process that has been consuming the last of the earth's open frontiers, with consequences that are yet uncertain.

◆ REALMS OF DEVELOPMENT

Finally, we turn to economic geography for a set of criteria that allow us to place countries, regions, and realms in regional groupings based on their level of **development**. The field of economic geography focuses on spatial aspects of the ways people make a living and thus deals with patterns of production, distribution, and consumption of goods and

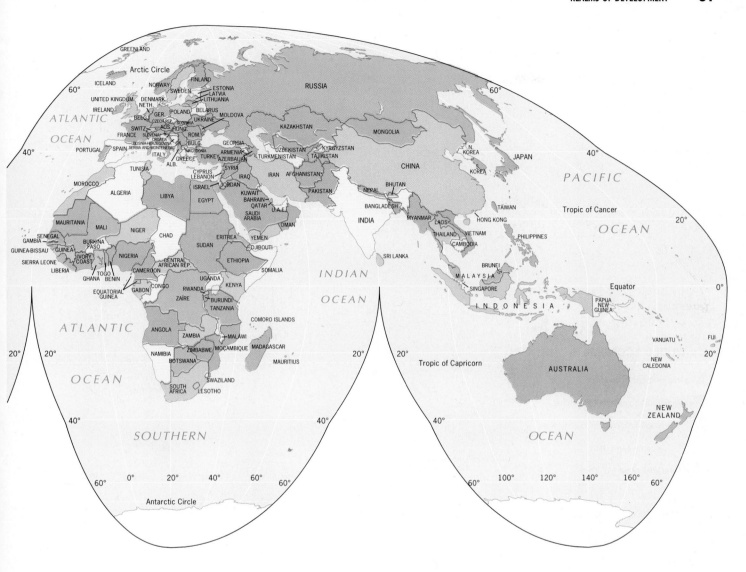

services. As with all else in this world, these patterns exhibit considerable variation. Individual states report their imports and exports, farm and factory production, and many additional economic data to the United Nations and other international agencies. From such information it is possible to determine the comparative economic well-being of the world's countries.

The World Bank, one of the agencies that monitor economic conditions, classifies countries into four groups: (1) high-income, (2) upper-middle-income, (3) lower-middle-income, and (4) low-income countries. These groupings display regional clustering when mapped (Fig. I-13). The high-income economies are concentrated in Europe, North America, and along the western Pacific Rim, including Japan and Australia. The low-income countries dominate in Africa and parts of Asia.

In this book the world's geographic realms are grouped into two main categories: developed realms (Part One) and developing realms (Part Two). The developed realms are dominated by developed countries (DCs), while the developing realms are mainly (though not exclusively) constituted by underdeveloped countries (UDCs). We use these terms because they have become commonplace, but they are synonyms for rich and poor, haves and have-nots, and, perhaps most appropriately, advantaged and disadvantaged. Not long ago, such categories as First (capitalist), Second (socialist), Third (underdeveloped), Fourth (severely underdeveloped), and Fifth (poorest) Worlds were routinely used. However, this kind of categorization has lost much of its relevance in the rapidly changing world of the 1990s. The most meaningful distinction we can presently make is based on the notion of advantage: there are

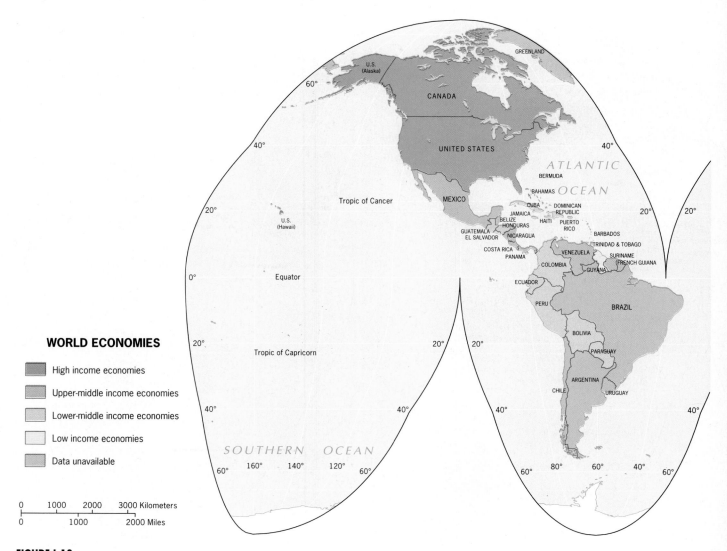

WORLD ECONOMIES

- High income economies
- Upper-middle income economies
- Lower-middle income economies
- Low income economies
- Data unavailable

0   1000   2000   3000 Kilometers
0   1000   2000 Miles

FIGURE I-13

countries that have it and many others that do not. As the gap between these two camps widens (and it *is* widening), the political stability of the world is at risk, and with it the advantages the "haves" have secured.

In Part Two we will have repeated occasion to focus on problems of underdevelopment and on the ways in which some countries manage to improve their situation and achieve what economists call "takeoff"—a pace of economic growth that, if sustained, will lift such countries out of their stagnation and into a higher level of income and development. In certain countries, small nodes or concentrations of rapid growth are creating "islands" of development within otherwise stagnant economies. This, as we note in Chapter 4, is a sign of the times, and it is especially evident today along the Asian edge of what has become known as the Pacific Rim.

Although we routinely (and arbitrarily) divide our world into developed and developing components, no universally applicable criteria exist to measure development accurately. Leading indexes are grouped under seven headings (see the box entitled "Measures of Development"), but they remain arbitrary and subject to debate. Moreover, what these numerical indexes fail to convey is the time factor. The DCs are the advantaged countries—but why can't the UDCs and "emerging" countries catch up? The issue here is not, as is sometimes suggested, simply a matter of environment, resource distribution, or cultural heritage (a resistance to innovation, for example).

The sequence of events that led to the present division of our world began long before the Industrial Revolution: Europe, even by the middle of the eighteenth century, had laid the foundations for its colonial expansion. The Industrial Revolution then magnified Europe's demands for raw materials, and its manufactures increased the efficiency of its imperial control. Western countries thereby gained an enormous head start, while colonial dependencies remained

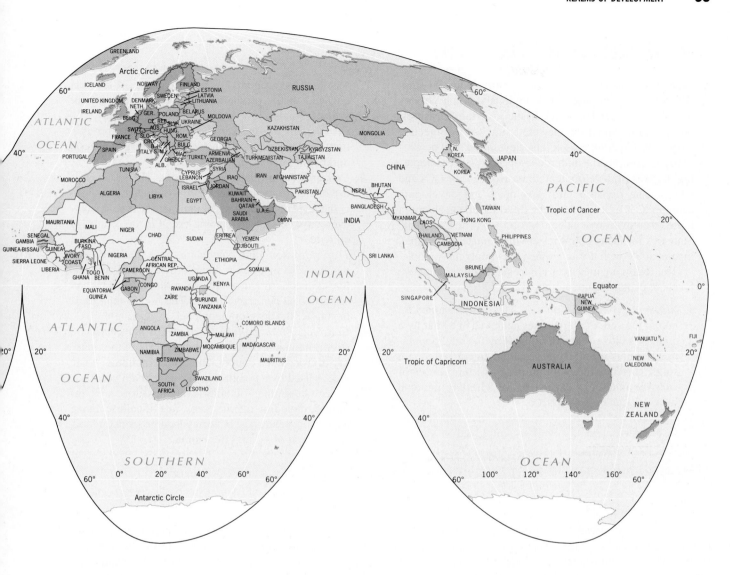

suppliers of resources and consumers of the products of the Western industries. Thus was born a system of international exchange and capital flow that really changed very little when the colonial period came to an end. Developing countries, well aware of their predicament, accused the developed world of perpetuating its long-term advantage through neocolonialism—the entrenchment of the old system under a new guise.

SYMPTOMS OF UNDERDEVELOPMENT

The disadvantaged countries suffer from numerous demographic, economic, and social ills. Their populations tend to have high birth rates and moderate to high death rates; life expectancy at birth is comparatively low (see Chapter 9 and Appendix A). A large percentage of the population, often as much as half, is 15 years of age or younger. Infant mortality is high. Nutrition is inadequate; diets are not well balanced; protein deficiency is a common problem. The incidence of disease is high; health-care facilities are inadequate; there is an excessively high number of persons per available doctor; and hospital beds are too few in number. Sanitation is poor. Substantial numbers of school-age children do not go to school; illiteracy rates are high.

Rural areas are overcrowded and suffer from poor surface communications. Men and women do not share fairly in the work that must be done; women's workloads are much heavier, and children are pressed into the labor force at a tender age. Landholdings are often excessively fragmented, and the small plots are farmed with outdated, inefficient tools and equipment. The main crops tend to be cereals and roots; protein output is low because its demand on the available land is higher. There is little production for the local market because distribution systems are poorly organized and demand is weak. On the farms, yields per

MEASURES OF DEVELOPMENT

What distinguishes a developed economy from an underdeveloped (developing) one? Obviously, it is necessary to compare countries on the basis of certain measures; the question cannot be answered simply by subjective judgment. No country is totally developed, and no economy is completely underdeveloped. We are comparing degrees of development when we identify DCs and UDCs. The division into developed and developing economies is arbitrary, and the dividing line is always a topic of controversy. There is also the problem of data. Statistics for many countries are inadequate, unreliable, incompatible with those of others, or simply unavailable.

The following measures are frequently used to gauge levels of economic development:

1. *National product per person.* This is determined by taking the sum of all incomes achieved in a year by a country's citizens and dividing it by the total population. Figures for all countries are then converted to a single currency index for purposes of comparison. In DCs the index can exceed (U.S.) $10,000; in some UDCs, it can be below $100.
2. *Occupational structure of the labor force.* This is given as the percentages of workers employed in various sectors of the economy. A high percentage of laborers engaged in the production of food staples, for instance, signals a low overall development level.
3. *Productivity per worker.* This is the sum of production over the period of a year divided by the total number of persons in the labor force.
4. *Consumption of energy per person.* The greater the use of electricity and other forms of power, the higher the level of national development. These data, however, must be viewed to some extent in the context of climate.
5. *Transportation and communications facilities per person.* This measure reduces railway, road, and airline connections and telephone, radio, television, and so forth to a per capita index. The higher the index, the higher the development level.
6. *Consumption of manufactured metals per person.* A strong indicator of development levels is the quantity of iron, steel, copper, aluminum, and other basic metals used by a population during a given year.
7. *Rates.* A number of additional measures are employed, including literacy rates, rate of caloric intake per person, percentage of family income spent on food, and amount of savings per capita.

---

unit area are low, subsistence modes of life prevail, and the specter of debt hangs constantly over the peasant family. These circumstances preclude investment in such luxuries as fertilizers and soil conservation techniques. As a result, soil erosion and land denudation scar the rural landscapes of many UDCs. Where areas of larger-scale modernized agriculture have developed, they produce for foreign markets with little resultant improvement of domestic conditions.

In the urban areas, appalling overcrowding, poor housing, inadequate sanitation, and a general lack of services prevail. Job opportunities are insufficient, and unemployment is always high. Per capita income is low, savings per person are minimal, and credit facilities are poor. Families spend a very large proportion of their income on food and basic necessities. The middle class remains small; often a substantial segment of that middle-income population consists of foreign immigrants.

These are some of the criteria that signal underdevelopment, and the list is not complete. For example, one of the geographic properties that mark UDCs is the problem of internal regional imbalance. But even in UDCs, there are local exceptions to the general economic situation. The capital city may appear as a skyscrapered model of urban modernization, with thriving farms in the immediate surroundings and factories on the outskirts. Road and rail may lead to a bustling port where luxury automobiles are unloaded for use by the privileged elite. Here in the country's core, the rush of "progress" is evident—but if you travel a few miles into the countryside (or even to the squatter "shacktowns" at the edge of the city), you will probably find that almost nothing has changed. And just as the rich countries become richer and leave the poorer ones farther behind, so the gap between progressing and stagnant regions within UDCs grows larger. This is a problem of global dimensions.

There can be no doubt that the world economic system works to the disadvantage of the UDCs, but sadly it is not the only obstacle that the less advantaged countries face. Political instability, corruptible leaderships and elites, misdirected priorities, misuses of aid, and traditionalism are among the conditions that commonly inhibit development. External interference by interests representing powerful DCs have also had negative impacts on the economic as well as political progress of many UDCs, especially during the period of the Cold War, when domestic strife was magnified into major conflict when the United States and the now-defunct Soviet Union took opposite sides. Angola, Ethiopia, Afghanistan, Vietnam, Nicaragua, and other UDCs suffered incalculably as a result of such superpower involvement.

Variations in the level of development of countries within geographic realms form an important criterion in the establishment of regions. One example is South America, where the three southernmost countries (Chile, Argentina, Uruguay) are well ahead of their northern neighbors Peru, Bolivia, Paraguay, and Brazil. As we note in Chapter 6, there are historic, cultural, and political reasons for delimiting a "southern cone" region in South America; economic geography confirms it. Development contrasts between the majority of countries in Eastern Europe as against those of Western Europe again help justify a regional boundary there (as we will see in Chapter 1).

It is now time to complete the map of world geographic realms (Fig. I-1) by delimiting their internal regions based on the criteria discussed in the preceding sections of this Introduction. That spatial framework will form the basis for the investigations that follow.

◆ THE REGIONAL FRAMEWORK

At the beginning of this Introduction, we outlined a map of the geographic realms of the world (Fig. I-1). We then addressed the task of dividing these realms into regions and enumerated several criteria to be used for that purpose. The result is Fig. I-14. This global framework is much more than a regionalization of cultural landscapes: it also reflects criteria of physical, political, economic, urban, and historical geography, and it is a spatial synthesis of human geography as a whole, not just cultural geography.

Thirteen geographic realms form the structure for our global regional survey. Five of these comprise the developed world (Europe, North America, Japan, Australia-New Zealand, and Russia), although there is uncertainty about the status and future of the Russian economy. The remaining eight realms consist of assemblages of developing countries.

DEVELOPED REALMS AND REGIONS

Europe ①

Europe merits recognition as a world realm despite the fact that it occupies only a small portion of the Eurasian landmass—a segment, moreover, that is largely made up of Eurasia's western peninsular extremities. But Europe's territorial size is no measure of its global significance. No other part of the world is (or ever has been) packed so full of the products of human achievement or has been the

source of so many innovations and revolutions that transformed the world far beyond its borders. Over the past several centuries, the evolution of global interaction consistently focused on European states and their capitals. Time and again—despite internal wars, the loss of colonial empires, and the threat and impact of external competition—Europe has proved to contain the human and natural resources needed to rebound and renew its progress.

Europe in the mid-1990s is embarked on an historic program of unification. Yet Europe's regional identities have actually been strengthened by events following the collapse of the Soviet Union and the end of communist domination over its former Eastern European satellites. Today, Europe displays five regions: the British Isles, Western Europe, Northern Europe, Southern (Mediterranean) Europe, and Eastern Europe.

Russia ②

Russia, the heart of the former Soviet Empire, still is the largest territorial state in the world, even after the loss of its 14 colonies in 1991. The Russian domain was achieved by the tsars, forged into an empire by the Soviets, and ruined by communist mismanagement. World War II dealt the Soviet Union a devastating blow, but it galvanized Russian nationalism, and after the war the state rose to a position of world superpower with a mighty military and a pioneering space program. However, the Soviet Union was a colonial empire, and as Europe's colonial domains fell apart in the postwar era, Russia's increasingly began to show signs of breakdown as well. On December 25, 1991, the Union of Soviet Socialist Republics passed into history.

Russia's vast realm can be divided into many regions, but for our purposes we recognize four: the western Russian Core, centered on Moscow; the Eastern Frontier; Siberia; and the Far East. Each of these, given their dimensions and internal diversity, has subregions (a topic addressed in Chapter 2).

North America ③

The North American geographic realm consists of two countries, the United States and Canada. This postindustrial realm is characterized by high levels of urbanization, sophisticated technology, unmatched mobility, and massive consumption of the planet's resources and commodities. This is also a realm of pluralistic societies troubled by the reality that minorities tend to remain separate from the dominant culture.

North America is divided into 8 regions. The Continental Core is the urban-industrial heartland of the realm; today it is being transformed into a postindustrial complex whose functions are increasingly shared with the South, the Southwest, and the West Coast. The world's most productive grain and livestock farming dominate the Agricultural Heartland. French Canada and the New England/

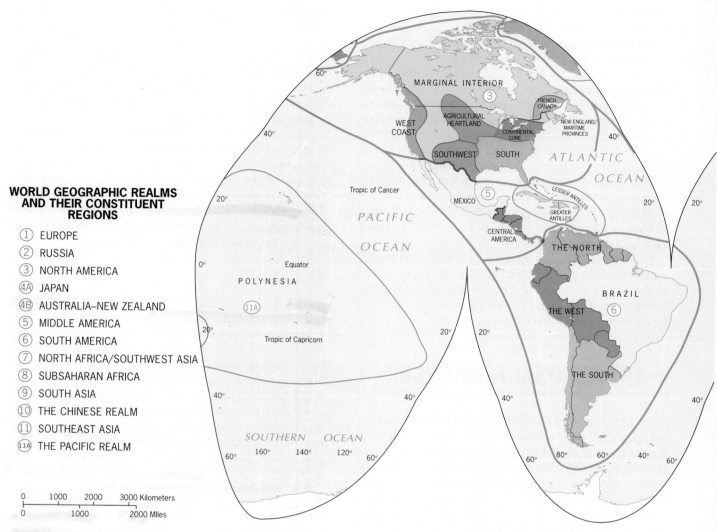

WORLD GEOGRAPHIC REALMS AND THEIR CONSTITUENT REGIONS

① EUROPE
② RUSSIA
③ NORTH AMERICA
④A JAPAN
④B AUSTRALIA–NEW ZEALAND
⑤ MIDDLE AMERICA
⑥ SOUTH AMERICA
⑦ NORTH AFRICA/SOUTHWEST ASIA
⑧ SUBSAHARAN AFRICA
⑨ SOUTH ASIA
⑩ THE CHINESE REALM
⑪ SOUTHEAST ASIA
⑪A THE PACIFIC REALM

FIGURE I-14

Maritime Provinces region are struggling to avoid being left behind as the economic spotlight continues to shift away from northeastern North America. The vast, resource-rich Marginal Interior awaits development in the twenty-first century. Note that many regions straddle the U.S.-Canada border—a harbinger of even greater economic interaction to come if the North American Free Trade zone takes effect in 1994.

Realms on the Western Pacific Rim ④A ④B

As we note in Chapter 4, regional change is underway all along the shores of the Pacific Ocean, especially on its western flank. Here lie two realms with developed economies: Japan ④A and Australia-New Zealand ④B.

Japan constitutes a geographic realm by virtue of its massive economic and financial power, technological capa-

bilities, and global trade linkages. Japan is an urban society, an industrial giant, and a modern hearth of innovation. Japan gained and lost a colonial empire, suffered a devastating defeat in World War II, and faced a postwar prospect of overpopulation and economic disadvantage. This country's rise to global (and realm) status is the story of the second half of the twentieth century.

As the map shows, Japan is not subdivided into regions. This is not merely a reflection of its small territorial size; rather, it reflects Japan's overwhelming culture homogeneity. Japan is a geographic realm *and* a culture region.

In many ways Australia and New Zealand are the reverse of Japan: their combined territory is huge, but their total population is very small. This realm's identity rests on its relative location, cultural heritage, and economic character. Australia's dominant regional division is twofold: the urbanized, oceanfront core area (fragmented into two

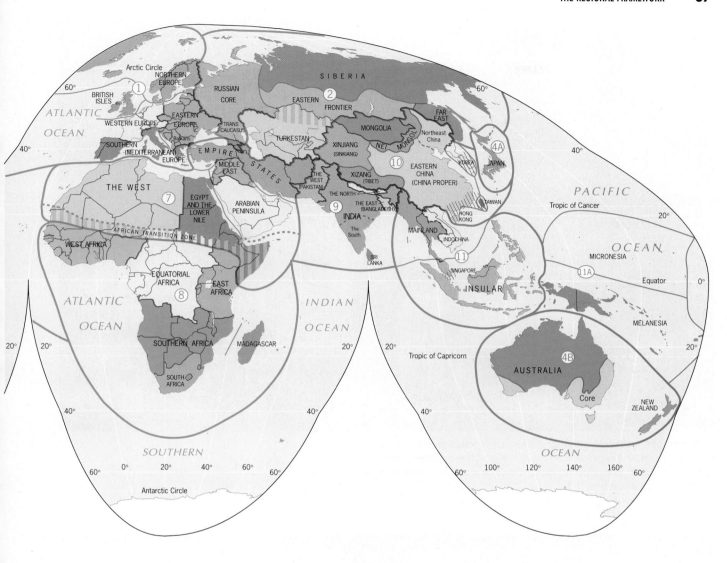

subregions) and the vast, mainly arid interior. As we note in Chapter 4, additional subregions emerge at a larger scale.

DEVELOPING REALMS AND REGIONS

Middle America ⑤

Middle and South America together are often called "Latin" America because of the Spanish and Portuguese imprint that resulted from the European expansion and the destruction of Amerindian cultures. This "Latin" imprint is far more pervasive in South America than in Middle America, however, and it represents one geographic basis for distinguishing between Middle and South American realms.

Middle America consists of four regions. Mexico, the realm's giant, by itself constitutes a region. The seven states of Central America form the second region. The larger islands of the Caribbean (the Greater Antilles) create the third Middle American region, and the arc of smaller islands (the Lesser Antilles) constitutes the fourth.

South America ⑥

The continent of South America also is a geographic realm, one shared by Portuguese-influenced Brazil and a former Spanish colonial domain now divided into nine separate countries. The Roman Catholic religion continues to dominate life and culture, and systems of land ownership, tenancy, taxation, and tribute also were transferred from the Old World to the New. Today, South America still carries the impress of its source region in the architecture of its cities, in the visual arts, in its music, and in other ways.

South America is divided into four regions. These are the Caribbean-facing countries of the North, the Amer-

Tokyo (Japan's capital), situated at the convergence of three tectonic plates in one of our planet's most dangerous earthquake zones, has become a global financial and economic center. The concentration of wealth and investment here in the world's largest city creates major risks.

Cultural landscape in the Muslim world: a minaret rises above the townscape of San'a, capital of Yemen on the Arabian Peninsula. Note the dryness of the countryside and the faults (fractures) in the otherwise horizontal rock layers here on the east wall of the Red Sea rift valley.

indian-influenced West, the comparatively developed South, and the giant Brazil. (In Chapter 6 Brazil is treated at some length in a separate vignette that acknowledges its special place among South America's regions.)

North Africa/Southwest Asia ⑦

This vast and sprawling geographic realm is known by several names, none of them completely satisfactory: the Islamic realm, the Arab world, the dry world. (Some geographers use the term *Naswasia*, and *Afrasia* is occasionally seen as well.) It is true that the Islamic (Muslim) religion is this realm's overriding cultural feature, often vividly expressed in the cultural landscape. It is also true that aridity dominates the natural environment, so that people are clustered where there is water—often in widely separated settlements. Such isolation perpetuates cultural discreteness, and we can distinguish regional contrasts within this realm quite clearly.

There are seven regions, three in North Africa and four in Southwest Asia. Egypt forms a region by itself, and to the west lies the North Africa dominated by the Maghreb (see Chapter 7). Southward across all of Africa lies the transition zone referred to earlier. In Southwest Asia the pivotal region is often called (for want of a more appropriate term) the Middle East, a crescent of five countries lying between the Mediterranean and the Persian Gulf. Southward lies the Arabian Peninsula, and to the north stretches a group of non-Arab (Empire) states in a region from Turkey to Afghanistan. East of the Caspian Sea, anchoring the realm's northeast, lies the group of five Muslim countries (Turkestan) now freed from Soviet communist rule, where Islam is reviving.

Subsaharan Africa ⑧

Between the southern margins of the Sahara and the southernmost cape of South Africa lies the geographic realm of Subsaharan Africa. Its border with the realm to the north is a broad transition zone with several countries straddling it (see Fig. I-14). Elsewhere, however, coastlines mark the limits of this realm, the cradle of humanity, a culturally very distinctive part of the world. European colonialism left its imprints, notably in the political boundary framework and in the export-oriented transport system. But African traditional cultures persisted; many hundreds of languages are spoken here, and numerous community religions endure despite Christian and Islamic penetrations. Africa also is the least developed and least urbanized of the major realms; this decidedly remains a realm of farmers.

The realm consists of four regions: West Africa, East Africa, Equatorial Africa, and Southern Africa. The last of these is dominated by South Africa, where a momentous political and social transformation is under way (the topic of a separate vignette at the end of Chapter 8).

South Asia ⑨

The familiar triangular shape of India outlines a subcontinent in itself—a clearly discernible physical realm bounded by mountain ranges and seas, and inhabited by a population that constitutes one of the greatest human concentrations on earth. The scene of ancient civilizations, this became the cornerstone of the vast British colonial empire. From the colonial period and its often violent aftermath emerged six states: India, Pakistan, Bangladesh, Sri Lanka, Nepal, and Bhutan.

This is a realm of many languages and great religions and consequently of deep cultural divisions. Although India is dominantly a Hindu society, the Muslim minority here is larger than the entire population of any single country in the North Africa/Southwest Asia realm. The South Asian realm consists of five regions: India's Ganges Plain-dominated heartland at its center, Pakistan to the west, the mountainous north comprised of Nepal and Bhutan, Bangladesh in the east, and India's Dravidian South and island Sri Lanka to the south.

The Chinese Realm ⑩

China lies at the heart of a vast East Asian sphere that encompasses the world's largest population agglomeration. Territorially, this realm extends from the Russian border not far from Siberia to the tropical coasts of the South China Sea, and from the westernmost Pacific islands to the deserts and rugged highlands of inner Asia. Demographically, China's 1.2 billion people dominate this realm; historically, they are heirs to one of the world's great ancient culture hearths. The ethnic Chinese still refer to themselves as the "People of Han," the eminent dynasty (207 B.C. to A.D. 220) that presided over the formative period in their national evolution. Yet the cultural individuality and continuity of China were established 2000 years before that time (and perhaps even earlier).

Long-term isolation, protected by desert and distance, was broken by the European invasions of the nineteenth century. In the first half of this century, civil wars, the Japanese intrusion, and the climactic postwar struggle between communists and nationalists kept China in turmoil. Since 1949 China has been a communist state; today, it is the last communist bulwark *and* the last great empire.

This realm's regional organization reflects China's sway. There are six regions: Eastern China (China Proper in Fig. I-14); Xizang (Tibet), under Chinese rule; far western Xinjiang (Sinkiang); Nei Mongol (Inner Mongolia), a part of China; Mongolia, an independent state between China and Russia; and the separate entities on the Pacific Rim—North Korea, South Korea, Taiwan, and Hong Kong. Taiwan, the island haven of the nationalists who were driven from the mainland in 1949, is not under communist Chinese control, but it also is not recognized as a sovereign state by most other countries. And Hong Kong, long a British colony, will be taken over by China in the middle of 1997.

Southeast Asia ⑪

Southeast Asia's corner of the world is a particularly varied mosaic of ethnic and linguistic groups. This realm has been the scene of countless contests for power and primacy and has been called "the Eastern Europe of Asia." During the colonial period, the term *Indochina* came into use to denote the eastern portion of mainland Southeast Asia. The term is a good one because it reflects the major sources of cultural influence that have affected the entire realm. Ethnic affinities are with China, but cultural imprints (Hindu, Buddhist, even Muslim) came from or via India.

Spatially, the realm's discontinuities are quite obvious: it consists of a peninsular mainland (where populations tend to be clustered in river basins) and thousands of islands forming the archipelagoes of Indonesia and the Philippines. The two regions, therefore, are based on this mainland-island distinction (Fig. I-14).

The Pacific Realm ⑪Ⓐ

Between Asia and Australia to the west and the Americas to the east lies the vast Pacific Ocean, larger than all the landmasses of the earth combined. In this great ocean lie tens of thousands of islands, large and small. This fragmented, culturally complex area is the Pacific realm (treated in the vignette that follows Chapter 11).

The Pacific realm has been divided into three regions as a matter of long-term custom. The most populous region, anchored by New Guinea, is named Melanesia (*melas* means black, in reference to the very dark skin, hair, and eyes of the majority of the people). Immediately to the north lies Micronesia (*micro* means small, an apparent reference to the smallness of many of this region's islands). And to the east lies Polynesia (*poly* means many), the huge central Pacific region of numerous islands extending from Hawaii southeastward to Easter Island and as far southwestward as New Zealand. In New Zealand, two realms meet: the majority population of European extraction justifies its inclusion in the Australian-dominated realm, but the native Maori population is one of the larger clusters of Polynesians in the Pacific.

We now have before us the comprehensive framework we need to enlarge our scale and deepen our perspective. Before we begin, one alert: do not let the regional boundaries in Fig. I-14 mislead you into assuming that the realms and regions of the world exist in isolation from one another. We observed previously that regions change, and change results from interaction. A few places in the world have been totally or nearly completely isolated over the past half-century—Cuba, North Korea, and, more significantly, China (between 1949 and 1972). But consider how even these deliberate self-isolations have affected other regions of the world. Cuba generated an exodus of a million refugees from communist rule who transformed the social and cultural geography of metropolitan Miami in nearby South Florida (see photo p. 3). North Korea's militant communism drew an American response along the border with South Korea, where tens of thousands of troops have been stationed as a defensive measure throughout the four decades since the end of the Korean War; the presence of these foreign forces in turn has had a major impact on politics in South Korea. And China's isolation from Western influences did not exclude, at first, a large cadre of Soviet advisers, technicians, and workers, who brought to China equipment ranging from railroad locomotives to bicycles, who helped strengthen China's military, and who bequeathed to the skylines of China's cities the drab architecture that is the hallmark of Russian and other former Soviet cities. So the signs of interaction are everywhere, and in this world isolation no longer works. And yet, as we will see, realms and regions retain their **iconographies**, their unifying symbols of culture and tradition, of ways of doing and building things. Our interconnected world still has its regional neighborhoods.

◆ REGIONAL AND SYSTEMATIC GEOGRAPHY

As this introductory chapter demonstrates, our world regional survey is no mere description of places and areas. We have combined the study of realms and regions with a look at geography's ideas and concepts—the notions, generalizations, and basic theories that make the discipline what it is. We continue this method in the chapters ahead so that we will become better acquainted with the world *and* geography.

By now you will be aware that geography is a wide-ranging, multifaceted discipline. Geography is often described as being a social science, but this tells only half the story: in fact, geography uniquely straddles the divide between the social and the physical (natural) sciences. Many of the ideas and concepts you will encounter relate to the interactions between human societies and natural environments.

To bring order to these diverse interests, geographers have established several *fields*. Among these, the key one is **regional geography**, the unifying field. As Fig. I-15 shows, regional geography lies at the heart of the discipline.

When geographers concentrate on certain aspects of the world—say, cities, economic activities, or politics—they work within one of many topical or **systematic fields**. The geography of cities, for instance, is urban geography. Economic activities are studied under economic geography,

THE RELATIONSHIP BETWEEN
REGIONAL AND SYSTEMATIC GEOGRAPHY

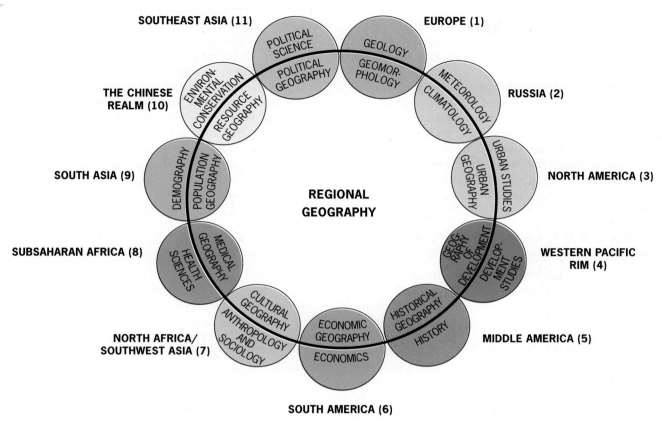

FIGURE I-15

WHAT DO GEOGRAPHERS DO?

A systematic spatial perspective and an interest in regional study are the unifying themes and enthusiasms of geography. Geography's practitioners include physical geographers, whose principal interests are the study of geomorphology (land surfaces), research on climate and weather, vegetation and soils, and the management of water and other natural resources. There also are geographers who concentrate their research and teaching on the ecological interrelationships between the physical and human worlds; they study the impact of humankind on our globe's natural environments and the influences of the environment (including such artificial contents as air and water pollution) on human individuals and societies.

Other geographers are regional specialists, often concentrating their work for governments, planning agencies, and multinational corporations on a particular region of the world. Still other geographers—who now constitute the largest group of practitioners—are devoted to certain topical or systematic subfields such as urban geography, economic geography, cultural geography, and many others (see Fig. I-15); they perform numerous tasks associated with the identification and resolution (through policy-making and planning) of spatial problems in their specialized areas. And, as in the past, there remain many geographers who combine their fascination for spatial questions with technical know-how.

Computerized cartography, geographic information systems, remote sensing, and even environmental engineering are among specializations listed by the 10,000-plus professional geographers of North America. In Appendix C (pp. A-8–A-14), *Opportunities in Geography*, you will find considerable information on the discipline, how one trains to become a geographer, and the many exciting career options that are open to the young professional.

politics under political geography (see the box entitled "What Do Geographers Do?" for an overview of the geography profession and Appendix C for a discussion of career opportunities). As long ago as 1919, geographer Nevin Fenneman drew a diagram that depicted the relationships among geography's various fields, including their linkage to regional geography; Fig. I-15 is a modification of his scheme.

We will highlight 11 of these topical fields of geography in brief essays—each entitled *Focus on a Systematic Field*—near the beginning of each chapter (as indicated along the circumference of Fig. I-15). This will strengthen our understanding not only of the realm under study but also of a systematic field that is especially relevant to it.

Take, for example, the North American realm (Chapter 3): here the systematic field of urban geography is particularly relevant because the United States and Canada, over the past century and a half, each experienced cycles of urban development that transformed their human geography. To know North America's regional geography, therefore, we need to know some basics of urban geography. As the book unfolds, we will thus be building our arsenal of systematic as well as regional knowledge.

And much of what we learn will be found on maps that accompany the text. Reading a map with maximum benefit requires some background, which we provide in Appendix B (pp. A-6-A-7) for easy reference.

◆ P R O N U N C I A T I O N   G U I D E

Alluvial   (uh-LOO-vee-ull)
Antilles   (an-TILL-eeze)
Aquifer   (AK-kwuh-fer)
Bangladesh   (bang-gluh-DESH)
Beijing   (bay-ZHING)
Belarus   (bella-ROOSE)
Bhutan   (boo-TAHN)
Bosnia-Herzegovina   (BOZ-nee-uh hert-suh-goh-VEE-nuh)
Brahmaputra   (brahm-uh-POOH-truh)
Caribbean   (kuh-RIB-ee-un/karra-BEE-un)
Cenozoic   (senno-ZOH-ick)
Chang Jiang   (chahng-jee-AHNG)
Chile   (CHILLI/CHEE-lay)
Coniferous   (kuh-NIFF-uh-russ)
Cyprus   (SYE-pruss)
Dhaka   (DAHK-uh)
El Niño   (ell-NEEN-yoh)
Ethiopia   (eeth-ee-OH-pea-uh)
Fenneman   (FENN-uh-munn)
Fungi   (FUN-jye)
Ganges   (GAN-jeeze)
Han   (HAHN)
Hierarchical   (hire-ARK-uh-kull)

Himalayas   (him-AHL-yuzz/himma-LAY-uzz)
Holistic   (hoh-LISS-tick)
Holocene   (HOLLO-seen)
Homogeneity   (hoh-moh-juh-NAY-eh-tee)
Huang He   (HWAHNG-HUH)
Humus   (HUE-muss)
Islam   (iss-LAHM)
Jawa   (JAH-vuh)
Kalahari   (kalla-HAH-ree)
Kazakhstan   (kuzz-uck-STAHN)
Kikuyu   (kee-KOO-yoo)
Köppen   (KER-pun)
Laos   (LAUSS)
Maghreb   (Mahg-GRAHB)
Maori   (MAH-aw-ree/MAU-ree)
Megalopolis   (meh-guh-LOPP-uh-liss)
Montreal   (mun-tree-AWL)
Muslim   (MUZZ-lim)
Myanmar   (mee-ahn-MAH)
Nei Mongol   (nay-MUNG-goal)
Nepal   (nuh-PAHL)
New Guinea   (noo-GHINNY)
Nicaragua   (nick-uh-RAH-gwuh)
Nigeria   (nye-JEERY-uh)

Niger   (nee-ZHAIR)
Pakistan   (PAH-kih-stahn)
Pangaea   (pan-GAY-uh)
Paraguay   (PAHRA-gwye)
Philippines   (FILL-uh-peenz)
Physiography   (fizzy-OGG-ruh-fee)
Pleistocene   (PLY-stoh-seen)
Quebec   (kwuh-BECK)
San'a   (suh-NAH)
Sauer   (SOUR)
Shanghai   (shang-HYE)
Siberia   (sye-BEERY-uh)
Sisal   (SYE-sull)
Slav   (SLAHV)
Spatial   (SPAY-shull)
Sri Lanka   (sree-LAHNG-kuh)
Steppe   (STEP)
Taiwan   (tye-WAHN)
Ukraine   (yoo-CRANE)
Uruguay   (OO-rah-gwye)
Wegener   (VAY-ghenner)
Xinjiang   (shin-jee-AHNG)
Xizang   (sheedz-AHNG)
Yangzi   (YANG-dzee)
Yemen   (YEMMON)
Zaïre   (zah-EAR)
Zimbabwe   (zim-BAHB-way)

PART ONE
DEVELOPED REALMS

Scale 1: 16 000 000; one inch to 250 miles. Conic Projection
Elevations and depressions are given in feet

CHAPTER 1

RESILIENT EUROPE: CONFRONTING NEW CHALLENGES

IDEAS & CONCEPTS

Relative location	Primate city
Geomorphology	Central business district (CBD)
Infrastructure	Devolution
Areal functional specialization	Supranationalism
Model	Conurbation
Von Thünen's Isolated State	Site
Industrial location	Situation
Centrifugal forces	Acid rain
Centripetal forces	Shatter belt
Nation-state	Balkanization
Spatial interaction principles	Irredentism
Complementarity	
Transferability	
Intervening opportunity	

REGIONS

The British Isles	Mediterranean (Southern) Europe
Western Europe	Eastern Europe
Nordic Europe (Norden)	

For centuries Europe has been the heart of the world. European empires spanned the globe and transformed societies far and near. European capitals were the focal points of trade networks that controlled distant resources. Millions of Europeans migrated from their homelands to the New World as well as to newly settled parts of the Old, creating new societies from North America to Australia.

In agriculture, in industry, and in politics, Europe went through revolutions—and then exported those revolutions throughout the world, serving to consolidate the European advantage. Yet during the twentieth century, Europe twice plunged the world into war. In the aftermath of World War II (1939–1945), Europe's weakened powers lost the colonial possessions that for so long had provided wealth and influence, and the continent was divided by an ideological Iron Curtain. Resilient Western Europe's recovery and Eastern Europe's rejection of communism have been the dominant events of the past few decades.

These internal regional groupings constitute major elements of Europe's human geography, whose overall framework is an intricate mosaic of 37 countries (see map p. 44). *Western Europe* includes the large countries of Germany and France; the so-called Low Countries of Belgium, the Netherlands, and Luxembourg; and the Alpine countries of Switzerland, Austria, and tiny Liechtenstein. *Eastern Europe* consists of the former communist satellites of Poland, the Czech Republic, Slovakia, Hungary, Romania, Bulgaria, Albania, and the remnants of former Yugoslavia—Serbia-Montenegro, Slovenia, Croatia, Bosnia-Herzegovina, and Macedonia; to these 12 countries, we must now also add the five republics of the former Soviet Union that joined Eastern Europe upon independence in 1991—Latvia, Lithuania, Belarus, Moldova, and Ukraine. Three additional regional groupings comprise the remainder of Europe: (1) the offshore *British Isles* (the United Kingdom and Ireland); (2) Mediterranean *Southern Europe* (Spain, Portugal, Italy, and Greece); and (3) Norden or *Northern Europe* (Denmark, Norway, Sweden, Finland, Estonia, and the island country of Iceland). This long list of states notwithstanding, when we add everything together, Europe is still a world geographic realm of quite modest proportions on the peninsular margin of western Eurasia (see Fig. I-14). Yet despite its comparatively small size, for more than 2,000 years the European realm has been a leading focus of human achievement, a hearth of innovation and invention.

Europe's human resources have been matched by its large and varied raw material base; whenever the opportunity or the need arose, the realm's physical geography proved to contain what was required. In fact, for so limited an area (slightly less than two-thirds the size of the United States), Europe's internal natural diversity is probably unmatched. From the warm shores of the Mediterranean to the frigid Scandinavian Arctic, from the flat coastlands of the North Sea to the grandeur of the Alps, and from the moist woodlands and moors of the Atlantic fringe to the semiarid prairies north of the Black Sea, Europe presents an almost infinite range of natural environments. The insular and peninsular west contrasts strongly against a more interior, continental east. A raw-material-laden backbone extends across the middle of Europe from England eastward to Ukraine, yielding

TEN MAJOR GEOGRAPHIC QUALITIES OF EUROPE

1. The European realm consists of the western extremity of the Eurasian landmass, a locale of maximum efficiency for contact with the rest of the world.
2. Europe's lingering and resurgent world influence results largely from advantages accrued over centuries of global political and economic domination.
3. The European natural environment displays a wide range of topographic, climatic, vegetative, and soil conditions and is endowed with many industrial resources.
4. Europe is marked by strong internal regional differentiation (cultural as well as physical), exhibits a high degree of functional specialization, and provides multiple exchange opportunities.
5. European economies are dominated by manufacturing, and the level of productivity has been high; levels of development generally decline from west to east.
6. Europe's nation-states emerged from durable power cores that formed the headquarters of world colonial empires. A number of those states are now bedeviled by internal separatist movements.
7. Europe's population is generally well off, highly urbanized, well educated, enjoys long life expectancies, and constitutes one of the world's three largest population clusters.
8. Europe is served by efficient transport and communications networks that promote extensive trade and other forms of international spatial interaction.
9. Europe has made important progress toward international economic integration. The push toward still stronger and broader coordination continues.
10. The European realm was enlarged on its eastern flank in 1991 after the collapse of the Soviet Union, absorbing the six westernmost republics of the former U.S.S.R.

coal, iron ore, and other valuable minerals. And this diversity is not confined to the physical makeup of the continent. The European realm contains peoples of many different cultural-linguistic stocks, including not only Latins, Germanics, and

RELATIVE LOCATION: EUROPE IN THE LAND HEMISPHERE

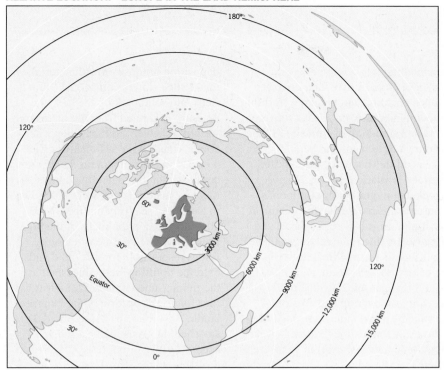

Azimuthal equidistant projection centered on Hamburg

FIGURE 1-1

Slavs but also numerous minorities such as Finns, Hungarians, and various Celtic-speaking groups (as Fig. 1-6 shows).

This diversity of physical and human content alone, of course, is not Europe's greatest asset. If differences in habitat and culture automatically led to rapid human progress, Europe would have had many more competitors for its world position than it did. But in Europe, there were virtually unmatched advantages of scale and geographic proximity.

Globally, Europe's **relative location**—at the heart of the *land hemisphere*—is one of maximum efficiency for contact with the rest of the world (Fig. 1-1). Regionally, Europe is also far more than a mere western extremity of the Eurasian landmass. Almost nowhere is Europe far from that essential ingredient of European development—the sea—and the water interdigitates with the land here as it does nowhere else on earth. Southern and Western Europe consist almost entirely of peninsulas and islands, from Greece, Italy, France, and the Iberian Peninsula (Spain and Portugal) to the British Isles, Denmark, and the Scandinavian Peninsula (Norway and Sweden). Southern Europe faces the Mediterranean, and Western Europe virtually surrounds the North Sea as it looks out over the Atlantic Ocean. Beyond the Mediterranean lies Africa, and across the At-

lantic are the Americas. Europe has long been a place of contact between peoples and cultures, of circulation of goods and ideas. The hundreds of miles of navigable waterways; the easily traversed bays, straits, and channels between numerous islands and peninsulas and the mainland; and the highly accessible Mediterranean, North, and Baltic seas all provided the avenues for these exchanges. Later, even the oceans became routes of long-distance spatial interaction.

This historic advantage of moderate distances applies on the mainland as well. Europe's Alps may form a transcontinental divider, but what they separate still lies in close juxtaposition (Alpine passes have for centuries provided several corridors for overland contact). Consider Rome and Paris: the distance between these long-time control points of Mediterranean and northwestern Europe is less than that between New York and Chicago or Miami and Atlanta. No place in Europe is very far from anyplace else on the continent, although nearby places are often sharply different from each other in terms of economy and outlook. Short distances and large differences make for much interaction—and that has marked the geography of Europe for over a millennium.

GEOMORPHOLOGY

The geographic study of the configuration of the surface of the earth is called **geomorphology**. The term *physiography* also appears in connection with this study, but it has a wider connotation than geomorphology. The contents of physiographic regions (or *provinces*) include not only terrain but also climate and weather, soils and vegetation. Geomorphology, in contrast, concentrates on landscapes and landforms alone.

As in other fields of geography, the study of the earth's surface features can be approached at various levels of detail. Some geomorphologists focus on one particular *landform*, trying to learn what forces and processes have shaped it. Thus, a landform is a single feature; a volcanic mountain, a segment of a river valley, a deposit laid down by a glacier, and a sinkhole are all landforms. Other physical geographers take a broader view and study entire landscapes in an effort to discover what led to their formation and present appearance. A *landscape*, therefore, is an assemblage of landforms (for a discussion of Europe's four landscape regions, see pp. 50–53). Some knowledge of landform formation may help in the analysis of landscapes, but it is not the entire story.

Obviously, geomorphology is a complicated field. To understand why landscapes appear the way they do, it is necessary to know the forces that bend and break the earth's crust *and* the agents of weathering and erosion that constantly alter the exposed surface. The internal crustal movements are caused by *tectonic* forces, and anyone who has experienced even a mild earthquake will know their power. Whole continents are slowly moved by these forces in response to a greater global process, whereby crust is "made" along mid-oceanic ridges and "recycled" where the pieces (*plates*) collide and are pushed under. The con-

tinental landmasses are dragged along as these plates move, their contact zones buckling and breaking (forming mountain belts from Alaska to Argentina and across all of Eurasia) as their trailing edges warp and bend.

As all this happens, the exposed surface is attacked by *weathering* (rock disintegration by continuous temperature change and by chemical action in the moist atmosphere) and by *erosion* (the removal of weathered rock and its further breakdown by streams, glaciers, wind, and waves). But this is not all. The material loosened by weathering and erosion is carried away and then deposited elsewhere, so that although there is wearing down (*degradation*) in some areas, there is accumulation (*aggradation*) in others. A river system such as the Mississippi and its tributaries thus erodes or degrades the landscapes of its interior basin and aggrades the landscape in its lower valley and delta. To keep track of all the forces—tectonic, erosional, depositional—involves much scientific detective work.

Tectonic geomorphic processes work at different rates, accentuating surface relief at any given time. Each process is associated with its own assemblage of landforms and thus leaves a distinct imprint on the landscape. A good example is volcanism, which can produce conical mountains built by the violent ejection and sudden cooling of molten rock on the surface. Volcanoes, like all landforms, are composed of crustal materials that are sculpted by tectonic and erosional forces. The type of rock associated with volcanism is known as *igneous*, which is formed by the cooling and solidification of molten material (called *magma* if it hardens underground, *lava* if it solidifies on the surface).

The second major type of rock, and the most widespread, is known as *sedimentary*, because it is formed by the

deposition of loose eroded material that is then slowly compacted (lithified) into solid rock by the heat and pressure generated from the increasing weight of newer sediments deposited above. South Asia's Himalayas, the world's greatest mountain range, are composed of geologically young sedimentary materials that were laid down in a shallow sea (tiny marine fossils are embedded in the rocks of the highest peaks). The enormous tectonic forces responsible for this rapid and extreme uplifting were unleashed as the Indian subcontinent (cast adrift as the Gondwana portion of the Pangaea landmass) broke apart (see Fig. 8-3), smashed into Asia, and caused an accordion-like crumpling of the crust, thrusting up the Himalayas in a process that still continues as the Australian-Indian Plate grinds against the Eurasian Plate (Fig. I-3).

Such transformations of the surface help create the final type of rock, appropriately called *metamorphic* because it is altered from a preexisting igneous or sedimentary rock by the reintroduction of great heat and pressure. Metamorphism is often associated with horizontal deformation of the crust, which can make sedimentary rock formations plastic, producing warping or *folding* of parallel rock strata. When these forces are too great, however, fracturing or *faulting* occurs, producing formations such as East Africa's rift valleys (see Fig. 8-2). Besides vertical faulting, horizontal faulting can also take place (California's San Andreas Fault is a famous example)—a dislocation that can trigger sudden and devastating earthquake activity.

Degradational processes wear down the land surface through weathering and erosion, reshaping landscapes and creating new ones where significant quantities of eroded materials are deposited. The progressive actions of erosional agents produce an orderly sequence of landforms. The recently uplifted Himalayas, for instance, will eventually be reduced by erosion to a low rounded mountain chain (much like the U.S. Appalachians), bordered on the south by vast depositional plains and plateaus composed of the eroded Himalayan materials transported by *running water* in streams—the most widespread and effective erosional agent (even in deserts).

The steepness of the slope determines in part the work of the river, dragging rock fragments along, carrying others in suspension, and even dissolving some along the way. Eventually the stream's velocity slows down as it approaches its mouth, and aggradation becomes the dominant process. Valleys are filled by sediment, and a delta extends from the river mouth into the sea. Where rock strata are soluble, as in limestone areas, the water not only erodes but also dissolves, producing a unique landscape (known as *karst*) pocked by sinkholes and caves. Water in a different form fashions shorelines as *waves* perform degradational functions. Where the coastline coincides with one of those crunching, mountain-creating plate contacts (as along much of the west coast of North and South America), the deep-water waves attack with great power and sculpt a spectacular high-relief landscape. Where the coastal landscape is plain-like (as in the southeastern United States), the waves deposit sediment and prove that they also can be aggradational agents, forming wide sandy beaches and offshore barrier islands.

Landscapes and landforms associated with *glaciation* usually have complex histories. Where the glaciers formed huge icesheets during the most recent (Pleistocene) glaciation, they scraped the underlying topography into a flat plain, carrying off the rub-

The Alps form Europe's mountain backbone, the realm's divider between north and south. Glaciated during the Pleistocene and snowcapped today, the magnificent scenery of Switzerland's high-relief Alpine landscapes attracts large numbers of visitors.

ble, grinding it down, and depositing it in thick layers far to the south, thereby burying much of the U.S. Midwest. Where the deepening cold caused mountain glaciers to form, as in the Rocky Mountains and European Alps, the ice descended into valleys that were formerly carved by rivers, deepening and widening them.

The erosive power of *wind* is at its greatest in the drier climates, where it picks up and hurls loose surface particles against obstacles, wearing them away by abrasion. Although thick sand deposits are shaped by the wind into various dune formations in certain deserts, the majority of the world's desert surfaces exhibit rocky landscapes. Wind also creates depositional features, the most important of which

is *loess*, a highly fertile windblown silt that can accumulate to great depths (as we will see in Chapter 10).

Geomorphology, a vital inquiry in and of itself, must also be linked to the other elements of physical geography because all together form a dynamically interacting environmental system. Climate is the ultimate generator of degradational agents; soils are intimately related to both geomorphic processes and climatic influences, and to vegetation and hydrography in varying degrees as well. Finally, human influences must also be considered because agricultural land use, urban development, water diversion, surface mining, and the like dominate an ever-expanding artificial landscape.

◆ LANDSCAPES AND OPPORTUNITIES

Europe may be small in areal size, but its physical landscapes are varied and complex. It would be easy to identify a large number of physiographic regions, but in doing so we might lose sight of the broader regional pattern—a pattern that has much to do with the way the European human drama unfolded. Accordingly, Europe's landscapes can be grouped regionally into four units: the Central Uplands, the Alpine Mountains in the south, the Western Uplands, and the great North European Lowland (Fig. 1-2).

The very heart of Europe is occupied by an area of hills and small plateaus, with forest-clad slopes and fertile valleys. These *Central Uplands* also contain the majority of Europe's productive coalfields. When the region emerged from its long medieval quiescence and stirred with the stimuli of the Industrial Revolution, towns on the Uplands' flanks grew into cities, and farms gave way to mines and factories.

The Central Uplands are flanked on the south by the much higher Alpine Mountains (and to the west and north by the North European Lowland). The *Alpine Mountains* include not only the famous Alps themselves (see photo p. 49) but also other ranges that belong to this great mountain system. The Pyrenees between Spain and France (one of Europe's few true barriers), Italy's Appennines, former Yugoslavia's Dinaric ranges, and the Carpathians of Eastern Europe are all part of this Alpine system, which extends even into North Africa (as the Atlas Mountains) and eastward into Turkey and beyond. Although the Alps are rugged and imposing, they have not been a serious obstacle to communication: traders have operated through their mountain passes for many centuries.

Europe's western margins are also quite rugged, but the *Western Uplands* of Scandinavia, Scotland, Ireland, France's Brittany, Portugal, and Spain are not part of the Alpine system (maximum elevations are markedly lower than those in the Alps). This western arc of highlands represents older geologic mountain building, contrasting sharply with the relatively young, still active, earthquake-prone Alpine mountains. Scandinavia's uplands form part of an ancient geologic shield underlain by old crystalline rocks now bearing the marks of the Pleistocene glaciation. Spain's central plateau or *Meseta* is also supported by comparatively old rocks, now worn down to a tableland.

The last of Europe's landform regions is also its most densely populated. The *North European Lowland* (also known as the Great European Plain) extends in a gigantic arc from southwestern France through the Low Countries across northern Germany and then eastward through Poland deep into southern Russia. Southeastern England, Denmark, and the southern tip of Sweden also are part of this region,

EUROPE: THE EASTERN BOUNDARY

The European realm is bounded on the west, north, and south by Atlantic, Arctic, and Mediterranean waters, respectively. Europe's eastern boundary, however, has always been a matter for debate. Some scholars place this boundary at the Ural Mountains (deep inside Russia), thereby recognizing a "European" Russia and, presumably, an "Asian" one as well. Others argue that because there is a continuous transition from west to east (which continues into Russia), there is no point in trying to define any boundary.

Still, the boundary used in this chapter—marking Europe's eastern boundary as the border with Russia—has geographic justifications. Eastern Europe shares with Western Europe its fragmentation into several states possessing distinct cultural geographies, a political condition that sets it apart from the Russian giant to the east. Historical and cultural contrasts also mark large segments of this boundary (see Fig. I-6). The former Soviet Union's postwar hegemony in Eastern Europe—a dominance lasting well back into tsarist times for several western republics of the U.S.S.R. that proclaimed independence in 1991—did not wipe out certain differences in ideology and nationalist loyalties between the two regions. Eastern Europe retains its discrete nationalisms, latent and actual conflicts, and pressures, qualities most vividly demonstrated by the upheavals that have swept through the region since 1989. Despite its own internal disunity, Russia remains Eurasia's territorial giant, several times as large as Europe, historically consolidated by force, and a potentially destabilizing influence on the emerging order of post-communist Europe.

which forms a continuous belt on the mainland from southern France to the plains northeast of the Black Sea (Fig. 1-2). Most of the North European Lowland lies below 500 feet (150 m) in elevation, and local relief rarely exceeds 100 feet (30 m). But make no mistake: this region may be topographically low-lying and flat or gently rolling; however, there its uniformity ends. Beyond this single topographic factor there is much to differentiate it internally. In France, this region includes the basins of three major rivers—the Garonne, Loire, and Seine. In the Netherlands, for a good part it is made up of land reclaimed from the sea, enclosed by dikes and lying below sea level. In southeastern England,

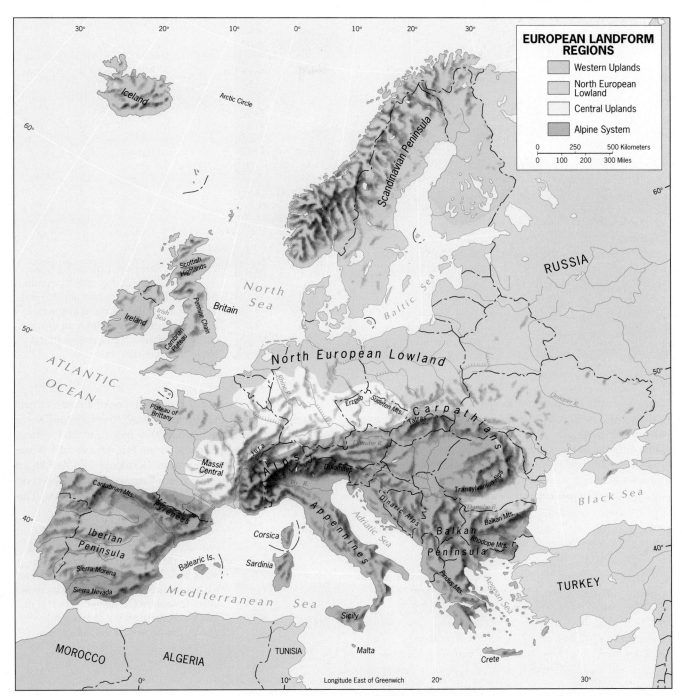

FIGURE 1-2

the higher areas of the Netherlands, northern Germany and Denmark, southern Sweden, and farther eastward, it bears the marks of the Pleistocene glaciation that withdrew only a few thousand years ago (see Fig. I-5). Each of these particular areas affords its own opportunities as soils and climates vary, giving rise to some of the world's most productive (and prestigious) agricultural pursuits.

The North European Lowland has also been one of Europe's major avenues of human contact. Entire peoples have migrated across it; armies have repeatedly marched through it. As settlement took place, agricultural diversity became a hallmark, and land use came to be dominated by intensive farming organized around a myriad of villages from which farmers commuted to their nearby fields. Today, centuries

The North European Lowland's low-relief topography allows intensive cultivation from France to Poland and beyond. This is part of the Lowland in southern Sweden, overlooking a lake. In this generally mountainous country, such flatlands are especially valuable.

later, it is still not possible to speak of a "dairy belt" or "wheat belt" in Europe like those in North America; even where one particular crop dominates the farming scene, some different crops stand just a few hundred yards away.

Finally, Europe's great lowland possesses yet another crucial advantage: its multitude of navigable rivers, emerging from higher adjacent areas and wending their way to the sea. In addition to the three rivers of France already mentioned, the Rhine-Meuse (Maas) river system serves one of Europe's most productive industrial and agricultural areas, reaching the sea via the Netherlands; the Weser, Elbe, and Oder cross northern Germany; the Vistula traverses Poland. In southeastern Europe, the Danube rivals the Rhine in regularity of flow and navigability. Thus

Europeans for centuries have been improving their system of natural waterways by linking them with canals. Here a barge passes vineyards on a waterway in southwestern Germany. Note that the vineyards stand on the slopes while vegetables are grown on the floodplain.

north of the Alpine Mountain system, Europe's major rivers create a radial pattern outward from the continent's interior highlands. In this way, the natural waterways as well as the land surface of the North European Lowland favor traffic and trade. Over many centuries, the Europeans have improved the situation still further by connecting navigable stretches of rivers with artificial canals. These waterways, and the roads and railroads that followed later, combined to bring tends of thousands of localities into contact with one another. Thus new techniques and innovations could spread rapidly, and trade connections and activity intensified continuously.

◆ HERITAGE OF ORDER

Modern Europe was peopled in the wake of the Pleistocene's most recent glacial retreat—a gradual withdrawal that caused cold tundra to turn into deciduous forest and ice-filled valleys into grassy vales. On Mediterranean shores, Europe witnessed the rise of its first great civilizations—on the islands and peninsulas of Greece, and later in Italy. Greece lay exposed to the influences radiating from the advanced civilizations of Mesopotamia and the Nile valley (see map p. 360), and the intervening eastern Mediterranean was crisscrossed by maritime trade routes.

ANCIENT GREECE

As the ancient Greeks forged their city-states and intercity leagues, they made impressive intellectual achievements as well (which peaked during the fourth century B.C.). Their political philosophy and political science became important products of their culture, and the writings they left behind have influenced politics and government ever since. Yet there was more to ancient Greece than politics; great accomplishments were also recorded in such fields as architecture, sculpture, literature, and education. Because of the fragmentation of their habitat, there was local experimentation and success, followed by active exchanges of ideas and innovations.

Although individualism and localism were elements that the Greeks turned to their advantage, internal discord was always present, and in the end, it got the better of them. The constant struggle between the two major cities, Athens and Sparta, ensured their decline, and by 147 B.C. the Romans had defeated the last sovereign Greek intercity league. Nevertheless, what the ancient Greeks had accomplished was not undone: they had transformed the eastern Mediterranean into one of the cultural cores of the world, and Greek culture became a major component of Roman civilization.

THE ROMAN EMPIRE

The Roman successors to ancient Greece also made their own essential contributions. The Greeks never achieved politico-territorial organization on the scale accomplished by Imperial Rome, and much additional progress was made in such spheres as land and sea communications, military organization, law, and governmental administration. During its greatest expansion (in the second century A.D.), the Roman Empire extended from Britain to the Persian Gulf and from the Black Sea to Egypt. Facing little opposition, the vast empire could organize internally without interference, and in Europe it evolved into the continent's first truly interregional political unit.

Given the variety of cultures that had been brought under Roman control and the resulting exchange of ideas and innovations, there were many opportunities for regional interaction. This process of economic development (for such it really was) had a profound impact on the whole structure of Mediterranean and Western Europe. Areas that had hitherto supported only subsistence modes of life were drawn into the greater economic framework of the state, and suddenly there were distant markets for products that had never found even local markets before. In turn, these areas received the farming know-how of the heart of the Roman state so that they could increase their yields and benefit even further. Foodstuffs now flowed into Rome from the entire Mediterranean Basin; with a population of perhaps a quarter-million, the city itself was the greatest single marketplace of the Empire and the first metropolitan-scale urban center in Europe.

This urban tradition came to characterize Roman culture throughout the Empire, and many cities and towns founded by the Romans continue to prosper today. Roman urban centers were also connected by an unparalleled network of highway and water routes, facilities that all formed part of an **infrastructure** needed to support economic growth and development. (Today, a modern state's infrastructure would include railroads, airports, energy-distribution systems, telecommunications networks, and the like.) More than anything else, though, the Roman Empire left Europe a legacy of ideas—concepts that long lay dormant but eventually played their part when Europe again discovered the path of progress. In political and military organization, effective administration, and long-term stability, the Empire was centuries ahead of its time. Moreover, never was a larger part of Europe unified by acquiescence than it was under the Romans, and at no time did Europe come closer to obtaining a *lingua franca* (common language) than it did during the age of Rome.

Finally, Europe's transformation under Roman rule heavily involved the geographic principle of **areal functional specialization**. Before the Romans brought order and connectivity to their vast domain, much of Europe was inhabited by tribal peoples whose livelihoods were on a subsistence level.

The ancient Romans permanently influenced cultural landscapes throughout their empire, building towns, roads, aqueducts, and other structures. This spectacular Roman aqueduct has stood the test of nearly 2,000 years near Arles in the Provence region of southeastern France.

Many of these groups lived in virtual isolation, traded little, and fought over territory when encroachment occurred. Peoples under Rome's sway, however, were brought into Roman economic as well as political spheres, and farmlands, irrigation systems, mines, and workshops appeared. Thus Roman-dominated areas began to take on a characteristic that has marked Europe ever since: *particular peoples and particular places concentrated on the production of particular goods.* Parts of North Africa became granaries for urbanizing (European) Rome; Elba, a Mediterranean island, produced iron ore; the Cartagena area of southeastern Spain mined and exported silver and lead. Many other locales in the Roman Empire specialized in the production of particular farm commodities, manufactured goods, or minerals. The Romans knew how to exploit their natural resources; at the same time, they also learned to use the diversified productive talents of their subjects.

◆ DECLINE AND REBIRTH

The eventual breakdown and collapse of the Empire in the fifth century A.D. could not undo what the Romans had forged in the spreading of their language, in the dissemination of Christianity (in some ways the sole strand of permanence through the ensuing Dark Ages), in education, the arts, and countless other spheres. But ancient Rome's decline was attended by a momentous stirring of Europe's peoples as Germanic and Slavic populations moved to their present positions on the European stage. The Anglo-Saxons invaded Britain from Danish shores, the Franks moved into France, the Allemanni traversed the North European Lowland and settled in Germany. Capitalizing on the disintegration of Roman power, numerous kings,

dukes, barons, and counts established themselves as local rulers. Europe was in turmoil, and its weakness invited invasion from North Africa and Southwest Asia. In Iberia, the Arab-Berber Moors conquered a large area; in Eastern Europe, the Ottoman Turks extended their Islamic empire. The townscapes of southern Spain and the Balkans still carry the cultural imprints of these Muslim invasions.

After nearly a thousand years of feudal fragmentation during the Dark and Middle Ages, modern Europe began its emergence in the second half of the fifteenth century. (Some date this rebirth from 1492, the year of Columbus's first arrival in the New World.) At home, monarchies strengthened at the expense of feudal lords and landed aristocracies and, in the process, forged the beginnings of nation-states. Abroad, Western Europe's developing states were on the threshold of discovery—the discovery of continents and riches across the oceans. Europe's emerging powers were fired by a new national consciousness and pride, and there was renewed interest in Greek and Roman achievements in science and government. Appropriately, this period is referred to as Europe's *Renaissance*.

The new age of progress and rising prosperity was centered in Western Europe, whose countries lay open to the new pathways to wealth—the oceans. Now the highly competitive monarchies of Western Europe engaged in economic nationalism that operated in the form of *mercantilism*. The objectives of this policy were the accumulation of as large a quantity of gold and silver as possible, and the use of foreign trade and colonial acquisition to achieve that end. Mercantilism was promoted and sustained by the state; precious metals could be obtained either by the conquest of peoples in possession of them or indirectly by achieving a favorable balance of international trade. Thus there was stimulus not only to seek new territories where such metals might lie, but also to produce goods at home that could be sold profitably abroad. The matrix of modern states in Western Europe was beginning to take shape, and the spiral had been entered that was to lead to great empires and a period of world domination.

◆ THE REVOLUTIONS OF MODERNIZING EUROPE

Strife and dislocation punctuated Europe's march to world domination. Much of what was achieved during the Renaissance was destroyed again as powerful monarchies struggled for primacy; religious conflicts dealt death and misery; and the beginnings of parliamentary government fell under new tyrannies. Nevertheless, revolutions—in several spheres—were in the making. Economic developments in Western Europe ultimately proved to be the undoing of absolute monarchs and their privileged, land-owning nobilities. The city-based merchant was gaining wealth and prestige, and the traditional measure of affluence—land—began to lose its status in these changing times. The merchants and businesspeople of Europe were soon able to demand political recognition on grounds the nobles could not match. Urban industries were thriving; Europe's population, more or less stable at about 100 million since the mid-sixteenth century, was on the increase.

THE AGRARIAN REVOLUTION

This transformation was heightened by an ongoing *agrarian revolution*—the significant metamorphosis of European farming that preceded the Industrial Revolution and helped make possible a sustained population increase during the seventeenth and eighteenth centuries. The Netherlands, Belgium, and northern Italy paved the way with their successes in commerce and manufacturing. The stimulus provided by expanding urbanization and markets led to improved organization of landownership and agriculture. Some of the new practices spread to England and France, where traditional communal landownership began to give way to individual landholding by small farmers. Land parcels were marked off by fences and hedges, and the new owners readily adopted innovations to improve crop yields and increase profits.

At the same time, methods of soil preparation, crop rotation, cultivation, harvesting, and livestock feeding improved. More effective farm equipment was adopted; there was better planning and experimentation; storage and distribution systems became more efficient. In the growing cities and towns, farm products fetched higher prices. New crops were introduced, especially from the Americas; the potato now became a European staple. More and more of

MODELS IN GEOGRAPHY

A current, widely used approach to generalization in both human and physical regional geography is the development of **models**. Peter Haggett, in his book *Locational Analysis in Human Geography*, offers an especially lucid definition: *"in model-building we create an idealized representation of reality in order to demonstrate its most important properties."*

The use of models by geographers is necessitated by the complexity of reality: to understand how things work, we must first filter out the main spatial processes and their responses from the myriad details with which they are embedded in a highly complicated world. Models therefore provide a simplified picture of reality in order to convey, if not the entire truth, then at least a useful and essential part of it. The theory-based derivation of the Von Thünen model (Fig. 1-3) and its empirical or real-world application to contemporary Europe (Fig. 1-4) offers a classical demonstration of this geographic method.

Europe's farmers were drawn from subsistence into profit-driven market economies. Later, the manufactured products of the Industrial Revolution further stimulated the transformation of the realm's agriculture.

As new forces and processes began to reshape the economic geography of Europe, certain scholars tried to interpret the new spatial patterns they produced. In 1826, the economist (and farmer) Johann Heinrich von Thünen (1783–1850) fashioned one of the world's first geographical *models* (see the box above). For four decades Von Thünen, who owned a large farming estate in northeastern Germany, studied the effects of distance and transportation costs on the location of productive activity. Eventually, he published a work entitled *The Isolated State*, and his methods in many ways constitute the foundations of modern location theory.

Von Thünen's Isolated State model was so named because he wanted to establish, for purposes of theoretical analysis, a self-contained country devoid of outside influences that would disturb the internal workings of the economy. Thus he created a sort of regional laboratory within which he could identify the factors that influence the locational distribution of farms around a single urban center. To do this, he made a number of limiting assumptions. First, he stipulated that the soil and climate would be uniform throughout the region. Second, no river valleys or mountains would interrupt a completely flat land surface. Third, there would be a single centrally positioned city in the Isolated State, and the latter would be surrounded by an empty, unoccupied wilderness. Fourth, the farmers in the Isolated State would transport their own products to market by oxcart, directly overland and straight to the central city. This, of course, is the same as assuming a system of radially converging roads of equal and constant quality; with such a system, transport costs would be directly proportional to distance.

Von Thünen integrated these assumptions with what he had learned from the actual data collected while running his estate, and he now asked himself: What would be the ideal spatial arrangement of agricultural activities within the Isolated State? He concluded that farm products would be raised in a series of concentric zones outward from the central market city. Nearest to the city would be grown those crops that perished easily and/or yielded the highest returns (such as vegetables), because this readily accessible farmland was in great demand and therefore quite expensive; dairying would also be carried on in this innermost zone. Farther away would be potatoes and grains. And eventually, since transport costs to the city increased with distance, there would come a line beyond which it would be uneconomical to produce crops. There the wilderness would begin.

Von Thünen's model incorporated four zones or rings of agricultural land use surrounding the market center (Fig. 1-3). The first and innermost belt would be a zone of intensive farming and dairying. The second zone, according to Von Thünen, would be an area of forest used for firewood and timber (still important as building material in his time). Next, there would be a third ring of increasingly extensive field crops. The fourth and outermost zone would be occupied by ranching and animal products beyond which would begin the wilderness that isolated the region from the rest of the world.

Von Thünen knew, of course, that the real Europe (or world) did not present idealized situations exactly as he had postulated them. Transport routes serve certain areas more efficiently than others. Physical barriers can impede the most modern of surface communications. External economic influences invade every area. But Von Thünen wanted to eliminate these disruptive conditions in order to discern the fundamental processes that shaped the spatial layout of the agricultural economy. Later, the distorting factors could be introduced one by one and their influence measured. First, however, he developed his model in theoretical isolation, basing it on total regional uniformity.

It is a great tribute to Von Thünen that his work still commands the attention of geographers. The economic-geographic landscape of Europe has changed enormously since his time, but geographers still compare present-day patterns of economic activity to the Thünian model. Such a comparison was made by Samuel van Valkenburg and Colbert Held, whose map of Europe's agricultural intensity reveals a striking ring-like concentricity (Fig. 1-4). The overriding spatial

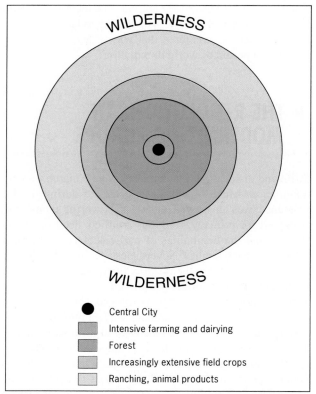

FIGURE 1-3

- ● Central City
- Intensive farming and dairying
- Forest
- Increasingly extensive field crops
- Ranching, animal products

change since Von Thünen's time is the improvement in transportation technology, which permitted the Isolated State to expand from *micro-* to *macro-scale*. Thus the model is no longer centered on a single city but rather on the vast urbanized area lining the southern coasts of the North Sea, which now commands a continent-wide Thünian agricultural system.

THE INDUSTRIAL REVOLUTION

Alongside its advances in agriculture, Europe also had developed significant industries before the *Industrial Revolution* began. In Flanders and England, specialization had been achieved in the manufacturing of woolen and linen textiles; in Saxony (in what is now eastern Germany), iron ore was mined and smelted. European manufacturers produced a wide range of goods for local markets, but their quality was often surpassed by textiles and other wares from India and China. This gave European entrepreneurs the incentive to refine and mass-produce their products. Because raw materials could be shipped home in virtually unlimited quantity, if they could find ways to mass-produce these commodities into finished goods, they could bury the Asian industries under growing volumes and declining prices.

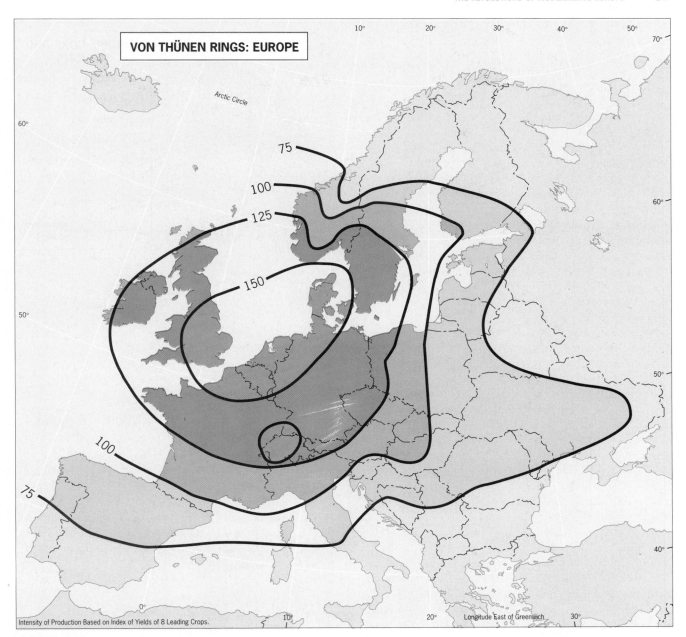

VON THÜNEN RINGS: EUROPE

Intensity of Production Based on Index of Yields of 8 Leading Crops.

Longitude East of Greenwich

FIGURE 1-4

Now the search for better machinery was on, especially improved spinning and weaving equipment. The first steps in the Industrial Revolution were not all that revolutionary, for the larger spinning and weaving machines that were built were still driven by the old source of power—water running downslope. But then, in the 1780s, James Watt and others succeeded in devising a steam-driven engine, and soon this invention was adapted for various uses. About the same time, it was realized that coal (converted into carbon-rich coke) was a vastly superior substitute for charcoal in smelting iron. These momentous innovations had a rapid effect. The power loom revolutionized the

weaving industry. Iron smelters, long dependent on Europe's dwindling forests for fuel, could now be concentrated near coalfields. Engines could move locomotives as well as power looms. Ocean shipping entered a new age.

England had an enormous advantage, for the Industrial Revolution occurred when British influence reigned worldwide and the significant innovations were achieved in Britain itself. The British controlled the flow of raw materials, they held a monopoly over products that were in global demand, and they alone possessed the skills necessary to make the machines that manufactured the products. Soon the fruits of the Industrial Revolution were being exported,

FIGURE 1-5

and the modern industrial spatial organization of Europe began to take shape (Fig. 1-5). In Britain, manufacturing regions, densely populated and heavily urbanized, developed near coalfields in the English Midlands, at Newcastle to the northeast, in southern Wales, and along the Clyde River in Scotland.

In mainland Europe, a belt of major coalfields extends from west to east, roughly along the southern margins of the North European Lowland, due eastward from southern England across northern France and Belgium, Germany (the Ruhr), western Bohemia in the Czech Republic, Silesia in southern Poland, and the Donets Basin in eastern Ukraine. Iron ore is found in a broadly similar belt, and the industrial map of Europe reflects the resulting concentrations of economic activity (Fig. 1-5). Another set of manufacturing regions emerged in and near the growing urban centers of Europe, as the same map demonstrates. London—already Europe's leading urban focus and Britain's richest domestic market—was typical of these developments. Many local industries were established here, taking advantage of the large supply of labor, the ready availability of capital, and the proximity of so great a number of potential buyers. Although the Industrial Revolution thrust other places into prominence, London did not lose its primacy: industries in and around the British capital multiplied.

Industrial and Urban Intensification

It is not surprising that the industrialization of Europe—or, rather, its industrial *intensification* following the Industrial Revolution—also became a focus of geographic research. What influences affected industrial location? How was Europe's industrialization channeled? Again, the first important studies were conducted by German scholars, mostly during the second half of the nineteenth century. Much of this work was incorporated in a volume by Alfred Weber (1868–1958), published in 1909 and entitled *Concerning the Location of Industries*. Like Von Thünen, Weber began with a set of limiting assumptions in order to minimize the complexities of the real Europe. But unlike Von Thünen, Weber dealt with activities that take place at particular *points* rather than across large areas. Manufacturing plants, mines, and markets are located at specific places, and so Weber created a model region marked by sets of points where these activities would occur. He eliminated labor mobility and varying wage rates, and thereby could calculate the "pulls" exerted on each point in his theoretical region.

In the process, Weber discerned various factors that affect **industrial location**. He recognized what he called "general" factors that would affect all industries—dominated by transport costs for raw materials and finished products—and "special" factors (such as perishability of foods). He also differentiated between "regional" factors (transport and labor costs) and "local" factors. The latter, Weber argued, involve *agglomerative* (concentrating) and *deglomerative* (deconcentrating) forces. Take the case of London discussed above: industries located there, in large part, because of the advantages of locating together. The availability of specialized equipment, a technologically sophisticated labor force, and a large-scale market made London (as well as Paris and other big cities not positioned on rich deposits of natural resources) an attractive site for many manufacturing plants that could benefit from agglomeration. On the other hand, such concentration may over time create strong disadvantages—chiefly competition for space, rising land prices, congestion, and environmental pollution. Eventually, an industry might move away and deglomerative forces would set in.

Europe's industrialization also speeded the growth of many of its cities and towns. In Britain in the year 1800,

only about 9 percent of the population lived in urban areas; but by 1900, some 62 percent resided in cities and towns (today that proportion surpasses 90 percent). All this was happening while the total population skyrocketed as well. As industrial modernization came to Belgium and Germany, to France and the Netherlands, and to other parts of Western Europe, the entire urban pattern changed. The nature of this process—the growth and strengthening of towns and cities—and related questions also became topics of geographic study, just as agriculture and industrialization had (principles of urban geography are discussed in the "Focus on a Systematic Field" at the outset of Chapter 3).

POLITICAL REVOLUTIONS

Europe had had long experience with experiments in democratic government, but the *political revolution* that swept the realm after 1780 brought transformation on an unprecedented scale. Overshadowing these events was the French Revolution (1789–1795), but France's—and Europe's—political catharsis lasted into the twentieth century as the rising tide of nationalism eventually affected every monarchy on the continent.

In France, the popular revolution had plunged the country into years of chaos and destruction. Only when Napoleon took control in 1799 was stability restored. Napoleon personified the new French republic, and he reorganized France so completely that he laid the foundations of the modern nation-state (a concept treated below). He also built an empire that extended from German Prussia to Spain and from the Netherlands to Italy. Although his armies were ultimately repulsed, Napoleon had forever changed the political spatial structure of Europe. His French forces had been joined by nationalist revolutionaries from all over the continent; one monarchy after another had been toppled. Even after his final defeat at Waterloo in 1815, there were popular uprisings in Spain, Portugal, Italy, and Greece. By then Europe had had its first real taste of democracy and nationalist power—and it would not revert to its old ways.

The Rise of the Nation-State

As Europe went through its periods of rebirth and revolutionary change, the realm's politico-geographical map was transformed. Smaller entities were absorbed into larger units, conflicts resolved (by force as well as negotiation), boundaries defined, and internal divisions reorganized. European nation-states were in the making.

But what is a nation-state and what is not? The question centers in part on the definition of the term *nation*. That definition usually involves measures of homogeneity: a nation should comprise a group of tightly knit people who speak a single language, have a common history, share the

CENTRIPETAL AND CENTRIFUGAL FORCES

Political geographers use the terms *centripetal* and *centrifugal* to identify forces within a state that tend, respectively, to bind that political system together and to pull it apart.

Centripetal forces tend to tie the state together, to unify and strengthen it. A real or perceived external threat can be a powerful centripetal force, but more important and lasting is a sense of commitment to the governmental system, a recognition that it constitutes the best option. This commitment is sometimes focused on the strong charismatic qualities of one individual—a leader who personifies the state, who captures the population's imagination. (The origin of the word *charisma* lies in the Greek expression that means "divine gift.") At times such charismatic qualities can submerge nearly everything else. Juan Peron's lasting popularity in Argentina is a case in point. Jomo Kenyatta in Kenya, Josip Broz Tito in Yugoslavia, Charles De Gaulle in France, Mao Zedong in China, and Jawaharlal Nehru in India all possessed similar charisma and individually played dominant roles (extending well beyond their lifetimes) in binding their states.

Centrifugal forces are disunifying or divisive. They can cause deteriorating internal relationships. Religious conflict, racial strife, linguistic cleavages, and contrasting regional outlooks are among the major centrifugal forces. During the 1960s, the Vietnam (Indochina) War became a major centrifugal force in the United States, the aftermath of which is still felt three decades later. Newly independent countries often find tribalism a leading centrifugal force, sometimes strong enough to threaten the very survival of the whole state system (as the Biafra conflict did in Nigeria).

The degree of strength and cohesion of the state depends on a surplus of centripetal forces over divisive centrifugal forces. It is difficult to measure such intangible qualities, but some attempts have been made in this direction—for example, by determining attitudes among minorities and by evaluating the strength of regionalism as expressed in political campaigns and voter preferences. When the centrifugal forces become excessively strong and cannot be checked, even by external imposition, the state breaks up (as the Soviet Union and Yugoslavia did in 1991); or it undergoes revolutionary internal change that makes it, in effect, a new entity, as Nicaragua became after the ouster of Somoza, and as Cuba became after the victory of Castro's forces.

---

same cultural background, and are united by common political institutions. Accepted definitions of the term suggest that many states are not nation-states because their populations are divided in one or more important ways. But cultural homogeneity may not be as important as a more intangible "national spirit" or emotional commitment to the state and what it stands for. One of Europe's oldest states, Switzerland, has a population that is divided along linguistic, religious, and historical lines—but Switzerland has proved to be a most durable nation-state nonetheless, marking its 700th anniversary in 1991.

A **nation-state**, therefore, may be defined as a political unit comprising a clearly defined territory and inhabited by a substantial population, sufficiently well organized to possess a certain measure of power, with the people considering themselves to be a nation having certain emotional and other ties that are expressed in their most tangible form in the state's legal institutions and political system. Supporting all this, of course, is a government that constantly works to ensure that the forces unifying the state prevail over those that would drive it apart (see the box entitled "Centripetal and Centrifugal Forces"). This definition of the nation-state essentially identifies the European model that emerged in the course of the realm's long period of evolution and revolutionary change. France is often cited as the best example among Europe's nation-states, but the United Kingdom, Poland, Hungary, Sweden, and newly reunified Germany are also among countries that satisfy the terms of the definition to a great extent. Belgium is an example of a European state that cannot at present be designated as a true nation-state. The former Yugoslavia and Czechoslovakia never were nation-states; their actual and latent centrifugal (disunifying) forces were too strong to allow long-term stability.

◆ THE EUROPEAN REALM TODAY

The nation-states of Europe are among the world's oldest, and the colonial empires of European powers were among the most durable. Despite many disruptive wars and revolutions, European nations survived, and from their long-term stability they forged a confidence—based on strong individual identities—that is a hallmark of European culture. Although Europe quite clearly constitutes a geographic realm, it exhibits little geographic homogeneity.

It is sometimes postulated that Europe may be viewed as a regional unit because its peoples share Indo-European languages (Fig. 1-6), Christian religious traditions (see

opposing forces at the international level continue to build toward stronger ties among the realm's states. Thus integration and disintegration proceed simultaneously. The year 1994 marks the fiftieth anniversary of the *Benelux* agreement, which established an economic union that linked the three small Western European countries of *Bel*gium, the *Ne*therlands, and *Lux*embourg. More importantly, it simultaneously ignited a still-ongoing experiment in international cooperation and association that is culminating in the formation of a multi-country European Union which could, some hope and believe, eventually lead to a United States of Europe.

European countries are not alone in trying to forge a mutually beneficial union, nor is this the first time that the Europeans have tried to achieve such a goal. But this particular European experiment of the late twentieth century is the greatest accomplishment of its kind to date. Political geographers define the voluntary association of three or more countries as **supranationalism**. Such association may involve economic, cultural, and/or political spheres. The key factor involves sovereignty: are the participant countries willing to give up some sovereignty for the betterment of their union?

In the case of Europe, that is indeed the situation. It all began in 1944, just before the end of World War II, when the Benelux organization was formed. Then, soon after the war, a far-sighted U.S. Secretary of State, George Marshall, proposed a massive infusion of American aid to the struggling countries of postwar Western Europe. The Marshall Plan (1948–1952) not only boosted national economies: it also led directly to the need for a multinational European economic-administrative structure. This, Europeans realized, was no time for national boundaries and troublesome tariffs to interfere with the flow of raw materials and finished products. Also, Europeans worried that future misunderstandings might lead to renewed conflict, so they established the Council of Europe, a deliberative body that met in Strasbourg, France (a city that has been nominated for the capital of a future united Europe).

These two initiatives in the economic and political spheres produced one success after another, although traditional European rivalries did not simply end. In 1957, six Western European countries—France, (then-West) Germany, Italy, and the Benelux group—signed the far-reaching Treaty of Rome. This established, as of 1958, the so-called Common Market—also known as the European Economic Community (or EEC for short) and later simply the European Community (EC). Not to be outdone, the British, who did not join the Common Market, in 1959 led six other countries to create the European Free Trade Association (EFTA). But this group of seven (the United Kingdom, Sweden, Norway, Denmark, Switzerland, Austria, and Portugal) was no match for the EEC in terms of productive capacity, markets, or raw materials. Soon EFTA lost its relevance as some of its members (including the British) began to join the expansion-minded EC (see the

SUPRANATIONALISM IN EUROPE

1944 Benelux Agreement signed.

1947 Marshall Plan proposed.

1948 Organization for European Economic Cooperation (OEEC) established.

1949 Council of Europe created.

1951 European Coal and Steel Community (ECSC) Agreement signed (effective 1952).

1957 Treaty of Rome signed, establishing European Economic Community (EEC) (effective 1958), also known as the Common Market.
European Atomic Energy Community (EURATOM) Treaty signed (effective 1958).

1959 European Free Trade Association (EFTA) Treaty signed (effective 1960).

1965 EEC-ECSC-EURATOM Merger Treaty signed (effective 1967).

1968 All customs duties removed for intra-EEC trade; common external tariff established.

1973 United Kingdom, Denmark, and Ireland admitted as members of EEC, creating "The Nine."

1979 First general elections for a European Parliament held; new 410-member legislature meets in Strasbourg.

1981 Greece admitted as member of EC (EEC has been shortened), creating "The Ten."

1986 Spain and Portugal admitted as members of EC, creating "The Twelve."
Single European Act ratified, targeting a functioning European Union in the 1990s.

1990 Charter of Paris signed by 34 members of the Conference on Security and Cooperation in Europe (CSCE).
Former East Germany, as part of newly reunified Germany, incorporated into EC.

1991 Maastricht meeting charts European Union course for the 1990s.

1993 Single European Market goes into effect.
Modified European Union Treaty scheduled for ratification.

1995 Projected date for admission of Austria, Finland, and Sweden into EC, which would create "The Fifteen."

box entitled "Supranationalism in Europe"). Now the Community began to grow, first to 9 members, then to 10, and, by 1986, to 12 (Fig. 1-8). This group of 12 countries was to become the vanguard of the European Union.

FIGURE 1-8

In the political sphere, too, progress was made that many believed impossible. The Council of Europe *did* evolve into an embryonic European Parliament, and in 1979 the first Europe-wide elections were held to send 410 representatives—now elected by the people and their parties rather than appointed by governments—to Strasbourg. The European Parliament began to reflect more accurately the political mainstream of what at the time was still Europe west of the Iron Curtain. Elections now are held at regular five-year intervals; those of 1984 and 1989 enlarged the parliament to 518 members. The 1994 elections will have special significance, being the first to come after the consolidation of the European Union.

While all this was happening, supranational cooperation in other spheres also took place. EURATOM (European Atomic Energy Community) was founded to develop the peaceful uses of nuclear energy, in which France is a world leader. ESRO (European Space Research Organization) coordinates Europe's successful space programs. And NATO (North Atlantic Treaty Organization) forms the umbrella for the military activities of member countries.

But the spotlight today is on "The Twelve," the countries forming the European Union. (To remember which countries these are, use the *3-4-5* method: the *three* giants—Germany, France, and the United Kingdom; the *four* southern countries—Spain, Portugal, Italy, and Greece; and the *five* small countries—Ireland, Denmark, Belgium, the Netherlands, and Luxembourg.) In 1986 these states signed the momentous Single European Act, promising to work toward the achievement and implementation of some 279 specific goals by the end of 1992.

In late 1991, The Twelve met at Maastricht in the Netherlands to confront the inevitable difficulties these goals raised. The British, especially, had reservations about giving more power to the European Parliament; they also continued to have doubts about eventual political union (which is one of the 279 goals). The prospect of a common currency, to be in place at the end of 1999, further worried several of The Twelve: imagine Britain without the pound, France without the franc, Germany without its mark! Nonetheless, a central European banking system with a single European currency was approved through the Economic and Monetary Union (EMU).

These are just a few of the still-unresolved items before the members of the emerging European Union; common foreign and defense policies, policies on immigrant-workers' rights, and many other matters also remain to be worked out. In the meantime, the Charter of Paris was signed by The Twelve plus 22 other European states in 1990. This charter acknowledged and approved Germany's reunification, ratified Europe's present boundaries, and established the first greater-European mechanism for settling disputes among member countries. In fact, the Charter itself may be regarded as less important than the list of signatories, now including Eastern European countries. As time goes on, The Twelve may well be seen as merely a stage in the still-evolving supranational process.

In the mid-1990s, that process is proving to be one of considerable difficulty in the aftermath of the signing of the Treaty on European Union at Maastricht. For all its recent convergence, the Union still (and for a very long time to come) will be a patchwork of nations with ethnic traditions and histories of conflict and competition. Economic success and growing well-being tend to submerge such differences, but should the EU face troubled economic or social times, these divisive forces will swiftly reemerge. That is precisely what happened during 1992 as Europe's economic recession deepened and resurgent nationalism prompted many member-states to reexamine the concessions they made at Maastricht. Their leaders quickly assembled and explained that the intention at Maastricht was to respect the separate national identity and diversity of each country—reassuring perhaps, but an un-

mistakable shift away from the path leading toward a "United States of Europe." In any case, the Union Treaty could not go into effect until each of The Twelve ratified it. Misgivings among the French and Danish electorates resulted in a bare majority voting approval—and then only after changes were made in the Treaty's provisions; nonetheless, full ratification was expected in late 1993, with the modified Treaty in effect by early 1994. The immediate prospects are brighter for the Single European Market: with more than 95 percent of the 279 goals attained, it began operating on January 1, 1993, allowing the unrestricted flow of goods, services, and capital among The Twelve (although problems with illegal aliens and terrorists delayed implementation of the planned passport-free flow of EU citizens).

If the European Union succeeds, Europe will become an even more formidable presence in the economic and political geography of the world than it has been in the past. The Twelve contain just under 350 million people, constitute one of the world's richest markets, and produce about 40 percent of all global exports. Already, other European countries want to join: in 1993 negotiations were under way to admit Austria, Sweden, and Finland in 1995 (applications from Switzerland, Turkey, Malta, and Cyprus are also pending, with the first two a possibility for membership by 1996). And in the east, Hungary, Poland, the Czech Republic, and Slovakia have publicly stated their intention to seek admission to the European Union. In 1993, these four countries plus Romania and Bulgaria were formally invited to apply for membership "when their economic and political conditions permit" (perhaps as soon as 2000).

This raises a key question. Can the system devised by (and for) The Twelve be adapted to encompass a larger number of member states? In 1993, Portugal, Greece, and Ireland were the poorest members of the EU, and at Maastricht the question of richer members' support for poorer ones was addressed but not resolved. Enlarging the Union to encompass parts of Eastern Europe (and later perhaps even fragments of the former Soviet Union) could endanger what has been achieved. After Maastricht and the new problems that arose there, it is likely that the tempo of integration will slow down, as will the rate of post-1994 admissions. Already, the deepening economic recession and growing antipathy toward some of Maastricht's stipulations are slowing Europe's progress toward unification.

◆ REGIONS OF EUROPE

Earlier in this chapter, on the regional map of Europe (see Fig. 1-5 on p. 58), the realm's core area was delineated. That boundary encloses Europe's heartland, historically a large-scale functional region of dynamic growth, dense population, burgeoning urbanization, and great industrial productivity. We now turn to a more traditional overview of the regions shown on that map, which were derived by grouping sets of European countries together (as listed on p. 46 at the opening of this chapter). Each regional grouping reflects those countries' proximity, historical-geographical association, cultural similarities, common habitats, and shared economic pursuits. On these bases, we can identify five regions of Europe: (1) the British Isles; (2) Western Europe; (3) Northern (Nordic) Europe; (4) Mediterranean Europe; and (5) Eastern Europe.

THE BRITISH ISLES

Off the coast of mainland Western Europe lie two major islands, surrounded by a constellation of tiny ones, that constitute the British Isles (Fig. 1-9). The larger of the two major islands, which also lies nearest to the mainland (a mere 21 miles or 34 km at the closest point), is the island of *Britain*; its smaller neighbor to the west is *Ireland*. Although Britain and Ireland continue to be known as the British Isles, the British government no longer rules over the state that occupies most of Ireland, the Republic of Ireland or *Eire*. After a very unhappy period of domination, the London government in 1921 gave up control over that part of the island, which remains overwhelmingly Roman Catholic. But the northeastern corner of Ireland, where English and Scottish Protestants had settled, was retained by the British and is called *Northern Ireland*.

Although all of Britain is part of the United Kingdom, political divisions exist here as well. *England* is the largest of these units, the center of power from which the rest of the region was originally brought under unified control. The English conquered *Wales* in the Middle Ages, and *Scotland* was tied to England in the seventeenth century when a Scottish king ascended the English throne. Thus, England, Wales, Scotland, and Northern Ireland became the *United Kingdom (UK)*.

Highland and Lowland Britain

Britain's natural environment is varied, and the relative location of different subareas matters much. The most simple and meaningful division of Britain is into a *lowland* and a *highland* region. Most of England, with the exception of the Pennine Mountains and the peninsular southwest, lies in lowland Britain; Highland Britain consists mainly of Scotland and Wales. Lowland Britain, therefore, lies opposite the western periphery of the North European Lowland on the continent, and it is essentially a geological continuation of it (see Fig. 1-2). In this part of England, the relief is low, soils are generally good, rainfall is ample, and agricultural productivity is relatively high.

Lowland Britain, particularly the accessible Thames Basin around London in southeastern England, has long been the Isles' power center. Before the birth of the British nation and empire following the eleventh century, numerous peoples had penetrated and fought over this subregion, among them the Celts, the Romans (who founded London and introduced Christianity), the Germanic Angles and Saxons, the Vikings, and the Normans (who had invaded from France under William the Conqueror in A.D. 1066). In its political development, Britain benefited from its insular separation from Europe, and England became the focus of regional power and economic organization. Centered on London, England steadily evolved into the core area of the British Isles. Ultimately, the British achieved a system of parliamentary government that had no peer in the Western world, and England rose from a regional core area into the headquarters of a global empire.

The economic development that to a considerable extent precipitated this political progress came in stages that could be accommodated without disrupting the whole society. Britain's original subsistence economy gave way to the commercial era of mercantilism, and the Industrial Revolution was foreshadowed by the early development of manufacturing based on the water power provided by streams flowing off the Pennines, Lowland Britain's mountain backbone. Then, when the Industrial Revolution came, coalfields in the Midlands (the crescent surrounding the southern Pennines) provided the needed energy source, which supported the budding industries of Sheffield, Leeds, and Manchester (Fig. 1-9). Other important coalfields were also exploited around nearby Birmingham and along the northeast coast in the vicinity of Newcastle.

The impact of the Industrial Revolution in England was felt most strongly in these central and northeastern industrial regions. Despite poor working and living conditions in these manufacturing centers, population totals soared, and Britain prospered. Nowhere was the principle of areal functional specialization better illustrated than here. Woolen-producing cities lay east of the Pennines, centered on Leeds and Bradford; the cotton textile industries of Manchester clustered on the western side of the range. Birmingham and its nearby satellite cities concentrated on the production of steel and metal products; Nottingham came to specialize in hosiery; boots and shoes were made in Leicester and Northampton. In northeastern England, shipbuilding and the manufacture of chemicals were the leading industries along with coal-mining and iron and steel production.

Today, the industrial specialization of many British cities continues, but the resource situation has changed drastically as local raw materials became exhausted. A one-time exporter of coal and iron ore, England has in recent years been forced to import these commodities as domestic supplies dwindle. Of course, Britain's industries always did use certain raw materials from other countries, such as wool from

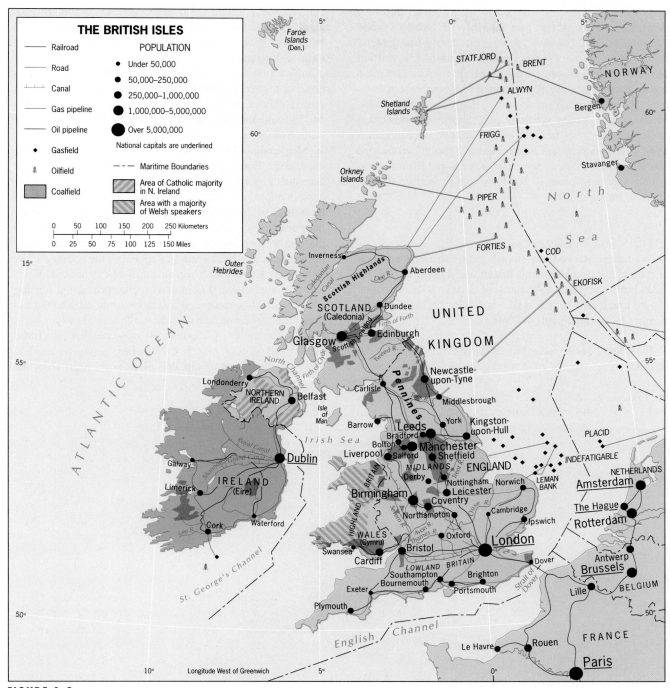

FIGURE 1-9

Australia and cotton from India. But with the dissolution of the British Empire and unrelenting competition from rival industrial powers, Britain today struggles to maintain its position as a major Western industrial power. The overall energy situation is much more favorable, and Britain is currently self-sufficient in oil and natural gas thanks to discoveries made beneath the North Sea since the 1950s. European countries facing the North Sea have allocated sectors of seafloor to themselves (see Fig. 1-9). The British sector has proven to contain major reserves of petroleum and natural gas, but it is estimated that today's self-sufficiency will not endure far into the next century.

Lowland Britain, in addition to its industrial development, also contains the vast majority of Britain's good agricultural land. However, there is not nearly enough land to feed the people of the United Kingdom (58 million of them) at present standards of caloric intake. The good arable land is heavily concentrated in southeastern England. Where the soils are adequate, there is intensive cultivation of grain crops, potatoes, and sugar beets. But much of Lowland Britain, because of soil quality, cool temperatures, and excessive moisture, cannot support field crops and is suitable only for pasture. Therefore, Britain must import a large quantity of food, enough to feed up to two-thirds of its population.

Regions of Britain

As we have observed, the British Isles constitute a region of the European realm. Both Britain and Ireland can be divided into subregions. In the case of Britain, four such units can be identified: (1) the Affluent South; (2) the Stagnant North; (3) Scotland; and (4) Wales.

The Affluent South   As industrial decline accelerates in the manufacturing regions of central and northern England, a

dual economy has emerged in recent years; its major geographic division lies roughly along a line connecting Bristol in the west with Norwich in the east (Fig. 1-9). South of that line, high-technology and service industries are thriving, dominated by financial, banking, engineering, communications, and energy-related activities. Nowhere is this boom greater than in the Southeast, the area anchored by London and its immediate hinterland, which is home to more than 20 million of the United Kingdom's 58 million inhabitants.

Metropolitan London itself contains 7.3 million people; its classic European-city layout of built-up zones and greenbelt is mapped in Fig. 1-10. At the same time, London centers one of the world's great **conurbations**, the general term used by geographers to describe vast multimetropolitan complexes (such as Megalopolis, discussed on p. 23) that are formed by the coalescence of two or more major urban areas. The central city, of course, is the country's historic focus, the seat of government, the headquarters of numerous industrial and commercial enterprises, the leading port, the most concentrated and richest market, and the national transportation hub—in short, the primate city.

The remainder of southern Britain contains two other areas worth noting. One is the tapering peninsula that extends southwestward from Bristol, a world apart from the

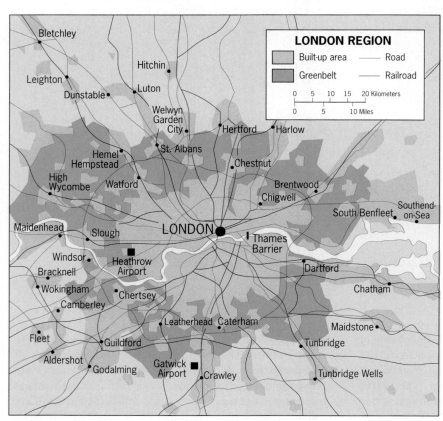

FIGURE 1-10

All is not well in many of Europe's urban centers: industrial decline is taking its toll. Liverpool, England shows the results.

crowded Southeast. This area belongs to Highland Britain; it offers scenic countrysides that attract many tourists but otherwise is an economic backwater that has long been awaiting development. And to the northeast of metropolitan London you will note a round bulge of land that constitutes the third component of the British South—East Anglia. Here is one of England's oldest settlement areas, which attracted Germanic invaders and eventually developed a major fishing industry. Today, the sea remains important (offshore gasfields and still-productive fishing)—but so is agriculture, because East Anglia contains some of Britain's best soils.

The Stagnant North   To the north of the Bristol-Norwich dividing line is a very different England, one dominated by the social and economic decline associated with the aging and disintegration of the once-prodigious regional industrial complex. Unemployment is one of the most significant measures of this north-south cleavage in Britain's economic geography: in the early 1990s, the unemployment rate in the North was nearly twice the rate for England as a whole. Every other economic indicator tells the same story. Even in car ownership rates, the North lags so far behind the rest of England (214 per 1,000 population versus the national average of 319) that we seem to be making an underdeveloped/developed country comparison.

Within the northern subregion of Britain, conditions are bleakest in the old industrial cities of the Midlands, including Manchester (2.3 million), Sheffield (560,000), and Leeds (1.5 million). The woes of nearby Birmingham (2.3 million) have been compounded by racial tensions as its large low-income black and Asian populations (immigrants from former British colonies) compete grimly with long-time

residents for the shrinking job opportunities. The old port city of Liverpool (485,000) offers particularly distressing prospects to its young generation: the docks today are all but totally shut down; over 50,000 jobs vanished during the 1980s; and the unemployment rate exceeds 50 percent for the inner city's youths, most of whom will never find a job (and will be forced to live out their lives on welfare).

Seeking to alleviate these severe problems of the North, the British government and the private sector have launched some new revitalization programs in the Tyneside area surrounding Newcastle (310,000). New facilities, especially industrial parks, have been successful, but local unemployment has not dropped significantly, and the longer-term impact of these investments remains uncertain. Another discouraging development is the intensified suburbanization of population and economic activities since the mid-1980s; throughout Britain, central cities are now being drained of their employment opportunities as office companies, factories, and stores relocate to sleek new business centers in growing, automobile-based suburbs.

Scotland and Wales   Scotland and Wales are distinct subregions of Britain because they have political as well as cultural, economic, and physiographic identities. Moreover, their relationships with long-dominant England are changing. By European standards of scale, these are by no means small provinces: Scotland is twice the areal size of the Netherlands and has 5.1 million inhabitants; Wales is one-fourth the size of Scotland and has 2.8 million residents. Here in Highland Britain's most rugged and remote territories, ancient Celtic peoples found refuge from encroaching invaders, and their languages still survive after centuries of domination by the English (Fig. 1-6).

Wales and Scotland shared in Britain's Industrial Revolution but with mixed results. Initially, the coalfields of southern Wales attracted industries, but when the better deposits were exhausted and production costs went up, competition from manufacturing centers in England proved too strong. Soon, Wales's countryside was ravaged by strip mines, its now-underemployed people dislocated, its cities slum-ridden—and many Welsh citizens left their country for new opportunities elsewhere.

Scotland was more fortunate. Along the narrow "waist" of the Clyde valley (Clydeside) and Firth of Forth, Scotland possessed an extensive coalfield and nearby iron ore deposits, which formed the basis for long-term manufacturing development (notably in shipbuilding). This lowland corridor became the region's core area, with industrial Glasgow (1.9 million) at one end and the cultural focus of Edinburgh (715,000) at the other. Although Clydeside shares with the English North much of the industrial decline caused by obsolescence, Glasgow has revived and is now a budding center of industrial research, especially in electronics. In addition, Scotland on the east now fronts a North Sea that produces lucrative oil and natural gas. Wales and Scotland also share a resurgent nationalism. The government in London (where Scottish and Welsh representatives are far outnumbered by their English counterparts) has always pursued economic policies that favor England, often to the clear disadvantage of Scotland and Wales. When an economic recession in the 1970s brought wide new support for longstanding Scottish and Welsh nationalist movements, strong demands for regional autonomy began to be heard. Given Scotland's importance to the successful new energy industries of the North Sea, particularly in the wake of the crisis in Northern Ireland, London swiftly learned that it must adjust at an early stage to regional pressures.

As a result, Parliament has in recent years engaged in formal discussions on the restructuring of the United Kingdom's political framework. We have already noted Scotland as one of Europe's leading candidates for devolution (see p. 65); in the 1990s the buzz-word has become *home rule*—the shifting of some significant powers from London to a new Scottish legislature. Although the latest (1992) national election did not accelerate this movement, a huge majority of Scots voted against the status quo. The outlook today is that Scotland will probably get its own parliament—but not just yet.

Fractured Ireland

Northern Ireland   The smallest of the United Kingdom's four units continues to be London's most serious domestic problem. Northern Ireland, with a population of 1.7 million, occupies the northeastern one-sixth of the island of Ireland, a legacy of England's colonial control. Despite a poor resource base, a diversified economy has emerged here as a result of infusions of British capital and raw materials. In Belfast (330,000), the shipbuilding industry grew to significant proportions; so did textile (mostly linen) manufacturing in Londonderry (105,000). But unemployment has been an erosive reality, and Northern Ireland has not fared well since 1970—economically or socially.

Northern Ireland, severed from the Republic of Ireland when the latter became independent from London in 1921, has always been different from the rest of Ireland because of its substantial Protestant population. About two-thirds of its people trace their ancestry to Scotland or England, from which their predecessors came across the Irish Sea to settle. Only about 35 percent of Northern Ireland's people are Catholics, in contrast to the Republic, which is overwhelmingly Roman Catholic. No clear spatial separation marks the religious geography of Northern Ireland: Protestants and Catholics mostly live in clusters throughout its territory.

For the past quarter-century, Northern Ireland has remained on the brink of civil war as Protestants and Catholics attacked each other in terrorist campaigns. Catholics have long protested what they perceive to be discrimination by the Protestant-dominated local administration, especially in housing and employment; Protestants accuse Catholics of seeking union with the Republic. For more than 20 years, the territory has been ruled directly from London, but even a large British military force sent to maintain order has been unable to end the persistent sectarian violence.

Republic of Ireland   The republic to the south (Eire) is one of Europe's younger independent states, having fought itself free from Britain only 73 years ago. Without protective mountains or industrial resources, the Irish—a nation of peasants—faced their English adversaries across the narrowest of seas. During the seventeenth century the British conquered the surprisingly resistant Irish, expropriated their good farmland, and turned it over to (mostly absentee) landlords. The English also placed restrictions on every facet of life, especially in Catholic southern Ireland though somewhat less in the Protestant north.

The island on which Eire is located is shaped like a saucer with a wide rim, which is broken toward the east where lowlands open to the sea. A large plainland is thereby enclosed, but this topographic advantage does not result in better agricultural opportunities than in Scotland or Wales. Ireland is wetter than England (as Fig. I-6 shows), and excessive moisture is the great inhibiting element in farming here. Hence, pastoralism is once again the dominant agricultural pursuit, and a good deal of potential cropland is turned over to fodder.

An important exception is the potato, which quickly took the place of other crops as a staple when it was introduced to Ireland from America in the 1600s; it does particularly well in this cool, moist environment. But even the potato could not withstand the effects of the heightened precipitation that fell during several successive years in the 1840s. Having long been the nutritive basis for Ireland's population (which by 1830 had reached 8 million), the potato crop failed repeatedly; ravaged by blight and soaking, potatoes by the millions simply rotted in the ground. Now the Irish faced famine: over a million people died, and nearly twice that number left the country within 10 years of this calamity. This started a pattern as emigration offset and even exceeded natural increase.

Together, Ireland and Northern Ireland now have a population of 5.3 million (3.6 million in Ireland proper), barely two-thirds of what it was when the famine of the 1840s struck. Not many places in the world can point to a decline in population since the days of the Industrial Revolution. Moreover, the exodus continued through 1990, particularly among the educated younger Irish who fled by the thousands each year. Economic causes were the main driving force in this emigration (Eire's unemployment rate still ranks among the highest in Europe), but also cited frequently was the conservatism of Irish cultural and religious life.

In the 1990s this outflow has (at least temporarily) leveled off. Efforts to speed the country's development and economic diversification during the 1980s met with considerable success. Foreign companies—lured by low labor costs, tax breaks, and the anticipation of easy access to a single-market EC—arrived by the hundreds, including a sizeable contingent from the United States that created more than 50,000 jobs. Dublin (925,000), the capital, became the center for a growing number of light industries. In the trading arena, long-dominant farm exports were increasingly balanced by manufactures that by the late 1980s accounted for over half the annual foreign sales. Unfortunately for the Irish, the economic boom did not continue after 1990; today, mired in recession, Ireland struggles again with rising unemployment as factories lay off personnel, recent emigrants return (unable to find jobs in other economically pressed countries), and thousands of new workers enter the domestic labor force each year.

WESTERN EUROPE

The essential criteria for identifying a Western European region are those that define a continental core: this is the Europe of industry and commerce, of great cities and bustling interaction, of functional specialization and areal interdependence (Fig. 1-5). This is dynamic Europe, whose countries founded empires while forging democratic governments and whose economies staged an astounding recovery from the devastation of World War II and now form the cornerstone of the EC.

Yet if it is possible to recognize some semblance of historical and cultural unity in British, Scandinavian, and Mediterranean Europe, that quality is absent here in the melting pot of Western Europe. Europe's western coreland may be a contiguous area, but in every other way it is as divided as any part of the continent is or ever was. Unlike the British Isles, there is no *lingua franca* here; unlike Scandinavia, there is considerable religious fragmentation even within individual countries; unlike Mediterranean Europe, there is no common cultural heritage. Indeed, the core region we think we can recognize today on the basis of economic realities could be shattered at almost any time by political developments. It has happened before.

The two leading states of Western Europe are Germany and France (Fig. 1-11); for most of the past half-century (1945–1990), the former was partitioned into West and East Germany, but the collapse of Eastern Europe's communist system in 1989 led to a quick reunification. In addition, there are the three Low Countries of Belgium, the Netherlands, and Luxembourg (collectively called *Benelux*) and the two Alpine states of Switzerland and Austria. Northern Italy, by virtue of its industrialization and its interconnections with Western Europe, was identified as part of the continental core (Fig. 1-5); however, for the purposes of the present regionalization of Europe, all of Italy will be treated as part of Mediterranean Europe.

Germany

Territorially, Western Europe's largest state is France, but demographically and economically, Germany is paramount. In many other ways these two mainland powers present interesting and sometimes enlightening contrasts. France is an old state, by most measures the oldest in Western Europe. Germany is a young country, created in 1871 after a loose association of German-speaking states had fought a successful war against . . . the French! Modern France bears the imprint of Napoleon, who died just a few years after the political architect of Germany, Bismarck, was born.

On the map of Western Europe, it would seem that Germany, smaller than France in area, also has a less advantageous geographic position than its neighbor (Fig. 1-11). France has a window on the Mediterranean, the North Atlantic Ocean, the English Channel, and, at Calais, even a corner on the North Sea. Germany, on the other hand, has only short coastlines on the North Sea and the icy Baltic; the rest of its territory seems tightly landlocked by the Netherlands and Belgium to the west, by the Alps to the south, and by Poland in the east. But such appearances can be deceptive. Actually, France is at a disadvantage when it comes to foreign trade, because none of its natural harbors is particularly good; moreover, most of France's rivers and inland waterways are not navigable by large ocean-going ships.

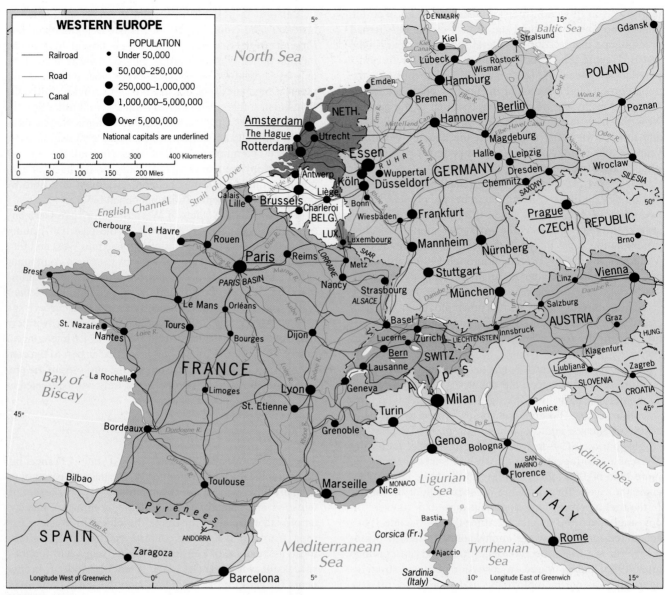

FIGURE 1-11

Germany is much better off. Although the mouth of the important Rhine River lies in the Netherlands, western Germany contains most of its course, which runs past one of Europe's major heavy industrial complexes, the Ruhr. Here, the Rhine is almost as good a connection with the North Sea and the Atlantic as a domestic coast would be. The Dutch port of Rotterdam, located at the mouth of the Rhine, is Europe's largest and serves as a more effective outlet for Germany than any French port is for France.

Another sharp contrast between Germany and France lies in the degree of urbanization in the two countries. Unlike moderately urbanized (73 percent) France, an overwhelming 90 percent of the German population reside in

towns and cities. Germany is also more industrialized than France, and the majority of its large urban centers lie along the zone of contact between the North European Lowland and the Central Uplands (see Fig. 1-2). This, as we know, is also the leading zone of coal deposits, and they gave rise to prewar Germany's three leading industrial agglomerations: the *Ruhr* near the Dutch border in the west; the *Saxony* area along the Czech border in the east; and *Silesia*, now in southern Poland (Fig. 1-11). Postwar West Germany was left with only the Ruhr, but it became the greatest heavy manufacturing complex in Europe. Today, it remains an advantaged industrial region, but many of its aging factories are experiencing the same economic-geo-

Continental Europe, too, has its rust-belts. Here in Duisburg, near the confluence of the Ruhr and Rhine rivers, aging facilities, idle smokestacks, and a half-empty worker parking lot mark a large steel mill. The once-dominant Ruhr industrial complex is in decline, with severe regional consequences.

graphic problems that bedevil the British North; despite a major effort to diversify and modernize its industries, the Ruhr's future is now an uncertain one.

The geography of Germany in the mid-1990s is dominated by the absorption of former East Germany into the economic system and infrastructure that was developed in postwar West Germany. Reunification of the country in 1990 created a new Germany that today contains 80.5 million residents (a population more than 20 million larger than that of any other European state), reinforced former West Germany's standing as the world's fourth biggest economy, and restored a formidable political unit in the center of Europe that borders no less than nine neighboring countries. The internal administration of reunified Germany is spatially organized within a framework of 16 *Länder* or federal states of recent origin, each of which reflects a traditional cultural or political region (or city-state) in the experience of the German nation (Fig. 1-12). Such subnational administrative restructuring is fully in line with the EC's desire to forge a more open Europe, one dominated less by self-centered nation-states and more by the direct interactions of their key component regions. This regionalization process is now well advanced in a number of countries, as we shall see when we discuss France and Spain.

On the new political map of Germany (Fig. 1-12), the old East Germany has been transformed into six *Länder*. The most important of these is the city-state of Berlin (3.3 million) (itself reunified in 1990 following the removal of its infamous wall), Germany's prewar capital, which in 1991 was once again designated the national headquarters. Not far behind Berlin in importance is resource-rich Saxony, eastern Germany's old industrial heartland anchored by the cities of Leipzig (585,000) and Dresden (555,000), which is now in the painful process of reinventing itself to fit into the country's capitalist system. The four remaining eastern *Länder*—

Mecklenburg-Western Pomerania, Brandenburg, Saxony-Anhalt, and Thüringia—are more rural in character, containing large areas of hilly, forested land suitable for recreational development that are interspersed with locally significant pockets of agricultural, raw-material, and specialized industrial production.

The ten *Länder* of former West Germany (Fig. 1-12) are dominated by the six states that contain the country's leading concentrations of economic activity: (1) North Rhine-Westphalia, home to the Ruhr manufacturing complex as well as nearby Bonn (315,000), seat of the old West German government now seeing most of its capital-city functions transferred to Berlin; (2) Hesse, with its major Rhine-Main industrial region centered on Frankfurt (3.6 million), Germany's financial headquarters and air-transport hub; (3) Baden-Württemberg and Bavaria, southern Germany's booming, high-technology industrial heartland, focused on the cities of München (Munich) (2.2 million) and Stuttgart (2.6 million), which is often labeled the country's "Sunbelt"; and (4) the city-states of Bremen (865,000) and Hamburg (2.6 million), Germany's leading seaports, now benefitting from the restoration of their hinterlands as the Iron Curtain no longer severs the basins of the Elbe River and other important northern German waterways.

The remaining four *Länder* are well known for their specialties: Rhineland-Palatinate (industrial clusters as well as the world-class wines of the Moselle and central Rhine valleys), Lower Saxony (Germany's leading agricultural *land*), Schleswig-Holstein (beef and hog production), and tiny Saarland (coal-mining and steel products).

The successful redrawing of the political map should not mask the enormous problems that attend the ongoing economic integration of former East and West Germany. At the outset of the 1990s, it was optimistically thought that within a year or two, the West's robust capitalist econ-

STATES (*LÄNDER*) OF REUNIFIED GERMANY

POPULATION

- Under 50,000
- 50,000–250,000
- 250,000–1,000,000
- 1,000,000–5,000,000
- Over 5,000,000

National capital is underlined

FIGURE 1-12

omy would be able to incorporate the defunct socialist system of the East. Experience has proven otherwise, however; the challenge is now regarded to be so much more difficult than the planners first realized that the process is expected to continue beyond the turn of the century and cost a staggering U.S. $750 billion.

There are several reasons for this. To begin with, the entire economic spatial organization of the East must be reoriented from its former Soviet fuel supplies, raw-material sources, and markets so that it meshes with the infrastructure of western Germany and the EC beyond. Physical facilities in the East, from power plants to farm buildings to transport and communications lines, were in shockingly poor shape when communism collapsed in 1989; the industrial base was so badly neglected and outmoded that half of it may have to be abandoned while much of the rest must be overhauled. Unfortunately, the East's service sector is so small that it cannot begin to absorb the hordes of displaced

When the countries under communist control were freed and opened to the world, the severity of industrial pollution there became evident. Air, waterways, soils, and groundwater were afflicted. This scene of what used to be East Berlin reflects the problem.

blue-collar and agricultural workers. Unemployment, driven ever higher as new private owners streamline inefficient operations, has reached epic proportions, widening the already severe imbalance in living standards (the gross domestic product of the East is still less than 20 percent of that of the West). Many desperate job seekers migrate west (over a million in 1990 and 1991 alone); more than 300,000 still live in the East but commute to work in the West. Social tensions run high as well: impatient *Ossies* often complain about the frustratingly slow improvement in their lifestyles at the hands of arrogant *Wessies*, and everyone grumbles about the huge influx of foreign refugees that arrived during the early 1990s thanks to the government's liberal attitude toward asylum seekers. These circumstances notwithstanding, the landscapes of the East are slowly transforming, opportunities there are rising, and Germany is undoubtedly making steady (though painful) progress toward its goal of truly uniting its long-divided halves.

Beyond Germany's borders, reunification has raised some politico-geographical questions that are certain to linger into the foreseeable future. Because Germany has twice plunged the world into war since 1914, there are millions who fear that "the third time never fails." Thus there is pressure for a demilitarized Germany, just as Japan was demilitarized in 1945. Another concern is united Germany's domination of Europe's economic geography; the United Kingdom and France, in particular, see themselves disadvantaged by the new consolidation of German economic power, and that may threaten the functioning of the EC. A third issue involves Germany's eastern boundary with Poland (the Oder-Neisse Line), superimposed at the end of World War II when Soviet-dominated East Germany lost a wide zone of territory to the Poles as part of the reconstruction of the postwar European map. West Germany's recognition of this boundary in 1990 as a precondition for reunification was decidedly reluctant, and many non-Germans can readily envision the rise of a renewed German nationalism driven by the cause of regaining "lost territories." The relocation of the capital from Bonn to Berlin—now peripherally positioned as a *forward capital* near the Polish border—could further spur such territorial demands.

France

France, Western Europe's other leading state, as we have already noted, is not as heavily urbanized as Germany. An important spatial expression of that distinction can be seen in Fig. 1-13, which shows the French population rather evenly distributed across the country. In fact, the only population concentration that really stands out on this map is the one centered on Paris in the region of Ile de France. True, Paris is without rival in France (and perhaps mainland Europe), but there is a huge gap between this cultural focus and primate city of 9.5 million and the second-ranking French city, Lyon (1.3 million). Why should Paris, without major raw materials in its immediate vicinity, be so dominantly large?

Paris owes its origins to advantages of *site* and its later development to a fortuitous *situation*. Whenever an urban center is studied, these are two highly significant locational qualities to consider. A city's **site** refers to the local physical attributes of the place it occupies—whether the land is flat or hilly, whether it lies on a river or coast, and whether

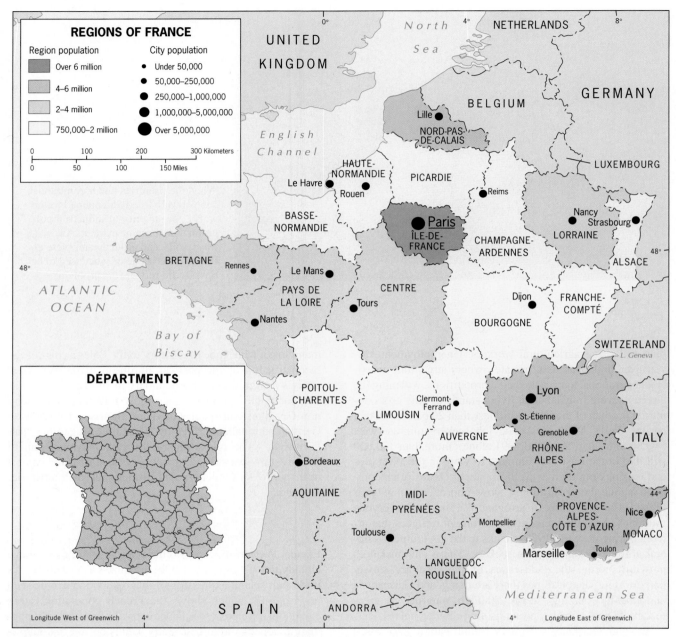

FIGURE 1-13 **Not shown on this map is the island region of Corse (Corsica).**

there are any obstacles to future expansion (such as ridges or marshes). By **situation** is meant the geographic position of the city with reference to surrounding areas of productive capacity, the size of its hinterland, the location of nearby competing towns—in short, the greater regional framework within which the city finds itself.

Paris was founded on an island in the Seine River, a place of easy defense where the waterway was often crossed. Exactly when pre-Roman settlement on this site, the *Ile de la Cité*, actually began is not known, but it functioned as a

Roman outpost some 2,000 years ago. For many centuries this defensibility of the site continued to be important, but then the island proved to be too small and the city began to expand along the banks of the river (Fig. 1-14).

Soon, the complementary advantages of Paris's situation were revealed. The city lay near the center of a large and prosperous agricultural area, and as a growing market its focality increased steadily. Significantly, the Seine is joined near Paris by several navigable tributaries providing access to various parts of the Paris Basin. Via these rivers

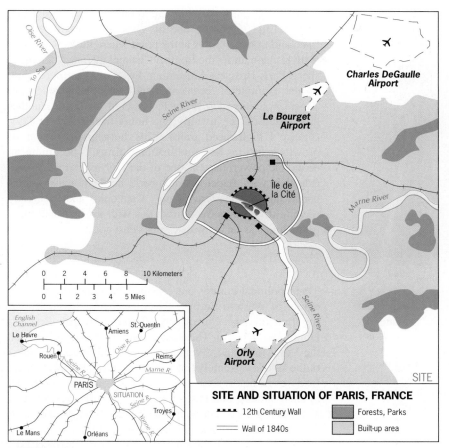

FIGURE 1-14

(the Oise, Marne, and Yonne), and via the canals that later extended them even farther, Paris can be reached from the Loire Valley, the Rhône-Saône Basin, the Lorraine industrial area, and from the Franco-Belgian border zone of coal-based manufacturing. Of course, there are land connections as well. The political reorganization that Napoleon brought to France included the construction of a radial system of roads—followed later by railroads—that focused directly on Paris (Fig. 1-14, inset map). To stay on top, of course, a city must always protect its situational advantages; therefore, to meet the movement challenges of the twenty-first century, a bold new plan is being implemented to provide the now-sprawling Paris region with an ultramodern, high-speed public transport system.

If it is not difficult to account for the greatness of Paris, it is quite another matter to account for the relatively limited development of French industrial centers because they possess coal as well as good-quality iron ore. What was lacking? For one thing, large supplies of high-quality coal; for another, the juxtaposition of such coal with cheap transport facilities and sizeable population clusters. So French

manufacturers did what Europeans have done almost everywhere: unable to compete in volume, they specialized—in high-quality textiles, precision equipment, and, of course, wines and cheeses.

France achieved greater strength in agriculture (it is still Europe's leading producer) with a much larger area of arable land than Britain or Germany and with the benefits of a moist, temperate climate free of extremes. Even so, a wide variety of local conditions exist, each with its special opportunities and limitations. Wheat is grown on the best soils and is France's leading crop; sugar beets and barley are raised on soils of lesser quality. Spatially, French agriculture is marked by an enormous diversity of production, so much so that it is impossible to delineate land-use regions—except in the vineyards of the Rhône-Saône, Garonne, and Loire valleys.

In the 1990s, the French are seeking to overcome past deficiencies in their economic geography by investing heavily in new high-tech industries, especially transportation and telecommunications. This effort is going well, and France has already earned a world-class reputation in the

The new Channel Tunnel (popularly called the "Chunnel") between Britain and France creates a direct link between island and continent for the first time in history. This view shows the portal on the British side, whose construction lagged well over a year behind its counterpart in France. Beneath the English Channel, the rail cars switch sides as the British keep left and the French (as well as all other Europeans) keep right.

manufacture of high-speed trains, aircraft, fiber-optic communications systems, and various space-related technologies. And in the all-important energy sphere, France is also developing, despite the well-known risks, a nationwide and state-of-the-art nuclear power grid that today supplies more than 75 percent of the country's electricity (whose prices are about two-thirds the Western European average because the French have cut their oil imports by half over the past two decades).

This ongoing development is taking place within a new subnational framework of 22 historically significant, province-level *regions* (analogous to Germany's *Länder*). These regions, which consist of groupings of the 96 smaller *départments* that date back to Napoleonic times, first appeared on France's political map in 1982 as part of a major decentralization of government control from Paris (Fig. 1-13). By the late 1980s the new regions had acquired substantial powers, among them the election of governing regional councils, the establishment of taxation levels, and considerable autonomy in borrowing and spending for local development. This administrative reorganization soon achieved its desired effects and sparked growth booms all over France, particularly in the cities and towns that centered each region. In the far north in Nord-Pas de Calais, for in-

stance, plans are well underway to make the old "Rustbelt" city of Lille (965,000) one of Western Europe's leading trade and transport hubs at the point where the new rail line from London under the English Channel intersects the high-speed Paris-Brussels-Amsterdam line. An even more spectacular example of recent regional development is Rhône-Alpes in southeastern France, centered on Lyon, another high-speed rail nexus only two hours from Paris by TGV (France's high-speed train). This burgeoning region of growth industries and multinational firms is becoming an economic powerhouse as one of the *four motors of Europe* (together with Baden-Württemberg, northern Italy's Lombardy, and northeastern Spain's Catalonia) and is already seeking to forge its own direct ties with international business centers as far away as China's east coast.

Benelux

Three political entities are crowded into the northwestern corner of Western Europe: Belgium, the Netherlands, and tiny Luxembourg, collectively referred to by their first syllables (Be-Ne-Lux). These states are also frequently called the Low Countries, a very appropriate label because most of the land is extremely flat and lies near (and in the Netherlands even below) sea level. Only toward the southeast, in Luxembourg and eastern Belgium's Ardennes, is there a hill-and-plateau landscape with elevations in excess of 1,000 feet (300 m). As with Germany and France, there are major differences between Belgium and the Netherlands. Indeed, there are such contrasts that the two countries find themselves in a position of complementarity.

Belgium is marked by two industrial corridors. One is the coal-based east-west axis through Charleroi and Liège, where there are heavy industries. The second corridor of lighter and more varied manufacturing extends north from Charleroi through Brussels to the major port of Antwerp. The diversified industrial products of these areas include metals, chemicals, furniture, and specialties such as pianos, soaps, and cutlery. The Netherlands, in contrast, has a large agricultural base (along with its vitally important transport functions); it can export dairy products, meats, vegetables, and other foods. Hence, there was mutual advantage in the Benelux economic union of the 1940s because it facilitated the reciprocal flow of needed imports to both countries, and it doubled the domestic market.

The Benelux countries are among the most densely populated on earth, and space is truly at a premium. Some 25 million people inhabit an area about the size of Maine (home to 1.2 million). For centuries the Dutch have been expanding their living space—not at the expense of their neighbors, but by wresting it from the sea. The greatest project so far is the draining of almost the entire Zuider Zee, begun in 1932 and scheduled for completion after the turn of the century. In the southwest the islands of Zeeland

THE CHANGING SOCIAL GEOGRAPHY OF WESTERN EUROPE

Europe in modern times has sent millions of its inhabitants to populate the Americas, Australia, and other overseas realms. Where they overwhelmed local communities, the white settlers from Europe created new societies in the European mold. Even where they remained in the minority, as in South Africa and Algeria, they drastically changed their new homelands.

But now Europe is experiencing an immigrant invasion itself. In recent decades, immigrants have come to Europe—especially to Western Europe—in large numbers. By the mid-1990s, about 15 million immigrants were living in Western Europe, nearly 9 percent of the region's population. The great majority came from the Third World, with many arriving from comparatively nearby countries (such as Turkey and Algeria) to take jobs that became available during Europe's industrial boom of the 1960s and 1970s. Others came from more distant one-time colonies, such as Indonesia, Angola, and Suriname, exercising their right to do so as subjects of the colonial power. And since the fall of communism in 1989, hundreds of thousands of migrants have entered the region from Eastern Europe. Almost all, whatever their origin, settled in Europe's great urban areas, where jobs and other opportunities were located.

When Europe boomed, these immigrants found employment and were generally welcomed. But when Western Europe's economies slowed after 1980, the newcomers were less welcome. Opposition to uncontrolled immigration arose. Competition between the immigrants and Europeans for scarcer jobs increased. Social problems in the cities intensified. For the Third World immigrants, Europe did not prove to be the melting pot that Europeans found in America. Charges of discrimination were made against European governments, and many European cities with ethnic communities (where the immigrants mostly cluster) found themselves with problems for which they were not prepared.

Major European cities, including capitals such as Paris and Amsterdam, contain large cohesive neighborhoods where the immigrants have implanted some of their own culture. France's leading cities (not only Paris but also Lyon and Marseille) have many suburbs where street signs are in Arabic, where Islam rules, and where the atmosphere is that of urban Morocco, Tunisia, or Algeria. The Turkish imprint on many German cities is similarly strong, although of 2-million-plus Turks only a few thousand have been permitted to become German citizens. In Amsterdam, hundreds of thousands of Surinamese immigrants have changed the face of the city, whose social geography has altered rapidly as large numbers of white residents have moved to outer suburbs. Such changes cannot occur without difficulty, and Amsterdam suffers from increased crime and related problems.

Thus Western Europe, itself a patchwork of societies and traditions, faces the need for new adjustments and the reality of a new social map. The transition is not going smoothly in many quarters, however, as reactionary political movements gain strength, agitation to sharply limit immigration intensifies (particularly in Germany), and acts of violence against newcomers multiply. Europe has been described as the most international of realms; almost overnight, it has also become the most intercultural.

---

are being connected by dikes and the water pumped out, creating additional *polders* (reclaimed lands). Another future project involves the islands that curve around northern Holland, which may be connected by dikes to dry and reclaim the intervening Wadden Sea.

Three cities—Amsterdam, Rotterdam, and The Hague—anchor the Netherlands' triangular core area. Amsterdam (1.1 million), the constitutional capital, remains very much the focus of the Netherlands, with a bustling commercial center, a busy port, a variety of light manufactures—and an increasingly heterogeneous social complexion (see box above). Rotterdam (1.1 million), Europe's busiest port, is the shipping gateway to Western Europe, commanding the entries to both the Rhine and Meuse (Maas) rivers. Its modern development mirrors that of Germany's Ruhr industrial region and the adjacent Rhine valley; thus ongoing manufacturing decline in the Rhine hinterland as well as increased competition from other European ports is eroding Rotterdam's historic situational advantage. The third city in the triangle, The Hague (455,000), is the seat of the Dutch government and the home of the United Nations' World Court.

Collectively, these three cities of the triangular core have spawned a conurbation called *Randstad*, and their coalescence has created a ring-shaped, multi-metropolitan complex that surrounds a still-rural center (the literal translation of *rand* is edge or margin). A more precise labeling of the conurbation, however, would be Randstad-Holland, because *Holland* (meaning "hollow country") refers specifically to the Dutch heartland that faces the North Sea in these lowest-lying western provinces of the Netherlands.

In contrast to Belgium, the Dutch resource base has always been heavily agricultural. With a premium on space, rural population densities are very high, and practically every square foot of available soil is in some form of productive use. Only in the southeast does the Netherlands share the Campine belt of good coal deposits that extends across northern Belgium. Natural gas reserves discovered in the postwar era have been opened up in the northeast, adding to Dutch energy supplies.

With the existing limitations of space and raw materials, the Dutch and Belgians have also turned to managing international trade as a leading economic activity. It is hard to think of any other countries that could be better positioned for this. Not only do the Rhine and Meuse form primary arteries that begin in the interior and terminate in the Low Countries, but Benelux itself is also surrounded by the most productive countries in Europe. Belgium's capital city, Brussels (1.3 million), is also a global city of consequence. This historic royal headquarters, positioned awkwardly astride the Flemish-French (Walloon) linguistic dividing line across Belgium (see Fig. 1-6), has become a major international administrative center. Hundreds of multinational corporations with European interests have their central offices here, enhancing the city's role as a financial center and commercial-industrial complex. And Brussels has also become the administrative headquarters for international economic and military organizations, including the European Community and NATO (North Atlantic Treaty Organization).

It was pointed out earlier that Belgium is now facing the threat of devolution as Flemish-Walloon ethnic divisions refuse to moderate. Moreover, these centrifugal forces are intensifying as the economic gap widens between more prosperous Flanders and heavy-industry-dependent Wallonia—and it is no longer difficult to envision a Czechoslovakian-style breakup of the Belgian federation.

Switzerland and Austria

Switzerland and Austria share a landlocked, increasingly peripheral location and the mountainous topography of the Alps—and little else. On the face of it, Austria would seem to have the advantage over its western neighbor; it is twice as large in area and has a bigger population than Switzerland. A sizeable portion of the upper Danube valley lies in Austria, and the Danube is to Eastern Europe what the Rhine is to Western Europe. Moreover, no Swiss metropolis can boast of a population even half as large as that of the famous Austrian capital, Vienna (2.2 million). Austria also has considerably more land that is relatively flat and cultivable, a prize possession in this part of Europe. And that is not all. From what is known of the resources buried within the Alpine topography, Austria is again the winner, with deposits of iron ore, coal, lead, zinc, and graphite; Switzerland has hardly any exploitable mineral deposits.

From all this, we might infer that Austria should be the leading country in this part of Europe, and that impression would probably be strengthened by a look at some cultural geography. Take the map of European languages (Fig. 1-6). Austria is a unilingual state in which only one language (German) is spoken throughout the country. But in Switzerland no less than four languages are in use: German is spoken in the largest (northern) part; French over the western quarter; Italian in the extreme southeast; and in the mountains of the central southeast, a small remnant of Romansch usage. Similarly (and adding to the picture of disunity), Swiss religious preferences are evenly split between Protestantism and Roman Catholicism, whereas Austria is 85 percent Catholic.

These hindrances notwithstanding, it is the Swiss people who have forged for themselves a superior standard of living, and it is the Swiss state, not Austria, that has achieved greater stability, security, and progress. The world geomorphological map (Fig. I-4) sometimes seems to suggest that mountainous countries share certain limitations on development. It is therefore tempting to generalize about the impact of mountainous terrain—and its frequent corollary, *landlocked location*—as preventing productive agriculture, obstructing the flow of raw materials, and hampering the dissemination of new ideas and innovations. Tibet, Afghanistan, and the Andean portions of South American states seem to prove the point. That is why Switzerland is such an important lesson in human geography: all the tangible evidence suggests that here is a European area that will be economically deprived and lack internal cohesion—but the actual situation is exactly the opposite!

The Swiss, through their skills and abilities, have overcome a seemingly restrictive environment; they have made it into an asset that has permitted them to keep pace with industrializing Europe. First they took advantage of their Alpine passes to act as middlemen in interregional trade; then they used the water cascading from those mountains to produce the hydroelectric power that spawned a high-quality, specialized industrial base; and finally they learned to accommodate with professional excellence the tourists who came to visit their beautiful mountain country. Simultaneously, Swiss farmers were learning how to get the most out of their efforts on the limited non-highland rural space available, and the country today successfully specializes in dairy farming and the worldwide export of cheeses and chocolate products.

The majority of the country's population is concentrated in the central plateau, where the land is at lower elevations, and here lie the major cities. Zürich (990,000) is the largest and functions as a banking center of global importance. Geneva (445,000) is the famous city of international organizations and conferences, and between them lies Bern (330,000), the quaint Swiss capital. Here, too, are found many of the industrial centers, which must import virtually all their raw materials; but their products (precision machinery, instruments, tools, fine watches) have so prestigious a reputation that these Swiss exports are guaranteed a prominent place on the world market. Thus a strong case can be made that Switzerland is part and parcel of the European core.

After more than seven centuries of strict neutrality vis-à-vis its neighbors, Switzerland in the mid-1990s is finally relaxing that tradition of isolation by considering the formalization of its linkages to Europe's coreland through membership in the EC. This shift was prompted by the growing fear that the Swiss would lose their international competitive edge in the vital domains of trade, education, and research and development if they continued to remain outside the arena of European economic integration. Switzerland's centrality on the map of Europe, despite its location astride the rugged Alps, virtually guarantees that it can swiftly become the geographic crossroads of the EC. As many as 100 million people per year already cross the Alps by road and rail, and plans are underway to build more tunnels and enlarge existing ones. Not surprisingly, all this activity is wreaking havoc on the Alpine environment. Air pollution generated by highway traffic has decimated trees and wildlife; roadbuilding and ski-resort construction has destabilized countless slopes and affected the flows of water courses; ubiquitous noise and trash are ruining the once-pristine landscapes of all but the least accessible uplands. In response, environmental activist groups and "green" political parties are banding together to try to halt the spread of this blight, and a major confrontation is shaping up that could yet curtail plans to open the Alps to further development.

Austria is a much younger state than Switzerland, and its history is far less stable. Modern Austria is a remnant of the Austro-Hungarian Empire that fell during World War I; subsequently, the country suffered through convulsions far more reminiscent of Eastern Europe than Switzerland's peaceful Alpine isolation. Stability finally returned in 1955 when the last foreign occupiers withdrew, but ever since then Austria's most difficult problem has been its reorientation to Western Europe. Even Austria's physical geography seems to demand that the country look eastward: it is at its widest, lowest, and most productive in the east; the Danube flows eastward; and Vienna also lies closer to the eastern perimeter. These obstacles notwithstanding, Austria in 1993 began negotiations (alongside Sweden and Finland) to enter the EC, a move that should at last bind the country to the prosperous spatial economy of the European heartland.

NORDIC EUROPE (NORDEN)

In three directions from its core area, Europe changes quite drastically. To the south, Mediterranean Europe is dominated by Greek and Latin influences and by the special habitat produced from the combination of Alpine topography and a Mediterranean climatic regime. To the east lies truly continental Europe, which is less industrialized and urbanized than the coreland. And to the north lies the Nordic Europe of Scandinavia (Sweden and Norway), Denmark, Finland, Estonia, and Iceland (Fig. 1-15), almost all of it separated by water from what we have defined as the European core.

Despite its peripheral location, Nordic Europe is not "underdeveloped" Europe. To the contrary, a great deal has been achieved here. In terms of resources and in a realmwide context, however, the northern region of Europe is not particularly rich (although the exploitation of the North Sea oilfields has lately boosted Norway's economy). Northern Europe possesses the realm's most difficult environments: cold climates, poorly developed soils, steep slopes, and sparse mineral resources mark much of the region, with only small Denmark offering better opportunities thanks to fertile soils and lower relief. This region's conditions are reflected by its comparatively small population—its six countries contain just 25.2 million, a total less than that of Benelux. The overall land area, on the other hand, is almost the size of the entire European core. People go where there is a living to be made, but the living in most of Scandinavia is not easy.

Several aspects of Nordic Europe's location have much to do with this. First, this is the world's northernmost group of states; although Russia, Canada, and the United States possess territory at similar latitudes, each of these much larger countries has its national core area in a more southerly position. The Northern Europeans themselves call their region *Norden*, an appropriate term indeed. Second, Norden, as viewed from Western Europe, is on the way to nowhere. How different would the relative location of the Norwegian coast be if important world shipping routes paralleled the shoreline on their way to and from the European core? Third, except for relatively small Denmark and Estonia, all of Norden is separated by water from the rest of Europe. As we know, water has often been an ally rather than an enemy in the development of Europe—but mostly where it could be used for the interchange of goods. Norden, however, lies separated from Europe *and* relatively isolated in its northwestern corner. The only major exception to the bleak Scandinavian rule is found in Denmark, southern Sweden, and Estonia, which are really extensions of the North European Lowland.

Nordic Europe's relative isolation did have some positive consequences as well. Its countries have a great deal in common. They were not repeatedly overrun by different European peoples. The three major languages—Danish,

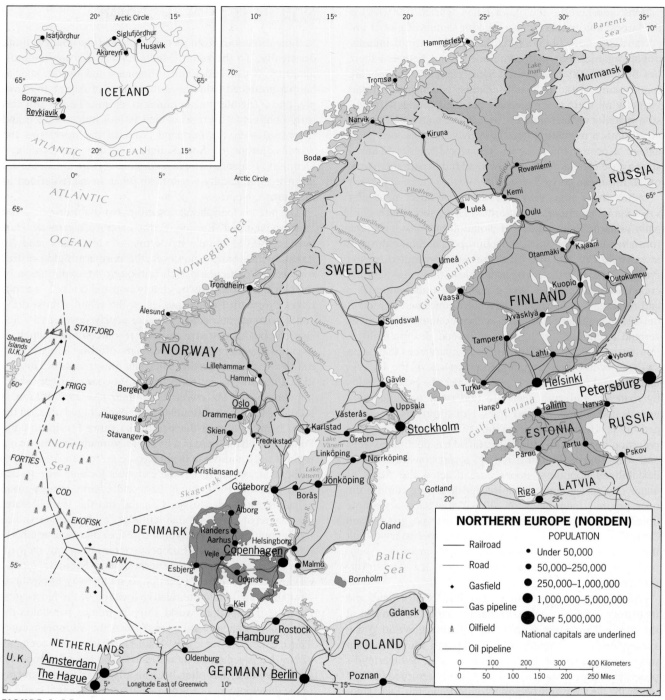

FIGURE 1-15

Swedish, and Norwegian—are mutually intelligible, allowing people to converse without requiring an interpreter. Icelandic belongs to the same Germanic language subfamily (see Fig. 1-6); only Finnish and closely related Estonian are of totally different origin, but Finland's long period of contact with Sweden has helped to overcome this linguistic barrier. Furthermore, in each of the Scandinavian countries, there is overwhelming adherence to the same Lutheran church, and in each it is the recognized state religion. Finally, there is considerable similarity in the political evolution of the Scandinavian states: democratic representative parliaments emerged early, and individual rights have long been carefully protected.

Because of their higher-latitude location, the six Nordic states also share generally common environmental conditions. Understandably, populations tend to be concentrated in the more southerly parts of each country. Although the warm waters of the Atlantic Ocean temper the Arctic cold and keep Norway's ports open year-round, their effect diminishes both northward and landward. Most important is Scandinavia's high mountain backbone that blocks air moving eastward across Norway and into Sweden. Not only does this highland limit the milder maritime belt to a narrow strip along the Norwegian cost, but by its elevation it also allows Arctic conditions to penetrate southward into the heart of the Scandinavian Peninsula. One cannot help speculating on what Western Europe's environment might have been if that Scandinavian backbone had continued southward through the Low Countries into France and Spain. At the same time, the lack of a physiographic barrier between Scandinavia and Western Europe has permitted airborne industrial pollution to become a major environmental hazard (see the box entitled "Acid Rain").

Norway

Norway, with its long Atlantic coastline, consists almost entirely of mountains whose valley soils have been stripped away by glaciation. Only in the southeast around the capital of Oslo (490,000), in the southwest below Bergen (235,000), and on the west coast near Trondheim (150,000) are there limited areas of good soil and agriculture (apart from tiny bottomland patches in fjorded valleys near the coast). Yet even though Norway came off second best in the division of land resources on the Scandinavian Peninsula, it has turned to the sea, which more than makes up for this deficiency. The Norwegian fishing industry is a sizeable one, and its fleets ply all the oceans, especially the highly productive fishing grounds that lie close to Norway's own waters. Another major maritime pursuit is the Norwegian merchant marine, which is mostly hired to transport goods between other trading countries. But what proved to be the most important benefit of the sea remained unrealized until the 1970s: enormous petroleum and natural gas deposits lying below Norway's share of the North Sea floor. Large-scale extraction of these energy resources has entered its third decade, annually earning about half the country's income and making now-wealthy Norway the "Sheikdom of Europe."

Sweden

Norway's eastern neighbor on the Scandinavian Peninsula is also favored in many respects. Of Sweden's two agricultural zones, the leading one lies at the southernmost end of the country just across the Kattegat Straits from Denmark. This area resembles Denmark agriculturally except that more grain crops, especially wheat, are raised here; Malmö (690,000) is the area's chief metropolis and agricultural service center. Sweden's other farming zone lies astride a

The rugged, highland, snowy countryside of Scandinavia stands in sharp contrast to the rest of Europe. Hammerfest, Norway represents the comparatively barren scenery of the far north, beyond the forest-clad slopes of the south.

ACID RAIN

One of the most discussed environmental perils of recent years is **acid rain**, caused by the considerable quantities of sulfur dioxide and nitrogen oxides released into the atmosphere as fossil fuels (coal, oil, natural gas) are burned. These pollutants combine with water vapor contained in the air to form dilute solutions of sulfuric and nitric acids, which are subsequently washed out of the atmosphere by rain or other types of precipitation such as fog and snow.

Although acid rain usually consists of relatively mild acids, they are sufficiently caustic to do great harm to certain natural *ecosystems* (the mutual interactions between groups of plant or animal organisms and their habitat). There is much evidence that this deposition of acid is causing lakes and streams to acidify (resulting in fish kills), forests to become stunted in their growth, and acid-sensitive crops to die in affected areas. In cities, the corrosion of buildings and monuments is both exacerbated and accelerated. To some extent, acid rain has always been present in certain humid environments, originating from such natural events as volcanic eruptions, forest fires, and even the bacterial decomposition of dead organisms. However, during the past century as the global Industrial Revolution has spread ever more widely, the destructive capabilities of natural acid rain have been greatly enhanced by human activities.

The geography of acid rain is most closely associated with patterns of industrial agglomeration and middle- to long-distance wind flows. The highest densities of coal and oil burning are associated with major concentrations of heavy manufacturing, such as those already discussed for Britain and Western Europe. As these industrial areas began to experience increasingly severe air pollution problems in the second half of the twentieth cen-

tury, many nations (including the United States in 1970) enacted environmental legislation to establish minimal clean-air standards. For industry, the easiest solution has often been the construction of very tall smokestacks (1,000 feet [300 m] or higher is now quite common) that disperse pollutants from source areas via higher-level winds. These longer-distance winds have been effective as transporters—with the result, of course, that more distant areas have become dumping grounds for sulfur- and nitrogen-oxide wastes. Regional wind flows all too frequently steer these acid rain ingredients to wilderness areas, where livelihoods depend heavily on tourism, agriculture, fishing, and forestry.

The spatial distribution of acid rain within Europe offers a classical demonstration of all this. As Fig. 1-16 shows, high-sulfur-emission sources are located in the leading manufacturing complexes of England, France, Belgium, Germany, Poland, and the Czech Republic. The map also indicates that prevailing winds from these areas converge northward toward Scandinavia, with a particularly severe acid rain crisis occurring in southern Norway. Lake acidity there is already in the moderately caustic 4-to-5 range on the pH scale of 0 to 14 (7 is neutral), and most fish species, the phytoplankton they feed on, and numerous aquatic plants have been obliterated.

Scandinavia is not alone in experiencing these environmental problems: at least 25 percent of Europe's forests have become affected. In Central Europe, forest ecosystems are now so badly damaged by acid rain that vast woodland areas are diseased and dying. More than one-third of Austria's forests are threatened; in Switzerland, the proportion nears 50 percent. The worst devastation, however, has occurred in Germany, where over two-thirds of the woodlands are afflicted and entire

line drawn southwest from the capital, Stockholm (1.8 million), to Göteborg (885,000). Here dairying is the main activity, but farming is overshadowed by industry. Swedish manufacturing, in contrast to that of certain Western European countries, is scattered through dozens of small and medium-sized towns.

Unlike Denmark and Norway, Sweden has the resources to sustain such industries. For a long time, Sweden served as an exporter of raw or semi-finished materials to industrializing countries; but increasingly, the Swedes are making finished products themselves, honing skills and specializing much the way the Swiss have done. The list of

Sweden's well-known products is now quite long and includes furniture, stainless steel, automobiles, electronic goods, and glassware. And there is a great deal more, much of it based on relatively minor local resources. Apart from iron, of which there is an ample supply in the area around Kiruna north of the Arctic Circle, the Swedes mine copper, lead, zinc, manganese, and even some silver and gold.

Finland

In Finland, the alternatives are fewer. Most of the country is too cold and its glacial soils too thin to sustain permanent

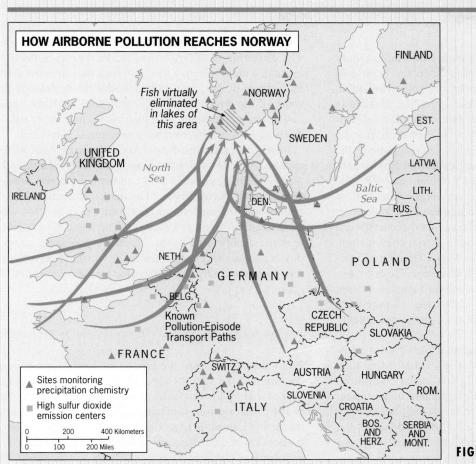

HOW AIRBORNE POLLUTION REACHES NORWAY

FINLAND

NORWAY

Fish virtually eliminated in lakes of this area

EST.

SWEDEN

LATVIA

UNITED KINGDOM

North Sea

Baltic Sea

LITH.

IRELAND

DEN.

RUS.

POLAND

NETH.

GERMANY

BELG.

Known Pollution-Episode Transport Paths

CZECH REPUBLIC

SLOVAKIA

FRANCE

AUSTRIA

HUNGARY

SWITZ.

SLOVENIA

ROM.

▲ Sites monitoring precipitation chemistry

■ High sulfur dioxide emission centers

ITALY

CROATIA

BOS. AND HERZ.

SERBIA AND MONT.

| 0 | 200 | 400 Kilometers |
| 0 | 100 | 200 Miles |

FIGURE 1-16

forests (such as the once-lovely Black Forest in the country's southwestern corner) are all but decimated. Government agencies have belatedly recognized this ecological disaster, but their proposed remedies may well prove to be a case of too little too late.

agriculture. Where farming is possible, mostly along the warmer coasts of the south and southwest, the objective is self-sufficiency rather than export. Known mineral deposits are few, but the Finns still have succeeded in translating their limited opportunities into a healthy economic situation; once again, the skills of the productive population play a major role. The domestic market sustains a textile industry centered at Tampere (335,000) and metal industries for shipbuilding and locally needed machinery at Turku (315,000) and the capital, Helsinki (1.0 million). Elsewhere, Finland's extensive forests have long supported pulp, paper, and timber production (an activity also wide-spread in Norway and Sweden); wood and wood products annually account for over half of the country's exports.

Finland today is accommodating itself to a changed politico-geographical reality. The demise of the Soviet Union in 1991 abruptly ended more than 40 years of *Finlandization*, an arrangement under which the powerful former neighbor to the east guaranteed security in exchange for the right to dictate the Finnish government's foreign and defense policies. Economically, however, reorientation away from the east is causing some problems because each year large quantities of Finnish goods were bartered to the Soviet Union for badly needed fuel supplies. The Finns are not looking backward, however; they are now enthusiastically turning toward Western Europe, where, as we have already noted, they are negotiating for full membership in the EC.

Estonia

Estonia became independent after more than a half-century of domination by the Soviet Union when the latter collapsed at the end of 1991. However, unlike its two Baltic neighbors to the south, Latvia and Lithuania, Estonia is now designated a part of Norden rather than Eastern Europe. If you refer back to Fig. 1-6, you will observe an important cultural basis for this regional assignment: the Estonian language belongs to the Finnic (not Baltic) subfamily. On ethnic grounds, too, the Estonians claim Nordic ties because their country was part of the Kingdom of Sweden when it was incorporated into the Russian Empire after 1710. The Estonians tried to maintain their northward connections even during the period of Soviet colonization, and Finland's independence on the Soviet Union's perimeter kept alive Estonia's dream of liberation. Today these linkages are being fully restored (Estonia's capital, Tallinn [540,000], lies almost directly across the 35-mile-wide Gulf of Finland from Helsinki and is easily and quickly reached by hydrofoil).

Estonia, about the size of Vermont plus New Hampshire, consists of a coastal plain of low relief. More than one-quarter of the country's small population (1.6 million) is Russian, which may prove to be a problem in the future. Estonia is industrialized, thanks to ample supplies of electric power derived from substantial oil shale deposits, but its overall economic opportunities are limited. Like the other republics of the former U.S.S.R., Estonia was tightly enmeshed in the planned Soviet economy, and extracting itself from that web is a slow and difficult task.

Denmark

Denmark is territorially the smallest of the countries of Northern Europe, but its population of 5.2 million ranks it second in the region after Sweden. Consisting of the Jutland Peninsula and adjacent islands between Scandinavia and Western Europe, Denmark has a comparatively mild, moist climate. It also has level land and soils good enough to sustain intensive agriculture over 75 percent of its area. Denmark exports dairy products, meats, poultry, and eggs, mainly to its chief trading partners, Germany, Sweden, and the United Kingdom.

Denmark's capital, Copenhagen (1.3 million), has long been a place where large quantities of goods are collected, stored, and transshipped. This is because it lies at the *break-of-bulk* point where many oceangoing vessels are prevented from entering the shallow Baltic Sea; conversely, ships with smaller tonnages ply the Baltic and bring their cargoes to this pivotal collecting station. Thus Copenhagen is an *entrepôt* whose transfer functions maintain the city's position as the lower Baltic's leading port.

Iceland

The westernmost of Norden's (and Europe's) countries is Iceland, a hunk of active volcanic rock that emerges above the surface of the frigid waters of the North Atlantic just south of the Arctic Circle. Its population (267,000) gives Iceland the dimensions of a microstate, and about half of its residents are concentrated in the capital, Reykjavik.

Iceland shares with Scandinavia its difficulties of terrain and climate (which are even more severe here), and it also has ethnic affinities with continental Norden (its settlers came from Norway and Denmark). Not until 1918 did Iceland gain its modern-era autonomy; only in 1944 were all remaining political ties to Denmark renounced and a republic finally established. Icelandic economic geography is almost entirely oriented to the surrounding sea, with fish products yearly accounting for over 80 percent of the country's exports.

MEDITERRANEAN (SOUTHERN) EUROPE

From Northern Europe we turn to the four Mediterranean countries of the south: Greece, Italy, Spain, and Portugal (Fig. 1-17). With this shift from near-polar Europe into near-tropical Europe, it is reasonable to expect strong contrasts, and indeed there are many—but there also are similarities. Once again, we are dealing with peninsulas: two are occupied singly by Greece and Italy, and the third—the Iberian Peninsula—jointly by Spain and Portugal. Moreover, there is effective separation from the Western European core. Greece lies at the southern end of Eastern Europe and has the sea and the Balkans between it and the rest of the realm; Iberia lies separated from France by the Pyrenees, which through history have proven to be an imposing mountain barrier; and then there is Italy. Southern Italy lies far removed from Western Europe, but the north is situated very close to it, pressing against France, Switzerland, and Austria. For many centuries northern Italy was in close contact with Western Europe and developed less as a Mediterranean area than as a part of the core area of Western Europe (see Fig. 1-5).

Just as the Nordic countries share many cultural bonds, so do the countries of Mediterranean Europe. Firm interconnections were established early by the Greeks and Romans. When this unity did not last, new political arrangements replaced the old, and the Romanic language differentiated into Portuguese, Spanish, and Italian. In Greece the Roman tide was resisted to a surprising extent, but the underlying shared legacy remains strong to this day.

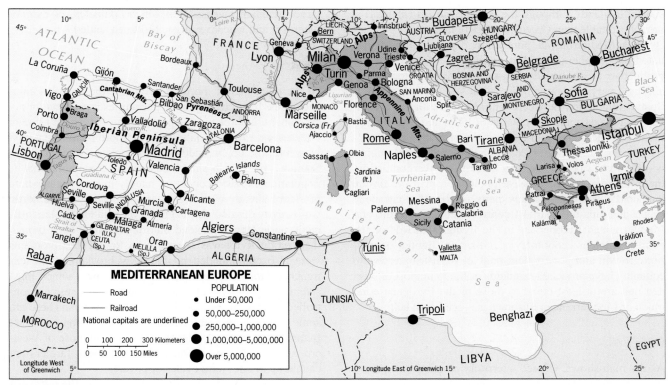

FIGURE 1-17

Like Norden, Mediterranean Europe lies largely within a single climatic zone, and from one end to the other the environmental opportunities and problems are similar. The opportunities lie in the warmth of the near-tropical location, and the problems are largely related to the moisture supply, particularly its limited quantity and the dryness of the warmest months. In topography and relief, too, the Mediterranean countries share similar conditions—conditions that would look quite familiar to a Norwegian or a Swede. Much of Mediterranean Europe is upland country with steep slopes and poor, thin, rocky soils. For its agricultural productivity, the region largely depends on river valleys and coastal lowlands. As with most rules there are exceptions (as in northern Italy and northern and interior Spain), but generally the typical Mediterranean environment prevails.

Neither is Mediterranean Europe better endowed with natural resources than Norden. Both Greece and Italy are deficient in coal and iron ore (Italian industry relies on massive imports); only in northwestern Iberia are sizeable deposits of these commodities positioned close to each other. Yet Spain, best endowed of all Mediterranean countries, has only moderately industrialized, exporting most of its minerals to Europe's coreland. Woodland resources in Southern Europe also offer a discouraging picture: unlike Scandinavia and Finland, the Mediterranean region stands largely denuded, its once extensive forests obliterated to supply fuel, building materials, and additional agricultural space. With regional fuels in generally short supply, hydroelectric power opportunities have been developed, but to a lesser extent than in Nordic Europe. Only northern Italy, near the Alps and their dependable precipitation, has had any real success; elsewhere, given the seasonal and rather low rainfall, water shortages are all too frequent when streams run dry and water levels behind dams recede.

From what we know of Mediterranean dependence on agriculture, the historical geography of trade routes and urban growth, and the general topography of Southern Europe, we can fairly well predict the distribution of the region's 118 million inhabitants. In Italy, more than 25 million of the country's 58 million people are concentrated in and near the basin of the Po River in the far north; to the south on the Italian Peninsula, large population clusters exist in coastal lowlands and river valleys and on both flanks of the Appennine mountain spine all the way from Genoa to Sicily. In Greece, today as in ancient times, densely settled coastal plains are separated from one another by relatively empty highlands; the largest agglomerations are in the lowland dominated by Athens and on the western side of the Peloponnesus Peninsula. Both Spain and Portugal also contain high-density settlements in their coastal lowlands, although the interior Meseta (plateau) is more hospitable than Greece's rocky uplands.

Thus Mediterranean population distribution is dominated by peripheral location, by heavy concentrations in productive areas (usually coastal and riverine lowlands), and by a varying degree of isolation within these clusters. Spatial interaction between central and southern Greece, between the east and west coasts of Italy, and between Atlantic and Mediterranean Iberia is therefore not always effective; terrain and distance are the obstacles. Although it is difficult to say exactly what constitutes overpopulation, there has long been excessive population pressure on the land and resources in many parts of the Mediterranean Basin. Perhaps the sharpest contrast between Nordic and Mediterranean Europe lies in the living standards of the people: whereas economic specialization, limited population growth, and government policies to distribute wealth equitably have produced standards of living in most of Norden that are at least equal to those of the European core, much of Mediterranean Europe has lagged well behind.

Greece

As the map shows, Greece's territory consists of a peninsular mainland flanked by numerous islands large and small (Crete is the largest). The country's barren terrain is rocky and rugged. Greece is perhaps the least favorably endowed of the Mediterranean countries: less than a third of its area is presently capable of supporting cultivation, under 1 acre (0.4 hectare) per person when Greece's extremely high rural population density is factored in. Thus, many Greek farmers are forced to engage in near-subsistence agriculture because their incomes are insufficient to buy modern farm machinery, fertilizer, and essential irrigation equipment. Yet agriculture is Greece's mainstay because other economic opportunities are still quite limited. Greek farmers raise wheat and corn for the home market, and tobacco, cotton, and such typical Mediterranean produce as olives, grapes, citrus fruits, and figs for export.

Reflecting the sizeable agricultural population, only 58 percent of Greece's inhabitants live in cities and towns (significantly below the Europe-wide average of 73 percent); the urban hierarchy that has evolved is dominated by primate Athens (3.6 million). This famous capital city is also the commercial, cultural, and historic focus of Greece, and despite its limited industry, Athens has grown to a size far beyond what would be expected for such a relatively poor, agrarian country. With its outport of Piraeus, it stands at the head of the Aegean Sea; Athens also has a major international airport. Moreover, thanks to its heritage of ancient structures, the Greek capital is one of the Mediterranean's major tourist attractions—despite a serious air pollution problem that threatens the fragile monuments of antiquity. Yet another source of national income is Greece's

large merchant marine fleet, which competes for cargoes wherever and whenever they need to be hauled.

Spain and Portugal

At the other end of the Mediterranean, the Iberian Peninsula is less restrictive in the opportunities it presents for development. Unbroken, compact Iberia is much larger than Greece, and raw materials are in far more plentiful supply. However, the countryside is overpopulated, and one price Spain has paid for its slow industrial development is that its population explosion occurred mainly in rural areas where pressures already were high. Land, especially in the north, was heavily divided; farms grew ever smaller and less efficient; agriculture expanded onto poorer soils even though their productivity was bound to be low. These problems were also compounded by the slow breakup of the entrenched *latifundia* land-tenure system, under which massive numbers of tenant farmers labored on huge estates controlled by wealthy (often absentee) landowners. These are some of the reasons why Southern Europe's crop yields are frequently more than 50 percent lower than those of Western Europe and why so many Mediterranean farmers are caught up in long-term cycles of poverty.

Spain's major industrial area is located in the northeast in Catalonia, centered on metropolitan Barcelona (3.5 million). Here it is neither a favorable location nor a rich local resource base that has stimulated industrialization; rather, the vigorous and progressive outlook of the Catalans has produced this development, aided by a strong regional identity that has enabled these people to forge ahead of the rest of Spain. The Basque provinces around the north-central city of Bilbao (430,000) constitute another manufacturing zone, particularly for metal and machine industries. To the west, people have agglomerated in Galicia near the coal-mining areas of the Cantabrian Mountains. Two other sizeable population clusters have developed in Andalusia in the south, where the lowland of the Guadalquivir River opens into the Atlantic, and in the center of the country around the capital of Madrid (5.5 million).

In Portugal, which possesses an Atlantic but no Mediterranean coast, the majority of the heavily agricultural population is distributed along the coastal lowlands rather than on the Iberian Plateau. Lisbon (1.7 million) and Porto (395,000) are the leading urban centers in a country that remains 70 percent rural.

Although Spain has lagged behind the economies of Western Europe, its entry into the EC in 1986 (along with Portugal) signalled a new determination to close the development gap between itself and the European core. The response from north of the Pyrenees was immediate: the country was inundated with so much foreign investment that during the late 1980s it recorded Europe's fastest economic growth rate. Among the major attractions to American and Japanese as well as European investors were Spain's newly stable democracy, market-oriented government policies, tax incentives, low wage levels, and an increasingly affluent domestic marketplace. Climatic amenities were important, too, not only in the rush to build new factories but also in the steady upsurge of tourism and the construction of recreational facilities, condominiums, and retirement communities. If all this sounds very much like recent growth in the U.S. Sunbelt, the analogy is appro-

priate—many Spaniards now talk about their country as the "Florida" and "California" of Europe, as resorts, high-technology industries, and ultramodern fruit and vegetable farms have multiplied.

The spatial pattern of this growth is also reminiscent of the Sunbelt: as in the United States, new development has been spotty, with many areas unaffected while others thrive. The regional disparities that have resulted are clearly visible on the map of per-capita gross domestic product (Fig. 1-18).

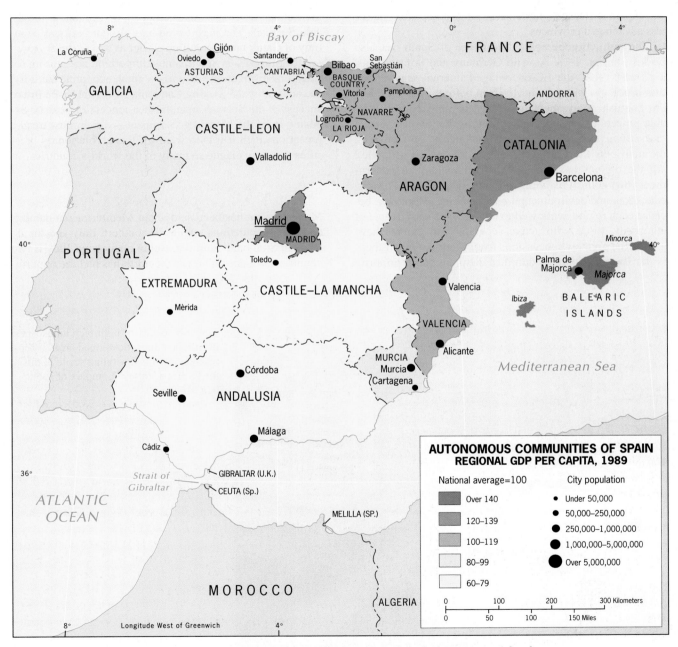

FIGURE 1-18 **Not shown on this map is the autonomous community of the Canary Islands.**

Led by Barcelona-focused Catalonia and the capital district of Madrid, northeastern Spain has burgeoned; but the Spanish northwest has been left relatively untouched, and the interior south and west lag far behind the rest of the country. In the 1990s this uneven development has been aggravated by the termination of the boom that accompanied the entry into the EC—a massive, six-year growth spurt that vaulted Spain to the enviable rank of the world's eleventh largest economy. Thus the likelihood of additional major investments and spread effects from the newly developed areas is diminished for the time being, and the Spanish government must confront serious problems of unemployment, labor unrest (Spain is Europe's leader in strikes), and poverty in the less advantaged provinces.

The politico-geographical structure of Spain can also be seen in Fig. 1-18. As with Germany and France (Figs. 1-12 and 1-13), the EC-encouraged internal regionalization process is well advanced. Even before joining the EC, the Spanish government in 1983 decentralized considerable powers to 17 subnational regions that are designated *autonomous communities* (these include the 16 shown on the map plus the Canary Islands in the nearby Atlantic). All the regions have their own parliaments and governments that control planning, public works, cultural affairs, education, and environmental policymaking. Moreover, because each region is allowed to negotiate its own degree of autonomy, those with active separatist movements have gained the greatest concessions from Madrid. One of these is the Basque Country, dominated by an ethnic minority whose passionate nationalism has been a centrifugal political force on the Spanish scene for decades; along with the powers listed above, the Basques have been granted rights to levy their own taxes, run their own police agencies, and control local television programming.

Perhaps the most powerful autonomous community is Catalonia, whose 6.5 million proud, fiercely nationalistic inhabitants have long yearned to loosen their bonds to the Spanish state and obtain full recognition of their historical identity and culture. When the Olympics were held in Barcelona in 1992, international advertisements proclaimed the "country" of Catalonia, and at the Games themselves, the Catalan language was even given official status alongside Spanish. The high-technology-driven regional economy of Catalonia—identified earlier as one of the *four motors* of Europe—is Spain's most important, accounting for fully one-fifth of the country's total economic activity. Because outward-looking Catalonia is so vital to the future success of the Spanish state, further concessions can be expected from the Madrid government—which must tread a precarious path if it is to ward off the devolutionary pressures that now plague so many of the world's countries.

Italy

Much of what has been said about Mediterranean habitats and rural populations applies to southern Italy and the islands of Sicily and Sardinia. But in many ways Italy is two countries, not one. Whereas the north has had the opportu-

The universality of high-tech landscapes in Europe's most developed areas is represented by this un-Italian scene in this, the "Silicon Valley" complex of suburban Milan.

nities—and the advantages of proximity to the European core—to sustain development in the style of Western Europe, the south or *Mezzogiorno* has for centuries been a lagging and stagnant region (its total income today is only 56 percent of that of the north). Together, north and south count over 58 million inhabitants (nearly equal to Spain, Portugal, and Greece combined), bound by the capital city of Rome (3.1 million), which lies astride the narrow transition zone between the two contrasting Italian halves. By every measure Italy is Mediterranean Europe's leading state—in the permanence of its contributions to Western culture, in the productivity of its industry and agriculture, and in its overall living standard.

The core area of Italy has now shifted northward from the vicinity of historic Rome to the region known as Lombardy in the central Po Basin. This broad lowland of the Po and its tributary rivers, lying between the Appennines and the Alps, is the largest in Mediterranean Europe. As the climatic map (Fig. I-7) shows, the Po Plain has an almost wholly non-Mediterranean regime, with a much more even rainfall distribution throughout the year. Certainly this area possesses superior advantages for agriculture, but what marks it today is the greatest development of manufacturing in Southern Europe. It is all a legacy of the early period of contact with northwestern Europe via nearby transalpine trading routes, an old exchange vigorously renewed with the stimulus of the Industrial Revolution. Hydroelectric power from Alpine and Appennine slopes is the only local resource other than a large and skilled labor force. But northern Italy imports huge quantities of iron ore, coal, and other raw materials that it needs. Italian industry seeks to create precision equipment, in which a minimum of metal and a maximum of skill produce the desired products and revenues. The metal industries are led by the manufacturing of automobiles, for which Turin (1.5 million) is the chief center. At nearby Genoa (1.0 million), Italy's leading port, there is a major shipbuilding industry.

The principal metropolis of northern Italy is Milan (5.3 million), located in the heart of Lombardy. This is not only Italy's financial, banking, and service-industry headquarters but also the leading manufacturing center in all of Mediterranean Europe (Lombardy is another of Europe's *four motors*). No Italian city rivals Milan's range of industries—from farm equipment to television sets, from fine silk to pharmaceuticals, from chinaware to shoes. Factories here tend to be small operations, but they are so well managed and so automated that they have acquired a global reputation for efficiency, adaptability, and speed. In fact, the enterprising Milanese are so productive that they account for fully one-third of Italy's national income with only 9 percent of its total population. This performance has not only made Milan one of Europe's wealthiest cities but lately has helped propel the Italian economy past that of the United Kingdom; by century's end, Italy will have surpassed France as well and is likely to be challenging Germany for the realm's top position.

These developments notwithstanding, many clouds loom on the Italian horizon in the mid-1990s. Decades of ineffective and corrupt national governments (there have been more than 50 of them in the last 50 years!) have produced major administrative problems that now directly threaten the country's ability to function within the single-market EC; among other things, Italy has already failed to implement more EC directives than any other member of the European Community and has accumulated a staggering national debt that the government has shown little inclination to reduce. More tragically, the lack of central political leadership has failed to narrow the ever-widening gap between the affluent north and the disadvantaged *Mezzogiorno*; most Italians now acknowledge the so-called "Ancona Wall," an invisible line connecting the cities of Rome and Ancona (Fig. 1-17) that separates their country's two halves. To make matters worse, a strong separatist movement has emerged in Lombardy, where discontent continues to rise over the tens of billions of (U.S.) dollars annually provided in public funds to the impoverished south—where conditions are unchanging and organized-crime clans increasingly control the countryside. In the absence of a respected and efficient government in Rome, such devolutionary pressures from the critically important north could intensify and eventually lead to the dissolution of the Italian state.

EASTERN EUROPE

The eyes of the world today are on Eastern Europe, a region in turmoil and change. Since 1990, civil war has raged in what was Yugoslavia, regional conflict has caused the division of Czechoslovakia, the Turkish minority has been under pressure in Bulgaria, and political instability has afflicted Romania. Many more actual and potential problems confront the peoples and governments of Eastern Europe—problems quite different from those facing British, Western, Northern, or Southern Europe.

The very outlines of Eastern Europe are changing in the 1990s. As recently as 1989, geographers defined Eastern Europe as that part of the European realm extending from Poland in the north to Albania in the south, and from Czechoslovakia and Yugoslavia in the west to Romania and Bulgaria in the east; at the heart of it lay Hungary, once the center of one of the many empires that have ruled over this region in the past. Outlined this way, Eastern Europe lay between Western Europe and the Soviet Union, so that its eastern borders faced the republics of the Soviet Empire.

Until the late 1980s, much of Eastern Europe (including, by some definitions, former East Germany) was under Soviet-communist control. The region lay behind what was called the Iron Curtain and according to some scholarly interpretations had become part of a greater communist realm. Soviet-communist domination in the region, it was believed, would permanently transform Eastern Europe's political culture and recast it in Moscow's mold.

Most geographers, however, did not subscribe to this viewpoint. More than 20 years ago, when the first edition of *Geography: Regions and Concepts* was published, its chapter on Eastern Europe contained the following commentary (pp. 112, 129):

> ... [N]*othing has ever succeeded in unifying Eastern Europe, and it is doubtful that even Soviet power can do it . . . the pendulum of power has swung across Eastern Europe many times, and it is likely to do so again.* [The Soviet] *phase is likely to lead to something else. . . . Eastern European nationalism is a potent force, and national pride and Soviet planning are not always compatible. Hence it may be that Eastern Europe will loosen its political and economic ties with the Soviet Union—though, one hopes, not to rekindle old quarrels and bring on another period of destructive instability.*

The prediction has come true, but the hope has been dashed. Once again, Eastern Europe is going through a traumatic aftermath of empire. This is the region's fate, and its geographic position with reference to the major cultural influences in this part of the world helps explain why (see Fig. 1-6). To the west lie Germanic and Latin cultures, represented politically by Germany and Italy. To the east looms the Slavic culture realm, dominated by Russia. To the south lie Turkey and Greece; the Turkish Ottoman Empire left durable legacies in this European region. Throughout Eastern Europe's modern history, these strong external forces have collided here, resulting in endless invasions, dislocations, territorial exchanges, and political fragmentation. Geographers call Eastern Europe a **shatter belt**, a zone of politico-geographical splintering and fracturing. There are other shatter belts in the world, but none as important as Eastern Europe.

The Region Old and New

Eastern Europe is not only changing, it is expanding. The breakup of the Soviet Union (discussed in geographic perspective in Chapter 2) has redefined Eastern Europe's outlines. No longer are the Baltic states (Estonia, Latvia, and Lithuania) part of the Soviet Empire. No longer do the other western republics of the former Soviet Union—Belarus (formerly called B[y]elorussia), Moldova (formerly known as Moldavia), and Ukraine—look only toward Moscow for trade, political and economic directives, and cultural guidance. As the boundaries between these former Soviet republics and their Eastern European neighbors become more porous, so does their common border with Russia take on stronger significance. Thus, changing Eastern Europe now extends from the German border to the Russian border.

How Many States?   The new Eastern Europe (Fig. 1-19) incorporates the territories of the seven states of the old (Poland, Czechoslovakia, Yugoslavia, Hungary, Romania, Bulgaria, and Albania), plus five new entities: the two Baltic states of Latvia and Lithuania, Belarus, Ukraine, and Moldova. That would suggest a total of 12 states, but note that we referred above to the *territories* of the seven countries of the old Eastern Europe. Already, Yugoslavia has fractured: in 1991 its northwesternmost internal "republic," Slovenia, seceded from the country, adding still another state to the region's lengthening list; moreover, Slovenia was quickly followed both by neighboring Croatia and *its* neighbor Bosnia-Herzegovina, whose proclamations of independence unleashed a full-scale war with *their* common neighbor, Serbia! And at the beginning of 1993, Czechoslovakia, bowing to a strong separatist movement in its eastern region, fragmented into the Czech Republic and Slovakia. In contrast, a campaign is now underway for the eventual reunification of Moldova with Romania. The politico-geographical instability that has been one of Eastern Europe's defining characteristics continues unabated.

However regionally defined, Eastern Europe is the European realm at its most continental and landlocked and at its most agrarian, least urbanized, and most underdeveloped. While British, Western, Northern, and Southern Europe lie open to the influences and opportunities of the Atlantic Ocean and/or the Mediterranean Sea, Eastern Europe's waters are the throttled Baltic and Black Seas and the rock-fringed Adriatic. Because of this remoteness, Eastern Europe was comparatively isolated from the innovations that accompanied the Industrial Revolution. The region suffered not only from economic stagnation but also from ethnic fragmentation, weak and ineffective government, and external interference. Europe's most severely underdeveloped country, Albania, lies in this part of the realm.

The present boundary framework of Eastern Europe is a legacy of negotiations that took place following the end of World War I during the Versailles Peace Conference (1919). Many modifications have been made in the decades that followed, but the intentions of the conference planners are still etched on the map. Poland, which did not exist when the war broke out in 1914 (it had been divided among Germany, Russia, and the Austro-Hungarian Empire), was reestablished. Czechoslovakia was newly created from the defeated Austro-Hungarian Empire, combining the Czechs and the Slovaks in one state. Hungary was separated from Austria. And in the south, the Serbs (who had thrown off the Turkish yoke and had established the Kingdom of Serbia), the Croats, and the Slovenes were

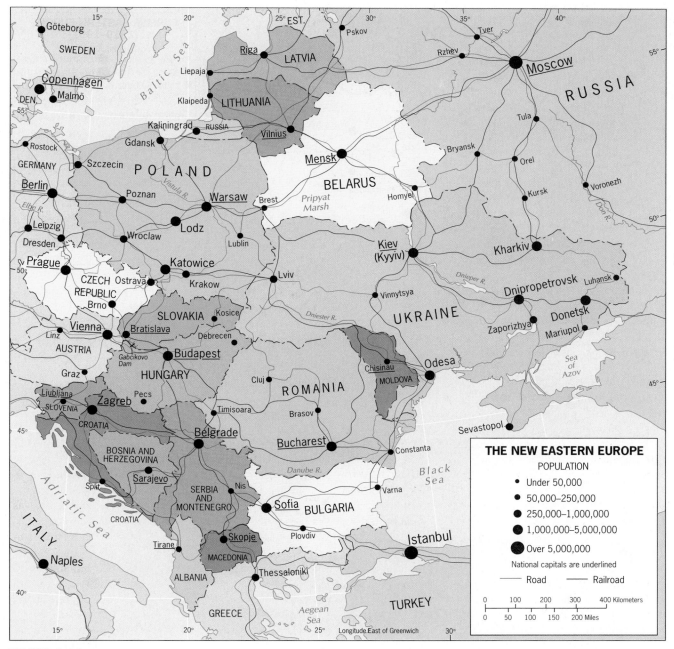

FIGURE 1-19

united in a single state named *Yugo*slavia, land of the South Slavs.

In the second quarter of the twentieth century, Russian expansionism and German aggression changed the Versailles map considerably. The map of the old Eastern Europe (Fig. 1-20) shows what happened to Poland. The entire country was moved westward (the gray lines mark the Versailles boundary), losing territory to the Russians in the east and later gaining it at the expense of the Germans in

the west (1945). The Soviets also annexed the eastern tip of Czechoslovakia (1945) and took Moldavia (Moldova) from the Romanians (1940). In addition, the three Baltic states, independent during the interwar period, were incorporated into the U.S.S.R. as Soviet Socialist "republics" (1940).

Yet even if external forces had not buffeted Eastern Europe's boundary framework, internal pressures would have modified it. The treatymakers at Versailles faced an insur-

FIGURE 1-20

mountable task: to strike a geographic compromise that would prevent Eastern Europe from turning into a jigsaw of tiny ethnic homelands (see Fig. 1-21) while simultaneously satisfying the nationalist aspirations of large populations. It seemed sensible to combine the Czechs and the Slovaks: both are West Slavic peoples, as are the Poles.

(The Russians and Ukrainians are East Slavs.) However, as Yugoslavia and then Czechoslovakia have proved, ethnic affinity did not guarantee social stability.

Nor was the physical geography of Eastern Europe helpful in reducing the region's divisiveness. An eastern arm of the Alpine Mountain system, the Carpathians, cuts

FIGURE 1-21

across Slovakia and northern Romania, separating Poland on the North European Lowland to the north from the higher-relief countries to the south (Fig. 1-2). In the middle of the region lies the Hungarian Basin, drained by the middle course of the Danube River and surrounded by the Carpathians to the north and east, the Transylvanian Alps to the southeast, the Austrian Alps to the west, and the highlands of (former) Yugoslavia to the southwest and south (Fig. 1-20). The whole southern part of Eastern Europe is

also quite mountainous. Climatic contrasts mark the region, too: the countries of the peninsular south enjoy the warmth associated with Mediterranean conditions, while the north has shorter growing seasons.

All of these physiographic divisions have long had the effect of impeding spatial interaction and of favoring the development of strong cultural regionalisms. When the boundaries of Eastern Europe were redrawn in 1919, different ethnic groups simply had to be joined together. At

BALKANIZATION

The southern half of Eastern Europe is sometimes referred to as the Balkans or the Balkan Peninsula. This refers to the triangular landmass whose points are at the southern tip of the Greek mainland, the head of the Adriatic Sea, and the northwestern corner of the Black Sea. The name itself comes from a mountain range in Bulgaria, but it has also become a concept, born of the reputation for division and fragmentation in this Eastern European subregion. Any good dictionary carries a definition of the verb *balkanize*: for example, "to break up (as a region) into smaller and often hostile units," to quote Webster.

the same time, people who shared ethnic and cultural backgrounds were sometimes separated by the postwar international boundaries. The result was that every country in Eastern Europe as constituted in 1919 found itself with sizeable minorities to govern. Such fragmentation and division was especially pronounced in the southern part of Eastern Europe, the area called the Balkan Peninsula (Fig. 1-21). Thus the term **balkanization** came into use to describe a politically fractured area (see box above). To our question, "How many states in Eastern Europe?" there has been no simple answer—as the unfolding events of the 1990s again underscore.

Governments and Minorities   Governments faced with the challenge of ethnic-minority groups within national boundaries must find ways to satisfy the aspirations of those minorities. During the Soviet-communist domination of Eastern Europe, dictatorial rule suppressed such minority desires. When Soviet control ended, the pent-up demands of minorities erupted into sometimes violent campaigns for identity and even independence.

Minority problems take another form as well. A close look at Fig. 1-20 reveals that an ethnic group that is a majority in one country is often a minority in neighboring countries. To keep this map relatively simple, we have only shown the ethnic mosaic of former Yugoslavia. Note that there is a substantial Hungarian minority in the subnational political entity of Vojvodina. Also note the large area in the south where Albanians form the majority, in Kosovo. Not shown on this map are the Hungarians who live in Transylvania, the Turks in Bulgaria, and many other minorities with ties to majorities across international boundaries (Fig.

1-21 displays the regionwide picture in greater complexity). Transylvania, the eastern part of the Hungarian Basin, was severed from Hungary and attached to Romania, complete with its Hungarian population. Hungary has not forgotten this and from time to time has openly laid claim to Romanian Transylvania. In 1991, while Yugoslavia's civil war between the Serbs and their neighboring republics gained intensity, Hungary warned the Serb-dominated government in Belgrade to do no harm to Hungarians living in Vojvodina.

Such behavior, when a state appeals to a regionally concentrated minority of ethnic cohorts in an adjacent state, is termed **irredentism**. This can range from veiled nationalist propaganda sent across the border by radio, television, and the press to outright demands for the acquisition of territory and people. The term derives from an old issue of this kind: a political pressure group in northern Italy sought to achieve the incorporation of an adjacent Italian-speaking part of Austria and called this area *Italia Irredenta* (Unredeemed Italy).

Irredentist problems now loom in the new Eastern Europe. During the existence of the Soviet Union, millions of Russians migrated into the neighboring "Soviet Socialist Republics." Today, none of these former republics is without its Russian minority. In the rush for freedom and independence, these Russian minorities have found themselves isolated; they represent the old and often despised Soviet Empire, are losing their power and privilege, and can expect only limited help from a disorganized and embattled Russia.

In the future some of these minorities will encounter discrimination in the new politico-geographical order, and they may be caught up in violent change. By that time, perhaps in the late 1990s, Russia will be better organized (either through economic growth or through renewed dictatorial rule) to look outward. Undoubtedly, Russian governments of the future will take an interest in the fate of ethnic Russian minorities outside Russia's borders, and a familiar pattern will reemerge. The potential for irredentism in the new Eastern Europe is as strong as it was in the old.

Two ways to alleviate irredentist pressures are to relocate boundaries and to resettle populations. Both have happened in Eastern Europe and will happen again. We have already noted the westward repositioning of Poland. When the Soviet communists and their Eastern European allies controlled most of the region, many hundreds of thousands of ethnic Germans were expelled from countries where they had formed sizeable minorities and sent to what was then East Germany. Many Hungarians moved from Czechoslovakia and Romania to Hungary. Ukrainians and Belarussians moved from Poland to their ethnic homelands, and Bulgars who had settled in Romania returned to Bulgaria. In the late 1980s the government of Bulgaria promulgated laws that were unacceptable to the country's large Turkish minority, and some 300,000 Turks left their homes and crossed the border into Turkey. Turkish irredentism toward southern Bulgaria, where the ethnic Turks are concentrated, was energized by these events. And in what was Yugoslavia, the resettlement of Serbs in villages formerly inhabited by Croatians and Bosnians began even while the civil war between Serbs and non-Serbs still raged. Neither boundary adjustment nor population resettlement has been enough, so far, to bring stability to Eastern Europe's volatile ethnic-cultural mosaic.

The Politico-Geographical Framework

The countries of Eastern Europe cluster in four relatively well-defined geographic areas (Fig. 1-19), as follows:

1. *Countries Facing the Baltic Sea.* These include Poland, Lithuania, and Latvia, and also the Belarus Republic. Although Belarus does not have a sea coast, it borders all three of the Baltic-coast states. (As noted on p. 90, the northernmost Baltic republic, Estonia, has far more in common with the Nordic countries than with Latvia and Lithuania, and therefore is now part of the region of Northern Europe.)
2. *The Landlocked Center.* These are the countries at the heart of the old Eastern Europe: Hungary, the Czech Republic, and Slovakia.
3. *Countries Facing the Adriatic Sea.* This is a complex group of countries consisting of former Yugoslavia (Serbia-Montenegro, Slovenia, Croatia, Bosnia-Herzegovina, [landlocked] Macedonia) and Albania.

4. *Countries Facing the Black Sea.* This cluster of countries includes Bulgaria, Romania, Moldova, and Ukraine. (Again, Moldova is actually landlocked [Fig. 1-19] but borders the Black Sea states.)

Countries Facing the Baltic Sea

Poland The northwestern part of Eastern Europe is dominated by Poland. Look again at Fig. 1-20 and you will note that Warsaw, the capital and primate city of Poland, once lay near the center of the national territory (outlined by the gray pre-war boundary) but today lies in its eastern quadrant. (Berlin, not far from present-day Poland's western border, once was much more central to a greater Germany too.) Situated on the North European Plain, Poland has traditionally been an agrarian country. Large-scale industrialization was introduced by the communist planners of the post-World War II period who exploited the substantial coal and mineral supplies contained in the south. Along the boundary with the Czech Republic lie the Sudeten Mountains, and just to the east of them, in Silesia, Poland's industrial heartland emerged (Fig. 1-22). Katowice, Wroclaw, and Krakow (combined population 5.4 million) came to symbolize the new era introduced by the communists. Unchecked environmental degradation, including the pollution of air, streams, and groundwater, accompanied this transformation of the south.

Poland's best farmlands also lie in the south, where farming is intensive and wheat is the leading crop. In central Poland, thin and often rocky soils that developed on glacial

Communist-era steel mills loom above the historic industrial landscape of the southern Polish city of Walbrzych (VAHLB-zhik), 40 miles south of Wroclaw (VRAUGHT-slahv). Here in the Silesian manufacturing region, residents are still forced to breathe some of the most polluted air on earth because Poland's economic resources are insufficient to tackle the country's severe environmental crisis.

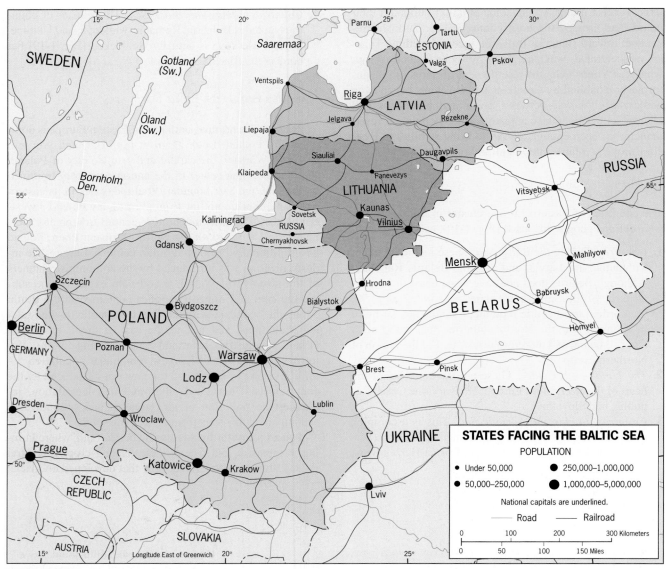

FIGURE 1-22

deposits support rye and potatoes; farther north, the Baltic rim has such poor soils that cultivation gives way to pasture-land and moors. Farming in Poland suffered severely during the communist period. The communist planners attempted to collectivize all farms but without making investments comparable to those made to state-owned factories. This created a severe economic-geographic imbalance, so that Polish farming today copes with many problems. Farmers work with tools and equipment that would be in an historical museum in any Western European country.

Warsaw (2.3 million) lies near the head of navigation on the Vistula River in a productive agricultural area. As Fig. 1-22 shows, this focus of Polish history and culture lies at the hub of a radiating network of transport routes that

reaches all parts of the country. One routeway links Warsaw to the port city of Gdansk (915,000) on the Baltic Sea coast. There, in the docks and shipyards, began the political events that were to put Poland in the vanguard of Eastern European liberation. The Solidarity labor union in the early 1980s became a beacon for the entire world, its calls for change resonating from the Baltics to Bulgaria. Today, Poland's nearly 40 million citizens are still struggling to transform their society and economy and to establish a new democratic political order.

Lithuania The Baltic state of Lithuania (3.8 million) is but a remnant of a once-powerful empire. The Grand Duchy of Lithuania dominated an area from the Baltic coast in the north to the Black Sea in the south. During the fourteenth century, this Lithuanian domain extended from Warsaw in the west to Tula in the east and included such major cities as Mensk (formerly Minsk) and Kiev (locally known as Kyyiv).

Around 1600, the Lithuanians, now allied with the Poles, still held sway over a vast region. But the Russians were on the rise, and their state of Muscovy whittled away at Lithuanian power. Eventually, the Lithuanians found themselves wedged between the Poles (allies no longer) and the Germans to the south, the Latvians to the north, and the Russians to the east in a small coastal territory (Fig. 1-22).

Things were to get still worse. In 1940, the Soviet Union annexed Lithuania, along with Latvia and Estonia. Then, after World War II, Lithuania's window on the Baltic Sea became even smaller when, at the Potsdam Conference in 1945, the victorious Soviets were awarded the former German base at Königsberg plus a Connecticut-sized territory in its hinterland. Renaming the place Kaliningrad (see box at right), the Russians proceeded to make the port into a major military naval base. As Fig. 1-22 shows, this left Lithuania with a mere 50 miles (80 km) of coastline, including the small port of Klaipeda. But Klaipeda would be of little importance: it was not even connected by rail to the interior. As a Soviet Socialist Republic, Lithuania would look inward toward Moscow, not outward to the Baltic Sea and beyond. As if to confirm this orientation, the capital, Vilnius (620,000), lies in the interior near the border with Belarus.

Lithuania is a country with limited agricultural and industrial potential. It does contain a notable resource in the form of amber, a golden-hued, translucent fossil resin that can be fashioned into ornamental objects such as beads and rings. But modernization here has come slowly. The Russian minority in Lithuania constitutes less than 10 percent of the population (80 percent are ethnic Lithuanians). When Moscow acknowledged Lithuania's independence in 1991, it set adrift a country facing enormous economic and political odds.

Latvia The "middle" Baltic republic in every way—in terms of relative location, population, and territorial size—Latvia also is Eastern Europe's northernmost country. (Estonia to the north is now part of Northern Europe.) About the size of West Virginia, Latvia has 2.7 million inhabitants. Many more Russian settlers were attracted to this country than to Lithuania; about one-third of the population is ethnic Russian. Latvians make up a bare majority (52 percent) in their own country.

Latvians, descendants of the ancient Balts in this area, speak a distinctive language and mostly belong to the Lutheran Church. They have seen their country ruled by Germans, Poles, Swedes, and Russians, but independence was restored during World War I. In 1940, that independence was lost again as the Soviets annexed the country.

KALININGRAD: ANOTHER BALTIC STATE?

The Russian *exclave* of Kaliningrad lies wedged between Lithuania and Poland, facing the Baltic Sea through its gigantic naval port (Fig. 1-22). Soon after the Soviets acquired this German base in 1945, virtually all ethnic Germans were expelled. Not much was left of their cultural landscape: relentless British bombing had devastated the place.

Russians replaced the departed Germans, and today Kaliningrad's population approaches 1 million people, 90 percent of whom are Russian. In the Soviet political scheme of things, Kaliningrad had the status of an *oblast* (see Chapter 2), and it was made part of the Russian Republic—as it remains today.

Uncertainty now prevails in Kaliningrad. Independence is not an issue, but that could change. Despite the destruction of much of seven centuries of German heritage, the German past still can be felt: the renowned philosopher Immanuel Kant lived and worked here and lies buried in the ruins of the German cathedral. The city's German name, Königsberg, is used frequently.

In 1990, Moscow decided to resettle some 20,000 ethnic Germans from Kazakhstan to Kaliningrad. Some political leaders in the oblast worry that the German influence in their territory will threaten the Russian majority. There is also talk that Kaliningrad might become a Russian free-trade zone, revitalizing the commercial port. Could Kaliningrad become the Hong Kong of the Baltic?

During the Soviet period, Latvia's economic development was considerable—tied, of course, into the Soviet centrally-planned system. Scientific equipment, machinery, and quite a number of consumer goods were manufactured from raw materials brought to Latvia from Soviet sources. When independence was regained in 1991, the challenge facing the Latvians was daunting: to reintegrate the country into the world at large without destroying the links to Russia. Like it or not, the future of the Baltic states is irrevocably tied to their hinterland—that is, Russia.

Belarus On the map of world geographic realms, certain countries lie in transition zones. In Africa, Chad and Sudan lie in such a zone, where Subsaharan Africa and the North Africa/Southwest Asia realms meet. In the 1990s, the country called Belarus (formerly Belorussia) is in a similar situation.

Given Belarus's relative location, this is not surprising. The country lies in the shatter belt that took the brunt of impact during World War II. Its compact territory, about the size of Kansas, lies centered on the capital and largest city, Mensk (1.8 million). A substantial part of what is today Belarus was eastern Poland during the period between the two world wars, and many of the urban centers in western Belarus began their existence as Polish towns. Many place-names were changed here after the Soviet intervention. Today, Belarus's neighbors are changing too: newly independent Latvia, Lithuania, and Ukraine, and reorganizing Poland. To the east Belarus adjoins Russia, and the country's relationship with Moscow is far from settled.

Belarus has a population of 10.3 million, of whom about 80 percent are Belarussians ("White" Russians). Like the Poles and the Slovaks, the Belarussians are a West Slavic people. Only 13 percent of the population is (East Slavic) Russian, a legacy of the Soviet period. Although Belarus was one of the four founding union republics when the Soviet Union was created in 1922, its role in the Soviet economy was overshadowed by the other republics, especially Ukraine, its powerful southern neighbor. The major Belarussian resources were peat and forest products (forests cover as much as one-third of its territory; marshes extend across the south). Already lagging, the republic's limited agricultural and industrial economies were shattered by World War II, when eastward-marching German armies made Belarus their first Soviet victim. One-quarter of the population was killed, and recovery was slow and painful.

Mensk, almost totally rebuilt after the war, has become an important industrial center because it was linked to the Soviet Union's expanding network of oil and gas pipelines, which provide energy and raw materials for a petrochemical complex. Some oil has been discovered in Belarus itself; a potash deposit is being developed for fertilizer production; and agriculture, though still inefficient in the familiar Soviet mode, is expanding. Nevertheless, Belarus remains comparatively isolated, its future uncertain between the Russian giant to the east and transforming Eastern Europe to the south and west.

The Landlocked Center

It is no overstatement to say that the Czech Republic and Slovakia (the two components of former Czechoslovakia) as well as Hungary are pivotal countries in the changing Eastern Europe. In 1956, Hungarians rose in revolt against Soviet domination, and Russian tanks left no doubt regarding the true relationship between Moscow and this Eastern European colony. In Czechoslovakia in 1968, what began as a political movement toward greater freedom—the so-called "Prague Spring"—also was crushed by Soviet arms, again with much loss of life. These uprisings failed, but they revealed Soviet rule for what it was and contributed to the endurance of national identities and aspirations. In 1989 communist rule was swept away, and a new era opened for this heart of Eastern Europe.

These three countries are important for other reasons too. Czechoslovakia (1919–1993) was one of those artificial creations negotiated after the end of World War I, but during the interwar period this country kept democracy alive—alone among Eastern European states. And Hungary was part of the nearest thing Eastern Europe has had to a domestic empire: that of the Austro-Hungarians, which collapsed in the chaos of the First World War.

In the mid-1990s, the most pressing challenges in this crossroads area of Eastern Europe are faced by the fledgling states of the Czech Republic and Slovakia, which on January 1, 1993 dissolved their 74-year marriage as Czechoslovakia. Bordered by Poland to the north and Hungary to the southeast, wedge-shaped Czechoslovakia had been a country of strong regionalism and local identity. It consisted of Bohemia in the west, Moravia in the north-center, and Slovakia in the east (Fig. 1-20). Today, Bohemia and Moravia constitute the Czech Republic, whose population totals 10.4 million; the eastern portion of former Czechoslovakia has become independent Slovakia, with a population of 5.4 million.

The Czech Republic   Bohemia, the new republic's western sector, lies centered in the mountain-enclosed Bohemian Basin. This always has been an important Eastern European core area, cosmopolitan in character and Western in its exposure, outlook, and development. The Basin's focus remains Prague (1.2 million), the capital and primate city of both former Czechoslovakia and the new Czech Republic. A major manufacturing center, Prague is located near the upper Elbe River (Bohemia's historic outlet to northern Germany and the North Sea) at the heart of the country's greatest concentration of wealth and productive capacity. The surrounding mountains contain many valleys in which lie small industrial towns that specialize, Swiss-style, in fabricating numerous high-quality goods. In Eastern Europe, the Czechs always have led in technology and engineering skills, with their products finding their way to foreign countries near and far.

The eastern portion of the Czech Republic is even more strongly marked by economic-geographic factors. Here, the country shares with Poland the Silesian industrial region, and local sources of raw materials lie astride the gap between the Sudeten Mountains and the Tatra extension of the Carpathians (Fig. 1-20). This sector of the Czech Republic is called Moravia, and the gap is known as the Moravian Gate because of its vital importance as a passageway between the Danube valley to the south and the North European Plain to the north. During the communist period, Moravia grew in importance as its industries produced steel, other metals, and chemicals.

Czechoslovakia's breakup originated in the schism that emerged following the demise of communism in 1989. The more affluent, West-oriented Czechs wanted to convert the country to capitalism as soon as possible and begin the process of integrating the national economy with those of Western Europe; but the Slovaks, simultaneously experiencing a resurgence of ethnic identity and their own nationalism, were generally opposed to economic reform and rapid development. With neither the Czechs nor the Slovaks strongly desiring to hold the Czechoslovakian federation together, cordial "divorce" negotiations began in the early 1990s, progressed steadily, and culminated in the division of the country at the outset of 1993. The immediate aftermath of this separation, however, has gone less smoothly: arguments erupted over the distribution of a number of Czechoslovakia's assets, a currency conversion deal fell through, a planned customs union failed to materialize, and trade between Bohemia-Moravia and Slovakia fell to less than a third of its pre-1993 level. Souring relations, of course, signal a potential new source of instability in this part of Eastern Europe, but it seemed unlikely in 1993 that any Czech-Slovak dispute would escalate into the kind of confrontation that has plagued former Yugoslavia.

The division of Czechoslovakia left the Czech Republic with far more than Slovakia: about two-thirds of the population and territory in addition to a substantially greater share of the former country's natural resources and productive capacity. Moreover, the Czechs no longer carry the financial burden of subsidizing the poorer, Slovak portion of Czechoslovakia. Given their established reputation for high-quality Bohemian manufactures and their still-important industrial operations in Moravian Silesia, the Czechs would seem to have a bright economic future. But these assets must be reevaluated in the context of the Republic's aspirations to compete in the high-technology arena of Western Europe. As we have seen, that poses difficult challenges even for Western Europe's own manufacturing complexes (such as Germany's once-mighty but now declining Ruhr). At the moment the Czechs are on the right track, trying to swiftly transfer the properties and enterprises of the former communist state into the hands of capable new private owners. Privatization, however, is a double-edged sword because the new companies must soon turn a profit in order to survive; the shaking-out process will almost certainly be painful. Two events in the months after independence serve to underscore the new realities. The Czechs had hoped to terminate their well-known weapons manufacturing at the end of 1992, but only six months later were quietly resuming operations to curtail rising unemployment and keep badly-needed foreign revenues flowing into the country. The second event was the takeover of the huge Skoda autoworks by Germany's Volkswagen—mainly because low labor costs at this Bohemian facility made it the most preferable site in Europe to mass produce low-priced cars for sale in countries throughout the realm.

Slovakia   This eastern component of former Czechoslovakia, the land of the Slovaks, is mountainous and disjointed.

Its capital is Bratislava (470,000), located in Slovakia's southwestern corner on the Danube River, just 50 miles (80 km) downstream from Vienna, Austria. The country's only other sizeable city is Kosice (255,000), situated in the far east, close to the Ukrainian (former Soviet) border.

The demise of communism here was followed, in the early 1990s, by a swift rise in Slovak nationalism (the Slovaks had been ruled by Czechs or Hungarians for most of the past 1000 years), which hastened the breakup of Czechoslovakia. But the creation of an independent Slovakia in 1993 has set into motion another separatist movement. More than 600,000 ethnic Hungarians (comprising 11 percent of the country's population) live in a part of southern (Hungary-adjoining) Slovakia that once was part of Hungary, and they have declared their intention to secede from the new Slovak state. Hungarian-Slovak tensions have also been heightened by the 1993 completion of Gabcikovo Dam on the Danube (the border between the two countries), which diverts much of the river's water into a new artificial channel lying inside Slovakia—from where it will supply much of the country's electricity. Hungary asserts this diversion constitutes a change in its international border—negotiated in a 1919 treaty with now-nonexistent Czechoslovakia—and that a new boundary will have to be delimited. Further exacerbating the situation is the dam itself: environmental activists as well as governments of other Danube-Basin countries complain it will lower both water availability and quality, flood valuable farmland, damage underground water supplies, disrupt river transport, and harm wildlife.

Slovakia was always the least developed sector of Czechoslovakia, and the new state's economic prospects, therefore, lag well behind those of the Czech Republic. During the period of Soviet hegemony, industrialization first arrived in heavily rural Slovakia, but most operations involved the manufacturing of semi-finished goods destined for Soviet and Czech markets that are rapidly disappearing. The manufacture of weapons is one of the few successful industries left, but will probably deteriorate because its products are almost exclusively tied to the arms specifications of the reorienting military forces of the former Soviet bloc. New Western investment remains a possibility, but Slovakia's aging, inefficient, and high-polluting industrial base is a major (and costly) impediment. Thus the outlook for the go-it-alone Slovaks is not encouraging, with massive unemployment and below-average living standards likely to bedevil Slovakia through the foreseeable future.

Hungary   On the threshold of the Balkan Peninsula lies Hungary, land of the middle Danube Basin, historic center of regional power, ally of Germany during World War II, and restive Soviet colony until 1989. Today, Hungary is reasserting itself in regional affairs, and relationships with neighboring countries where large Hungarian minorities reside are proving problematic. In a recent, reliable survey, more than two-thirds of Hungarians expressed the belief that parts of

Modern times on the riverfront of an ancient city—Budapest, Hungary. The transition to market economies is accompanied by the transformation of urban landscapes from Berlin to Bucharest.

adjacent countries rightfully belong to Hungary. Not surprisingly, Hungarians outside Hungary's boundaries again hear the siren song of irredentism.

The Danube River is Eastern Europe's major waterway. It originates in southern Germany, traverses Austria, and then crosses the heart of Eastern Europe, first forming the boundary between Slovakia and Hungary; then, after crossing Hungary, it forms part of the border between Serbia-Montenegro and Romania; finally, it marks the Romania-Bulgaria boundary. It is ironic that such a great transport artery, which could form the focus for a large region, serves mostly as a dividing line. After it emerges from the Austrian Alps, the Danube flows across the Hungarian Basin, which is larger than the state of Hungary itself; Slovakia, Croatia, Bosnia-Herzegovina, Serbia-Montenegro, and Romania also occupy parts of this lowland. Leaving the basin, the Danube squeezes through the Transylvanian Alps at the Iron Gate (near the town of Orsova) and next flows into the basin of its lower course, shared by Romania and Bulgaria (Fig. 1-20). No other river in the world touches so many countries—but the Danube has not been a regional bond.

The Danube and its basin do serve as unifiers for the Hungarians (or Magyars), who arrived in the Hungarian Basin in the ninth century A.D. A people of Asian origin, the Hungarians have neither Slavic nor Germanic roots. For more than a thousand years, they have held onto their fertile lowland, retaining their cultural and linguistic identity and vying for power in their turbulent region. Their twin-cities capital on the Danube, Buda and Pest (or Budapest, 2.2 million), is a primate city nearly 10 times as populous as the next largest town, a reflection of the rural character of the country. With its river port and its large industries, its cultural distinctiveness, and its nodal location within the state, Budapest typifies Eastern Europe generally: urbanization has been slow, and the capital city is the only urban center of any magnitude.

Hungary's rural economy is the region's most productive, but it has had its problems. After the country was delimited in 1919, its large estates were carved into smaller holdings. Productivity rose, but soon World War II and its destruction, especially of livestock, set farming back severely. In the postwar era, the Soviets coveted fertile Hungary's potential as a breadbasket for Eastern Europe, but they tried to collectivize farming—not the best way to increase output. Through it all, the country remained agriculturally self-sufficient (and was the region's only food exporter) with its harvests of wheat, corn, barley, oats, and rye. Today, with the constraints of communism removed, Hungary finally is on its way toward full realization of its agricultural opportunities.

Industrially, too, Hungary has not yet been able to take full advantage of its potential. Some local coal is available, and iron ore can be brought in via the Danube for steel production; this has been done in quantity only recently. For a long time, Hungary, like so many countries in the developing world, was exporting millions of tons of raw materials—particularly bauxite (aluminum ore), with which it is abundantly endowed—to other areas.

Despite these limitations, Hungary (with a population of 10.3 million) has become Eastern Europe's most successful country, enjoying a living standard well above the region's norm. Ironically, the economic growth sustaining this trend originated in the aftermath of the second Soviet invasion (the first occurred in 1944) that crushed the short-lived revolution of 1956. By convincing fellow citizens to think of the advantages of accommodating to renewed Soviet domination, Hungary's leaders were quickly able to stabilize the country; the grateful Soviets soon quietly began to grant Hungary permission to experiment in free markets, private initiatives, and greater personal freedoms.

This balancing of communism and capitalism provided incentives for the resourceful Hungarians to make substantial progress and set the stage for the 1989 breakthrough when Soviet domination ended. But perhaps their most consequential action occurred in 1988, when Hungary decided to open its border with Austria, thereby cracking the Iron Curtain. It then became possible for ethnic Germans to go from Czechoslovakia into Hungary, from Hungary into Austria, and from there to West Germany. This turning point in the Cold War was soon followed by the breakdown of the Berlin Wall and the elimination of the border between West and East Germany.

Today, Hungary is reconstructing its economy, restructuring its political system, and reconsidering its relationships with neighboring states that contain large Hungarian minorities. As the Soviet military presence ends, Hungary also is reestablishing its armed forces (the news media report substantial weapons purchases by the government). Hungary's stability is crucial to Eastern Europe's future, and it is to be hoped that no need will arise for the use of those armaments.

Countries Facing the Adriatic Sea

As recently as 1990, only two Eastern European countries faced the Adriatic Sea: Yugoslavia and Albania. By late 1993, Albania survived, but Yugoslavia had split into five new countries.

The breakup of Yugoslavia is the great human tragedy of Europe during the second half of the 20th century. Long-dormant, long-contained centrifugal forces broke apart a multinational, multicultural state that had survived seven decades of turmoil and war. Yugoslavia ("Land of the South Slavs") lay between the Adriatic Sea to the west and Romania to the east, and between Austria and Hungary to the north and Bulgaria and Greece to the south. This was a country thrown together on maps after World War I, a land of seven major and 17 smaller ethnic and cultural groups. The north, where Slovenes and Croats prevailed, was Roman Catholic; the south, Serbian Orthodox. Several million Muslims lived in Christian-surrounded enclaves. Two alphabets were in use.

No Yugoslav nation existed except in the legal sense. This was a zone of ancient animosities first held together by the Royal House of Serbia and later by communist dictatorship personified by one man, the war hero Marshal Tito. But when they got the chance, the Yugoslavs fought each other: Nazi-supporting Croats against anti-Hitler Serbs during World War II, Muslims against non-Muslims after the communist system collapsed. Here in what used to be Yugoslavia, people think of themselves first and foremost as Serbs, Croats, Slovenes, Muslims, Macedonians, or as members of other, smaller cultural groups (Fig. 1-23).

As a result, Yugoslavia is no more. When the communists ruled the country, they divided it up into six internal "republics" on the Soviet model, each dominated (except Bosnia) by one major group. Now these republics are independent countries, and we must learn the outlines of a new map still taking shape. Out of the collapse of Yugoslavia have come the newly recognized states of Slovenia, Croatia, Bosnia (whose survival remained in doubt in late 1993), Serbia-Montenegro, and Macedonia. The list may grow longer still. The outcome of Yugoslavia's disintegration is far from clear.

As we study the still-evolving political geography of this area, we should take note of the implications. Twice during the twentieth century, Europe has plunged the planet into world wars; each time Europeans emerged from their conflicts resolved to build a better world. In the former Yugoslavia in the 1990s, that resolve faced its first major test at a time when grandiose Euro-unification schemes were underway. Europe failed the test, with consequences yet unforeseeable. Europe's major powers stood by as more than 300,000 people were killed, hundreds of thousands more were injured, refugees streamed into neighboring countries, and historic treasures were demolished. In the meantime, the conferees in Maastricht and Brussels poured champagne to herald the "Community" movement's progress.

Serbia-Montenegro How to approach this difficult politico-geographical jigsaw? It is appropriate to begin with the country of the Serbs, the largest nation (10.1 million out of the area's 26 million including Albania). The

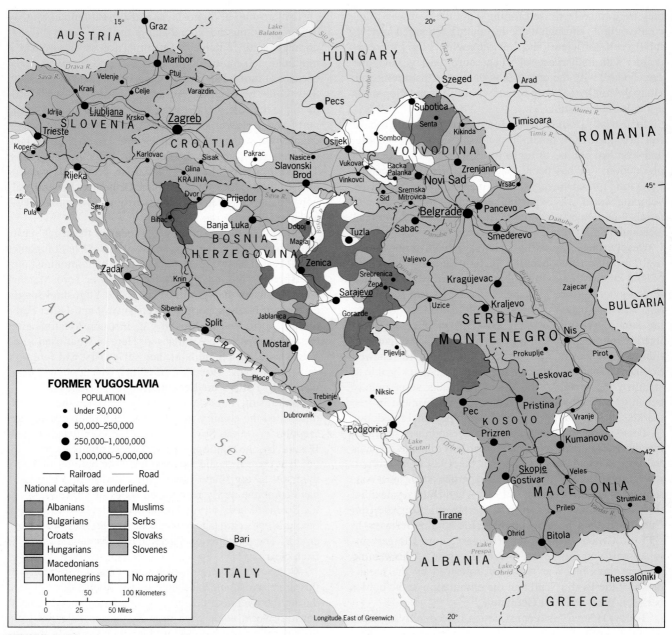

FIGURE 1-23

Serbs continue to call their country "Yugoslavia," but in fact it is an uneasy amalgam of cultures dominated by Serbs. Figure 1-23 shows what this country is made of: Serbia, Montenegro (the coastal territory inhabited by about 700,000 Montenegrins, allies of the Serbs), Vojvodina (on the border with Hungary and home to a large Hungarian minority), and Kosovo (population over 2 million, dominantly Muslim and ruled tightly by Serbian authorities).

Serbia-Montenegro, as this country probably should be called, is centered on Yugoslavia's old capital, Belgrade (1.8 million). In the north, the Danube River crosses it; the north also contains oil and natural gas reserves. These reserves, and the Danube's supply line, have helped the Serbs wage a war against their non-Serbian neighbors with devastating effect.

About six million of the area's 9 million Serbs live in Serbia-Montenegro. Another 3 million occupy parts of Bosnia and Croatia (Fig. 1-23). The notion of a Greater Serbia, to unite these Serb-held areas, has propelled a war of domination that has pitted Serb against Slovene, Serb against Croat, and Serb against Bosnian Muslim. Initially, the Serbian objective was to prevent the secession of any former Yugoslav "republic." When that policy failed and Slovenia became independent, Serbian tactics changed. Unable to prevent the proclamation of independence by Croatia and Bosnia, the Serbs took to consolidating and expanding their strongholds. The result was a devastating civil war that raged unabated when this book went to press.

Slovenia Looking back, it should not surprise us that Slovenia, the northwesternmost part of former Yugoslavia, was the first and most successful state to secede. The map shows why: Slovenia is the most remote area from Serbia. It is small, but compact and very homogeneous ethnically. It had only 8 percent of former Yugoslavia's population (just under 2 million), but its economy contributed about 20 percent of Yugoslavia's gross national product. Slovenia is the most developed, most Western-appearing part of Adriatic Eastern Europe, a nation with its own language, cultural landscape (and primate city, Ljubljana [410,000]), and social atmosphere.

Slovenia's alpine topography helped the country withstand the Serbian onslaught when independence was declared in 1991. For the first time since World War II, warplanes bombed Slovenian targets and tanks attacked local strongholds. Some civilians died, but Croatians and Bosnians in Yugoslavia's "federal" army had little stomach for fighting in Slovenia. Soon the conflict ended, Slovenia's secession was recognized, and briefly there was hope that a similar sequence of events would occur in neighboring Croatia.

Croatia When the communists laid out the internal republics of Yugoslavia, they did not, of course, foresee that these territories would some day become sovereign states. This is evident from the outlines of several of them, especially Croatia. Geographers often commented on the unusual layout of this "republic." It looked like a wide arc, with two prongs along the Hungarian border and the Adriatic coast (Fig. 1-23).

The ethnic map of the area shows why: the territory's shape reflects the general distribution of the 5 million Croats. But the Croats are not alone in their country. Large parts of it (notably the area known as Krajina in the west) have Serbian majorities.

Shortly after Croatia followed Slovenia's example and declared its independence, in mid-1991, the long-feared conflict between Serbs and Croats erupted into full-scale war. Prospects of a settlement receded when, later that year, the Germans declared their intention to recognize not only Slovenia's sovereignty but their old ally Croatia's as well; in doing so, the Germans broke ranks with other European powers who believed that withholding such recognition might help counter the violence.

As the land war swept back and forth across Croatia's mountainous countryside and losses mounted, a new specter arose: war in neighboring Bosnia. In Croatia, the battle lines were clearly drawn and the adversaries were well defined. "Ethnic cleansing" became a commonly used term to describe the ouster of Croats from Serb-conquered territory and vice versa. Damage to historic buildings and bridges was extensive: the picturesque old port of Dubrovnik will never be the same again. Dreadful as this war was, worse lay ahead.

Bosnia When you study the map of the former Yugoslavia's internal republics, one will stand out for its bulk—and for the shortness of its coastline. That country is Bosnia, wedged between the twin prongs of Croatia.

A late 1993 view of central Mostar, the former capital and chief city of Herzegovina, situated in the southern corner of Bosnia. This historic bridge across the Neretva River, which connected the Muslim area to the rest of the city, was destroyed in the massive shelling of Mostar by Croatian forces.

Although Bosnia often is officially entitled Bosnia-Herzegovina, this addendum has no ethnic or cultural significance. The very south of Bosnia, centuries ago, was the domain of *herzogs* (the German word for feudal lords or dukes). This "Herzog-land" became corrupted into Herzegovina, but Herzegovinians are not different from Bosnians the way Montenegrins differ from Serbians. So it is reasonable to call this country, simply, Bosnia.

As the map shows, Bosnia's population of 4.3 million has three components: Muslims, the largest group (about 40 percent of the total), Serbs, and, especially in the south, Croatians. Once again the early recognition of Bosnia as an independent country contributed to conflict, but there would have been war in any case. Here, contiguous to Serbia, the Serbs saw an opportunity to expand their domain. They laid siege to the captial, Sarajevo (465,000), and drove the Muslims (whose Islamic allegiance harks back to the days of the Ottoman Empire) into ever-smaller pockets of territory. The Croats within Bosnia seized the opportunity and pushed the Muslims back as well, so that by the autumn of 1993 the Muslims, though 40 percent of the population, held a mere (estimated) 15 percent of their country's land.

The carnage in Bosnia went on despite attempted United Nations intervention, and in late 1993 it appeared doubtful that a Bosnian state could emerge from the war. Various mediators proposed maps of a culturally divided but territorially-whole Bosnia. Islamic countries in Asia and Africa called for armed support of the beleaguered Bosnian Muslims. Western countries, unable or unwilling to stop the flow of arms and fuel to the Serbs, argued that arming the Muslims would "raise the death toll." As the wounded and the refugees streamed out of the country, the word Bosnia grew synonymous with European shame.

Macedonia   As if the Bosnian crisis were not enough, another problem looms: the proclaimed independence of Macedonia, a country whose neighbor, Greece, will not even permit the use of its name.

The name Macedonia has been associated with Greece—and with such renowned Greeks as Aristotle and Alexander the Great—since antiquity. But after another round of Balkan war, in 1912–13, the Macedonian region was partitioned among Greece, Serbia, and Bulgaria. Macedonia was made a Yugoslav republic in 1946, and Greeks did not object then. But in 1992, when it declared sovereignty, they protested. Macedonia is Greek, they argued, and no external area will bear that name.

Macedonia (the Greeks call it the Republic of Skopje, after its capital) is poor, landlocked, and incapable of threatening neighbors. Its 2.0 million inhabitants are dominantly Slavic Macedonians, but there is a strong Muslim minority of Albanians and ethnic Turks. The question is: will the Serbs' attempt to extend their power over this country as

well? If the Serb-Muslim conflict were to spread to Kosovo, Macedonia would soon be involved. In mid-1993, a contingent of United States armed forces, under U.N. auspices, was sent to Macedonia to forestall just such a domino effect. What started in Slovenia may be far from over.

Albania   If our generalization about declining development and southerly location in the Balkans holds true, then Albania should be even less developed than the states that used to constitute Yugoslavia. And so it is. Albania ranks lowest in Europe on many indices of development. A majority of its 3.4 million people subsist on livestock herding and farming on the one-seventh of this mountainous country that can be cultivated. There is some petroleum; tobacco and some foodstuffs have been exported; and a chromite deposit is mined. Otherwise, however, Albania is impoverished and stagnant.

Albania is an anachronism even in Eastern Europe. Alone among remnants of the Ottoman Empire, it became independent with a large (70 percent) Muslim majority. Alone among postwar Eastern European countries, it turned to China, not the Soviet Union, for its communist ideology. Alone among communist states, it remained tightly closed and retained its communist dogma long after its neighbors had opened their doors and altered their course. When change finally did come to Albania in 1991, thousands clambered aboard boats to flee across the Adriatic to Italy; others tried to cross into Greece.

Albania is reorganizing politically and economically, but the country's opportunities are severely limited. Across the northeastern border in Kosovo, an autonomous province of Serbia-Montenegro, lies a serious challenge: there, nearly two million ethnic Albanians live under Serbian rule (Fig. 1-23). Dozens have died in clashes with their Serbian adversaries, and the potential for further conflict remains significant; here, yet again, is one of those ethnic geographies that give meaning to the term balkanization.

Countries Facing the Black Sea

Four countries lie in Eastern Europe's Black Sea quadrant: Bulgaria, Romania, Moldova, and Ukraine (Fig. 1-19). All except Moldova have coastlines on the Black Sea, but none of the national capitals lies on the coast. The Ukrainian port of Odesa is the largest and most important city along the entire Black Sea coastline of Eastern Europe. The Black Sea is a sea, not a lake, and it does have navigable—if narrow and vulnerable—straits (lying in Turkey) that connect to the Mediterranean Sea and thence the open ocean. Yet the Black Sea countries have long looked inward, not to the world at large. The situation of their core areas, capitals, major urban centers, and transport networks reflects this orientation.

This is a multifaceted part of Eastern Europe. In the south, Bulgaria epitomizes the Balkan condition. In the east, Ukraine is the region's largest and most productive country, its potential enormous, its progress impeded by Stalin's murderous policies, by incalculable war damage during Hitler's aggression, and by postwar entanglement in the Soviet communist economic system. Today, Ukraine again is an independent state, looking westward for new directions.

Bulgaria A Bulgarian state did not appear until 1878, when the tsar's Russian armies drove the Turkish rulers out of the area. The Slavic Bulgarians have not forgotten who their liberators were. Nor have they always treated kindly the Turks and other Muslims who remained in the country (about 8 percent of the total population of 8.9 million, clustered in the east and south).

Bulgaria is a mountainous country, except for the northern Danube lowland and the plains of the Maritsa River, which to the south forms the boundary between Greece and Turkey. Even the capital, Sofia (1.4 million), is located in the far interior, away from the plains, the rivers, and the Black Sea; the highlands were used as protection for the headquarters of the weak embryonic state, and the capital has remained here ever since. The east-west range that forms Bulgaria's backbone and separates the Danube and Maritsa basins carried the subregion's name—the Balkan Mountains. Much of the rest of Bulgaria is mountainous as well, and its large rural population lives clustered in basins and valleys separated by rough terrain.

The Turkish period eliminated the wealthy landowners so that Bulgarian farms tend to be small gardens, carefully cultivated and quite productive. These garden plots survive today, although collectivization has been carried further here than in any other Eastern European country. On the bigger collective farms, the domestic staples of wheat, corn, barley, and rye are grown; but the smaller gardens produce vegetables, olives, plums, grapes, and other fruits (reminding us that we are again approaching the Mediterranean).

At the time of transition to Soviet domination in the mid-1940s, Bulgaria was in a different position from its Eastern European neighbors, Romania and Yugoslavia. It was poorer than either, and it had still less in terms of agricultural and industrial resources, but there was a lengthy history of positive association with the Russians. The Bulgars are also more Slavicized than the Romanians, and in the Cold War era they proved to be the Soviet Union's most loyal Eastern European satellite.

Under those geographic and historic circumstances it is not surprising that Bulgaria responded slowly when political and economic reforms swept across Eastern Europe at the end of the 1980s. Typically, its preoccupations were internal: in 1974 the government had issued a decree pro-

hibiting the use of the Turkish language in teaching, and closing the country's 1,300 mosques. Then, in 1984, Bulgarian Turks and Muslims were forced to Slavicize both their family names and given names. This led to hunger strikes by ethnic Turks, followed by clashes and the deaths of Muslims opposed to Bulgaria's "assimilation" policy. In 1989, Turkey opened its border with Bulgaria, and more than 300,000 Bulgarian Turks became refugees in Turkish camps. Later that year, Sofia rescinded its discriminatory policies, but the damage had been done. Another latent ethnic conflict had been activated; another fertile field for irredentism was being plowed.

Romania Bulgaria's northern neighbor, Romania, is both richer and larger, more than twice as large in terms of territory as well as population. But Romania's 23.2 million inhabitants have not benefited in general from the country's advantages. Here, communist totalitarianism reached its destructive extreme in the person (and personality cult) of the executed dictator Nicolae Ceausescu, who squandered Romania's wealth, destroyed much of the country's architectural heritage, and plunged the state into a social abyss from which it will take generations to extract itself.

Romania is part of the old Eastern Europe, and (as Fig. 1-20 shows) it contains a varied physiography. The resource-laden Carpathian Mountains sweep into the country from the northwest in a broad arch. The Transylvanian Alps mark the limit of the lower Danube River basin, shared with Bulgaria to the south. Productive and tourist-attracting Transylvania lies in the northwest. Along the southern flank of the Transylvanian Alps lie several oilfields, and Romania was long an energy exporter.

Like other Eastern European countries, Romania incorporates minorities. The Romanians trace their ancestry to the time, 2,000 years ago, when their country was a Roman province; as Fig. 1-6 shows, Romania is an outlier of the Romance subfamily of Indo-European languages, encircled by Slavic languages. This map (as well as Fig. 1-21) also reveals an ethnic distribution: in Transylvania live some 2 million Hungarians, Magyars who were detached from the Hungarian homeland when Eastern Europe was carved into states in 1919. As we noted earlier, many Hungarians regard this arrangement as unacceptable and temporary. (The German ethnic minorities shown in Fig. 1-6 have dwindled considerably.)

Together with former Yugoslavia, Romania is Eastern Europe's most bitter tragedy. The capital, Bucharest (2.3 million), once was known as the Paris of the Balkans; in 1857 it had the world's first street-lighting system. Today its townscape is dominated by huge buildings in Ceausescu's personally approved "monumental" style, including Republic House, reputedly the largest building in all of Europe (see photo p. 112). Such structures rise above a city still in disrepair and decay. Endless rows of gray, featureless

Bucharest's massive Republic House and its ceremonial promenade. This still-unfinished complex is the epitome of the Ceausescu style, and stands as a stark reminder of the executed dictator's relentless looting of communist Romainia's limited resources in order to pursue his personal whims.

apartment blocks represent the communist ideal; remaining single houses have become hovels, with farm animals and patches of crops attesting to the people's need to survive under any circumstances. Tens of thousands of unwanted orphans languish with minimal care in institutions, products of Ceausescu's repugnant population policies that required all women to bear five children. Romania's future in the new Europe is not bright.

Moldova If things were different in Romania, it is likely that a reunion between Romania and Moldova (Moldavia) would already be implemented. On the map, Moldova looks like an extension of Romania, and in fact it is just that (Fig. 1-19). Two-thirds of Moldova's population of 4.4 million have Romanian ancestry. Moldova, a landlocked territory about twice the size of New Jersey, was taken from Romania by the Soviets in 1940 and made a Soviet Socialist Republic of the U.S.S.R. A half-century later, when self-determination for the republics became a possibility, Moldovans of Romanian heritage, encouraged by Romanian nationalists across the border, rekindled the old ethnic ties. A merger seemed a likely prospect, but then it became clear what conditions prevailed in Romania. Moldova declared its independence in 1991, but the fervor for formal reunification with Romania has faded.

Bessarabia, as this area between the Prut and the Dnestr Rivers was formerly called, has changed hands many times. It was part of the Ottoman Empire, and then the Turks ceded it to the Russians in 1812. After World War I, Moldova (then called Moldavia) was reattached to Romania, but only until 1940. Following the Soviet annexation, the Russians imposed their culture, including the use of the

Cyrillic alphabet to spell the Romanian dialect spoken here (see Fig. 1-6). The Moldovans never stopped campaigning to get Roman characters reintroduced, and in 1989 they finally succeeded. But by then, events were set to overtake this Soviet "republic."

Soviet political planners deprived Moldova of a Black Sea coast by leaving a narrow strip of land in Ukrainian hands (Fig. 1-24). Thus landlocked, the country lies in the hinterland of the port of Odesa. The capital, Chisinau (formerly called Kishinev) (760,000), is the center for its modest industrialization; Moldova remains primarily an agricultural country. Best known for its very good wines, Moldova also produces corn and wheat as well as fruits and vegetables. Soils are fertile and pastures are rich, so meat and dairy products also figure among exports.

Is all this enough to sustain Moldova as an independent state? Analysts differ on the country's prospects. Only 65 percent of the population has Romanian ancestry; 14 percent are Ukrainians, and 13 percent are Russians. Add to this the small minorities of Bulgarians and Gagauz, and the familiar Eastern European pattern emerges. The prospect of reunification with Romania may destabilize a territory that first needs to consolidate its own gains.

Ukraine Ukraine, as part of the new Eastern Europe, is the region's largest and most populous country (Fig. 1-19). Its territory of 232,000 square miles (604,000 sq km), nearly the size of Texas, is the largest in the entire European realm. Its population of 52.2 million is nearly the same as that of the United Kingdom, France, or Italy. The capital, Kiev (Kyyiv) (2.8 million residents), is a major historic, political, and cultural center.

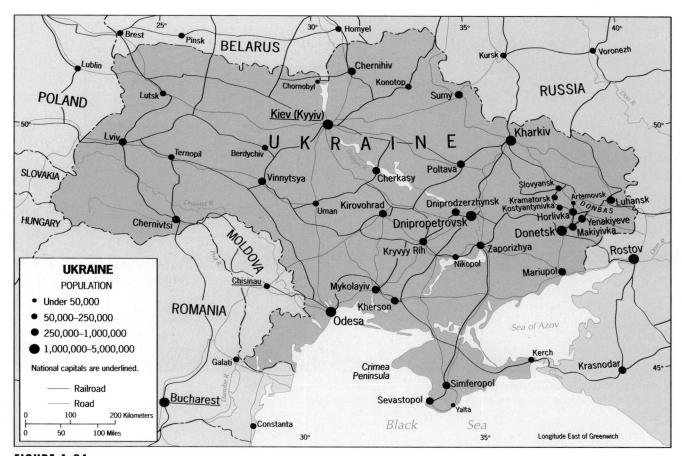

FIGURE 1-24

As Fig. 1-24 shows, Ukraine has seven neighboring countries, including Hungary and Slovakia, and it has a lengthy Black Sea coastline that includes the Crimea Peninsula. This peninsula once was home to the Crimean Tatars and had the status of Soviet Autonomous Republic (see Chapter 2). But the Crimean Tatars were accused of collaborating with the Nazi invaders during World War II, and in 1945 they were ruthlessly expelled. Russians entered the area to farm the land and settle in the towns, but in 1954 the peninsula was transferred to Ukrainian control. Today, the population remains dominantly Russian, and Russian leaders have suggested that a "revision" of the boundary may be sought.

Other territorial and boundary issues loom elsewhere along Ukraine's lengthy border, but more urgent problems confront the new Ukraine internally. The population is about 73 percent Ukrainian and 22 percent Russian, and the Russian minority is strongly concentrated in the urban-industrial east and the coastal south. The more agrarian west is almost exclusively Ukrainian. The Russian minority, furthermore, is not only more urbanized but generally better educated and has been disproportionately influential in Ukraine's affairs.

Nevertheless, Russians joined Ukrainians in voting in favor of independence; overall, the approval vote was 90 percent, and even in the Crimea Peninsula Russians voted 56 to 44 percent in favor of sovereignty. Still, the future of Ukrainian-Russian relations is uncertain. Economic hard times will test this relationship in the future.

Ukraine, an economic-geographic cornerstone of the former Soviet Union, is rich in raw materials for industry and in good soils for farming. In the past, Soviet strength was built on a mineral wealth of great variety and volume; the U.S.S.R.'s huge territory presented an almost limitless range of raw materials for industry. In Ukraine, major industrial resources lay in relatively close proximity. The eastern part of the country began to emerge as a leading industrial region toward the end of the nineteenth century, and one major reason for this was the Donets Basin, one of the world's largest coalfields. This area, known as the *Donbas* for short, initially yielded as much as 90 percent of the tsarist empire's high-grade coal; by the late 1980s, its supplies dwindling, it still produced about 25 percent of the total Soviet output.

Ukraine's Donets Basin (Donbas) anchors an industrial region that extends into Russia, and which was a crucial manufacturing complex for the former Soviet Union. Shown here is the heavy industrial landscape that dominates the city of Donetsk.

Less than 200 miles (320 km) from all this Donbas coal lay vast deposits of high-quality iron ore at Krivoy Rog (740,000), which has now been renamed Kryvyy Rih (Fig. 1-24). Not surprisingly, major metallurgical industries arose in both places: around the Donets coalfields and near the Krivoy Rog iron mines. Donetsk (1.2 million) and its satellite Makeyevka (now renamed Makiyivka) (565,000) dominate the Donbas area, whereas Dnipropetrovsk (1.2 million) is the major center of the Krivoy Rog cluster.

Like Germany's Ruhr industrial complex, Ukraine's heavy manufacturing developed in an area of dense population (offering a substantial labor pool) near large and productive farmlands, and these industries were served by adequate transport systems. Moreover, when the better-grade ores were exhausted, extensive lower-grade ores could sustain production. Later, additional iron ores were discovered near Kerch on the eastern tip of the Crimean Peninsula, close enough for use in Ukraine's established plants. Nonetheless, Ukrainian steel mills have begun to depend to some extent on nearby Russian ores, notably those of the Kursk Magnetic Anomaly, one of the world's richest and largest known iron ore reserves.

Whatever the political relationships between Ukraine and Russia in the future, their economic interdependence will continue. This, in fact, will be a major challenge in years to come, because Ukraine supplied the former Soviet Union with nearly one-half of total farm output by value, plus between one-quarter and one-third of all manufactures. Ukraine's farms produce wheat, corn, vegetables, tobacco, sugar beets, and a host of other crops; its pastures raise one-quarter of all of the former Soviet Union's cattle; its factories make televisions, refrigerators, chemicals, canned foods, and much more. Ukraine even exports raw materials, including manganese (a ferroalloy vital in steel production) from a deposit near Krivoy Rog and, of course, coal. If Ukraine has a natural-resource weakness, it lies in oil and natural gas. Comparatively small reserves in the north near Sumy (335,000) and in the far west, south of Lviv (860,000), cannot keep up with demand, and Ukraine has long imported oil from Azerbaijan's reserves near Baku. There is also a gas pipeline from Ploesti in Romania to Odesa (1.1 million) that helps satisfy Ukraine's energy demand.

Now Ukrainian leaders have visions of European and world markets—and world prices—for their country's wide range of products, at a time when Russia's need for them is greater than ever. There has even been talk of Ukraine applying for admission to the European Community. The transition will be difficult, and Ukraine's relative location and existing linkages will not help. As the map reminds us, Ukraine lies farthest (overland) of all the Eastern European countries from Europe's western core; it is not a major maritime trading nation as yet.

In the meantime, Ukraine faces some internal challenges as well. The much-dammed Dnieper River (Fig. 1-24) is a significant divide between the agrarian, rural-Ukrainian, Catholic west and the industrialized, urban-Russian, Orthodox east. Entanglement with Russia will continue, involving the sensitive question of former Soviet military forces and nuclear armaments on Ukrainian soil.

The specter of Chornobyl (as the Ukrainians have renamed Chernobyl) also continues to hang over the country. In 1986, a serious accident at the Chernobyl nuclear power plant in northern Ukraine killed dozens, made thousands ill with radiation poisoning, and caused the dislocation of many thousands more. Prevailing winds wafted the deadly radiation away from the nearby capital of Kiev but poisoned soils, plants, and animals in Belarus, Poland, and places far beyond (even northern Norway's reindeer were affected, and with them the people who depend on this resource). Moscow's stonewalling and the Soviets' incompetent and deceitful handling of the Chernobyl crisis galvanized anti-Soviet feelings in Ukraine. The Soviet Union is no more, but in Ukrainian eyes that entity and Russia always were one and the same. Overcoming the past to shape a new Ukrainian-Russian association while setting its compass westward will be a massive challenge as Ukraine enters its first decade as a European country.

Throughout its modern history, beginning with the revolutions that emerged from the Renaissance, Europe has been a changeable realm. As such, Europe's progress was in no small way propelled by the resilience and resourcefulness of the Europeans in successfully overcoming an unrelenting series of new and recurring challenges.

As we have seen, further challenges loom in the mid-1990s. At the national scale, many countries are threatened by devolution, which has already led to political disintegration and brutal warfare in former Yugoslavia, the bifurcation of Czechoslovakia, and the weakening of the state system in the United Kingdom. At the regional level, Mediterranean Europe's lagging economies are marked by a widening gap between wealthy and poverty-stricken areas; in the new Eastern Europe, five former Soviet republics—together with the entire tier of erstwhile satellites of the defunct Soviet Union—struggle to keep afloat during the difficult transition from communism to capitalism. And most importantly, at the realm scale, the EC confronts forces that would derail its mission to consolidate recent gains and achieve ever greater European unification. Europe's past successes suggest that these challenges, too, will eventually be overcome.

THE REALM IN TRANSITION

1. Europe's global connections may be the strongest of any of the world's geographic realms; despite Europe's economic and political problems, these connections continue to grow.
2. Supranationalism is weakening; the EC Union Treaty is being watered down, and prospects for Eastern European and Turkish participation are uncertain.
3. Centrifugal forces are threatening the survival of Belgium as a single state, and even Italy's fragmentation is possible.
4. Cultural conflict in Eastern Europe is creating Europe's worst foreign policy crisis since World War II.
5. Countries on the Russian border continue to suffer from poor integration into the European framework. Ukraine and Belarus in particular are disconnected.

◆ PRONUNCIATION GUIDE

Adriatic   (aydree-ATTIC)
Aegean   (uh-JEE-un)
Allemanni   (alla-MAH-nee)
Ancona   (ahng-KOH-nuh)
Andalusia   (unda-looh-SEE-uh)
Appennines   (APP-uh-nines)
Arc de Triomphe   (arc duh tree-
   AWMFF)
Ardennes   (ar-DEN)
Arles   (ARL)
Azerbaijan   (ah-zer-bye-ZHAHN)
Baden-Württemberg   (BAHDEN
   VEERT-um-bairg)
Baku   (bah-KOO)
Balkan   (BAWL-kun)
Barcelona   (bar-suh-LOH-nuh)
Basque   (BASK)
Bavaria   (buh-VAIRY-uh)
Belarus   (bella-ROOSE)
Belarussia   (bella-RUSH-uh)
Bern   (BAIRN)
Bessarabia   (bess-uh-RAY-bee-uh)
Bilbao   (bil-BAU)
Bohemia   (boh-HEE-mee-uh)
Bosnia-Herzegovina   (BOZ-nee-uh
   hert-suh-go-VEE-nuh)
Bratislava   (BRUDDIS-lahva)
Bremen   (BRAY-mun)
Bucharest   (BOO-kuh-rest)
Budapest   (BOODA-pest)
Byzantine   (BIZZ-un-teen)
Calais   (kah-LAY)
Campine   (kam-PEEN)
Cantabrian   (kan-TAY-bree-un)
Carpathians   (kar-PAY-theons)
Cartagena   (karta-HAY-nuh)
Catalan   (katta-LAHN)
Catalonia   (katta-LOH-nee-uh)
Caucasian   (kaw-KAY-zhun)
Ceausescu, Nicolae   (chow-SHESS-
   koo, NICK-oh-lye)
Celt(ic)   (KELT [ick])
Centrifugal   (sen-TRIFFA-gull)
Centripetal   (sen-TRIPPA-tull)
Champs d'Elysées   (SHAWZ elly-
   ZAY)
Charleroi   (SHARL-rwah)
Chisinau   (kee-shih-NAU)
Chornobyl   (CHAIR-noh-beel)
Conurbation   (konner-BAY-shun)
Copenhagen   (koh-pen-HAHGEN)

Crimea   (cry-MEE-uh)
Croat   (KROH-aht/KROH-aht)
Croatia   (kroh-AY-shuh)
Cyrillic   (suh-RILL-ick)
Czech   (CHECK)
Czechoslovakia   (check-uh-sloh-
   VAH-kee-uh)
Dalmatia   (dall-MAY-shuh)
Danube   (DAN-yoob)
Défense, La   (day-FAWSS, lah)
Départment   (day-pahrt-MAW)
Devolution   (DEE-voh-looh-shun)
Dnestr   (duh-NYESS-truh)
Dnieper   (duh-NYEPP-er)
Dnipropetrovsk   (duh-nep-roh-puh-
   TRAWFSSK)
Donbas   (DAHN-bass)
Donets   (duh-NETTS)
Donetsk   (duh-NETTSK)
Dubrovnik   (doo-BROV-nik)
Duisburg   (DYOOSE-boorg)
Edinburgh   (EDDIN-burruh)
Eire   (AIR)
Elbe   (ELB)
Entrepôt   (AHNTRA-poh)
Erzgebirge   (AIRTSS-guh-beer-guh)
Fjord   (FYORD)
Frankfurt   (FRUNK-foort)
Gabcikovo   (gahb-CHEE-kuh-voh)
Gaelic   (GALE-ick)
Gagauz   (guh-GOOSH)
Galicia   (guh-LEE-see-uh)
Garonne   (guh-RON)
Gdansk   (guh-DAHNSK)
Genoa   (JENNO-uh)
Göteborg   (GOAT-uh-borg)
Grande Arche   (GRAWND-ARSH)
Guadalquivir   (gwahddle-kee-VEER)
Hamburg   (HAHM-boorg)
Hegemony   (heh-JEH-muh-nee)
Herzog   (HAIR-tsog)
Iberia   (eye-BEERY-uh)
Igneous   (IG-nee-uss)
Ile de France   (EEL duh-FRAWSS)
Ile de la Cité   (EEL duh-la-see-TAY)
Irrendentism   (irruh-DEN-tism)
Jutland   (JUT-lund)
Kaliningrad   (kuh-LEEN-in-grahd)
Katowice   (kott-uh-VEET-suh)
Kattegat   (KAT-ih-gat)
Kenyatta   (ken-YUTTA)

Kerch   (KERTCH)
Kiev   (KEE-yeff)
Kiruna   (kih-ROONA)
Kishinev   (KISH-un-neff)
Klaipeda   (KLYE-puh-duh)
Königsberg   (KAY-nix-bairk)
Kosice   (KOH-shuh-tsay)
Kosova   (KAW-suh-voh)
Krajina   (CRY-eena)
Krakow   (KRAH-kow)
Krivoy Rog   (krih-voy-ROAG)
Kursk   (KOORSK)
Kyyiv   (KAY-eff)
Land   (LUNT)
Länder   (LEN-derr)
Latifundia   (latta-FOON-dee-uh)
Le Mans   (luh-MAW)
Leicester   (LESS-ter)
Leipzig   (LYPE-sik)
Liechtenstein   (LIK-ten-shtine)
Liège   (lee-EZH)
Lille   (LEEL)
Lingua franca   (LEEN-gwuh
   FRUNK-uh)
Ljubljana   (lee-oo-blee-AHNA)
Loess   (LERSS)
Loire   (luh-WAHR)
Louvre   (LOOV)
Luxembourg   (LUX-em-borg)
Lviv   (luh-VEEF)
Lyon   (lee-AW)
Maastricht   (mah-STRICT)
Macedonia   (massa-DOH-nee-uh)
Madrid   (muh-DRID)
Main   (MINE)
Makeyevka   (muh-KAY-uff-kuh)
Malmö   (MAL-muh)
Mao Zedong   (MAU zee-DUNG)
Maritsa   (muh-REET-suh)
Marseille   (mar-SAY)
Mecklenburg   (MEK-lun-boorg)
Meseta   (meh-SAY-tuh)
Meuse/Maas   (MERZZ/MAHSS)
Mezzogiorno   (met-soh-JORR-noh)
Milan   (muh-LAHN)
Moldavia   (moal-DAY-vee-uh)
Moldova   (moal-DOH-vuh)
Montpellier   (maw-pell-YAY)
Moravia   (more-RAY-vee-uh)
Moselle   (moh-ZELL)
München   (MIN-shun)

Munich   (MYOO-nik)
Muscovy   (muh-SKOH-vee)
Muslim   (MUZZ-lim)
Nehru   (NAY-rooh)
Nord-Pas de Calais   (NORD pah-duh-kah-LAY)
Norwich   (NORRITCH)
Oder-Neisse   (OH-der NICE)
Odesa   (oh-DESSA)
Oise   (WAHZ)
Oslo   (OZ-loh)
Ossi   (OH-see)
Palatinate   (puh-LATTEN-utt)
Pangaea   (pan-GAY-uh)
Peloponnesus   (pelloh-puh-NEEZE-uss)
Peron   (puh-ROAN)
Piraeus   (puh-RAY-uss)
Ploesti   (ploh-ESS-tee)
Pomerania   (pahm-uh-RAY-nee-uh)
Provence   (pro-VAHNSS)
Prague   (PRAHG)
Prut   (PRROOT)
Pyrenees   (PEER-unease)
Randstad   (RUND-stud)
Reykjavik   (RAKE-yah-veek)
Rhône-Saône   (ROAN say-OAN)
Romania   (roh-MAIN-yuh)
Romansch   (roh-MAHN-ssh)

Ruhr   (ROOR)
Saarland   (ZAHR-lunt)
Sarajevo   (sahra-YAY-voh)
Saxony-Anhalt   (SAX-uh-nee UN-hultt)
Schleswig-Holstein   (SHLESS-vik HOAL-shtine)
Seine   (SENN)
Sicily   (SISS-uh-lee)
Silesia   (sye-LEE-zhuh)
Skopje   (SKAWP-yay)
Slav(ic)   (SLAHV [ick])
Slavicize   (SLAH-vih-size)
Slovakia   (slow-VAH-kee-uh)
Slovenia   (slow-VEE-nee-uh)
Sofia   (SOH-fee-uh)
Strasbourg   (STRAHSS-boorg)
Stuttgart   (SHTOOT-gart)
Sudeten   (soo-DAYTEN)
Sumy   (SOO-mee)
Suriname   (soor-uh-NAHM-uh)
Tallinn   (TALLEN)
Tampere   (TAHM-puh-ray)
Tatar   (TAHT-uh)
Tatra   (TAHT-truh)
Thames   (TEMZ)
Thüringia   (tyoor-RIN-jee-uh)
Train à grande vitesse   (TRAN ah-grawnd-vee-TESS)

Trondheim   (TRAHN-hame)
Tula   (TOO-luh)
Tunisia   (too-NEE-zhuh)
Turin   (TOOR-rin)
Turku   (TOOR-koo)
Tyrol   (tih-ROLL)
Ukraine   (yoo-CRANE)
Ural   (YOOR-ull)
Versailles   (vair-SYE)
Vienna   (vee-ENNA)
Vilnius   (VILL-nee-uss)
Vistula   (VIST-yulluh)
Vojvodina   (VOY-vuh-deena)
Volkswagen   (VOAKS-vah-gun)
Von Thünen, Johann Heinrich   (fon-TOO-nun, YOH-hahn HINE-rish)
Walbrzych   (VAHLB-zhik)
Walloon   (wah-LOON)
Weber   (VAY-buh)
Weser   (VAY-zuh)
Wessi   (VESS-see)
Wroclaw   (VRAUGHT-slahv)
Yonne   (YAHN)
Yugoslavia   (yoo-goh-SLAH-vee-uh)
Zeeland   (ZAY-lund)
Zuider Zee   (ZYDER ZEE)
Zürich   (ZOOR-ick)

CHAPTER 2

RUSSIA'S FRACTURING FEDERATION

Relief

Meters		Feet
3050		10 000
1525		5000
610		2000
305		1000
152.5		500
0	Sea Level	0
152.5		500
1525		5000
3050		10 000
		Below Sea Level

IDEAS & CONCEPTS

Climatology   Core area
Imperialism   Centrality
Forward capital   Inaccessibility
Colonialism   Heartland theory
Federation   Frontier
Economic planning

REGIONS

Russian Core
Eastern Frontier
Siberia
Far East

The name *Russia* evokes geographic images of the past: terrifying tsars, conquering Cossacks, relentlessly advancing armies, poverty-stricken peasants, Napoleon's defeat, the communist revolution, a Byzantine church, a checkered culture. Russia is a land of vast distances, bitter cold, impenetrable forests, treacherous mountains, isolated outposts, and remote frontiers. Pre-communist Russian culture was a culture of strong nationalism, resistance to change, political despotism, bejeweled aristocrats, desperate serfs. Great authors lamented the plight of the poor; major composers celebrated the indomitable Russian people and, as Tchaikovsky did in his *1812 Overture*, commemorated their victories over foreign foes.

Under the tsars, Russia grew from nation into empire. The insatiable demands of these rulers for wealth, territory, and power sent Russian armies across the plains of Siberia, through the deserts of interior Asia, and into the mountains along Russia's rim. Russian pioneers ventured even farther, entering Alaska, traveling down the Pacific coast of North America, and planting the Russian flag near San Francisco

THE SOVIET UNION, 1924–1991

For 67 years Russia was the cornerstone of the *Soviet Union,* the so-called Union of Soviet Socialist Republics (U.S.S.R.). The Soviet Union was the product of the Revolution of 1917, when more than a decade of rebellion against the rule of Nicholas II led to the tsar's abdication. Russian revolutionary groups were called *soviets* ("councils"), and they had been active since the first workers' uprising in 1905. In that crucial year, thousands of Russian workers marched on the tsar's palace in St. Petersburg in protest, and soldiers opened fire on them. Hundreds were killed and wounded. Russia descended into chaos.

The tsar's abdication in 1917 was forced by a coalition of military and professional men. After the tsar, Russia was ruled briefly by a Provisional Government. In November 1917, the country held its first democratic election ever—and, as it turned out, the last for more than 70 years to come.

The Provisional Government allowed the return to Russia of exiled activists in the Bolshevik camp (there were divisions among the revolutionaries): Lenin from Switzerland, Trotsky from New York, and Stalin from internal exile in Siberia. In the political struggle that ensued, Lenin's Bolsheviks gained control over the revolutionary soviets, and this ushered in the era of communism. In 1924, the new communist empire was formally renamed the Union of Soviet Socialist Republics, or Soviet Union in shorthand.

Lenin the organizer (who died in 1924) was succeeded by Stalin the tyrant, and many of the peoples under Moscow's control suffered unimaginably. In pursuit of communist reconstruction, Stalin and his elite starved millions of Ukrainian peasants to death, forcibly relocated entire ethnic groups, and exterminated "uncooperative" or "disloyal" peoples. The full extent of these horrors will never be known. Many of the country's most creative people were executed.

On December 25, 1991, the inevitable occurred: the Soviet Union ceased to exist, its economy a shambles, its political system shattered, the communist experiment a failure. The last Soviet President, Mikhail Gorbachev, resigned, and the Soviet hammer-and-sickle flying on the flagstaff atop the Kremlin was lowered for the last time, immediately replaced by the Russian tricolor. Eleven former Soviet republics proclaimed the formation of a new union, the *Commonwealth of Independent States (CIS),* to be headquartered at Mensk in Belarus. Political geographers reasoned that weak centripetal and strong centrifugal forces would soon erode the CIS into ineffectiveness. By mid-1993, this post-Soviet union had run its course.

in that year of triumph, 1812. As Russia's empire expanded, its internal weaknesses gnawed at the power of the tsars. Serfs rebelled. Unpaid (and poorly fed) armies mutinied. When the tsars tried to initiate reforms, the aristocracy objected. The empire at the beginning of the twentieth century was ripe for revolution, which began in 1905 (see the box entitled "The Soviet Union, 1924–1991").

The tsar was overthrown in 1917, and a struggle for control followed. The victorious communists of V. I. Lenin soon swept away many vestiges of the Russia of the past. The Russian flag disappeared. The last tsar and his entire family were executed. The old capital of Russia, St. Petersburg, was renamed Leningrad in honor of the revolutionary leader. The interior city of Moscow was chosen as the new capital for a country with a new name, *the Soviet Union*. Eventually, this Union came to consist of 15 political entities, each a Soviet Socialist Republic. Russia was just one of these republics, and so the name Russia disappeared from the international map.

Russia remained first among equals, however. Not for nothing had the communist revolution been known as the *Russian* Revolution. The Soviet Empire was the legacy of the tsars' expansionism, and the new communist rulers were Russians first and foremost. As the new Union was laid out, Russians moved by the millions to the fringes of the empire—the fringes where the "republics" formed from the tsars' colonies (or conquered later by the Red Army) lay. The Russification of the Soviet Empire proceeded just as the British and French and Belgians and Portuguese were also moving in large numbers to their colonies. The Soviet Union was a Russian colonial empire.

The world's great colonial empires did not endure, and (as we predicted in earlier editions of this book) neither did the Soviet Union. Non-Russian peoples stirred in opposition to Moscow, their nationalism or ethnic consciousness mobilized by memories of a long-suppressed past. Lithuanians, Ukrainians, Georgians, and other peoples enmeshed in the Soviet Empire moved to throw off the communist yoke. And in Russia itself, nationalism stirred as well—not in opposition to other peoples of the empire but in opposition to the communist system that for nearly 70 years had bound them together.

The last Soviet (Russian) leader, Mikhail Gorbachev, responded to those pressures in new and unfamiliar ways: unlike his predecessors, he initiated changes to accommodate and guide them. Two Russian words became part of the international vocabulary: *glasnost* (openness) and *perestroika* (restructuring). Gorbachev also ended decades of secrecy and denial by allowing news and information to flow, permitting the expression of views his predecessors would have regarded as treasonous, and encouraging debate. At the same time, he began an effort to transform the Soviet Union while holding the old empire together, an effort that in some ways resembled earlier attempts made by Western European powers to retain some form of union in their collapsing empires. As we now know, that effort was unsuccessful: the Soviet

TEN MAJOR GEOGRAPHIC QUALITIES OF RUSSIA

1. Russia is by far the largest territorial state in the world. Its area is nearly twice as large as the next-ranking country (Canada).

2. Russia is the northernmost large and populous country in the world; much of it is very cold and/or very dry. Extensive rugged mountain zones separate Russia from warmer subtropical air, and the country lies open to Arctic air masses.

3. Russia was one of the world's major colonial powers. Under the tsars, the Russians forged the world's largest contiguous empire; this empire was taken over and expanded by the Soviet rulers who succeeded the tsars.

4. For so large an area, Russia's population of 150 million is comparatively small. The population remains heavily concentrated in the westernmost one-fifth of the country.

5. Development in Russia is concentrated in the western part of the country, west of the Ural Mountains; here lie the major cities, leading industrial regions, densest transport networks, and most productive farming areas. National integration and economic development east of the Urals extend mainly along a narrow corridor that stretches from the southern Urals region to the southern Far East around Vladivostok.

6. Russia is a multicultural state with a complex domestic political geography. Twenty-one internal republics, originally based on ethnic clusters, continue to function as politico-geographical entities.

7. Its large territorial size notwithstanding, Russia suffers from land encirclement within Eurasia; it has few good and suitably located ports.

8. Regions long part of the Russian and Soviet empires are realigning themselves in the post-communist era. Eastern Europe and the heavily Muslim Southwest Asia realm are encroaching on Russia's imperial borders.

9. The failure of the Soviet-communist system left Russia in economic disarray. Many of the long-term components described in this chapter (food-producing areas, railroad links, pipeline connections) broke down in the transition to the post-communist order.

10. Russia long has been a source of raw materials but not a manufacturer of export products, except weaponry. Very few Russian (or Soviet) automobiles, televisions, cameras, or other consumer goods reached world markets.

◆ F O C U S   O N   A   S Y S T E M A T I C   F I E L D

CLIMATOLOGY

Climatology is one of the main branches of physical geography. Climatologists analyze the distribution of climatic conditions over the earth's surface. They study the causes behind this distribution and the processes that change it. Such topics as the earth's long-term cooling (glaciation) and short-term warming (the so-called "greenhouse effect"), desertification, and hurricane tracking are part of this field.

The term *climate* implies an average, a long-term record of *weather* conditions at a certain place or across a region. Weather conditions are thus recorded in specifics for any given moment in time: the temperature, the amount of rainfall, percentage of humidity, wind speed and direction, and other data. Climate, on the other hand, is described in more general terms, as shown in Fig. I-7 (pp. 16-17). *Humid equatorial* climates, *dry* climates, and *cold polar* climates are marked by certain prevailing characteristics that can be mapped, resulting in the regional framework in Fig. I-7.

It is important to remember that Fig. I-7 is a kind of cartographic snapshot of the earth's changing climatic environment. Just 10,000 years ago, the map would have looked quite different; the most recently developed icesheets were just receding. Even only one century ago, we would, at a larger scale, be able to detect differences. In West Africa, for example, the Sahara (desert) has spread southward far enough that we would be able to see the evidence on a 100-year-old map.

To interpret the global climatic map, we must understand the processes and conditions that underlie both weather and climate. Our planet orbits the sun in such a way that the tropics receive the maximum radiation (heat) and the polar areas the minimum. Warmth is then redistributed by the atmosphere and by the oceans. The whole planet is enveloped by a layer of atmosphere; about 70 percent of the earth's crust lies under water. The earth's rotation, coupled with the differential heating between equatorial and polar areas, sets up a system of redistribution of warmth from the tropics to the higher latitudes. Without the Gulf Stream current and its offshoot, the North Atlantic Drift, Europe would be as cold and barren as Canada's Labrador. Without the moisture brought from the oceans by humid air masses, the U.S. Midwest would be a desert rather than the nation's breadbasket.

In the earth's great ocean basins, water circulates in giant cells or *gyres* that, because of the earth's rotation, move slowly in a clockwise direction in the Northern Hemisphere and in a counterclockwise flow south of the equator. This places warm water, moving from low toward high latitudes, along east coasts, and cool water, flowing from polar latitudes toward the tropics, along west coasts. The Europe-warming North Atlantic Drift is actually a deviation from this pattern, caused by the configuration of North Atlantic coastlines.

Flowing over land and sea are currents of air, or *winds*. Again, the earth's rotation and the sun's heating generate the pattern, but atmospheric circulation is further complicated by the sizes, shapes, and topographies of the landmasses. We know some of these wind belts by familiar names: the easterly *trade winds* that form persistent belts in the northern and southern tropics and that propelled the earliest transoceanic sailing ships; and the *westerlies*, the mid-latitude wind belt that carries weather across much of the United States from west to east. Other wind and air-pressure belts may be less familiar, but we have learned to pay attention to the behavior of high-speed, upper-atmosphere winds called *jet streams*. These tube-like streams of air snake around the globe in the middle latitudes, alternately forming northward and southward loops. By doing so, they interrupt average conditions, sometimes steering frigid polar air into lower latitudes and at other times wafting tropical air poleward to soften a winter's cold.

The complex climatic pattern in Fig. I-7 results from these and other variables. Just as the ocean circulates in gyres, so the atmosphere is patterned into circulating cells called *highs* or *lows*, depending on the pressure or weight of the air they contain. High-pressure cells are associated with cold, heavy, "stable" air. Low-pressure cells tend to contain warmer, moister, less stable air which, if caused to rise, quickly condenses and precipitates its moisture.

The capacity of an air mass to carry moisture is limited, however. When warm moist air is carried by the westerlies across Europe, it may rain—because the air rises against a cool mountain side; because it rides up against a colder, stabler air mass; or because it crosses warm terrain and rises through convection. By the time that air mass reaches Poland, it may already have lost much of its moisture; thus when it reaches Russia, it may be dry. Countries that lie in the deep interiors of continents, perhaps separated from the ocean by moun-

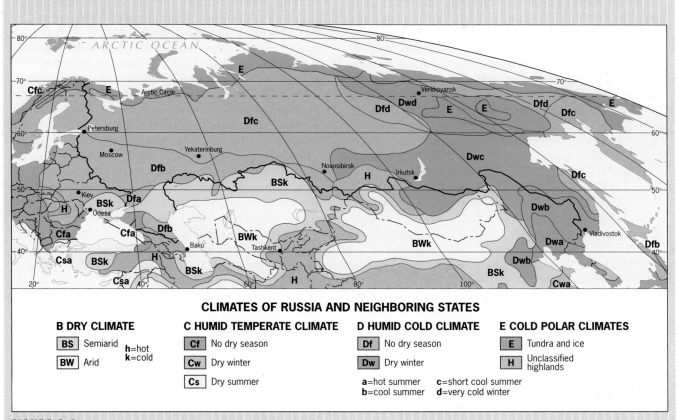

CLIMATES OF RUSSIA AND NEIGHBORING STATES

B DRY CLIMATE

| BS | Semiarid | **h**=hot |
| BW | Arid | **k**=cold |

C HUMID TEMPERATE CLIMATE

Cf	No dry season
Cw	Dry winter
Cs	Dry summer

D HUMID COLD CLIMATE

| Df | No dry season |
| Dw | Dry winter |

a=hot summer   **c**=short cool summer
b=cool summer   **d**=very cold winter

E COLD POLAR CLIMATES

| E | Tundra and ice |
| H | Unclassified highlands |

FIGURE 2-1

tains, are at a disadvantage when it comes to the essence of life—water.

Russia lies remote from the warm North Atlantic Drift current, and moisture-laden air masses must cross Europe before they reach its farmlands. As Fig. 2-1 shows, not only is Russia a high-latitude country, but it also lies exposed to polar coldness. No protective mountain barriers stretch across the north, and Arctic air has free reign from the Kola to the Kamchatka Peninsula. But in the south, from where warm, moist air

might have come, Russia and its former-Soviet-republic neighbors are virtually ringed by high mountains. As a result, the entire southern region, from the Caspian Sea to the mountains on the border with China, is desert or semiarid steppe—climatic regions that do not yield (as the map shows) until conditions become those of a cool-summer continental climate.

The climatic map (Fig. 2-1) goes far to explain why the majority of the Russian population today remains concentrated in the western one-fifth

of this gigantic country, why agriculture remains a critical problem, and why areas that are known to contain valuable raw materials remain virtually uninhabited. Russian tsars sought warm-water ports because the country's higher-latitude harbors froze up for several months each year; later, Soviet planners diverted entire rivers to bring cultivation to desert lands. Russia is a storehouse of mineral wealth, but it faces a host of environmental problems that can be summarized under one rubric: climate.

Union no longer exists. But Gorbachev's leadership from 1985 to 1991 was crucial in constraining the process, even as he slowly lost control. Had an old-style communist hard-liner been at the helm, the breakdown of the Soviet Union might have been calamitous for the entire world.

And so, today, Russia's national flag flies once again above the Kremlin in Moscow, the city of St. Petersburg has its old name back (it is now called *Petersburg* by most Russians), the tsar's name may be mentioned again, and the empire is no more. Church bells peal anew as the Russian Orthodox Church, suppressed under communism's official atheism, is in revival.

The six westernmost Soviet Socialist Republics (S.S.R.s) of the former U.S.S.R. are now part of changing Europe (see Fig. 1-19), looking westward as they seek a new place in the outside world. Similarly, the five south-central S.S.R.s that made up Soviet Central Asia today constitute the Muslim-dominated *Turkestan* region of the North Africa/Southwest Asia realm (see Fig. 7-18). Russia's relationships with these former colonies are in flux. Russia itself is a country in transition, a state of many nations, and a newly diminished realm of still uncertain margins.

In this chapter we examine what is likely to be the Russia of the foreseeable future. In the vignette that follows (pp. 165-169), we treat the three remaining former S.S.R.s of *Transcaucasia*, a region of uncertain prospects on Russia's southern rim in which the legacies of empire still hang heavily.

◆ RUSSIA'S DIMENSIONS

Even without its 14 former Soviet co-republics, territorially Russia is by far the largest country in the world. From the Bering Sea near Alaska to the Gulf of Finland, Russia stretches across *eleven* time zones. From well inside the Arctic Circle, Russia reaches southward to the latitude of Salt Lake City. It is a land of almost unimaginable dimensions, more than twice as large as the United States or China.

Russia also is a land of vast empty spaces. The country's population of 150 million is far outnumbered by China's 1.2 billion and India's 900-plus million. And, as Fig. I-9 reveals, the Russian population still is strongly concentrated in what is sometimes called "European" Russia—Russia west of the Ural Mountains, which divide the west from Siberia (see map p. 118). The name Siberia means "sleeping land," appropriate for a vast area in which the population still lives in isolated clusters and along discontinuous ribbons. If the Russia of the tsars and the Soviet Union of the communists had something in common, it was their desire to populate the east, to strengthen the Russian presence in that remote frontier. Over time an eastward vanguard of settlements did emerge, the largest ones along two railroads: one built by

the tsars before the revolution and the other constructed by the communists afterward. The suitably named city of Novosibirsk is such a center, one of many trans-Ural places whose names start with *novo* ("new").

Russia is the northernmost large and populous country on earth, and its land lies unprotected by natural barriers against the onslaught of Arctic air masses. Moscow lies farther north than Edmonton, Canada, and Petersburg (the former Leningrad) lies at latitude 60° North—the latitude of the southern tip of Greenland. Winters are long, dark, and bitterly cold in most of Russia; summers are short and growing seasons limited. To make things worse, rainfall ranges from modest to minimal because the warm, moist air carried from the North Atlantic Ocean across Europe loses most of its warmth and much of its moisture by the time it reaches western Russia. (The term *continentality* is used to describe such an inland climatic environment remote from moderating maritime influences.) Drought, variable rainfall, and temperature extremes, therefore, have plagued Russian farmers as long as they have tilled the land (Fig. 2-1).

The tsars embarked on their imperial conquests in part because of Russia's relative location: Russia always lacked warm-water ports. Their southward push might have reached the Persian Gulf or even the Mediterranean Sea, had the Revolution not intervened. Tsar Peter the Great envisaged a Russia open to and trading with the entire world; he developed St. Petersburg into Russia's leading port. But in truth, Russia's historical geography is one of remoteness from the mainstreams of change and progress, and of self-imposed isolation. Not even a string of warm-water ports would have been likely to transform Russia into an outward-looking, trading state. The tsars' objectives were primarily strategic, not economic.

◆ AN IMPERIAL MULTINATIONAL STATE

Centuries of Russian expansionism did not confine itself to empty land or unclaimed frontiers. The Russian state became an imperial power that annexed and incorporated peoples of many nationalities and cultures. This was done by force of arms, by the overthrow of uncooperative rulers, by annexation, and by stoking the fires of ethnic conflict. By the time the ruthless Russian regime began to face revolution among its own people, the tsars held sway over the largest contiguous empire on earth. Tsarist Russia was a hearth of **imperialism**, and its empire contained peoples representing more than 100 nationalities. The winners in the ensuing revolutionary struggle—the communists who forged the Soviet Union—did not liberate these subjugated peoples. Rather, they changed the empire's framework, binding the peoples colonized by the tsars into a new system that would in theory give them autonomy and identity. In practice, it doomed those peoples to bondage and, in some cases, extinction.

A EURASIAN HERITAGE

The historical geography of this turbulent realm focuses on Russia west of the Urals and also involves Ukraine and other neighboring areas. Although our knowledge of Russia before the Middle Ages is only fragmentary, it is clear that peoples moved in great migratory waves across the plains on which the modern state eventually was to emerge. The dominant direction seems to have been from east to west; many groups came from interior Asia and left their imprints on the makeup of the population. Scythians, Sarmatians, Goths, and Huns came, settled, fought, and were absorbed or driven off. Eventually, the Slavs emerged as the dominant people in what is today Ukraine; they were peasants, farming the good soils of the plains north of the Black Sea. Their first leadership came not from their own midst but from a Scandinavian people to the northwest they called the Varangians (better known to us as the Vikings), who had for some time played an important role in the fortified trading towns (*gorods*) of the area.

The first Slavic state (ninth century A.D.) came about for reasons that will be familiar to anyone who remembers the importance of the transalpine routes across the middle of Europe. The objective was to render stable and secure an eastern crossing of the continent, from the Baltic Sea and Scandinavia in the northwest to Byzantine Europe and Constantinople in the southeast. This route ran southward from the Gulf of Finland to Novgorod (positioned on Lake Ilmen) and then crossed what is today Belarus to Kiev (Kyyiv). From there, it proceeded along the Dnieper River to the shores of the Black Sea.

Novgorod, near Scandinavia and close to the Baltic coast, was the European center on this long route; it had a cosmopolitan population and trade connections with the Hanseatic League. It benefited from its position and became the capital of its area, a distinctive princedom known as a *Rus*. Kiev, in contrast, lay in the heart of the land of the Slavs. Its situation was near an important confluence of the Dnieper, and it was located near the zone of contact between the forests of middle Russia and the grassland steppes of the south. Kiev also had centrality: it served as a meeting place for Scandinavian and Mediterranean Europe. Though distinctly different from northern Novgorod, Kiev became a Europeanized urban center as well. It, too, was the center of a Rus, the Kievan Rus. During the eleventh and twelfth centuries, Kiev was the political and cultural focus for a large region. Briefly, Novgorod and Kiev even united, and regional stability brought still greater prosperity.

Mongols

Prosperity, however, attracts competition, and the Kievan Rus suffered internal division and external invasion. The external threat came from the Mongol Empire far to the east in interior Asia, which had been building under Genghis Khan. The Tatar hordes rode into the Kievan Rus, and the city as well as the state fell in 1240. Many Russians fled into the forests, where the horsemen of the open steppes were far less threatening. What remained of the western Rus, then, lay in the forest between the Baltic and the steppes, and in that area a number of weak feudal states arose. Many of these were ruled by princes who paid tribute to the Tatars in order to be left in relative peace.

From among these feudal states, the one centered on Moscow (Muscovy) possessed superior locational advantages. It was positioned on a river that formed a route to Novgorod, and it was centrally situated with respect to other Russian trading sites. While the Russians managed to hold off their Asian enemies and the westward drive of the Tatars' "Golden Horde" slowed, a major state emerged where the old Kievan Rus had lain. This was the Grand Duchy of Lithuania, which for a time extended from near the Baltic Sea to the vicinity of the Black Sea. When Moscow became the dominant Russian center of power, the Lithuanians, allied with the Poles, controlled a region that included present-day Poland, Belarus, Ukraine, and adjacent areas. Russians and Lithuanians competed for regional power.

Soon Moscow's geographic advantages began to be felt. During the fifteenth century, Moscow's ruler was able to take control of Novgorod. But Muscovy became a real tsardom under the reign of Ivan (IV) the Terrible (1547–1584), who began Russia's advances into non-Russian territory—a campaign that did not end until well after the tsarist regime was terminated by the Revolution of 1917. Ivan first established control over the entire basin of the important Volga River to the east, inflicted heavy defeats on the Tatars, and then pushed eastward across the Urals into Siberia.

Cossacks

This eastward expansion of Russia was spearheaded by a relatively small group of semi-nomadic peoples who came to be known as Cossacks and whose original home was in present-day Ukraine. Opportunists and pioneers, they sought the riches of the eastern frontier, chiefly fur-bearing animals, as early as the sixteenth century. By the middle of the seventeenth century they reached the Pacific Ocean, defeating Tatars in their path and consolidating their gains by constructing *ostrogs*, strategic fortified way-stations along river courses. Before the eastward expansion halted in 1812 (Fig. 2-2), the Russians had moved across the Bering Strait to Alaska and down the western coast of North America into what is now northern California (see box on p. 126).

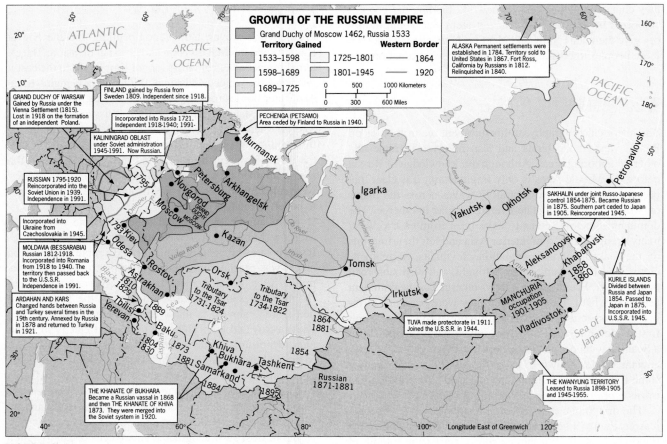

GROWTH OF THE RUSSIAN EMPIRE

Grand Duchy of Moscow 1462, Russia 1533

Territory Gained   **Western Border**

1533–1598   1725–1801   ——— 1864

1598–1689   1801–1945   ——— 1920

1689–1725

0   500   1000 Kilometers

0   300   600 Miles

ALASKA Permanent settlements were established in 1784. Territory sold to United States in 1867. Fort Ross, California by Russians in 1812. Relinquished in 1840.

FINLAND gained by Russia from Sweden 1809. Independent since 1918.

PECHENGA (PETSAMO) Area ceded by Finland to Russia in 1940.

GRAND DUCHY OF WARSAW Gained by Russia under the Vienna Settlement (1815). Lost in 1918 on the formation of an independent Poland.

Incorporated into Russia 1721. Independent 1918-1940; 1991-

KALININGRAD OBLAST under Soviet administration 1945-1991. Now Russian.

RUSSIAN 1795-1920 Reincorporated into the Soviet Union in 1939. Independence in 1991.

Incorporated into Ukraine from Czechoslovakia in 1945.

MOLDAVIA (BESSARABIA) Russian 1812-1918. Incorporated into Romania from 1918 to 1940. The territory then passed back to the U.S.S.R. Independence in 1991.

ARDAHAN AND KARS Changed hands between Russia and Turkey several times in the 19th century. Annexed by Russia in 1878 and returned to Turkey in 1921.

THE KHANATE OF BUKHARA Became a Russian vassal in 1868 and then THE KHANATE OF KHIVA 1873. They were merged into the Soviet system in 1920.

SAKHALIN under joint Russo-Japanese control 1854-1875. Became Russian in 1875. Southern part ceded to Japan in 1905. Reincorporated 1945.

KURILE ISLANDS Divided between Russia and Japan 1854. Passed to Japan in 1875. Incorporated into U.S.S.R. 1945.

TUVA made protectorate in 1911. Joined the U.S.S.R. in 1944.

THE KWANYUNG TERRITORY Leased to Russia 1898-1905 and 1945-1955.

Tributary to the Tsar 1731-1824

Tributary to the Tsar 1734-1822

MANCHURIA occupation 1901-1905

Russian 1871-1881

FIGURE 2-2

RUSSIANS IN NORTH AMERICA

The first white settlers in Alaska were Russians, not Western Europeans, and they came across Siberia and the Bering Strait, not across the Atlantic and North America. Russian hunters of the sea otter, valued for its high-priced pelt, established their first Alaskan settlement at Kodiak Island in 1784. Moving southward along the North American coast, the Russians founded additional villages and forts to protect their tenuous holdings until they reached as far as the area just north of San Francisco Bay, where they built Fort Ross in 1812.

But the Russian settlements were isolated and vulnerable. European fur traders began to put pressure on their Russian competitors, and Moscow found the distant settlements a burden and a risk. In any case, American,

British, and Canadian hunters were decimating the sea otter population, and profits declined. When U.S. Secretary of State William Seward offered to purchase Russia's holdings in 1867, St. Petersburg quickly agreed—for $7.2 million. Thus Alaska, including its lengthy southward coastal extension, became American territory. Although Seward was ridiculed for his decision—Alaska was called "Seward's Folly" and "Seward's Icebox"—his reputation was redeemed when gold was discovered there in the 1890s. The twentieth century has proved Seward's action one of great wisdom, strategically as well as economically. At Prudhoe Bay off Alaska's northern Arctic slope, large oil reserves are still being exploited. And like Siberia, Alaska probably contains other yet-unknown riches.

Tsar Peter the Great

By the time Peter the Great took over the leadership of Russia (he reigned from 1682 to 1725), Moscow lay at the center of a great empire—great at least in terms of the territories under its hegemony. As such, emergent Russia had many enemies. The Mongols had ceased to be a menace, but the Swedes, no longer allies in trade, threatened from the northwest, as (still) did the Lithuanians. To the west was a continuing conflict with the Germans and the Poles. And to the southwest, the Ottoman Turks were heirs to the Byzantine Empire and also posed a threat.

Peter consolidated Russia's gains and did much to make a modern European-style state out of the loosely knit country. He wanted to reorient the empire to the Baltic, to give it a window on the sea, and to make it a maritime as well as a land power. In 1703, following his orders, the building of St. Petersburg began. It was built by Italian architects at the tsar's behest; they designed numerous ornate buildings arranged around the grandiose city's many waterways in a manner reminiscent of Venice.

Petersburg is positioned at the head of the Gulf of Finland, which opens into the Baltic Sea. The city not only provided Russia with an important maritime outlet but also was designed to function as a **forward capital**: it lay on the doorstep of Finland, at that time a Swedish possession, and thus represented the Russian determination to maintain its presence in this strategic area. In 1709, Peter's armies defeated the Swedes, confirming Russian power on the Baltic coast. Four years later, Peter took the momentous step of moving the Russian capital from Moscow to the new Baltic headquarters, where it remained until 1918.

Peter was an extraordinary leader, in many ways the founder of modern Russia. In his desire to remake Russia—to pull it from the forests of the interior to the western coast, to open it to outside influences, to end its comparative isolation—he left no stone unturned. Not only did he move the capital, but he himself, aware that the future of Russia as a major force lay in strength at sea as well as power on land, went to Holland to work as a laborer in the famed Dutch shipyards to learn how ships were most efficiently built. Peter wanted a European Russia, a maritime Russia, a cosmopolitan Russia. He developed St. Petersburg into a leading seat of power as well as one of the most magnificent cities in the world, and it remains to this day a European-style city apart from all others in Russia.

Tsarina Catherine the Great

During the eighteenth century, Tsarina Catherine the Great, who ruled from 1762 to 1796, continued to build Russian power but on another coast and in another area: the Black Sea in the south. Here the Russians confronted the Turks, who had taken the initiative from the Greeks; the Byzantine Empire had been succeeded by the Turkish Ottoman Empire. The Turks were no match for the Russians. The Crimean Peninsula soon fell, as did the old and important trading city of Odesa (Odessa). Before long, the whole northern coast of the Black Sea was in Russian hands. Soon afterward, the Russians penetrated the area of the Caucasus to the southeast, and in due course they took Tbilisi, Baku, and Yerevan. Yet as they pushed farther into the corridor between the Black and the Caspian seas, the Russians faced growing opposition from the British—who held sway in Persia (modern-day Iran)—as well as from the Turks. Their advance was halted short of its probable ultimate goal: controlling a coast on the Indian Ocean.

The Nineteenth Century

Russian expansionism was not yet satisfied. While extending the empire southward, the Russians also took on the Poles, old enemies to the west, and succeeded in taking most of what is today the Polish state, including the capital of Warsaw. To the northwest, Russia took over Finland from

Although Moscow is Russia's capital, Petersburg in many ways is Russia's prominent city, exceptionally expressive of national traditions and feelings. Peter the Great hired the architects to remake this most European of Russian cities, its many waterways giving Petersburg a Venetian quality. Gold-domed St. Isaac's Cathedral is flanked to the right by a Russian Orthodox-style church. Farther to the right is the golden spire of the Admiralty Building. The Neva River (foreground) bisects the city.

the Swedes in 1809. During most of the nineteenth century, however, the Russian preoccupation was with Central Asia—the region between the Caspian Sea and western China—where Tashkent and Samarkand came under St. Petersburg's control (Fig. 2-2). The Russians here were still bothered by raids of nomadic horsemen, and they sought to establish their authority over the Central Asian steppe country as far as the edges of the high mountains that lay to the south. Thus Russia gained a considerable number of Muslim subjects, for this was Islamic Asia they were penetrating. Under tsarist rule, these people acquired a sort of ill-defined protectorate status while retaining some autonomy. Much farther to the east, a combination of Japanese expansionism and a decline of Chinese influence led Russia to annex from China several provinces to the east of the Amur River. Soon thereafter, in 1860, the port of Vladivostok on the Pacific was founded.

Now began the course of events that was to lead, after five centuries of almost uninterrupted expansion and consolidation, to the first setback in the Russian drive for territory. In 1892, the Russians began building the Trans-Siberian Railroad in an effort to connect the distant frontier more effectively to the western core. As the map shows (Fig. 2-2), the most direct route to Vladivostok was across northeastern China (Manchuria). The Russians wanted China to permit the construction of the last link of the railway across their territory, but the Chinese resisted. Taking advantage of the 1900 Boxer Rebellion in China (see Chapter 10), Russia responded by annexing Manchuria and occupying it. This brought on the Russo-Japanese War of 1905, in which the Russians were disastrously defeated; Japan even took possession of southern Sakhalin Island (which they called Karafuto). For the first time in nearly five centuries, Russia sustained a setback that resulted in a territorial loss.

COLONIAL LEGACY

Thus Russia—recipient of British and European innovations in common with Germany, France, and Italy—expanded by **colonialism** too. Yet where other European powers traveled by sea, Russian influence traveled overland into Central Asia, Siberia, China, and the Pacific coastlands of the Far East. What emerged was not the greatest empire but the largest *territorially contiguous* empire in the world. It is tempting to speculate what would have happened to this sprawling realm had European Russia (for such it still was) developed politically and economically in the manner of the other European power cores. At the time of the Japanese war, the Russian tsar still ruled over more than 8.5 million square miles (22 million sq km), just a tiny fraction less than the area of the Soviet Union after the 1917 Revolution. Thus the communist empire, to a very large extent, was the legacy of St. Petersburg and European Russia, not the product of Moscow and the socialist revolution.

◆ THE PHYSICAL STAGE

Russia's climatic and biotic environments, we noted earlier, reflect the country's interior location, its high-latitude position, its exposure to Arctic air masses, and its encirclement by mountains on its landward side. Low temperatures, short growing seasons, and limited water supplies combine to challenge Russian farmers.

While the Soviet Union existed and control was exercised from Moscow, the communist planners initiated major irrigation projects to bring water to the warmer areas then under their control, especially the arid Muslim republics in Central Asia east of the Caspian Sea. Some of these schemes were successful, but others led to environmental disaster. Surface streams were diverted, groundwater supplies dwindled, pesticides caused widespread chemical pollution, and social costs (especially deterioration of public health) were high. The Aral Sea, on the border between the former Kazakh and Uzbek republics, has lost more than half its water surface since 1960 (see photo) because streams feeding it were diverted for irrigation.

Climate is a major reason why the independence of Ukraine has been of such great concern to the Russians. Ukraine has precious areas of *Cfa* and *Dfa* climate (the *f* signifying year-round precipitation and the *a* meaning comparatively warm summers), and it also contains the most southerly portion of the still-moderate *Dfb* climate (see Fig. 2-1). In an average year, Ukraine produced as much as 45 percent of all farm produce harvested in the entire Soviet Union. Much of that production was consumed in Russia—at prices set by the communist system, not by world markets. The prospect that this bounty would be lost to Russia following the dissolution of the Soviet Union was a major stimulus for the creation of the successor "Commonwealth of Independent States" in December 1991 (see the box entitled "The Soviet Union, 1924–1991"). Russia's President Boris Yeltsin took the lead; to his Russian Republic, keeping Ukraine in the fold was a most urgent matter. Ultimately this campaign failed, and Ukraine not only is no longer in the Russian fold: relations between the two countries have become difficult.

Whatever the political manipulations, however, the climatic environment (and the biotic and soil conditions that go with it) will for many years continue to make Russia dependent on outside food sources. Expensive grain imports were a major problem for the old Soviet Union, Ukraine's productivity notwithstanding. The new Russia will have it no easier.

PHYSIOGRAPHIC REGIONS

Some of the forces shaping Russia's harsh environments are revealed in the map of its physiography (Fig. 2-3). South of the Russian state lie mountains—the Caucasus ⑧ in the

This is what has become of the former shoreline of the rapidly receding Aral Sea, scene of one of the greatest ecological disasters of the century.

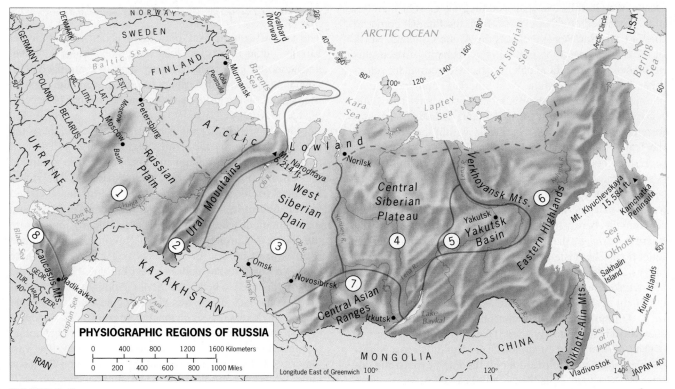

FIGURE 2-3

west, the Central Asian Ranges ⑦ in the center, and the Eastern Highlands ⑥ facing the Pacific from the Bering Strait to the North Korean border. Warm subtropical air has little or no opportunity to penetrate here, while cold Arctic air masses can sweep southward without interruption. Russia's Arctic fringe is a lowland sloping gently toward the frigid Arctic Ocean. Overall, we can discern eight physiographic regions in Fig. 2-3.

The *Russian Plain* ① is the eastward continuation of the North European Lowland, and here the Russian state formed its first core area. At its heart lies the Moscow Basin. Travel northward from Moscow, and the countryside soon is covered by needleleaf forests like those of Canada; to the south lie the grain fields of southern Russia and, beyond, those of Ukraine. Note the Kola Peninsula and the Barents Sea in the far north: warm water from the North Atlantic comes around northern Norway and keeps the port of Murmansk open most of the year. The Russian Plain is bounded on the east by the *Ural Mountains* ②, not a very high range but topographically prominent because it separates two extensive plains. The range is more than 2,000 miles (3,200 km) long and reaches from the shores of the Kara Sea to the border with Kazakhstan. It is no barrier to east-west transportation, and its southern end is quite densely populated. Here the Urals yield a variety of minerals as well as fossil fuels.

East of the Urals lies Siberia. The *West Siberian Plain* ③ has been described as the world's largest unbroken lowland; this is the vast basin of the Ob and Irtysh Rivers. Over the last 1,000 miles (1,600 km) of its course toward the Arctic Ocean, the Ob falls less than 300 feet (90 m). On Fig. 2-3, note the dashed line that extends from the Finnish border to the East Siberian Sea and offsets the Arctic Lowland. North of the line, water in the ground is permanently frozen; this *permafrost* creates yet another obstacle to permanent settlement. Looking again at the West Siberian Plain, the north is permafrost-ridden, and the central zone is marshy. The south, however, carries significant settlement, including such major cities as Omsk and Novosibirsk, within the corridor of the Trans-Siberian Railroad.

East of the West Siberian Plain the country begins to rise, first into the *Central Siberian Plateau* ④, another sparsely settled, remote, permafrost-afflicted region. Here winters are long and extremely cold, and summers are short; the area still remains barely touched by human activity. Beyond the *Yakutsk Basin* ⑤, the terrain becomes mountainous and the relief high. The *Eastern Highlands* ⑥ are a jumbled mass of ranges and ridges, precipitous valleys, and volcanic mountains. Lake Baykal lies in a trough that is over 5,000 feet (1,500 m) deep—the deepest rift lake in the world (see Chapter 8). On the Kamchatka Peninsula, volcanic Mount Klyuchevskaya reaches nearly 15,600 feet (4,750 m).

The northern part of region ⑥ is Russia's most inhospitable zone, but southward along the Pacific coast the climate is less severe. Nonetheless, this is a true frontier region. The forests provide opportunities for lumbering, a fur trade

exists, and there are gold and diamond deposits. To help develop this promising area, the Soviets constructed the new Baykal-Amur Mainline (BAM) Railroad, a 2,200-mile-long (3,540 km) route that roughly parallels the aging Trans-Siberian line to the south (see Fig. 2-6).

The southern margins of Russia are also marked by mountains: the *Central Asian Ranges* ⑦ from the Kazakh border in the west to Lake Baykal in the east, and the *Caucasus* ⑧ in the land corridor between the Black and Caspian seas. The Central Asian Ranges rise above the snow line and contain extensive mountain glaciers. The annual melting brings down alluvium to enrich the soils on the lower slopes and water to irrigate them. The Caucasus form an extension of Europe's Alpine mountain system and exhibit a similar topography, but they do not provide convenient passes. Here Russia's southern border is sharply defined by topography.

As the physiographic map suggests, the more habitable terrain in Russia becomes latitudinally narrower from west to east; beyond the southern Urals, settlement becomes a discontinuous ribbon. As we will note later in this chapter, isolated towns did develop in interior Siberia (such as Yakutsk on the Lena River and Norilsk near the Yenisey River). Yet as in Canada, the population is clustered markedly in the southern, most livable zone of the country, a narrow belt that widens only on Russia's Far Eastern, Pacific rim.

◆ THE SOVIET LEGACY

The era of communism may have ended in the Soviet Empire, but its effects will long remain on Russia's political and economic geography. Seventy years of centralized planning and implementation cannot be erased overnight; regional reorganization toward a market economy cannot be accomplished in a day.

While a world of capitalism celebrates the failure of the communist system in the former Soviet realm, it will do well to remember why communism found such fertile ground in the Russia of the 1910s and 1920s. In those days Russia was infamous for the wretched serfdom of its peasants, the cruel exploitation of its workers, the excesses of its nobility, and the ostentatious palaces and riches of its rulers. Ripples from the Western European Industrial Revolution introduced a new age of misery for those laboring in factories. There were workers' strikes and ugly retributions, but when the tsars finally tried to better the lot of the poor, it was a case of too little too late. There was no democracy, and the people had no way to express or channel their grievances. Europe's democratic revolution passed Russia by, and its economic revolution touched the tsars' domain only slightly. The vast majority of Russians, and tens of millions of non-Russians under the tsars' control, faced exploitation, corruption, starvation, and harsh subjugation. When the people began to rebel in 1905, there was no hint of what lay in store; even after the full-scale Revolution of 1917, Russia's political future hung in the balance.

The Russian Revolution was no unified uprising. There were factions and cliques; the Bolsheviks ("Majority") took their ideological lead from Lenin, while the Mensheviks ("Minority") saw a different, more liberal future for their country. The so-called "Red" army factions fought against the "Whites," while both battled the forces of the tsar. The country stopped functioning; terrible deprivations visited the people in the countryside as well as in the cities. Most Russians (and other nationalities within the empire, too) were ready for radical change.

That change came when the Revolution succeeded and the Bolsheviks bested the Mensheviks, most of whom were exiled. In 1918, the capital was moved from Petrograd (as St. Petersburg had been renamed in 1914, to remove its German appellation) to Moscow. This was a symbolic move, the opposite of the forward-capital principle: Moscow lay deep in the Russian interior, not even on a major navigable waterway (let alone a coast), amid the same forests that much earlier had afforded the Russians protection from their enemies. The new Soviet Union would look inward, and the communist system would achieve with Soviet resources and labor the goals that had for so long eluded the country. The chief political and economic architect for this effort was the revolutionary leader who prevailed in the power struggle: V. I. Lenin (born Vladimir Ilyich Ulyanov).

THE POLITICAL FRAMEWORK

Russia's great expansion had brought a large number of nationalities under tsarist control; now it was the turn of the revolutionary government to seek the best method of organizing this heterogeneous ethnic mosaic into a smoothly functioning state. The tsars had conquered, but they had done little to bring Russian culture to the peoples they ruled. The Georgians, Armenians, Tatars, and residents of the Muslim khanates of Central Asia were among dozens of individual cultural, linguistic, and religious groups that had not been "Russified." The Russians themselves in 1917, however, constituted only about one-half of the population of the entire country. Thus, it was impossible to establish a Russian state instantly over the whole of this vast political region, and these diverse national groups had to be accommodated.

The question of the nationalities became a major issue in the young Soviet state after 1917. Lenin, who brought the philosophies of Karl Marx to Russia, talked from the beginning about the "right of self-determination for the nationalities." The first response by many of Russia's subject peoples was to proclaim independent republics, as was done in Ukraine, Georgia, Armenia, Azerbaijan, and even in Central Asia. But Lenin had no intention of permitting any breakup of the state. In 1923, when his blueprint for the new Soviet Union went into effect, the last of these briefly independent units was fully absorbed into the sphere of the Moscow regime. Ukraine, for example, declared itself independent in 1917, and until 1919 managed to sustain this initiative; but in that year the Bolsheviks set up a provisional government in Kiev, thereby ensuring the incorporation of the country into Lenin's Soviet framework.

The Bolsheviks' political framework for their Soviet Union was based on the ethnic identities of its numerous incorporated peoples. Given the size and cultural complexity of the empire, it was impossible to allocate territory of equal political standing to all the nationalities; the communists controlled the destinies of well over 100 peoples, including large nations as well as small isolated groups. It was decided to divide the vast realm into *Soviet Socialist Republics (S.S.R.s)*, each of which was delimited to correspond broadly to one of the major nationalities. At the time, Russians constituted about half of the developing Soviet Union's population, and, as Fig. 2-4 shows, they also were (and still are) the most widely dispersed ethnic group in the realm. The Russian Republic, therefore, was by far the largest designated S.S.R., comprising just under 77 percent of total Soviet territory.

Within the S.S.R.s, smaller minorities were assigned political units of lesser rank. These were called *Autonomous Soviet Socialist Republics (A.S.S.R.s)*, which in effect were republics within republics; other areas were designated *Autonomous Regions* or other nationality-based units. It was a complicated, cumbersome, often poorly designed framework, but in 1924 it was launched officially under the banner of the *Union of Soviet Socialist Republics (U.S.S.R.)*.

Eventually, the Soviet Union came to consist of 15 S.S.R.s (shown in Fig. 2-5), including not only the original Republics of 1924 but also such later acquisitions as Moldova (formerly Moldavia), Estonia, Latvia, and Lithuania. The internal political layout often was changed, sometimes at the whim of the communist empire's dictators. But no communist *apartheid*-like system of segregation could accommodate the shifting multinational mosaic of the Soviet realm. The republics quarreled among themselves over boundaries and territory. Demographic changes, migrations, war, and economic factors soon made much of the layout of the 1920s obsolete. Moreover, the communist planners made it Soviet policy to relocate entire peoples from their homelands to better fit the grand design, and to reward or to punish— sometimes, it would appear, capriciously. The overall effect, however, was to move minority peoples eastward, and to replace them with Russians. And, as time went on, the Russification of the Soviet Empire also produced a map of substantial ethnic Russian minorities in all the non-Russian republics (Fig. 2-5).

The Russian Republic, although only one of the 15 S.S.R.s, was the Soviet Union's dominant entity—the centerpiece of a tightly controlled **federation** (see p. 134). With half the population, the capital city, the realm's core area, and over three-quarters of the Soviet Union's territory, Russia was

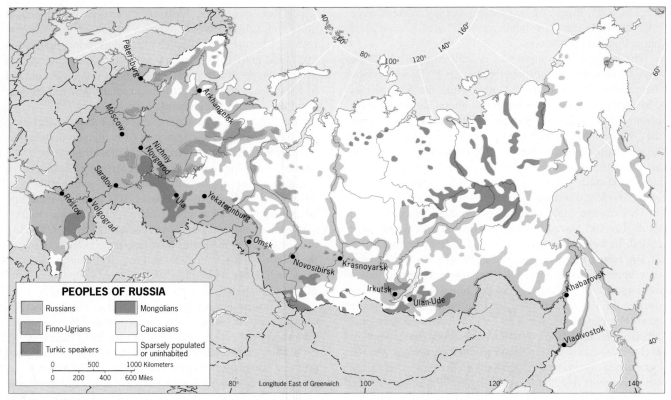

FIGURE 2-4

the empire's nucleus. In other republics, "Soviet" often was simply equated with "Russian"—it was the reality with which the lesser republics lived. Russians came to the other republics to teach (Russian was taught in the colonial schools), to organize (and often dominate) the local Communist Party, and to implement Moscow's economic decisions. This was colonialism, but somehow the communist guise and the contiguous spatial nature of the empire made it appear to the rest of the world as something else. Indeed, on the world stage, the Soviet Union became a champion of oppressed peoples, a force in the decolonization process. It was an astonishing contradiction that would, in time, be fully exposed.

THE SOVIET ECONOMIC FRAMEWORK

The geopolitical changes that resulted from the founding of the Soviet Union were accompanied by a gigantic economic experiment: the conversion of the empire from a capitalist system to communism. From the early 1920s onward, the country's economy would be *centrally planned*—that is, all decisions regarding **economic planning** and development were made by the communist leadership in Moscow. Soviet planners had two principal objectives: (1) to speed industrialization and (2) to collectivize agriculture. To accomplish this, the entire country was mobilized, with a national planning commission (*Gosplan*) at the helm. For the first time ever on such a scale, and for the first time in accordance

with Marxist-Leninist principles, an entire country was organized to work toward national goals prescribed by a central government.

The Soviet planners believed that agriculture could be made more productive by organizing it into huge state-run enterprises. The holdings of large landowners were expropriated, private farms were taken away from the farmers, and land was consolidated into collective farms. Initially, it was intended that all such land would be part of a *sovkhoz*, literally a grain-and-meat factory in which agricultural efficiency, through maximum mechanization and minimum labor requirements, would be at its peak. But the Soviets ran into opposition from many farmers who tried to sabotage the program in various ways, hoping to retain their land.

The fate of farmers and peasants who obstructed the communists' grand design was dreadful; it is one of the factors that have galvanized anti-Russian nationalism in the now-independent republics. In the 1930s, for instance, Stalin confiscated Ukraine's agricultural output and then ordered part of the border between the Russian and Ukrainian republics sealed—thereby creating a famine that killed several million farmers and their families. In the Soviet Union under communist totalitarianism, the ends justified any means, and untold hardship came to millions who had earlier suffered under the tsars. (In his book *Lenin's Tomb*, author David Remnick states that between 30 and 60 million people lost their lives from imposed starvation, purges, Siberian exile, and other causes.)

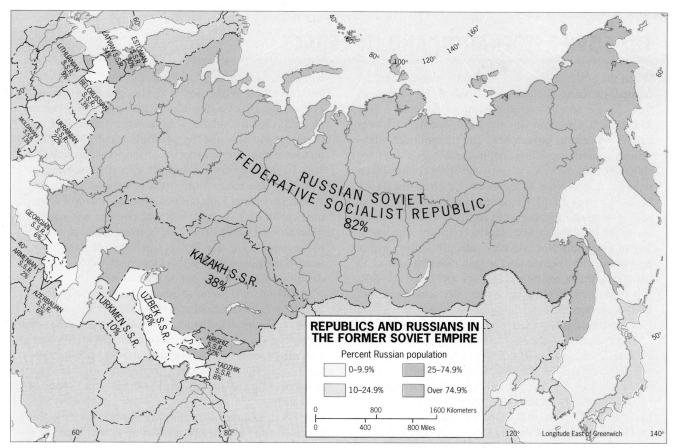

FIGURE 2-5

It was a human tragedy of incalculable dimensions, but it was hidden from the world by the secretive and covert character of Soviet officialdom. Only after the last Soviet president, Mikhail Gorbachev, initiated his *glasnost* campaign did meaningful data on such events begin to emerge. One result has been the rehabilitation of the reputations of many prominent Russians and others who protested these actions and who were executed during Stalin's dictatorial rule (1929–1953).

While dissidents were ruthlessly eliminated, Soviet planners did realize that a smaller-scale collective farm, the *kolkhoz*, would be more acceptable and more efficient than the *sovkhoz*. After 15 years of reform, just before the outbreak of World War II in 1939, about 90 percent of all farms, ranging from large estates to peasant holdings, had been collectivized.

Productivity, however, did not increase as the Soviet planners had hoped. Farmers like to tend their own land, and they take better care of it than they do the land of the state. Resentment over the harsh imposition of the program undoubtedly played a role also, but ultimately the two persistent problems were poor management and weak incentives for farmers to do their best. Add to this the recurrent weather problems caused by the country's disadvantageous relative location, and it is not surprising that Soviet farm yields were below expectations—and below requirements. Through-

out the three generations of communist rule, the Soviets had to turn to the outside world for food supplies.

Even as collectivization proceeded, the Soviets brought millions of acres of new land into cultivation through ambitious irrigation schemes. During the 1950s they launched the Virgin and Idle Lands Program in Kazakhstan, turning pasturelands into wheatfields. The program transformed parts of the Kazakh S.S.R., but success was linked to environmental disaster as diverted streams dried up, groundwater was poisoned by pesticides, and large numbers of people became ill. Again, the grand design created new problems as it solved old ones.

The Soviet planners hoped that collectivized and mechanized agriculture would free hundreds of thousands of workers to labor in the industries they wanted to establish. Enormous amounts of money were allocated to the development of manufacturing; the Soviets knew that national power would be based on the country's factories. The transport networks were extended; a second rail route to the Far East (the BAM) was built in the 1970s and 80s, and such remote places as Alma-Ata and Karaganda were connected to the system (Fig. 2-6). Energy development was given priority; compared to agriculture, the U.S.S.R.'s industrialization program had good results. Productivity rose rapidly, and when World War II engulfed the empire, the Soviet manufacturing

THE SOVIET FEDERATION AND ITS LEGACY

When the Soviet Union was originally planned and delimited, its name implied what it was to be—a federation. As we note in our discussion of federalism (pp. 251–253), a *federation* involves the sharing of power between a central government and the political subdivisions (provinces, States, or, in the Soviet case, "Socialist Republics") of a country. In theory, any Soviet republic that wanted to leave the Union could do so. Study the map of the former Soviet Union (Fig. 2-5), and an interesting geographic corollary emerges: every one of the 15 Soviet republics had a boundary with a non-Soviet neighbor. Not one was spatially locked within the others. This seemed to give geographic substance to the notion that a republic could opt out of the U.S.S.R. if it so desired. Reality, of course, was something else: Moscow's control over the republics made the Soviet Union a federation in theory only.

Nevertheless, the term "federal" appeared even on official Soviet maps. The former Russian Republic's official name was the Russian Soviet *Federative* Socialist Republic (R.S.F.S.R.), as any pre-1992 political map should show. The other 14 republics, however, did not have "federal" or "federative" in their names. Why not?

The answer lies not only in Russia's central and paramount position among the republics of the former Soviet Union but also in the enormous politico-geographical complexity of the Russian Republic itself. Today, after the devolution of the Soviet Union, the term *Russian Federation* is in common use in Russia, signifying (again theoretically) the relationship between the central authority of Moscow and the governments of 21 republics that lie within the boundaries of the Russian state. As we note in the profiles of these 21 internal republics (pp. 143–150), this relationship often is strained. When the Russian president took the initiative in creating a Commonwealth of Independent States (see the box entitled "The Soviet Union, 1924–1991"), the leaders of several of the republics within Russia announced that they, too, wanted the option to join this Commonwealth—not under Moscow's umbrella but on their own.

A federation is not established simply by proclaiming it. After a half-century of supranational negotiation, there still is no European federation. Federation involves the voluntary sharing of power, the willing association of sometimes disparate peoples who understand that yielding some sovereignty will bring rewards to all concerned. The years of Soviet rule did little to spread the ideas and ideals of federalism to the former empire's diverse peoples. Present-day Russia needs those lessons if it is to be a true federal state, but they may be lost on those forging the new system.

sector was able to generate the equipment and weapons needed to repel the German invaders.

Yet even in this context, the Soviet grand design held liabilities for the future. Because they could ignore market pressures and certain cost factors, Soviet planners assigned the production of particular manufactured goods to particular places, often disregarding the locational considerations of economic geography. For example, the manufacture of railroad cars might be assigned (as indeed it was) to a factory in Latvia. No other factory anywhere else would be permitted to produce this particular equipment—even if supplies of raw materials would make it much cheaper to build them near, say, Volgograd 1,200 miles away. Such practices made manufacturing in the U.S.S.R. extremely expensive, and the absence of competition made managers complacent and workers less productive than they could be.

The Soviet system tightly bound the economic geography of the republics, however. Each part of the Soviet Union depended for raw materials, energy, or other needs on another part. When the political system collapsed, this areal interdependence created a major impediment to reform.

The dramatic changes arising from the Soviet Union's devolution are reflected in Fig. 2-7. Even before the reform-minded Gorbachev took office in 1985, Soviet economic, political, and social systems were showing signs of failure. A costly military campaign in Afghanistan had gained nothing. Expensive food imports were necessary. Stirrings of democracy were being felt in communist-dominated Eastern Europe. The Soviet weapons buildup was becoming unaffordable. The country's infrastructure was breaking down. Gorbachev knew that the great Soviet communist experiment was ending in failure, but he worked to guide the processes of political and economic transformation. That the dissolution of the old order occurred with minimal loss of life will be a permanent monument to the leader who opened Soviet society to the world.

As Fig. 2-7 shows, Soviet devolution shrank the Russian sphere of influence in huge areas of Eurasia. The Iron Curtain was lifted, and countries from Eastern Europe to Mongolia were freed from Moscow's hegemony. The Soviet Empire's borders lost their functions and Russia's boundaries gained new ones. Ethnic Russian minorities in the 14

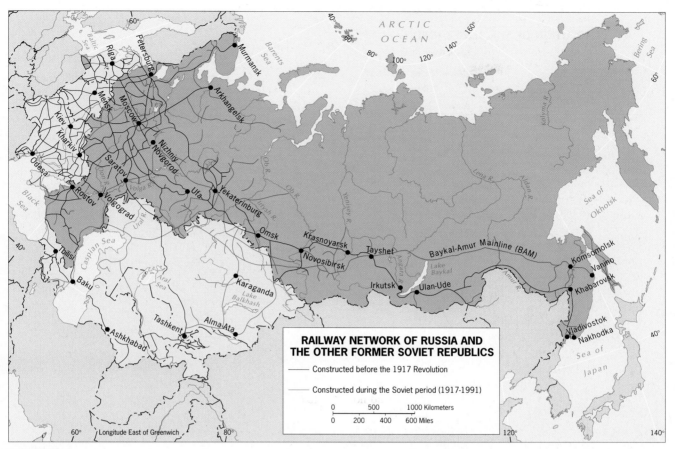

FIGURE 2-6

neighboring republics, once secure under Moscow's protection, now faced uncertain postcolonial futures. With remarkable speed, the former Soviet republics reoriented themselves: the western ones toward Europe, the southern ones toward the Islamic world of which they had been part. What was left of the Soviet Empire was Russia—itself an amalgam of political and cultural entities.

◆ RUSSIA'S CHANGING POLITICAL GEOGRAPHY

Even without its empire, Russia remains a vast and varied country—territorially still the largest on earth but ranking well down the list in terms of population (below such countries as Indonesia and Brazil). Russia's livable space is latitudinally widest in the west but narrows eastward to a corridor between frigid Siberia to the north and deserts and mountains to the south, as reflected by its urban system and transport networks (Fig. 2-8).

When the Soviet Union went out of existence on December 25, 1991, Russia entered a new era as a republic, with an elected President (Boris Yeltsin) and, for the first time in its history, a representative government. Not all Russians approved of this new order, and from the beginning a conservative minority made itself heard in the parliament and sometimes in the streets of Moscow and other cities. These conservatives, many of them so-called *apparatchiks* (people who served in and benefited from the old communist machine) also obstructed the plans of Russia's economic reformers. Change could easily be slowed in the vast reaches of this enormous country merely by continuing old practices of corruption and slow production. Many apparatchiks still remained in charge of farms and factories.

Political maps sometimes show a "European" Russia (bounded on the east by the Urals) and an "Asian" Russia. No geographic justification exists for this. True, as Fig. 2-8 reveals, the great majority of Russia's population of 150 million is concentrated in the western one-fifth of the national territory. But as the map of ethnic groups (Fig. 2-4) shows, Russians—who make up about 83 percent of the population—are as dominant in the vast eastern regions of the country as they are in the western heartland. The Far Eastern cities of Komsomolsk and Khabarovsk are no less "European" Russian than Rostov or Nizhniy Novgorod (formerly Gorkiy). In short, the Russian geographic realm extends from Petersburg to Vladivostok and from Murmansk to the Georgian border.

DEVOLUTION OF THE SOVIET UNION

Former Soviet Sphere

Republics in Russia

Proclaimed republics

Major Russian minorities

0   500   1000   1500 Kilometers
0   250   500   750 Miles

ARCTIC OCEAN

NORWAY

SWEDEN

FINLAND

Arctic Circle

Barents Sea

Kara Sea

KARELIA

KOMI

RUSSIAN FEDERATION

Russian Boundary, 1992

KHAKASSIA

ALTAYA

TUVA

Iron Curtain, 1945–1990

E. GER.

CZECH. REP.

SLOVAKIA

HUNG.

ROMANIA

BULG.

POLAND

KALININGRAD (Russia)

LITH.

LAT.

EST.

BELARUS

UKRAINE

MOLDOVA

Russian Boundary, 1992

MORDVINIA

CHUVASHIA

MARI-EL

UDMURTIA

TARTARSTAN

BASHKORTOSTAN

URALS

KALMYKIA

TURKEY

Black Sea

ARMENIA

GEORGIA

DAGESTAN

AZERBAIJAN

SYRIA

IRAQ

IRAN

Caspian Sea

Aral Sea

KAZAKHSTAN

Lake Balkhash

UZBEKISTAN

TURKMENISTAN

KYRGYZSTAN

TAJIKISTAN

Soviet Union Boundary, 1989

AFGHANISTAN

SOUTHERN REPUBLICS

1 Adygeya

2 Karachayevo-Cherkessia

3 Kabardino-Balkaria

4 North Ossetia

5 Ingushetia

6 Chechenya

FIGURE 2-7

East Siberian Sea

Laptev Sea

SAKHA

Sea of Okhotsk

SAKHALIN

NORTHERN TERRITORIES

BURYATIA

Lake Baykal

MARITIME

MONGOLIA

Sea of Japan

Soviet Sphere Boundary, 1989

N. KOREA

JAPAN

S. KOREA

C H I N A

Longitude East of Greenwich

RUSSIA: MAJOR URBAN CENTERS AND SURFACE COMMUNICATIONS

POPULATION

- Under 50,000
- 50,000–250,000
- 250,000–1,000,000
- 1,000,000–5,000,000
- Over 5,000,000

—— Railroad
—— Road

National capitals are underlined

0   200   400   600   800 Kilometers
0   100   200   300   400   500 Miles

FIGURE 2-8

138

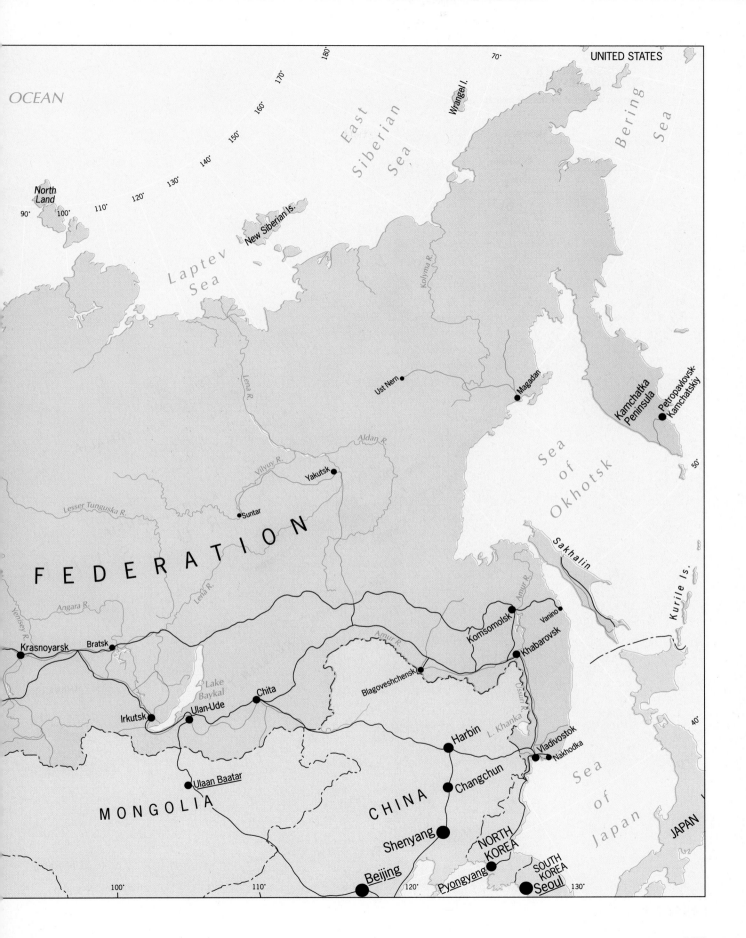

OCEAN

North
Land

90° 100° 110° 120° 130° 140° 150° 160° 170° 180° 70°

UNITED STATES

East
Siberian
Sea

Wrangel I.

Laptev
Sea

New Siberian Is.

Kolyma R.

Bering
Sea

Lena R.

Aldan R.

Ust Nern

Magadan

Kamchatka
Peninsula

Petropavlovsk-
Kamchatskiy

50°

Vilvuy R.

Yakutsk

Sea
of
Okhotsk

Lesser Tunguska R.

Suntar

F E D E R A T I O N

Sakhalin

Angara R.

Lena R.

Amur R.

Komsomolsk

Vanino

Kurile Is.

Yenisey R.

Krasnoyarsk

Bratsk

Amur R.

Blagoveshchensk

Khabarovsk

Ussuri R.

40°

Lake
Baykal

Chita

Irkutsk

Ulan-Ude

L. Khanka

Harbin

Vladivostok

Nakhodka

Ulaan Baatar

Changchun

Sea
of
Japan

M O N G O L I A

C H I N A

Shenyang

NORTH
KOREA

JAPAN

Beijing

100° 110° 120° Pyongyang

SOUTH
KOREA
Seoul

130°

139

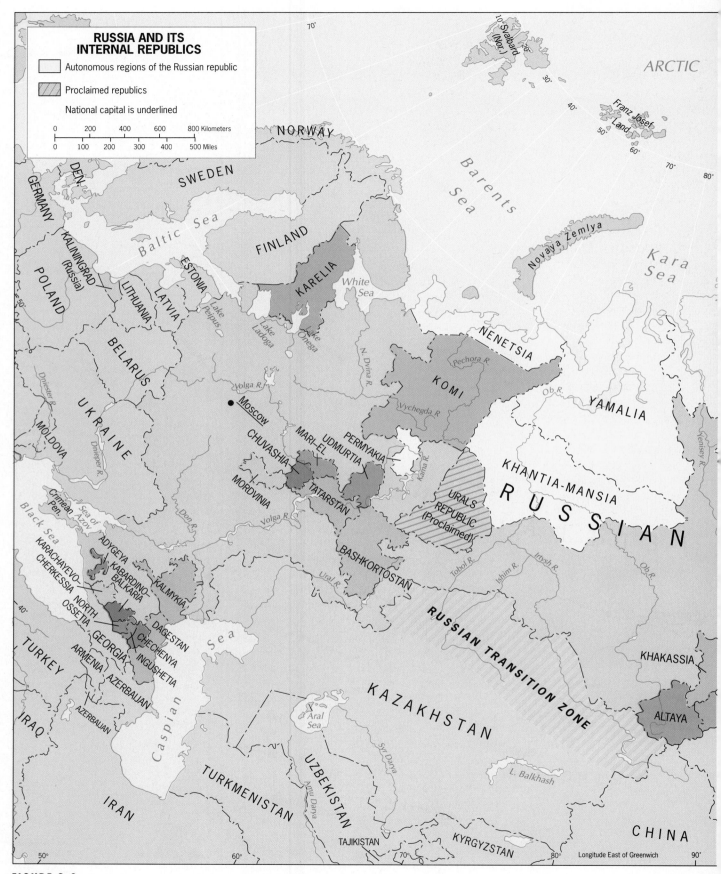

RUSSIA AND ITS INTERNAL REPUBLICS

Autonomous regions of the Russian republic

Proclaimed republics

National capital is underlined

| 0 | 200 | 400 | 600 | 800 Kilometers |
| 0 | 100 | 200 | 300 | 400 | 500 Miles |

FIGURE 2-9

140

OCEAN

North
Land

90° 100° 110° 120° 130° 140° 150° 160° 170° 180° 70°

UNITED STATES

Laptev
Sea

New Siberian Is.

East
Siberian
Sea

Wrangel I.

Bering
Sea

CHUKOTKA

KORYAKIA

TAYMYRIA

Lena R.

SAKHA
(YAKUTIA)

Aldan R.

Sea
of
Okhotsk

Kamchatka
Peninsula

50°

EVENKIA

Vilyuy R.

Lesser Tunguska R.

F E D E R A T I O N

Sakhalin

Angara R.

Lena R.

Amur R.

BIROBIJAN
(JEWISH)

Kurile Is.

Yenisey R.

BURYATIA

Amur R.

UST-ORDA

Lake
Baykal

AGA

Ussuri R.

MARITIME
REPUBLIC
(Proclaimed)

40°

L. Khanka

TUVA

Sea
of
Japan

JAPAN

M O N G O L I A

C H I N A

NORTH
KOREA

SOUTH
KOREA

100° 110° 120° 130°

141

Russia is nonetheless a patchwork of nations and ethnic groups, and even after nearly 75 years of communist domination and enforced Russification, many non-Russian cultures remain vibrant and strong. As discussed earlier, when the Soviet planners were confronted by the U.S.S.R.'s cultural diversity, they created a politico-geographical framework that would, they hoped, satisfy ethnic demands and pressures by establishing smaller political units within the Soviet Socialist Republics. Within the Transcaucasus republic of Georgia, for instance, they laid out two so-called Autonomous Soviet Socialist Republics (A.S.S.R.s) and one Autonomous Region (see the vignette on Transcaucasia that follows this chapter).

The most complicated part of this framework lay in Russia itself. The Russian Republic was administratively and politically divided into 6 Territories, 49 Regions, 16 A.S.S.R.s, 5 Autonomous Regions, and 10 Autonomous Areas. The 6 Territories and 49 Regions were designed as administrative units, mainly for the purposes of national economic planning, census taking, and so forth. However, the original 16 A.S.S.R.s and 5 Autonomous Regions were designated to award recognition to some of the nearly 40 substantial minorities on Russian soil. One of these Regions, for example, was the Jewish Autonomous Region in the remote Soviet Far East, a rugged and near-isolated corner of the country not much larger than Vermont (and never populated by more than 30,000 Russians of Jewish descent).

Even before the Soviet Union fell apart, this unwieldy politico-geographical framework was failing. While the 5 Autonomous Regions held a higher status than the 10 Autonomous Areas, the peoples and leaders of several of the Autonomous Areas had begun to play a more prominent role in national affairs than those of some of the regions (the Jewish Region, for instance, was never a factor). But these issues were submerged when the devolution of the whole Soviet Union gathered momentum. When Russia emerged as a newly independent republic at the end of 1991, it took the official name *Russian Federation* in acknowledgment of its inherited, multi-tiered political geography and the diversity of its cultural geography.

The Russian Federation inherited the complicated administrative structure the Soviets had created, and naturally it was impossible to change that structure overnight. Yet something had to be done, because some of the republics *within* Russia, and even some other administrative units, were themselves agitating for independence or greater autonomy.

On March 31, 1992, most of Russia's internal republics, oblasts, krays, and autonomous regions (all part of the administrative hierarchy) signed the Russian Federation Treaty. Four of the oblasts were awarded the higher status of republic, making a total of 20 republics inside Russia. Two republics refused to sign the treaty: Tatarstan, whose president insisted that his country should be as independent as Latvia or Georgia, and Checheno-Ingushetia, a republic in the troubled Caucasus region of the southwest. Soon afterward, Checheno-Ingushetia experienced a political crisis and split into two separate republics, Chechenya and Ingushetia. By mid-1993, Chechenya had ratified the Russian Federation Treaty, but Ingushetia had not. Further changes in Russia's federal map (Fig. 2-9) will undoubtedly occur, but in mid-1993 the number of republics stood at 21.* Also note that Fig. 2-9 refers to all the communist-era autonomous territories as Autonomous Regions, whether they were "regions" or "areas."

Undoubtedly the most prominent point made by Fig. 2-9 relates to Russia's own territory. Subtract the internal republics and the Autonomous Regions, and what is left of Russia proper is an elongated, fragmented, perforated strip of territory that constitutes less than half the total area of the Russian Federation, including islands in the Arctic Ocean, the tip of the Kamchatka Peninsula, and parts of Siberia. Most of the country's western heartland is Russian, but east of Moscow lies a cluster of six republics (including troublesome Tatarstan) that extends into Russia's very core area. Several of these republics contain energy installations, weapons factories, and crucial railroad lines. These are not mere administrative divisions.

As the map shows, the Republics and Autonomous Regions lie dispersed across the country, from the Caucasus in the southwest to the Arctic shores of the far northeast. On the next eight pages we group them as follows, their order based on relative location:

(1) *The Republics of the Heartland*
(Mordvinia, Chuvashia, Mari-el, Tatarstan, Udmurtia, Bashkortostan (plus the newly proclaimed Urals Republic not covered here)

(2) *The Republics of the Caucasian Periphery*
Dagestan, Chechenya, Ingushetia, North Ossetia, Kabardino-Balkaria, Karachayevo-Cherkessia,† Adygeya,† Kalmykia

(3) *The Republics of the North*
Karelia, Komi, Sakha (Yakutia)

(4) *The Republics of the Southeast*
Altaya,† Khakassia,† Buryatia, Tuva (plus the newly proclaimed Maritime Republic not covered here)

(5) *The Autonomous Regions*

---

*In July 1993, two additional regions proclaimed themselves republics: the *Urals Republic*, centered on the city of Yekaterinburg, and the *Maritime Republic*, whose capital is Vladivostok. This brings the number of internal republics to 23.

†Formerly Autonomous Regions or Autonomous Areas.

REPUBLICS OF THE HEARTLAND

MORDVINIAN REPUBLIC (MORDVINIA)

Travel southeastward from Moscow, and the first republic you will reach is Mordvinia (Fig. 2-10). Once under the sway of the Tatars of Kazan, Mordvinia came under Russian control in the sixteenth century and was proclaimed a Soviet A.S.S.R. in 1934. Once known as Mordovia, its population grew beyond 1 million only recently.

The Mordvinians, a people with Finnish and Ugric (Hungarian) ancestries, speak a language belonging to the Ural-Altaic family, which has not only Finnish and Hungarian but also Persian roots. Although Mordvinian culture has been fading, more than one-third of the population remains ethnic Mordvinian (Russians make up about 60 percent). In recent years, Mordvinian national costumes, elaborately embroidered tunics and dresses, have made a comeback in the cultural landscape.

FIGURE 2-10

Unlike many other republics, Mordvinia was never given a key role in the former Soviet economic system. The local economy remains chiefly agricultural; beekeeping is an old and continuing tradition here. Manufacturing, most of it of the light variety, concentrates in and near the capital, Saransk (338,000).

CHUVASH REPUBLIC (CHUVASHIA)

Western Russia's great Volga River flows eastward north of Moscow until, in the vicinity of the Tatar city of Kazan (1.2 million), it bends southward on its journey to the Caspian Sea. Before the bend, on its right bank lies the small, New Jersey-sized Chuvash Republic (Fig. 2-10).

The Chuvash are a Turkic people, sedentary farmers who inhabited this part of the middle Volga valley before the tsar's armies swept across their land in the sixteenth century. Farming and livestock raising continue to be the dominant way of life here; to this day, the Chuvash outnumber Russians in their republic by 70 to 25 percent.

Chuvashia became an A.S.S.R. in 1925. Under Soviet administration, some industrialization took place, and the capital, Cheboksary (457,000 today), situated on the Volga River, became a minor manufacturing center. The Chuvash, who are distantly related to the Bulgars, largely converted from Islam to the Russian Orthodox faith before the 1917 Revolution; religion has not been an issue in Chuvash-Russian relations. During the collapse of the U.S.S.R. in 1991, Chuvash leaders reaffirmed their republic's "sovereignty."

MARI REPUBLIC (MARI-EL)

Across the Volga River from the Chuvash Republic, about 400 miles (650 km) east of Moscow, lies the small Mari Republic (Fig. 2-10), a Massachusetts-sized enclave whose historical geography has been closely linked to that of its neighbors. The Mari, like the Komi and other minorities in east-central and northern Russia, have ancient ties to the Finns, the Hungarians, and even the Persians. At one time, they were subjugated under the Islamic Tatar khanate, but when Tsar Ivan the Terrible conquered the Tatar state in the sixteenth century, the Mari domain became a Russian colony. After the Revolution it became an *oblast*, and in 1936 it was proclaimed an A.S.S.R.

Today, about 45 percent of the population of nearly 800,000 has Komi ethnic ancestry. Nearly half of the people are Russians. Tatars (7 percent) still make up a noticeable minority.

Timber-based industries are important here, but under Soviet economic planning the Mari Republic was assigned machine building as its principal role. Still, logs from Mari forests continue to float down the Volga, to be made into furniture, prefabricated homes, and paper in the factories located in large Russian industrial cities.

Mari ethnic consciousness revived during the Soviet Union's disintegration, and the republic's leaders are demanding greater autonomy. The course of events in the neighboring Tatar Republic will be a key to Mari's future—as was the case in the past.

TATAR REPUBLIC (TATARSTAN)

Russia's Tatars are scattered remnants of the Mongol invaders, the "Golden Horde" that swept across the realm nine centuries ago. Between 5 and 6 million ethnic Tatars survive, more than half of them clustered in the middle Volga valley (Fig. 2-10). Here lies the Tatar Republic, established in 1920.

The Tatar Republic lies where the Kazan khanate once did, a powerful Muslim state centered on the city of Kazan. In 1552, Tsar Ivan the Terrible conquered the khanate, destroying some 400 mosques (even as he ordered the construction of St. Basil's Cathedral in Moscow). Sporadic Muslim uprisings followed, always crushed with great loss of life; but Islam did not die here. When the Revolution erupted, the Kazan Tatars joined it, hoping for a better life after the tsars' overthrow. They were rewarded with their republic even before the U.S.S.R. was formally constituted.

Under Soviet rule, atheism was state policy, and Islam continued to suffer. In addition, the republic (somewhat smaller than Maine) lay in the path of communist-planned industrial expansion. Agricultural and pastoral Tatarstan was "endowed" with truck factories, chemical industries, and engineering plants. For a time it was the former Soviet Union's largest oil producer; the first well was drilled in 1943, and growth was rapid. All this attracted Russian workers, and the population balance shifted. The Tatars began to object, their anger already aroused by cultural Russification.

Today, ethnic Tatars constitute around 48 percent of the population of about 3.7 million, Russians 44 percent. The Tatars in the republic are trying to revive their remaining mosques, rekindle their Arabic language, and stem the tide of Slavic culture. In 1991, the republic's president demanded the right to sign any new treaties as an equal partner with former Soviet Socialist Republics, and in 1992 he refused to sign the Russian Federation Treaty. Tatar nationalism remains a potent force here.

UDMURT REPUBLIC (UDMURTIA)

Udmurtia on its south adjoins the Tatar Republic and is the fifth of the cluster of republics situated east of Moscow (Fig. 2-10). About the size of Estonia, the Udmurt Republic is the ancestral home of still another Finnic people, the Udmurts. Their autonomy was proclaimed in 1934. Today, ethnic Udmurts constitute barely over 30 percent of the population; they are far outnumbered by Russians (nearly 60 percent). A Tatar minority (7 percent) also remains here, as well as some Mari.

During the Soviet era, Udmurtia became heavily industrialized. Weapons, steel, machines, tools, engines, and vehicles were produced in factories in Izhevsk (683,000 today), the capital, and other cities. Lumbering, sawmilling, leatherworking, and food processing also take place; grain cultivation and livestock raising further contribute to a varied economic geography. All this explains the large Russian population in this republic. The overall population now totals 1.7 million.

Undoubtedly reflecting the large Russian majority, the political scene in Udmurtia has been relatively calm, although local leaders have given notice that they wanted to strengthen their ability to negotiate with Moscow over "exports."

BASHKORT REPUBLIC (BASHKORTOSTAN)

Located north of Kazakhstan and extending westward from the slopes of the southern Ural Mountains (Fig. 2-10), Illinois-sized Bashkortostan has a population of 4.1 million and an economy based on rich mineral resources (including oil and natural gas) and productive agriculture. Of major significance is Bashkortostan's oil refining capacity because huge facilities have been built, making this republic one of the most important and centralized nodes in Russia's oil pipeline network.

The Bashkorts, a Turkic people, settled here during the Mongol invasions. The Russians conquered the area during the rule of Ivan the Terrible (1552), and the capital and largest city, Ufa (1.2 million today), was founded shortly thereafter.

In 1919, Muslim Bashkort leaders, enticed by Bolshevik promises of autonomy, joined the "Reds" in the Revolution. When the acreage of the large estates was redistributed, however, Russians—not Bashkorts or Tatars also living here—got most of it. A local rebellion was ruthlessly put down after the Bashkort A.S.S.R. was established (also in 1919).

Today, Bashkorts (24 percent) are outnumbered by Russians (40 percent) and Tatars (25 percent) in their own republic. Major industrialization has occurred in the Urals portion of the republic, based not only on oil production and refining but also on natural gas, timber, iron, manganese, and copper ores.

Non-Russians remain in the majority, and Islam and Orthodox Christianity are reviving. Bashkortostan's leaders, using the importance of the republic's economy to Russia as leverage, have demanded greater autonomy in the new Russian framework.

REPUBLICS OF THE CAUCASIAN PERIPHERY

DAGESTAN REPUBLIC (DAGESTAN)

At the eastern end of the Caucasus Mountains, between the crestline of the main range and the Caspian shore, lies the Dagestan Republic (Fig. 2-11), about half the size of Virginia. Dagestan—the name means "mountain country"—contains some 30 distinct nationalities among its 2 million people, many of them confined to isolated valleys. The Russians penetrated this territory in the fifteenth century, forcing the Persians (whose frontier it was) to yield in 1723. Formal annexation occurred in 1813, but the locals continued to fight back until 1877. In 1921, Dagestan was made an Autonomous Republic, a congregation of minorities in which the largest ethnic group is the Avartsy (27 percent) and the second largest the Dargintsy (16 percent). Russians account for only 11 percent.

Dagestan lies in a sensitive area, between Shi'ite Muslim Azerbaijan to the south and the Kalmyk Republic to the north, and adjacent to contentious Chechenya in the west. To Russia, the republic is important because of its sizeable oil and gas deposits on the Caspian coastal plain near the capital, Makhachkala (348,000).

CHECHEN REPUBLIC (CHECHENYA)

Until mid-1992, Chechenya was part of the Chechen-Ingush Republic. Old Muslim enemies, the Chechens and the Ingush heroically stood together and repelled the Cossacks fighting for the tsars. By 1870, however, the Russians prevailed. Forcibly united by the Soviet-communist planners in 1936, the two ethnic groups were awarded an Autonomous Republic in 1957.

In 1991, Chechen separatists moved to install their leader as president of the Republic. Russian President Yeltsin then signed a decree imposing a state of emergency and

FIGURE 2-11

voiding the election that had brought the Chechen leader to office. The Russian Parliament, in turn, voided Yeltsin's decree. Before long, the Chechens (who made up more than half of the population of 2.1 million) and the Ingush (12 percent) were in conflict. The strife was worsened by the situation in neighboring North Ossetia to the west, where 50,000 Ingush were persecuted by the Ossetian majority.

The solution was to divide the Chechen-Ingush Republic into two. The eastern (larger) part, dominated by Chechens, now forms Chechenya (Fig. 2-11). It includes the capital, Groznyy (436,000), one of the Russian Federation's major oil-refining centers, where most of the Republic's

Russians (30 percent of the population) live and work. Chechenya contains major oil reserves, and pipelines radiate as far afield as the Donets Basin in eastern Ukraine. Natural gas is also found in quantity, and the Chechen Republic has become an important entity in Russia's economic geography.

INGUSH REPUBLIC (INGUSHETIA)

Caucasus peaks over 14,000 feet (4,200 m) high tower over this tiny new republic on Russia's Transcaucasian rim

(Fig. 2-11). The Ingush, one of Russia's many small ethnic groups (perhaps numbering 250,000 people), have occupied mountain valleys here for centuries—and suffered dreadfully for it. In the 1940s, Stalin placed part of the historic Muslim Ingush domain under the control of neighboring non-Muslim, pro-Soviet North Ossetia. The other part of their territory lay in a Chechen-dominated Autonomous Republic.

In 1991, Chechens seized political control of the eastern part of Ingush territory, and in 1992 North Ossetian soldiers, helped by Russian forces, began a campaign of "ethnic cleansing" that destroyed hundreds of Muslim villages and drove thousands of Ingush families into the icy mountains, where many perished. The Russian Parliament's response was to declare a separate republic in Ingushetia, extending over the western part of the former Chechen-Ingush Republic but *not* including any Ingush land in North Ossetia.

Moscow's contradictory actions—creating an Ingush Republic on the one hand while abetting the killing of Ingush citizens in North Ossetia on the other—probably arise from a new regional concern here: the Muslim Confederation of Caucasian Mountain Peoples, an organization that seeks to unite Islamic groups on the Russian side of the Caucasus with the principal aim of breaking away from the Russian Federation.

NORTH OSSETIAN REPUBLIC (NORTH OSSETIA)

Territorially the smallest of all of Russia's internal republics, North (Severo) Ossetia's 3,000 square miles (8,000 sq km) of mountainous terrain lie on the northern flank of the Caucasus Mountains (Fig. 2-11). Immediately to the south lies the South (Yugo) Ossetian Autonomous Region of Georgia (see map p. 165). Important road and rail links cross the Caucasus through the republic.

Ossetian historical geography is long and turbulent. The Ossetians lived in this area as long as 2,500 years ago. The Tatars incorporated Ossetia into their empire in the thirteenth century; the Turks contested the area; and then the Russians colonized it in the late 1700s. Today, Ossetians still constitute half the population of about 700,000; Muslim Ingush number about 50,000. Russians make up 35 percent. The Republic was proclaimed in 1936.

Ossetian ethnic consciousness was raised by the Soviet collapse and by Georgia's push for independence, seen by the South Ossetians as a threat. Voices favoring reunification of the Ossetian domain were heard, and on the Georgian side of the border an insurrection was put down with armed force. On the Russian side, the Ingush were mercilessly persecuted. In 1993, these issues were far from resolution; one tangible change was the renaming of the North Ossetian capital from Ordzhonikidze to Vladikavkaz (population today: 333,000), its pre-Soviet appellation.

KABARDINO-BALKAR REPUBLIC (KABARDINO-BALKARIA)

Located in the central segment of the Caucasus cluster of internal republics (Fig. 2-11), Kabardino-Balkaria has some of Russia's most spectacular scenery, including Mount Elbrus (18,510 feet/5,640 m) in its southwestern corner. Glaciers, swift-running streams, alpine meadows, forests, and lower plains form a highly varied topography, despite the republic's small area (about the size of Connecticut).

About half of the 800,000 inhabitants are Kabardinians, people of the lower plains who allied themselves with the Russians as early as the sixteenth century. One-third of the population is Russian, many of Cossack descent. The Balkars of the mountains (10 percent) fought the Russian incursions for a long time. During World War II, the communists alleged, the Balkars sympathized with the German invaders of the U.S.S.R., and Stalin ordered their deportation as punishment. Many later returned to the republic, which was officially constituted in 1936.

Small but important deposits of valuable minerals (gold, chromium, nickel, tungsten, molybdenum) make mining an important industry. Soviet economic planners assigned the manufacture of oil-drilling equipment to this republic as part of their grand design—although Kabardino-Balkaria has no oil.

KARACHAYEVO-CHERKESS REPUBLIC (KARACHAYEVO-CHERKESSIA)

The spectacular scenery of this small, Connecticut-sized republic is reminiscent of Switzerland: jagged peaks of rock and ice tower over forested slopes and alpine meadows. The southern boundary of Karachayevo-Cherkessia (Fig. 2-11) coincides with the crestline of the Caucasus Mountains, and here elevations exceed 13,000 feet (4,000 m).

This republic exemplifies the Soviet administration of minorities. A combined Autonomous Region for the Turkic Karachay and Christian Cherkess peoples was created in 1922, but a few years later this unit was divided to give each group an oblast. In 1943, this arrangement was dissolved because the Karachay were accused of supporting the Germans during World War II; many Karachay were exiled. In 1957, the present territory was re-established as an Autonomous Region to accommodate both ethnic groups; today it is identified officially as a Republic of the Russian Federation.

Livestock herding continues on the northern plains, but mining, lumbering, and some manufacturing have developed. The total population remains around 500,000. The capital, Cherkessk (146,000), lies in the non-Turkic part of the republic.

ADYGEY REPUBLIC (ADYGEYA)

The Adyghian people are related to the Cherkess, and their republic was first created as an Autonomous Region in 1922. Somewhat smaller than Puerto Rico, Adygeya lies in the hinterland of the city of Krasnodar (669,000) in the Russian Federation's southwestern corner (Fig. 2-11), where the plains to the north merge into the foothills of the Caucasus to the south. Today, Adyghians make up about one-quarter of the total population of 480,000. Agriculture dominates here, and a wide variety of farm produce is exported; the area is famous for its flowers, especially its "Crimean" roses.

KALMYK REPUBLIC (KALMYKIA)

Kalmykia does not lie on the slopes of the Caucasus, but it adjoins Dagestan to the south. The Kalmyks, a nomadic people of Mongol ancestry, migrated to their present home in the lowlands northwest of the Caspian Sea during the seventeenth century (Fig. 2-11). They brought with them their Tibetan Buddhist faith (some Kalmyks are Muslims, however); their yurts (tents of felt fabric on a lattice frame, easily moved); and their livestock (horses, cattle, goats, sheep, and camels as beasts of burden). Overrun by the Russians in the eighteenth century, they found themselves in a Soviet oblast (administrative area) in 1920; in 1935 they were awarded an Autonomous Republic. Stalin abolished Kalmykia in 1943, alleging that the Kalmyks were collaborating with the invading Germans; after the war he exiled the entire population (with much loss of life) to Soviet Central Asia. After Stalin's death in 1953, the Kalmyks were allowed to return to Kalmykia, and in 1958 their Autonomous Republic was restored.

About as large as South Carolina, Kalmykia flanks the lower Volga River as well as the Caspian Sea, leaving a narrow Russian corridor to the Caspian shore (Fig. 2-11). It was Soviet policy to permanently settle the traditionally nomadic Kalmyks, who continue to depend on their livestock. Sheep breeding is a major source of income. Russian farmers moved into the arable lands of Kalmykia, and today the population of 400,000 is about 43 percent Kalmyk, 42 percent Russian, and 15 percent other ethnic minorities.

In 1993, Kalmykia's high-profile president, touting his republic as "Europe's only Buddhist state," demanded greater autonomy for his country and "international recognition" from abroad. Russians show little enthusiasm for such notions.

REPUBLICS OF THE NORTH

KARELIAN REPUBLIC (KARELIA)

In the far northwestern corner of Russia lies a republic larger in area than California but smaller in total population than San Francisco: Karelia (Fig. 2-10). To get there, you go north from Petersburg or south from Murmansk, but no matter what your route, expect cold weather, clear lakes, dense forests, and very little surface relief. The glaciers scoured the north of Karelia, and they deposited their till (unconsolidated materials) over the south as they melted back.

One would expect that an area as remote and forbidding as this would not have a checkered politico-geographical past, but Karelia has been a changeable frontier. The east has been a part of Russia since the early fourteenth century, but the west has seen many boundary changes involving Russia's neighbor, Finland. The Karelian A.S.S.R. was proclaimed in 1923, and after the war between the Soviets and Finland in 1940, territory gained by the U.S.S.R., including the port of Vyborg, was attached to Karelia. In the same year, Karelia was made a full-scale Soviet Republic. In 1946, however, Karelia's coastline was taken away and attached to the Russian Republic, and the status of the republic reverted to A.S.S.R. within the boundaries on the map today.

The Karelians are a Finnic people, not Slavs, but they constitute only 11 percent of the current population. Russians make up 70 percent, and Belarussians about 8 percent. Karelia's economic geography is based on its vast timber resources. It was a feared place of exile for criminals and political dissidents during the communist era.

KOMI REPUBLIC (KOMI)

The vast northeastern corner of Russia west of the Urals is the territory of Komi, land of the Arctic Circle (Fig. 2-10). From the crestline of the mountains in the east (sometimes rising above 6,000 feet), the surface drops into the basin of the Pechora River on its way to the Arctic Ocean. This is a country of tundra and taiga (snowforest), of frigid winters and short cool summers.

Komi became a Russian domain as early as the fourteenth century, and fur hunting drew settlers who made Syktyvkar (254,000 today) now the capital, a major base of operations. An A.S.S.R. was established in 1936 expressly to recognize the Komi ethnic group, a Finnic people who today constitute a mere 25 percent of the population of nearly 1.5 million.

In the 1940s the Pechora Railway was built diagonally across the entire republic from near Syktyvkar in the southwest to the extreme northeastern corner. The economic geography of Komi centers on its extensive forest resources but importantly also on its coal, natural gas, and oil. In terms of value, mining is the leading activity. A nuclear power plant near the town of Pechora provides electricity—and, being of Chernobyl vintage and construction, it worries local residents.

SAKHA REPUBLIC (YAKUTIA)

Territorially, the Yakut Republic (renamed *Sakha* in 1993) in Russia's far northeast (Fig. 2-9) is larger than all the other internal Russian republics combined—but its population barely exceeds 1 million. Sakha is nearly one-third the size of Canada, and its environments are even harsher: in fact, this republic has the harshest climate in all the inhabited world.

The Yakuts, traditionally nomadic pastoralists, moved northward from central Asia into the basin of the Lena River; their ancestral connections are Mongol and Turkic, but their history is uncertain. They arrived here more than a thousand years ago, and despite the difficult climate they continued to herd cattle as well as reindeer. In the first half of the seventeenth century, Russian pioneers arrived in their domain, and the present capital, Yakutsk, was founded in 1632. The Yakut A.S.S.R. was established in 1922.

Sakha is a permafrost-ridden, isolated, topographically varied land; surface communications remain primitive. But the republic's mineral potential is enormous. Oil and gas have been found as well as gold and diamonds. Timber resources are huge. Russians (52 percent) outnumber ethnic Yakuts (38 percent) in the population, but in 1991 the Yakuts presented a united front; demanding that any "foreigners" exploiting or exploring for Yakutia's raw materials pay Yakutsk for the right to do so—Russian outsiders included. Japan has expressed interest in Yakutia's resources, but who will strike the deal: Moscow or Yakutsk?

REPUBLICS OF THE SOUTHEAST

ALTAY REPUBLIC (ALTAYA)

Altaya is situated at the sensitive crossroads where Russia, Kazakhstan, China, and Mongolia meet (Fig. 2-9). About the size of Maine, it is home to a small population (roughly 250,000), most of whom still depend on cattle breeding for their livelihood; the mining of gold and coal also contribute to the economy.

The original designation of the Autonomous Region in 1922 was based on the local people's distinct language and traditional culture. This republic has had various names on older maps, including Oirot and Gorno-Altay.

KHAKASS REPUBLIC (KHAKASSIA)

Khakassia began its existence in 1930 as an oblast in Krasnoyarsk Territory, a West Virginia-sized area in the upper basin of the Yenisey River (Fig. 2-9). The heart of this republic, including the capital, Abakan (191,000), lies in the valley of the Abakan River, a mountain-encircled basin with an arid climate and steppe vegetation. The grasslands were converted into grain and potato fields during the Virgin Lands campaign of the 1950s, but livestock herding continues to be an important industry. Pine forests clothe the mountain slopes, and timber is an important export product.

The Khakass people who first occupied this area were Turkic nomads who drove their herds across the plains, but today they are greatly outnumbered by ethnic Russians. Copper ores first attracted the Russian immigrants; later, iron, gold, coal, and ferroalloys were discovered. The republic is linked by rail to the Trans-Siberian mainline.

The population remains about 600,000, but Khakassia lies in a sensitive zone along Mongolia's northern border. Some current Chinese texts describe Khakassia as part of a zone where territorial issues remain "unsettled."

BURYAT REPUBLIC (BURYATIA)

Montana-sized, mountainous Buryatia flanks Lake Baykal and extends westward to border the Tuva Republic (Fig. 2-9). As the map shows, a comparatively narrow Russian corridor separates Sakha (Yakutia) to the north from Buryatia on the Mongolian border.

Mineral-rich and scenically spectacular, Buryatia was settled by nomadic Buddhist Mongols who kept cattle as well as camels. The territory was annexed from China by the Russians during the seventeenth century, and colonization began. Gold and furs were the early objectives, but the Buryats offered strong opposition. Then the Trans-Siberian Railroad was built across Buryat territory during the late nineteenth century, and industrial and urban expansion accelerated. Valuable ferroalloys are found here as well as coal and iron. Mining and metallurgy dominate the economy; pastoralism is still important as well.

Reflecting the economic significance of Buryatia, many Russians settled in the area. The population of nearly 1.1 million is more than 70 percent Russian today. Less than one-quarter are ethnic Buryats in this important prize of tsarist expansion.

TUVA REPUBLIC (*FORMERLY* TANNU TUVA)

The Tuva Republic, about the size of North Dakota, lies wedged between Mongolia to the south, Buryatia to the east, Altaya to the west, and Siberian Russia to the north (Fig. 2-9). Its varied environments range from high snow-capped mountains and dry steppe-like basins to dense pine forests. Population is sparse (about 330,000).

This area was part of China's empire from the mid-eighteenth century until 1911, when the Russians incited an anti-Chinese uprising among the Turkic Tuvans. In 1914, the tsar established a "protectorate" for the area, but after the Revolution the Tuvans declared themselves independent from both China and Russia and formed the Tannu Tuva People's Republic. The Soviets annexed their small neighbor in 1944 and proclaimed an A.S.S.R. there in 1961. In 1971, the name was changed to the Tuva Republic.

Tuvans (or Tuvinians) still constitute about 60 percent of the population (Russians 35 percent), and pastoralism remains the chief economic activity. Some gold has been found, along with cobalt and asbestos.

There are signs that China has not forgotten Russia's takeover of Tuva. Some Chinese maps show this as disputed territory.

THE AUTONOMOUS REGIONS

In addition to its 21 Republics, the Russian Federation in mid-1993 included 11 Autonomous Regions (Fig. 2-9). One of these, Permyakia, is a sparsely peopled, densely forested expanse situated between the Komi and Udmurt republics near the Russian core. Five large A.R.s occupy most of Siberia west of Sakha; two others, Chukotka and Koryakia (formerly Kolyma, the most notorious and deathly prison-camp area of Stalinist Russia), lie in the (potentially) mineral-rich far northeast, facing the Bering Sea and Alaska. The three remaining A.R.s are comparatively small and are located along Russia's southeastern periphery; of these ethnic enclaves, the most distant is the Jewish Region.

The Russian Federation now contains 90 political-administrative divisions. Each of these has a governmental bureaucracy. All of them have expectations of greater political autonomy and better economic times. Their leaders have become an important pressure group in Russian national politics, and Russia's president must take their interests into consideration even as he deals with Russia's parliament.

What are Russia's prospects? In 1993, this gigantic state continued to struggle with three reformations: (1) a politico-geographical one, restructuring the country from a colonial power into a multicultural nation-state; (2) an economic one, reorienting productive activities to capitalist, market-based principles; and (3) a democratic one, introducing social and ideological notions not practiced here—ever. In China, coping with only one of these reforms (the second) is testing the very fabric of the state. Russia's challenge is much greater, and its capacity to survive the transition is far from assured.

◆ REGIONS OF THE RUSSIAN REALM

Russia is a geographic realm by itself, a territorial giant with unsteady neighbors, uncertain boundaries, and unstable structures. Always the Soviet Union—and therefore Russia—was a complex of contradictions. Great achievements in science and industry stood in stark contrast to the fear that stalked dissident scientists, authors, and artists. The treatment of the former Soviet Union's Jewish communities sparked a worldwide outcry. The communist state could boast of powerful armed forces, great universities, magnificent orchestras, and spectacular achievements in space—but its State Security Committee (KGB) spied on its own citizens, its judges sent nonconformists to "mental hospitals," and countless millions of its people died in Siberian exile. When, in 1991, a group of communist hardliners attempted a *coup d'état* against then-Soviet President Gorbachev, the specter of the totalitarian past was raised again. The coup failed, but it may not be the last challenge for power—Soviet or Russian. Therefore, when we undertake our survey of the regional map of Russia, we should keep in mind not only that the realm is being transformed, but also that long-entrenched patterns cannot change instantly.

THE RUSSIAN CORE

The heartland of a state is its **core area**. Here a large part of the population is concentrated, and leading cities, major industries, dense transport networks, intensively cultivated lands, and other essentials of the country cluster within a relatively small region. Core areas of long standing carry the imprints of culture and history especially strongly. The Russian core area, broadly defined, is the region that extends from the western border of the Russian realm to the Ural Mountains in the east (Fig. 2-12). This is the Russia of Moscow and Petersburg, of the Volga River and its industrial cities, of farms and forests. Here, the Muscovy Rus-

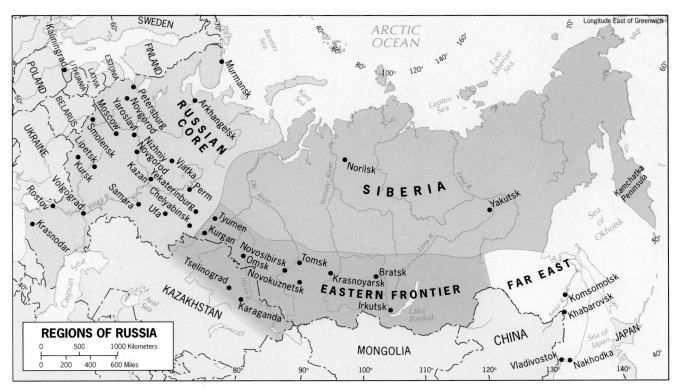

FIGURE 2-12

sians asserted their power and began the formation of the Russian Empire, forerunner to the now-defunct Soviet Union.

Central Industrial Region

At the heart of the Russian Core lies the Central Industrial Region (Fig. 2-13). The precise definition of this subregion varies, as all regional definitions are subject to debate. Some geographers prefer to call this the Moscow Region, thereby emphasizing that for over 250 miles (400 km) in all directions from the capital, regional orientations are toward this historic focus of the state. As the map shows, Moscow has maintained its decisive **centrality**: roads and railroads radiate in all directions to Ukraine in the south; to Mensk (Belarus) and the rest of Eastern Europe in the west; to Petersburg and the Baltic coast in the northwest; to Nizhniy Novgorod (formerly Gorkiy) and the Urals in the east; to the cities and waterways of the Volga Basin in the southeast (a canal links Moscow to the Volga, Russia's most important navigable river); and even to the subarctic northland that faces the Barents Sea (see box).

Moscow itself is a transforming metropolis (1994 population: approximately 9.4 million), as high-rise apartment complexes increasingly dominate the residential landscape (see photo). Although this helps to alleviate the capital's severe housing shortage, Moscow—like nearly all Russian cities—is horribly overcrowded, with most people forced to accept unreasonably cramped personal living spaces.

Moscow is also the focus of an area that includes some 50 million inhabitants (one-third of the country's total population), many of them concentrated in such major cities as Nizhniy Novgorod (1.5 million)—the automobile-producing "Soviet Detroit"; Yaroslavl (667,000)—the tire-producing center; Ivanovo (541,000)—the heart of the textile industry; and Tula (573,000)—the mining and metallurgical center where lignite (brown coal) deposits are worked.

Petersburg (the former Leningrad) remains Russia's second city, with a population of 5.2 million. There was a time when Petersburg was the focus of Russian political and cultural life and Moscow was a distant second city. Today, though, Petersburg has none of Moscow's locational advantages, at least not with respect to the domestic market. It lies well outside the Central Industrial Region near the northwestern corner of the country, 400 miles (650 km) from Moscow. Neither is it better off than Moscow in terms of resources: fuels, metals, and foodstuffs must all be brought in, mostly from far away. The former Soviet emphasis on self-sufficiency even reduced Petersburg's asset of coastal location, because some raw materials could have been imported much more cheaply across the Baltic Sea

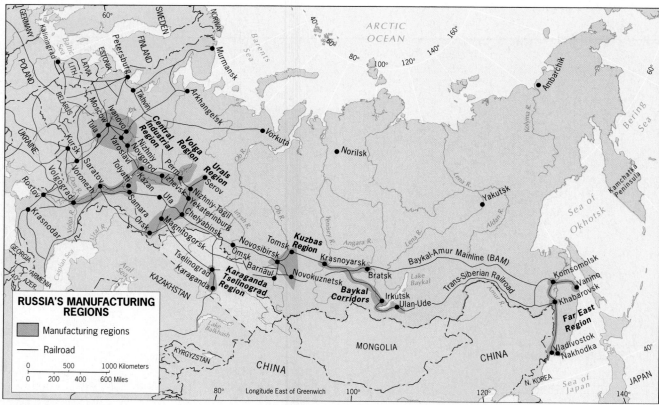

FIGURE 2-13

During the Soviet era, Russian cities acquired huge, drab, often poorly-built apartment complexes such as these in an outer neighborhood of Moscow.

FACING THE BARENTS SEA

North of the latitude of Petersburg, the region we have named the Russian Core takes on Siberian properties. However, the Russian presence in this remote northland is much stronger than it is in Siberia proper. Two substantial cities, Murmansk (487,000) and Arkhangelsk (456,000), are road- and rail-connected outposts in the shadow of the Arctic Circle.

Murmansk lies on the Kola Peninsula not far from the border with Finland (Fig. 2-10). In its hinterland lie a variety of mineral deposits, but Murmansk is particularly important as a naval base. During World War II, allied ships brought supplies to Murmansk; the city's remoteness shielded it from German occupation. After the war, it became a base for nuclear submarines. This city is also an important fishing port and a container facility for cargo ships.

Arkhangelsk is located near the mouth of the Northern Dvina River where it reaches an arm of the White Sea (Fig. 2-10). Its site was chosen by Ivan the Terrible during Muscovy's early expansion; the tsar wanted to make this the key port on a route to maritime Europe. Yet Arkhangelsk, mainly a port for lumber shipments, suffers from a more restricted ice-free season than Murmansk, whose port can be kept open with the help of the North Atlantic Drift ocean current (and by icebreakers when the need arises).

Nothing in Siberia east of the Urals rivals either of these cities—yet. But their very existence and growth prove that Siberian barriers to settlement can be overcome.

Although the former Soviet Union was a prodigious producer of equipment, almost all of this production was for domestic use; little was exported to the non-communist world. One exception was weapons, and Russia today continues to market arms worldwide. Here in the port of Petersburg, cannons and armored cars await transportation, reportedly to India. Also note the red farm tractors ready for transfer, the large container facility, and Petersburg's high-rise apartment buildings from the Soviet era in the background.

from foreign sources than from domestic sites in distant Central Asia (only bauxite deposits lie nearby, at Tikhvin).

Yet Petersburg was at the vanguard of the Industrial Revolution in Russia, and its specialization and skills have remained. Today, the city and its immediate environs contribute about 10 percent of the country's manufacturing, much of it through the building of high-quality machinery. In addition to the usual association of industries (metals, chemicals, textiles, and food processing), Petersburg has major shipbuilding plants and, of course, its port and naval station. This productive complex, although not enough to maintain its temporary pre-revolution advantage over Moscow, kept the city in the forefront of modern Soviet development and will continue to do so in the new Russia.

Povolzhye: The Volga Region

A second region lying within the Russian Core is the *Povolzhye*, the Russian name for an area that extends along the middle and lower valley of the Volga River. It would be appropriate to call this the Volga Region, for that greatest of Russia's rivers is its lifeline, and most of the cities that lie in the *Povolzhye* are situated on its banks (Fig. 2-13).

In the 1950s, a canal was completed to link the lower Volga with the lower Don River (and thereby the Black Sea), extending this region's waterway system still farther.

The Volga River was an important historic route in old Russia, but for a long time neighboring regions overshadowed this one. The Moscow area and Ukraine were far ahead in industry and agriculture. The Industrial Revolution that came late in the nineteenth century to the Moscow Region left the *Povolzhye* little affected. Its major function remained the transit of foodstuffs and raw materials to and from other regions.

This transport function is still important, but things have changed in the *Povolzhye*. First, World War II brought furious development because the Volga River, located east of Ukraine, was protected by distance from the German armies that invaded from the west. Second, in the postwar era the Volga-Urals region proved to be the greatest source of petroleum and natural gas in the entire Soviet Union. From near Volgograd (formerly Stalingrad) in the southwest to Perm (1.1 million) on the Urals' flank in the northeast lies a belt of major oilfields (Fig. 2-14). These deposits were once believed to constitute the former Soviet Union's largest reserve, but as the map shows, later discoveries in

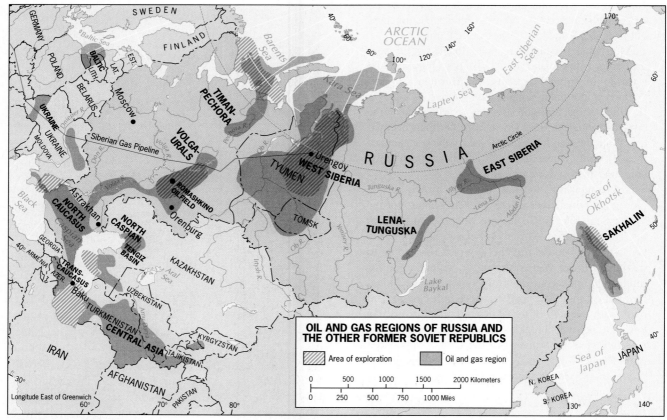

FIGURE 2-14

It is unlikely that you have ever seen a Russian car in the street, but Russia does have automobile factories. This view of part of the LADA auto assembly plant in Tolyatti, on the Volga River south of Kazan, shows vehicles ready for shipment.

western Siberia indicate that even more extensive oil- and gasfields lie beyond the Urals. Still, because of their location and size, the Volga Region's fossil-fuel reserves retain their importance.

Third, the transport system has been greatly expanded. The Volga-Don Canal directly connects the Volga waterway to the Black Sea; the Moscow Canal extends the northern navigability of this river system into the very heart of the Central Industrial Region; and the Mariinsk Canals provide a link to the Baltic Sea. Today, the Volga Region's population exceeds 25 million, and the cities of Samara (formerly Kuybyshev), Volgograd, Kazan, and Saratov all have populations in the range of 1.0 to 1.3 million. Manufacturing has also expanded into the middle Volga Basin, emphasizing more specialized engineering industries. The huge Fiat-built auto assembly plant in Tolyatti (693,000), for example, is one of the world's largest of its kind (see photo).

The Urals Region

The Ural Mountains form the eastern limit of the Russian Core. These mountains are not particularly high; in the north they consist of a single range, but southward they broaden into a zone of hilly topography. Nowhere do they form an obstacle to east-west transportation. An enormous storehouse of metallic mineral resources located in and near the Urals has made this area a natural place for industrial development. Today, the Urals Region, well connected to the Volga and Central Industrial regions, extends from Serov in the north to Orsk in the south (Fig. 2-13).

The Urals Region rose to prominence during World War II and its aftermath when it benefited from its remoteness from the German invaders, allowing its factories to support the war effort without threat of destruction. Coal had to be imported, but the problem of energy supply here has been considerably relieved by oilfields discovered in the zone lying between the Urals and the Volga Region (Fig. 2-14). More recent West Siberian oil and gas exploitation (from fields lying northeast of the Urals) increasingly supports development in the Urals as well.

The Central Industrial, Volga, and Urals regions form the anchors of the Russian core area. For decades they have been spatially expanding toward one another, their interactions ever more intensive. These regions of the Russian Core stand in sharp contrast against the comparatively less developed, forested, Arctic north and the remote upland south between the Black and Caspian seas; thus even within this Russian coreland, frontiers still await growth and development.

THE EASTERN FRONTIER

Karaganda-Tselinograd

The eastward march of Russian development made the Karaganda-Tselinograd area of the northern Kazakh Republic (Kazakhstan) one of its hubs of growth. (We will consider the regional geography of Russia's southern neighbor in detail in Chapter 7.) During the Soviet period, northern Kazakhstan became almost totally Russified. It lay directly in the path of Russia's eastward expansion (Fig. 2-12) and, from the Urals Region in the west to the Kuznetsk Basin in the east, transport routes were laid across it. This helps explain why Kazakhstan, a Soviet republic established for the Kazakhs and other Muslim minorities, today has almost as many Russians (38 percent) among its population of 17.4 million as Kazakhs (40 percent); it also explains why Kazakhstan contains a higher percentage of ethnic Russians than any other former S.S.R. (Fig. 2-5). The ethnic spatial pattern is such that Russians dominate in the modernized north, and Kazakhs and others dominate in the rest of the republic.

The Karaganda-Tselinograd Region (Fig. 2-13) reflects this sequence of events. Karaganda, once a small Kazakh town, has become a center for iron and steel production along with chemicals and other manufactures. With a population of more than 650,000, Karaganda is a Russified city that exchanges raw materials and finished products with other Russian production zones, especially the Urals Region. Tselinograd (324,000) emerged in the 1950s as an administrative and collection center for the troubled Virgin and Idle Lands Program; together, the two cities are linked by rail to the major developing areas of the Eastern Frontier (Fig. 2-13).

From the beginning of Soviet rule, the penetration and consolidation of the Eastern Frontier were high-priority planning goals. Concerns over effective national control along sensitive eastern boundaries were one reason for this; security in the event of renewed German invasion from the west was another. The exploitation of mineral resources was a further incentive.

The Kuznetsk Basin (Kuzbas)

Northeast of the Karaganda area, back in Russia proper and fully 1,200 miles (2,000 km) east of the Urals, lies another primary region of heavy manufacturing—the Kuznetsk Basin or *Kuzbas* (Fig. 2-13). In the 1930s, this area was first opened up as a supplier of raw materials (especially coal) to the Urals, but that function steadily diminished in importance as local industrialization accelerated. The original plan was to move coal from the Kuzbas west to the Urals and allow the returning trains to carry iron ore east to the coalfields; however, good-quality iron ore deposits were subsequently discovered in the vicinity of the Kuznetsk Basin itself. As the new resource-based Kuzbas industries grew, so did its urban centers. The leading city, located just outside the region, is Novosibirsk (1.5 million), which stands at the intersection of the Trans-Siberian Railroad and the Ob River as the very symbol of Russian enterprise in the vast eastern interior. To the northeast lies Tomsk (542,000), one of the oldest Russian towns in all of Siberia, founded three centuries before the Bolshevik uprising and now caught up in the modern development of the Kuzbas region. Southeast of Novosibirsk lies Novokuz-

The Kuzbas region of heavy manufacturing has become even more important than it was, because of the separation of Ukraine and its Donbas industrial complex (see pp. 113–115). These huge plants at Novokuznetsk produce steel.

netsk, a city of nearly 650,000 people that specializes in the production of steel for the region's machine and metalworking plants; aluminum products using Urals bauxite are also manufactured here.

Impressive as the concentration of coal, iron, and other resources may be in the Kuznetsk Basin, the industrial and urban development that has taken place here must in large measure be attributed once again to the ability of the communist state and its planners to promote this kind of expansion, notwithstanding what capitalists would see as excessive investment. In the absence of such financial constraints, the state planners were able to push the country vigorously ahead on the road toward industrialization, with the hope that certain areas would successfully reach "takeoff" levels. Then these areas would require progressively fewer direct investments because growth would to an evergreater extent become self-perpetuating. The Kuzbas, for instance, was expected to grow into one of the Soviet Union's leading industrial agglomerations, with its own important market and with a location, if not favorable to the Urals and points west, then at least fortuitous with reference to the developing markets of the Soviet Far East.

The Lake Baykal Area

East of the Kuzbas, development becomes more insular, and distance becomes a stronger adversary (Fig. 2-13). North of the Tuva Republic and eastward around Lake Baykal, larger and smaller settlements cluster along the two railroads to the Pacific coast (Fig. 2-6). West of the lake, these rail corridors lie in the headwater zone of the Yenisey River and its tributaries. A number of dams and hydroelectric projects serve the valley of the Angara River, particularly the city of Bratsk (298,000). Mining, lumbering, and some farming sustain life here, but isolation dominates it. The city of Irkutsk (677,000), near the southern end of Lake Baykal, is the principal service center for a vast Siberian region to the north and for a lengthy east-west stretch of southeastern Russia.

Eastward beyond Lake Baykal, the Eastern Frontier really lives up to its name: this is southern Russia's most rugged, remote, forbidding country. Settlements are few and far between; many of them are mere camps. The Buryat Republic (Fig. 2-9) is part of this zone; the territory bordering it to the east was taken from China by the tsars and may become an issue in the future. Where the Russian-Chinese boundary turns southward, along the Amur River, the region called the Eastern Frontier ends and Russia's Far East begins.

SIBERIA

Before we assess the potentials of Russia's Pacific Rim, we should remind ourselves that the ribbons of settlement just discussed hug the southern perimeter of this giant country, avoiding the vast Siberian region to the north (Fig. 2-12). Siberia extends from the Ural Mountains to the Kamchatka Peninsula; a vast, bleak, frigid, forbidding land. Larger than the conterminous United States but inhabited by only an estimated 15 million people, Siberia quintessentially symbolizes the Russian environmental plight: vast distances, cold temperatures worsened by strong Arctic winds, difficult terrain, poor soils, and limited options for survival.

But Siberia is also a region of resources. From the days of the first Russian explorers and Cossack adventurers, word of Siberia's riches filtered back to the west. Gold, diamonds, and other precious minerals were found. Later, metallic ores including iron and bauxite were discovered. Still more recently, the Siberian interior proved to contain sizeable quantities of oil and natural gas (Fig. 2-14) and began to contribute significantly to the country's energy supply.

As the physiographic map (Fig. 2-3) shows, several major rivers—the Ob, Yenisey, and Lena—flow gently northward across Siberia and the Arctic Lowland into the Arctic Ocean. Hydroelectric power development in the basins of these rivers has generated electricity used in the extraction and refinement of local ores, and in the lumber mills that have been set up to exploit the vast Siberian forests. Soviet planners even gave serious consideration to grandiose schemes to reverse the flow of Siberian streams (to divert Siberian water onto parched farmlands far to the southwest)—ideas their Russian successors are likely to discard.

The human geography of Siberia is fragmented, and much of the region is for all intents and purposes uninhabited (as shown in Fig. 2-4). Ribbons of Russian settlement have developed; the Yenisey River, for instance, can be traced on this map of Soviet peoples (a series of small settlements north of Krasnoyarsk [981,000]), and the upper Lena valley is similarly fringed by ethnic Russian settlement. Yet these ribbons and other islands of habitation are separated by hundreds of miles of empty territory. During the worst excesses of the Stalinist period (and long afterwards as well), dissidents and criminals were exiled to Siberian mining and lumbering camps to serve their terms at hard labor under the harshest of conditions; many never returned.

The political geography of eastern Siberia is marked by the growing identity of Sakha (the Yakut Republic) (see profile on p. 149). As additional resources are discovered here (including oil and natural gas), the importance of this republic, centered on the capital, Yakutsk (214,000), will increase.

Siberia, Russia's freezer, is stocked with goods that may become mainstays of future national development. Already, precious metals and mineral fuels contribute importantly to the Russian economy. As time passes, we may expect Siberian resources to sustain the development of the Eastern Frontier. As that happens, this southern zone of development will expand northward as well as eastward, bringing more of Siberia into the national sphere. One step in that

Russia's largest oil and gas region lies below northwestern Siberia as Fig. 2-14 indicates. Here, amid the vastness of the West Siberian Plain near the mouth of the Ob River at the latitude of the Arctic Circle, a new oil well begins to tap these rich fossil-fuel reserves.

direction has already been accomplished with the completion of the BAM (Baykal-Amur Mainline) Railroad during the 1980s. This route, lying north of and parallel to the Trans-Siberian Railroad, extends 2,200 miles (3,540 km) eastward from Tayshet (near Krasnoyarsk) directly to Komsomolsk (Fig. 2-6). To date, however, the BAM has faced problems, and only portions of it are in regular use; nonetheless, it is still expected to become the axis of a major new industrial corridor in the twenty-first century.

THE FAR EAST

Russia has about 5,000 miles (8,000 km) of Pacific coastline—more than the United States (including Alaska). However, most of its coastal margin lies north of the latitude of the State of Washington, and from the point of view of coldness, the Russian coastline lies on the "wrong" side of the Pacific. The port of Vladivostok (686,000), the original eastern terminus of the Trans-Siberian Railroad, lies at a latitude about midway between San Francisco and Seattle, but it must be kept open throughout the winter by icebreakers—something unheard of in these American ports. The climate, to say the least, is harsh. Winters are long and bitterly cold; summers are cool.

Nonetheless, the Russians have been determined to develop their Far East Region (Fig. 2-12) as intensively as possible. Their resolve has recently been spurred by ideological differences between democratizing Russia and hardline-communist China. In the Asian interior, high mountains and empty deserts mark most of the Russo-Chinese borderland zone; south of the Kuznetsk Basin and Lake Baykal, the Republic of Mongolia functions as a sort of buffer between Chinese and Russian interests. But in the Pacific area, the Russian Far East and Northeast China confront each other directly across an uneasy river boundary (along the Amur and its tributary, the Ussuri). Not only the Russian tsars but also the Soviet communists always wanted to consolidate their distant Pacific outposts. The Soviets even offered financial and other incentives to western residents willing to move east.

As the map of climates (Fig. 2-1) reminds us, the Russian Far East is no Garden of Eden. The proximity of the Pacific Ocean is not very effective in moderating the cold; summers are only slightly longer and milder than they are in Siberia. Most of the Far East, with its rugged terrain and vast wilderness, remains a sparsely peopled region. Lumbering centers and fishing villages only occasionally break the emptiness of the countryside; seafood is still the leading product here. From Vladivostok and points north, Russian fishing fleets sail the Sea of Okhotsk and the northern Pacific in search of salmon, herring, cod, and mackerel to be frozen, canned, and shipped via rail to far-off western markets.

The mineral potential of the Far East, like that of eastern Siberia, is still only partially known. A deposit of high-quality coal lies in the Bureya River valley (a tributary of the Amur), and there is iron ore near Komsomolsk (361,000), now the region's first steel-producing center. Tin deposits near Komsomolsk are especially important because they are Russia's main source.

The major axis of urban and industrial development in the Far East has emerged along the Amur-Ussuri river system, from Vladivostok in the south (with its huge military installations, shipbuilding, and fish-processing plants) to Komsomolsk in the north (Fig. 2-13). Near the confluence of the two rivers lies Khabarovsk (656,000), the city with the greatest advantages of centrality. At Khabarovsk, machine and metal-working industries process iron and steel from Komsomolsk; chemical industries depend on nearby Sakhalin's oil, and furniture factories use the area's ubiquitous timber.

Until now, very little development has taken place in Vanino, the coastal terminus of the new BAM Railroad. There is a ferry to Kholmsk on the island of Sakhalin, and Vanino and its neighbor, Gavan, are home to small fishing fleets. The more significant growth is taking place at the new coastal endpoint of the Trans-Siberian Railroad, which now goes 55 miles (90 km) beyond Vladivostok to Nakhodka. As Fig. 2-8 shows, this rapidly growing port (population 408,000) is the southernmost city of the Russian Far East and faces the Sea of Japan; across the water lies one of the world's most powerful—and most raw-material-deficient—economies.

Why did Soviet-Japanese economic interactions in this area fail to develop, given its regional geography? The circumstances are ironic. Japan and the former Soviet Union never signed a treaty ending the war between them (World War II) because the Soviets occupied and annexed four small Japanese islands in the Kurile chain, northeast of Hokkaido (see Fig. 4-3). The Japanese demanded their return, the Soviets refused, and there the matter stood, holding up not only the treaty but also economic relations between the two powers. In 1990, then-Soviet President Gorbachev met with Japanese leaders, hoping to resolve the issue. Japan reportedly offered a (U.S.) $26 billion deal for the islands, a proposal that included the intensified exploration of the Russian Far East and eastern Siberia, the purchase of raw materials, and assistance in developing the region's infrastructure. In fact, some Japanese help was given in the building of the port of Nakhodka. It was reported that Gorbachev reacted favorably to the Japanese proposal, but that Russian President Yeltsin refused to consider the transfer of "one square meter" of Russian territory. In 1992, however, the tables were turned. Now it was Yeltsin, president of an independent Russia, who had to negotiate with the

Nakhodka, about 50 miles (80 km) east of Vladivostok on the Sea of Japan, has an excellent natural harbor and developed rapidly as an alternative to the closed naval port. Now, however, Vladivostok is an open port again, and competes with Nakhodka for the overseas trade.

Japanese. Predictably, his earlier stand now made negotiations difficult. There the matter stood and still stands today—holding up the growth of an entire region and delaying Russia's entry into the economic sphere of the western Pacific Rim.

In its Far East, Russia faces problems even more serious than was the case during the days of communist rule. The farther eastward the country's developing areas lie, the greater is their **inaccessibility** and, in the new era, the larger their need for self-sufficiency in view of the ever-increasing costs of transportation. In the past, Soviet planners could counter these disadvantages by allocating funds and resources to such remote areas as a matter of national economic policy. Under the new capitalist system, however, development may actually slow down because the rules are now very different. What Russia needs is a breakthrough in its Far East, and Japan's Northern Territories form the opportunity. But until that opportunity is realized, the Far East will suffer: in the early 1990s, the region suffered from fuel and food shortages unknown even during the bleakest of communist times. The Far East's old enemies, distance and remoteness, again held sway.

◆ UNRESOLVED POLITICO-GEOGRAPHICAL ISSUES

HEARTLANDS AND RIMLANDS

In a recent article in a major geography journal, a prominent political geographer pointed to what he called "hopelessly out of date" and "obsolete" theories of power and influence in Eurasia. He argued that earlier ideas about Russia's challenge for global might are no longer relevant, given what has happened in the realm over the past several decades.

Specifically, he was referring to a debate that started nearly a century ago, at a time when Russian armies had been disastrously defeated by the Japanese and when tsarist Russia was in chaos on the eve of the 1917 Revolution. Who would predict that a national state centered on decaying Russia would become a world power in the twentieth century? The British geographer Sir Halford Mackinder (1861–1947) did just that. In an article published in 1904, he analyzed the potential strengths and weaknesses of Eurasian regions, concluding that one such region, which he called the *pivot area*, enjoyed a combination of natural protection and resource wealth that would propel its occupants to world power. This pivot area (Fig. 2-15, upper left) consisted of the Moscow region, the Volga valley, the Urals, Central Asia, and western and central Siberia.

Mackinder's article sparked a heated debate. In 1919, he published a revision of his idea, incorporating Eastern

Europe into the framework (Fig. 2-15, lower left). He now called the crucial (and expanded) strategic area the *heartland*, and argued that control over Eastern Europe would mean global power. This became known as Mackinder's **heartland theory**:

> *Who rules East Europe commands the Heartland;*
> *Who rules the Heartland commands the World Island;*
> *Who rules the World Island commands the World.*

Mackinder's theory seemed to become a prescription for national policy: the Germans during World War II sought to control Eastern Europe and pushed into the heartland, and then the Soviets tried to subjugate postwar Eastern Europe. The World Island, in Mackinder's parlance, consisted of Eurasia and Africa; this huge territory would fall to the heartland power, leading ultimately to world domination.

Other geographers took a different view of the capacities and limitations of the heartland, and some, notably Nicholas Spykman, suggested that the future of Eurasia might lie in a *rimland* rather than in Mackinder's heartland (Fig. 2-15, upper right). But no one could fail to note the Soviets' own reaction: David Hooson sketched the Soviet planners' attempt to move the Soviet core area eastward after World War II, away from the vulnerable west (Fig. 2-15, lower right). In an age of intercontinental missiles and nuclear weapons, a dispersed core area confers a certain degree of security upon a threatened state. Until very near the time of his death in 1947, Mackinder participated in the continuing discussion, convinced that he was as correct near mid-century as he had been in 1904.

Were they alive today, Mackinder and Spykman would argue that their respective positions are anything but outdated by the course of events. The Slavic heartland's strength was sapped by the failures of the communist system and is now (perhaps temporarily) undermined by political fragmentation. But the potential for another challenge for world power (if not domination) remains. Spykman's rimland notion is reaffirmed by what is happening in the non-Russian republics on Moscow's western and southern flanks. In short, what Mackinder saw in the strategic geography of Eurasia at the start of the century now ending has lost little of its relevance.

NEIGHBORS AND BOUNDARIES IN THE SENSITIVE EAST

Before its demise in 1991, the Soviet Union had borders with 12 neighboring countries. As the inheritor of a Russian empire, the U.S.S.R. throughout its existence faced unfinished boundary business with several of these adjoining states. When Soviet boundaries were defined, some affected countries, such as China, were in a weakened state. Other countries, such as Poland and Romania, lost territory to the Soviet Union in the aftermath of World War II.

EURASIAN HEARTLANDS

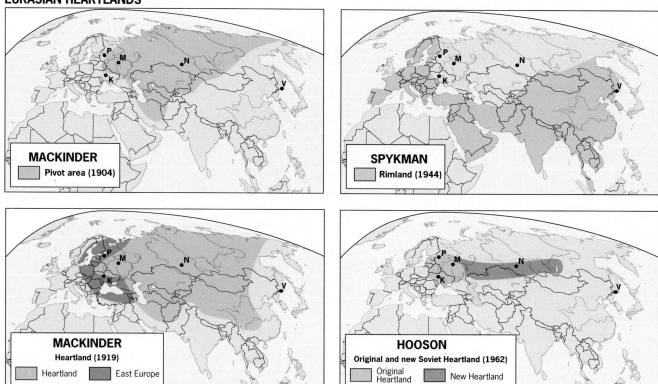

FIGURE 2-15

Russia's boundaries with its neighbors in many areas are sensitive and disputed. In response, these borders have been demarcated and reinforced. This electrified fence, running through the center of an off-limits border zone, marks the contact between Russia and China near the western corner of Mongolia.

FIGURE 2-16

When the Russian tsars sent their armies far afield to conquer distant territories, large areas on the margins of the empire were under uncertain authority. Just a century ago, such ill-defined **frontiers** still separated countries and colonizers. This enabled the stronger powers to impose borders on weaker ones. In a relatively short time, most frontiers disappeared from the world map, replaced by boundaries defined according to power relationships or by treaties that were not always fair. Understandably, these gave rise to some serious territorial issues, most of them in central and eastern Asia where they involve China.

When tsarist forces occupied and secured the Far East, China was in disarray. Repeatedly, after the Soviets consolidated the tsarist empire, Chinese governments demanded the "return" of areas on the Sino-Soviet boundary. As Fig. 2-16 shows, these included virtually all of the Russian Far East, a part of the border area northeast of Mongolia, a portion of what is now the Altay Republic northwest of Mongolia, and parts of what were then the Kazakh and Kirghiz S.S.R.s (Kazakhstan and Kyrgyzstan today).

More specific border issues also have arisen in several places. The tsars managed to impose boundary treaties on the Chinese during the period from 1858 to 1864 (see Fig. 2-2), but China has now repudiated those treaties. As Fig. 2-16 shows, five boundary segments (shown by heavy black lines) have caused particular trouble: three sections of the Russo-Chinese border in the Far East and Eastern Frontier regions, one section affecting Kazakhstan just west of Mongolia, and one involving Tajikistan and westernmost China.

In 1992, China and newly reconstituted Russia signed a series of agreements that settled disputes over more than a dozen boundary segments. But other issues along their joint border remain unresolved, and the larger question—over territory, not borders—looms as a source of future conflict.

THE REALM IN TRANSITION

1. Russia has not been willing to sign an agreement that would stipulate its boundary with Ukraine.
2. Russia's enormous oil and natural gas resources are being wasted by corruption, inefficiency, and outdated equipment, costing Moscow a huge opportunity as it tries to revive the Russian economy.
3. Not only the 21 republics but nearly 70 *other* internal entities are vying for enhanced stakes in the Russian Federation. In the mid-1990s, the Russian president had to deal with 90 regional and local leaders.
4. Environmental degradation, pollution, and continual hazards from poorly designed and constructed nuclear power plants pose serious problems for the Russian state.
5. Russia's old nemesis, variable weather and uncertain food supplies, still afflict this vast, northerly country; the drive to reconstruct the Russian economy will not soon overcome these problems.

◆ PRONUNCIATION GUIDE

Abakan   (ahb-uh-KAHN)
Adygeya   (ah-duh-GAY-uh)
Adyghian   (ah-duh-GAY-un)
Alma-Ata   (al-muh-uh-TAH)
Altaic (al-TAY-ik)
Altay   (AL-tye)
Altaya   (al-TYE-uh)
Amur   (uh-MOOR)
Angara   (ahng-guh-RAH)
Apartheid   (APART-hate)
Apparatchik   (appa-RAH-chick)
Aral   (ARREL)
Arkhangelsk   (ahr-KAN-jelsk)
Armenia   (ar-MEENY-uh)
Astrakhan   (ASTRA-kahn)
Avartsy   (ah-VAHR-tsee)
Azerbaijan   (ah-zer-bye-JAHN)
Baku   (bah-KOO)
Barents   (BARRENS)
Bashkort   (BAHSH-kort)
Bashkortostan   (bahsh-kort-oh-STAHN)
Baykal   (bye-KAHL)
Baykalsk (bye-KAHLSK)
Belarus   (bella-ROOSE)
Bolshevik   (BOAL-shuh-vick)
Bratsk   (BRAHTSK)
Bureya   (buh-RAY-yuh)
Buryat(ia)   (boor-YAHT [ee-uh])
Byzantine   (BIZZ-un-teen)
Caspian   (KASS-spee-un)
Caucasus   (KAW-kuh-zuss)
Cheboksary   (cheb-ahk-SAH-ree)
Chechen   (CHEH-chen)
Checheno-Ingushetia   (cheh-CHENNO in-goo-SHETTY-uh)
Chechenya   (cheh-CHEN-yuh)
Cherkessk   (cher-KESK)
Chernobyl   (CHAIR-nuh-beel)
Chukotka   (chuh-KAHT-kuh)
Chuvash(ia)   (choo-VAHSH [ee-uh])
Coup d'état   (koo-day-TAH)
Crimean   (cry-MEE-un)
Dagestan   (dag-uh-STAHN)
Dargintsy   (dar-GHINT-see)
Dnieper   (duh-NYEPPER)
Donets   (duh-NETTS)
Dvina   (duh-vee-NAH)
Finno-Ugric   (finno-YOO-grick)
Gavan   (guh-VAHN)

Georgia   (GEORGE-uh)
Glasnost   (GLUZZ-nost)
Gorbachev, Mikhail   (GOR-buh-choff, meek-HYLE)
Gorkiy   (GORE-kee)
Gorno-Altay   (gore-noh-AL-tye)
Gorod   (guh-RAHD)
Groznyy   (GRAWZ-nee)
Gyre   (JYER)
Hegemony   (heh-JEH-muh-nee)
Hokkaido   (hah-KYE-doh)
Ingush   (in-GOOSH)
Irkutsk   (ear-KOOTSK)
Ingushetia   (in-goo-SHETTY-uh)
Irtysh   (ear-TISH)
Ivanovo   (ee-VAH-nuh-voh)
Izhevsk   (EE-zheffsk)
Kabardinian   (kabber-DIN-ee-un)
Kabardino-Balkar(ia)   (kabber-DEE-noh bawl-KAR [ree-uh])
Kalmyk   (KAL-MIK)
Kalmykia   (kal-MIK-ee-uh)
Kamchatka   (kahm-CHUT-kuh)
Kara   (KAHR-ruh)
Karachay   (kah-ruh-CHYE)
Karachayevo-Cherkessia   (kahra-CHAH-yeh-vuh cheer-KESS-ee-uh)
Karaganda   (karra-gun-DAH)
Karelia   (kuh-REE-lee-uh)
Kazakh   (KUZZ-uck)
Kazakhstan   (kuzz-uck-STAHN/ KUZZ-uck-stahn)
Kazan   (kuh-ZAHN)
Khabarovsk   (kuh-BAHR-uffsk)
Khakassia   (kuh-KAHSS-ee-uh)
Khan, Genghis   (KAHN, JING-guss)
Khanate   (KAHN-ate)
Kholmsk   (KAWLMSK)
Kiev   (KEE-yeff)
Kievan   (kee-EVAN)
Kirghiz   (keer-GEEZE)
Klyuchevskaya   (klee-ooh-CHEFF-skuh-yuh)
Kodiak   (KOH-dee-ak)
Kolkhoz   (KOLL-koze)
Kolyma   (koh-LEE-mah)
Komi   (KOH-mee)
Komsomolsk   (kom-suh-MAWLSK)
Koryakia   (kor-YAH-kee-uh)

Krasnodar   (KRASS-nuh-dahr)
Krasnovodsk   (krass-nuh-VAUGHTSK)
Krasnoyarsk   (krass-nuh-YARSK)
Kura   (KOOR-uh)
Kurile   (CURE-reel)
Kuybyshev   (KWEE-buh-sheff)
Kuzbas   (kooz-BASS)
Kuznetsk   (kooz-NETSK)
Kyrgyzstan   (KEER-geeze-stahn)
Kyyiv   [*See* Kiev]
Lena   (LAY-nuh)
Lenin   (LENNIN)
Mackinder, Halford   (muh-KIN-der, HAL-ferd)
Makhachkala   (muh-kahtch-kuh-LAH)
Mari(-el)   (MAH-ree [el])
Mariinsk   (muh-ree-EENTSK)
Menshevik   (MEN-shuh-vick)
Meshketian   (mesh-KETTY-un)
Moldova   (moal-DOH-vuh)
Mordvinia   (mord-VINNY-uh)
Moscow   (MOSS-cau)
Murmansk   (moor-MAHNTSK)
Muscovy   (muh-SKOH-vee)
Muslim   (MUZZ-lim)
Nakhodka   (nuh-KAUGHT-kuh)
Neva   (NAY-vuh)
Nizhniy Novgorod   (NIZH-nee NAHV-guh-rahd)
Norilsk   (nuh-REELSK)
Novgorod   (NAHV-guh-rahd)
Novokuznetsk   (noh-voh-kooz-NETSK)
Novosibirsk   (noh-voh-suh-BEERSK)
Odesa   (oh-DESSA)
Oirot   (AW-ih-rut)
Okhotsk   (oh-KAHTSK)
Ordzhonikidze   (or-johnny-KIDD-zuh)
Ossetia   (oh-SEE-shuh)
Pechora   (peh-CHORE-ruh)
Perestroika   (perra-STROY-kuh)
Perm   (PAIRM)
Permyakia   (pairm-YAH-kee-uh)
Petrograd   (PETTRO-grahd)
Povolzhye   (puh-VOLL-zhuh)
Rus   (ROOSE)

Sakha (SAH-kuh)
Sakhalin (SOCK-uh-leen)
Samara (suh-MAH-ruh)
Saransk (sun-RAHN-tsk)
Saratov (suh-RAHT-uff)
Sarmatian (sahr-MAY-shee-un)
Scythian (SITH-ee-un)
Serov (SAIR-roff)
Shi'ite (SHEE-ite)
Siberia (sye-BEERY-uh)
Slavic (SLAH-vick)
Sovkhoz (SOV-koze)
Spykman (SPIKE-mun)
Stalin (STAH-lin)
Syktyvkar (sik-tiff-KAR)
Tajikistan (tah-JEEK-ih-stahn)
Tannu Tuva (tan-ooh-TOO-vuh)
Tatar (TAHT-uh)

Tatarstan (TAHT-uh-STAHN)
Tayshet (tye-SHET)
Tbilisi (tuh-BILL-uh-see)
Tchaikovsky (chye-KOFF-skee)
Tianshan (TYAHN-SHAHN)
Tikhvin (TIK-vun)
Tolyatti (tawl-YAH-tee)
Transcaucasia (tranz-kaw-KAY-zhuh)
Tsar (SAHR)
Tsarina (sah-REE-nuh)
Tselinograd (seh-LINN-uh-grahd)
Tula (TOO-luh)
Turkmenistan (terk-MEN-uh-stahn)
Tuva (TOO-vuh)
Tuvinian (too-VINNY-un)
Udmurt(ia) (ood-MOORT [ee-uh])
Ufa (oo-FAH)

Ukraine (yoo-CRANE)
Ulyanov (ool-YAH-noff)
Ural (YOOR-ull)
Ussuri (ooh-SOOR-ree)
Uzbekistan (ooze-BECK-ih-stahn)
Vanino (VAH-nih-noh)
Varangian (vuh-RANGE-ee-un)
Vladikavkaz (vlad-uh-kuff-KAHZ)
Vladivostok (vlad-uh-vuh-STAHK)
Volgograd (VOLL-guh-grahd)
Yakut(sk) (yuh-KOOT [sk])
Yakutia (yuh-KOOTY-uh)
Yaroslavl (yar-uh-SLAHV-ull)
Yekaterinburg (yeh-KAHTA-rin-berg)
Yenisey (yen-uh-SAY)
Yerevan (yair-uh-VAHN)
Yurt (YOORT)

TRANSCAUCASIA:
Cauldron of Conflict

Between the Black Sea to the west and the Caspian Sea
to the east lies a geographic region of enormous physiographic, historical, and cultural complexity. Physiographically, this region is dominated by the rugged terrain of the
Caucasus Mountains, in whose valleys countless struggles
have been waged. Historically, this has been a battleground
for Christians and Muslims, Russians and Turks, Armenians and Persians. Culturally, the region is a jigsaw of languages, religions, and traditions. It is, by many measures,
the Balkans of Asia.

IDEAS & CONCEPTS

Shatter belt (2)
Centrifugal forces (2)
Exclave

The geographic name for this turbulent region is *Transcaucasia,* signifying its situation astride the great ranges
that define its physiography. As Fig. T-1 shows, Transcaucasia at present consists of three political entities: Georgia,

FIGURE T-1

TEN MAJOR GEOGRAPHIC QUALITIES OF TRANSCAUCASIA

1. As a region, Transcaucasia has relief that is among the highest anywhere in the world.
2. The international political geography of Transcaucasia is complicated by enclaves, exclaves, and poorly defined and delimited boundaries.
3. The internal, national political geography of Transcaucasia is marked by centrifugal forces most strongly expressed in Georgia.
4. The complex political and cultural mosaic of Transcaucasia results from its location relative to imperial powers pursuing geopolitical objectives in the region: the Russians, the Turks, and the Persians.
5. Transcaucasia is a region of spatial-religious extremes. Shi'ite Islam prevails in the east (Azerbaijan); Christianity in various forms in the west (Armenia, Georgia).
6. Physiographic and historic factors have combined to keep population groups separated and divided. The region's internal circulation is minimal.
7. Strong nationalism has developed in Transcaucasia, much of it at the subnational, ethnic-group level.
8. Historic links connect parts of Transcaucasia to external areas. Ossetians occupy territory in Russia. Armenians have ties to Turkey. Azerbaijan is the northern end of a much larger area (of the same name) in adjacent Iran.
9. Energy resources are important in Transcaucasia: Azerbaijan is a major producer. But Armenia is energy-poor, and Georgia also depends on imports.
10. In the post-Soviet period, severe conflict has engulfed Transcaucasia. Territorial disputes have pitted Azerbaijan against Armenia; ethnic and political strife has affected Georgia.

Armenia, and Azerbaijan, all former Soviet Socialist Republics in the communist empire.

Before we examine the regional geography of Transcaucasia, we should consider its location relative to the states that border it. To the north lies the Russian Federation—but not Russia itself. Almost all the way across Transcaucaasia's northern flank, from the Caspian to the Black Sea, lies a tier of Russia's internal republics—some stable and quiet, others volatile and active (Fig. 2–11). To the south lie two Muslim countries, each with links to the region: Iran, with historic ties to Azerbaijan, and Turkey, with a long history of adversarial relationships with Armenians. These outside forces continue to influence the course of events in Transcaucasia—just as external powers have long done in Eastern Europe's Balkan **shatter belt**.

Where does Transcaucasia, as a region, belong in the global spatial framework? That question remains unanswered in the mid-1990s. Georgia, the country that faces the Black Sea, has proclaimed its "Europeanness," and its embattled government has stated its intention to seek closer ties to Europe. Armenia, landlocked and fragmented, is a Christian country nearly surrounded by Muslim domains. And Azerbaijan, as we see on Fig. T-1, is actually the northern half of a larger province, the southern part of which lies in Iran today. So pressing and numerous are the unresolved territorial issues in Transcaucasia that we discuss the region separately here, awaiting indications of its future disposition. Transcaucasia is in transition; its three republics are small and comparatively weak, and major contenders for influence in the region (and perhaps for control over parts of it) lie nearby.

◆ GEORGIA

Of the three republics of Transcaucasia, only Georgia has a Black Sea coast. Somewhat smaller than South Carolina, Georgia is a country of high mountains and fertile valleys, with a complicated political geography. Its population of 5.6 million people is more than 70 percent Georgian but also includes Armenians (8 percent), Russians (6 percent), Azerbaijanis or Azeris (6 percent), and smaller numbers of Ossetians (3 percent) and Abkhazians (2 percent). Within the republic lie three minority-based autonomous entities, the Abkhazian and Adjarian Autonomous Republics, and the South Ossetian Autonomous Region (Fig. T-1).

Sakartvelo, as the Georgians call their country, has a long and turbulent history. Tbilisi (1.4 million today), the capital for 15 centuries, lay at the core of a major empire around the turn of the thirteenth century, but the Mongol invasion ended this era. Next, the Christian Georgians found themselves in the path of Islamic wars between Turks and Persians. Turning northward for protection, the Georgians soon were annexed by the Russians, who were looking for warm-water ports. Like other peoples overpowered by the tsars, the Georgians seized on the Russian Revolution to reassert their independence; but the Soviets reincorporated Georgia in 1921 and proclaimed a Georgian Soviet Socialist Republic in 1936. Josef Stalin was a Georgian.

Georgia is renowned for its scenic beauty, warm and favorable climates, varied agricultural production (especially tea), timber, manganese, and other products. Georgian wines, tobacco, and citrus fruits are much in demand. The diversi-

The Georgian capital, Tbilisi, extends for over 30 miles along the banks of the Kura River, Transcaucasia's longest by far. This is one of the oldest cities in this part of Eurasia.

fied economy has the potential to support Georgia's consolidation as a viable state.

Unfortunately, Georgia's political geography is loaded with **centrifugal forces**. After Georgia declared its independence in 1991, factional fighting destroyed its first elected government. But worse was to come. In the Autonomous Region of South Ossetia, which borders the Russian internal republic of North Ossetia, conflict broke out over local demands for self-determination; many of Georgia's Ossetians preferred unification with Russia's Ossetians over minority status in the new Georgian state. Even as this costly civil strife continued, one of Georgia's two internal republics, Abkhazia (in the country's northwestern corner [Fig. T-1]) proclaimed its independence. Abkhazians, a Muslim minority placed under Georgian control during Soviet times, saw an opportunity to gain their own independence—although even in their own tiny republic, they constitute only about one-sixth of the population.

When Georgian forces intervened in Abkhazia, Muslims from the Russian side of the border rose in support of the Abkhazians. Georgia's leaders accused the Russians of failing to control the Muslims under their rule, and relations between Moscow and Tbilisi worsened. In the meantime, the economy suffered severely; tourism is a major source of income here, and the country's health resorts stood empty.

If there is a positive side to the political geography of Georgia, it is that the southwestern republic, Adjaria, is less likely than Abkhazia to attempt secession. More than 80 percent of the population there is Georgian, and of the remainder 10 percent are Russians and 5 percent Armenians. Turkic Adjarians, once dominant here, now form a miniscule minority.

◆ ARMENIA

Georgia's problems pale when compared to those of Armenia and its eastern neighbor, Azerbaijan. Armenia occupies some of the most rugged and earthquake-prone terrain in the Transcaucasus. Territorially the smallest of all the former Soviet republics, Maryland-sized Armenia has a prorupt shape (see p. 568) that has the effect of fragmenting its eastern neighbor, Azerbaijan, into two parts (Fig. T-1).

More than 90 percent of Armenia's predominantly Christian population of 3.6 million people is Armenian, and less than 2 percent is Russian. But Armenians also form a large majority in the Nagorno-Karabakh Autonomous Region, an Armenian **exclave** encircled by Muslim Azerbaijan (Fig. T-1).

This spatial recipe for trouble, engineered by Soviet sociopolitical planners in the 1920s, now destabilizes the region. Armenians, who adopted Christianity 17 centuries ago,

War in the Transcaucasus has destroyed much of the region's infrastructure. In the capital of Armenia, Yerevan, trees have been cut down for firewood and buildings have been damaged by people searching for fuel to heat their homes during the wartime blockade.

for more than a millennium have sought to establish a secure homeland here on the margin of the Muslim world. Their ancient domain was overrun by Turks, Persians, and later by Russians. During World War I, the Ottoman Turks decimated their Armenian adversaries on that empire's eastern flank. At the end of the war an independent Armenia briefly arose, but it lasted just two years. In 1920 it became a Soviet republic, and in 1936 it was proclaimed one of the 15 constituent republics of the Soviet Union.

Yerevan (1.3 million), the capital, lies within sight of the Turkish border. Mineral-rich and producing a variety of subtropical fruits, Armenia has opportunities for development; hydroelectric power is plentiful. But severe setbacks have afflicted the republic. A disastrous earthquake struck in 1988, killing tens of thousands. Continuing conflict with its eastern neighbor has ruined the economy and has cost

thousands of lives. Armenia's hopes at independence were almost immediately dashed in the early 1990s.

What does Armenia's future hold? In 1993, the war with Azerbaijan over Nagorno-Karabakh exacted ever higher costs in this already poverty-stricken country. Early during the campaign, the Armenians succeeded in establishing a corridor between their exclave and the main territory of the republic, but Muslim forces advanced into the Christian stronghold from the east. Eventually Armenia's geography proved to be its greatest adversary: its main source of energy, the oil pipeline from the Caspian shore to Yerevan, was cut off. No help could be expected from the Turks to the west. And Armenia's other external link, from Russia across Georgia, was undependable because of Georgia's own instability and Russia's economic troubles. Without powerful protectors and without external contacts, Armenia's future hangs in the balance.

◆ AZERBAIJAN

The 7.4 million people of Azerbaijan—the *Azeris*—occupy a corner of Transcaucasia that lies separated from Russia by mountains, from its western neighbors by religion and ethnicity, and from the Muslim republics to the east by the waters of the Caspian Sea (Fig. T-1). Small wonder that the Shi'ite Muslim Azeris look southward across the border to Iran for support. About 4 million Azeris live in northern Iran (and perhaps 15 million in Iran overall); they speak the language and practice the faith of their counterparts across the border.

Azerbaijan is a legacy of the northward march of Islam into the corridor between the Black and Caspian seas. The Tatars declared independence here in 1918, but the Red Army conquered them and made the area a Soviet republic in 1920. The complex political geography of Maine-sized Azerbaijan includes the Nagorno-Karabakh Autonomous Region (population 210,000; more than three-quarters Armenian) and the Nakhichevan A.S.S.R. (325,000), an Azeri exclave created by an intervening corridor of Armenian territory (Fig. T-1).

Baku (1.8 million), the coastal capital, has grown into Transcaucasia's largest city, based on major nearby oil reserves, associated industries, and an expanding port function. Many Armenians came to work here, and friction developed. In 1989, Azeris attacked the Armenian communities, killing many. About 250,000 Armenians fled, and although a few thousand returned when Soviet troops restored order, this remnant of the Armenian community was ousted after Azerbaijan declared its independence in 1991. When Azerbaijan in effect annexed the Armenian exclave of Nagorno-Karabakh, all hope for long-term stability disappeared.

The waterfront of Azerbaijan's capital, Baku, the former Soviet Union's fifth largest city. Since the late nineteenth century, Baku has been synonymous with petroleum production from a huge oilfield that extends beneath the Caspian Sea. A number of oil tankers can be seen in the background.

In the mid-1990s, Iran and Turkey were vying for influence in independent Azerbaijan. Iran tried to broker a peace between the Azeris and the Armenians, but this effort was rejected by radicals in Azerbaijan; the Turks never wavered in their support of the Azeris. In the process, the geopolitical trend lines of the future may have been revealed. Azerbaijan once was a large domain, divided between the (more powerful) Russian and Persian empires during the nineteenth century. More than twice as many Azeris live in Iran as in independent Azerbaijan (the Iranian province is also called Azerbaijan). Will sovereign Azerbaijan engage in irredentist policies toward its neighbors? Will Iranian Azeris wish to follow the example of their independent cohorts? The new map of Transcaucasia is still in the making.

◆ PRONUNCIATION GUIDE

Abkhazia (ahb-KAHZ-zee-uh)
Adjaria (uh-JAR-ree-uh)
Armenia (ahr-MEENY-uh)
Azerbaijan (ah-zer-bye-JAHN)
Azeri (ah-ZAIRY)
Baku (bah-KOO)
Caspian (KASS-spee-un)
Caucasus (KAW-kuh-zuss)
Georgia (GEORGE-uh)

Iran (ih-RAN/ih-RAHN)
Islam (iss-LAHM)
Kura (KOOR-uh)
Moscow (MOSS-cau)
Muslim (MUZZ-lim)
Nagorno-Karabakh (nuh-GORE-noh KAH-ruh-bahk)
Nakhichevan (nah-kee-chuh-VAHN)
Ossetia (oh-SEE-shuh)

Sakartvelo (sah-KART-vuh-loh)
Shi'ite (SHEE-ite)
Stalin (STAH-lin)
Tatar (TAHT-uh)
Tbilisi (tuh-BILL-uh-see)
Transcaucasia (tranz-kaw-KAY-zhuh)
Yerevan (yair-uh-VAHN)

Relief

Meters		Feet
3050		10 000
1525		5000
610		2000
305		1000
0	Sea Level	0
152.5		500 Below
		Sea Level
1525		5000
3050		10 000
6100		20 000

A-520000-76- -14
COPYRIGHT BY
RAND McNALLY & COMPANY
MADE IN U.S.A.

0   200   400   600   800   1000 Miles

0   400   800   1200   1600 Kilometers

Scale 1:40 000 000; one inch to 630 miles. Lambert's Azimuthal Equal Area Projection
Elevations and depressions are given in feet

CHAPTER 3

NORTH AMERICA: THE POSTINDUSTRIAL TRANSFORMATION

The North American realm consists of two countries that are alike in many ways. In the United States and in Canada, European cultural imprints dominate. Indeed, the realm is often called *Anglo-America*: English is the official language of the United States and shares equal status with French in Canada. The overwhelming majority of church-goers adhere to Christian faiths. Most (but not all) of the people trace their ancestries to various European countries. In the arts, architecture, and other modes of cultural expression, European norms prevail.

North American society is the most highly urbanized of the world's realms, and nothing symbolizes the New World quite as strongly as the skyscraper panoramas of New York, Toronto, Chicago, or San Francisco. North Americans are also hypermobile, with networks of superhighways, commercial air lanes, and railroads efficiently interconnecting the realm's far-flung cities and regions. Commuters stream into and out of suburban business centers and central-city downtowns by the millions each working day. A U.S. family changes its residence on average once every five and a half years.

In the 1990s North America has entered a new age, the third since the arrival of Columbus in the New World more than 500 years ago. The first four centuries were dominated by agriculture and rural life; the second age—industrial urbanization—has endured over the past century but is now ending. In its place, the United States and Canada are today experiencing the maturation of a **postindustrial society and economy**, which is dominated by the production and manipulation of information, skilled services, and high-technology manufactures, and operates within a global-scale framework of business interactions. As blue-collar rapidly yields to white-collar employment and as the automated office becomes the dominant workplace, fundamental dislocations are felt throughout North America. The

IDEAS & CONCEPTS

Postindustrial society and economy
Urban geography
Cultural pluralism
Time-space convergence
Physiographic province
Rain shadow effect
Pollution
Culture hearth

Epochs of metropolitan evolution
Megalopolitan growth
Eras of intraurban structural evolution
Suburban downtown
Urban realms model
Economies of scale
Historical inertia

REGIONS

Continental Core
New England/Maritime Provinces
French Canada
Agricultural Heartland

The South
The Southwest
Marginal Interior
West Coast

aging and declining Manufacturing Belt of the northeastern quadrant of the United States (and adjacent southeastern Canada) is now often referred to as the *Rustbelt*. But the U.S. southern-tier states (the so-called *Sunbelt*) contain growing numbers of glamorous, high-prestige locales, led by suburban San Francisco's Silicon Valley and its high-technology counterparts in Texas, North Carolina, Florida, Arizona, and Southern California.

◆ F O C U S   O N   A   S Y S T E M A T I C   F I E L D

URBAN GEOGRAPHY

Urban geography is concerned with the spatial interpretation of city-centered population concentrations that exhibit a high-density, continuously built-up settlement landscape. The significance of contemporary worldwide urbanization has already been discussed (p. 27), including definitions of such key terms as *urbanization, city, suburb,* and *metropolitan complex (metropolis)*; accompanying Table I-1 revealed that no world realm is more heavily urbanized than North America—a dimension of vital importance for understanding the human geography of the United States and Canada.

Contemporary urban geography is organized into four leading components. The first is an appreciation for the *historical evolution* of urban society. The second and third involve the study of contemporary spatial patterns at two distinct levels of generalization: macroscale or *interurban geography*, which treats cities as a system of interacting points that serve large surrounding areas, and microscale or *intraurban geography*, which focuses on the internal structure and functioning of individual metropolitan complexes. The fourth component deals with *planning and policy-making*, applications of urban-geographical knowledge to help define and solve spatial problems.

Urban evolution is concerned with the historical geography of cities, emphasizing the forces that produced urban growth during each of the major periods of human development since the city emerged in the Mesopotamian area

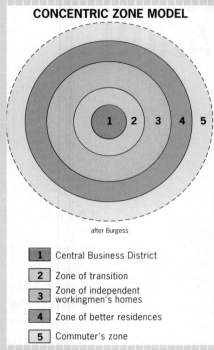

CONCENTRIC ZONE MODEL

after Burgess

1	Central Business District
2	Zone of transition
3	Zone of independent workingmen's homes
4	Zone of better residences
5	Commuter's zone

FIGURE 3-1A

of Southwest Asia about 7,000 years ago. As urbanization spread outward from its hearth, it was incorporated into the cultures of various realms—as we observe in each of this book's chapters.

Interurban geography deals with the broad scope of national-scale urban systems, viewing individual cities as points in a network that interact with one another and serve hinterlands whose territories are commensurate with the size and functional diversity of each city. The study of city types, functions, and spheres of influence within such

urban systems has led to the development of useful classifications and models. One is *central place theory*, which describes some of the basic rules that govern the locations of and relationships among settlements within an urban system.

Intraurban geography is concerned with the internal spatial organization of the metropolis, emphasizing structural form as well as the distributions of diverse population groups and activities. One of the earliest attempts at modeling urban structure, proposed in the 1920s, viewed the city as a set of *concentric zones* of declining intensity of land use and increasingly affluent residents as distance from the city center was increased (Fig. 3-1A). An alternative model soon argued that the differential influence of radial transport routes was so important that it shaped an intra-city structure based on contrasting *sectors* (Fig. 3-1B). By the 1940s, a third model claimed that automobile-based intraurban dispersal was creating a *multiple nuclei* structure (Fig. 3-1C) that loosened the pull of the downtown central business district (CBD). In combination, concentric zones, sectors, and nucleations still provide a useful generalization of the land-use organization of today's large central cities. But the contemporary metropolis has spilled out of its central-city confines in the final third of this century, and these models can no longer accommodate the new urban reality wherein the suburbs have become the essence of the American city. (A model of this metropolitan

Not surprisingly, the human geography of the United States and Canada is undergoing a parallel transformation as dynamic new locational forces surface. As new regions

emerge, the older ones struggle to reinvent themselves; simultaneously, at the intrametropolitan scale, the industrial cities turn inside out and provide major new opportunities

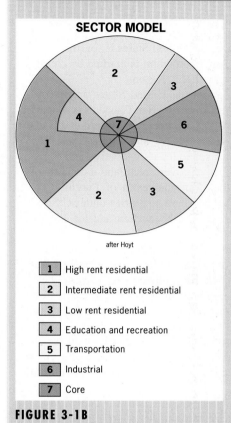

SECTOR MODEL

after Hoyt

1	High rent residential
2	Intermediate rent residential
3	Low rent residential
4	Education and recreation
5	Transportation
6	Industrial
7	Core

FIGURE 3-1B

transformation is discussed on pp. 191–192.)

The spatial arrangement of people within the metropolis is the domain of *urban social geography*. The residential territorial mosaic is the focus of attention and is studied from a number of perspectives, including ethnicity, race, socioeconomic status, community formation, housing-market operations, and the dynamics of intraurban migration. The intrametropolitan distribution of

nonresidential activities is the sphere of *urban economic geography*. Today, this involves major new suburban business centers as well as older central-city commercial districts.

The discipline's concern with the contemporary human condition is expressed in the deepening involvement of professional geographers with the fourth leading component, *planning and policy-making*, which provides the opportunity for employing spatial concepts and methods to help resolve social, economic, and environmental problems. Urban geographers have been in the forefront of such efforts, building on durable ties with the urban-planning community. *Planning* is defined by geographer Harold Mayer as a "process of decision-making [involving] the conscious intervention of people, individually or collectively, into the evolutionary process in order to facilitate progress toward predetermined goals." Since this process is fundamentally a spatial one, geographers have much to contribute through the practical application of such advanced skills as computer mapping, interpretation of satellite imagery, and the management of geographic information systems.

Geographers also participate in policy-making by demonstrating the likely outcomes of various policy options. Many practitioners have gone on to argue that the achievement of social justice means the elimination of spatial inequalities that still deeply divide the metropolitan population mosaic. In the United States these geographic disparities are intensifying as new forces

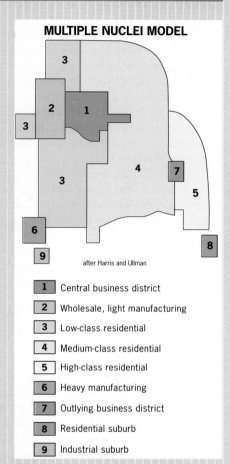

MULTIPLE NUCLEI MODEL

after Harris and Ullman

1	Central business district
2	Wholesale, light manufacturing
3	Low-class residential
4	Medium-class residential
5	High-class residential
6	Heavy manufacturing
7	Outlying business district
8	Residential suburb
9	Industrial suburb

FIGURE 3-1C

continue to transform the metropolis. Today, the aging industrial city, its once-dominant manufacturing function all but exhausted, is trying to carve out a new economic role for its downtown office complex—while its vast, beleaguered inner residential areas house the least fortunate people in urban America.

for their surrounding suburbs as well as the urbanizing countryside beyond. Whatever the outcome, the winners and losers in the current scramble to adapt to the changing spatial infrastructure will shape the geography of these two leading countries in the fast-maturing postindustrial age—which is certain to extend well into the next century.

◆ TWO HIGHLY ADVANCED COUNTRIES

Although it is true that Canada and the United States share a number of historical, cultural, and economic qualities, the two countries do differ in important ways. The differences, in fact, have firm spatial dimensions. The United States, somewhat smaller territorially than Canada, occupies the heart of the North American continent and, as a result, encompasses a greater environmental range. In the United States, we tend to think of an "East" and "West" as a major geographic breakdown, but in Canada it is more appropriate to identify a *Southern* Canada and a Canadian *North*. The overwhelming majority of Canadians live in Southern Canada, mostly within 200 miles (320 km) of the U.S. border. The United States also encompasses North America's northwestern extension, Alaska (offshore Hawaii, however, belongs to the Pacific Realm). Thus, unlike Canada, the United States is a *fragmented state*, a discontinuous country whose national territory consists of two or more individual parts separated by foreign territory and/or international waters.

POPULATION CONTRASTS

Differences also become apparent when population totals and composition are examined. The 1994 population of the United States was approximately 259.6 million; Canada's was approximately 27.8 million, just over one-tenth as large. Although comparatively small, Canada's population is divided by culture and tradition, and this division has a pronounced regional expression. English is the home language of 62 percent of Canada's citizens, French is spoken by 25 percent, and other languages are used by 13 percent of the population (Amerindians and Inuit peoples [formerly called Eskimos] make up only 1.5 percent of the total).

Canada's multilingual situation is accentuated by the strong spatial concentration of French speakers in the second-largest of the country's 10 provinces—Quebec (which also contains the second-biggest Canadian metropolis, Montreal [3.1 million]). More than 85 percent of Quebec's population is French Canadian, and Quebec is the historic, traditional, and emotional focus of French culture in Canada. Quebec straddles the central and lower St. Lawrence River valley, the leading route of access into North America for the early French settlers. In recent decades a strong nationalist movement has emerged in Quebec, at times demanding nothing less than outright separation from the rest of Canada.

Internal regionalism also affects Canada to the west of *Francophone* (French-speaking) Quebec. The country's core area lies mainly in Ontario, Canada's most populous pro-

TEN MAJOR GEOGRAPHIC QUALITIES OF NORTH AMERICA

1. North America encompasses two of the world's biggest states territorially (Canada is the second largest in size; the United States is fourth).
2. The North American realm is marked by clearly defined physiographic regions.
3. Both Canada and the United States are federal states, but their systems differ. Canada's is adapted from the British parliamentary system and is divided into 10 provinces and 2 territories. The United States separates its executive and legislative branches of government, and it consists of 50 states, the Commonwealth of Puerto Rico, and a number of island territories under U.S. jurisdiction in the Caribbean Sea and the Pacific Ocean.
4. Both Canada and the United States are plural societies. Although ethnicity is increasingly important, Canada's pluralism is most strongly expressed in regional bilingualism. In the United States, major divisions occur along racial/ethnic lines.
5. Despite North America's internal social cleavages and regional economic inequalities, the realm is unified by the prevalence of European cultural norms.
6. By world standards, this is a rich realm where high incomes and high rates of consumption prevail. North America possesses a highly diversified resource base, but nonrenewable raw materials are consumed prodigiously and domestic energy prospects remain uncertain.
7. North America's population, not large by international standards, is the most highly urbanized and mobile among the world's geographic realms.
8. North America is still home to one of the world's great manufacturing complexes. The realm's industrialization generated its unparalleled urban growth, but a new postindustrial society and economy are rapidly maturing in both countries.
9. Agriculture in North America employs less than 3 percent of the labor force; it is overwhelmingly commercial, mechanized, and specialized, and it normally produces a huge annual surplus for sale in overseas markets.
10. The North American Free Trade Agreement will link the economies of the United States, Canada, and Mexico ever more tightly; international barriers to trade and investment flow are being dismantled among the three countries.

vince, centered on metropolitan Toronto (3.7 million); Ontario's French-speaking minority, constituting only about 5 percent of the population, is clustered mostly in the east near the Quebec border. Farther west lie the interior Prairie Provinces of Manitoba, Saskatchewan, and Alberta, where the French join the Germans and Ukrainians to form small minorities ranging from 4 to 7 percent of each provincial population. The French cause is weakest in Canada's westernmost province, British Columbia, which is third in population and focuses on Greater Vancouver (1.6 million); here, over 80 percent of the population claims English ancestry, whereas only about 1.5 percent are ethnic French—well behind the province's rapidly growing Asian community.

No multilingual divisions affect the unity of the North American realm's other federation, but **cultural pluralism** of another kind prevails south of the border in the United States. More persistent in the U.S. cultural mosaic than language or ethnicity is the division between peoples of European descent (just over 80 percent of the population) and those of African origin (roughly 12 percent). Despite the significant progress of the modern civil rights movement, which decidedly weakened *de jure* racial segregation in public life, whites overwhelmingly refuse to share their immediate living space with blacks; thus *de facto* residential segregation is all but universal. Although separatist regional thinking on the provincial/state scale of Quebec does not exist, a compelling case can be made that persistent local racial segregation has, in effect, produced two societies—one white and one black—that are both separate and unequal.

ECONOMIC AFFLUENCE AND INFLUENCE

The United States and Canada rank among the most highly advanced countries of the world by every measure of national development, possessing two of the highest living standards on earth. Yet the good life is not shared equally by all of North America's residents (see photo below). Deprivation is surprisingly widespread, with notable spatial concentrations in the United States inside the inner-ring slums of big cities and on rural reservations containing Native Americans or Inuit. In the worst of these poverty pockets, malnutrition is commonplace, although its severity pales in comparison to the daily misery experienced by the poor of the developing realms.

The vast, lower-income, rowhouse-dominated inner city of South Philadelphia, looking north toward Center City. Overindustrialized Philadelphia, burdened by a deteriorated physical plant, has seen its share of metropolitan manufacturing employment drop by over 60 percent since 1970.

North America's highly developed societies have clearly achieved a global leadership role, which arose from a combination of history and geography. Presented with a rich abundance of natural and human resources over the past 200 years, Americans and Canadians have brilliantly converted these productive opportunities into continentwide affluence and worldwide influence as their booming Industrial Revolution surpassed even Europe's by the early decades of this century.

Perhaps the greatest triumph accomplished by human ingenuity in the North American realm was the hard-won victory over a vast and difficult natural environment. By persistently improving transportation and communications, the United States and Canada were finally able to spatially organize their entire countries across great distances (exceeding 2,500 miles/4,000 km) in the needed east-west direction across a terrain in which the grain of the land—particularly mountain barriers—is consistently oriented north-south. As breakthroughs in railroad, highway, and air transport technology succeeded each other over the past 150 years, geographic space-shortening—or **time-space convergence**—allowed distant places to become ever nearer to one another and constantly enabled people and activities to disperse more widely.

Although they continue to have their share of differences, the United States and Canada maintain close and cordial relations, and the border between them is by far the longest open international boundary on earth (about 80 million people cross it each year). Clearly, the two countries are firmly locked together in a mission of free-world leadership. What is now happening in the North American realm—and why—matters enormously to every other country. The United States and Canada constitute the most advanced realm in the world; wherever it goes in the new postindustrial age, many of the rest will eventually follow.

◆ NORTH AMERICA'S PHYSICAL GEOGRAPHY

Before we examine the human geography of the United States and Canada more closely, it is important to consider the physical setting in which they are rooted. The North American continent extends from the Arctic Ocean to Panama, but we will confine ourselves here to the territory north of Mexico—a geographic realm that still stretches from the near-tropical latitudes of southern Florida and Texas to the subpolar lands of Alaska and Canada's far-flung northern periphery. The remainder of the North American continent, which constitutes a separate realm (together with the island chains of the Caribbean Sea), comprises *Middle America* and will be treated in Chapter 5.

PHYSIOGRAPHY

North America's physiography is characterized by its clear, well-defined division into physically homogeneous regions called **physiographic provinces**. Each region is marked by a certain degree of uniformity in relief, climate, vegetation, soils, and other environmental conditions, resulting in a scenic sameness that comes readily to mind. For example, we identify such regions when we refer to the Rocky Mountains, the Great Plains, and the Appalachian Highlands. However, not all the physiographic provinces of North America are so easily delineated.

The complete layout of the continent's physiography is seen in Fig. 3-2. The most obvious aspect of this map of North America's physiographic provinces is the north-south alignment of the continent's great mountain backbone, the Rocky Mountains, whose rugged, often-snow-covered topography dominates the western segment of the continent from Alaska to New Mexico. The major feature of eastern North America is another, much lower chain of mountain ranges called the Appalachian Highlands; these uplands also trend approximately north-south and extend from Canada's Atlantic Maritime Provinces to Alabama. The orientation of the Rockies and Appalachians is important because unlike Europe's Alps, they do not form a topographic barrier to polar or tropical air masses flowing southward or northward, respectively, across the continent's interior.

Between the Rocky Mountains and the Appalachians lie North America's vast interior plains, which extend from the shores of Hudson Bay to the coast of the Gulf of Mexico. These flatlands can be subdivided into several provinces. The major regions are: (1) the great Canadian Shield, which is the geologic core area containing North America's oldest rocks; (2) the Interior Lowlands, covered largely by glacial debris laid down by meltwater and wind during the Late Cenozoic glaciation; and (3) the Great Plains, the extensive sedimentary surface that slowly rises westward toward the Rocky Mountains. Along the southern margin, these interior plainlands merge into the Gulf-Atlantic Coastal Plain, which extends from southern Texas along the seaward margin of the Appalachian Highlands and the neighboring Piedmont until it ends at New York's Long Island.

On the western side of the Rocky Mountains lies the zone of Intermontane Basins and Plateaus. This physiographic province (within the conterminous United States) includes: (1) the Colorado Plateau in the south, with its thick sediments and spectacular Grand Canyon; (2) the lava-covered Columbia Plateau in the north, which forms the watershed of the Columbia River; and (3) the central Basin-and-Range country (Great Basin) of Nevada and Utah, which contains several extinct lakes from the glacial period as well as the surviving Great Salt Lake.

The reason this province is called *intermontane* has to do with its position between the Rocky Mountains to the east and the Pacific Coast mountain system to the west. From the

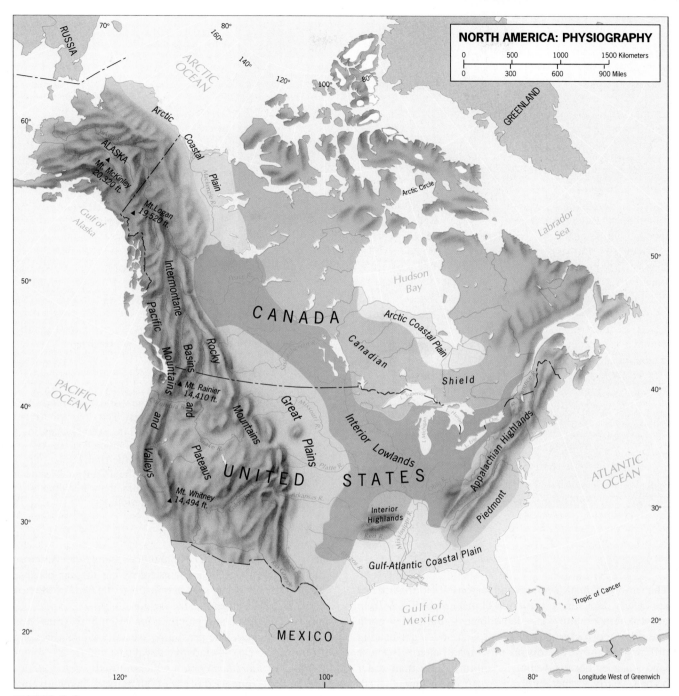

FIGURE 3-2

Alaskan Peninsula to Southern California, the west coast of North America is dominated by an almost unbroken corridor of high mountain ranges whose origins stem from the contact between the North American and Pacific Plates (Fig. I-3). The major components of this coastal mountain belt include California's Sierra Nevada, the Cascades of Oregon and Washington, and the long chain of highland massifs that line the British Columbian and southern Alaska coasts. Two broad valleys in the conterminous U.S. portion of the province—which contain dense populations—are the only noteworthy interruptions: California's Central (San Joaquin-Sacramento) Valley and the Puget Lowland of Washington State that extends southward into western Oregon's Willamette valley.

Water supply is a growing concern in many parts of the world, particularly the United States West. The Colorado River has been dammed in several places, including the Glen Canyon Dam in Arizona just south of the Utah border, upstream from the Grand Canyon. Despite such grand schemes, the water situation continues to deteriorate.

CLIMATE

The various climatic regimes and regions of North America are clearly depicted on the world climate map (Fig. I-7). In general, temperature varies latitudinally—the farther north one goes, the cooler it gets. Local land-and-water-heating differentials, however, distort this broad pattern. Because land surfaces heat and cool far more rapidly than water bodies, yearly temperature ranges are much larger where *continentality* (see p. 123) is greatest.

Precipitation generally tends to decline toward the west—except for the Pacific coastal strip itself—as a result of the **rain shadow effect**, whereby most Pacific Ocean moisture is effectively screened from the continental interior (see box at right). This broad division into Arid (western) and Humid (eastern) America, however, is marked by a fuzzy boundary that is best viewed as a wide transitional zone. Although the separating criterion of 20 inches (50 cm) of annual precipitation is easily mapped (see Fig. I-6), that generally north-south *isohyet* (the line connecting all places receiving ex-

actly 20 inches per year) can and does swing widely across the drought-prone Great Plains from year to year because highly variable warm-season rains from the Gulf of Mexico come and go in unpredictable fashion.

On the other hand, precipitation in Humid America is far more regular. The prevailing westerly winds (blowing from west to east—winds are always named for the direction *from which they come*), which normally come up dry for the large zone west of the 100th meridian, pick up considerable moisture over the Interior Lowlands and distribute it throughout eastern North America. A large number of storms develop here on the highly active weather front between tropical Gulf air to the south and polar air to the north. Even if major storms do not materialize, local weather disturbances created by sharply contrasting temperature differences are always a danger. There are more tornadoes (nature's most violent weather) in the central United States each year than anywhere else on earth. And in winter, the northern half of this region receives large amounts of snow, especially around the Great Lakes.

THE RAIN SHADOW EFFECT

The dryness of most of the western United States is a result of the relationship between climate and physiography. Despite the prevailing wind direction from west to east, moisture-laden air moving onshore from the Pacific is unable to penetrate the center of the continent because high mountains stand in the way. When eastward-moving Pacific air reaches the foot of the Cascade Mountains and other north-south-aligned ranges, it is forced to rise up the western (windward) slope in order to surmount the topographic barrier. As the air rises toward the summit level, it steadily cools. Since cooler air is less able to hold moisture, falling temperatures produce substantial rainfall (and snowfall), in effect "squeezing out" the Pacific moisture as if the air were a sponge filled with water. This altitude-induced precipitation is called *orographic* (mountain) rainfall.

Although precipitation totals are quite high on the upper windward slopes of these West Coast mountain ranges, the major impact of the orographic phenomenon is felt to the east. Robbed of most of its moisture content by the time it is pushed across the summit ridge, the eastward-moving air now rushes down the leeward (downwind) slopes of the mountain barrier. As the air warms, its capacity to hold moisture greatly increases, and the result is a warm dry wind that can blow strongly for hundreds of miles inland. This widespread existence of semiarid (and in places even truly arid) environmental conditions is known as the **rain shadow effect**—with mountains quite literally creating a leeward "shadow" of dryness. As air masses continue to move eastward across the Intermontane and Rocky Mountain provinces, they are again subjected to orographic uplift (and even further drying out). This reinforces the spreading of dryness, in most years, well to the east of the Rockies. Here in the Great Plains, with Pacific moisture sources usually blocked, most farmers depend on fickle south-to-north winds from the Gulf of Mexico to bring sufficient precipitation during the growing season; no wonder this region is susceptible to recurring droughts as well as dust-bowl environmental disasters.

Figure I-7 shows the absence of humid temperate (*C*) climates from Canada (except along the narrow Pacific coastal zone) and the prevalence of cold in Canadian environments. East of the Rocky Mountains, Canada's most *moderate* climates correspond to the *coldest* of the United States. Nonetheless, Southern Canada does share the environmental conditions that mark the Upper Midwest and Great Lakes areas of the United States, so that agricultural productivity in the Prairie Provinces and in Ontario is substantial. Canada is a leading food exporter (chiefly wheat), as is the United States, in spite of its comparatively short growing season.

The broad environmental partitioning into Humid and Arid America is also reflected in the distribution of the realm's soils and vegetation. For farming purposes there is usually sufficient soil moisture to support crops where annual precipitation exceeds the critical 20 inches; where the yearly total is less, soils may still be fertile (especially in the Great Plains) but irrigation is often necessary to achieve their full agricultural potential. Vegetation patterns are displayed on the world map (Fig. I-8), but the enormous human modification of the environment in North America's settled zones has all but reduced this regionalization scheme to the level of the hypothetical. Nonetheless, the Humid/Arid America dichotomy is again a valid generalization: the natural vegetation of areas receiving more than 20 inches of water annually is *forest*, whereas the drier climates give rise to a *grassland* cover.

HYDROGRAPHY (WATER)

Surface water patterns in North America are dominated by the two major drainage systems that lie between the Rockies and the Appalachians: (1) the five Great Lakes (Superior, Michigan, Huron, Erie, and Ontario) that drain into the St. Lawrence River, and (2) the mighty Mississippi-Missouri river network, supported by such major tributaries as the Ohio, Tennessee, and Arkansas rivers. Both are products of the last episode of Late Cenozoic glaciation, and together they amount to nothing less than the best natural inland waterway system in the world. Human intervention has further enhanced this network of navigability, mainly through the building of canals that link the two systems as well as the St. Lawrence Seaway (opened in 1959 but now of limited service because its locks and artificial channels cannot accommodate today's larger ships).

Elsewhere, the northern east coast of the continent is well served by a number of short rivers leading inland from the Atlantic. In fact, many of the major northeastern seaboard cities of the United States—such as Washington, D.C., Baltimore, and Philadelphia—are located at the waterfalls that marked the limit to tidewater navigation (hence their designation as *fall line cities*). Rivers in the Southeast and west of the Rockies at first offered little practical value owing to orientation and navigability problems. In the Far West, however, the Colorado and Columbia rivers have today become supremely important as suppliers of drinking and irrigation water as well as hydroelectric power as much of the Pacific coast continues its regional development.

HUMAN ENVIRONMENTAL IMPACTS

More than a century of advanced industrial technology has taken its toll on the natural environment of North America. For decades, the growing problems of air and water **pollution** were ignored until a public outcry in the United States during the 1960s led the federal government to establish the Environmental Protection Agency (EPA) and take an active role in enforcing new pro-environmental legislation. Although substantial progress was made in the 1970s, administrations between 1981 and 1993 were criticized for weakening the federal enforcement apparatus. Also during the 1970s, cancer surpassed heart disease as the leading killer in the United States. Since many cancers (especially lung cancer) are environmentally related, medical geographers have increased their study of the subject (this subdiscipline is reviewed in the "Focus on a Systematic Field"

essay in Chapter 8). The spatial distribution of cancer mortality, shown in Fig. 3-3, reveals that respiratory-system cancer coincides with a number of major manufacturing and refining centers.

One of the most severe *air pollution* problems of large metropolitan complexes is smog. Dozens of major cities experience this atmospheric hazard, with Los Angeles and Denver among the worst offenders on the continent. Smog is usually created by a temperature inversion in which a warm, dry layer of air, hundreds of feet above the ground, prevents cooler underlying air from rising; this causes the surface air to become stagnant, thereby trapping motor vehicle and industrial emissions that intensify and turn into chemical smog (the word is a contraction of "smoke" and "fog"). One of the realm's most serious *water pollution* problems is acid rain (see pp. 88-89). Although at least 75 percent of North America's sulfur and nitrogen emissions emanate from the U.S. Manufacturing Belt (particularly Illinois, Indiana, Ohio, and Michigan), much of the resultant acid rain falls in southeastern Canada as prevailing winds blow from the Midwest toward eastern Ontario and southern Quebec.

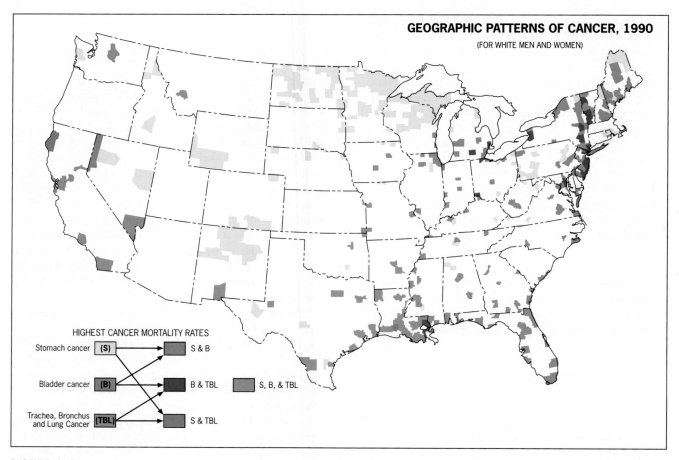

FIGURE 3-3

Los Angeles is infamous for its atmospheric pollution, underscored here in contrasting views of the air-quality extremes over the central business district. Smog shrouds this sprawling city on many days of the year, despite efforts to curb the production of pollutants from automobile exhaust, factories, and other sources.

◆ THE UNITED STATES

The broad outline of North America we have just sketched will be useful in developing the regionalization scheme that appears in the final section of this chapter. To fully appreciate the realm's internal regional organization, however, it is first necessary to examine in some detail the changing human geography of each country. We begin with the United States, and it may be helpful to take a few moments to review and further familiarize yourself with the map of its basic contents (Fig. 3-4). Our focus is on contemporary urban, cultural, and economic geography, and we start by tracing the evolution of that most essential of all human-geographical expressions—the map of population distribution.

POPULATION IN TIME AND SPACE

The current population distribution of the United States is shown in Fig. 3-5. It is important to note that this map is the latest "still" in a motion picture, one that has been unreeling for nearly four centuries since the founding of the first permanent European settlements on the northeastern coast. Slowly at first, then with accelerating speed after 1800, as one major technological breakthrough followed another, Americans (and Canadians) took charge of their remarkable continent and pushed the settlement frontier westward to the Pacific. The swiftness of this expansion was dramatic, but Americans have long been the world's most mobile people (about 18 percent of the U.S. population changes residence each year). In fact, migrations continue to reshape the United States in the 1990s, perhaps the most

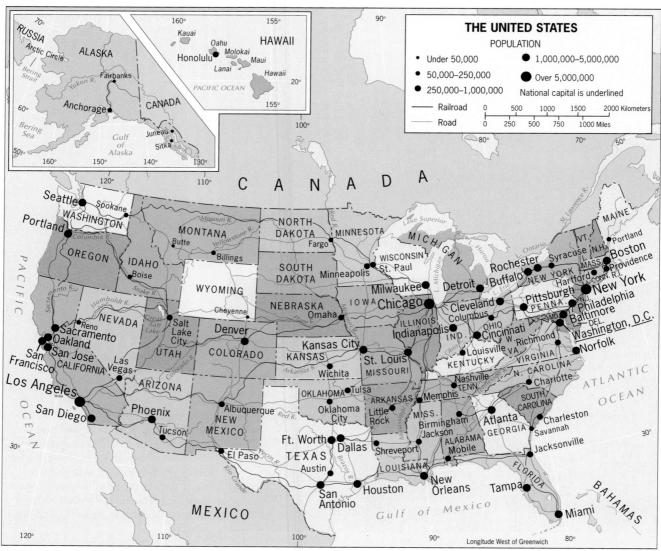

FIGURE 3-4

significant being the unrelenting interregional shift of people and livelihoods toward the south and west (the Sunbelt), away from the north and east. The 1980 census revealed that for the first time, the geographic center of the U.S. population had crossed west of the Mississippi River, capping a 200-year movement in which every decennial census reported a westward shift. The 1990 census found that this center had shifted an additional 40 miles (65 km) southwestward during the 1980s; undoubtedly, the 2000 census will report another movement in this direction.

To understand the contemporary population map, we need to review the major forces that have shaped, and continue to shape, the distribution of Americans and their activities. Since its earliest days, the United States has been perceived as the world's premier "land of opportunity," and it has attracted a steady influx of immigrants who were rapidly assimilated into the societal mainstream. (Despite

strict quotas, immigration continues, with well over 1 million new arrivals annually—at least two out of three of whom enter illegally.) Within the country, people have sorted themselves out to maximize their proximity to existing economic opportunities, and they have shown little resistance to relocating as the nation's changing economic geography has successively favored different sets of places over time.

During the past century these transformations have spawned a number of major migrations. Although the American frontier closed 100 years ago, the westward shift of population continues, but now with that pronounced deflection to the south. The explosive growth of cities, triggered by the Industrial Revolution in the final decades of the nineteenth century, launched a rural-to-urban migratory stream that lasted into the 1960s. The middle decades of this century also witnessed significant migration from

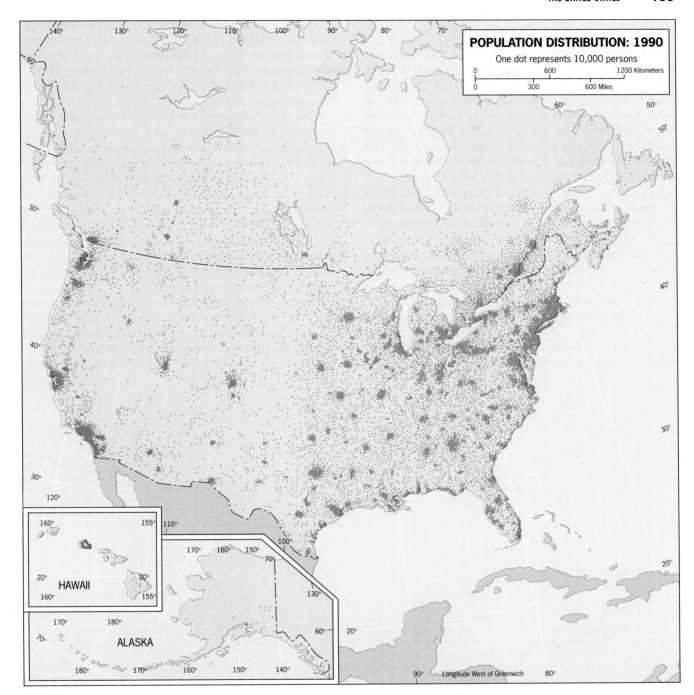

POPULATION DISTRIBUTION: 1990
One dot represents 10,000 persons

FIGURE 3-5

south to north—particularly by blacks—but that, too, has ended and even reversed as sizeable numbers of blacks today return south. This is part of the much larger north-to-south movement toward the Sunbelt mentioned above—a migratory stream that remains strong for the following reasons: (1) the U.S. economy and its higher-paying jobs are shifting in that direction; (2) the substantial retirement migration of the affluent elderly to such states as Florida and

Arizona, though slowing, shows little sign of ending; and (3) the new tide of immigration from Middle America is overwhelmingly directed toward the zone adjacent to the southern border.

Let us now look more closely at the historical geography of these changing population patterns, considering first the initial rural influence and then the decisive impacts of industrial urbanization and its postindustrial aftermath.

Populating Rural America Before 1900

The spatial distribution of the U.S. population is rooted in the colonial era of the seventeenth and eighteenth centuries that was dominated by England and France. The French were concerned with penetrating the continental interior in order to establish a lucrative fur-trading network. The other major (and larger) colonizing force, the English, concentrated their settlement efforts along the coast of what is today the northeastern U.S. seaboard. These British colonies quickly became differentiated in their local economies, a diversity that was to endure and later shape American cultural geography. The northern colony of New England (Massachusetts Bay and environs) specialized in commerce; the southern Chesapeake Bay colony (Tidewater Virginia and Maryland) emphasized the large-scale plantation farming of tobacco; the Middle Atlantic area lying in between (southeastern New York, New Jersey, eastern Pennsylvania) was home to a number of smaller, independent-farmer colonies.

As these three adjacent colonial regions thrived and yearned to expand after 1750, the British government kept the inland frontier closed; it also exerted tightening economic controls, thereby unifying the three groups of colonies in their growing dislike for the increasingly heavy-handed mother country. By 1775, escalating tensions produced open rebellion and the onset of the eight-year-long Revolutionary War. This resulted in a resounding English defeat and independence for the newly formed United States of America.

The western frontier of the fledgling nation now swung open, and the old British Northwest Territory (Ohio-Michigan-Indiana-Illinois-Wisconsin) was promptly settled. Propelling this rise of the Trans-Appalachian West was the discovery that the soils (and climate) of the Interior Lowlands were especially favorable for farming—a remarkable improvement over the relatively infertile seaboard soils. That sparked the rapid growth of agriculture and the widening of coastal-interior trading ties across ever greater distances. New interregional complementarities now emerged, a cornerstone of an economy whose spatial organization was assuming national-scale proportions.

By the time the westward-moving frontier swept across the Mississippi valley in the 1820s (Fig. 3-6), it was clear that the three former seaboard colonies had become separate **culture hearths**—primary source areas and innovation centers (**A**, **B**, and **C** on Fig. 3-6) from which migrants carried

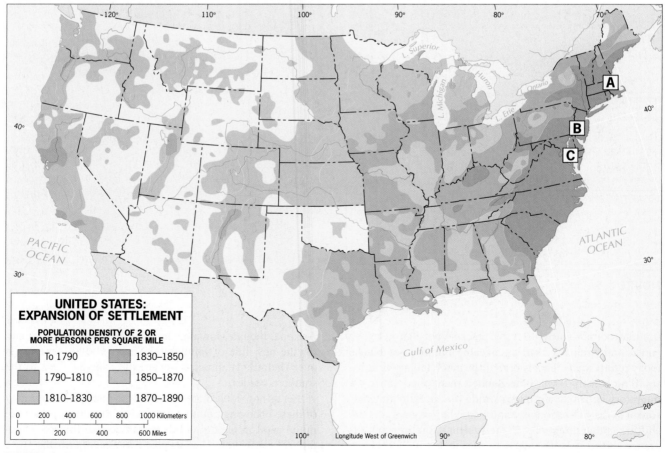

FIGURE 3-6

cultural traditions into the central United States. The New England region (**A**) influenced the southern and western margins of the Great Lakes, with settlers here creating a cultural landscape that reflected New England's architecture and village patterns. Immediately to the south in the Trans-Appalachian West lay a much larger area, focused on the Ohio valley but stretching as far south as southern Tennessee, that similarly resembled Middle-Atlantic Pennsylvania (**B**); and along the Atlantic and Gulf Coastal Plain, from Delaware to newly purchased Louisiana, lay an emerging culture area dominated by the tobacco- (and now cotton-) plantation tradition that had originated in the Tidewater Maryland/Virginia hearth (**C**).

The northern half of this vast interior space soon became well unified as its infrastructure constantly im-proved. By 1860 the railroad had replaced earlier plank roads and canals, providing a more efficient network that offered fast and cheap long-distance transportation, thereby producing significant time-space convergence among once-distant regions within the northeastern quadrant of the United States. The American South, however, did not wish to integrate itself economically with the North to any great degree, preferring instead to export cotton and tobacco to overseas markets. This divergent regionalism, together with its insistence on preserving slavery, soon led the South into secession and the ruinous Civil War (1861–1865). In the dismal aftermath, the South took decades to rebuild.

The second half of the nineteenth century also saw the frontier cross the western United States, although at first it was a *hollow frontier*: parts of the Pacific coast were settled before the dry steppelands of the Great Plains region far to the east (Fig. 3-6). As early as 1869, agriculturally booming California was linked to the rest of the nation by transcontinental railroad, and these same steel tracks also began to open up the bypassed Great Plains. Those semi-arid grasslands, it turned out, were quite fertile after all, particularly for raising wheat on a large scale. The Plains region was deficient only in water supplies, which could be stretched through careful farming methods or enhanced by irrigation near surface and underground water sources. Two new inventions—automated farm machinery for large wheatfields and barbed-wire fences to protect crops from grazing cattle—further supported the development of grain farming and the regional economy.

By the time the U.S. frontier closed in the 1890s, today's rural settlement pattern was firmly in place, anchored to a set of enduring national agricultural regions (discussed later in this chapter). The closure of the frontier also coincided with an accelerating shift in the distribution of the American population: the census of 1900 revealed that more people lived in newly developed metropolitan areas than in the countryside. An urban revolution was accompanying the Industrial Revolution that had taken hold after 1880. To be sure, cities had been vitally important since colonial times. Yet the degree and extent of urbanization by 1900 was approaching massive proportions; and it was clear, as the twentieth century opened, that the tilt from rural to urban America was just beginning.

American Industrial Urbanization, 1900–1970

In the United States the Industrial Revolution occurred almost a century later than in Europe, but when it finally did cross the Atlantic in the 1870s, it took hold so successfully and advanced so robustly that only 50 years later America was surpassing Europe as the world's mightiest industrial power. Thus the far-reaching social and economic changes that Europe's industrializing countries experienced were accelerated in the United States, heightened all the more by the arrival of nearly 25 million European immigrants—who overwhelmingly headed for jobs in the major manufacturing centers—between 1870 and 1914.

The impact of industrial urbanization occurred simultaneously at two levels of spatial generalization. At the national scale or *macroscale*, a system of new cities rapidly emerged, specializing in the collection, processing, and distribution of raw materials and manufactured goods, linked together by an efficient web of long-distance and local railroad lines. Within that urban network, at the *microscale*, individual cities prospered in their new roles as manufacturing centers, generating a new internal structure that still forms the geographic framework of most of the central cities of America's large metropolitan areas. We now examine the urban trend at both of these scales.

Macroscale Urbanization The rise of the national urban system in the late nineteenth century was based on the traditional external role of cities: providing goods and services for their hinterlands in exchange for raw materials. This function had been present since colonial times, with preindustrial cities located at the most accessible points on the existing transportation network so that movement costs could be minimized. Because handicrafts and commercial activities were already clustered in those cities, the emerging industrialization movement gravitated toward them. These urban centers also contained concentrations of labor and investment capital, provided a large market for finished goods, and possessed favorable transport and communications connections.

Importantly, these cities could also absorb the hordes of newcomers who would cluster by the thousands around new factories built within and just outside the municipal boundaries. Their constantly growing incomes, in turn, permitted industrially-intensifying cities to invest in a bigger local infrastructure of private and public services as well as housing—and thereby convert each round of industrial expansion into a new stage of urban development. Once generated, this whole process unfolded so quickly that planning was impossible; quite literally, America awoke one morning near the turn of the twentieth century to discover it had unexpectedly built a number of large cities.

The rise of the national urban system, unintended though it may have been, was a necessary by-product of industrialization, without which rapid U.S. economic development could not have taken place. Even though it emerged during the Industrial Revolution of the 1870–1910 period, the American urban system was in the process of formation for several decades preceding the Civil War. The evolutionary framework of the system from 1790 to 1970 is summarized within a model developed by John Borchert consisting of four **epochs of metropolitan evolution** based on transportation technology and industrial energy.

The preindustrial *Sail-Wagon Epoch* (1790–1830) was the first stage of development, marked by slow, primitive overland and waterway circulation. The leading cities were the northeastern ports of Boston, New York, and Philadelphia (none of which had yet emerged as the primate city), which were at least as oriented to the European overseas trade as they were to their still rather inaccessible western hinterlands—though the Erie Canal linking the coast and the Great Lakes was completed as the epoch came to a close.

Next came the *Iron Horse Epoch* (1830–1870), dominated by the arrival and spread of the steam-powered railroad, which steadily expanded its network from east to west until the transcontinental line was completed as the epoch ended. Accordingly, a nationwide transport system had been forged, coal-mining centers boomed (to keep locomotives running), and—aided by the easier movement of raw materials—small-scale urban manufacturing began to disperse outward from its New England hearth. The national urban system started to take shape: New York advanced to become the primate city by 1850, and the next level in the hierarchy was occupied by such booming new industrial cities as Pittsburgh, Detroit, and Chicago.

This economic/urban development process crystallized during the third stage—the *Steel-Rail Epoch* (1870–1920)—which encompassed the U.S. Industrial Revolution. Among the powerful forces now shaping the growth and full establishment of the national metropolitan system were the following: (1) the rise and swift dominance of the vital steel industry along the Chicago-Detroit-Pittsburgh axis (as well as its coal and iron ore supply areas in the northern Appalachians and Lake Superior district, respectively); (2) the increasing scale of manufacturing that necessitated greater agglomeration in the most favored raw-material and market locations for industry; and (3) the use of steel in railroad construction, which permitted significantly higher speeds,

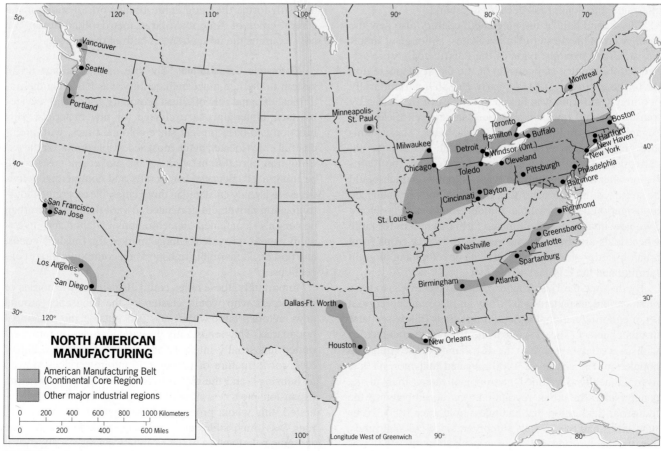

FIGURE 3-7

longer hauls of bulk commodities, and the time-space convergence of hitherto distant rail nodes.

The *Auto-Air-Amenity Epoch* (1920–1970) constituted the final stage of U.S. industrial urbanization and maturation of the national urban hierarchy. The key innovation was the gasoline-powered internal combustion engine, which permitted ever greater automobile- and truck-based regional and metropolitan dispersal. And as technological advances in manufacturing spawned the increasing automation of blue-collar jobs, the American labor force steadily shifted toward domination by white-collar personal and professional services to manage the industrial economy. This kind of productive activity responded less to traditional cost- and distance-based location forces and ever more strongly to the *amenities* or pleasant environments available in suburbia and the outlying Sunbelt states in a nation now fully interconnected by jet travel and long-distance communication networks.

The growth of the U.S. urban system and its industrial-based economy produced dramatic spatial changes as populations relocated to keep pace with shifting employment opportunities. The most notable regional transformation was the early-twentieth-century emergence of the continental *core area* or American Manufacturing Belt, which contained the lion's share of industrial activity in both the United States and Canada. As Fig. 3-7 shows, the geographic form of the core region—which includes Canada's southernmost Ontario—was a great rectangle whose four corners were Boston, Milwaukee, St. Louis, and Baltimore. However, because manufacturing is such a spatially concentrated activity, the core area should *not* be thought of as a continuous factory-dominated landscape. In truth, well under 1 percent of the territory of the Manufacturing Belt is actually devoted to industrial land use. Most of its mills and foundries are tightly clustered into a dozen districts that center on metropolitan Boston; Hartford-New Haven; New York-northern New Jersey; Philadelphia; Baltimore; Buffalo; Pittsburgh-Cleveland; Detroit-Toledo; Chicago-Milwaukee; Cincinnati-Dayton; St. Louis; and Ontario's Toronto-Hamilton-Windsor.

At the subregional scale, as transportation breakthroughs permitted progressive urban decentralization and **megalopolitan growth**, the expanding peripheries of major cities soon coalesced to form a number of conurbations (see p. 72). The most important of these by far was the *Atlantic Seaboard Megalopolis* (Fig. 3-8), the 600-mile (1,000-km) urbanized northeastern coastal strip extending from southern Maine to Virginia that contains metropolitan Boston,

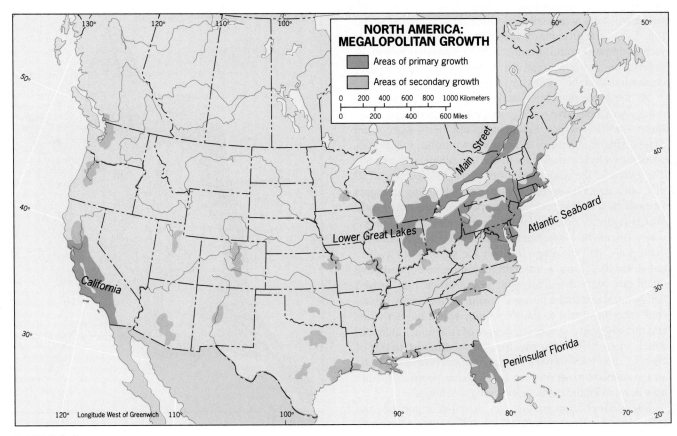

FIGURE 3-8

New York, Philadelphia, Baltimore, and Washington. This is the economic heartland of the U.S. core region, the seat of American government, business, and culture as well as the trading "hinge" between the United States and the Old World across the Atlantic. Three other primary conurbations also emerged—*Lower Great Lakes* (Chicago-Detroit-Pittsburgh), *California* (San Diego-Los Angeles-San Francisco), and *Peninsular Florida* (Jacksonville-Orlando-[Tampa]-Miami)—and are expected to continue growing, along with a number of secondary megalopolitan concentrations (Fig. 3-8). And note, for later reference, that north of the border the same forces created a nationally predominant conurbation linking Quebec City, Montreal, Toronto, and Windsor (called *Main Street*), capping a parallel episode of rapid industrial urbanization in Canada.

Microscale Urbanization The internal structure of the American metropolis reflected the same mixture of forces that shaped the national urban system. As an industrial city, however, it represented a departure from its European parentage. Whereas Europe's major cities were historically centers of political and military power—onto which industrialization was grafted almost as an afterthought—a large number of U.S. cities came into being as economic machines to produce the goods and services required to sustain an ongoing Industrial Revolution. Thus, right from the start, the performances of America's cities were judged mainly in terms of their profit-making abilities, and the less successful ones were callously discarded. The chief social function of the city was to receive and process foreign (as well as domestic rural) immigrants for assimilation into the mainstream American nativist culture, which increasingly concentrated in the rapidly growing suburbs after 1920.

At the intraurban microscale, too, transportation technology was a decisive force in shaping geographic patterns. Rails—in this instance, lighter street rail lines—once again shaped spatial structure, as horse-drawn trolleys were succeeded by electric streetcars in the late nineteenth century. The arrival of the automobile after World War I changed all that, and America began to turn from building compact cities to the widely dispersed metropolises of the post-World War II highway era. By 1970 the new intraurban expressway network had equalized location costs throughout the metropolis, and the stage was set for suburbia to swiftly transform itself from a residential preserve into a complete *outer city* with amenities and new prestige that proved highly attractive to the business world. As the newly urbanized suburbs started to capture major economic activities and thereby gain a surprising degree of functional independence, many large cities saw their status diminish to that of coequal, their once thriving central business districts (CBDs) all but reduced to serving the less affluent populations that now dominated the central city's neighborhoods.

Intrametropolitan growth can also be organized into **eras of intraurban structural evolution** based on trans-

portation, as decribed by John Adams (Fig. 3-9). The initial Stage I, the *Walking-Horsecar Era* (prior to 1888), entailed a pedestrian city, slightly augmented after 1850 when slow horse-drawn trolleys began to operate. Urban structure was dominated by compactness (everything had to be within a 30-minute walk), and very little land-use specialization could occur. The invention in 1888 of the electric traction motor, a device that could easily be attached to horsecars, launched the *Electric Streetcar Era* (1888–1920). Speeds of up to 20 miles (32 km) per hour enabled the 30-minute travel radius and the urbanized area to expand considerably along new outlying trolley corridors (Stage II in Fig. 3-9), spawning streetcar suburbs and helping to differentiate space within the older core city. In the latter, the CBD, industrial, and residential land uses emerged in their modern form (see Fig. 3-1).

Stage III was the *Recreational Automobile Era* (1920-1945), when the initial impact of cars and highways steadily improved the accessibility of the outer metropolitan ring, thereby launching a wave of mass suburbanization that

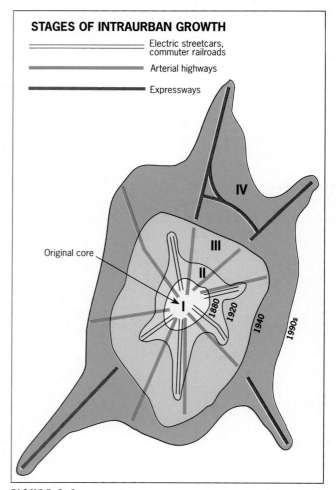

FIGURE 3-9

further extended the urban frontier (and the half-hour time-distance radial). Meanwhile, the still-dominant central city experienced its economic peak and the partitioning of its residential space into neighborhoods sharply defined by income level, ethnicity, and race. The final Stage IV, the *Freeway Era* (1945 to the present), saw the full impact of automobiles as high-speed expressways pushed the metropolitan-development and time-distance limits in certain sectors to more than 30 miles (50 km) from downtown—thereby propelling round after round of massive new suburbanization.

The social geography of the evolving industrial metropolis over the past 150 years has been marked by the development of a residential mosaic that exhibited the congregating of ever-more-specialized groups. Before the arrival of the electric streetcar, which introduced "mass" transit affordable by every social class, the heterogeneous city population had been unable to sort itself into ethnically uniform neighborhoods. Immigrants pouring into the industrializing cities were forced to reside within walking distance of their workplaces in crowded tenements and row houses, whose tiny apartments were literally filled in the order that their tenants arrived in town. Once the inner-city neighborhood pattern was able to form in the 1890s, the familiar residential rings, sectors, and nucleations quickly materialized (see Fig. 3-1).

When the United States essentially closed its doors to foreign immigration in the 1920s, industrial managers soon discovered the large African-American population of the rural Deep South—increasingly unemployed there as cotton-related agriculture declined—and began to recruit them by the thousands to work in the factories of Manufacturing Belt cities. This new migration had an immediate impact on the social geography of the industrial city because whites refused to share their living space with the racially different newcomers. The result was the involuntary segregation of these newest migrants, who were channeled into geographically separate, all-black areas that by 1945 became large expanding *ghettoes,* speeding the departure of many white central-city communities in the postwar era and helping to create a racially divided society.

The suburban component of the intraurban residential mosaic is generally home to the more affluent residents of the metropolis. The social makeup of its congregations is largely determined by income (i.e., housing prices), with the residential turf of the suburbs marked by an abundance of minor class- and status-related variables—perfectly suited to a society in which frequent upward social and spatial mobility go hand in hand.

Urbanization in Postindustrial America Since 1970

By 1970 the long arc of industrial urbanization had been traversed. In 1800 only 5 percent of the agriculture-dominated U.S. population lived in cities; today, less than 2 per-

cent live on the farms while over 90 percent reside within a two-hour drive of a metropolitan center. The rise of the postindustrial economy has coincided with a number of recent urban spatial changes. At the macroscale, while metropolitan growth has leveled off nationally, many Sunbelt and internationally connected cities—as well as newer suburbs everywhere—continue to expand. At the microscale, a new intraurban spatial structure has emerged as three decades of deconcentration have now turned the central city inside out, producing a metropolis so widely dispersed that large portions of it have subdivided into self-contained functional areas. We now examine current trends at each of these levels of generalization.

Macroscale Urbanization The highly complex current epoch that began in 1970 is defined by John Borchert as the *Satellite-Electronic-Jet Propulsion Epoch,* the fifth stage in his model of American metropolitan evolution. Propelled by advancements in communications, computer technologies, and long-distance travel, the key trends working to restructure the U.S. urban system today are: (1) increasing international interaction; (2) information- and management-related research and development; (3) the heightened lure of recreational amenities; (4) the constraints of more costly energy supplies; and (5) the cutback in military expenditures. The most successful metropolises are those with direct international air connections, overseas trade and business linkages, and thriving high-technology research communities; they include a number of urban regions along the West and East Coasts as well as Texas and the rest of the developed Mexican borderland zone. Among the now-disadvantaged, slow-growth metropolises are the older manufacturing centers, particularly those hitherto dependent on defense-industry contracts; not surprisingly, most of those cities are located in the American Manufacturing Belt.

Geographically, these changes of the past quarter-century—which appear to be hallmarks of our new postindustrial age—strongly reinforce the continuing rise of the Sunbelt. At the same time they signal that once-significant interregional differences in location costs are becoming equalized throughout the United States, especially for the white-collar, office-based industries. Noneconomic factors therefore now play a pivotal role in locational decision-making, with geographic prestige, local amenities, and proximity to recreational activities heading the list.

Perhaps no other place in the Sunbelt better symbolizes the totality of the post-1970 transformation than *Silicon Valley* (see photo p. 190). This is the headquarters of the U.S. microprocessor industry (which produces components for computers and electronic appliances), located near both San Francisco International Airport and Stanford University in the Bay Area's high-amenity Santa Clara valley. Silicon Valley continues to play a world leadership role, and its domestic success has also spawned a number of similar complexes in the outer cities of San Diego, Los Angeles,

The heart of the Silicon Valley research-and-development and manufacturing complex, the most important innovation center of its kind in North America.

Dallas, Miami-Ft. Lauderdale, Raleigh-Durham (N.C.), and Denver—as well as outside Boston, where the electronics industry was born in the 1940s.

Microscale Urbanization The most dramatic changes in postindustrial urban geography to date have occurred at the scale of the metropolis. The completion of the intraurban expressway system in effect destroyed the regionwide centrality advantage of the central city's CBD, making most

places on the freeway network just as accessible to the rest of the metropolis as only downtown had been before the 1970s. Industrial and commercial employers quickly realized that most of the advantages of being located in the CBD were now eliminated. Some companies in the office-based services sector chose to remain and even enlarge their presence downtown; but many others, together with myriad firms in every other sector, chose to respond to the new economic-geographic reality by voting with their

The landscape of suburbia is increasingly indoors. The Mall of America in Bloomington, Minnesota (opened in 1992) is a marriage of the shopping mall and theme park—and is by far the largest such mall in the United States. Why would such a leading facility be placed in the suburbs of the comparatively small Minneapolis-St. Paul metropolitan area? Because it serves not only the Twin Cities market, but a far wider population in the Upper Midwest, from the Great Plains to the Great Lakes.

South Coast Metro in Costa Mesa, California is one of the nation's largest suburban downtowns. This huge activity complex is located about 30 miles southeast of central Los Angeles in the heart of Orange County, which has been called the quintessential outer city of the late twentieth century. The office towers of neighboring Irvine, another sizeable suburban business center, can be seen in the background.

feet—or rubber tires—and headed for outlying sites. As early as 1973, the suburbs surpassed the central cities in total employment; by the mid-1990s, even Sunbelt metropolises were experiencing the suburbanization of a critical mass of jobs (greater than 50 percent of the urban-area total).

As the outer city grew rapidly, the volume and level of interaction with the central city constantly lessened and suburban self-sufficiency increased. This shift toward functional independence was heightened as new suburban nuclei sprang up to serve the new local economies, the major ones locating near key freeway interchanges. These multipurpose activity nodes developed around big regional shopping centers, whose prestigious images attracted scores of industrial parks, office campuses and high-rises, hotels, restaurants, entertainment facilities, and even major-league sports stadiums and arenas, which together formed burgeoning new **suburban downtowns** that are an automobile-age

version of the CBD (see photo above). As suburban downtowns flourished, they attracted tens of thousands of local residents to organize their lives around them—offering workplaces, shopping, leisure activities, and all the other elements of a complete urban environment—and thereby further loosened ties not only to the central city but also to other sectors of the suburban ring.

These spatial elements of the contemporary metropolis are assembled in the model displayed in Fig. 3-10, which should be regarded as an updating and extension of the classical models of intraurban structure (Fig. 3-1). The rise of the outer city has now produced a *multicentered* metropolis consisting of the traditional CBD plus a set of increasingly co-equal suburban downtowns, with each activity center serving a discrete and self-sufficient surrounding area. James Vance defined these new tributary areas as **urban realms**, recognizing in his studies that each such

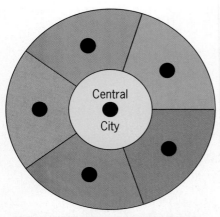

IDEAL FORM OF MULTICENTERED URBAN REALMS MODEL

FIGURE 3-10

realm maintains a separate, distinct economic, social, and political significance and strength. The urban realms of Los Angeles are mapped in Fig. 3-11; a similar regionalization scheme could easily be drawn for other large American metropolises.

The position of the U.S. central city within the new multinodal metropolis of realms is eroding. No longer the dominant metropolitanwide center for urban goods and services, the CBD increasingly serves the less affluent residents of the innermost realm and those working downtown. As inner-city manufacturing employment declined precipitously, many large cities adapted successfully by shifting toward the growing service industries. Accompanying this switch is downtown commercial revitalization, which has been widespread since 1970; but in many cities, for each shining new skyscraper that goes up, several old commercial buildings are abandoned. Moreover, in a number of Sunbelt cities, a whole forest of new office towers contains suburban commuters whose only contact with the CBD below is the short drive between the freeway exit and their buildings' parking garages.

Residential revitalization in and near the CBD has also occurred in many central cities since 1970. However, the number of people involved is not that significant; in fact, most reinvestment is undertaken by those already residing in the central city, so a "return-to-the-city" movement by suburbanites has not taken place. Such downtown-area neighborhood redevelopment involves *gentrification*—the upgrading of residential areas by new higher-income settlers. To succeed, however, this development usually requires the displacement of established lower-income residents—an emotional issue that has sparked many conflicts. Beyond the CBD zone, the vast inner city remains the problem-ridden domain of low- and moderate-income people, with most forced to reside in ghettoes. Financially ailing big-city governments are unable to fund adequate schools, crime-prevention programs, public housing, and sufficient social

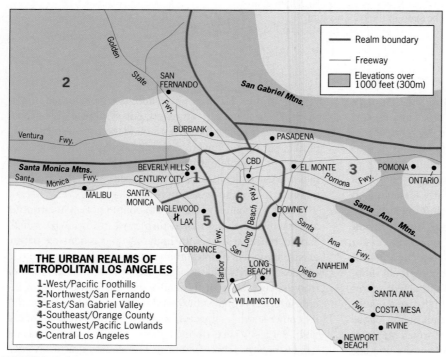

THE URBAN REALMS OF METROPOLITAN LOS ANGELES

1-West/Pacific Foothills
2-Northwest/San Fernando
3-East/San Gabriel Valley
4-Southeast/Orange County
5-Southwest/Pacific Lowlands
6-Central Los Angeles

FIGURE 3-11

The problems of America's older central cities are nowhere more vividly displayed than in Detroit. Renaissance Center in the background, built in the 1970s in the hope that CBD decline might be reversed, towers over streets where little remains of the vitality that once marked the urban life of the Motor City.

services, and so the downward spiral, including abandonment in the old industrial cities, continues unabated into the mid-1990s.

CULTURAL GEOGRAPHY

The United States belongs to one of the world's newer geographic realms, yet the contributions of a wide spectrum of immigrant groups over the past two centuries have shaped—and continue to shape—a rich and varied cultural mosaic. Great numbers of these newcomers were willing to set their original cultural baggage aside in favor of assimilation into the emerging culture of their adopted homeland, which itself was a hybrid nurtured by constant infusions of new influences. For most upwardly mobile immigrants, this plunge into the much-touted "melting pot" promised a ticket for acceptance into mainstream American society. Yet for millions of other would-be mainstreamers, the melting pot has proven to be a lumpy stew. Whether by choice or not, they continue to stick together in ethnic communities (especially within the inner cities of the Manufacturing Belt), with a large proportion of those neighborhoods inhabited by racial minorities who do not participate fully in the national culture.

American Cultural Bases

As the nativist culture of the United States matured, it came to develop a set of powerful values and beliefs: (1) love of newness; (2) a desire to be near nature; (3) freedom to move; (4) individualism; (5) societal acceptance; (6) aggressive pursuit of goals; and (7) a firm sense of destiny. Brian Berry

has discerned these cultural traits in the behavior of people throughout the evolution of urban America. A "rural ideal" has prevailed throughout U.S. history and is still expressed in a strong bias against residing in cities. When industrialization made urban living unavoidable, those able to afford it soon moved to the emerging suburbs (*newness*) where a form of country life (*close to nature*) was possible in a semi-urban setting. The fragmented metropolitan residential mosaic, composed of a myriad of slightly different neighborhoods, encouraged frequent *unencumbered mobility* as middle-class life revolved around the *individual* nuclear family's *aggressive pursuit* of its aspirations for *acceptance into the next higher stratum of society*. These accomplishments confirmed to most Americans that their goals could be attained through hard work and perseverance and that they possessed the ability to realize their *destiny* by achieving the "American Dream" of homeownership, affluence, and total satisfaction.

Language

Although linguistic variations play a far more important role in Canada, no less than one-eighth of the U.S. population spoke a primary language other than English in 1990. Differences in English usage are also evident at the subnational level in the United States, where regional variations (*dialects*) are still widespread, despite the recent trend toward a truly national society. The South and New England immediately come to mind as areas that still possess distinctive accents. An even closer connection between language and landscape is established through *toponymy,* the naming of places. U.S. place-name geography provides important clues to the past movements of cultural influences and national groups: for instance, the preponderance of such Welsh place names as Cynwyd, Bryn Mawr, and Uwchlan just to the west of Philadelphia is the final vestige of an erstwhile colony of Celtic-speaking settlers from Wales.

Religion

North America's Christian-dominated kaleidoscope of religious faiths contains important spatial variations. Many major Protestant denominations are clustered in particular regions, with Baptists localized in the southeastern quadrant of the United States, Lutherans in the Upper Midwest and northern Great Plains, and Mormons in Utah and southern Idaho. Roman Catholics are most visibly concentrated in two locations: (1) the Manufacturing Belt metropolises and nearby New England, which received huge infusions of Catholic Europeans over the past century; and (2) the entire Mexican borderland zone that is home to a burgeoning Hispanic-American population. Judaism is the nation's most highly agglomerated major religious group; its largest con-

gregations are clustered in the cities and suburbs of Megalopolis, Southern California, South Florida, and the Midwest.

Ethnicity

Ethnicity (which means nationality) was a decisive influence in the shaping of American culture, which in turn also reshaped the cultural traditions of newcomers who were assimilated into mainstream society. The current U.S. ethnic tapestry is, as always a complex mosaic (see box at right). The spatial distribution of ethnic minorities is mapped in Fig. 3-12, which also includes the Native American population of the conterminous United States. Because the latter largely occupy tribal lands on reservations ceded by the federal government, they are undergoing very little distributional change. The Hispanic population, on the other hand, is growing rapidly through in-place natural increase as well as in-migration from Middle America (much of it illegal). Spatially, Hispanics are both increasing in density along the southwestern border and fanning out toward large metropolitan areas to the north and east. Blacks are not as mobile

SPATIAL DISTRIBUTION OF ETHNIC MINORITIES

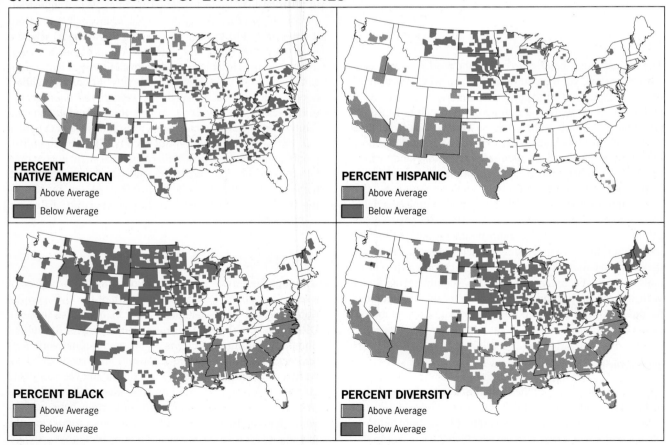

PERCENT NATIVE AMERICAN
- Above Average
- Below Average

PERCENT HISPANIC
- Above Average
- Below Average

PERCENT BLACK
- Above Average
- Below Average

PERCENT DIVERSITY
- Above Average
- Below Average

FIGURE 3-12

THE ETHNIC TAPESTRY OF THE UNITED STATES

The diversity and complexity of America's ancestral makeup was reported by the 1990 census, which showed that nearly 85 percent of the U.S. population identified with one of more than 140 different national backgrounds. There were more Americans of full or partial English descent than the total population of England, and more than one-half as many Americans of German stock as currently reside in Germany; the Irish-Americans outnumber the population of the Republic of Ireland by a ratio of almost 13 to 1. In addition, so many Americans listed themselves as African-Americans that only four Subsaharan African states recorded a larger national population total. Other significant ethnic ancestries include French, Italian, Scottish, Polish, and Mexican. Those of Native American background accounted for 0.8 percent of the U.S. total.

Geographically, the dominant European ancestries—French, Irish, and those originating in the United Kingdom—were dispersed throughout the United States. Most of the others showed an affinity for a particular region: Italians, Portuguese, and Russians clustered in the Northeast, whereas Scandinavians and Czechs were localized in the north-central states. California, the most populous state, best exhibited the nation's diverse ethnicity (see Fig. 3-12), with more Americans of English, German, Irish, French, Scottish, Dutch, Swedish, and Danish origin concentrated there than in any other state.

Although the African-American population remains fairly stable in relative size (about 12 percent of the U.S. total), two other minority groups—Asians and Hispanics—are substantially increasing their presence in the national ethnic tapestry through in-place growth and sizeable immigration. During the 1980s, the number of Asians and Pacific Islanders more than doubled, for the most part remaining clustered in the urban areas of the Far West. The biggest single Asian group today is the Chinese, followed by the Filipinos, Japanese, Asian Indians, Koreans, and Vietnamese. The Hispanic population, which is expected to surpass the black population to become the nation's largest minority by 2005, grew by nearly 55 percent between 1980 and 1990. Disaggregated by national background, the Mexican component accounts for 60 percent of the Hispanic population, Central and South Americans 15 percent, Puerto Ricans 12 percent, and Cubans 5 percent. As Fig. 3-12 indicates, Hispanics are most heavily concentrated in the southwestern quadrant of the United States.

as Hispanics, undoubtedly constrained by racially segregated housing markets. Nonetheless, upwardly mobile blacks are increasing their presence both in suburbia and, via return migration, in the more affluent areas of the South—although settling in communities that remain highly segregated. The newest elements in the ethnic tapestry (as indicated in the box above) are Asian-Americans, who heavily cluster in major West Coast metropolitan areas.

The Mosaic Culture

American cultural geography continues to evolve. What is now taking place is a new fragmentation into a *mosaic culture,* an increasingly heterogeneous complex of separate, uniform "tiles" that cater to more specialized groups than ever before. No longer based solely on such broad divisions as income, race, and ethnicity, today's residential communities of interest are also forming along the dimensions of age, occupational status, and especially lifestyle. Their success and steady proliferation reflect an obvious satisfaction on the part of a majority of Americans. Yet such balkanization—fueled by people choosing to interact only with others exactly like themselves—threatens the survival of important democratic values that have prevailed throughout the evolution of U.S. society.

THE CHANGING GEOGRAPHY OF ECONOMIC ACTIVITY

The economic geography of the United States in the 1990s is the product of all of the foregoing, as bountiful environmental, human, and technological resources have cumulatively blended together to create one of the world's most advanced economies. Perhaps the greatest triumph was overcoming the "tyranny" of distance, as people and activities were organized into a continentwide spatial economy that took maximum advantage of agricultural, industrial, and urban development opportunities. Yet despite these past achievements, American economic geography today is once again in upheaval as the transition is completed from industrial to postindustrial society.

Major Components of the Spatial Economy

Economic geography is mainly (though not exclusively) concerned with the locational analysis of productive activities. Four major sets may be identified:

1. **Primary activity**: the extractive sector of the economy in which workers and the environment come into direct contact, especially *mining* and *agriculture*.
2. **Secondary activity**: the *manufacturing* sector, in which raw materials are transformed into finished industrial products.
3. **Tertiary activity**: the *services* sector, including a wide range of activities—from retailing to finance to education to routine office-based jobs.
4. **Quaternary activity**: today's dominant sector, involving the collection, processing, and manipulation of information; a subset, sometimes referred to as **quinary activity**, is managerial or control-function activity associated with decision-making in large organizations.

Historically, each of these activities has successively dominated the American labor force for a time over the past 200 years, with the quaternary sector now dominating the economy for the foreseeable future. Agriculture was dominant until late in the nineteenth century, giving way to manufacturing by 1900. The steady growth of services after 1920 finally surpassed manufacturing in the 1950s but now shares a dwindling portion of the limelight with the still-rising quaternary sector. The approximate breakdown by major sector of employment in the U.S. labor force today is agriculture, 2 percent; manufacturing, 15 percent; services, 18 percent; and quaternary, 65 percent (with about 10 percent in the quinary sector). We now sequentially treat these major productive components of the spatial economy in the following coverage of resource use, agriculture, manufacturing, and the postindustrial revolution.

Resource Use

The United States (and Canada) was blessed with abundant deposits of industrial and energy resources. Fortunately, these were usually concentrated in sufficient quantities to make long-term extraction an economically feasible proposition, and most of the richest raw material sites are still the scene of major drilling or mining operations. Moreover, the continental and offshore mineral storehouse may yet contain outstanding resources for future exploitation.

Industrial Mineral Resources   North America's rich mineral deposits are localized in three zones: the Canadian Shield north of the Great Lakes, the Appalachian Highlands, and scattered areas throughout the mountain ranges of the West. The Shield's most noteworthy minerals are iron ore (Minnesota's Mesabi Range just west of Lake Superior and the eastern Shield in Quebec and Labrador); nickel (along Lake Huron's north shore); and gold, uranium, and copper (northwestern Canada). Besides vast deposits of soft (bituminous) coal, the Appalachian region also contains hard (anthracite) coal in northeastern Pennsylvania and iron ore in central Alabama. The western mountain zone contains significant deposits of coal, copper, lead, zinc, molybdenum, uranium, silver, and gold.

Fossil Fuel Energy Resources   The realm's most strategically important resources are its coal, petroleum (oil), and natural gas supplies—the *fossil fuels*, so named because they were formed by the geologic compression and transformation of plant and tiny animal organisms that lived hundreds of millions of years ago. These energy supplies, mapped in Fig. 3-13, reveal abundant deposits and distribution networks, particularly in the conterminous United States and Alaska.

The realm's *coal* reserves are among the greatest anywhere on earth, the U.S. portion alone containing at least a 400-year supply. Three coal regions are evident: (1) Appalachia, still the largest region but declining steadily because its high-sulfur coal must be expensively treated to meet federal clean-air standards; (2) the Western coal region, centered on the northwestern Great Plains (Wyoming is now the leading coal-producing state), which continues to expand because its vast, low-sulfur, near-surface coal beds can easily be strip-mined; and (3) the Midcontinent coalfields, an arc of high-sulfur, strippable deposits centered on southern Illinois and western Kentucky, which is also declining as the West surges.

The major *oil*-production areas of the United States are located along and offshore from the Texas-Louisiana Gulf Coast; in the Midcontinent district, extending through western Texas-Oklahoma-eastern Kansas; and along Alaska's central North Slope facing the Arctic Ocean. Lesser oilfields are also found in Southern California (major untapped reserves lie offshore), west-central Appalachia, the northern Great Plains, and the southern Rockies.

The distribution of *natural gas* deposits resembles the geography of oilfields because petroleum and energy gas are usually found in similar geologic formations (the floors of ancient shallow seas). Accordingly, major gasfields are located in the Gulf, Midcontinent, and Appalachian districts. However, when subsequent geologic pressures are exerted on underground oil, the liquid is converted into natural gas. This has happened frequently in mountainous zones so that western gas deposits in and around the Rockies tend to stand apart from oilfields.

Agriculture

Despite the twentieth-century emphasis on urbanization and the development of the nonprimary sectors of the spa-

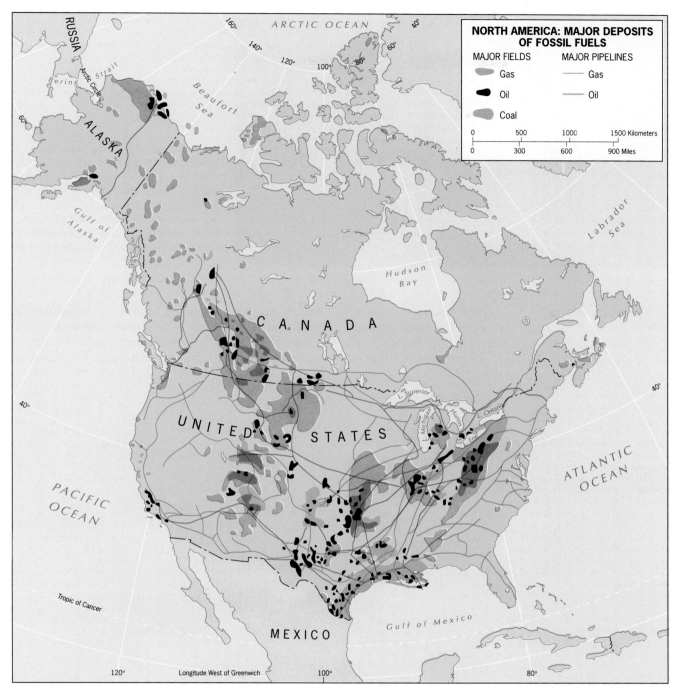

FIGURE 3-13

tial economy, agriculture remains an important element in America's human geography. Because it is the most extensive (space-consuming) economic activity, vast expanses of the U.S. (and Canadian) landscape are clothed with fields of grain. Moreover, great herds of livestock are sustained by pastures and fodder crops because this wealthy realm can afford the luxury of feeding animals from its farmlands—and demands huge quantities of red meat in its diet. The increasing application of high-technology mechanization to farming has steadily increased both the volume and value of total agricultural production and has been accompanied by a sharp reduction in the number of those actively engaged in agriculture (in 1990, only a dwindling 1.9 percent of the U.S. population still lived on farms).

Soybeans are of growing importance in what used to be the nearly exclusive Corn Belt. Here a new, no-till farming method involves the rotation of corn and soybeans: the corn is being sown directly into soybean stubble from the previous year's harvest. The soil is not ploughed; liquid fertilizer is injected as the corn is inserted.

The regionalization of U.S. agricultural production is shown in Fig. 3-14, its spatial organization developing largely within the framework of the Von Thünen model (see pp. 55–56). As in Europe (Fig. 1-4), the early nineteenth-century, original-scale model of town and hinterland—in a classical demonstration of time-space convergence—expanded outward (with constantly improving transportation technology) from a locally "isolated state" to encompass

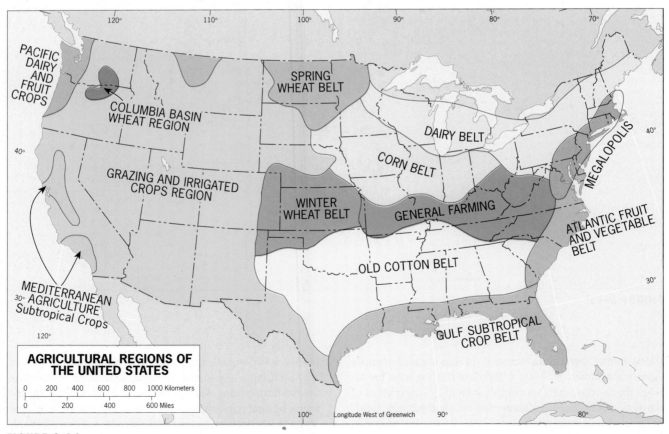

FIGURE 3-14

the entire continent by 1900. As this macrogeographic structure formed, the greatly enlarged original Thünian production zones (Fig. 1-3) were modified: (1) the first ring was now differentiated into an inner fruit/vegetable zone and a surrounding belt of dairying; (2) the forestry ring was displaced to the outermost limits of the regional system because railroads could now transport wood quite cheaply; (3) the field crops ring subdivided into an inner, mixed crop-and-livestock ring to produce meat (the Corn Belt, as it came to be called) and an outer zone that specialized in the mass production of wheat grains; and (4) the ranching area remained in its outermost position, a grazing zone that supplied young animals to be fattened in the meat-producing Corn Belt as well as supporting an indigenous sheep-raising industry. The "supercity" anchoring this macro-Thünian regional system was the northeastern Megalopolis, well on its way toward coalescence and already the dominant food market and transport focus of the entire country.

Although the circular rings of the model are not apparent in Fig. 3-14, many spatial regularities can be observed (remember that Von Thünen, too, applied his model to reality and thereby introduced several distortions of the theoretically ideal pattern). Most significant is the sequence of farming regions as distance from the national market increased, especially westward from Megalopolis toward central California, which was the main directional thrust of the historic inland penetration of the United States. The Atlantic Fruit and Vegetable Belt, Dairy Belt, Corn Belt, Wheat Belts, and Grazing Region are indeed consistent with the model's logical structure, each zone successively farther inland astride the main transcontinental routeway. Deviations from the scheme may be attributed to irregularities in the environment and/or to unique conditions. For example, the nearly year-round growing seasons of California and the Gulf Coast-Florida region permit those distant areas (with the help of efficient refrigerated transport) to produce fruits and vegetables in competition with the regional system's innermost zone.

Manufacturing

The geography of North America's industrial production has long been dominated by the Manufacturing Belt (Fig. 3-7). As noted earlier, the internal structure of the Belt is organized around a dozen urban-industrial districts interlinked by a dense transportation network. This highly agglomerated spatial pattern developed because of the economic advantages of clustering industries in cities, which contained the largest concentrations of labor and capital, constituted major markets for finished goods, and were the most accessible nodes on the national transport network for assembling raw materials and distributing manufactures. As these industrial centers expanded, they also achieved

economies of scale, savings accruing from large-scale production in which the cost of manufacturing a single item was further reduced as individual companies mechanized assembly lines, specialized their work forces, and purchased raw materials in huge quantities.

This efficient production pattern served the nation well throughout the remainder of the industrial age. Because of the effects of **historical inertia**—the need to continue using hugely expensive manufacturing facilities for their full, multiple-decade lifetimes to cover initial investments—the Manufacturing Belt is not about to disappear for a long time to come. However, as aging factories are retired, a process well underway, the distribution of American industry will change. As transportation costs even out among U.S. regions, as energy costs now favor the south-central oil- and gas-producing states, as high-technology manufacturing advances reduce the need for lesser-skilled labor, and as locational decision-making intensifies its attachment to noneconomic factors, industrial management has increasingly demonstrated its willingness to relocate to regions it perceives as more desirable in the South and West.

The Decline of Traditional Industry   The decline of traditional heavy industry has been a major trend in the Manufacturing Belt since 1980. One recent study estimates that smokestack-industry employment will plunge from 20 percent of the U.S. work force in 1980 to 8 percent by 1995. Within the Belt today, problems of obsolescence, economic decline, and foreign competition are widespread. The region's aging physical plant is seldom worth reinvesting in, and factory closures are regularly reported in the evening newscasts; on the blighted landscape of the inner industrial city, in particular, the abandoned manufacturing complex is a depressing, all-too-familiar sight. Competing high-quality industrial imports from abroad have tripled their share of the U.S. market during the past two decades. Once based exclusively on cheap labor, America's chief foreign competitors in the western Pacific Rim and Europe now possess superior technologies and productive efficiencies in a growing list of industries. (The Japanese, for instance, use five times as many robots than the Americans in running some of the world's most sophisticated factories.)

The reduced global position of the United States is well illustrated in the case of steel, one of the most basic heavy industries and the very foundation on which the Manufacturing Belt was built. The world leader in production as recently as the late 1960s, American steelmaking then fared so poorly that by 1986 sales had declined by over 40 percent and employment by nearly 60 percent. That decline was due to lowered worldwide demand and the steady rise of steel imports (whose lower prices found many U.S. customers). Yet the American steel industry also fell behind its competitors because it was too slow to modernize its old-fashioned manufacturing methods.

Factories in what is now called the Rust-belt have closed in large numbers, idling many thousands of workers and devastating hundreds of towns. This plant in Youngstown, Ohio is among the casualties of changing economic times—and places.

The Revitalization Effort Belatedly, U.S. manufacturers are learning from their pacesetting foreign rivals. A major effort has been underway since the late 1980s to create high-technology "factories of the future," with much of this activity concentrated in the Midwest portion of the Manufacturing Belt. High-skilled jobs have been created by the tens of thousands in research and development and new equipment manufacturing, emphasizing robotics and other state-of-the-art automation technologies. Significantly, the steel industry has been in the forefront, with many companies doubling their productivity over the last 10 years; among the most successful ventures have been nonunionized mini-mills built by smaller firms, which manufacture steel by using electric power to melt down scrap metal instead of beginning with ore and coal.

There is, however, a downside to this progress: the more successful high-tech manufacturing becomes, the fewer the factory jobs to be filled. Thus it is *not* the American manufacturing sector that is declining—industrial production in the early 1990s was more than 60 percent higher than in 1970. Rather, it is the rapid shrinkage of blue-collar employment opportunities that has caused most of the pain,

The trend toward modernization of the steel industry is not limited to the Manufacturing Belt: this is the North Star Steel Company's new mini-mill at Beaumont on the Gulf Coast of Texas.

THE LINEAMENTS OF POSTINDUSTRIAL SOCIETY

In his classic work, *The Coming of Postindustrial Society*, sociologist Daniel Bell sketched the distinguishing features of the emerging economy and its impact on American life. The new society and economy is marked by a fundamental change in the character of technology use—from fabricating to processing—in which telecommunications and computers are vital for the exchange of knowledge. Information becomes the basic product, the key to wealth and power, and is generated and manipulated by an intellectual rather than a machine technology. Yet postindustrial society does not displace industrial society, just as manufacturing did not obliterate agriculture. Instead, in Bell's words, "the new developments overlie the previous layers, erasing some features and thickening the texture of society as a whole."

Several hallmarks of postindustrialism can be identified. Knowledge is central to the functioning of the economy, which is led by science-based and high-tech industry. The technical/professional or "knowledge" class increasingly dominates the work force: quaternary and quinary activity already employed 40 percent of the U.S. labor force in 1975, a proportion that has surpassed 65 percent in the 1990s. The nature of work in the information society focuses on person-to-person interaction rather than person-product or person-environment contacts. With the eclipsing of manufacturing, more and more women participate in today's labor force. A meritocracy now prevails in which individual advancement is based on education and acquired job skills.

Among the many spatial implications of this socio-economic transformation, Bell observed that postindustrial occupations would gravitate toward five major types of "situses" or locations: economic enterprises, government, universities, social-service complexes, and the military. Central North Carolina's well-known *Research Triangle Park*, located near the center of the "triangle" circumscribed by the nearby cities of Raleigh, Durham, and Chapel Hill, is an outstanding example of a prestigious Sunbelt high-technology manufacturing and research complex. Even a small sampling of the list of tenants fully supports Bell's hypothesis: (1) among its business enterprises are such corporate giants as IBM, TRW, and the headquarters of Burroughs Wellcome Pharmaceuticals; (2) government is represented by the U.S. Environmental Protection Agency, the Southern Growth Policies Board, and North Carolina's Board of Science and Technology; (3) three local universities are prime movers in the Park's research operations—Duke, North Carolina, and North Carolina State—which have also attracted the National Humanities Center and the Sigma Xi Scientific Research Society; (4) social services are the business of the National Center for Health Statistics Laboratory and the International Fertility Research Program; and (5) the military presence is embodied by the U.S. Army Research Office.

a still-intensifying human problem because the dynamics driving the postindustrial revolution continue to mandate change in the structure and composition of the U.S. labor force.

The Postindustrial Revolution

The signs of postindustrialism are visible throughout the United States (and Canada) today, and they are popularly grouped under such names as "the information society" and "the electronic era." The term *postindustrial* by itself, of course, tells us mainly what the theme of the American economy is no longer; yet the term is also used by many social scientists to refer to a specific set of societal traits that signal an historic break with the recent past (see box above). Many of the urban spatial expressions of the currently transforming society have already been highlighted; we focus here on some broader economic-geographical patterns.

High-technology activities are the leading growth industries of the postindustrial economy. Most are spatially footloose and therefore quite sensitive to the noneconomic locational forces discussed on p. 189. Increasingly, their location decision-making is dominated by the same concerns that would govern a person's choice of residence if he or she had vast financial resources to draw upon: finding a site and building that maximizes geographic prestige, environmental amenities, commuting convenience, and access to recreational activities.

Northern California's Silicon Valley epitomizes the blend of locational qualities that attract a critical mass of high-tech companies to a given locality. According to a recent study there, they include: (1) a nearby major university that offers an excellent graduate engineering program; (2) close proximity to a cosmopolitan urban center; (3) a large local pool of skilled and semi-skilled labor; (4) year-round pleasant weather and recreational opportunities; (5) affordable nearby housing as well as luxury housing for executives; and (6) a local economic climate especially conducive to corporate prosperity. Their glamour notwithstanding, there are no guarantees concerning the stability of certain high-tech industries, which appear to be vulnerable to economic downturns and foreign competition. Silicon Valley experienced such a "shaking out" in the late 1980s, when overexpansion and fierce competition from Japanese and South Korean semiconductor manufacturers resulted in numerous business failures and the loss of thousands of jobs.

The geographic impact of the postindustrial revolution has been uneven over the past quarter-century. States offering high-amenity environments have generally fared best, especially Florida, Texas, California, North Carolina, Virginia, and those of the Pacific Northwest. Not surprisingly, the Manufacturing Belt states as a group performed weakly, although the Midwest has enjoyed a resurgence since 1985. Most interesting has been the overall performance of the highly touted Sunbelt.

The notion that the southern tier of states is a uniformly booming area is a myth, and stark contrasts in development abound. Even at the state level, significant internal variations are obvious, and all too often the burgeoning economy of one county leaves a dormant neighbor completely untouched. It is worth pointing out, too, that even prosperous corners of the Sunbelt are not immune to chang-

ing economic currents—as Texas, Oklahoma, and Louisiana discovered when the oil industry entered a long slump in the early 1980s. Yet however spotty the overall economic development pattern, it is still very likely that many Sunbelt metropolises will continue to capture an expanding share of major U.S. business activity.

◆ CANADA

Like the United States, Canada is a federal state, but it is organized differently. Canada is divided administratively into 10 provinces and two federal territories (Fig. 3-15). The two latter, the Yukon and the Northwest Territories,* together occupy a massive area half the size of Australia—but are inhabited by fewer than 90,000 people. The 10 provinces range in territorial size from tiny, Delaware-sized Prince Edward Island to sprawling Quebec, more than twice

---

*As Fig. 3-15 indicates, a third Territory, Nunavut, lies to the east of the Northwest Territories. In 1992, Nunavut, comprising one-fifth of Canada's land, was carved out of the much bigger, pre-existing Northwest Territories; it is now in the process of becoming a full self-governing entity, homeland to an Inuit (formerly Eskimo) Amerindian population of less than 20,000.

Vancouver, leading city of Canada's British Columbia, has been favored by its location on the Pacific Rim. Large numbers of Asian immigrants (many recently from Hong Kong) have come to the Vancouver area to settle, strengthening the city's ties to its cross-Pacific trading partners.

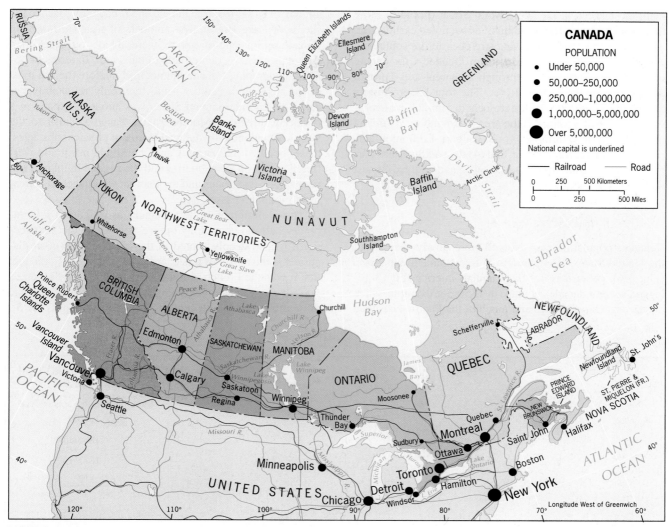

FIGURE 3-15

the area of Texas. Beginning in the east, the four provinces that face the Atlantic—Nova Scotia, New Brunswick, Prince Edward Island, and Newfoundland—are known as the Maritime Provinces. To their west lies Quebec, and to *its* west Ontario, Canada's two biggest provinces. Most of central Canada is covered by the three Prairie Provinces—Manitoba, Saskatchewan, and Alberta. In the far west, beyond the Canadian Rockies and facing the Pacific, lies the tenth province, British Columbia.

In terms of population (in descending order), Ontario (10.2 million) and Quebec (7.0 million) are again the leaders; British Columbia ranks third with 3.4 million; next come the three Prairie Provinces of Alberta, Manitoba, and Saskatchewan, with a combined total of 4.7 million; the Maritime Provinces of Nova Scotia, New Brunswick, Newfoundland, and Prince Edward Island are the four smallest, together containing 2.4 million. The sparsely populated federal territories in the far north contain only about

88,000 residents. Canada's total population of 27.8 million is only slightly larger than one-tenth the size of the U.S. population. Spatially, as we noted earlier in the chapter, most Canadians reside within 200 miles (325 km) of the United States border—reminding us that north-south distinctions are of far greater importance than an east-west geographical breakdown of Canada.

POPULATION IN TIME AND SPACE

The map showing the distribution of Canada's population (Fig. 3-5, p. 183) immediately tells us that only a fraction of this enormous country is effectively settled—about one-eighth, to be more precise. Even within that most heavily populated strip along the southern border, people tend to cluster in a pattern that resembles a strung-out island chain in a vast ocean. Five such clusters can be identified on the map, the largest by far being Main Street (home to more than 6 out of every 10 Canadians); as we noted earlier (Fig. 3-8), this is the conurbation that stretches across southeasternmost Quebec and Ontario, from Quebec City on the lower St. Lawrence River through Montreal and Toronto to Windsor on the Detroit River. The four lesser clusters are: (1) the Saint John-Halifax crescent in central New Brunswick and Nova Scotia; (2) the area surrounding Winnipeg in southeastern Manitoba; (3) the Calgary-Edmonton corridor in south-central Alberta; and (4) the southwestern corner of British Columbia, focused on Canada's third-largest metropolis, Vancouver.

Pre-Twentieth-Century Canada

Canada's population map evolved more slowly than that of the United States. Compared to U.S. settlement expansion (Fig. 3-6), penetration of the Candian interior (Fig. 3-16) lagged several decades behind; in 1850, for example, when the eastern United States and much of the Pacific coast had

already been settled, Canada's frontier had only reached the shores of Lake Huron. The numbers of westward-moving Canadian pioneers, of course, were always far fewer than the totals of their counterparts to the south. In terms of political geography, Canada did not unify before the last third of the nineteenth century—and then only reluctantly because of fears the United States was about to expand in a northerly direction. Moreover, the Canadian push to the west was undoubtedly delayed by major physical obstacles; these included the barren Canadian Shield north and west of Lakes Huron and Superior and the rugged highlands between the Great Plains and the Pacific, which lacked convenient mountain passes.

The evolution of modern Canada, as well as its contemporary cultural geography (as we will see in the next section), is also deeply rooted in the bicultural division discussed at the outset of this chapter. Its origin lies in the fact that it was the French, not the British, who first colonized present-day Canada, beginning in the 1530s. *New France*, during the seventeenth century, grew to encompass the St.

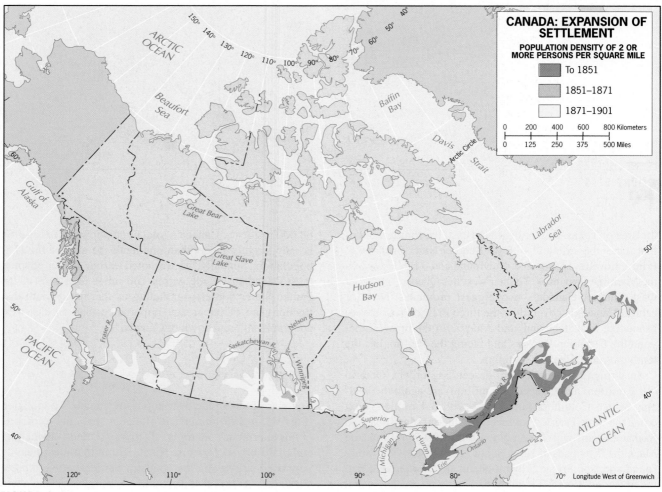

FIGURE 3-16

Lawrence Basin, the Great Lakes region, and the Mississippi valley. In the 1680s a series of wars between the English and French began, ending with France's defeat and the cession of New France to Britain in 1763. By the time London took control of this new possession, the French had made considerable progress in their North American domain. French laws, the French land-tenure system, and the Roman Catholic church prevailed, and substantial settlements (including Montreal on the St. Lawrence) had been established. The British, anxious to avoid a war of suppression and preoccupied with problems in their American colonies to the south, gave former French Quebec—the region extending from the Great Lakes to the mouth of the St. Lawrence—the right to retain its legal and land-tenure systems as well as freedom of religion.

After the American War for Independence, London was left with a region it called British North America (the name Canada—derived from an aboriginal word meaning settlement—was not yet in use), whose cultural imprint still was decidedly French. The war drove many thousands of English refugees northward, and soon there were difficulties between them and the French in British North America. In 1791, heeding appeals by these new settlers, the British Parliament divided Quebec into two provinces: Upper Canada, the region upstream from Montreal centered on the north shore of Lake Ontario, and Lower Canada, the valley of the St. Lawrence. Upper and Lower Canada became, respectively, the provinces of Ontario and Quebec (Fig. 3-17). By parliamentary plan, Ontario would become English-speaking and Quebec would remain French-speaking.

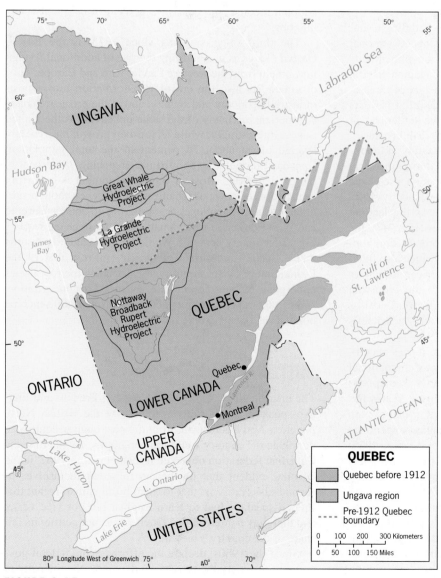

FIGURE 3-17

This earliest politico-cultural partitioning did not work well, and in 1840 the British Parliament tried again. This time it reunited the two provinces in a so-called Act of Union under which Upper and Lower Canada would have equal representation in the provincial legislature. That, too, was a failure, and efforts to find a better system finally led to the 1867 British North America Act that established the Canadian federation (consisting initially of Upper and Lower Canada, New Brunswick, and Nova Scotia, later to be joined [between 1870 and 1949] by the other provinces and territories). Under the 1867 Act, Ontario and Quebec were once again separated, but this time Quebec was given important guarantees: the French civil code was left unchanged, and the French language was protected in parliament and in the courts.

Twentieth-Century Canada

By the close of the nineteenth century, the fledgling Canadian federation was making major strides toward regional development and the spatial integration of a continentwide economy. In 1886, the transcontinental Canadian Pacific Railway had finally been completed to Vancouver, a condition of British Columbia's entry into the federation 15 years earlier. This new lifeline soon spawned the settlement of not only the far west but also the fertile Prairie Provinces, whose wheat-raising economy expanded steadily as immigrants from the east and abroad poured in during the early years of this century (Fig. 3-16). Industrialization also began to increasingly affect Canada at this time, and by 1920 manufacturing had surpassed agriculture as the leading source of national income. As noted earlier (Fig. 3-7), the dominant zone of industrial activity is the Toronto-Hamilton-Windsor corridor of southernmost Ontario (which also functions as one of the 12 districts of the American Manufacturing Belt); that has been the case since the Industrial Revolution took hold in Canada during World War I (1914–1918), when the country was heavily involved in Britain's war effort on the European mainland.

As industrial intensification proceeded, it was accompanied by the expected parallel maturation of the national urban system. Along the lines of Borchert's five epochs of American metropolitan evolution, Maurice Yeates has constructed a similar multistage model of urban-system development that divides the briefer Canadian experience into three eras. The initial *Frontier-Staples Era* (prior to 1935) encompasses the century-long transition from a frontier-mercantile economy to one oriented to staples (production of raw materials and agricultural goods for export), with increasing manufacturing activity in the budding industrial heartland. By 1930, Montreal and Toronto, reflecting their different cultural constituencies, had emerged as the two leading cities atop the national urban hierarchy (thus Canada has no single primate city).

Next came the *Era of Industrial Capitalism* (1935–1975), during which Canada achieved U.S.-style prosperity. However, most of this development—involving the tremendous growth of manufacturing, tertiary activities, and urbanization—took place after 1950 because the Great Depression lasted until World War II, which was followed by a long recession. A major stimulus was the investment of U.S. corporations in Canadian branch-plant construction, especially in the automobile industry in Ontario near the automakers' Detroit-area headquarters. In western Canada, the rapid growth of oil and natural gas production fueled Alberta's urban development, but new agricultural technologies reduced farm labor needs and sparked a rural outmigration in neighboring Saskatchewan and Manitoba. The postwar period also saw the ascent of Main Street, which on less than 2 percent of Canada's land quickly came to contain more than 60 percent of its people, two-thirds of its national income, and nearly 75 percent of its manufacturing jobs.

The third stage, ongoing since 1975, is the *Era of Global Capitalism*, signifying the rise of additional foreign investment from the western Pacific Rim and Europe. This, of course, is also the era of transformation into a postindustrial economy and society, and in the process Canada is experiencing many of the same upheavals as the United States. Interestingly, while U.S. urban growth has recently leveled off at about 75 percent of the total population, Canada still continues to urbanize (reaching 78 percent in the early 1990s). Most of this development is occurring in the form of new suburbanization. Yet even though suburban business complexes are springing up outside Toronto and West Edmonton's megamall continues to be the world's biggest shopping center, it would be a mistake to presume that Canada's intraurban geography mirrors that of the United States. On the contrary, the Canadian metropolis has undergone far less automobile-generated deconcentration, and the central city has retained most of its middle-income population and economic base.

CULTURAL GEOGRAPHY

The historic cleavage between Canada's French- and English-speakers, supposedly resolved by the British North America Act, has resurfaced in the past quarter-century to dominate the country's cultural geography. By the time the Canadian federation observed its centennial in 1967, it had become evident that Quebecers regarded themselves as second-class citizens; that to them bilingualism meant that French-speakers had to learn English but not vice versa; and that they perceived that Quebec was not getting its fair share of the country's wealth.

Since the 1960s, despite the efforts of the federal government to satisfy the province's demands, the intensity of

Canada's French-speaking population is not confined to the Province of Quebec. This pair of signs in Edmundston, New Brunswick (a town 10 miles beyond Quebec's border) presents an interesting contrast. The official sign has English on top, but the commercial establishment is obviously French-owned.

devise another plan. They responded in 1992 with a complicated new proposal, but it was voted down in a national referendum by Quebec as well as five other provinces—thereby exhausting Canada's interest in constitutional reform.

As has become obvious in the 1990s, the Quebec controversy is also stirring the ethnic feelings of the 1.5 million Amerindians, who belong to more than 600 distinct bands. Their leaders were quick to point out that the "distinct society" clause in the 1987 Meech Lake Accord, if it could apply to French-speaking Quebecers, must apply to their communities as well. Such ethnic assertions—which are not falling on deaf ears in Ottawa—could have a major impact on the future map of Canada, as the creation of Nunavut in 1992 demonstrates (see the footnote on p. 202). Most of all, the Amerindians want their rights protected by the federal government against the provinces.

Quebec itself is a good example of this concern. The Crees of the Ungava region in northwestern Quebec (Fig. 3-17), whose administration was assigned to the provincial government in Quebec City in 1912, would probably be empowered to seek independence, should Quebec secede from Canada—thereby leaving Quebec with only about 45 percent of its present area. As the map shows, the Crees' domain is no unproductive wilderness; it contains vital segments of the James Bay Hydroelectric Project, a massive scheme of dikes, dams, and artificial lakes that will soon transform much of western Quebec and yield electric power for a huge market within and outside the province. Clearly, the stakes of independence are very high and the risks great. At the moment, Canada's centripetal forces still prevail; yet even though the country has entered a new period of uncertainty as to the resolution of its multicultural issues, stronger economic ties could well prove to be the glue that keeps the centrifugal forces from achieving a critical mass.

ECONOMIC GEOGRAPHY

As in the United States, the growth of Canada's spatial economy has been supported by a diversified, high-quality *resource base*. We noted earlier that the Canadian Shield is endowed with rich mineral deposits, and all three fossil fuels are extracted in substantial quantities throughout the Interior Lowlands zone to the west of the Shield (see Fig. 3-13). Canada has long been a leading *agricultural* producer and exporter, especially of wheat and other grains from its breadbasket in the fertile Prairie Provinces; new technologies keep productivity high, but labor requirements are diminishing to the point where farm workers now account for less than 3 percent of the national work force. Postindustrialization has caused massive recent employment decline in the *manufacturing sector*, including the loss of at least 300,000 factory jobs during the deep

ethnic feelings in Quebec has risen in surges. During the 1970s, while a separatist political party came to power in Quebec, a new federal constitution was drawn up in Ottawa. In 1980, Quebec's voters rejected sovereignty when given that choice in a referendum; but the new constitution did not satisfy Quebec, and polls soon indicated that given another opportunity, the province's voters would indeed support separation from Canada.

To avoid a crisis, Quebec was asked to give its terms for remaining in the Canadian federation. It did so, the key demand being its recognition as a "distinct society" within Canada. At a meeting at Meech Lake in 1987, the leaders of all ten of Canada's provinces agreed to those terms, which were to be ratified by their parliaments. But when two failed to ratify by the 1990 deadline, Quebec had reason to feel rejected. Demands for a new referendum on secession arose again, and Canada's leaders promised to

NORTH AMERICAN FREE TRADE AGREEMENT (NAFTA)

The economies of Canada, the United States, and Mexico are becoming ever more intertwined, and their interactions are now formalizing under the North American Free Trade Agreement (NAFTA) as it takes effect in 1994. This grand economic alliance is set to forge the world's largest trade bloc, a (U.S.) $6.5 trillion market of 380 million consumers (surpassing the European Community's 348 million). NAFTA will bind the three economies together over the next decade through a number of steps. Thousands of individual tariffs, quotas, and import licenses are scheduled to be removed or rolled back to eliminate most barriers to trading agricultural and manufactured goods as well as tertiary and quaternary services. Restrictions to the flow of investment capital across international borders will be lifted. And regulations concerning foreign ownership of productive facilities will also be liberalized.

The architects of NAFTA cite the opportunities the pact will bring. Business investment in all three countries will surge as unimpeded exports flow throughout the vast new marketplace; at the same time, intensified competition will give consumers access to a wider variety of quality products at reduced prices. Canada—which joined NAFTA mainly to safeguard the concessions it received in its 1989 Free Trade Agreement with the United States—faces the most skeptical citizenry, because opinion polls show that a majority of Canadians believe that the 1989 pact caused them more economic harm than good. Despite these feelings, the Ottawa government and many economists insist that free trade is resolving serious internal economic problems and is about to make Canada the fastest-growing industrial country. Specifically, they argue that restructuring Canadian firms have been forced to become more global in their operations, are making the country more competitive than the United States in costs and prices of key goods and services, are reversing the decline in industrial productivity, and will be creating large numbers of jobs in the tertiary and quaternary sectors.

If NAFTA realizes its goals, it will require each country to concentrate on the production of those goods and services for which it has a comparative advantage (in research and development, technological skills, managerial know-how, and the like). As workers move from less productive to more productive industries, jobs will be redistributed—a process that would be occurring anyway as the three countries continue to adapt to continental and global economic change.

NAFTA's supporters argue that the pact will boost trade, investment, and living standards. Its opponents insist NAFTA will destroy domestic industries as jobs relocate permanently to one of the other countries where wages and/or other economic conditions make production more profitable. Following a climactic debate, the U.S. Congress voted in favor of NAFTA in late 1993, and the trade agreement went into effect in 1994.

---

recession of the early 1990s. These sharp cutbacks have most severely affected the Southern Ontario industrial heartland, whose regional economy will continue to be bedeviled by unemployment and retraining problems as manufacturing plunges from 14 percent of the Canadian labor force (in 1990) to an expected mere 8 percent by 2000. As Canada in the mid-1990s emerges from its harsh economic downturn, it will be paced by its already robust *tertiary and quaternary sectors* (which employed an expanding 67 percent of the total work force in 1990).

Today, concerns about the economic future are heightening throughout Canada, coming to the forefront after the anticlimactic 1992 referendum on constitutional reform. In the aftermath of that vote, the Quebecer publisher of an influential French-language newspaper captured his province's mood this way: "Let's face the obvious—if after 30 years of [an active independence movement], nine years of [separatist-party provincial] government, one [secessionist] referendum, and 10 years of sterile threats, Quebecers have not yet separated, it's because they don't want to leave Canada." Going farther, the mayor of a once staunchly separatist Quebec city detected a shifting agenda: "Quebec sovereignty is all well and good, but we're a little tired of that battle when jobs are what really matter."

Those latter sentiments are on the minds of many Canadians these days because along with the postindustrial transformation, their country is beginning to feel major economic effects triggered by the 1989 U.S.-Canada Free Trade Agreement. That pact started the clock ticking on an ambitious 10-year timetable, wherein the two countries will phase out all tariffs and investment restrictions between them. This will indeed amount to a monumental achievement, given that Canada and the United States already engage in the world's biggest two-nation trade relationship (annually, about three-quarters of Canada's exports now go to the United States, while at least two-thirds of its imports originate there). Moreover, in 1994 these tightening economic linkages were further reinforced by the implemention of the

North American Free Trade Agreement (NAFTA), which consolidates the gains of the 1989 pact and opens new opportunities for both countries by adding Mexico as a third partner (see box at left).

At a 1992 conference of Canadian scholars and public officials, participants were asked to focus on the present and near-future consequences of the Free Trade pacts for their nation's economy. Most predicted a swift spatial reorientation of Canada, whereby traditional east-west linkages would weaken as north-south ties strengthened within a framework of transnational regions straddling the U.S.-Canadian border. Thus several parallel north-south relationships—many already well developed because of geographical and historical commonalities—can be expected to intensify: the Maritimes with neighboring New England, Quebec with New York State, Ontario with Michigan and surrounding Midwestern states, the Prairie Provinces with the Upper Midwest, and British Columbia with the (U.S.) Pacific Northwest. These international interactions are also well established in the overall regional configuration of North America, to which we now turn in the concluding section of this chapter.

◆ REGIONS OF THE NORTH AMERICAN REALM

The ongoing transformation of North America's human geography is fully reflected in its internal regional organization. As new forces uproot and redistribute people and activities, a number of old locational rules no longer apply; however, the varied character of the realm's physical, cultural, and economic landscapes assures that meaningful regional differences will persist. The current areal arrangement of the United States and Canada will now be examined within a framework of eight major regions (Fig. 3-18).

THE CONTINENTAL CORE

The Continental Core region (Fig. 3-7)—synonymous with the American Manufacturing Belt—has been discussed earlier. Serving as the historic workshop for the linked spatial economies of the United States and Canada, this region was the unquestioned leader and centerpiece during the century between the Civil War and the close of the industrial age (1865–ca. 1970). With the onset of postindustrialization, that linchpin regional role is unraveling today as the Continental Core increasingly shares a growing number of major functions with fast-rising areas to the west and south.

Unfortunately for the Manufacturing Belt, its continuing decline in blue-collar employment is not being matched by the creation of sufficient replacement jobs in other sectors. Nor is there yet much retraining of the hundreds of thousands of displaced factory workers who will probably never find another job in heavy industry. Thus parts of the region are becoming home to the ever-larger ranks of the permanently unemployed—a problem most apparent in the inner cities and innermost suburbs of Chicago (8.2 million), Detroit (4.7 million), Philadelphia (5.9 million), Pittsburgh (2.3 million), Cleveland (2.8 million), and Baltimore (2.5 million).

Manufacturing remains a highly important activity within the transformed American economy, but the productivity and obsolescence problems that elevate production costs in the Core region present obstacles in an era when its traditional industries are better able to respond to locational forces that operate in favor of newer regions. To remain competitive, a key concern for the Manufacturing Belt is the attraction of new investment capital. As we saw, parts of the Midwest are succeeding in pursuing the high-tech upgrading of an aged industrial base; how quickly that effort spreads will be an important indicator as the Belt struggles to forge a new role for itself in a wholly new age.

In certain parts of the Core region, where manufacturing has traditionally played a lesser role, postindustrial development has produced a number of new growth centers. The major metropolitan complexes of the northeastern Megalopolis—already the scene of much quaternary and quinary economic activity—are adjusting fairly well. Though still mired in the aftermath of recession, the Boston area (4.1 million), well endowed with research facilities, has attracted innovative high-tech businesses that are focused on the Route 128 freeway corridor that girdles the central city. New York City (7.3 million) remains the national leader in finance and advertising, and it houses the broadcast media; but even here the city increasingly shares its once-exclusive decision-making leadership with its own outer suburban city (10.8 million), which during the 1970s surpassed Manhattan in total number of corporate headquarters facilities.

Perhaps the Core region metropolis that has gained the most from postindustrialization is Washington, D.C. (4.0 million). As the information and control-function sectors have blossomed and as the U.S. federal government extends its connections ever deeper into America's business operations, the District of Columbia (585,000) together with its surrounding outer city of hyperaffluent Maryland and Virginia suburbs (3.4 million) has amassed an enormous complex of office, research, trade-organization, lobbying, and consulting firms. Suburbanization of facilities has been heightened by a lack of space in the District's center, avoidance of poor inner-city neighborhoods near downtown, and particularly the lure of the Capital Beltway—a 66-mile (105-km) freeway that encircles Washington and connects its most prestigious suburbs. Indeed, this thriving curvilinear outer city underscores the fact that nonresiden-

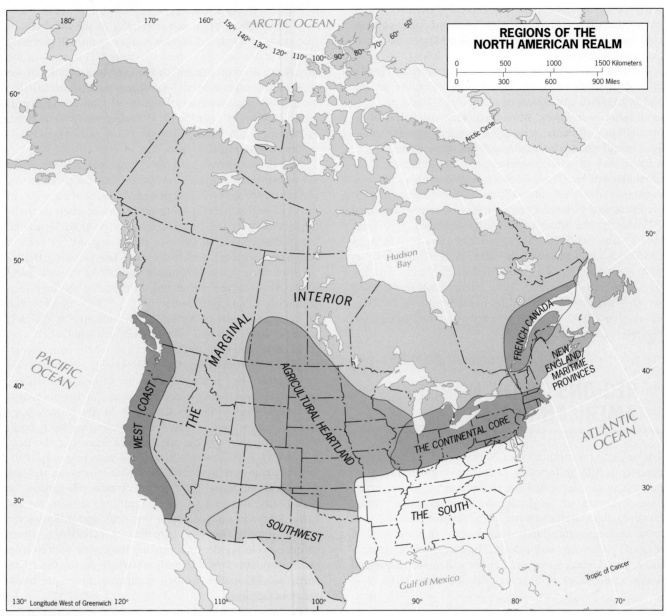

FIGURE 3-18

tial suburbia constitutes the healthiest category of economic subareas throughout the Continental Core.

NEW ENGLAND/MARITIME PROVINCES

New England, one of the realm's historic culture hearths, has retained a powerful regional identity for almost 400 years. Although the urbanized southern half of New England has been the northeastern anchor of the Continental Core since the mid-nineteenth century—where it must be regionally classified—the six New England states (Maine, New Hampshire, Vermont, Rhode Island, Massachusetts, and Connecticut) still share many common characteristics. Besides this overlap with the Manufacturing Belt (and Megalopolis) in its south, the New England region also extends northeastward across the Canadian border to encompass most of the four Maritime Provinces of New Brunswick, Nova Scotia, Prince Edward Island, and Newfoundland (Fig. 3-18).

A long association based on economic and cultural similarities has tied northern New England to Maritime Canada. Both are rural in character, possess rather difficult environments in which land resources are limited, and were historically bypassed in favor of more fertile inland areas. Thus economic growth here has always lagged be-

New Englanders made the most of their high-relief, thin-soil, limited-resource, harsh-winter environment, creating one of America's most characterful and agreeable cultural landscapes, represented here by East Corinth, Vermont.

hind the rest of the realm. Development has centered on primary activities, mainly fishing the rich offshore banks of the nearby North Atlantic, forestry in the uplands, and farming in the few fertile valleys available. Recreation and tourism have boosted the regional economy in recent times, with scenic coasts and mountains attracting millions from the neighboring Core region; the growth of skiing has also helped, extending the tourist season through the harsh winter months.

A rich sense of history and tradition permeates the close cultural affinity of New England and the Maritimes. Both possess homogeneous English-based cultures that have long withstood persistent Francophone incursions from adjacent Quebec, and both have given rise to a staunchly self-sufficient, pragmatic, and conservative population. Village settlement is overwhelmingly preferred, in sharp contrast to the dispersed rural population of the North American interior that clings to its individual farmsteads.

In the past decade, New England's economy has experienced a swing from one extreme to the other. The middle and late 1980s were years of remarkable growth, fueled by the computer industry and surging federal expenditures on military research. But without warning, the region's fortunes nose-dived during the early 1990s as smaller computers (mainly manufactured outside New England) quickly captured the market and the end of the Cold War signalled substantial cutbacks in defense spending. Moreover, these developments coincided with a nationwide recession—felt so strongly here that in 1991 New England was declared to be the U.S. region with the most severe economic down-

turn since the Great Depression of the 1930s. To make matters even worse, *every* economic sector was hard hit, and the region's business exodus and unemployment totals reached unprecedented levels (also practically unheard of was the Democratic sweep of the staunchly Republican upper New England states in the 1992 presidential election). By the mid-1990s, recovery had barely begun, a painfully slow process that could last into the next century.

FRENCH CANADA

Francophone Canada constitutes the effectively settled, southern portion of Quebec, which straddles the lower St. Lawrence valley from where that river crosses the Ontario-Quebec boundary just upstream from Montreal to its mouth in the Gulf of St. Lawrence. Also included are sizeable concentrations of French speakers who reside just across the provincial border in New Brunswick and the U.S. border in northernmost Maine (Fig. 3-18). The old-world charm of Quebec's cities is matched by an equally unique rural settlement landscape introduced by the French—narrow rectangular farms, known as *long lots*, are laid out in sequence perpendicular to the St. Lawrence and other rivers, allowing each farm access to the waterway.

The economy of French Canada is no longer rural (although dairying remains a leading agricultural pursuit), exhibiting urbanization rates similar to those of the rest of the country. Industrialization is widespread, supported by cheap hydroelectric energy generated at huge dams in northern

Quebec. Tertiary and postindustrial commercial activities are centered in Montreal (3.1 million); tourism and recreation are also important.

As we have seen, Quebec's heightened nationalism has had a major impact on the Canadian federation since 1970. Many important changes have also occurred within the province. During the 1980s, the provincial government enacted new laws that strengthened the French language and culture throughout Quebec. With English domination ended, the French Canadians have been channeling their energies into developing Quebec's economy (a drive that stalled when the Canadian economy went into deep recession in the early 1990s). A by-product of that effort was the rapid urbanization of young families and a loosening of their ties to Roman Catholicism; that produced a sharp decline in the provincial birth rate, which the Quebec government now tries to offset by encouraging (with cash payments) large families and Francophone immigration (mostly from former colonies of France around the world).

THE AGRICULTURAL HEARTLAND

In the heart of North America, agriculture becomes the predominant feature of the landscape. Whereas the innermost fruit-and-vegetable and dairying belts in the U.S. macro-Thünian regional system (Fig. 3-14) are in competition with other economic activities in the Continental Core, by the time one reaches the Mississippi valley, meat and grain production prevail for much of the 1,000 miles (1,600 km) west to the base of the Rocky Mountains. Because the eastern half of the Agricultural Heartland lies in Humid America and closer to the national food market on the northeastern seaboard, mixed crop-and-livestock farming wins out over less competitive wheat raising, which is relegated to the fertile but semiarid environment of the central and western Great Plains on the dry side of the 100th meridian. The latter area also contains Canada's agricultural heartland north of the 49th-parallel border; however, although these fertile Prairie Provinces are somewhat less subject to serious drought than the U.S. high plains to the south, they are situated at a more northerly latitude that makes for a shorter growing season.

The distribution of North American corn and wheat production is shown in Fig. 3-19, which rather neatly defines the boundaries of the Agricultural Heartland region. The Corn Belt to the east is focused on Illinois and Iowa, with extensions into neighboring states (see Fig. 3-14). The area of north-central Illinois, which displays a major cluster of corn farming, is a classic example of transition along a regional boundary: the Agricultural Heartland/Continental Core dividing line connecting Milwaukee (1.6 million)

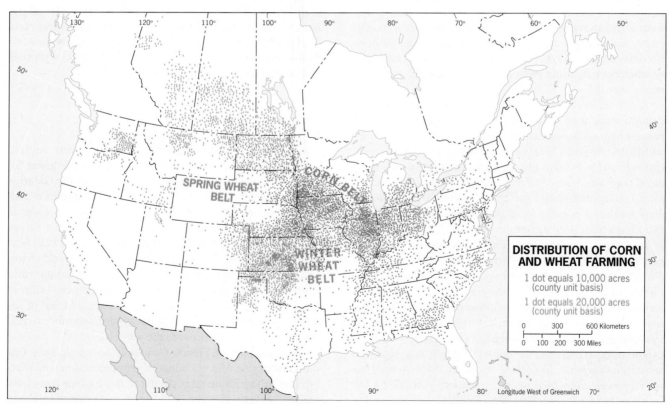

DISTRIBUTION OF CORN AND WHEAT FARMING

1 dot equals 10,000 acres (county unit basis)

1 dot equals 20,000 acres (county unit basis)

0   300   600 Kilometers

0   100   200   300 Miles

Longitude West of Greenwich

FIGURE 3-19

and St. Louis (2.5 million) passes right through this zone (Fig. 3-18), which contains a number of sizeable urban-industrial concentrations interspersed by some of the most productive corn-producing land on earth. On its western margins, the Corn Belt quickly yields to the dryland Wheat Belts. The growth of foreign wheat-sale opportunities in recent years has prompted the intensification of production in the Winter Wheat Belt through the use of center-pivot irrigation, a means of overcoming the moisture deficit as long as groundwater supplies hold up.

Throughout the Agricultural Heartland, nearly everything is oriented to farming. Its leading metropolises—Kansas City (1.6 million), Minneapolis-St. Paul (2.5 million), Winnipeg (680,000), Omaha (625,000), and even Denver (1.9 million)—are major processing and marketing points for beef packing, flour milling, and pork production. People in this region are generally of northern European ancestry, and conservative. Yet rapid technological advances require that farmers keep abreast of increasingly scientific agricultural techniques and business methods if they are to survive in a hypercompetitive atmosphere.

THE SOUTH

Of the realm's eight regions, none has undergone more change, during the past quarter-century, than the American South. Choosing to pursue its own sectional interests practically from the advent of nationhood, this region's economic and cultural isolation from the rest of the United States deepened during the long and bitter aftermath of the ruinous Civil War. For over a century, the South languished in economic stagnation; but the 1970s witnessed a reassessment in the nation's perception of the region that launched a still-ongoing wave of growth and change unparalleled in Southern history.

Propelled by the forces that created the Sunbelt phenomenon, people and activities streamed into the urban South. Cities such as Atlanta (3.1 million), Houston (3.7 million), Miami (2.1 million), Tampa (2.2 million), and New Orleans (1.3 million) turned into booming metropolises practically overnight, and conurbations swiftly formed in such places as southern and central Florida, the Carolina Piedmont, and along the Gulf Coast from Houston to Mobile, Alabama (505,000). This "bulldozer revolution" was matched in the most favored rural areas by an agricultural renaissance that stressed such higher-value commodities as beef, soybeans, poultry, and lumber. On the social front, institutionalized racial segregation was dismantled. Although certain problems in minority relations persist, inequalities in the South today are no worse than in other parts of the country; in fact, the opening of the 1990s saw African-American mayors in Atlanta, Birmingham (925,000), and New Orleans, and women in Houston and Tampa.

Yet for all the growth that has taken place since the 1960s, the South remains a region beset by many economic problems because the geography of its development has been decidedly uneven. Although certain urban and farming areas have benefited, many others containing large populations have not. Even places adjacent to boom areas have often been left unaffected, and the juxtaposition of progress and backwardness is frequently encountered in the Southern landscape. The geographic selectivity of non-agricultural growth is noteworthy in another regard: with the exception of centrally located Atlanta, the most important development has occurred near the region's periphery in the Washington suburbs of northern Virginia, the North Carolina Piedmont, the Houston area, and Florida's central

Jarring contrasts mark the landscape of today's South, where the old and the new often stand side-by-side—even here in Houston's CBD.

and southern-coast corridors. And as the checkerboard development pattern intensifies, even places within boom areas are now falling behind—especially large Southern central cities, whose shrinking prospects are converging with those of their northern counterparts as burgeoning outer suburban cities around Atlanta, Houston, Miami, Tampa, Orlando (1.2 million), Memphis (1.0 million), and New Orleans capture ever larger shares of total metropolitan employment.

In retrospect, recent Southern development has produced much beneficial change and has helped to remove the region's longstanding inferiority complex. But one must keep in mind, too, that the South had—and still has—a long way to go, and that even today much of it lags behind the times. One of its foremost students, the historian C. Vann Woodward, offers this revealing insight about the so-called New South:

> *The old Southern distinction of being a people of poverty among a people of plenty lingers on. There is little prospect of closing the gap overnight.*

THE SOUTHWEST

As recently as the 1960s, North America geography textbooks did not identify a distinct southwestern region, classifying the Mexican borderland zone into separate southward extensions of the Great Plains, Rocky Mountains, and Intermontane Plateaus. Today, however, this area must be recognized as a major regional entity, albeit the realm's youngest. The emerging Southwest is also unique in the United States because it is a bicultural regional complex, peopled by in-migrating Anglo-Americans atop the crest of the Sunbelt wave and by a rapidly expanding Mexican-American population anchored to the sizeable long-time resident Hispanic population, which traces its origins to the Spanish colonial territory that once extended from Texas's Gulf Coast to San Francisco. In fact, if one counts the large Native American population, then the Southwest is truly a *tricultural* region.

Recent rapid development in Texas, Arizona, and New Mexico is essentially built on a three-pronged foundation: (1) availability of huge amounts of *electricity* to power air conditioners through the long, brutally hot summers; (2) sufficient *water* to supply everything from swimming pools to irrigated crops so that large numbers may reside in this dry environment; and (3) the *automobile*, so that affluent newcomers may spread themselves out at much-desired low densities. The first and third of these foundation prongs have been rather easily attained because the eastern flank of the Southwest is abundantly endowed with oil and natural gas (Fig. 3-13). The future of water supplies, however, is far more problematical; for instance, Phoenix (2.2 mil-

lion) and Tucson (730,000) could not keep growing at their current swift pace without the massive new Central Arizona Project canal, the state's last major source (the Colorado River) of drinking water.

So far, the generation of new wealth in this growing region makes it an undeniable success. However, since much of this achievement is linked to the fortunes of the oil business, the Southwest may well be entering a future beset by unpredictable economic turns if the experiences of the energy industry since 1970 are an indicator. Much of Texas was staggered by the falling fortunes of oil in the 1980s, rebounding only slowly during the 1990s. But the postindustrial revolution is also prominently represented in this region—with a specialized activity complex of electronic and space-technology facilities located in the eastern Texas triangle formed by connecting Houston, San Antonio (1.3 million), and Dallas-Fort Worth (3.9 million). Moreover, the state capital of Austin (825,000) near the triangle's center is becoming a leading high-tech research complex in a partnership between manufacturers and the University of Texas.

THE MARGINAL INTERIOR

The Marginal Interior is an appropriate name for North America's largest region by far. It covers most of Canada, all of Alaska, the northern salients of Minnesota-Wisconsin-Michigan, New York's Adirondacks, and the inland U.S. West between (and including) the Sierra Nevada-Cascades and the Rocky Mountains. There can be no doubt about the interior position of this vast slice of the North American realm; its marginal nature stems from its isolation and rugged environment, which have attracted only the sparsest of populations relative to the other seven regions. In the U.S. portion, for example, even after counting substantial recent growth, population density is still only 12 per square mile (5 per sq km) in contrast to 71/28 overall. Yet, its disadvantages notwithstanding, the Marginal Interior contains great riches because it is one of the earth's major storehouses of mineral and energy resources.

Not surprisingly, this region's history has been a frustrating one of boom-and-bust cycles, determined by technological developments, economic fluctuations, and corporate and government decisions made in the Continental Core. The events of the past several years underscore this bittersweet development process. Following the oil shortage of the early 1970s, there was a virtual invasion in the search for additional petroleum and natural gas supplies, with efforts concentrating on the Alaskan North Slope as well as southwestern Wyoming and northwestern Colorado. With the opening of the Trans-Alaska Pipeline in 1977, optimism reached an all-time high, and western Wyoming and Colorado were inundated by tens of thousands of migrants who taxed local community facilities to the breaking point. But, as has happened so many times before, the boom soon fizzled out (when the oil shortage ended and fuel prices plunged).

Elsewhere in the region, recent economic recessions negatively affected the mining industry: parts of Minnesota's Iron Range were closed down, and once-productive copper mines in Montana, Utah, and Arizona were faced with unprecedented slowdowns. Nevertheless, the future still beckons brightly. Rocky Mountain coal- and uranium-extraction operations were relatively untouched by these economic misfortunes, and the mining of silver, lead, zinc, and nickel endured at scattered sites throughout western North America and on the Canadian Shield to the east. And surely, with the steady decline of fossil-fuel supplies worldwide, oil and gas drillers will one day return in force to the central Rockies and its vast deposits of oil shale.

The recent sharp swings of the extractive economy also mask a steadier influx of population and nonprimary activities to other parts of the Marginal Interior. In fact, from 1970 to 1990, the U.S. segment grew by more than 60 percent, thereby advancing its relative position from 4 to 7 percent of the national population. The guiding force was the search for high-amenity locations, with the clean, stress-free, wide-open spaces of certain parts of the Rockies and Intermontane Plateaus a particular attraction. Accordingly, Nevada, Alaska, Arizona, Utah, Wyoming, Colorado, and Idaho all ranked among the 10 fastest-growing states between 1970 and 1995.

THE WEST COAST

The Pacific coast of the conterminous United States and southwesternmost Canada has been a powerful lure to migrants since the Oregon Trail was pioneered more than 150 years ago. Unlike the remainder of the North American west, the narrow strip of land between the Sierra Nevada-Cascade mountain wall and the sea receives adequate moisture. It also possesses a far more hospitable environment, with generally delightful weather south of San Francisco, highly productive farmlands in California's Central Valley, and such scenic glories as the Big Sur coast, Washington State's Olympic Mountains, and the spectacular waters surrounding San Francisco-Oakland (6.3 million), Seattle-Tacoma (2.7 million), and Vancouver (1.6 million). Most major development here took place during the post-World War II era, accommodating enormous population and economic growth, and the West Coast is just now beginning to face the less pleasant consequences of regional maturity.

Fifty years of unrelenting growth have especially taken their toll in California, because the massive development of America's most populated state (31.8 million in 1994) has been overwhelmingly concentrated in the teeming conurbation extending south from San Francisco through San Jose (1.6 million), the San Joaquin valley, the Los Angeles Basin, and the southwestern coast into San Diego (2.6 million) at the Mexican border. Environmental hazards bedevil this entire corridor, including inland droughts, coastal-zone

San Francisco, site of the 1994 meeting of the Association of American Geographers, constitutes the core of a large metropolis that encircles one of the West's best natural harbors, San Francisco Bay. Seen here are the CBD, the Bay Bridge, part of the port, Alcatraz Island, and the Golden Gate Bridge (upper left corner).

flooding, mudslides, brush fires, and earthquakes—with the ominous San Andreas Fault practically the axis of megalopolitan coalescence. To all this, humans have added their own abuses of California's fragile habitat, from overuse of water supplies (requiring vast aqueduct systems to import water from hundreds of miles away) to the incredible air pollution of Los Angeles (see the photos on p. 181) caused by the emissions of almost 10 million motor vehicles, which are vital to movement within this particularly dispersed metropolis of 14.6 million.

Lost somewhere in the shuffle is yesterday's glamorous image of Southern California—communicated so stunningly in the motion pictures of the postwar period—which is based on relaxed outdoor living in luxurious horticultural suburbs in one of the most agreeable climates to be found anywhere on earth. Undeniably, economic development has brought California prosperity, and the state has established its leadership as a national innovator. But this affluence has also been subject to sudden shifts because much of the California economy is tied to aerospace, computer, and other volatile industries. (The state's deep recession of the early 1990s is but the latest cyclical economic downturn.) Clearly, the years of unbridled optimism are over, and the future will bring new struggles to maintain the state's still-considerable advantages in the face of growing competition from would-be new Californias elsewhere on the continent.

The northern portion of the West Coast region includes the Pacific Northwest—focused on Oregon's Willamette valley and the nearby Cowlitz-Puget Sound lowland of western Washington—and the British Columbia coastal zone of southwesternmost Canada. Originally built on timber and fishing (primary activities that still thrive here), this area found its impetus for industrialization in the massive Columbia River dam projects of the 1930s and 1950s, which generated cheap hydroelectricity. This in turn attracted aluminum and aircraft manufacturers (the huge Boeing aerospace complex around Seattle makes that metropolis one of the world's biggest company towns).

Unique environmental amenities—zealously safeguarded here—have lured hundreds of growth companies, and the Pacific Northwest is adjusting smoothly to the maturing postindustrial economy. Perhaps one of its greatest advantages, shared with urban California to the south, is its location as a gateway to the western Pacific Rim—now emerging as one of the world's most dynamic economic arenas (as we will see in Chapter 4). As North America becomes increasingly enmeshed in the global economy, the eastern

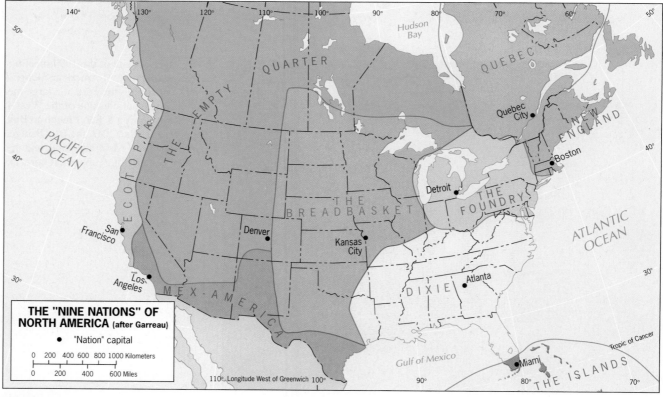

FIGURE 3-20

edge of Asia will be a critically important marketplace; Portland (1.6 million) and Seattle, as well as Vancouver north of the Canadian border (which has attracted major investment from Hong Kong in recent years), are bound to benefit because their airports are the closest to East Asia of any in the conterminous United States.

◆ ON TO THE TWENTY-FIRST CENTURY

As the postindustrial transformation continues to run its course, North America's regional system remains in the throes of change—a dynamism highlighted in the profile of each region we surveyed. A fascinating overview of the direction and significance of all this change is contained in geographer-commentator Joel Garreau's 1981 book, *The Nine Nations of North America*. His hypothesis is that the realm is reorganizing into nine separate "nations":

> *Each has its capital and its distinctive web of power and influence. A few are allies, but many are adversaries . . . Some are close to being raw frontiers; others have four centuries of history. Each has a peculiar economy; each commands a certain emotional allegiance from its citizens. These nations look different, feel different, and sound different from each other, and few of their boundaries match the political lines drawn on current maps . . . Each nation has its own list of desires. Each nation knows how it plans to get what it needs from whoever's got it. Most important, each nation has a distinct prism through which it views the world.*

Garreau's thought-provoking regionalization scheme is mapped in Fig. 3-20 (which should be compared to Fig. 3-18) and covers not only the United States and Canada but the Caribbean Basin as well. Although it needs a brief updating and probably the addition of Mexico, the nine-nations scheme is a valuable contribution to the captivating business of monitoring the geographical restructuring of this remarkable realm; moreover, it may well prove to be correct in its central argument, which points toward a more fragmented future North America. Time will tell.

THE REALM IN TRANSITION

1. The economic geography of North America will be substantially modified as NAFTA takes effect in the mid-1990s; interconnections between North and Middle America will be strengthened, but the long-range effects are uncertain.
2. U.S. dependence on foreign energy sources has returned to the levels of two decades ago, placing the country's economy at even greater risk.
3. North America is not immune from separatist movements. A group of counties has proclaimed its desire to split from Kansas. A plan to divide California has been seriously promoted. The notion of a U.S.-Canadian "Pacifica" on the Pacific Rim has some support.
4. Emigration by English speakers and very low birth rates are clouding the future of Canadian Quebec; together with improving economic ties to the rest of the country, they are defusing (perhaps temporarily) the province's sovereignty drive.
5. Soviet-era nuclear and chemical pollution threaten Canadian-Arctic waters; exposed are Inuit inhabitants of the realm's far north.

◆ P R O N U N C I A T I O N   G U I D E

Bryn Mawr (BRINN-mar)
Calgary (KAL-guh-ree)
Caribbean (kuh-RIB-ee-un/karra-BEE-un)
Cenozoic (senno-ZOH-ik)
Cynwyd (KIN-widd)
Francophone (FRANK-uh-foan)
Garreau (GARROW)
Inuit (IN-yoo-it)
Isohyet (EYE-so-hyatt)
Megalopolis (mega-LOPP-uh-liss)
Mesabi (meh-SAH-bee)

Montreal (mun-tree-AWL)
Newfoundland (NYOO-fun-lund)
Nova Scotia (nova-SKOH-shuh)
Nunavut (NOON-uh-voot)
Ottawa (OTTA-wuh)
Puerto Rico (pwair-toh-REE-koh)
Puget (PYOO-jet)
Quaternary (kwuh-TER-nuh-ree)
Quebec (kwih-BECK)
Quebecer (kwih-BECK-er)
Quinary (KWY-nuh-ree)
San Joaquin (san-wah-KEEN)

Saskatchewan (suss-KATCH-uh-wunn)
Tertiary (TER-shuh-ree)
Toponymy (toh-PONN-uh-mee)
Ungava (ung-GAH-vuh)
Uwchlan (YOO-klan)
Valdez (val-DEEZE)
Von Thünen (fon-TOO-nun)
Willamette (wuh-LAMM-ut)
Winnipeg (WIN-uh-peg)
Yeates (YATES)

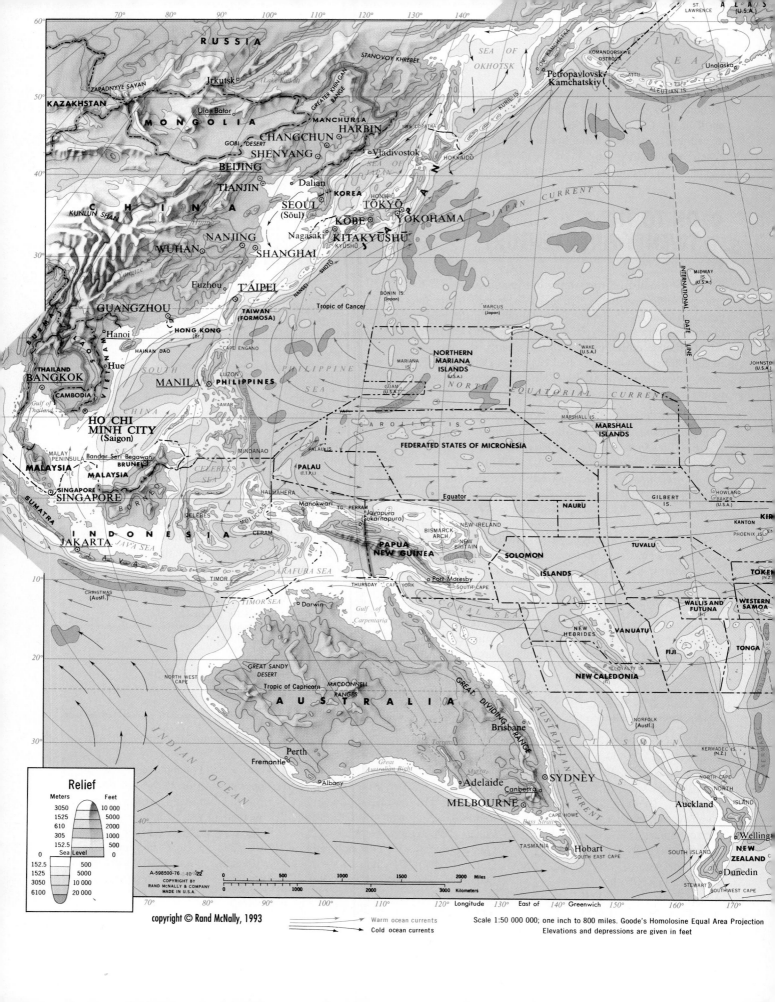

CHAPTER 4

THE PACIFIC RIM OF AUSTRASIA

I t is a sign of our times that countries are fragmenting internally—not only politically and culturally but economically as well. Economic growth is propelling parts of countries toward full-development status while other parts of those same countries are left behind. Nodes of spectacular development then stand in stark contrast against expanses of stagnation. Regional inequalities, disparities, and contrasts between technologically throbbing urban-industrial cores and comparatively changeless rural-agricultural peripheries become even sharper.

This phenomenon occurs worldwide, but it is a hallmark of a newly emerging world region: the **Pacific Rim**. Along the shores of countries that face the Pacific Ocean, hearths of industrial and technological activity are creating a discontinuous yet interconnected series of economic giants. These are the crucibles of their national economies, the hubs of development, the guideposts toward what governments hope will be prosperity and security. From Korea to Malaysia, such beehives of activity are transforming the economic geography of the zone that surrounds the Pacific.

IDEAS & CONCEPTS

Development (2)
Economic tiger
Modernization
Relative location (2)
Areal functional organization
Regional complementarity (2)

State capitalism
Special Economic Zone (SEZ)
Federalism
Migration
Import-substitution industry

REGIONS

Japan
South Korea
Taiwan
Hong Kong

Shenzhen (China)
Singapore
Australia
New Zealand

◆ WHERE IS THE PACIFIC RIM?

The regional term *Pacific Rim* has come into general use to combine a far-flung group of countries and parts of countries that share certain spatial and economic characteristics. First, they all adjoin the world's largest ocean. Second, they are comparatively highly urbanized and industrialized, reflecting a relatively high level of economic development. Third, they consist of island countries (Japan, Taiwan), peninsulas (Korea, Hong Kong) and other coast-hugging clusters and ribbons of burgeoning economic activity. Fourth, most of their imports and exports move across Pacific waters.

Defined this way, the Pacific Rim would include not only Japan and South Korea but also Canada's British Columbia, the U.S. states of Alaska, Washington, Oregon, and California, and parts of Middle and South America as far south as central Chile. Certainly these areas participate in and are influenced by the dramatic developments within the Pacific Rim as a whole. But the most spectacular of

AUSTRASIA

In regional geography, as in other fields of study, it is sometimes necessary to introduce new terminology to designate changing realities or new discoveries. The landmass of which the "continents" of Europe and Asia are part of has long been known as *Eurasia*, a term heard more frequently these days as Europe and Russia are ever more closely involved with each other. Along the rim of the western Pacific Ocean, a significant reorientation also is taking place. Australia, the southern continent (*austral* means "southern" in Latin), is becoming less a European outpost and more a multicultural society connected to eastern Asia, especially Japan. So we introduce the term **Austrasia** to recognize the regional realignment of the western Pacific Rim from Hokkaido to Tasmania.

◆ F O C U S   O N   A   S Y S T E M A T I C   F I E L D

GEOGRAPHY OF DEVELOPMENT

Scholars from many academic disciplines study **development**—the economic, social, and institutional growth of national states. The introduction to this topic in the opening chapter differentiated between developed countries (*DCs*) and underdeveloped countries (*UDCs*). It showed that the widening development gap that separates DCs and UDCs is deeply rooted in *colonialism*. The rapidly advancing Western countries, the colonizers, gained a decisive head start economically while their dependencies remained suppliers of raw materials and consumers of manufactured goods. The inequalities of wealth, advantage, and human well-being that resulted from this schism constitute a continuing crisis, because today just under 80 percent of the world's population resides in the UDCs of the developing geographic realms.

Underdevelopment, as we noted in the Introduction, is characterized by several specific symptoms. *Demograhically*, UDCs exhibit steadily increasing populations, comparatively short life expectancies, and mostly overcrowded living conditions. *Health and nutrition standards* are modest at best, with widespread hunger and disease, high rates of infant mortality

and dietary deficiencies, poor sanitation, and inadequate health care services. Among the symptomatic *social ills* are rigid, elite-dominated class structures that inhibit upward mobility, inferior education systems, and high illiteracy rates. *Economic problems* are pervasive, including far-reaching poverty and unemployment, inferior circulation systems, internal regional imbalances, and labor forces in which women and children perform grueling physical work. *Rural areas* are often dominated by low-yield subsistence agriculture as well as inefficient land-tenure practices and farming techniques; *urban centers* are marked by massive overcrowding, a chronic lack of sufficient jobs and human-services provision, and abysmal housing conditions in both city slums and outlying shantytowns.

In the search for meaningful generalizations, various attempts have been made to improve our understanding of the development process. A useful global model, still much discussed, was formulated by the economist Walt Rostow in the 1960s; his model suggested that all developing countries follow an essentially similar path through five interrelated growth stages. In the earliest of these stages, a *tradi-*

tional society engages mainly in subsistence farming, is locked in a rigid social structure, and resists technological change. When Stage 2—*preconditions for takeoff*—is reached, a progressive group of leaders moves the country toward greater flexibility, openness, and diversification. Old ways are abandoned by substantial numbers of people; birth rates decline; more products than farm-related goods are made and sold; transportation improves. A sense of national unity and purpose may emerge to help nurture this progress. If that happens, Stage 3—*takeoff*—is reached.

Now the country experiences something akin to an industrial revolution, and sustained growth takes hold. Industrial urbanization proceeds, and technological and mass-production breakthroughs occur. If the economy continues to develop, it enters Stage 4—*drive to maturity*. Technologies diffuse nationwide, sophisticated industrial specialization occurs, international trade expands. Some countries reach a still more advanced Stage 5, that of *high mass consumption*, marked by high incomes, the robust production of consumer goods and services, and a majority of workers in the tertiary and quaternary sectors. In the

these developments have taken place along the western shores of the Pacific Ocean. Here is where the world's regional geography is being transformed. And that is why our focus in this chapter is the Pacific Rim of *Australasia* (see box and map on pages 218-219).

A word of caution: because what is happening in eastern Asia (and Australia to the southeast) is so impressive, we might be tempted to redraw the map of world geographic realms to introduce a globe-girdling Pacific Rim realm into our macro-spatial framework. Realms and regions do

change, but for the time being the Pacific Rim remains an assemblage of other regions and parts thereof. A century ago we might have been tempted to speak of an "Atlantic Rim," when the diffusion of the Industrial Revolution was creating the urban-industrial complexes in Western Europe and eastern North America that made the North Atlantic Ocean the busiest theater of long-distance trade in the world. Today, vestiges of that era still exist, but Europe and North America remain distinct and discrete geographic realms. And so it is likely to be along the Pacific Rim. In the next cen-

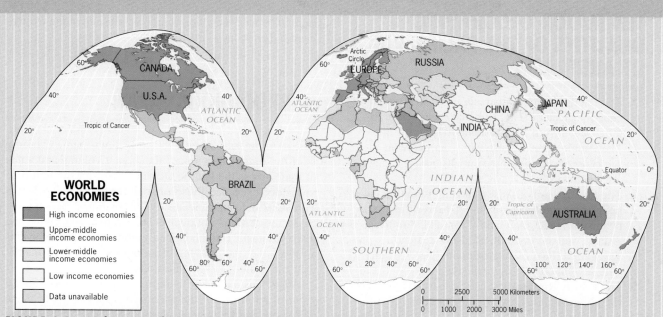

WORLD ECONOMIES

- High income economies
- Upper-middle income economies
- Lower-middle income economies
- Low income economies
- Data unavailable

FIGURE 4-1 **(A larger-scale version of this map appears in Fig. 1-13)**

1990s we might add a further stage to Rostow's model: the *postindustrial* stage of ultra-sophisticated high technology.

On today's world map (Fig. 4-1), the low income economies of Stages 1 and 2 are heavily concentrated in tropical Africa and southern and eastern Asia; the most advanced DCs associated with Stages 4 and 5 (high income economies) lie mainly in North America and Western Europe. Any assessment of the stage of development of a country involves difficult problems. For instance, by some measures countries in the "takeoff" stage include Mexico, Brazil, Venezuela, and Turkey. Yet while some of these (and other) countries in the takeoff stage seem to be headed for the next upward stage, others are stagnating or sliding backward. One example is Nigeria, which during its 1970s oil boom seemed poised for takeoff; it now has most of the characteristics of a Stage 2, not a Stage 3 society.

In the 1990s, the prognosis for UDCs is generally not good. The UDCs have been thrust into a global system of exchange and capital flow over which they have little or no control. Unable in many instances to stem the tide of population growth, UDCs see the prices of their products fall while interest rates rise; they must borrow, and are plunged into debt that deepens every year. Thus the poorest of the world's countries owe billions of dollars to the richest, a situation that reminds the peoples and leaders of the UDCs that the geography of development is the geography of inequality. While UDCs must get their economic and political houses in order (corruption is yet another obstacle to development), the DCs must modify a discriminatory system of international commerce that aborts many a takeoff.

tury, mainland East Asia and Japan may have converged enough to allow us to combine them into an East Asian geographic realm. Hong Kong may be just another major urban-industrial node on the mainland, perhaps overshadowed by Shanghai and by the growing complexes of Taiwan-Fuzhou and others in Korea. Yet we are likely still to distinguish geographically between East Asia, Southeast Asia, and Australia. Cultural geographies are durable and resilient, no matter how strong the economic forces in their midst.

In this chapter we focus on the Pacific Rim from Japan to Australia. We concentrate here on the modern developments that have thrust not only Japan but also South Korea, Taiwan, Hong Kong, and Singapore onto the world economic stage. We investigate China's recent efforts to stimulate rapid economic development in several of its oceanfront cities. And we examine the transformation of Australia, less and less a European outpost and more and more a disadvantaged participant in the affairs of the Pacific Rim. Later in this book, we will encounter these regions

TEN MAJOR GEOGRAPHIC QUALITIES OF THE PACIFIC RIM OF AUSTRASIA

1. The Pacific Rim of Austrasia is the western anchor of a greater Pacific Rim that encircles the Pacific Ocean from New Zealand to central Chile.
2. The Pacific Rim of Austrasia is an evolving assemblage of regions whose dimensions are changing continuously.
3. The Pacific Rim of Austrasia is a discontinuous geographic entity consisting of parts of insular, peninsular, and mainland East and Southeast Asia as well as Australia and New Zealand.
4. The Pacific Rim of Austrasia overlaps spatially with several other geographic realms.
5. The Pacific Rim of Austrasia is marked by high rates of economic growth and high levels of productive capacity.
6. In general, the Pacific Rim of Austrasia is not well endowed with natural resources. Australia is its richest component in these terms.
7. The economic geography of the Pacific Rim of Austrasia is characterized by industries that use raw materials from other, often distant locales and sell finished products all over the world.
8. The cultural geography of the Pacific Rim of Austrasia is marked by the contact (sometimes collision) of modern and traditional societies.
9. Much of the economic success of the economic tigers on the Pacific Rim is due to high skills and low wages.
10. The political geography of the Pacific Rim of Austrasia is fraught with actual and potential conflicts ranging from Japan's Kurile Islands and the embattled Spratly Islands to Maori and Aboriginal land issues in New Zealand and Australia.

and places again, but in different contexts. South Korea is emerging as one of the economic powers (or **economic tigers**, as economic geographers like to call them) on the Pacific Rim, but South Korea's relationships with communist North Korea overshadow much of its achievement. Taiwan, another tiger, faces a China whose leaders insist that it is a wayward province of the motherland, not an independent nation. Hong Kong will revert to the Chinese in 1997. Singapore lies off the tip of the Malay Peninsula,

and mainland Malaysia itself is showing signs of economic takeoff. Such issues will be discussed in subsequent chapters. Here, the focus is on the geographic makeup of the western Pacific Rim today—and tomorrow.

◆ PRODIGIOUS JAPAN

One morning late in the 1980s, the news media carried a brief but remarkable report: the annual volume and value of trade across the Pacific Ocean, it said, for the first time in history exceeded those of trade crossing the Atlantic. The story did not generate major headlines, perhaps because momentous political events were transforming the map of Eurasia. But it heralded a momentous turn of events all the same: the growing economic power of the Pacific Rim had passed another milestone.

All of this started with Japan. As long ago as 1868, a group of reform-minded modernizers seized power from the old guard, and by the end of the nineteenth century Japan was a military as well as an economic force, in the process of aggressively forging a large colonial empire in East Asia (Fig. 4-2). From the factories in and around Tokyo and from urban-industrial complexes elsewhere poured forth a stream of weapons and equipment that gave Japan primacy over a Pacific-Asian domain once monopolized by European colonizers. Not even a disastrous earthquake in 1923 could stop the Japanese; much of Tokyo was destroyed, but Japan barely broke stride. In 1941, Japanese-built aircraft carriers moved Tokyo's warplanes within striking range of Hawaii, and their December 7th surprise attack on the United States fleet in Pearl Harbor underscored Japan's confidence in its war machine. When, nearly four years later, at the conclusion of World War II, American nuclear bombs devastated two Japanese cities, the country lay in ruins. But once again Japan, aided this time by an enlightened U.S. postwar administration, surmounted disaster.

Japan's economic recovery and its rise to the status of world economic superpower has been the story of the second half of the twentieth century. Japan lost the war and its empire, but it scored many economic victories in a new global arena. Japan became an industrial giant, a technological pacesetter, a fully urbanized society, a political power, and a thriving affluent nation. No city in the world today is without Japanese cars in its streets; few photography stores lack Japanese cameras and film; laboratories the world over use Japanese optical equipment. From microwave ovens to VCRs, from oceangoing ships to camcorders, Japanese goods flood the world's markets.

Japan's brief colonial adventure helped lay the groundwork for other economic successes along the western Pacific Rim. The Japanese ruthlessly exploited Korean and Formosan (Taiwanese) natural and human resources, but

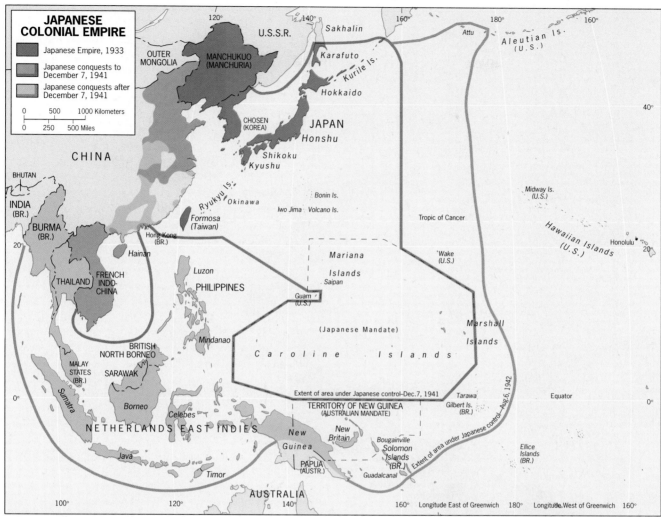

FIGURE 4-2

they also installed a new economic order there. After World War II, this infrastructure facilitated the economic transition—and soon made both Taiwan and South Korea competitors on world markets.

SPATIAL LIMITATIONS

In discussing the economic geography of Europe, we noted how the comparatively small island of Britain became the crucible of an Industrial Revolution that gave it a huge head start and advantage over the mainland, across which the Industrial Revolution spread decades later. Britain was not large, but its insular location, its local raw materials, the skills of its engineers, the presence of a large labor force, and the nature of its social organization combined to endow it not only with industrial strength but also with a vast overseas empire.

After the Japanese modernizers took over control of the country in 1868—an event known as the *Meiji Restoration* (the return of "enlightened rule" centered on the Emperor Meiji)—they turned to Britain for guidance in their efforts to reform their nation and its economy. The British, in the decades that followed, made many contributions. They advised the Japanese on the layout of cities and the construction of a railroad network, on the location of industrial plants and the organization of education. The British influence still is visible in the Japanese cultural landscape today: the Japanese, like the British, drive on the left side of the road.

The Japanese reformers of the late nineteenth century undoubtedly saw many geographic similarities between Britain and Japan. At that time, most of what mattered in Japan was concentrated on the country's largest island, Honshu (literally, "mainland"). The ancient capital, Kyoto, lay in the interior, but the modernizers wanted a coastal, outward-looking

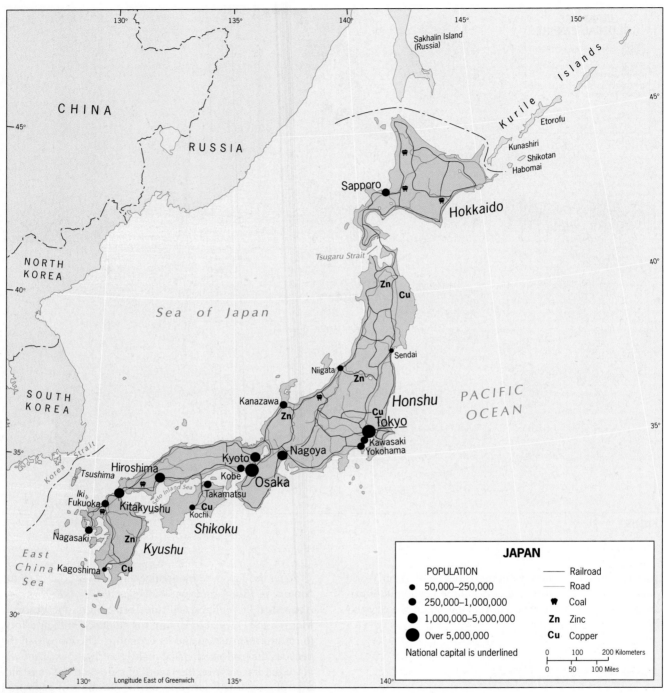

FIGURE 4-3

headquarters. So they chose the town of Edo, on a large bay where Honshu's eastern coastline turns sharply (Fig. 4-3). They renamed the place Tokyo ("eastern capital"), and little more than a century later it was the largest conurbation in the world. From Honshu's coasts, it was no great distance to the shores of mainland Asia, where sources of raw materials and potential markets for Japanese products lay. The notion

of a greater Japanese empire followed naturally from the British example.

But in other ways, the British and Japanese archipelagos, at opposite ends of the Eurasian landmass, differed considerably. In terms of total area, Japan is larger. In addition to Honshu, Japan has three other large islands—Hokkaido to the north and Shikoku and Kyushu to the south—plus nu-

Japan lies at the eastern margin of the Eurasian tectonic plate, where it moves against the Pacific and Philippine plates. Volcanism occurs throughout much of Japan, and the Japanese fight back by trying to control lava flows and mudslides. Here, on the slope of Mt. Sakurajima on Kyushu Island, barriers have been built across the path of the lava so that the next flow will be contained.

merous small islands and islets, for a total land area of about 146,000 square miles (377,000 sq km). Much of this area is mountainous and steep-sloped, geologically young, earthquake-prone, and studded with volcanoes. Britain has lower relief, is older geologically, does not suffer from severe earthquakes, and has no active volcanoes. And in terms of the raw materials for industry, Britain was much better endowed than Japan. Self-sufficiency in iron ore and high-quality coal gave Britain the head start that lasted for a century.

Japan's high-relief topography has been an ever-present challenge. All of Japan's major cities, except the ancient capital of Kyoto, are perched along the coast, and virtually all lie partly on artificial land claimed from the sea. Sail into Kobe harbor, and you will pass artificial islands designed for high-volume shipping and connected to the mainland by automatic space-age trains. Enter Tokyo Bay, and the refineries and factories to your east and west stand on huge expanses of landfill that have pushed the bay's shoreline outward. With 125 million people, the vast majority (77 percent) living in towns and cities, Japan uses its

habitable living space very intensively—and expands it wherever possible.

As Fig. 4-4 shows, farmland in Japan is both limited and regionally fragmented. Urban sprawl has invaded much cultivable land. In the hinterland of Tokyo lies the Kanto Plain; around Nagoya, the Nobi Plain; and surrounding Osaka, the Kansai District—each a major farming zone under relentless urban pressure. All three of these plains lie within Japan's fragmented but well-defined core area (delimited by the gray line on the map), the heart of Japan's prodigious manufacturing complex.

A GEOGRAPHIC REALM?

The boundaries and contents of geographic realms and regions often provide issues for debate. Should Japan be designated as a discrete geographic realm? If the British Isles are defined as one of the regions of the European realm, why is Japan not merely one region of an East Asian realm?

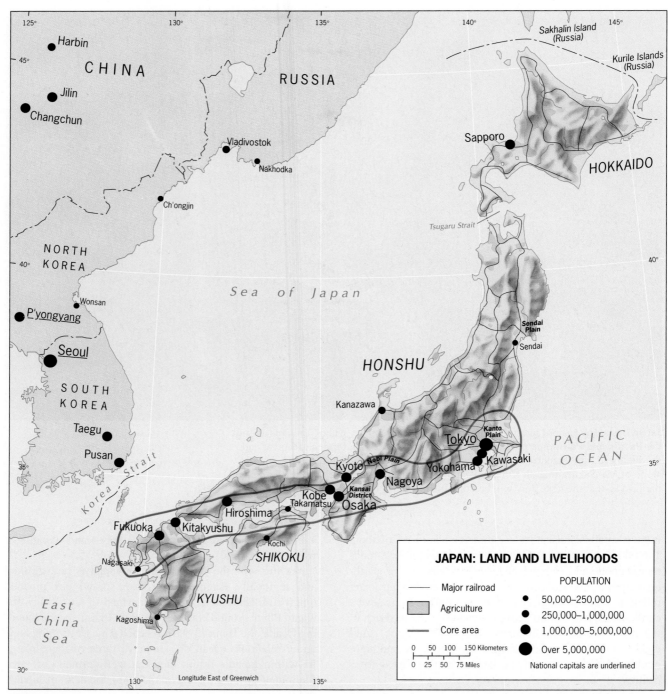

FIGURE 4-4

In time, the convergence of Japan and East Asia may justify such incorporation. At present, however, the Japanese geographic realm is a reality despite its comparatively small territorial size and relatively modest population numbers. Not the numbers, but the remarkable homogeneity of that population constitutes one of the main reasons why Japan is a separate realm. Centuries of insulation and isola-tion, coupled with more recent government policies de-signed to perpetuate the country's ethnic uniformity, have given Japan the most unvaried population of its size in the world. While other countries in the 1970s and 80s accepted Vietnamese refugees fleeing by the hundreds of thousands from the communist regime in their country, Japan allowed fewer than 200 to enter. When a labor shortage developed

Japan remains a cohesive, conformist society despite its modernization and Westernization. These youngsters (all in blue; two "teams," red hats and green hats) have just visited Heian Buddhist shrine in Gotemba near Tokyo (as revealed by a remarkably clear Mount Fuji).

recently, the Japanese looked for immigrant workers . . . among the large Japanese-emigrant community in Brazil. In this world of circulation and interaction, the Japanese have so far succeeded in maintaining a unique ethnic homogeneity.

Closely linked with this ethnic sameness is a cultural uniformity that confirms Japan's geographic identity. From Hokkaido in the north to Kyushu in the south, from one end of Honshu to the other, Japan exhibits a cultural constancy that is pervasive and even baffling. Japan does have its huge modern cities and its small traditional villages—in that sense there is contrast. Yet so do most other societies. In Japan, however, uniformity is ensured by history and custom, by inherited pattern and through the control of behavior. From the universal language to the Shinto national belief system, from the strength of family ties to the role of education and the schools (everyone in Japan wears the same approved school uniform), the Japanese exhibit a cultural uniformity unparalleled in modern times.

A further confirmation of Japan's status as a discrete geographic realm lies in its unmatched technological development, and hence its economic geography. Using raw materials from virtually every corner of the globe, the Japanese have created a manufacturing complex whose production far outstrips that of giant China, the two Koreas, Taiwan, and Hong Kong combined. On that basis alone, Japan cannot be delimited as a mere corner of an East Asian geographic realm. Japan is one of the hearths of the modern world, a truly distinct realm on the Pacific Rim.

EARLY DIRECTIONS

On the map, the Korean Peninsula seems to extend like an unfinished bridge from the Asian mainland toward southern Japan. Two Japanese-owned islands, Tsushima and Iki, in the Korea Strait come close to completing the connection. In fact, Korea *was* linked to Japan in prehistoric times, when sea levels were lower. Peoples and cultural infusions reached Japan from Asia time and again, even after the land bridge between them was inundated by rising waters.

The Japanese islands were inhabited even before the ancestors of the modern Japanese arrived from Asia. The Ainu, a people of Caucasian ancestry, had established themselves on all four of the major islands thousands of years ago. But the Ainu could not hold their ground against the better organized and armed invaders, and they were steadily driven northward until they retained a mere foothold on Hokkaido. Today, the last vestiges of Ainu ancestry and culture are disappearing.

By the sixteenth century, a highly individualistic culture had evolved in Japan, characterized by social practices and personal motivations based on the Shinto belief system. People worshipped many gods, venerated ancestors, and glorified their emperor as a divine figure, infallible and omnipotent. The emperor's military ruler or *shogun* led the campaign against the Ainu and controlled the nation. The capital, Kyoto, became an assemblage of hundreds of magnificent temples, shrines, and gardens. In their local architecture, modes of dress, festivals, theater, music, and other cultural expressions, the Japanese forged a unique society. Handicraft industries, based on local raw materials, abounded.

In the mid-nineteenth century, nonetheless, Japan hardly seemed destined to become Asia's leading power. For 300 years the country had been closed to outside influences, and Japanese society was stagnant and tradition-bound. When the European colonizers first appeared on Asia's shores, their merchants and missionaries were tolerated, even welcomed, in Japan. But as the European presence in East Asia grew stronger, the Japanese began to shut their doors. Toward the end of the sixteenth century, the emperor decreed that all foreign traders and missionaries should be expelled. Christianity (especially Roman Catholicism) had gained a considerable foothold in Japan, but now this was feared as a prelude to colonial conquest. Early in the seventeenth century, the *shogun* launched a massive and bloody campaign that practically stamped out Christianity.

Determined not to suffer the same fate as the Philippines, which had by then fallen to Spain, the Japanese henceforth allowed only the most minimal contact with Europeans. A few Dutch traders, confined to a little island near the city of Nagasaki, were for many decades the sole representatives of Europe in all of Japan. And thus Japan entered a long period of isolation, lasting past the middle of the nineteenth century.

Japan could maintain its aloofness while other areas were falling victim to the colonial tide because of its strong central authority and well-organized military, because of the natural protection provided by its islands, and because of its remoteness along East Asia's difficult northern coast. Also, Japan's isolation was far less splendid than that of China, whose exquisite silks, prized teas, and skillfully made wares attracted traders and usurpers.

When Japan finally came face to face with the new weaponry of its old adversaries, it had no answers. In the 1850s, the steel-hulled "black ships" of the American fleet sailed into Japanese harbors and the Americans extracted one-sided trade agreements. Soon the British, French, and Dutch were also on the scene, seeking similar treaties. When there was local resistance to their show of strength, the Americans quickly demonstrated their superiority by shelling parts of the Japanese coast. By the late 1860s, even as Japan's modernizers were about to overturn the old order, no doubt remained that Japan's protracted isolation had come to an end.

A JAPANESE EMPIRE

When the architects of the Meiji Restoration confronted the challenge to build a Japan capable of competing against powerful adversaries in a changing world, they took stock of the country's assets and liabilities. Material assets, they found, were limited. To achieve industrialization, coal and iron ore were needed. In terms of coal, there was enough to support initial industrialization. Coalfields in Hokkaido and Kyushu were located near the coast; since the new industries also were on the coast, cheap water transportation was possible. The shores of the Seto Inland Sea (Fig. 4-3) became the sites of many factories as a result. But there was little iron ore, certainly nothing like what was available domestically in Britain, and not enough to sustain the massive industrialization the reformers had in mind. This commodity would have to be purchased overseas and imported.

On the positive side, manufacturing—light manufacturing of the handicraft type—already was widespread in Japan. In cottage industries and in community workshops, the Japanese produced textiles, porcelain, wood products, and metal goods. The small ore deposits in the country were enough to supply these local industries. Power came from human arms and legs and from wheels driven by water; the chief source of fuel was charcoal. Importantly, Japan did have an industrial tradition and an experienced labor force that possessed appropriate manufacturing skills. All this was not enough to lead directly to industrial modernization, but it did hold promise for capital formation. The community and home workshops were integrated into larger units, hydroelectric plants were built, and some thermal (coal-fired) power stations were constructed in critical areas.

Now for the first time, Japanese goods (albeit of the light manufacture variety) began to compete with Western products on international markets. The Japanese planners resisted any infusion of Western capital, however: this would have accelerated the industrialization process, but would have cost Japan its economic autonomy. Instead, farmers were more heavily taxed, and the money thus earned was poured into the industrialization effort.

Another advantage for Japan's planners lay in the country's military tradition. Although the *shogun*'s forces had been unable to repel the invasions of the 1850s, this was the result of outdated equipment, not lack of manpower or discipline. So while the economic transformation gathered momentum, military forces, too, were modernized. Barely more than a decade after the Meiji Restoration, Japan laid claim to its first Pacific prize: the Ryukyu Islands (1879). A Japanese colonial empire was in the making.

The Japanese success is written on the map. As Japan's industries grew and diversified, they produced armaments as well as consumer goods. Japanese-made naval ships carrying Japanese troops armed with Japanese weapons challenged and broke China's hold over Formosa (Taiwan) in 1895, confronted and defeated the tsar's Russian army on Sakhalin Island (Karafuto) in 1904, occupied Korea (Chosen) in 1910, and then established a sphere of influence in Northeast China (Manchukuo). Various archipelagos in the Pacific Ocean were acquired by annexation, conquest, or mandate (Fig. 4-2). By the late 1930s, Japan had penetrated and occupied a large part of eastern China, and the Japanese began to propose the establishment and recognition of a so-called Greater East Asia Co-Prosperity Sphere, a *de facto* colonial realm that would combine all of China, Southeast Asia, and numerous Pacific territories under Tokyo's flag.

Japan's imperial designs got a boost during World War II (1941–1945), when the European powers' hold over their colonial realms weakened as a result of the conflict with Germany. As Fig. 4-2 shows, the Japanese now extended their domain in China, Southeast Asia, and the Pacific, including the Philippines. The surprise attack on Pearl Harbor, a severe blow to U.S. military installations there, was a major success—and is still admired in Japanese literature and folklore.

When Japan gained control over its first major East Asian colonies, Taiwan and Korea, its domestic raw-material problems were essentially solved, and a huge labor pool fell under its sway as well. High-grade coal, high-quality iron ore, and other resources were shipped to the factories of Japan; and in Taiwan as well as in Korea, the Japanese built additional manufacturing plants to augment production. This, in turn, provided the equipment to sustain the subsequent drive into China and Southeast Asia.

TEN MAJOR GEOGRAPHIC QUALITIES OF JAPAN

1. Japan is an archipelago. In common with most of the lands of the western Pacific Rim, it is dominated physiographically by high relief.
2. Japan buys raw materials and sells finished products the world over. It lies remote from most of the locations with which it has close economic contact.
3. Japan is the prime example in the world of modernization in a non-Western society.
4. Japan was the non-Western world's major modern colonial power. During the colonial era, it strongly influenced the development of the western Pacific.
5. Although recent interaction has been limited, Japan lies poised for a major role in the development of East Asia, including eastern Siberia as well as China.
6. Japan's modern industrialization depends on the acquisition of raw materials (including energy resources) from distant locales. Japan's own mineral-resource base is limited.
7. Modern and traditional society exist intertwined in Japan.
8. Japan's economy in the mid-1990s is buffeted by growing competition from economic tigers on Australia's Pacific Rim.
9. Japan's low population growth (which may soon turn into negative growth) is posing unprecedented problems for Japanese social programs and employment.
10. Japan's areal functional organization, based on regional specialization, is maturely developed.

MODERNIZATION

The reformers, who in 1868 set Japan on a new course, probably did not anticipate that three generations later, their country would lie at the heart of a major empire sustained by massive military might. They set into motion a process of **modernization**, but they managed to build on, not replace, Japanese cultural traditions. We in the Western world tend to equate modernization with Westernization: urbanization, the spread of transport and communications facilities, the establishment of a market (money) economy, the breakdown of local traditional communities, the proliferation of formal schooling, the acceptance and adoption of foreign

innovations. In the non-Western world, the process often is viewed differently. There, "modernization" is seen as an outgrowth of colonialism, the perpetuation of a system of wealth accumulation introduced by foreigners driven by greed. The local elites who replaced the colonizers in the newly independent states, in that view, only continue the disruption of traditional societies, not their true modernization. Traditional societies, they argue, can be modernized without being Westernized.

Japan's modernization, in this context, is unique in many ways. Having long resisted foreign intrusion, the Japanese did not achieve the transformation of their society by importing a Trojan horse; it was done by Japanese planners, building on the existing Japanese infrastructure, to fulfill Japanese objectives. Certainly Japan imported foreign technologies and adopted innovations from the British and others; but the Japan that was built, a unique combination of modern and traditional elements, was basically an indigenous achievement.

RELATIVE LOCATION

Japan's changing fortunes over the past century reveal the influence of relative location in the country's development. Until well after the Meiji Restoration took place, Britain, on the other side of the Eurasian landmass, lay at the center of a global empire. The colonization and Europeanization of the world were in full swing. The United States was still a developing country, and the Pacific Ocean was an avenue for European imperial competition. Japan, even while it was conquering and consolidating its first East Asian colonies (the Ryukyus, Taiwan, Korea), lay remote from the mainstreams of global change.

Then Japan became embroiled in the Second World War and dealt severe blows to Asia's European colonial armies. The Europeans never recovered: the French lost Indochina, and the Dutch were forced to abandon their East Indies (now Indonesia). When World War II ended, Japan was a defeated and devastated country, but the Japanese had done much to diminish the European presence in the Pacific basin. Moreover, the global situation had changed dramatically. The United States, Japan's trans-Pacific neighbor, had become the world's most powerful and wealthiest country, whereas Britain and its global empire were fading. Suddenly Japan was no longer remote from the mainstreams of global action: now the Pacific was becoming the avenue to the world's richest markets. Japan's **relative location**—its situation relative to the economic and political foci of the world—had changed. Therein lay much of the opportunity the Japanese seized after the postwar rebuilding of their country.

JAPAN'S SPATIAL ORGANIZATION

Imagine this: 125 million people crowded into a territory the size of the state of Montana (pop. 850,000), most of it mountainous, subject to frequent earthquakes and volcanism, with no domestic oilfields, very little coal and few raw materials for industry, and not much level land for farming. If Japan today were an underdeveloped country in need of food relief and foreign aid, explanations would abound: overpopulation; inefficient farming; energy shortages.

True, only an estimated 18 percent of Japan's national territory is designated as habitable. And Japan's large population is crowded into some of the world's largest cities. Moreover, Japan's agriculture *is* not especially efficient.

But Japan defeated the odds by calling on old Japanese virtues: organizational efficacy, massive productivity, dedication to quality, and adherence to common goals. Even before the Meiji Restoration, Japan was a tightly organized, populous country with as many as 30 million citizens; historians suggest that Edo, even before it was selected as the capital and renamed Tokyo, may have been the world's largest urban center. The Japanese were no strangers to urban life; they knew manufacturing, and they prized social and economic order.

Areal Functional Organization

All this proved invaluable to the modernizers when they set Japan on its new course. The new industrial growth of the country could be based on the urban and manufacturing development that was already taking place. As we noted, Japan does not possess major domestic raw material sources, so no substantial internal reorganization was necessary. However, some cities did possess advantages over others in terms of their sites and their situations relative to those limited local resources and, more importantly, relative to external sources of raw materials. As Japan's regional organization took shape, a hierarchy of cities developed; Tokyo took and kept the lead, but other cities also grew rapidly as industrial centers. This process was governed by a geographic principle that Allen Philbrick called **areal functional organization**, a set of five interrelated tenets that help explain the evolution of regional organization (not only in Japan but everywhere else).

First, human activity has spatial focus in that it is concentrated in some locale—a farm or factory or store. Second, such "focal" activity is carried on in certain particular places. Obviously, no two establishments can occupy exactly the same spot on the earth's surface, so every one of them has a finite or absolute location; but what is more relevant is that every establishment has a location relative to other establishments and activities. Since no human activity is carried on in complete isolation, the third idea is that interconnections develop among the various establishments. Farmers send crops to markets and buy equipment at service centers. Mining companies buy gasoline from oil companies and lumber from sawmills, and they send ores to refineries. Thus a system of interconnections emerges and grows more complex as human capacities and demands expand, and they are expressed spatially as units of areal organization. Philbrick's fourth idea is that these units of areal organization—or regions—evolve as a result of human "creative imagination" as people apply their total cultural experience as well as technological know-how when they decide how to organize and rearrange their living space. Finally, it is possible to recognize levels of development in areal organization, a ranking or hierarchy based on type, extent, and intensity of exchange. As Philbrick put it: "The progression of [areal]

Kyoto is the old capital of Japan, the city of 1,600 religious shrines. The serenity of Shinto (the Japanese ethnic religion closely related to Buddhism), with its marvelous temples and meticulously maintained gardens, pervades this historic city.

units from individual establishments to world regions and the world as a whole are formed into a hierarchy of regions of human organization."

In the broadest sense, regions of human organization can be categorized under subsistence, transitional, and exchange types, with Japan's areal organization reflecting the last of these. Within each type of areal organization however—and especially within the complex exchange type of unit—individual places can also be ordered or ranked on the basis of the number and kinds of activities and interconnections they generate. A map of Japan showing its resources, urban settlements, and surface communications can tell us a great deal about the kind of economy the country has (Fig. 4-4). It looks just like maps of other parts of the world where an exchange type of areal organization has developed—a hierarchy of urban centers ranging from the largest cities to the tiniest hamlets, a dense network of railways and roads connecting these places, and productive agricultural areas near and between the urban centers.

In Japan's case, though, Fig. 4-5 shows us something else: it reflects the country's external orientation, its dependence on foreign trade. All primary and secondary regions lie on the coast; of all the cities in the million-size class, only Kyoto (1.7 million) lies in the interior. If we were to deduce that Kyoto does not match Tokyo-Yokohama, Osaka-Kobe (10.6 million), or Nagoya (3.2 million) in terms of industrial development, that would be correct—the old capital remains a center of small-scale light manufacturing. Actually, Kyoto's ancient character has been deliberately preserved, and large-scale industries have been discouraged. With its old temples and shrines, its magnificent gardens, and its many workshop and cottage industries, Kyoto remains a link with Japan's premodern past (see photo above).

Leading Economic Regions

As Fig. 4-5 shows, Japan's dominant region of urbanization and industry (along with very productive agriculture) is the Kanto Plain, which contains about one-third of the Japanese population and is focused on the Tokyo-Yokohama-Kawasaki metropolitan area (26.2 million). This gigantic cluster of cities and suburbs (the world's largest urban agglomeration), interspersed with intensively cultivated farmlands, forms the eastern anchor of the country's elongated and fragmented core area (Fig. 4-5). Besides its flatness, the Kanto Plain possesses other advantages: its fine natural harbor at Yokohama, its relatively mild and moist climate, and its central location with respect to the country as a whole. (The region's only disadvantage is its vulnerability to earthquakes—see box entitled "When the Big One Strikes".) It has also benefited enormously from Tokyo's designation as the modern capital, which coincided with Japan's embarkation on its planned course of economic development. Many industries and businesses chose Tokyo as their headquarters in view of the advantages of proximity to the government's decision-makers.

The Tokyo-Yokohama-Kawasaki conurbation has become Japan's leading manufacturing complex, producing more than 20 percent of the country's annual output. The raw materials for all this industry, however, come from far away. For example, the Tokyo area is among the chief steel producers in Japan, using iron ores from the Philippines, Malaysia, Australia, India, and even Africa; most of the coal is imported from Australia and North America and the petroleum from Southwest Asia and Indonesia. The Kanto Plain cannot produce nearly enough food for its massive resident population. Imports must come from Canada, the United States, and Australia as well as from other areas in Japan. Thus Tokyo depends completely on

FIGURE 4-5

its external trading ties for food, raw materials, and markets for its wide variety of products, which run the gamut from children's toys to high-precision optical equipment to the world's largest oceangoing ships.

The second-ranking economic region in Japan's core area is the Osaka-Kobe-Kyoto triangle—also known as the Kansai (formerly Kinki) District—located at the eastern end of the Seto Inland Sea. The position of the Osaka-

Kobe conurbation, with respect to the old Manchurian Empire (Manchukuo) created by Japan, was advantageous. Situated at the head of the Seto Inland Sea, Osaka was the major Japanese base for the China trade and the exploitation of Manchuria, but it suffered when the empire was destroyed and its trade connections with China were lost after World War II. Kobe (like Yokohama, which is Japan's chief shipbuilding center) has remained one of the

WHEN THE BIG ONE STRIKES

The Tokyo-Yokohama-Kawasaki urban area is the biggest metropolis on earth. But Tokyo is more than a large city: it constitutes the most densely concentrated financial and industrial complex in the world. In 1993, 12 of the world's 13 largest banks were headquartered in Tokyo. Two-thirds of Japan's businesses worth more than $50 million are also clustered here, many with vast overseas holdings. More than half of Japan's huge industrial profits (averaging about U.S. $100 billion annually since the late 1980s) are generated in the factories of this gigantic metropolitan agglomeration.

But Tokyo has a worrisome environmental history because three active tectonic plates are converging here (Fig. 4-6). All Japanese know about the "70-year" rule: over the past three-and-a-half centuries, the Tokyo area has been struck by major earthquakes roughly every 70 years—in 1633, 1703, 1782, 1853, and 1923. The Great Kanto Earthquake of 1923 set off a firestorm that swept over the city and killed an estimated 140,000 people. Moreover, Tokyo Bay virtually emptied of water; then a *tsunami* (seismic sea wave) roared back in, sweeping homes, factories, and all else before it. That Japan could overcome this disaster was evidence of the strength of its ongoing economic miracle.

Today, Tokyo is more than a national capital. It is a global financial and manufacturing center in

which so much of the world's wealth and productive capacity are concentrated that an earthquake comparable to the one of 1923 would have a calamitous effect worldwide. Ominously, the Tokyo of the 1990s is a much more vulnerable place than the Tokyo of the 1920s. True, building regulations are stricter and civilian preparedness is better. But whole expanses of industries have been built on landfill that will liquefy; the city is honeycombed by underground gas lines that will rupture and stoke countless fires; congestion in the area's maze of narrow streets will hamper rescue operations; and many older high-rise buildings do not have the structural integrity that has lately emboldened builders to build skyscrapers of 50 stories and more. Add to this the burgeoning population of the Kanto Plain—approaching 30 million on what may well be the most dangerous 4 percent of Japan's territory—and we realize that the next big earthquake in the Tokyo area will not be a remote, local news story.

The Japanese would need cash to rebuild, which would force them to sell many of their worldwide holdings. This would quickly precipitate a global financial crisis. It is a measure of the interconnectedness of our world that we should all hope that nature will break its "70-year" rule in the mid-1990s and prolong Tokyo's stability.

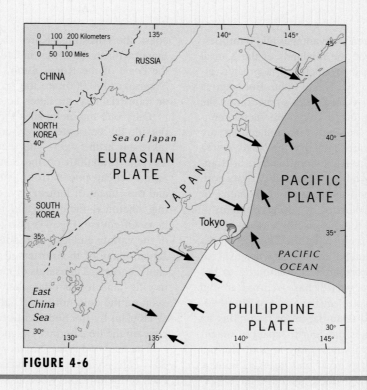

FIGURE 4-6

country's busiest ports, handling inland-sea traffic as well as extensive overseas linkages. Kyoto, as we noted, remains much as it was before Japan's great leap forward, a city of small workshop industries. The Kansai District is also an important farming area. Rice, of course, is the most intensively grown crop in the warm moist lowlands, but this agricultural zone is not as large as the one on the

Kanto Plain. Here is another huge concentration of people that requires a large infusion of foodstuffs every year.

The Kanto Plain and Kansai District, as Fig. 4-5 indicates, are the two leading primary regions within Japan's core. Between them lies the Nobi Plain (also called the Chubu District), focused on the industrial metropolis of Nagoya, Japan's leading textile producer. The map indi-

This satellite view of the Tokyo metropolitan area shows Tokyo Bay (A) (beneath which three tectonic plates converge), Tokyo proper (B), Kawasaki (C) and Yokohama (D); the latter two are both built substantially on landfill; and are subject to earthquake-induced liquefaction. Nearly one-third of Japan's 125 million people live in the area shown here.

cates some of the Nagoya area's advantages and liabilities. The Nobi Plain is larger than the lowlands of the Kansai District; thus, its agricultural productivity is greater, although not as great as that of the Kanto Plain. But Nagoya has neither Tokyo's centrality nor Osaka's position on the Seto Inland Sea: its connections to Tokyo are via the tenuous Sen-en Coastal Strip. Its westward connections are somewhat better, and there are signs that the Nagoya area is coalescing with the Osaka-Kobe conurbation. Still another disadvantage of Nagoya lies in the quality of its port, which is not nearly as good as Tokyo's Yokohama or Osaka's Kobe and has been plagued by silting problems.

Westward from the three regions just discussed—which together comprise what is often called the *Tokaido* megalopolis—extends the Seto Inland Sea, along whose shores the remainder of Japan's core area is continuing to develop. The most impressive growth has occurred around the western entry to the Inland Sea (the Strait of Shimonoseki), where Kitakyushu—a conurbation of five cities on northern Kyushu—constitutes the fourth Japanese manufacturing complex and primary economic region. Honshu and Kyushu are connected by road and railway tunnels, but the northern Kyushu area does not have an urban-industrial equivalent on the Honshu side of the strait. The Kitakyushu conurbation (2.7 million) includes Yawata, site of northwest Kyushu's (rapidly declining) coal mines, and it was on the basis of this coal that the first steel plant in Japan was built there; for many years it was the country's largest. The advantages of transportation here at the western end of the Seto Inland Sea are obvious: no place in Japan is better situated to do business with Korea and China. As relations with mainland Asia expand, this area will reap many of the benefits. Elsewhere on the inland-sea

coast, the Hiroshima-Kure urban area (1.0 million) has a manufacturing base that includes heavy industry. And on the coast of the Korea Strait, Fukuoka (1.3 million) and Nagasaki (490,000) are the principal centers—the former an industrial city, the latter a center of large shipyards.

Only one major Japanese manufacturing complex lies outside the belt extending from Tokyo in the east to Kitakyushu in the west—the secondary region centered on Toyama (360,000) on the Sea of Japan. The advantage here is cheap power from nearby hydroelectric stations, and the cluster of industries reflects it: paper manufacturing, chemical factories, and textile plants have located here. Of course, Fig. 4-5 also gives an inadequate picture of the variety and range of industries that exist throughout Japan, many of them oriented to local (and not insignificant) markets. Thousands of manufacturing plants operate in cities and towns other than those shown on this map, even on the cold northern island of Hokkaido, connected to Honshu by the Seikan rail tunnel, the world's longest, beneath the treacherous Tsugaru Strait.

The map of Japan's areal organization shows the country's core area to be dominated by four primary regions, each of them primary because they duplicate to some degree the contents of the others. Each contains iron and steel plants, each is served by one major port, and each lies in or near a large, productive farming area. What the map does not show is that each also has its own external connections for the overseas acquisition of raw materials and the sale of finished products. These linkages may even be stronger than those among the four internal regions of the core area; only in the case of Kyushu and its coal have domestic raw materials played much of a role in shaping the nature and location of heavy manufacturing, and these resources are all but depleted today. In the structuring of the country's areal functional organization, therefore, more than just the contents of Japan itself is involved. In this respect Japan is not unique: all countries that have exchange-type organizations, must, to some degree, adjust their spatial forms and functions to the external interconnections required for progress. But it would be difficult to find a country in which this is truer than in Japan.

FOOD: PRODUCTION AND PRICE

Japan's economic modernization so occupies the center stage that it is easy to forget that considerable achievements have been made in the venerable field of agriculture. Japan's planners, no less interested today in closing the food gap than in expanding industries, have created extensive networks of experiment stations to promote mechanization, optimal seed selection and fertilizer use, and information services to distribute to farmers as rapidly as possible knowledge that is useful for enhancing crop yields. Although this program has been very successful, Japan faces the unalterable reality of its stubborn topography: there simply is insufficient land to farm.

Less than one-fifth of the country's total area is in cultivation; although there may still be some land that can be brought into production, it is so mountainous or cold that its contribution to the total harvest would be minimal anyway. Japanese agriculture may resemble Asian agriculture in general, but nowhere else do so many people depend on so little land. Japan's overall population density is 859 per square mile (338 per sq km) to begin with; but when the arable land alone is considered, it turns out that 6,936 people depend on the average square mile of farmland (2,678 per sq km)! This measure of persons per unit of cultivable land, as we note on p. 462, is the *physiologic* density, and Japan's is one of the highest on earth. Even Bangladesh (3,263 per square mile [1,260 per sq km]) and China (3,207 [1,238]), notoriously crowded in confined farmlands, exhibit far lower physiologic densities than Japan.

To augment Japan's meager land resources, farmers have long been required to cultivate steep slopes. This aerial view near the Seto Inland Sea shows the intricate terracing of hillsides that is necessary to create level surfaces and safeguard against ruinous soil erosion.

The Japanese go to extraordinary lengths to maximize their limited farming opportunities. Rice yields per unit area are among the highest in the world, and the country manages to produce about 70 percent of its annual food requirements. This is achieved through the use of higher-yielding hybrid rice varieties, by assigning more than 90 percent of all farmland to food crops, mechanizing every possible operation, irrigating more than half of Japan's agriculture, painstakingly terracing steep slopes, and practicing multiple cropping, intercropping, intensive fertilization, and transplanting (i.e., the use of seedbeds to raise the young plants of the next crop while its predecessor matures in the field). Hilly terrain is given over to cash crops such as tea and grapes (Japan has a thriving wine industry). Vegetable gardens ring the outskirts of all the major cities.

But all this costs money, and Japan's food prices are high. In a country where the opportunities of the cities and industries beckon, it is difficult to keep farmers on the land; in 1920, more than 50 percent of the labor force were full-time farmers, but today under 8 percent remain on the farms—many of them part-timers who derive much of their income from non-agricultural sources. To induce them to stay there, the Japanese governmental system assigns considerable political power to these underpopulated rural areas, and it keeps the prices of farm products artificially high. During the early 1990s, when the Americans and the Japanese got into a quarrel over the flow of products to their respective markets, it was pointed out that American rice farmers could provide Japanese consumers with rice at *one-sixth* the price Japanese rice was costing them. But the Japanese market was not as open to American rice as the American market was to Japanese-made automobiles.

With their national diet so rich in such starchy foods as rice, wheat, barley, and potatoes, the Japanese need protein to balance it. Fortunately, this can be secured in sufficient quantities, not through foreign purchase but by harvesting it from rich fishing grounds near the Japanese islands. With customary thoroughness, the Japanese have developed a fishing industry that is larger than that of the United States or any of the long-time fishing nations of northwestern Europe, and it now supplies the domestic market with a second staple after rice. Although mention of the Japanese fishing industry brings to mind a fleet of ships scouring the oceans and seas far from Japan, the fact is that most of this huge catch (about one-seventh of the world's annual total) comes from waters within a few dozen miles of Japan itself. Where the warm Kuroshio and Tsushima currents meet colder water off Japan's coasts, a rich fishing ground exists that yields sardines, herring, tuna, and mackerel in the warmer waters and cod, halibut, and salmon in the seas to the north. Along Japan's coasts there are about 4,000 fishing villages, and tens of thousands of small boats ply the waters offshore to bring home catches that are distributed both locally and in city markets. The Japanese also practice aquaculture—the "farming" of freshwater fish in artificial ponds and flooded paddy (rice) fields, seaweeds in home aquariums, and oysters, prawns, and shrimp in shallow bays; they are even experimenting with the cultivation of algae for their food potential.

When you walk the streets of Japanese cities, you will notice something that Japanese vital statistics are reporting: the Japanese are getting taller and heavier. Coupled with this is evidence that heart disease and cancer are rising as causes of mortality. The reason seems to lie in the changing diets of many Japanese, especially the younger people. When rice and fish were the staples for virtually all, and calories available were limited but adequate, the Japanese as a

Shikoku is the smallest of Japan's four largest islands, the southern flank of the Seto Inland Sea. The south shore of this mountainous island is lined with prosperous fishing villages such as this one. Shikoku is now linked to Honshu by the world's longest set of suspension bridges, and development is speeding up.

nation were among the world's healthiest and longest-lived. But as fast-food establishments diffused throughout Japan, and Western (especially American) tastes for red meat and fried food spread among youths and young adults, their combined impact on the population was soon evident. Coupled with the heavy cigarette-smoking that prevails in all the Asian countries of the Pacific Rim, this is changing the region's medical geography.

JAPAN'S FUTURE

In the mid-1990s, Japan remains the economic giant on Austrasia's Pacific Rim, an industrial powerhouse whose products dominate world markets and whose investments span the globe. Nevertheless, Japan faces problems. Competition from other Pacific Rim countries, including South Korea, Taiwan, and Hong Kong, is undercutting Japanese products on world markets, such as Japan's own products once undercut more expensive Western goods. Many of Japan's overseas investments have lost much of their value. And always, Japan's dependence on foreign oil is a potential weakness: rising costs and/or supply interruptions would have grave consequences for the country's energy-dependent economy.

Nonetheless, Japan still has enormous further potential, much of it arising from its relative location. As we noted in Chapter 2, the Russian Far East, lying opposite northern Honshu and Hokkaido, is a storehouse of raw materials awaiting exploitation, and the Japanese are ideally located to help achieve this (a major obstacle, however, has been a group of small islands northeast of Hokkaido—see box at right). Moreover, Japan's technological prowess is finding an ever more sizeable market in developing China, and again the Japanese are situated on the doorstep of growing opportunity.

All this should be viewed against the backdrop of a changing social and cultural Japan. The population of 125 million is aging, and demographers report that it will soon stabilize and then begin a slow decline. That is why, as we noted earlier, Japan faces a labor shortage—a situation previously unheard of. And the growing number of elderly people is increasing the costs the state must pay in the form of health care, retirement, and other services for the aged. This also comes at a time when modernization is putting strains on the family. The Japanese have a strong tradition of family cohesion and veneration of its oldest members, but there is much evidence that this custom is breaking down. Younger people want more privacy and self-determination, and they are less willing than their parents and grandparents to accept overcrowded housing and limited comforts. Yet in Japan, for all its material wealth, family homes tend to be small, cramped, often flimsily built, and sometimes without basic amenities considered indispensable in other developed countries. With

THE WAGES OF WAR

Japan and the former Soviet Union never signed a peace treaty to end their World War II conflict. The reason? There are actually four reasons, all of them on the map just to the northeast of Japan's northernmost large island, Hokkaido (Fig. 4-3). Their names are Habomai, Shikotan, Kunashiri, and Etorofu. The Japanese call these four rocky specks of the Kurile Island chain their "Northern Territories." The Soviets occupied them late in the war, but they never gave them back to Japan. Now they are part of Russia, and the Russians have not turned them over, either.

The islands themselves are no great prize. During the Second World War, the Japanese brought 40,000 forced laborers, most of them Koreans, to the islands to mine the minerals found there. When the Red Army overran them in 1945, the Japanese were ordered out, and most of the Koreans fled. Today the population of about 50,000 is mostly Russian, many of them members of the military based on the islands, and their families. At their closest point the islands are only 3 miles (5 km) from Japanese soil, a constant and visible reminder of Japan's defeat and the loss of Japanese land. Moreover, their territorial waters bring Russia even closer, so that the islands' geostrategic importance far exceeds their economic potential.

Attempts to settle the issue have failed. In 1956, Moscow offered to return the tiniest two, Shikotan and Habomai, but the Japanese refused, demanding all four islands back. In 1989, then-Soviet President Gorbachev visited Tokyo in the hope of securing an agreement. The Japanese, it was widely reported, offered an aid-and-development package worth U.S. $26 billion to develop Russia's eastern zone—its Pacific Rim and the vast resources of the eastern Siberian interior. This would have begun the transformation of Russia's Far East, stimulated the ports of Nakhodka and Vladivostok, and made Russia a participant in the spectacular growth of the Austrasian Pacific Rim.

But it was not to be. Subsequently, Russian President Boris Yeltsin also was unable to come to terms with Japan on this issue, facing opposition from the islands' inhabitants and from his own government in Moscow. And so World War II, a half-century after its conclusion, continues to cast a shadow over this northernmost segment of the western Pacific Rim.

Japan today is experiencing a North-American-style postindustrial revolution. The rise of an ultramodern, information-based economy is evident across the country, and is symbolized by Tsukuba Science City (pop. 275,000) in suburban Tokyo. This 70,000-acre community, reminiscent of North Carolina's Research Triangle Park (box, p. 201), is a complex of high-tech university and government facilities that perform cutting-edge research in science and engineering.

their high incomes, the Japanese have come to rank among the world's most-traveled tourists, and, having seen how Americans and Europeans of similar income levels live, they have returned dissatisfied.

This dissatisfaction also spills over into the workplace and into the schools. Japan's work ethic is world renowned, but it is sustained at a high cost to the individual and often involves enormous demands of time and dedication to the corporate good. In education the requirements are perhaps the most rigorous anywhere, and the competition is fierce. Now some leaders are beginning to question the appropriateness of such standards and regulations in the new, economic-world-power Japan.

For the countries and economic clusters along Asia's Pacific Rim, Japan remains the model—a model to be emulated but also challenged. Workers in the factories of Taiwan are paid less than in Japan, and in South Korea they are paid less than in Taiwan. And so their cheaper products compete with those of the Japanese; but from other places (China, Malaysia) come cheaper goods still. Japan stands at a crossroads: of great opportunity on East Asia's threshold, of growing rivalry from its Pacific Rim neighbors to the south.

◆ SOUTH KOREA

On the Asian mainland, directly across the Sea of Japan, lies the peninsula of Korea (Fig. 4-3), a territory about the size of the state of Idaho, much of it mountainous and rugged, and containing a population of 68.3 million. Unlike Japan, however, Korea has long been a divided country, and the Koreans a divided nation. For uncounted centuries Korea has been a pawn in the struggles of more powerful neighbors. It has been a dependency of China and a colony of Japan; when it was freed from Japan's oppressive rule at the end of World War II (1945), Korea was divided for administrative purposes by the victorious allied powers. That division gave North Korea (north of the 38th parallel) to the forces of the Soviet Union, and South Korea to those of the United States. In effect, Korea traded one master for two new ones. The country was not reunited for the rest of the century because North Korea immediately fell under the communist ideological sphere and evolved as a dictatorship in the familiar (but in this case extreme) pattern. South Korea, with massive American aid, became part of East Asia's capitalist perimeter. Once again, it was the will of external powers that prevailed over the desires of the Korean people.

In 1950, North Korea sought to reunite the country by force and invaded South Korea across the 38th parallel. This attack drew a United Nations military response led by the United States. Thus began the devastating Korean War (1950–1953), in which North Korea's forces pushed far to the south only to be driven back across their own half of Korea almost to the Chinese border. Then China's Red Army entered the war and drove the U. N. troops southward again. A ceasefire was arranged in 1953, but not before the people and the land had been ravaged in a way that was unprecedented even in Korea's violent past. The cease-fire line shown in Fig. 4-7 became a heavily fortified *de facto* boundary.

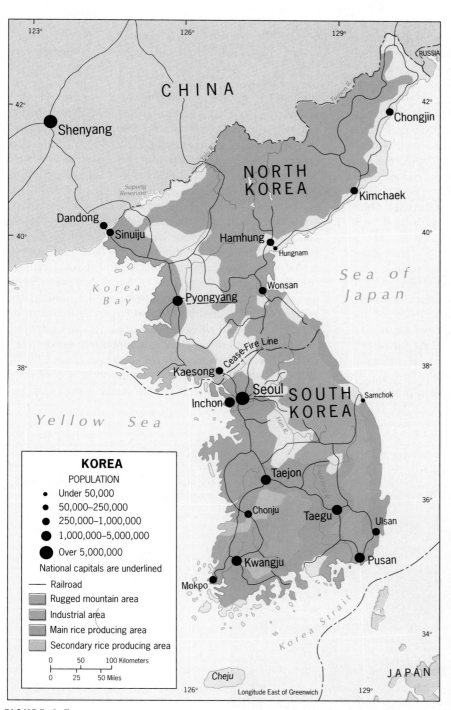

FIGURE 4-7

For decades there was no contact of any kind across this border, but in the early 1990s there were signs of mutual interest in an approach toward accommodation.

Our focus here is on South Korea, the country that, with 45 percent of Korea's land area but with two-thirds of the Korean population, has emerged as one of the economic tigers on the Pacific Rim. There is no telling what Korea might have achieved as an undivided country, because North and South Korea are in a situation that geographers call **regional complementarity**. Such a condition arises when two adjacent regions complement each other in economic-geographic terms. In this case, North Korea has raw materials that the industries of South Korea need; South Korea produces food required by the North; North Korea produces chemical fertilizers needed on farms in the South.

Under the political circumstances of the past half-century, however, the two Koreas were cut off from each other, and they developed in opposite directions. North Korea, whose large coal and iron ore deposits attracted the Japanese and whose hydroelectric plants produce ample electricity (which could be used in the South), traded mainly with the neighboring Chinese and the Soviet Russians. South Korea's external trade connections were with the United States, Japan, and Western Europe.

In the early postwar years, there were few indications that South Korea would emerge as a major economic force on the Pacific Rim and, indeed, on the world stage. Over 70 percent of all workers were farmers; agriculture was inefficient; the country was stagnant. But huge infusions of aid, first from the United States and later also from Japan, coupled with the reorganization of farming and the stimulation of industries (even if they lost money), produced a dramatic turnaround. Large feudal estates were parcelled out to farm families in 3-hectare (7.5-acre) plots, and a program of massive fertilizer importation was begun. Production rose to meet domestic needs, and some years yielded surpluses.

The industrialization program was based on modest local raw materials, plentiful and capable labor, ready overseas (especially American) markets, and continuing foreign assistance. Despite recurrent political instability and social unrest, successive South Korean governments managed to sustain a rapid rate of economic growth that placed the country, by the late 1980s, among the world's top 10 trading powers. To get the job done, the government borrowed heavily from overseas, and it controlled banks and large industries. South Korea's growth resulted from **state capitalism** rather than free-enterprise capitalism, but it had impressive results. South Korea became the world's largest shipbuilding nation; its automobile industry grew rapidly; and iron and steel and chemical industries thrive. There has been less emphasis, however, on smaller, high-technology industries, and these are the strengths of other Pacific Rim economic tigers. So the future remains uncertain.

Today, however, South Korea is prospering, and its economic prowess can be seen on the map (Fig. 4-7). The capital, Seoul, with 12.4 million inhabitants, is the world's ninth largest metropolis and the anchor of a huge industrial complex facing the Yellow Sea at the waist of the Korean Peninsula. Hundreds of thousands of farm families migrated to the Seoul area after the end of the Korean War (today only about 25 percent of South Koreans remain on the land). They also moved to Pusan (4.3 million), the nucleus of the country's second-largest manufacturing zone, located on the Korea Strait opposite the western tip of Honshu. And the government-supported, urban-industrial drive here continues. Just 25 years ago, Ulsan City, 40 miles (60 km) north of Pusan along the coast, was a fishing center with perhaps 50,000 inhabitants; today its popu-

On Ulsan City's docks, a scene indicative of South Korea's industrial prowess: freshly-assembled Hyundai automobiles await shipment to the United States.

lation approaches 700,000, nearly half of them the families of workers in the Hyundai automobile factories and the local shipyards. The third industrial area shown on Fig. 4-7, focused on the city of Kwangju (1.3 million) at the southwestern tip of the peninsula, does not yet match the Seoul and Pusan complexes, but it has an advantageous relative location and is bound to develop further.

South Korea's success inevitably has brought pressure from its competitors, including its capitalist allies. Cheap South Korean textiles undercut the Japanese textile industry, and Japanese textile makers demanded protection. Low-priced Korean automobiles captured a share of the American and European markets, and Japanese as well as American automakers wanted to limit the Koreans' access to these markets. But South Korea diversified is production and continued its climb on the ladder already topped by Japan.

◆ TAIWAN

The island of Taiwan, China's government in Beijing insists, is a wayward part of the Chinese state, some day to be brought back into the fold. We will defer a discussion of the background to that claim until Chapter 10. Here we observe an undeniable geographic reality: as a separate society, Taiwan has become one of the most potent economic powers on the Pacific Rim, second only to Japan by many measures of economic success.

Taiwan, formerly named Formosa, experienced a half-century of Japanese colonialism beginning in 1895 and ending at the close of the Second World War. The Japanese viewed Taiwan not only as a source of food and raw materials but also as a market for Japanese products. To ensure the latter, Japan launched a prodigious development program, involving road and railroad construction, irrigation projects, hydroelectric schemes, mines (mainly for coal), and factories. Farmlands were expanded, and farming methods improved.

After Japan's defeat and ouster in 1945, Taiwan was invaded four years later by another group of outsiders—the defeated Chinese "nationalists" who had been fighting the communists on the mainland. Led by Chiang Kai-shek, these invaders in 1950 accounted for about one-fifth of the island's population, and with their military might and weapons they took control of the place. Being anti-communist, they were supported by the United States, which (along with Chiang's regime) began the reconstruction of the war-damaged infrastructure.

Taiwan, as Fig. 4-8 shows, is not a large island. It is smaller than Switzerland but has a population of 21.3 million. Most of this population is clustered in an arc lining the western and northern coasts; the eastern half of Taiwan is dominated by the Chungyang Mountains, an area of high elevations (some over 10,000 feet [3,000 m]), steep slopes,

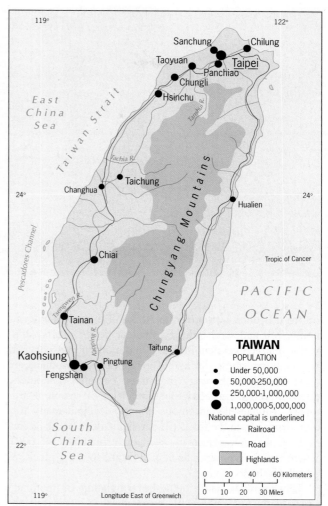

FIGURE 4-8

and dense forests. Westward, these mountains yield to a zone of hilly topography, and, facing the Taiwan Strait, a substantial coastal plain. Streams from the mountains irrigate the paddy fields, and farm production has more than doubled since 1950 even as hundreds of thousands of farmers left the fields to seek work in Taiwan's mushrooming industries.

Today, the lowland urban-industrial corridor of western Taiwan is anchored by the capital, Taipei (Taibei) (3.4 million), at the island's northern end and rapidly growing Kaohsiung (Gaoxiong) (1.7 million) in the far south.* Chilung (Jilong) (380,000), Taipei's outport, was developed by the Japanese to export nearby coal, but now the raw materials flow the other way. Taiwan imports raw cotton for its textile industry, bauxite (for aluminum) from Indonesia,

---

*The Taiwanese have retained the old Wade-Giles spelling of place names; China now uses the Pinyin system (see Chapter 10). Names in parentheses are written according to the Pinyin system.

The bustling heart of Kaohsiung (pronounced gau-shee-AHNG), Taiwan's second city. This rapidly-growing metropolis is the urban anchor of southern Taiwan, a busy seaport, and the site of the country's first export processing zone.

oil from Brunei, iron ore from Africa. Taiwan has a developing iron and steel industry, nuclear power plants, shipyards, a large chemical industry, and modern transport networks. Increasingly, though, Taiwan's export products are those of high-technology industries: personal computers, telecommunications equipment, precision electronic instruments. Taiwan has enormous brainpower, and many foreign firms join in the R&D (research and development) carried on in such places as Hsinchu (Xinzhu) (345,000), where the government has helped establish a high-tech research facility.

Taiwan, with one-fiftieth the population of mainland China, in the early 1990s still ranked ahead of its giant neighbor as a trading nation (and thirteenth in the world!). Two significant geographic aspects should be noted. First, Taiwan's trade with the People's Republic of China, all of it via Hong Kong, was reaching significant proportions in the early 1990s, annually exceeding U.S. $6 billion in value. And for the future, note that directly opposite Taiwan's capital and core area, across the Taiwan Strait, lies one of China's new Open Coastal Cities, Fuzhou, in Fujian Province (see Chapter 10). Should Taiwan and China manage to overcome their political estrangement, the benefits to both could be enormous.

In some ways Taiwan's emergence as an economic tiger on the Pacific Rim is more spectacular even than Japan's. True, Taiwan received much Western assistance, but it got out of debt faster than could have been expected. Today, income per capita is well over U.S. $10,000 per year, which is higher than in many European countries. Profits are so high that Taiwan plans to invest hundreds of billions of dollars in its own development projects over the remainder of the 1990s; and where Taiwanese investment is welcome outside the country, additional billions are being spent. Taiwan, for instance, is the largest foreign inves-

tor in the rebuilding of Vietnam and may thereby have a foothold in a future economic tiger.

Taiwan's greatest problem is political, not economic. We will discuss this issue in Chapter 10; suffice it to say here that Taiwan is not represented in the United Nations (except by the claimant state, China); that Taiwan, which stood firm against communism in the 1960s and 1970s, is now off limits to U.S. cabinet members at the insistence of Beijing; and that Taiwan's own democracy movement culminated in 1991—not in a massacre, but in free elections for a new parliament. Now Taiwan's political options are as limited as its economic horizons are endless, and a single misstep, such as a move perceived in Beijing as a move toward independence and nationhood, could precipitate an invasion and war. Thus the much-championed rights of the Lithuanians, Armenians, and Slovenians do not apply to Taiwan.

◆ HONG KONG

The economic tigers of the Pacific Rim often are described as economic miracles, and Hong Kong certainly qualifies. Here, on the left bank of the estuary of South China's Pearl (Zhu) River, nearly 6 million people (97 percent of Chinese ancestry) crowded onto a fragmented, hilly patch of territory covering just 400 square miles (1,000 sq km) under the tropical sun have created an economy that is larger than those of more than a hundred countries.

Hong Kong consists of three parts, of which the island named Hong Kong (32 square miles/82 sq km) is one. Other islands, including Lan Tao, fringe the mainland parts of Hong Kong: the Kowloon Peninsula and the New Territories (Fig. 4-9). The capital, Victoria, lies on Hong

Kong Island and overlooks Victoria Harbor, one of the busiest in the world. As these place names suggest, Hong Kong has long been a British colony. The islands and the Kowloon Peninsula were ceded permanently by China to Britain in 1841 and 1860, respectively, but the rest, the New Territories, were leased on a 99-year basis in 1898. That lease expires in 1997, and the British and Chinese have agreed on the transfer of authority over all three of the dependency's components from London to Beijing on July 1 of that year. East Asia's economic tigers seem to share uncertain futures.

With its excellent deep-water harbor, Hong Kong long has been the major *entrepôt* of the western Pacific between Shanghai and Singapore. Certainly the Chinese could have recaptured the colony during the late 1940s, when the communist forces were victorious throughout China. But both Beijing and London saw advantages in maintaining the status quo. During the period of isolation and communist reorganization, Hong Kong provided the People's Republic of China with a convenient place of contact with the Western world without any need for long-distance entanglement. Even when hundreds of thousands of Chinese from Guangdong and Fujian provinces fled to Hong Kong, and later when the city was a place of rest and recuperation for American forces during the Indochina War, the Chinese government continued to tolerate the British colonial presence. Indeed, the colony depends on China for vital supplies, including fresh water and food. It could easily have been choked into submission in a very short time.

Until the 1950s, Hong Kong was just another trading colony—a busy port, but little more. Then the Korean War and the United Nations embargo in trade with China cut the colony's connections with its hinterland, and an economic reorientation was necessary. Hong Kong possessed no mineral resources, but it did have human resources in virtually unlimited numbers—people in often desperate need to earn a wage, no matter how small. All that was needed was a supply of raw materials, investment in equipment, and efficient organization. Within just a few years, a huge textile industry developed, along with many light

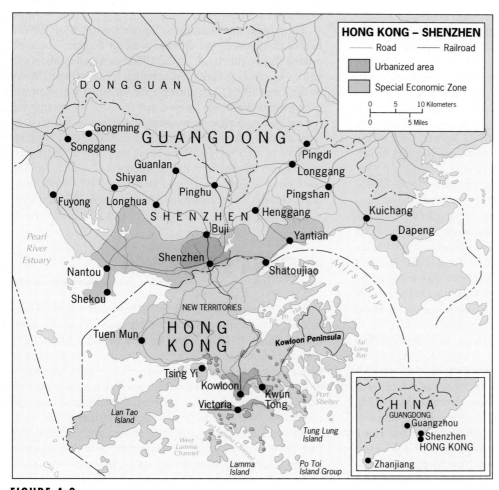

FIGURE 4-9

This forest of skyscrapers constitutes the Central District of Victoria, capital city of Hong Kong. Note how the buildings are shoehorned onto the flat land between bustling Victoria Harbor and the upland (capped by Victoria Peak at the upper left) that dominates this part of Hong Kong Island.

manufacturing industries of other kinds. Products were made at very low cost; they found ready markets throughout the world. Accumulated capital was then used to establish factories making electrical equipment, appliances, and literally countless other consumer goods. As the flood of immigrants continued, labor costs remained low, and products remained competitive. Today, textiles and fabrics still account for about 40 percent of the colony's total export revenues by value, but the range of products continues to expand. Simultaneously, Hong Kong became one of the world's leading financial centers, with a strong and broad-based stock market.

Hong Kong's consumer goods are exported to the United States, China, Germany, the United Kingdom, Japan, and many other countries. In the case of neighboring China, Hong Kong's capitalists account for about 75 percent of all foreign investment in the People's Republic and handle around 25 percent of all the external trade of the People's Republic (including trade with Taiwan). Hong Kong remains the major economic gateway for China to the outside world. China, however, also is investing heavily in Hong Kong, in anticipation of its takeover in 1997.

This, of course, is the crucial issue now facing Hong Kong. When the transfer plan was approved, real estate prices in the colony declined, the stock market crashed, and the HK dollar plunged. Confidence, however, was soon restored. Not only did China promise that Hong Kong's way of life (and economy) would be allowed to continue unchanged for 50 years after 1997, but Hong Kong seemed to recognize its own strength once again. Visions of Hong Kong as the future New York of China—its banking, financial, and trading center—created a new boom. Property values soared above pre-crash levels; the

stock market climbed to record heights; the HK dollar rose to its highest value ever.

Then came the Tiananmen Square massacre of 1989, a loss of confidence, and another cycle of decline. The emigration stream that began in the 1980s gathered momentum, and as many as 60,000 residents per year left the colony in the early 1990s. Yet during a day of street interviews in the spring of 1992, we learned that many of those applying for residency permits elsewhere had already returned or were planning to return to Hong Kong. Rather than emigrating, tens of thousands of Hong Kong residents were creating a foreign base—in Singapore, Australia, Canada, Britain, the United States—to be used in case the post-1997 years prove difficult. In the meantime, Hong Kong in 1992 and 1993 was enjoying another of its booms, a tiger still on the prowl.

◆ CHINA'S SHENZHEN

There is a spot on a hillside at the northern edge of Hong Kong's New Territories from which you can see into communist China. In 1970, when we first visited it, going there required a long wait for a permit. Once there, the glimpse of China's closed society was disappointing. Past the foot of the hill ran a tall, barbed-wire-topped fence studded with floodlights. Armed guards patrolled the tarmac strip on the communist side of this border. In the distance was rural China—a landscape laced with waterways, the strips of land occupied by duck farms and bush. Only the engine of an armored car, policing the boundary, broke the silence.

In 1992, a second visit produced a stunning contrast. Now you can drive right up to the place, just a short walk away from the vantage point. The fence is still there, but the armed guards are gone. The real contrast, though, is in the distance. Where farms and waterways once lay, skyscrapers stand. Where boats floated slowly toward the Pearl River, trains, trucks, buses, and automobiles move on newly built tracks and roadways. Factories abound. Cranes and scaffolding rise above the townscape. Bridges and overpasses are under construction. The noise never ceases: bells, whistles, blasts. We are witness to the rise of Shenzhen, China's most dramatic experiment with market-oriented economic reforms.

In 1979 the Chinese government initiated a new "open-door" policy to attract technology transfers and substantial investment from overseas. A key element in this policy was the creation of five **Special Economic Zones (SEZs)** and 14 Open Coastal Cities. Three of the SEZs lie in the province of Guangdong, whose population is approaching 70 million. And the most successful one is Shenzhen, right across the border from Hong Kong (Fig. 4-9).

In 1980, Shenzhen was just a sleepy fishing and duck-farming village of 20,000; in 1994, the population had surpassed 2.5 million—perhaps the fastest growth of an urban center anywhere on earth. Visit Shenzhen, and it is hard to believe you are in China: a raucous stock exchange, gleaming hotels, pet stores, street vendors, luxury shops, fancy restaurants. The Hong Kong dollar circulates freely, and foreign symbols are everywhere.

The map reveals why Shenzhen has been the most successful of China's SEZs. Not only does it lie adjacent to Hong Kong and across an increasingly porous border, but it is situated centrally vis-à-vis the entire province of Guangdong, economically the most advanced of all of China's provinces. Not far upstream along the Pearl River lies Guangzhou (Canton), with a population of 4.1 million.

It has advantages Hong Kong once had: low start-up costs for factories (perhaps one-quarter of what it would now cost in Hong Kong); labor much cheaper than elsewhere, particularly Hong Kong; and few regulations to contend with. As a result, Hong Kong companies are investing heavily in Guangdong, especially in Shenzhen; and while job opportunities are dropping in the British colony, they are growing by the tens of thousands in Shenzhen.

Is Shenzhen the newest economic tiger on the Pacific Rim? It is too early to rank it with the four established ones, and in any case it lies in the immediate hinterland of Hong Kong, without which its development would be much less spectacular. But we should take note of the enormous potential not just of Shenzhen but of the whole Hong Kong-Shenzhen-Guangzhou complex, soon to be unified politically. This may become the focus not only of a Pacific Rim economic giant but possibly also of a political maverick in the stifling political system of the People's Republic of China. While Hong Kong frets about being taken over politically, Hong Kong is in the process of taking over much of Guangdong economically. China may find that owning and controlling Hong Kong is akin to riding a tiger.

◆ SINGAPORE

The fourth established economic powerhouse along the western Pacific Rim lies in Southeast, not East, Asia: Singapore (Fig. 4-10). Situated at the southern tip of the Malay Peninsula and sited on a small island, Singapore has overcome, with its geography and human resources, the limitations of space and the absence of local raw materials.

The skyline of central Shenzhen, the booming city just across the Chinese border from Hong Kong. Besides its burgeoning retail trade and tourist industry, more than a thousand foreign-owned manufacturing companies are propelling Shenzhen's rise into the ranks of the leading economic centers of the western Pacific Rim.

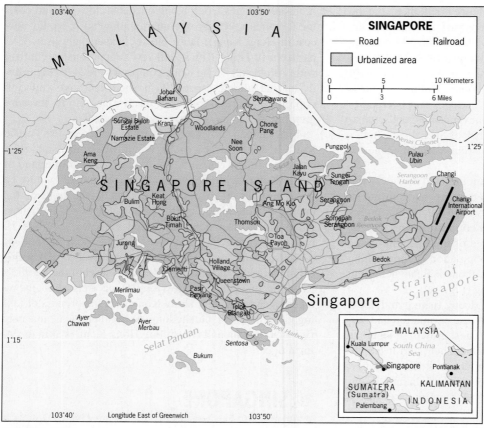

FIGURE 4-10

With a mere 240 square miles (600 sq km) of area, space is at a premium in Singapore, and this is a constant worry for the government. Singapore's only local spatial advantage over Hong Kong is that its small territory is less fragmented (there are just a few small islands in addition to the compact main island). With a population of 2.8 million and an expanding economy, Singapore must develop space-conserving, high-tech industries. In this it has been very successful, and the urban area's forest of high-rise buildings reflects the city-state's prosperity. You will find no slums in Singapore; there are dilapidated streets in the old Chinatown, but these are left to conserve a fragment of the city's past. Mostly, Singapore impresses with its newness, its modernity, and with the order of daily life.

As the map shows, Singapore lies where the Strait of Malacca (leading westward to India) opens into the South China Sea and the waters of Indonesia (Fig. 4-10). The port developed as part of the British Southeast Asian empire, and when independence came in 1963, it was made a part of the Malaysian Federation. However, two years later Singapore seceded from Malaysia and became a genuine city-state on the Pacific Rim. No reunification, merger, or annexation looms here.

Benefiting from its relative location, the old port of Singapore had become the world's fourth busiest (by number of ships served) even before independence. It thrived as an *entrepôt* between the Malay Peninsula, Southeast Asia, Japan, and other emerging economic powers on the Pacific Rim and beyond. Crude oil from Southwest Asia still is unloaded and refined at Singapore, then shipped to Asian destinations. Raw rubber from the adjacent peninsula and from Indonesia's island of Sumatera is shipped to Japan, the United States, China, and other countries. Timber from Malaysia, rice, spices, and other foodstuffs are processed and forwarded via Singapore. In return, automobiles, machinery, and equipment are imported for Southeast Asia via Singapore.

But that is the old pattern. Singapore's leaders want to redirect the city's economy and move toward high-tech industries for the future. In Singapore the government tightly controls business as well as other aspects of life (some newspapers and magazines have been banned for criticizing the regime, and there are even fines for such things as eating on the subway and failing to flush a public toilet). Its overall success after secession has tended to keep the critics quiet: while personal incomes from 1965 to 1992

Singapore is the product of geography, proof positive that location is itself a resource. Once a restive part of Malaysia, Singapore now is a city-state, an industrial power, a financial center, and one of the economic tigers of the Pacific Rim.

increased tenfold—to over U.S. $10,000 per year—those of neighboring Malaysia reached just U.S. $2,000. Among other things, Singapore became (and for many years remained) the world's largest producer of disk drives for small computers.

As multinational corporations settled in, Singapore's political stability on the volatile Pacific Rim was another advantage. Nonetheless, a problem emerged: unlike Taiwan, where brainpower is plentiful, Singapore's technical capacities were limited. In 1991, only 6 percent of the population was university educated. So while the government reduced the incentives for low-technology manufacturers to come here, and wages rose, the high-tech industries found themselves thinking about moving to nearby Malaysia and Thailand. That slowed Singapore's boom and challenged the government to find ways to stimulate a revival.

To accomplish this, Singapore has moved in several directions. First, it will focus on three growth areas: information technology, automation, and biotechnology. Second, there are notions of a "Growth Triangle" involving Singapore's developing neighbors, Malaysia and Indonesia; those two countries would supply the raw materials and cheap labor, and Singapore the capital and technical know-

how. Third, Singapore has opened its doors to capitalists of Chinese ancestry who may be wanting to leave Hong Kong and who wish to relocate their enterprises here. Singapore's population is 78 percent Chinese, 15 percent Malay, and 6 percent South Asian. The government is Chinese-dominated, and its policies have been to sustain Chinese control. Hong Kong's uncertain future seemed to present an opportunity to reinvigorate Singapore's Chinese community. Nevertheless, this smallest of the economic tigers may also be the most vulnerable. Singapore faces competition near and far; over the long term its relative location, which endowed it with its early wealth, may turn out to be its most enduring guarantee of sustenance—if not prosperity.

◆ AUSTRALIA

Imagine a country on the Pacific Rim, 10 times the size of Texas, rich in mineral resources, well endowed with farmlands and vast pastures, major rivers, ample underground

TEN MAJOR GEOGRAPHIC QUALITIES OF AUSTRALIA AND NEW ZEALAND

1. Australia and New Zealand constitute a geographic realm by virtue of territorial dimensions, relative location, and cultural distinctiveness—not population size.
2. Despite their inclusion within a single geographic realm, Australia and New Zealand differ physiographically. Australia is marked by a vast, dry, low-relief interior; New Zealand is mountainous.
3. Australia and New Zealand are marked by peripheral development—Australia because of its aridity, New Zealand because of its topography.
4. The populations of Australia and New Zealand are not only peripherally distributed but also highly clustered in urban agglomerations.
5. Australia and New Zealand lie remote from the places with which they have the strongest cultural ties.
6. In Australia, immigration is changing the ethnic makeup. In New Zealand, minority awareness is posing a growing challenge.
7. Australia's Aboriginal population was almost completely submerged by the European invasion, remains numerically small, and participates only slightly in modern society.
8. Australia and New Zealand are being integrated into the economic framework of the western Pacific Rim, principally as suppliers of raw materials.
9. Australia continues to yield newly discovered mineral deposits, proving its substantial untapped potential.
10. Australian agriculture is highly mechanized and produces large surpluses for sale on foreign markets. Huge livestock herds feed on vast pasturelands.

water sources, served by good natural harbors, and populated by about 18 million mostly well-educated people who have enjoyed stable government since the beginning of this century. Surely such a country would be a leading economic tiger in the region—a tough competitor for the others where population pressures, labor problems, scarce domestic resources, political uncertainties, and other obstacles must be overcome?

The answer is—no. Australia today is what locals wryly call an *NDC*, a newly declining country; a seller of raw materials, not finished ones; a purveyor of livestock, meat, and wheat on undependable world markets; a society deeply in debt; an economy in the doldrums. Spatially, Australia anchors the Pacific Rim in the south. Functionally, Australia is an open-pit mine for the industrial economies to the north. Its manufactures are no threat to those of Singapore, Hong Kong, Shenzhen, Taiwan, or South Korea, let alone Japan.

For most of the twentieth century, Australians could afford to ignore the obvious. Australia was a British progeny, a European outpost, and with nearby New Zealand it enjoyed standards and ways of living that were the envy of much of the world. With its wide range of environments, magnificent scenery, vast open spaces, and seemingly endless opportunities, Australia appeared invulnerable to the problems of the rest of the world. Europeans peopled the continent; the small remnant Aboriginal population (under 300,000 today) posed no threat. When the Japanese empire expanded, Australia's remoteness saved the day. When immigration became an issue, Australia adopted an all-white admission policy (not terminated until 1976). Australia-New Zealand became an outlier of Europe where, it seemed, time had stopped, a geographic realm distinct from its Pacific surroundings and remote from its cultural sources.

Today, that geographic realm still persists—but things are changing. It is estimated that at the beginning of the century, Australia's productivity per person (GNP) may have been the highest in the world. In 1990, it ranked seventeenth among the world's countries, but others are now overtaking it. Australia's share of global trade has fallen, over the past four decades, by more than half. And its internal cultural mosaic is being transformed as well. More than 600,000 Asians now live permanently in Australia, and their number continues to grow steadily. By a wide margin, Japan has become Australia's leading trade partner, and Australian schools are today teaching Japanese to tens of thousands of children. Unmistakably, Australia is beginning to become part of the Asian world it adjoins.

AUSTRALIAN RIMLAND

Australia is a large landmass, but the country's population is heavily concentrated in a core area that lies in the east and southeast, most of which faces the Pacific Ocean (here named the Tasman Sea between Australia and New Zealand). As Fig. 4-11 shows, this crescent-like Australian heartland extends from north of the city of Brisbane to the vicinity of Adelaide and includes the largest city, Sydney, the capital, Canberra, and the second-largest city, Melbourne. A secondary core area has developed in the far southwest, centered on Perth and its outport, Fremantle.

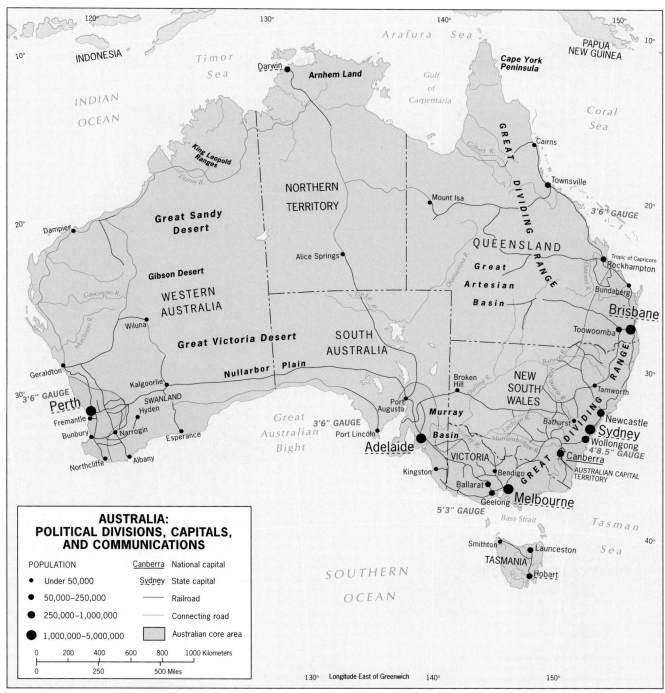

FIGURE 4-11

To better understand the evolution of this spatial arrangement, it helps to refer again to the map of world climates (Fig. I-7, p. 17). Note that the bulk of Australia's dry interior is desert (*BWh*) or steppe (*BSh/BSk*), inhospitable but containing many valuable mineral deposits. The core area described above coincides rather closely with the hu-

mid temperate climatic zone mapped as *Cfa/Cfb*, extending between the spine of the Great Dividing Range (Fig. 4-12) and the east coast, and from near the Tropic of Capricorn to Tasmania in the south. Around Adelaide and in the far southwest (including Perth), a dry-summer, Mediterranean regime (*Csa/Csb*) prevails. And the far north lies under

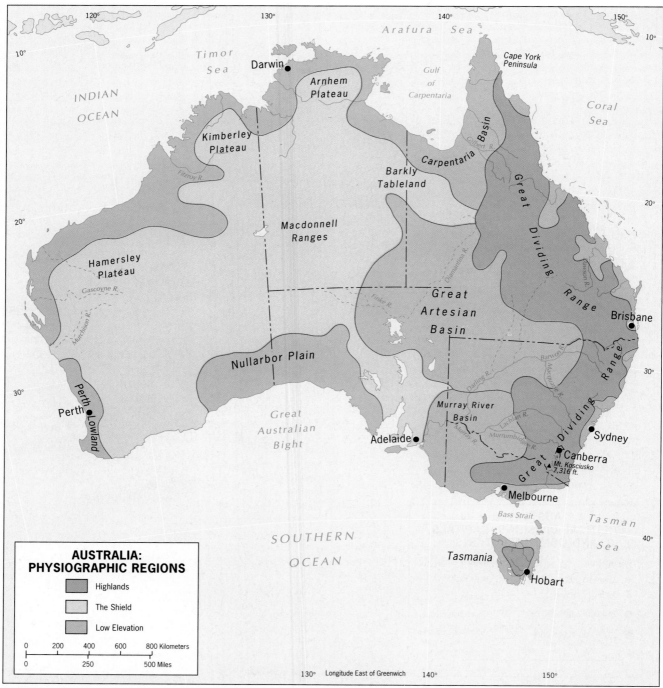

FIGURE 4-12

tropical, wet-and-dry savanna (*Aw*) conditions, with a small monsoonal strip (*Am*) along the neck of the Cape York Peninsula.

Environmentally, Australia's most favored strips face the Pacific and Southern oceans, and they are not large. The country may be described as a coastal rimland with cities, towns, farms, and forested slopes giving way to a vast, arid interior Australians call the *Outback*. On the western flanks of the Great Dividing Range lie the extensive grassland pastures that catapulted Australia into its first commercial age—and on which still graze one of the largest sheep herds on earth (over 160 million sheep, producing more than one-

The Australian *Outback* is a vast, arid, sparsely-vegetated expanse. On the Köppen climate-region map, this appears as *BW* (desert), but the landscape seems greener than the usual image of a desert. This nevertheless is a fragile ecology, easily damaged by human intervention.

fifth of all the wool sold in the world). Where it is moister, to the north and east, cattle by the millions are ranched. This is frontier Australia, over which livestock have ranged for nearly two centuries.

Aboriginal Australians reached this landmass as long as 50 to 60 thousand years ago, crossed the Bass Strait into Tasmania, and had developed a patchwork of indigenous cultures by the time the British immigration began in 1788 (see the box entitled "The Migration Process"). European settlement started on the coasts, where both penal colonies and free towns were situated. Distance protected the Aboriginal communities of the northern interior longer than elsewhere; in Tasmania, the indigenous Australians were exterminated.

Eventually, the major coastal settlements became the centers of separate, quite different colonies. There were seven of these, each with their own hinterlands, and by 1861 Australia was delimited by its now-familiar pattern of straight-line boundaries (Fig. 4-11). Sydney was the focus for New South Wales; Melbourne, Sydney's rival, anchored Victoria. Adelaide was the heart of South Australia; and Perth lay at the core of Western Australia. Brisbane was the nucleus of Queensland; and in Tasmania, the town of Hobart was the seat of government. The largest clusters of surviving Aboriginal people were in the so-called Northern Territory, with Darwin, on Australia's tropical north coast, its colonial city. Notwithstanding their shared cultural heritage, the Australian colonies were at odds not only with London over colonial policies but also with each other over economic and political issues. The building of an Australian nation during the late nineteenth century was a slow and difficult process.

A Federal State

On January 1, 1901, years of difficult negotiations finally produced the Australia we know today: the Commonwealth of Australia, consisting of six States and two Federal Territories (Table 4-1; Fig. 4-11). The special status of Federal Territory was assigned to the Northern Territory to protect the interests of the Aboriginal population there. A second Federal Territory was created in southern New South Wales for the establishment of a new federal capital (Canberra). Both Sydney and Melbourne had fiercely competed for this honor, but it was decided that a completely new seat of government, in a specially designated area, was the best solution. In 1927, the government buildings were ready for use and the administration moved into its new home, a city planned from scratch to serve a nation built from a group of quarrelsome colonies.

Today Canberra, Australia's only large city not situated on the coast, has a population of 323,000, less than one-tenth of Sydney's. As a growth pole, Canberra has been no great success. But as a symbol of Australian **federalism**, it has been a triumph. Australia's unification was made possible through the adoption of an idea with ancient Greek and Roman roots, one also familiar to Americans and Cana-

THE MIGRATION PROCESS

Australia is the product of **migration**. After tens of thousands of years of Aboriginal settlement, Europeans reached these shores during the sixteenth century. James Cook, the British sea captain, landed on the east coast several times during the 1770s, and his reports convinced the British that this *terra australis incognita* should become a British outpost. In 1788, Captain Arthur Phillip sailed into what is today Sydney Harbor and established the beginnings of modern Australia. The Europeanization of Australia doomed the continent's Aboriginal societies, but this was only one of many areas around the world where local cultures and foreign invaders came face to face. Between 1835 and 1935, perhaps as many as 75 million Europeans departed for distant shores—for the Americas, for Africa, and for Australia (Fig. 4-13). Some sought religious freedom; others escaped famine; still others simply hoped for a better life.

Studies of the *migration decision* indicate that migration flows—then as now—vary in size with: (1) the perceived degree of difference between one's home, or source, and the destination; (2) the effectiveness of the information flow, that is, the news about the destination sent back by those who migrated to those who stayed behind waiting to decide; and (3) the distance between the source and the destination (shorter moves attracting many more migrants than longer ones). A century ago, the British social scientist Ernst Georg Ravenstein studied the migration process; many of his conclusions remain valid today. For example, every migration stream from source to destination produces a counter-stream of returning migrants who are unable to adjust, unsuccessful, or otherwise persuaded to go back home.

Studies of migration also conclude that several factors are at work in the migration process. *Push factors* motivate people to move away; *pull factors* attract them to new destinations. To those early Europeans, Australia was a new frontier, a place where one might acquire a piece of land, some livestock. The skies (it was reported) were clear, the air fresh, sunshine plentiful.

Many of the first Europeans who arrived in Australia were not free but in bondage. *Voluntary migrants*, such as the millions who came to America, could make choices. But European governments also used their overseas domains as distant prisons for those convicted of some offense and sentenced to deportation—or "transportation," as the practice was called by the British. These *forced migrants* were sent to Australia by the thousands, often under dreadful circumstances on overcrowded boats. Countless prisoners died on the way and were thrown overboard; many who survived the journey arrived sick and unable to fend for themselves in the penal colonies of New South Wales, Tasmania, or the far west. Captain Phillip's flotilla of boats carried both free settlers and prisoners, and Australia's reputation as a penal colony was established.

By the middle of the nineteenth century, as many as 165,000 deported offenders had reached Australian shores. Large numbers of free colonists also were arriving, and prisoners who had served out their terms entered society as well. Meanwhile, the practice of

TABLE 4–1
States and Territories of Federal Australia

State	Area (Sq Mi)	Population	Capital	Population
New South Wales	309,500	6,085,000	Sydney	3,895,000
Queensland	666,900	3,115,000	Brisbane	1,437,000
South Australia	379,900	1,585,000	Adelaide	1,102,000
Tasmania	26,200	475,000	Hobart	191,000
Victoria	87,900	4,580,000	Melbourne	3,236,000
Western Australia	975,100	1,775,000	Perth	1,306,000
Territory				
Australian Capital Territory	900	323,000	*Canberra*	323,000
Northern Territory	519,800	205,000	Darwin	76,000

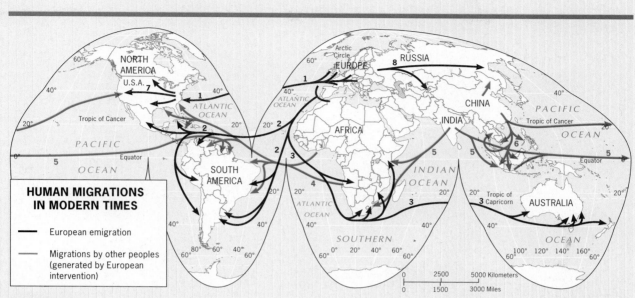

FIGURE 4-13

transportation came under heavy criticism in England, and in 1849 the last convict ship made the trip from Britain to Australia. Soon the penal colonies were phased out, and Australia's assets attracted an ever-growing stream of voluntary migrants.

The end of transportation did not signal the end of forced migration as a process on the world stage. During the twentieth century, a time of many voluntary migration streams, there also were major forced migrations. Josef Stalin used forced migration to punish dissidents during his murderous communist regime, exiling them to distant, unfamiliar, and often unsuitable lands in remote areas of the Soviet Union, with calamitous consequences for uncounted millions. Adolf Hitler, during World War II, organized the forced migration of millions of Jews to the concentration camps of the Holocaust. Ugandan dictator Idi Amin in 1972 ordered the immediate expulsion of 75,000 Asian residents from his country, producing a forced emigration of major proportions.

dians: the notion of *association*. The term to describe it comes from the Latin *foederis*; in practice, it means alliance and coexistence, a union of consensus and common interest—a federation. It stands in contrast to the idea that states should be centralized, or *unitary*. For this, too, the ancient Romans had a term: *unitas*, meaning unity. Most European countries are unitary states, including the United Kingdom of Great Britain and Northern Ireland. Although the majority of Australians came from that tradition (a kingdom, no less) they managed to overcome their differences and establish a Commonwealth that was, in effect, a federation of states with different viewpoints, economies, and objectives, separated by vast distances along the rim of an island continent. And yet, the experiment succeeded: in a few years, Australia will celebrate 100 years of federation.

An Urban Culture

In this century of federal association, the Australians have developed an urban culture. For all those vast open spaces and romantic notions of frontier and Outback, fully 85 percent of all Australians live in cities and towns. On the map, Australia's areal functional organization is somewhat similar to Japan's: large cities lie along the coast, the centers of manufacturing complexes as well as the focii of agricultural areas. In Japan, mountainous topography contributed to this situation; in Australia, an arid interior. There, however, the similarity ends. Australia's territory is 20 times larger than that of Japan, and Japan's population is seven times the size of Australia's. Japan's port cities are built to receive raw materials and to export finished products. Aus-

Sydney, the original—and still the largest—Australian city, has one of the world's best and most spectacular harbors. This southward view shows the Harbour Bridge, Opera House, and ferry terminal (at the foot of the downtown). Sydney will be the site of the Olympic Games in the year 2000.

tralia's cities forward minerals and farm products from the Outback to foreign markets and import manufactures from overseas. Distances in Australia are much greater, and spatial interaction (which tends to decrease with increasing distance), is less. In comparatively small, tightly organized Japan, you can travel from one end of the country to the other along highways, through tunnels, and over bridges with utmost speed and efficiency. In Australia, the overland trip from Sydney to Perth, or from Darwin to Adelaide, is time-consuming and slow. Nothing in Australia compares to Japan's high-speed bullet train.

For all its vastness and youth, Australia nonetheless developed a remarkable cultural identity, a sameness of urban and rural landscapes that persists from one end of the continent to the other. Sydney (3.9 million), often called the New York of Australia, lies on a spectacular estuarine site, its compact, high-rise central business district overlooking a port bustling with ferry and freighter traffic (see photo above). Sydney is a vast, sprawling metropolis with multiple outlying centers studding its far-flung suburbs; brash modernity and reserved British ways blend here. Melbourne (3.2 million), sometimes regarded as the Boston of Australia, prides itself on its more interesting architecture and more cultured ways. Brisbane (1.4 million), the capital of Queensland, which also anchors Australia's Gold Coast and adjoins the Great Barrier Reef, is the Miami of Australia; unlike Miami, however, its residents can find nearby relief from the summer heat in the mountains of its immediate hinterland (as well as at its beaches). Perth (1.3 million), Australia's San Diego, is one of the world's most isolated cities, separated from its nearest Australian neighbor by two-thirds of a continent and from Southeast Asia and Africa by thousands of miles of ocean.

And yet, each of these cities—plus the capitals of South Australia (Adelaide [1.1 million]), Tasmania (Hobart [191,000]), and, to a lesser extent, the Northern Territory (Darwin [76,000])—exhibit an Australian character of unmistakable quality. Life is orderly and unhurried. Streets are clean, slums are few, graffiti rarely seen. By American and even European standards, violent crime (though rising) is uncommon. Standards of public transportation, city schools, and health care provision are high. Spacious parks, pleasing waterfronts, and plentiful sunshine make Australia's urban life more acceptable than almost anywhere else in the world. Critics of Australia's way of life say that this very pleasant state of affairs has persuaded Australians that hard work is not really necessary. But Australia's privileges are being eroded away, and standards of living are in danger of serious decline. The country's cultural geography evolved as that of a European outpost, prosperous and secure in its isolation. Now Australia must reinvent itself as a major link in an Austrasian chain, a Pacific partner in a transformed regional economic geography.

ECONOMIC PERSPECTIVES

Distance has been an ally as well as an enemy to Australia. Its remoteness helped save the day when Japan's empire expanded over the western Pacific. From the very beginning, though, goods imported from Britain (and later from the United States) were expensive, largely because of transport costs. This encouraged local entrepreneurs to set up their own industries in and near the developing cities—industries economic geographers call **import-substitution industries**.

When the prices of foreign goods became lower because transportation was more efficient and therefore cheaper, the local businesses demanded protection from the colonial governments. This was done by enacting high tariffs against imported goods. The local products now could continue to be made inefficiently because their market was guaranteed. Japan could not afford this—how could Australia? The answer can be seen on the map (Fig. 4-14). Even before federation in 1901, all the colonies could export valuable minerals whose earnings were used to shore up those inefficient, uncompetitive local industries. By the time the colonies unified, the income from the pastoral industries contributed as well. So the miners and the farmers paid for those imports Australians could not reproduce themselves, plus the products made in the cities. No wonder the cities grew: here were secure manufacturing jobs, jobs in state-run service enterprises, and jobs generated by a growing government bureaucracy. When we

noted earlier that Australians once achieved the highest per capita GNP in the world, this was achieved in the mines and on the farms—not in the cities.

But the good times had to come to an end. The prices of farm products fluctuated, and international market competition increased. The cost of mining ores and minerals, transporting, and shipping them also rose. Australians like to drive (there are more road miles per person in Australia than in the U.S.), and expensive petroleum imports were needed. Meanwhile, the government-protected industries had been further fortified by strong labor unions. Not surprisingly, as the economy declined the national debt rose, inflation grew, and unemployment crept upward; an Australian leader said that his country was on its way toward becoming a banana republic.

Agriculture

And yet, Australia has material assets of which other countries on the Pacific Rim can only dream. In terms of agriculture, the raising of sheep was the earliest of commercial ventures, but it was the technology of refrigeration that brought world markets within reach of Australian beef producers. Wool, meat, and wheat have long been the country's big three income earners; Fig. 4-14 displays the vast pastures in the east, north, and west that constitute the ranges of Australia's huge herd. The zone of commercial grain farming forms a broad crescent extending from northeastern New South Wales through Victoria into South Australia, and covers a large area in the hinterland of Perth. Keep in mind the scale of this map: Australia is only slightly smaller than the 48 contiguous states of the United States! Commercial grain farming in Australia is big business. As the climatic map would suggest, sugarcane grows along most of the warm, humid coastal strip of Queensland, and Mediterranean crops (including grapes for Australia's wines) cluster in the hinterlands of Adelaide and Perth. Mixed horticulture concentrates in the basin of the Murray River system, including rice, grapes, and citrus fruits, all under irrigation. And, as elsewhere in the world, dairying has developed near the large urban areas. With its considerable range of environments, Australia yields a diversity of crops.

Mineral Resources

Australia's mineral resources, as Fig. 4-14 shows, also are diverse. Major gold discoveries in Victoria and New South Wales produced a 10-year gold rush starting in 1851 and ushered in a new economic era. By the middle of that decade, Australia was producing 40 percent of the world's gold. Subsequently, the search for more gold led to the discoveries of other minerals. New finds are still being made today, and even oil and natural gas have been found both

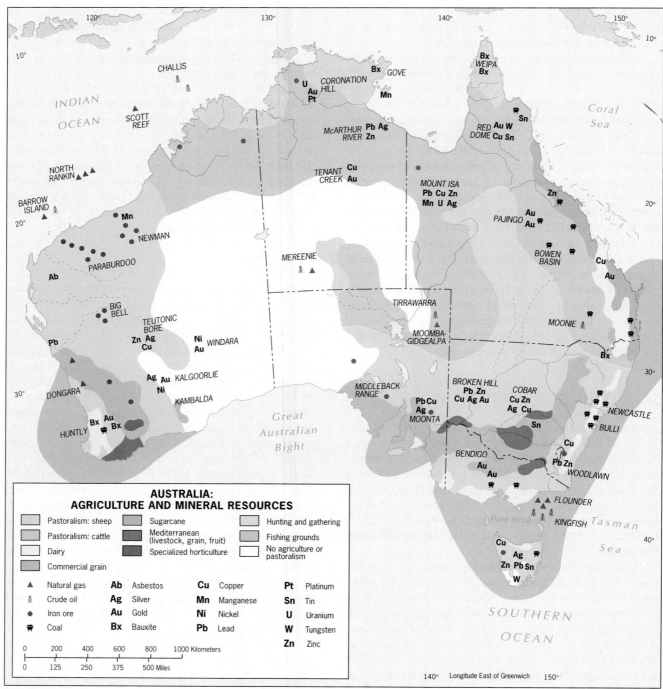

FIGURE 4-14

inland and offshore (see the symbols in Fig. 4-14 in the Bass Strait between Tasmania and the mainland, and off the northwest coast of Western Australia). Coal is mined at numerous locations, notably in the east near Sydney and Brisbane but also in Western Australia and even in Tasmania; before coal prices fell, this was a very valuable export.

In terms of metallic and nonmetallic minerals, major deposits abound—from the complex at Broken Hill and the mix of minerals at Mount Isa to the huge nickel deposits at Kalgoorlie and Kambalda, the copper of Tasmania, the tungsten and bauxite of northern Queensland, and the asbestos of Western Australia. A glance at the map reveals the wide distribution of iron ore (the red dots), and for this as for many other raw materials, Japan has been Australia's best customer in recent years. In the mid-1990s, Japan was buying more than one-third of all Australian mineral exports.

Manufacturing

Australian manufacturing, as we noted earlier, remains oriented to local domestic markets. Do not expect to find Australian automobiles, electronic equipment, or cameras challenging the Pacific Rim's economic tigers for a place on world markets—not yet, at any rate. Australian manufacturing is quite diversified, producing some machinery and equipment made of locally produced steel as well as textiles, chemicals, paper, and many other items. These industries cluster in and near the major urban areas where the markets are. The domestic market in Australia is not large, but despite declining real incomes, it remains a relatively affluent one. This makes it attractive to foreign producers, and Australia's shops are full of goods from Japan, South Korea, Taiwan, and Hong Kong. Indeed, despite its long-term protectionist practices, Australia still does not produce many goods that could be manufactured at home. Overall, the economy continues to display symptoms of a still-developing rather than a fully developed country.

AUSTRALIA'S FUTURE

A new age is dawning in Australia, a country that just a few years ago (1988) celebrated its bicentennial—two centuries as a European outpost facing the Pacific Ocean. Now, as Australia enters its third century, its European bonds are weakening and its Asian ties are strengthening (see box at right).

On the face of it, Australia and its northern neighbors on the Pacific Rim would seem to exhibit a geographic complementarity: Australia has excess food, metals, and minerals needed by Japan and the economic tigers, and Australia needs the cheap manufactures Asia produces. But it is not that simple. Australia still has tariff barriers against imported goods, and the Asian countries on the Pacific Rim maintain import barriers against processed foods and minerals, thus discouraging Australia from refining its exports and earning more from them. Asia's economic tigers, furthermore, are more interested in the potential of large markets (such as Indonesia, with 190 million people) than Australia. It is difficult for Australia to open its economy, to lower its protective tariffs, without reciprocation from its Asian trading partners.

In the mid-1990s, Australia faced other challenges as well. The Aboriginal population of under 300,000 (including many of mixed ancestry) has gained much influence in the country's affairs and has successfully obstructed further exploration and mining on its ancestral and sacred lands. A notable case involved the platinum, gold, and palladium deposits at Coronation Hill (see Fig. 4-14), near the famous Kakadu National Park in the Northern Territory. At first the Canberra government authorized operations there, but then, bowing to Aboriginal leaders, it reversed its decision because the project would violate an Aboriginal community's sacred lands. This was a significant victory, and it put companies, investors, and prospectors on notice that the days of freewheeling exploitation over much of the country's most promising frontier have ended. Business leaders warn that this could adversely affect the economy over the long term.

In fact, only a few Aboriginal communities still pursue their original life of hunting and gathering and fishing in the remotest corners of the Northern Territory, the north of Western Australia, and northern Queensland. The remainder are scattered across the continent, working on cattle stations, subsisting on reserves set aside for them by the government, or performing menial jobs in cities and towns. Lately, the Australian conscience has been aroused, and an effort has begun to undo centuries of persecution and neglect—but these attempts are met, understandably, with suspicion and distrust. In comparatively affluent Australia,

REPUBLIC OF AUSTRALIA?

Australians in 1993 were facing an emotional decision: whether to end their ties to the British Crown and make their country a republic. While the Australian economy deteriorated, politicians spent critical time arguing over an issue that, many agreed, would ultimately make little difference to the nation. But the question did raise fears in some quarters about Australia's future. State governments insisted that the matter must be put to a referendum—not nationally, but State by State. In Western Australia, public opinion seemed to favor the monarchy, and the suggestion was even raised that a "no" vote in that State might lead to secession. Other observers feared that republic status would eventually change the State system, marking the beginning of the end of a framework that has served Australia well. Thus even Australia is not immune to the politico-geographical stirrings that afflict today's world.

the Aboriginal peoples remain victims of poverty, disease, inadequate education, and even malnutrition. The Aboriginal question will loom large in Australia's future.

Another growing issue involves environment and conservation. Australia's wealth was derived not only from its ores and minerals and from soils and pastures: its ecology, too, paid a heavy price. Great stands of magnificent forest were destroyed. In Western Australia, centuries-old trees were simply "ringed" and left to die so that the sun could penetrate through their leafless crowns to nurture the grass below. Then the sheep could be driven into these new pastures. In Tasmania, where Australia's native eucalyptus tree reaches its greatest dimensions (comparable to North American redwood stands), tens of thousands of acres of this irreplaceable treasure have been lost to chain saws and pulp mills. Australia's unique marsupial fauna have suffered the loss of numerous species, and many more are endangered or threatened. "Never have so few people wreaked so much havoc on the ecology of so large an area in so short a time," observed a geographer in Australia recently; but awareness of this environmental degradation is growing. In Tasmania, the "Green" environmentalist political party has become a force in state affairs, and its activism has slowed deforestation, dam-building, and other "development" projects. Still, many Australians fear the environmental movement as an obstacle to economic growth at a time when the economy needs stimulation. This, too, is an issue for the future.

The population question has preoccupied Australia as long as Australia has existed—and even longer. Fifty years ago, when Australia had less than half the population it has today, 95 percent of the people were of European ancestry, and more than three-quarters of them came from the British Isles. Eugenic (race-specific) immigration policies until the 1970s maintained this situation. Today, the picture is dramatically different: of 18 million Australians, only about one-third have British-Irish origins, and Asian immigrants outnumber both Europeans and natural increase each year. During the late 1980s, nearly 150,000 legal immigrants arrived in Australia annually, the majority from Hong Kong, Vietnam, China, the Philippines, Malaysia, India, and Sri Lanka. (Immigration from Britain and Ireland continues at around 20-25,000 annually, and between 5-10,000 New Zealanders also emigrate to Australia each year.)

All this is happening at a time of economic difficulty, and the government is under pressure again to limit immigration. The quota was set at 80,000 for the next few years, but such restrictions touch sensitive nerves among many Australians. Some want to reduce the numbers still more: the Australian Nationalist Movement (headquartered in Perth) and the League of Rights, two extremist groups, oppose further immigration from Asia. So the population issue roils Australia once again, and as the ethnic makeup of Australia changes, multiculturalism and its challenges emerge as a national problem for a transforming society.

In the meantime, thousands of Japanese tourists stream through Australia's many attractions. Bilingual signs, maps, and menus are everywhere; Japanese newspapers sit on city newsstands; Japanese is taught in hundreds of Australian schools. The two ends of the Austrasian Pacific Rim are sharply different in countless ways, but their economic meeting ground grows every year. East Asian freighters in Australian ports symbolize the interconnections and interactions that are forging a new geographic region, of which Australia will be an integral part.

◆ NEW ZEALAND

Fifteen hundred miles east-southeast of Australia, in the Pacific Ocean across the Tasman Sea, lies New Zealand. In an earlier age, New Zealand would have been part of the Pacific geographic realm because its population was Maori, a people with Polynesian roots. But New Zealand, like Australia, was invaded and occupied by Europeans. Today, the country's population of 3.5 million is 85 percent European, and the Maori form a minority of less than a half million—with many of mixed Euro-Polynesian ancestry.

New Zealand consists of two large mountainous islands and numerous scattered smaller islands (Fig. 4-15). The two large islands, with South Island somewhat larger than North Island, look diminutive in the great Pacific Ocean, but together they are larger than Britain. In sharp contrast to Australia, the two islands are mainly mountainous or hilly, with several peaks rising far higher than any on the Australian landmass. South Island has a spectacular snow-capped range appropriately called the Southern Alps, with peaks reaching beyond 11,700 feet (3,500 m). Smaller North Island has proportionately more land under low relief, but it also has an area of central highlands along whose lower slopes lie the pastures of New Zealand's chief dairying district. Hence, while Australia's land lies relatively low in elevation and exhibits much low relief, New Zealand's is on the average quite high and evinces mostly rugged relief.

Thus the most promising areas must be the lower-lying slopes and lowland fringes on both islands. On North Island, the largest urban area, Auckland (900,000), occupies a comparatively low-lying peninsula. On South Island, the largest lowland is the agricultural Canterbury Plain, centered on Christchurch (325,000). What makes these lower areas so attractive, apart from their availability as cropland, is their magnificent pastures. Such is the range of soils and pasture plants that both summer and winter grazing can be carried on. Moreover, a wide variety of vegetables, cereals, and fruits can be produced in the Canterbury

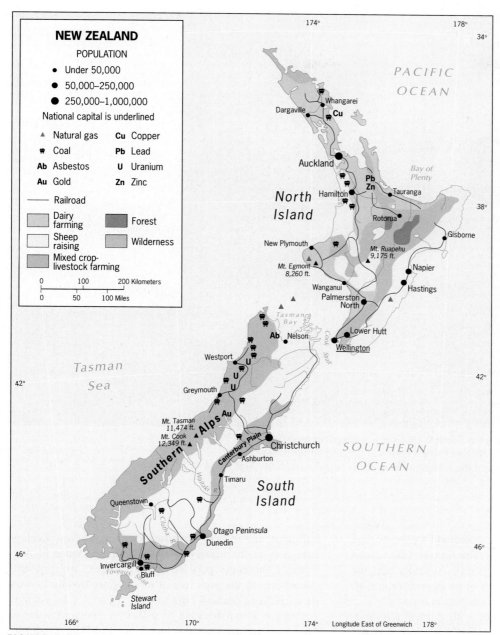

NEW ZEALAND

POPULATION
- Under 50,000
- 50,000–250,000
- 250,000–1,000,000

National capital is underlined

▲ Natural gas   **Cu** Copper
⚒ Coal   **Pb** Lead
Ab Asbestos   **U** Uranium
Au Gold   **Zn** Zinc

— Railroad

Dairy farming   Forest

Sheep raising   Wilderness

Mixed crop-livestock farming

0   100   200 Kilometers
0   50   100 Miles

FIGURE 4-15

Plain, the chief farming region. About half of all New Zealand is pasture land, and much of the farming is done to supplement the pastoral industry in order to provide fodder when needed. Sheep and cattle dominate these livestock-raising activities, with wool, meat, and dairy products providing over one-third of the islands' export revenues.

Despite their contrasts in size, shape, physiography, and history, New Zealand and Australia have a great deal in common. Apart from their joint British heritage, they share a sizeable pastoral economy, a small local market, the problem of great distances to world markets, and a desire to stimulate (through protection) domestic manufacturing. The high degree of urbanization in New Zealand (84 percent of the total population) again resembles Australia: substantial employment in city-based industries, mostly the processing and packing of livestock and farm products, and government jobs.

More remote even than Australia, New Zealand also has been affected by developments on the western Pacific Rim. In the mid-1990s, Japan and other countries on the Pacific

New Zealand's Canterbury Plain is the South Island's leading agricultural area. Here, at the innermost margin of this productive lowland, animal grazing abruptly gives way to the foothills that quickly lead up to the soaring ranges of the rugged Southern Alps.

Rim together bought more of New Zealand's exports than did Australia or the United States (Britain takes only about 6 percent of New Zealand's exports). Australia and the U. S. still send New Zealand most of its imports (about 40 percent combined), but Japan's contribution is rising.

Spatially, New Zealand shares with Australia its pattern of peripheral development (Fig. I-9), imposed not by desert but by high rugged mountains. The country's major cities—Auckland and the capital of Wellington (together with its satellite of Hutt, 345,000) on North Island, and Christchurch and Dunedin (120,000) on South Island—are all located on the coast, and the entire railway and road system is peripheral in its configuration (Fig. 4-15). This is more pronounced on South Island than in the north because the Southern Alps are New Zealand's most formidable barrier to surface communication.

Compared to brash, progressive, and modernizing Australia, New Zealand seems quieter. A slight regional contrast might perhaps be discerned between more forward-looking North Island and conservative South Island, but the distinction fades in the light of Australia's historic inter-

nal urban and regional rivalries. Nothing in New Zealand compares to the variety of the urban experience in Australia; a sameness pervades New Zealand and probably contributes to the high rate of emigration, mainly to Australia. If New Zealand fails to satisfy many of its younger citizens, it is not because of a lack of personal security. The government has developed an elaborate cradle-to-grave system of welfare programs that are affordable, given the country's high incomes and standards of living (taxes also are high). Whether this has suppressed entre-preneurship and initiative is a matter for debate, but New Zealand is a country with few excesses and much stability. In any case, the government in the 1980s began a major restructuring of the economy into a free-market business environment—with wide approval by the electorate.

Other significant change can be discerned as well. Auck-land not only is New Zealand's largest city, it also may be the largest Polynesian city of all, with over 100,000 Maoris, Samoans, Cook Islanders, Tongans, and others making up about one-sixth of the metropolitan population. The twenti-eth century has witnessed a revival of Maori culture—and a persistently slow pace of integration of the Maori minority into New Zealand society (nearly all the Maoris reside on North Island, where three-quarters of all New Zealanders live). In the 1990s, however, the Maori presence has become the leading national issue, because the Maoris have launched a vigorous effort to get the courts to enforce the terms of the 1840 Treaty of Waitangi that granted the British sovereignty over New Zealand. In particular, the Maoris have laid claim to at least half of the national territory, asserting that their tribal lands were illegally wrested away by British settlers. Judicial rulings in recent years have begun to support the Maori position, and the government may soon be required to return land that Maori tribes can prove once belonged to them. The Waitangi treaty was intended to be the framework for a partnership with the Maoris. If that promise can finally be fulfilled, New Zealand's future could be shaped by a harmonious bicultural society. If it fails, the specter of serious racial polarization looms because the more-rapidly-growing Maori population could double to 25 percent of New Zealand's total by the year 2010.

AUSTRASIA IN TRANSITION

1. Japan's economy was in deep recession in 1993; its long-entrenched political leadership was embroiled in turmoil. But the underlying stability of the Japanese system held firm, auguring an early recovery from these troubles and continued economic leadership on the global scene.
2. By some measures, China's national economy in 1993 had become the world's third largest, after the United States and Japan.
3. China's armed forces, already the largest in the world, are being modernized even as China's weapons industries are producing and selling ever more sophisticated armaments.
4. North Korea, called by some political geographers the "Iraq of East Asia," challenges China as well as its other neighbors by pursuing an independent nuclear-weapons policy.
5. For the first time in many years, representatives from China and Taiwan met in 1993 to discuss "matters of mutual concern," raising hopes for a softening of differences between "the Two Chinas."

◆ PRONUNCIATION GUIDE

Aboriginal   (abb-uh-RIDGE-uh-null)
Adelaide   (ADDLE-ade)
Ainu   (EYE-noo)
Algae   (AL-jee)
Auckland   (AWK-lund)
Austral   (AW-strull)
Austrasia   (aw-STRAY-zhuh)
Bauxite   (BAWKS-ite)
Beijing   (bay-ZHING)
Brisbane   (BRIZZ-bun)
Brunei   (broo-NYE)
Canberra   (KAN-burruh)
Chiang Kai-shek   (jee-AHNG kye-SHECK)
Chilung [Jilong]   (JEE-LOONG)
Chosen   (CHOH-SEN)
Chubu   (CHOO-BOO)
Chungyang   (joong-YAHNG)
Dunedin   (duh-NEED-nn)
Edo   (EDD-oh)

Etorofu   (etta-ROH-foo)
Formosa   (for-MOH-suh)
Fuji   (FOODGY)
Fujian   (foo-jee-ENN)
Fukuoka   (foo-kuh-WOH-kuh)
Fuzhou   (foo-ZHOH)
Gotemba   (got-TEM-buh)
Guangdong   (gwahng-DUNG)
Guangzhou   (gwahng-JOH)
Habomai   (HAH-boh-mye)
Heian   (HAY-ahn)
Hiroshima   (hirra-SHEE-muh/huh-ROH-shuh-muh)
Hokkaido   (hah-KYE-doh)
Honshu   (HONN-shoo)
Hsinchu [Xinzhu]   (shin-JOO)
Hyundai   (HUN-dye)
Iki   (EE-kee)
Indigenous   (in-DIDGE-uh-nuss)
Jilong   (jee-LUNG)

Kakadu   (KAH-kuh-doo)
Kalgoorlie   (kal-GHOOR-lee)
Kambalda   (kahm-BAHL-duh)
Kansai   (KAHN-SYE)
Kanto   (KAN-toh)
Kaohsiung [Gaoxiong]   (gau-shee-AHNG)
Karafuto   (kahra-FOO-toh)
Kawasaki   (kah-wah-SAH-kee)
Kinki   (kin-KEE)
Kitakyushu   (kee-TAH-KYOO-shoo)
Kobe   (KOH-bay)
Korea   (kuh-REE-uh)
Kunashiri   (koo-NAH-shuh-ree)
Kure   (KOOH-ray)
Kurile   (CURE-reel)
Kuroshio   (koo-roh-SHEE-oh)
Kwangju   (GWONG-JOO)
Kyoto   (kee-YOH-toh)
Kyushu   (kee-YOO-shoo)

Lan Tao (LAHN-DAU)
Malacca (muh-LAHK-uh)
Malay (muh-LAY)
Malaysia (muh-LAY-zhuh)
Manchukuo (mahn-JOH-kwoh)
Maori (MAU-ree/MAH-aw-ree)
Marsupial (mar-SOOPY-ull)
Meiji (may-EE-jee)
Melbourne (MEL-bun)
Mount Isa (mount EYE-suh)
Nagasaki (nah-guh-SAHKEE)
Nagoya (nuh-GOYA)
Nakhodka (nuh-KAUGHT-kuh)
Nobi (NOH-bee)
Osaka (oh-SAH-kuh)
Progeny (PRODGE-uh-nee)
Pusan (POO-sahn)
Ravenstein (RAVVEN-steen)

Ryukyu (ree-YOO-kyoo)
Sakhalin (SOCK-uh-leen)
Sakurajimi (suh-koora-JEE-muh)
Samoan (su-MOH-un)
Seikan (say-KAHN)
Sen-en (sen-NENN)
Seoul (SOAL)
Seto (SET-oh)
Shanghai (shang-HYE)
Shenzhen (shun-ZHEN)
Shikoku (shick-KOH-koo)
Shikotan (shee-koh-TAHN)
Shimonoseki (shim-uh-noh-
    SECKEE)
Shogun (SHOH-goon)
Singapore (SING-uh-poar)
Sri Lanka (sree-LAHNG-kuh)
Sumatera (suh-MAH-tuh-ruh)

Taipei [Taibei] (tye-BAY)
Taiwan (tye-WAHN)
Tasman (TAZZ-mun)
Tasmania (tazz-MAY-nee-uh)
Tiananmen (TYAHN-un-men)
Tokaido (toh-KYE-doh)
Tokyo (TOH-kee-oh)
Tsugaru (tsoo-GAH-roo)
Tsukuba (tsoo-KOOB-uh)
Tsunami (tsoo-NAH-mee)
Tsushima (tsoo-SHEE-muh)
Ulsan (OOL-SAHN)
Vietnam (vee-et-NAHM)
Vladivostok (vlad-uh-vuh-STAHK)
Waitangi (WYE-tonggy)
Yawata (yuh-WAH-tuh)
Yokohama (yoh-kuh-HAH-muh)
Zhu (JOO)

PART TWO

DEVELOPING REALMS

Scale 1:1 000 000

Scale 1:16 000 000; one inch to 250 miles. Polyconic Projection
Elevations and depressions are given in feet

copyright © Rand McNally, 1993

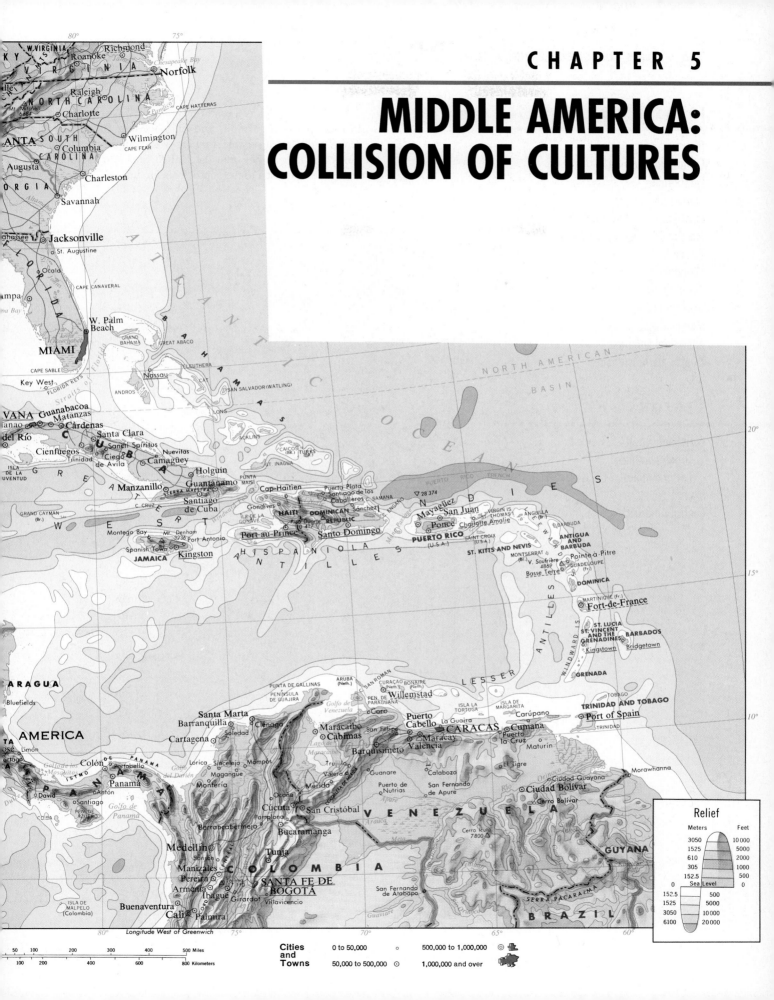

CHAPTER 5

MIDDLE AMERICA: COLLISION OF CULTURES

Relief

Meters		Feet
3050		10 000
1525		5000
610		2000
305		1000
152.5		500
0	Sea Level	0
152.5		500
1525		5000
3050		10 000
6100		20 000

Cities and Towns

0 to 50,000 ○
50,000 to 500,000 ⊙
500,000 to 1,000,000 ◎
1,000,000 and over

Longitude West of Greenwich

50   100   200   300   400   500 Miles
100   200   400   600   800 Kilometers

IDEAS & CONCEPTS

Historical geography
Cultural landscape (2)
Land bridge
Culture hearth (2)
Environmental determinism
Mainland-Rimland framework
Hacienda
Plantation
Plural society
Transculturation
Maquiladora
Altitudinal zonation
Tropical deforestation

REGIONS

Caribbean Basin
   Greater Antilles
   Lesser Antilles
Mexico
Central America

Middle America is a realm of vivid contrasts, turbulent history, political turmoil, and an uncertain future. The realm's diversity, visible in Fig. 5-1, comprises all the lands and islands between the United States to the north and South America to the south. This includes the substan-

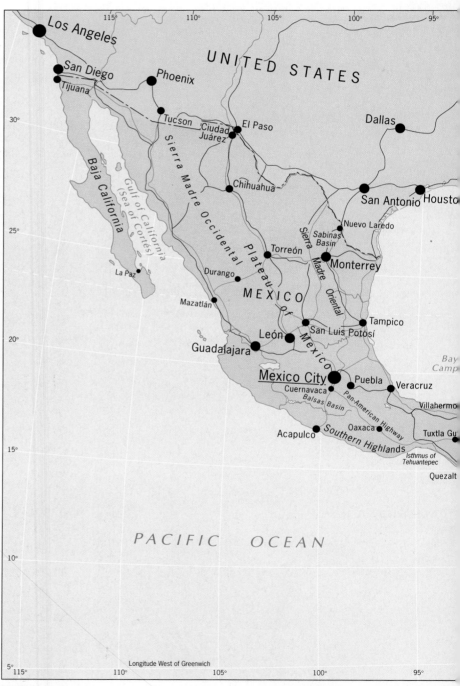

FIGURE 5-1

tial landmass of Mexico, the narrowing strip of land to its southeast that extends from Guatemala to Panama, and the many large and small islands of the Caribbean Sea to the east. Middle America is a realm of soaring volcanoes and forested plains, of mountainous islands and flat coral cays. Moist tropical winds sweep in from the sea, watering windward coasts while leaving leeward areas dry. Soils vary from fertile volcanic to desert barren. Spectacular scenery abounds, and tourism is one of the realm's leading industries.

Culturally, Middle America's contrasts are stark. Teeming, poverty-afflicted cities grow ever larger. People in search of a better existence cross into the United States by the hundreds of thousands, transforming the social geography

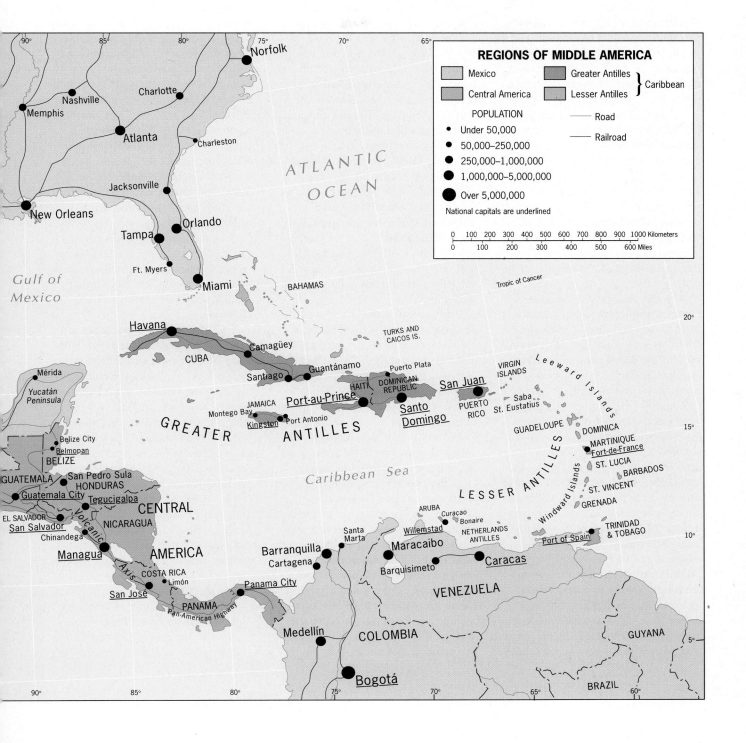

REGIONS OF MIDDLE AMERICA

- Mexico
- Central America
- Greater Antilles
- Lesser Antilles } Caribbean

POPULATION
- Under 50,000
- 50,000–250,000
- 250,000–1,000,000
- 1,000,000–5,000,000
- Over 5,000,000

National capitals are underlined

— Road
— Railroad

0 100 200 300 400 500 600 700 800 900 1000 Kilometers
0 100 200 300 400 500 600 Miles

HISTORICAL GEOGRAPHY

Although our survey of world realms is less than half complete, it has become evident that the geographic present is a product of the past. This is just as true of developing realms as it was of those in the developed world already covered. The study of how a region evolves and continues to change is the domain of **historical geography**. Because it treats the all-encompassing dimension of time, the concepts and methods of historical geography—like those of regional geography—are applicable to every branch of the discipline. Historical geography, however, is clearly distinct from the field of history: it is dominated by the work of geographers whose main task is the interpretation of spatial change on the earth's surface.

The rich historical dimension of regional geography is indispensable to an understanding of contemporary landscapes and societies. For example, the shifting distribution of economic activities inside U.S. metropolitan areas, as we saw in Chapter 3, is a response to changing geographic forces as the ending industrial age gives way to its postindustrial successor. Those dynamics remind us of the central role of *spatial processes*, the causal forces that act and unfold over time to shape the spatial distributions we observe today. In turn, these geographic patterns may be regarded as individual frames in a relentlessly advancing film. Thus spatial organization at any given moment of time represents a point in a long process and is clearly related to what went before (as well as to what will develop in the future). Emphasizing this central concern, historical geographers today

approach their subdiscipline from a number of different, although not mutually exclusive, perspectives.

The first of these is the study of the *geographic past*, wherein earlier spatial patterns are reconstructed and compared for time periods that were crucial in the formation of the present regional structure. The raw materials for assembling these snapshots of the past come from many sources: historical maps, books, censuses, archives, and interviews of long-time residents of the study area. This approach can also be used to build evolutionary models of spatial organization: the four-stage model of U.S. intraurban growth (Fig. 3-9) is a good example, with the structural pattern of each era shaped by its prevailing transportation technology.

The second perspective focuses on *landscape evolution*, the historical analysis and interpretation of existing **cultural landscapes**. This search for "the past in the present" takes a number of paths. The study of individual relics, such as Maya pyramids, is one of them. Another is comparative regional analysis of landscape artifacts such as house types, which facilitates the mapping of cultural change.

A third approach to historical geography can be called *perception of the past*, the application of modern behavioral concepts to recreate and study the attitudes of previous generations toward their environments. The study of migration decision-making, for instance, is of particular interest where it involves the acquisition of accurate spatial information on the "mental maps" of potential migrants. For instance, before the U.S. Civil

War, the Great Plains area was widely perceived as part of a "Great American Desert," a myth that helped delay its settlement until late in the nineteenth century. Abstractions are valuable, too; a recent example is Donald Meinig's analysis of symbolic historical landscapes, from which he derived three idealized American community types that still exert a powerful hold on residential preferences throughout the United States: the New England village, the small-town Main Street, and the California horticultural suburb. More formal modeling approaches exist as well, such as the *time-geography* perspective developed by the Swedish geographer Torsten Hägerstrand, in which allocations of time for performing tasks in space are measured and regionally analyzed.

One of the most active areas of historico-geographical research is the study of *settlement geography*, the facilities people build while occupying a region. Because these facilities usually survive beyond the era of their original functions, they often provide some of the clearest expressions of the past in the contemporary landscape. The colonial landscapes of the New World have received considerable attention, especially the layout of Spanish colonial towns in Middle America. Shown in Fig. 5-2, this layout's historico-geographical significance is described by Charles Sargent:

The morphology, or form, of a city reflects its past, and the clearest reflections are found in the street pattern, the size of city blocks, the dimensions of urban lots, and the

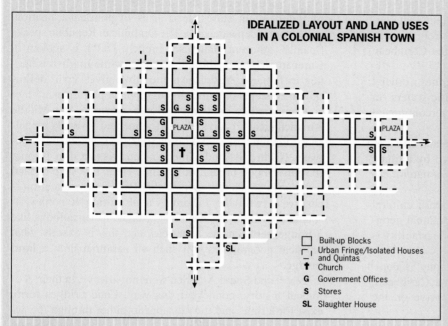

IDEALIZED LAYOUT AND LAND USES IN A COLONIAL SPANISH TOWN

Built-up Blocks
Urban Fringe/Isolated Houses and Quintas
✝ Church
G Government Offices
S Stores
SL Slaughter House

FIGURE 5-2

surviving colonial architecture. In the typical Spanish American city, the colonial past is on view in what is now the city core, which is commonly laid out in a gridiron of square or rectangular blocks subdivided into long, narrow lots along equally narrow streets.

This particular spatial pattern hardly came about by accident—it was decreed by royal regulations down to the smallest details of street, building, and central-plaza construction—and was repeated throughout Hispanic America (including former Spanish-controlled areas of the southwestern United States, such as New Mexico). Because most Spaniards were urban dwellers, it was hardly surprising that

they chose to build towns in their colonial territories. But why this particular town form? The answer rests with that other major characteristic of an urban settlement—function.

The Spanish colonial town possessed several functions that were best suited to the compact, gridiron-street-pattern layout. It is often said that the threefold aims of the rulers of New Spain were "God, glory, and gold." Because the main function of the settlement was administrative (a point developed in this chapter), town sites were chosen to maximize accessibility to regional trade routes and sources of tribute from local Amerindians. This control function also extended to the internal structure of the town: everything was tightly focused on the cen-

tral *plaza*, or market square, under the watchful eye of government authorities in adjacent buildings (**G** symbols in Fig. 5-2). Another leading town function was expressed in the central role of the church, which always faced the plaza as well. The Roman Catholic Church sought to convert as many Amerindians to its faith as possible, and the easiest way to do this was to forcibly resettle the dispersed aboriginal population in Spanish towns, where the collection of tribute, the recruitment of mine workers, and the farming of land surrounding the town were also facilitated. Because of the gridiron street plan, any insurrections by the resettled Amerindians could be contained by having a small military force seal off the affected blocks and root out the troublemakers.

The Greeks and Romans learned this lesson when they established far-flung empires, and the grid-plan tradition was passed down to their Mediterranean European successors. This idea even has modern applications: the battle for Vietnam's capital, Saigon (a non-gridiron city), in the 1960s and 1970s favored guerrillas who could move at will through twisting streets and alleyways, whereas the inner-city riots of 1967 in gridded Newark and Detroit were quickly squelched by the U.S. Army, which systematically surrounded and pacified block after block.

Historical geography will continue to be a prominent theme in upcoming chapters. Two additional important concepts that involve the meshing of time and space will also be discussed: *spatial diffusion* in Chapter 7 and *sequent occupance* in Chapter 8.

TEN MAJOR GEOGRAPHIC QUALITIES OF MIDDLE AMERICA

1. Middle America is a fragmented realm that consists of all the mainland countries from Mexico to Panama and all the islands of the Caribbean Sea to the east.

2. Middle America's mainland constitutes a crucial barrier between Atlantic and Pacific waters. In physiographic terms, this is an intercontinental land bridge.

3. Middle America's tropical location and climates are in important places ameliorated by altitude and the resulting vertical zonation of natural environments.

4. Middle America is a realm of intense cultural and political fragmentation. The political geography defies unification efforts, and instability is a persistent problem.

5. Middle America's cultural geography is complex. African influences dominate the Caribbean, whereas Amerindian traditions survive on the mainland.

6. Middle America's historical geography is replete with involvement by its powerful neighbor, the United States.

7. Underdevelopment is endemic in Middle America; the realm contains the Americas' least-developed territories.

8. In terms of area, population, and economic strength, Mexico dominates the realm.

9. The realm (notably Mexico) contains major actual and potential reserves of fossil fuels.

10. Mexico in the 1990s is reforming its economy and experiencing major industrial growth. Its hopes for continuing this development are heavily tied to expanding trade with the United States under the North American Free Trade Agreement.

ian (Spanish and Portuguese) heritage and the prevalence of the Roman Catholic religion. But these criteria apply far more strongly to South America. In Middle America, large populations exhibit African and Asian as well as European ancestries. Nowhere in South America has native, Amerindian culture contributed to modern civilization as strongly as it has in Mexico. The Caribbean Basin is a patchwork of independent states, territories in political transition, and colonial dependencies. The Dominican Republic speaks Spanish, adjacent Haiti uses French; Dutch is spoken in Curaçao and on neighboring islands, while English is spoken in Jamaica. Middle America thus gives vivid definition to concepts of cultural-geographical pluralism.

Compared to continental South America, the Middle American realm is divided and fragmented. Even its funnel-shaped mainland, a 3,800-mile (6,000-km) connection between North and South America, narrows to a slim 40-mile (65-km) ribbon of land in Panama. Here, this strip of land or *isthmus* bends eastward so that Panama's orientation is east-west (with the Panama Canal cutting it northwest-southeast). On the map, mainland Middle America looks like a bridge between the Americas, and this is exactly what physical geographers call such an isthmian link: a **land bridge**.

North and South America were not always, in their geological history, connected this way. Land bridges form, exist for a time, and then disappear, either through the action of geologic processes or because sea levels rise and submerge them. If you examine a globe, you can see other present and former land bridges: the Sinai Peninsula between Asia and Africa, the (now-broken) Bering land bridge between northeasternmost Asia and Alaska, and the former connection between the island chain off Southeast Asia and Australia. Such land bridges have played crucial roles in the dispersal of animals and humans across the globe. Many scholars believe that the Bering land bridge was the gateway for the first humans to enter the Americas. Migration routes into South America probably lay along the Middle American land bridge, and if dense forests made that passage difficult, the coast afforded a frame of reference for ancient migrants.

Even though mainland Middle America forms a land bridge, its fragmentation inhibits movement to this day. Mountain ranges, volcanoes, swampy coastlands, and dense rainforests make contact and interaction difficult, especially where they all combine in much of Central America (see the box entitled "Middle America and Central America").

The islands of the Caribbean Sea stretch in a vast arc from Cuba to Trinidad, with numerous outliers outside (such as Barbados) and inside (for example, the Cayman Islands) the main chain. The large islands (Cuba, Hispaniola, Puerto Rico, Jamaica) are called the *Greater Antilles*; the smaller islands are referred to as the *Lesser Antilles*. Again, the map gives a hint of the physiography: the island

of States from California to Texas to Florida (see Fig. 3-12). Northern South America, too, carries the imprints of Caribbean cultures and economies (see Fig. 6-6); those imprints range from Afro-Caribbean to Hispanic-American and include European, Amerindian, and even Asian elements.

Is Middle America a discrete geographic realm? Some geographers combine Middle and South America into a realm they call "Latin" America, citing the dominant Iber-

MIDDLE AMERICA AND CENTRAL AMERICA

Middle America, as we define it, includes all the mainland and island countries and territories that lie between the United States and the continent of South America. Sometimes the term is used to identify the same realm, but Central America is actually a region within Middle America. Central America comprises the republics that occupy the strip of mainland between Mexico and Panama: Guatemala, Belize, Honduras, El Salvador, Nicaragua, and Costa Rica. Panama itself is regarded here as belonging to Central America as well; however, it should be noted that many Central Americans do not consider Panama to be part of their region because that country was, for most of its history, a part of South America's Colombia.

arc consists of the crests and tops of mountain chains that rise from the floor of the Caribbean. Some of these crests are relatively stable; elsewhere, however, they consist of active volcanoes. Almost everywhere, earthquakes are a hazard; in the islands as well as on the mainland, the crust is unstable where the Caribbean, Cocos, and North American plates come together (Fig. I-3). Add to this the seasonal risk of hurricanes spawned and nurtured by the realm's abundant tropical waters, and the Middle American environment ranks among the world's most difficult and dangerous.

◆ LEGACY OF MESOAMERICA

Mainland Middle America was the scene of the emergence of a major ancient civilization. Here lay one of the world's true **culture hearths** (see Fig. 7-4), a source area from which new ideas radiated and whose population could increase and make significant advancements. Agricultural specialization, urbanization, and transport networks developed, and writing, science, art, architecture, religion, and other spheres of achievement saw major progress. Anthropologists refer to the Middle American culture hearth as *Mesoamerica*. What is especially remarkable about its development is that it occurred in very different geographic environments.

In the low-lying tropical plains of Honduras, Guatemala, Belize, and Mexico's Yucatán Peninsula, the Maya civilization arose; later, on the high plateau of present-day

Mexico, the Aztecs founded their well-organized civilization. In the process, the Maya and the Aztecs overcame some serious environmental obstacles. Mayan Yucatán may not have been as hot and humid as it is today, but the integration of so large an area was a huge accomplishment. The Aztecs also solved problems of distance and managed to unify people over a wide area despite the topographic barriers of the interior upland (Fig. 5-3).

In the 1910s and 1920s, a school of thought developed in American geography that favored the view that human cultural and economic progress could occur only under certain environmental conditions. Pointing to the present concentration of wealth and power in the middle latitudes of the Northern Hemisphere, these geographers postulated that the tropics were simply not conducive to human productivity. They based their argument largely on the role of climate. The monotonous heat and humidity of tropical climates were supposed to inhibit cultural progress. Mid-latitude climates, in contrast, with their variable weather, were alleged to stimulate human achievement.

This school of **environmental determinism** (or simply *environmentalism*) held that the natural environment to a large extent dictated the course of civilization. The leading proponent and popularizer of this hypothesis was Ellsworth Huntington (1876–1947), who believed that human progress rested on three bases: climate, heredity, and culture. Not surprisingly, his controversial conclusions came under attack as ill-founded and supportive of "master-race" philosophies. Although he may have generalized carelessly, Huntington did pose crucial questions that are still unanswered today; approaching them from another angle, sociobiologists now face similar obstacles. Human societies and natural environments do interact, but how? When does a combination of particular environmental circumstances, inherited capacities, and cultural transmissions stimulate a new cultural explosion?

THE LOWLAND MAYA

Certainly, few environmentalists could easily account for the rise of Maya civilization in the lowland tropics of Middle America. Some scholars have suggested that cultural stimuli from ancient Egypt actually reached Middle American shores and that the pyramid-like stone structures of Mayan cities represent imitations and variations of Egyptian achievements. It seems more likely that Maya civilization arose spontaneously and independently. It experienced successive periods of glory and decline, reaching its zenith in present-day Guatemala from the fourth to the tenth centuries A.D.

The Maya civilization unified an area larger than any of the modern Middle American states except Mexico. Its population was probably somewhere between 2 and 3 mil-

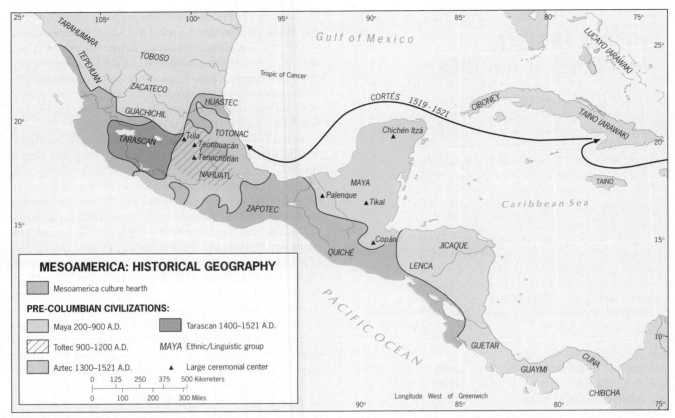

FIGURE 5-3

lion; the Maya language—the local *lingua franca*—and some of its related languages remain in use in the area to this day. The Maya state was a theocracy (ruled by religious leaders) with a complex religious hierarchy, and the great cities that today lie in ruins served in the first instance as ceremonial centers. Their structures, which testify to the architectural capabilities of the Maya, include huge pyramids and magnificent palaces. We also know that Maya culture boasted of skilled artists, writers, mathematicians, and astronomers. No doubt the Maya civilization also had poets and philosophers; but it had a practical side too. In agriculture and trade, these people achieved a great deal. They grew cotton, created a rudimentary textile industry, and even exported finished cotton cloth by seagoing canoes to other parts of Middle America in return for, among other things, the cacao (raw material of chocolate) they prized so highly.

THE HIGHLAND AZTECS

About the same time that the Maya civilization emerged in the tropical lowland forests, the Mexican highland also witnessed the rise of a great Amerindian culture, similarly

focused on ceremonial centers and marked as well by major developments in agriculture, architecture, religion, and the arts. Although these achievements were comparable to those of the Maya, they were to be overshadowed by what followed. An important successor of the early highland civilization were the Toltecs, who moved into this area from the north, conquered and absorbed the local Amerindian peoples, and formed a powerful state centered on one of the first true cities of Middle America: Tula. The Toltecs' period of hegemony was relatively brief, lasting for less than three centuries after their rise to power around A.D. 900, but they conquered parts of the Maya domain, absorbed many Mayan innovations and customs, and introduced them on the plateau. When the Toltec state was in turn penetrated by new elements from the north, it was already in decay, but its technology was readily adopted and developed by the conquering Aztecs (or Mexicas).

The Aztec state, the pinnacle of organization and power in Middle America, is thought to have originated in the early fourteenth century when a community of Nahuatl- (or Mexicano-) speaking Amerindians founded a settlement on an island in one of the many lakes that lay in the Valley of Mexico (the area surrounding present-day Mexico City). This ceremonial center, named Tenochtitlán, was soon to become the greatest city in the Americas and the capital of a large and powerful state. Through a series of alliances with neighboring peoples, the early Aztecs gained control over the entire Valley of Mexico, the pivotal geographic feature of Middle America that is still the heart of the modern state of Mexico. This 30-by-40-mile (50-by-65-km) region is, in fact, a mountain-encircled basin positioned about 8000 feet (nearly 2,500 m) above sea level. Elevation and interior location both affect its climate; for a tropical area, it is quite dry and very cool. The region's lakes formed a valuable means of internal communication for the Aztec state. The native peoples of Middle America never developed wheeled vehicles; thus they relied heavily on portage and, where possible, the canoe for the water transportation of goods and people. The Aztecs connected several of the Mexican lakes by canals and maintained a busy canoe traffic on their waterways, bringing agricultural produce to the cities and tribute paid by their many subjects to the headquarters of the ruling nobility.

Throughout the fourteenth century, the Aztec state strengthened its position by developing a strong military force, and by the early fifteenth century the conquest of neighboring peoples had begun. The Aztec drive to expand the empire was directed primarily eastward and southward. To the sparsely settled north, the rugged and unproductive land quickly became drier; to the west lay the powerful competing state of the Tarascans, with whom the Aztecs sought no quarrel, given the opportunities to the east and south. Here, then, they penetrated and conquered almost at will. The Aztec objective was neither the acquisition of territory nor the spreading of their language and religion, but rather the subjugation of peoples and towns in order to extract taxes and tribute. The Aztecs also carried off thousands of people for purposes of human sacrifice in the ceremonial centers of the Valley of Mexico, which provoked a constant state of enmity with weaker peoples.

As Aztec influence spread throughout Middle America, the goods streaming back to the Valley of Mexico included gold, cacao beans, cotton and cotton cloth, and the skins of wild animals. The state grew ever richer, its population mushroomed, and its cities expanded. At its peak Tenochtitlán had over 100,000 inhabitants or more—perhaps as many as a quarter of a million. These cities were not just ceremonial centers but true cities with a variety of economic and political functions and large populations, including specialized, skilled labor forces.

The Aztecs produced a wide range of impressive accomplishments, although they were better borrowers and refiners than they were innovators. They practiced irrigation by diverting water from streams to farmlands, and they built elaborate walls to terrace slopes where soil erosion threatened. Indeed, when it comes to measuring the legacy of Mesoamerica's Amerindians to their successors (and to humankind) the greatest contributions surely came from the agricultural sphere. Corn (maize), the sweet potato, various kinds of beans, the tomato, squash, cacao tree, and tobacco are just a few of the crops that grew in Mesoamerica when the Europeans first made contact.

◆ COLLISION OF CULTURES

We in the Western world all too often are under the impression that history began when the Europeans arrived in some area of the world, and that the Europeans brought such superior power to the other continents that whatever existed there previously had little significance. Middle America bears out this perception: the great, feared Aztec state fell before a relatively small band of Spanish invaders in an incredibly short time (1519–1521). Let us not lose sight, however, of a few facts. At first the Spaniards were considered to be "White Gods," whose arrival was predicted by Aztec prophecy. Having entered Aztec territory, the earliest Spanish visitors could see that great wealth had been amassed in Aztec cities. And Hernán Cortés, for all his 508 soldiers, did not single-handedly overthrow the Aztec authority.

What Cortés brought on was a revolt, a rebellion by peoples who had fallen under Aztec domination and who had seen their relatives carried off for human sacrifice to Aztec gods. Led by Cortés with his horses and artillery, these indigenous peoples rose against their Aztec oppressors and followed the band of Spaniards toward Tenochtitlán, where thousands of them died in combat against the Aztec warriors. They fed and guarded the Spanish soldiers, maintained connections for them with the coast, carried supplies from the shores of the Gulf of Mexico to the point Cortés had reached, and secured and held captured territory while the Europeans moved on. Cortés started a civil war; he got all the credit for the results. Yet it is reasonable to say that Tenochtitlán would not have fallen so easily to the Spaniards without the sacrifice of many thousands of Amerindian lives.

Actually, in the Americas as well as in Africa, the Spanish, Portuguese, British, and other European visitors considered many of the peoples they confronted to be equals—equals to be invaded, attacked, and, if possible, defeated—but equals nonetheless. The cities and farms of Middle America, the urban centers of West Africa, the great Inca roads of highland South America all reminded the Europeans that technologically they were only a few steps ahead of their new contacts. Thus the gap that developed between the European powers and the indigenous peoples of many other parts of the world emerged clearly only when the Industrial Revolution came to Europe—centuries after Vasco da Gama, Columbus, and Cortés.

EFFECTS OF THE CONQUEST

In Middle America the confrontation between Hispanic and native cultures spelled disaster for the Amerindians in every conceivable way: a drastic decline in population, rapid deforestation, pressure on vegetation from grazing animals, substitution of Spanish wheat for maize (Indian corn) on cropland, and construction of new Spanish towns. The quick defeat of the Aztec state was followed by a catastrophic decline in population. Of the 15 or 25 million native inhabitants of Middle America when the Spaniards arrived (estimates vary), only a century later just 2.5 million survived.

The Spanish were ruthless colonizers, but not much more so than other European powers that subjugated other cultures. True, the Spanish first enslaved the Amerindians and were determined to destroy the strength of aboriginal society. But biology accomplished what ruthlessness could not have achieved in so short a time. Nowhere in the Americas did the native peoples have immunity to the diseases the Spaniards brought: smallpox, typhoid fever, measles, influenza, and mumps. Nor did they have any protection against the tropical diseases that the Europeans introduced through their African slaves, such as malaria and yellow fever, which took enormous tolls on human life in the hot, humid lowlands of Middle America.

Middle America's cultural landscape—its great cities, its terraced fields, its dispersed aboriginal villages—was thus drastically modified. The Amerindian cities' functions ceased as the Spanish brought in new traditions and innovations in urbanization, agriculture, religion, and other pursuits. Having destroyed Tenochtitlán, the Spaniards did recognize the attributes of its site and situation and chose to rebuild it as their mainland headquarters. Whereas the natives had used stone almost exclusively as their building material, the Spaniards employed great quantities of wood and used charcoal for heating, cooking, and smelting metal. The onslaught on the forests was immediate, and expanding rings of deforestation quickly formed around the Spanish towns. Soon, the Amerindians also adopted Spanish methods of wooden construction and charcoal use, further accelerating forest depletion and the erosional scarring of the land.

The Amerindians had been planters but had no domestic livestock that made demands on the original vegetative cover. Only the turkey, the dog, and the bee (for honey and wax) had been domesticated in Mesoamerica. The Spaniards, however, brought with them cattle and sheep, whose numbers multiplied rapidly and made increasing demands not only on the existing grasslands but on the cultivated crops as well. Again, the natives adopted these Spanish practices, putting further pressure on the land. Cattle and sheep became avenues to wealth, and the owners of the herds benefited. However, the livestock now competed with the people for available food (requiring the opening up of vast areas of marginal land in higher and drier locations), thereby contributing to a major disruption of the region's food-production balance. Hunger quickly became a significant problem in Middle America during the sixteenth century, heightening the susceptibility of the Amerindian population to many diseases.

The Spaniards also introduced their own crops (notably wheat) and farming equipment, of which the plow was the most important. Thus, large fields of wheat began to make their appearance beside the small plots of corn that the natives cultivated. The encroaching wheatfields soon further reduced the natives' lands. Moreover, because this grain was raised by and for the Spanish, what the Amerindians lost in farmland was not made up in additional available food. Neither were their irrigation systems spared. The Spaniards needed water for their fields and hydropower for their mills, and they had the technological know-how to take over and modify regional drainage and irrigation systems. This they did, leaving the natives' fields insufficiently watered, thereby diminishing even further these peoples' chances for an adequate food supply.

The most far-reaching changes in the cultural landscape introduced by the Spanish had to do with their traditions as town dwellers. To facilitate control, they relocated the Amerindians from their land into nucleated villages and towns that the Spanish established and laid out. In these settlements, the Spanish could exercise the kind of government and administration to which they were accustomed. The focus of each town was the Catholic church (see photo). Each town was located near what was thought to be good agricultural land so that the Amerindians could go out each day and work in the fields. Unfortunately, the selection was not always good, and the land surrounding numerous villages was not suitable for native farming practices. Here food shortages and even famine resulted—but only rarely was a settlement abandoned in favor of a better opportunity.

DOMINATION

In the towns and villages, the Amerindians came face to face with Spanish culture. Here they learned the white invaders' religion and paid their taxes and tribute to a new master—or found themselves in prison or in a labor gang according to European regulations. Packed tightly in a concentrated settlement, they were rendered even more vulnerable to the diseases that regularly ravaged the population. Despite all this, the nucleated indigenous village survived. Its administration was taken over by the Spaniards (and later by their post-colonial successors), and today it is still a key landscape feature of the Amerindian areas of Mexico and Guatemala. However, anyone who wants to see remnants of the dispersed native dwellings and hamlets must travel into the remotest parts of mainland Middle America, where Amerindian languages still prevail (Fig. 5-4).

Spain's urban imprint on Middle America, the crucible in which the colonial-era acculturation of the Amerindians under their new masters took place. This is the small city of Taxco in Mexico's State of Guerrero. We have used earlier versions of this photo in previous editions of the book; the satellite dish in the center first appeared in the 1993 image, evidence of the ever-expanding reach of the communications age. But the Santa Prisca Church continues to dominate the townscape.

FIGURE 5-4

Once the indigenous population was conquered and resettled, the Spanish were able to pursue another primary goal in their New World territory: the exploitation of its wealth for their own benefit. Lucrative trade, commercial agriculture, livestock ranching, and especially mining were the avenues to affluence. The mining of gold held the greatest initial promise, and the Spaniards simply took over the Amerindian miner workforce already in operation. But the small-scale *placering* or "washing" of gold from streams carrying gold dust and nuggets yielded a diminishing supply of the precious metal, and with the abolition of Amerindian slavery, Spanish prospectors began searching for other valuable minerals. They were quickly successful, finding enormously profitable silver and copper deposits, particularly in a wide zone north of the Valley of Mexico (see Fig. 5-7).

The development of these resources set into motion a host of changes in this part of Middle America. Mining towns drew laborers by the thousands; the mines required equipment, timber, mules; the people in the towns needed food. Because most of these settlements were located in dry country, irrigated fields were laid out wherever possible. Mule trains connected farm-supply areas to the mining towns and these towns to the coastal ports. Thus was born a new urban system, one that not only integrated and organized the Spanish domain in Middle America but also extended effective economic control over some far-flung parts of New Spain. Mining truly was the mainstay of colonial Middle America.

◆ MAINLAND AND RIMLAND

In Middle America outside Mexico, only Panama, with its twin attractions of interoceanic transit and gold deposits, became an early focus of Spanish activity (the Spaniards founded Panama City in 1519). Apart from their use of the

corridor of the modern Panama Canal as an Atlantic-Pacific link, their main interest lay on the Pacific side of the isthmus. From here, Spanish influence began to extend northwestward into Central America. Amerindian slaves were taken in large numbers from the densely peopled Pacific lowlands of Nicaragua and shipped to South America via Panama. The highlands, too, fell under Hispanic control; before the middle of the sixteenth century, Spanish exploration parties based in Panama met those moving southeastward from Mesoamerica.

However, the leading center of Spanish activity remained in what is today central and southern Mexico, and the major arena of international competition in Middle America lay not on the Pacific side but on the islands and coasts of the Caribbean Sea. (Only the British gained a foothold on the mainland, controlling a narrow, low-lying coastal strip that extended from Yucatán to what is now Costa Rica.) As the colonial-era map (Fig. 5-5) shows, in the Caribbean the

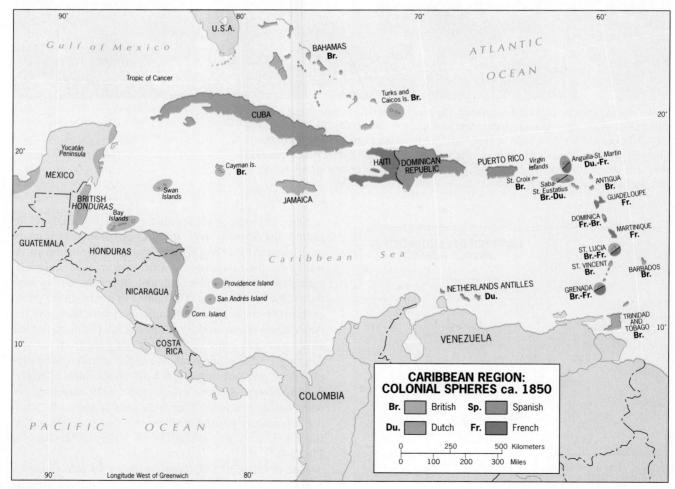

FIGURE 5-5

INDIGENOUS CARIBBEAN PEOPLES

Mesoamerican aboriginal cultures have in some measure survived the European invasion. Amerindian communities remain, and indigenous languages continue to be spoken by about 3 million people in southern Mexico and Yucatán and another million in Guatemala (Fig. 5-4). On the islands of the Caribbean, however, the Amerindian communities were smaller and more vulnerable. At about the time of Columbus's arrival, there probably were 3 to 4 million natives living in the Caribbean, a majority of them in Cuba, Jamaica, Puerto Rico, and Hispaniola (the island containing modern-day Haiti and the Dominican Republic). These larger islands of the Greater Antilles were peopled by the Arawaks, whose farming communities raised root crops, tobacco, and cotton. In the eastern Caribbean, the smaller islands of the Lesser Antilles (Guadeloupe, Martinique, Dominica, and dozens of others) had more recently been peopled by the adventurous Caribs, who traversed the waters of the Caribbean in huge canoes that carried several dozen persons. When the European sailing ships began to arrive, the Caribs were in the process of challenging the Arawaks for their land, just as the Arawaks centuries earlier had ousted the Ciboney.

The Europeans quickly laid out their sugar plantations and forced the Amerindians into service. But Arawaks and Caribs alike failed to adapt to the extreme rigors of the labor to which they were subjected, and they perished by the thousands. Some fled to smaller islands where the European arrival was somewhat delayed. Yet after only half a century, just a few hundred survived in the dense interior forests of some of the islands. There they were soon joined by runaway African slaves (brought to the Caribbean to replace the dwindling aboriginal labor force), and, with their mixture, the last pure Caribbean Amerindian strain disappeared forever.

Spaniards faced the British, French, and Dutch, all interested in the lucrative sugar trade, all searching for instant wealth, and all seeking to expand their empires. Later, after centuries of European colonial rivalry in the Caribbean Basin, the United States entered the picture and made its influence felt in the coastal areas of the mainland—not through colonial conquest but through the introduction of widespread, large-scale plantation agriculture.

The effects of these plantations were as far-reaching as the impact of colonialism on the Caribbean islands. The economic geography of the Caribbean coastal zone was transformed as hitherto unused alluvial soils in the many river lowlands were planted with thousands of acres of banana trees. Because the diseases the Europeans had brought to the New World had been most rampant in these hot and humid areas, the Amerindian population that survived was too small to provide a sufficient labor force; the native peoples of the Caribbean islands had faced an even harsher fate (see box at left). Consequently, tens of thousands of black laborers were brought to the mainland coast from Jamaica and other islands (many more came when the Panama Canal was dug between 1904 and 1914), completely altering the demographic mix. Physically, in many ways, the coastal belt already resembled the islands more than the Middle American plateau, and now the economic and cultural geographies of the islands were also extended to the coastal belt.

These contrasts between the Middle American highlands on the one hand, and the coastal areas and Caribbean islands on the other, were conceptualized by John Augelli into a **Mainland-Rimland framework** (Fig. 5-6). Augelli recognized (1) a Euro-Amerindian *Mainland*, consisting of mainland Middle America from Mexico to Panama, with the exception of the Caribbean coast from mid-Yucatán southeastward; and (2) a Euro-African *Rimland*, which included this coastal zone and the islands of the Caribbean. The terms *Euro-Amerindian* and *Euro-African* underscore the cultural heritage of each region: on the Mainland, European (Spanish) and Amerindian influences are paramount; in the Rimland, the heritage is European and African.

As Fig. 5-6 shows, the Mainland is subdivided into several areas on the basis of the strength of the Amerindian legacy. In southern Mexico and Guatemala, Amerindian influences are prominent; in northern Mexico and parts of Costa Rica, those influences are limited; between these areas lie sectors with moderate Amerindian influence. The Rimland, too, is subdivided. The most obvious division is between the mainland-coastal plantation zone and the islands. But the islands themselves can be classified according to their cultural heritage. Thus there is a group of islands with Spanish influence (Cuba, Puerto Rico, and the Dominican Republic on old Hispaniola) and another group with other European influences, including the former British West Indies, the various French islands, and the Netherlands Antilles.

These contrasts of human habitat are supplemented by regional differences in outlook and orientation. The Rimland was an area of sugar and banana plantations, of high accessibility, of seaward exposure, and of maximum cultural contact and mixture. The Mainland, being farther re-

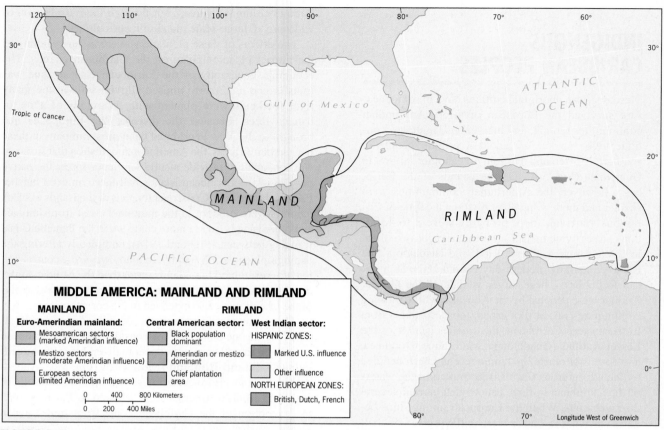

FIGURE 5-6

moved from these contacts, was an area of greater isolation. The Rimland was the region of the great plantation, and its commercial economy was therefore susceptible to fluctuating world markets and tied to overseas investment capital. The Mainland was the region of the hacienda, more self-sufficient and considerably less dependent on external markets.

THE HACIENDA

In fact, this contrast between plantation and hacienda land tenure in itself constitutes strong evidence for a Rimland-Mainland division. The hacienda was a Spanish institution, but the modern plantation, Augelli argued, was the concept of Europeans of more northerly origin. In the **hacienda**, Spanish landowners possessed a domain whose productivity they might never push to its limits: the very possession of such a vast estate brought with it the social prestige and comfortable lifestyle they sought. The native workers lived on the land—which may once have been *their* land—and had plots where they could grow their own subsistence crops. Traditions survived: in the Amerindian villages incorporated into the early haciendas, in the methods of farming, and in

the means of transporting produce to markets. All this is written as though it is mostly in the past, but the legacy of the hacienda, with its inefficient use of land and labor, is still visible throughout mainland Middle America (as well as parts of South America).

THE PLANTATION

The **plantation**, in contrast, was conceived as something entirely different. In *Middle America: Its Lands and Peoples*, Robert West and John Augelli listed five characteristics of Middle American plantations that clearly illustrate the differences between hacienda and plantation: (1) plantations are located in the humid tropical coastal lowlands of the realm; (2) plantations produce for export almost exclusively—usually a single crop; (3) capital and skills are often imported so that foreign ownership and an outflow of profits occur; (4) labor is seasonal—needed in large numbers during the harvest period but often idle at other times—and such labor has been imported because of the scarcity of Amerindian workers; and (5) with its "factory-in-the-field" operation, the plantation is more efficient in its use of land and labor than the hacienda. The objective was not self-sufficiency but profit, and wealth rather than social prestige is a dominant motive for the plantation's establishment and operation.

During the past century, both systems of land tenure have changed a great deal. The massive U.S. investment in the Caribbean coastal belt of Guatemala, Honduras, Nicaragua, Costa Rica, and Panama transformed that area and brought a new concept of plantation agriculture to the region. On the Mainland, the hacienda has been under increasing pressure from national governments that view it as an economic, political, and social liability. Indeed, some haciendas have been parceled out to small landholders, and others have been pressed into greater specialization and productivity. In Mexico, still other land has been placed in *ejidos*, where it is communally owned by groups of families. Both the hacienda and the plantation have for centuries contributed to the different social and economic directions that gave the Mainland and Rimland their respective regional personalities.

◆ POLITICAL DIFFERENTIATION

Mainland Middle America today is fragmented into eight different countries, all but one of which (Belize, the former British Honduras) have Hispanic origins. Largest of them all—the giant of Middle America—is Mexico, whose 762,000 square miles (1,972,500 sq km) constitute over 70 percent of the entire land area of Middle America (the Caribbean Basin included) and whose 91.8 million people outnumber those of all the other countries and islands of Middle America combined.

The cultural variety in Caribbean Middle America is much greater. Here Cuba dominates: its area is almost as large as that of all the other islands put together, and its population of 11.1 million is well ahead of the next-ranking country (the Dominican Republic, with 7.8 million). But, as we have seen, the Caribbean is hardly an area of exclusive Spanish influence: whereas Cuba has an Iberian heritage, its southern neighbor, Jamaica (population 2.6 million, mostly black), has a legacy of British involvement, and in nearby Haiti (6.8 million, overwhelmingly black) the strongest imprints have been African and French. The crowded island of Hispaniola (population 14.6 million) is shared between Haiti and the Dominican Republic, where Spanish influence prevails (it also predominates in neighboring Puerto Rico).

In the Lesser Antilles, too, there is great cultural diversity. There are the (once Danish) U.S. Virgin Islands; French Guadeloupe and Martinique; a group of British-influenced islands, including Barbados, St. Lucia, St. Vincent, and Grenada; and Dutch St. Maarten (shared with the French), Saba, St. Eustatius, and the A-B-C islands—Aruba, Bonaire, Curaçao—off the northwestern Venezuelan coast. Standing apart from the Antillean arc of islands (off northeastern Venezuela) is Trinidad, another former British dependency that, along with its smaller neighbor Tobago, became a sovereign state in 1962.

Spain was the first colonial power in Middle America and also the first to be forced to yield power in the face of independence movements. Independence came to the mainland colonies much earlier than in the Caribbean, however, because the islands were easier to hold. Mexico proclaimed sovereignty in 1810; the Central American republics emerged during the 1820s and 1830s. In the Greater Antilles, however, Spain held Cuba and Puerto Rico until the Spanish-American War of 1898. By that time, these two islands had come to resemble the mainland republics in the composition of their population and in their cultural imprint. Spain's colonial archrival, Britain, also had a share of the Greater Antilles in Jamaica, and it gained several footholds in the Lesser Antilles to the southeast. The sugar boom and the strategic character of the Caribbean Sea had attracted the European competitors to the West Indies. By the opening of the twentieth century, the United States was making its presence felt as well—although the plantation crop was now the banana and the American strategic interest even more immediate than that of the European powers.

In parts of Caribbean Middle America, the period of colonial control is only just ending. Jamaica attained full independence from Britain in 1962, as did Trinidad and Tobago. An attempt by the British to organize a Caribbean-wide West Indies Federation failed, but other long-time British dependencies, including Barbados, St. Lucia, Dominica, and Grenada, were steered toward a precarious independence nonetheless (with disastrous results in Grenada and a U.S. invasion in 1983). France, on the other hand, has made no moves to end the status of Martinique and Guadeloupe as overseas *départments* of the French national state; and the Dutch A-B-C islands remain within the Netherlands Empire (although Aruba now holds semiautonomous status while on the road to full independence in 1996—a goal the Arubans were seriously reconsidering in 1993).

◆ CARIBBEAN REGIONAL PATTERNS

Caribbean America today is a land crowded with so many people that, as a region (encompassing the Greater and Lesser Antilles), it is the most densely populated part of the Americas. It is also a place of grinding poverty and, in all too many localities, unrelenting misery with little chance for escape. In some respects U.S.-affiliated Puerto Rico (see box next page) constitutes an exception to any such generalization made about Caribbean America—as did communist Cuba before the 1991 demise of its benefactor, the

PUERTO RICO: CLOUDED FUTURE?

The largest, most populous U.S. domain in Middle America is Puerto Rico, the easternmost and smallest island of the Greater Antilles chain. The 3,400-square-mile (8,800-sq km) island, now home to 3.8 million people, fell to the United States during the 1898 Spanish-American War; at the same time, the United States also acquired the Philippines and Guam in the western Pacific.

Thus Puerto Rico's struggle for independence from Spain ended, but American administration was at first difficult. Not until 1948 were Puerto Ricans allowed to elect their own governor. In 1952, following a referendum, the island became the autonomous Commonwealth of Puerto Rico; San Juan and Washington share governmental responsibilities in a complicated arrangement. Puerto Ricans have American citizenship but pay no federal taxes on local incomes. The political situation fails to satisfy everyone: there still is a small pro-independence movement, a sizeable number of voters who favor statehood, and—according to the most recent plebiscite in 1967—a majority who want to continue the commonwealth status but with certain modifications. Another plebiscite took place in late 1993; the status quo was again endorsed as commonwealth supporters narrowly outvoted those favoring statehood.

Puerto Rico stands in sharp contrast to Hispaniola and Jamaica. Long dependent on a single-crop economy (sugar), Puerto Rico during the 1950s and 1960s industrialized rapidly as a result of tax breaks for corporations, comparatively cheap labor, governmental incentives of various kinds, political stability, and special access to the U.S. market. Today, chemicals—not bananas or sugar—rank as the leading export. Puerto Rico does not have substantial mineral resources, and it lies far from the U.S. core area (San Juan is about 1,600 miles [2,600 km] from New York), but these disadvantages have largely been overcome by a development program that began a half-century ago.

Nonetheless, the Puerto Rican economy responds to and reflects the mainland-U.S. economic picture. Times of recession and high unemployment generate an exodus of islanders to the mainland, where they can qualify for welfare support even if they cannot find work. Since 1950, the Puerto Rican population in New York City (the major destination of out-migrants) has grown from about 250,000 to nearly 1 million today—which has markedly contributed to the low population growth rate on the island (1.2 percent annually in the early 1990s, half of Middle America's average). Association with the United States has undoubtedly speeded Puerto Rico's development. The question now is whether the island's social order and political system will be able to withstand the pressures that lie ahead.

Soviet Union. On most of the other islands, however, life for the average person is difficult, often hopeless, and tragically short.

All this is in jarring contrast to the early period of riches based on the sugar trade. Yet that initial wealth was gained while an entire ethnic group (the Amerindians) was being wiped off the Caribbean map and while another (the Africans) was being imported in bondage. The sugar revenues, of course, always went to the planters, not the laborers. Subsequently, the regional economy faced rising competition from other tropical sugar-producing areas; it soon lost its monopoly of the European market, and difficult times prevailed. Meanwhile, just as they did in other parts of the world, the Europeans helped stimulate the rapid growth of the island population. Death rates were lowered, but birth rates remained high and explosive population increases resulted.

With the decline of the sugar trade, millions of people were pushed into a life of subsistence, malnutrition, and hunger. Many sought work elsewhere. Tens of thousands of Jamaican laborers went to the plantations of the Rim-land coast; large numbers of British West Indians went to England in search of a better life; Puerto Ricans and (more recently) Dominicans streamed to New York. But this outflow has failed to stem the tide of regional population growth: today there are over 35 million people on the Caribbean islands, a total expected to increase by 30 percent over the next three decades.

OPPORTUNITIES AT HOME

The Caribbean islanders simply have not had many alternatives in their search for betterment. Their dispersed habitat is fragmented by both water and mountains; the total amount of flat cultivable land is only a small fraction of the far-flung Antillean *archipelago* (island chain). Although some diversification of production has occurred in the region, agriculture remains a significant economic activity. Sugar is still the leading product and continues to head the export lists of the Dominican Republic and Cuba; in Haiti, coffee has become the leader among exported primary products. Trinidad is fortunate in possessing sizeable oil-fields, but the global oil depression produced hard times during the 1980s for a country in which petroleum and petrochemicals regularly account for at least 80 percent of annual exports. In the Lesser Antilles, sugar has retained a somewhat less prominent position, having been supplanted by such crops as bananas, sea-island cotton, limes, and nut-megs. Even here, however, sugar still dominates in places, particularly in Guadeloupe, St. Kitts, and Barbados.

All the crops grown in the Caribbean—Haiti's coffee, Jamaica's bananas, the Dominican Republic's cacao, the Lesser Antilles' fruits as well as the pivotal sugar indus-try—constantly face severe competition from other parts of the world and have still not become established at a scale that could begin to have an effect on improving standards of living. Those minerals that do exist in this region—Jamai-ca's bauxite (aluminum ore), Cuba's iron and chromium, Trinidad's oil—do not support any significant industrial-ization within the Caribbean Basin itself. As in other parts of the developing world, these resources are exported for use elsewhere.

PROBLEMS OF WIDESPREAD POVERTY

Given these conditions, the vast majority of the people in this region continue to eke out a precarious living from small plots of ground, mired in poverty and threatened by disease. Food is not always adequate on Caribbean islands, and some countries—most notably Haiti—are chronically food-deficient. What foodstuffs there are tend to be sold in small markets whose produce is dominated by a few local staples that hardly offer a balanced diet. Farm tools are still primitive, and cultivation methods have undergone little change over the generations. Land inheritance customs have divided and redivided peasant families' plots until they have become so small that the owner must share-crop some other land or seek work on a plantation in order to supple-ment the meager harvest. Bad years of drought, winter cold waves, or hurricanes can spell disaster for the peasant family.

Furthermore, soil erosion constantly threatens. Much of the Jamaican countryside is scarred by gulleys and ravines, and Haiti's land has become so ravaged by spectacular

Along the border between the two countries, the ravaged landscape of Haiti (left) stands in stark contrast to the thick forests of the neighboring Dominican Republic (right). Hungry Haitians have been stripping the land of its trees for centuries to clear space for raising crops. Trees, however, bind the soil; with the country 99 percent de-forested, tropical rains are now steadily washing away the soils, leaving behind a completely denuded land surface.

erosion (frequently down to bedrock level) that the entire country may soon become an ecological wasteland (see the photo on p. 281). Where soils are not eroded, their nutrients are depleted, and only the barest yields are extracted. The good land lies under cash crops for the export trade, not in food crops for local consumption. Those expanses of sugarcane and banana trees symbolize the persistently disadvantaged position of the Caribbean countries, dependent on uncertain markets for their revenues and trapped in an international economic order they cannot change.

With such problems, it would be unlikely for Caribbean America to have many large cities; after all, these countries have little basis for major industry, little capital, little local purchasing power. Indeed, the figures reflect this situation, with a modest 58 percent of the region's population classified as urban. (Moreover, this proportion is rising slowly across the region; only the handful of large cities are growing at more than a moderate pace.) Cuba and Puerto Rico, however, have nearly 75 percent of their populations in urban areas; but the Caribbean's second- and third-largest cities, Cuba's Havana (2.2 million) and Puerto Rico's San Juan (1.5 million), owe a great deal of their development to earlier U.S. influences. The region's largest city is the Dominican Republic's Santo Domingo (2.6 million); next on the list are the capitals of Haiti and Jamaica—respectively, Port-au-Prince (1.3 million) and Kingston (717,000). Cities on these less fortunate islands, however, often exhibit even more miserable living conditions than the poorest rural areas. The slums of Port-au-Prince are among the developing world's worst (see photo below); no wonder such abysmal conditions drive away the most desperate Haitians as "boat people" in search of a better life elsewhere.

TOURISM: THE IRRITANT INDUSTRY

The cities of Caribbean America constitute a potential source of income as tourist attractions and ports of call for cruise ships. The thriving cruise-ship industry (based in Miami and Ft. Lauderdale) has experienced spectacular growth since 1980, and places such as Ocho Rios (Jamaica) and Puerto Plata (Dominican Republic) were added to cruise itineraries that already included San Juan, Port-au-Prince, and Nassau in the Bahamas. The Caribbean has long been known for its magnificent beaches and beautiful island landscapes, but visitors also are attracted by the night life and gambling of San Juan, the cuisine and shopping of Martinique's Fort-de-France, and the picturesque colonial architecture of Curaçao's Willemstad.

Certainly, Caribbean tourism is a prospective money-earner, and it already ranks at or near the top on many of the islands of the Lesser Antilles. But tourism has serious drawbacks. The invasion of overtly poor communities by wealthier visitors at times leads to hostility, even actual antagonism on the part of the hosts. For some island residents, tourists have a "demonstration effect," which leads locals to behave in ways that may please or interest the visitors but are disapproved by the larger community. Free-spending, sometimes raucous tourists contribute to a rising sense of local anger and resentment. Moreover, tourism has the effect of debasing local culture, which is adapted to suit the visitors' tastes. Anyone who has attended hotel-staged "culture" shows has witnessed this process. And many workers say that employment in the tourist industry is dehumanizing; expatriate hotel and restaurant managers demand displays of friendliness and servitude that locals find difficult to sustain. Nonetheless, the Caribbean's tour-

The Caribbean is a region of stark, even searing contrasts. Haiti is the Western Hemisphere's poorest country, its city slums (here in the capital, Port-au-Prince) the most desperate of human environments. Tourism and wealthy waterfront living create another face of the region, generating landscapes of conspicuous consumption—but producing some of the limited economic opportunities in the Caribbean Basin.

The Sunbelt phenomenon is not unique to North America. By the hundreds of thousands, tourists from Britain, Germany, and Scandinavia seek the warmth of Mediterranean Europe. A virtual navy of sea ferries allows them to bring their cars and mobile homes. In Santander, Spain, and elsewhere along the Iberian coast, the issue is not cultural: here, no clash of tourist wealth and local poverty clouds the horizon. Rather, the issue is sustainability: tourists strain water resources, clog the roads, crowd the facilities.

ist trade has stimulated several island industries to produce for the visitors. A typical handicraft "industry" that has mushroomed with the tourist business is Haiti's visual arts: when cruise ships are in port, dockside areas are bedecked with countless paintings and carvings for sale by local artists.

Tourism does generate income in the Caribbean where alternatives are few, but the flood of North American tourists cannot be said to have a beneficial effect on the great majority of Caribbean residents. In the popular tourist areas, the intervention of island governments and multinational corporations has removed opportunities from local entrepreneurs in favor of large operators and major resorts; tourists are channeled on prearranged trips in isolation from the local society. There are some cultural advantages to this because many tourists do not enhance international understanding when they invade Caribbean communities; but such practices undoubtedly deprive local small establishments and street vendors of potential income.

Tourism, then, is a mixed blessing for the developing Caribbean Basin. Given the region's limited options, it provides revenues and jobs where otherwise there would be none. Yet there is a negative cumulative effect that intensifies contrasts and disparities: gleaming hotels tower over substandard housing, luxury liners glide past poverty-stricken villages, opulent meals are served in places where, down the street, children suffer from malnutrition. Clearly, the

tourist industry contributes positively to island economies but strains the fabric of the local communities involved.

AFRICAN HERITAGE

The Caribbean region is also a legacy of Africa, and there are places where its cultural landscapes strongly resemble those of West and Equatorial Africa. In the construction of village dwellings, the operation of rural markets, the role of women in rural life, the preparation of certain kinds of food, methods of cultivation, the nature of the family, artistic expression, and in an abundance of other traditions, the African heritage can be read throughout the Caribbean-American scene.

Nevertheless, in general terms it is still possible to argue that the European or white person is in the best position in this island chain, politically and economically; the *mulatto* (mixed white-black) ranks next; and the black person ranks lowest. In Haiti, for instance, where 95 percent of the population is "pure" black and only 5 percent mulatto, this mulatto minority holds a disproportionate share of power. On neighboring Jamaica, the 15 percent "mixed" sector of the population plays a role of prominence in island politics far out of proportion to its numbers. In the Dominican Republic, the pyramid of power puts the white sector (15 percent) at the top, the mixed group (75 percent) next, and the black population (10 percent) at the bottom; this country clings tenaciously to its Spanish-European legacy in the face of a century and a half of hostility from neighboring Afro-Caribbean Haiti. In Puerto Rico, Spanish values persist despite American cultural involvement and a non-Hispanic sector accounting for about one-tenth of the island's 3.8 million people. In Cuba, too, the 12 percent of the population that is black has found itself less favored than the white sector (33 percent), the mulatto sector (32 percent), and the *mestizo* (mixed white-Amerindian) population (22 percent).

The composition of the population of the islands is further complicated by the presence of Asians from both China and India. During the nineteenth century the emancipation of slaves and ensuing local labor shortages brought some far-reaching solutions. Some 100,000 Chinese emigrated to Cuba as indentured laborers (today they still constitute almost 1 percent of the island's population); and Jamaica, Guadeloupe, Martinique, and especially Trinidad saw nearly 250,000 East Indians arrive for similar purposes. To the African-modified forms of English and French heard in the Caribbean, therefore, can be added several Asian languages; Hindi is particularly strong in Trinidad, whose overall population is now 41 percent South Asian. The ethnic and cultural variety of the **plural societies** of Caribbean America is indeed endless.

◆ THE MAINLAND MOSAIC

Mainland Middle America consists of two regions, Mexico and Central America, the former constituted by a single country and the latter by seven. Mexico is Middle America's giant in virtually every way, a region by virtue of its physical size, population, cultural qualities, resource base, and relative location. Mexico's geographic position adjacent to the United States has been a blessing as well as a curse—a blessing because it has facilitated economic interaction, a curse because it has led to neighborly friction over massive illegal emigration. Much of that cross-border population outflow was stimulated by economic conditions in Mexico, which now may be changing for the better. In the mid-1990s Mexico is reforming its economy, and certain areas are enjoying substantial industrial growth; the Mexicans are hoping that the new North American Free Trade Agreement (NAFTA), which takes effect in 1994, will broaden and accelerate this development—thereby providing expanding economic opportunities for millions of its citizens.

The seven republics of Central America cannot match Mexico in terms of total population or territory, but they constitute a geographic region nonetheless—a region plagued by conflict (with recent big-power involvement), economic stagnation, major refugee flows, military coups, and en-

vironmental crises. We focus first on Mexico and then turn to Central America.

MEXICO: LAND OF PROGRESS AND UNFINISHED REVOLUTION

Mexico is the colossus of Middle America, with a 1994 population of 91.8 million—exceeding the combined total of all the other countries and islands of the realm by 24 million—and a territory more than twice as large (Fig. 5-7). Indeed, in all of Spanish-influenced Middle and South America, no country has even half as large a population as Mexico (Colombia and Argentina are next). Moreover, Mexico has grown so rapidly that its population has *doubled* since the early 1970s.

The United Mexican States—the country's official name—consists of 31 States and the Federal District of Mexico City, the capital. Urbanization (as noted in the introductory chapter) has rapidly expanded in the developing world, and no less than 71 percent of the Mexican people now reside in towns and cities (only 50 percent were urbanites as recently as the 1960s). Almost one-fourth of the national total (22.4 million people) are crammed into the conurbation centered on Mexico City alone, which is growing at the astonishing rate of about 750,000 per year (see the box on pp. 286-287). The country's second city, Guadalajara, located approximately 300 miles (470 km) northwest of the capital, has 3.1 million inhabitants.

The physiography of Mexico is reminiscent of the western United States, although environments are more tropical. Figure 5-1 shows several prominent features: the elongated Baja (Lower) California Peninsula in the northwest, an extension of California's Coastal Ranges, separated from the mainland by the Gulf of California (which Mexicans call the Sea of Cortés); the Yucatán Peninsula in the southeast, jutting out into the Gulf of Mexico; and the Isthmus of Tehuantepec, southeast of Veracruz, where Mexico's landmass becomes narrowest. Here in the southeast, Mexico most resembles Central America physiographically; a mountain backbone forms the isthmus and extends northwest toward Mexico City. Shortly before reaching the capital, this mountain range divides into two chains, the Sierra Madre Occidental (in the west) and Sierra Madre Oriental (in the east). These diverging ranges frame the funnel-shaped Mexican heartland, the center of which consists of the extensive Plateau of Mexico (the Valley of Mexico lies near its southeastern end). This rugged tableland is some 1,500 miles (2,400 km) in length and up to 500 miles (800 km) wide. The plateau is highest in the south, near Mexico City, where it is about 8,000 feet (2,450 m) in elevation; from there it gently declines to the northwest toward the Rio Grande. As Fig. I-7 reveals, Mexico's climates are marked by dryness, particularly in the broad, mountain-flanked north. It has been estimated that only 12 percent of the country receives adequate rainfall throughout the year.

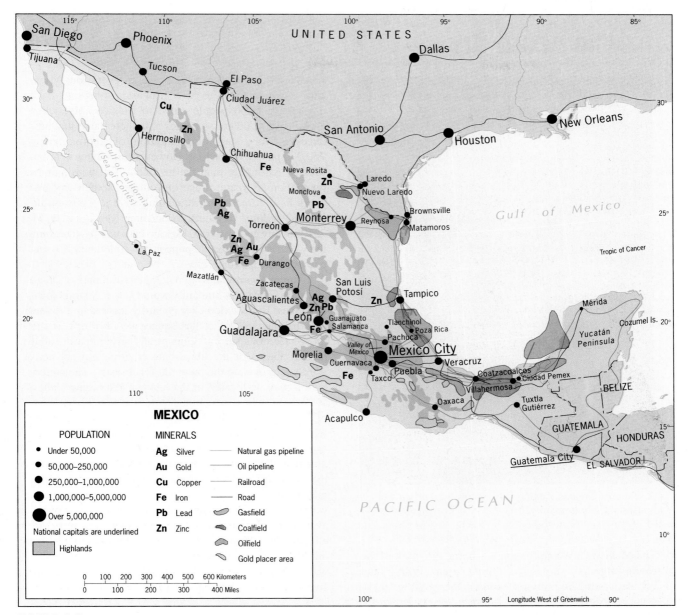

FIGURE 5-7

Most of the better-watered areas lie in the south and east—the areas where Mesoamerica's great indigenous civilizations arose.

The Amerindian imprint on Mexican culture remains extraordinarily strong. Today, 60 percent of all Mexicans are mestizos and 29 percent are Amerindians; only about 9 percent are Europeans. There is a Mexican saying that Mexicans who do not have Amerindian blood in their veins nevertheless have the Amerindian spirit in their minds. Certainly, the Mexican Amerindian has been Europeanized, but the Amerindianization of modern Mexican society is so powerful that it would be inappropriate here to speak of one-way, European-dominated *acculturation*. Clearly, what took place in Mexico is **transculturation**—the two-way exchange of culture traits between societies in close contact. Several hundred thousand Mexicans still speak only an Amerindian language, and an additional 5 million still use these languages in everyday conversation even though they also speak Mexican Spanish (Fig. 5-4). Mexican Spanish itself has been strongly influenced by Amerindian languages, but this is only one aspect of Mexican culture that has received an Amerindian impress. Distinctive Mexican modes of dress, foods and cuisine, sculpture and painting, architectural styles, and folkways also vividly reflect the Amerindian contribution.

Mexico is a large country, but its population is substantially concentrated in a zone that centers on Mexico City and extends across the southern "waist" from Veracruz

TEEMING MEXICO CITY

A sector of Greater Mexico City, whose vast urban web houses a sea of humanity.

The Mexico City conurbation has borne the brunt of the recent migratory surge toward urban areas. With a total population of 22.4 million (second in the world in 1994),

it is already home to nearly one out of every four Mexicans, and it continues to grow at an astonishing rate. Each day, about 1,000 people move to Mexico City; when added to the 1,000 or so babies born there daily, that produces a truly staggering addition of approximately 750,000 people every year. However, birth rates in urban Mexico are higher than the national level; with half its population 18 years of age or younger, some demographers have forecast an astounding total of between *40 and 50 million* residents for greater Mexico City by 2010! In any case, by 2000 Mexico City will have become the world's largest single population agglomeration, surpassing metropolitan Tokyo.

Even in a well-endowed natural environment, such an enormous cluster of humanity would severely strain local resources. But Mexico City is hardly located in a favorable habitat; in fact, it lies squarely within one of the most hazardous surroundings of any city on earth—and human abuses of the immediate environment are constantly aggravating the potential for disaster. The conurbation may be located in the heart of the scenic Valley of Mexico, whose situational virtues led to the building of Tenochtitlán by the Aztecs, but serious geologic problems loom: the vulnerability of the basin to major volcanic and seismic activity (such as the devastating earthquake of 1985), and the overall instability resulting from the weak, dry-lakebed surface that underlies much of the metropolis (aggravated by land subsidence as groundwater supplies are pumped out in vast quantities). Water availability presents another problem in this semiarid climate: dwindling local supplies must be augmented by increases in the long-distance transportation of drinking water from across the mountains, an enormously expensive undertaking that requires the accompanying growth

(367,000) in the east through Guadalajara in the west. This zone contains over half the Mexican people, as is shown on the world population distribution map (Fig. I-9). Just under 30 percent of Mexico's population lives in rural areas associated with the country's better farmlands. Although some notable development has recently occurred in agriculture, there still are enormous difficulties to be overcome.

After achieving independence from Spain in the early 1800s, Mexico failed for nearly a century to come to grips with the problems of land distribution, which were a legacy of the colonial period. By the opening of the twentieth century, the situation had worsened to the point where 8,245 haciendas covered nearly 40 percent of Mexico's en-

tire area; moreover, about 96 percent of all rural families owned no land whatsoever and worked as *peones* (landless, constantly indebted serfs) on the haciendas. There was deprivation and hunger, and the few remaining Amerindian lands and small holdings owned by mestizos or whites could not produce enough food to satisfy the country's needs. Meanwhile, thousands of acres of arable land lay idle on the haciendas, which blanketed just about all the good farmland in Mexico.

Revolution

Not surprisingly, a revolution began in 1910 and set into motion a sequence of events that is still unfolding today.

of a parallel network to pipe sewage out of the waste-choked basin.

Mexico City's appalling air pollution, however, poses the greatest health hazard and is conceded to be the world's most serious, exacerbated by the thin air that contains 30 percent less oxygen than at sea level (the city's elevation is 7,350 feet/2,240 m). The conurbation's 3-million-plus cars and 7,500 diesel buses produce about 75 percent of the smog, with the remainder caused by the daily spewing of 15,000 tons of chemical pollutants into the atmosphere by the area's 37,000 factories. Despite new legislation aimed at improving air quality, on any given day the pollution of Mexico City's air can approach *100 times* the acceptable level.

For the affluent and tourists, Mexico City is undoubtedly one of the hemisphere's most spectacular primate cities, with its grand boulevards, magnificent palaces and museums, vibrant cultural activities and night life, and luxury shops. But most of its residents dwell in a world apart from the glitter of the *Paseo de la Reforma*. An increasing majority of them are forced to live in the miserable poverty and squalor of the conurbation's 500 slums as well as the innumerable squatter shacktowns that form the burgeoning metropolitan fringe (the notorious *ciudades perdidas*, or "lost cities"). This is the domain of the newcomers, the peasant families who have abandoned the hard life of the difficult countryside, lured to the urban giant in search of a better life. With Mexico's underemployment rate hovering above 30 percent in recent years, decent jobs and upward mobility quickly become elusive goals for most of the new arrivals. Yet, despite the overwhelming odds, a surprising number of migrants eventually do enjoy some economic success by becoming part of the so-called *informal sector*. This is a primitive form of capitalism that is now common in many developing countries; it takes place beyond the control—and especially the taxation—of the government. Participants are unlicensed sellers of homemade goods (such as arts and crafts, clothing, food specialties) and services (auto repair, odd jobs, and the like), and their willingness to engage in this hard work has transformed many a slum into a beehive of activity that can propel resourceful residents toward a middle-class existence. For its part, although officially discouraging the growth of squatter settlements, the government has recently made life on the Mexico City outskirts more comfortable by improving schools, roads, and other municipal services; moreover, it still permits squatters who settle on public lands to gain free title to those properties after a period of five years.

The ongoing growth of Mexico City (and its problems of overcrowding and environmental degradation) is not a unique phenomenon. According to the United Nations, in 1992 there were 37 metropolitan areas worldwide whose populations exceeded 5 million; by 2010, that total is predicted to rise to 60. Of these 23 additions to the 5-million-plus category, all are located in developing realms. Moreover, 15 of these "megacities" will contain at least 15 million people by 2010: Mexico City; Tokyo, Japan; São Paulo, Brazil; Bombay, India; Shanghai, China; Lagos, Nigeria; Beijing, China; Dhaka, Bangladesh; Jakarta, Indonesia; New York, U.S.A.; Karachi, Pakistan; Manila, the Philippines; Calcutta, India; Tianjin, China; and Delhi, India. These growth leaders, of course, are simply the tip of the iceberg, because urbanization rates are skyrocketing throughout the developing world. This is a topic to be explored further in Chapter 6, with emphasis on the contemporary urban geography of Middle and South America.

One of its major objectives was the redistribution of Mexico's land, and a program of expropriation and parceling out of the haciendas was made law by the Constitution of 1917. Since then, about half the cultivated land of Mexico has been redistributed, mostly to peasant communities consisting of 20 families or more. Such lands are called *ejidos*; the government holds title to the land, and use rights are parceled out to villages and then individuals for cultivation. Most of the *ejido* lands carved out of haciendas lie in central and southern Mexico, where Amerindian traditions of landownership and cultivation survived and where the adjustments were most successfully made.

With such a far-reaching program, it is understandable that agricultural productivity temporarily declined. The miracle is that land reform has been carried off without a major death toll and that the power of the wealthy landowning aristocracy could be broken without ruin to the state. Mexico alone among the region's countries with large Amerindian populations has made major strides toward solving the land question, although there is still widespread malnutrition and poverty in the countryside. But the revolution that began in 1910 did more than that; it also resurrected the Amerindian contribution to Mexican life and blended Spanish and Amerind heritages in the country's social and cultural spheres. It brought to Mexico the distinctiveness that it alone possesses in Middle and South America.

The revolution could change the distribution of land, but it could not change the land itself or the methods by which it was farmed. Corn (maize), beans, and squash continue to form the subsistence food of most Mexicans, with corn the chief staple (still occupying over half the cultivated land). Yet corn is grown all too often where the conditions are not right for it, so that yields are low; wheat might do better, but the people's preference, not soil suitability, determines the crop. And if the people's preference has not changed a great deal, neither have farming methods over much of the country.

Economic Geography

Commercial agriculture in Mexico has diversified and made major strides in recent decades with respect to both the home market and export. The greatest productivity still emanates from private cultivators, although much of the land involved has been subdivided into *ejidos*. The central plateau is geared mainly to the domestic production of food crops, but in the arid north, major irrigation projects have been built on streams flowing down from the interior highlands. Along the booming northwest coast, mechanized large-scale cotton production supplies the domestic market as well as the thriving export trade. For the home market, wheat and winter vegetables are grown; but fruit and vegetable cultivation is also interesting foreigners, who are soon expected to increase significantly their investments in such crops as bananas and sugarcane. Cattle raising is another leading pursuit and is continuing to expand onto the Gulf Coast lowlands from its long-time base in the northern interior.

Mexico's metal mining industries are less important today than they once were. The country still exports a major share of the world's silver; other important commodities include copper, zinc, and lead. Mining activity is scattered throughout northern and central Mexico (Fig. 5-7), but many of the mines that were important in the colonial period—and near which urban centers of some size developed—have been exhausted.

More recently, Mexico has enjoyed the advantages—and suffered the problems—of a large and productive petroleum industry. Centered on the southern Gulf Coast's Bay of Campeche around Villahermosa (445,000), these oilfields brought Mexico huge revenues when the world oil price was high and serious economic difficulties when the price fell. Discoveries of huge oil and natural gas reserves in this area since 1970 have made Mexico self-sufficient in these energy resources, adding to already substantial reserves located in oilfields lying along the Gulf Coast between Veracruz and Tampico (Fig. 5-7).

As the world oil price rose to unprecedented heights during the late 1970s, Mexico's economic geography began to be transformed. The government had great plans for the economy, based on anticipated future income from oil and natural gas (much of it to be sold to the United States). Mexico borrowed and spent as never before, running up a huge debt to foreign countries and banks. But then, after 1980, the price of oil plunged—and with it Mexico's income. The government could not pay the interest on the loans it had taken, and a long economic crisis ensued. This experience, of course, is hardly unique; many developing countries, counting on huge future incomes, strapped themselves with enormous debts. But in Mexico's case, serious efforts were made by the government to reduce this burden, and by the early 1990s these economic reforms were ushering in a new wave of prosperity.

The rising importance of manufacturing in the Mexican economy has much to do with this latest episode of national economic progress. To be sure, Mexico possesses a wide range of raw materials, many of which are located in the north (Fig. 5-7). As far back as 1903, an iron and steel plant (now abandoned) was built in Monterrey in the northeast,

Over the border from Southern California—down Tijuana way. These are businesses that attract hordes of customers from the United States (only a few blocks from here). In fact, there's more English in view on this street than in Miami's Little Havana (cf. photo on p. 3)!

using iron ore located near Durango and coking coal from the Sabinas Basin just north of Monterrey; a second steel complex was developed in the 1950s at Monclova near the source of coking coal. Most of Mexico's industrial production takes place in cities, which means there is a particularly heavy concentration of manufacturing—and its accompanying pollution—in and around Mexico City.

The most significant recent development in Mexico's manufacturing geography is the steady growth of **maquiladora** plants in the northern border zone. The maquiladoras are foreign-owned factories (mainly owned by large U.S. companies) that assemble imported, duty-free components and/or raw materials into finished industrial products. At least 80 percent of these goods are then re-exported to the United States, whose import tariffs are limited to the value added to the products during their Mexican fabrication stage.

All parties benefit from this industrial system: the Mexicans gain tens of thousands of jobs, and the foreign owners benefit from Mexico's wage rates, which are well below those north of the border. Although this development program was initiated in the 1960s, the number of maquiladoras grew only to a modest 588 (with 122,000 employees) as recently as 1982. Then this phenomenon suddenly took flight: by 1992, nearly 1,800 assembly plants were employing over 500,000 people and the maquiladoras were accounting for a robust 20 percent of Mexico's industrial labor force and more than 5 percent of its gross domestic product.

Among the goods being assembled were electronic equipment, electrical appliances, auto parts, clothing, plastics, and furniture. Today these are being joined by white-collar services, especially routine data-processing operations, as myriad back-office jobs shift southward from Southern California. Two "twin" borderland metropolises have enjoyed the greatest growth surge: Tijuana/San Diego on the Pacific coast and Ciudad Juarez/El Paso on the Rio Grande in westernmost Texas. Tijuana (948,000), in particular, has experienced a swift transformation from a honky-tonk border enclave into one of the developing world's more prosperous cities. It is now the world's busiest border-crossing point, admitting millions of free-spending U.S. tourists each year; its 600-plus maquiladoras employ over 80,000 workers, who already assemble more television sets than anywhere else on earth (Japanese manufacturers have arrived in droves).

The Mexican government is now trying to capitalize on the success of the maquiladora program by encouraging the industrial development of other parts of the country. It would most like to see industrial firms create complete manufacturing facilities within Mexico rather than limit their investments to assembly plants that hug the U.S. border. During the early 1990s, several major multinational corporations did undertake such ventures, particularly in the automobile and electronics industries. A major boost in this activity may occur when NAFTA takes effect in the mid-

MEXICO'S NAFTA CONNECTION

The box on p. 208 outlined the provisions and potentials of the North American Free Trade Agreement (NAFTA), which is taking effect in the mid-1990s. Basically, NAFTA will bind together the economies of Mexico, the United States, and Canada over the next decade by eliminating most tariff barriers, liberalizing the foreign ownership of businesses, and removing restrictions on the flow of investment capital across the trading bloc's international borders.

Mexico's President Salinas (1988–1994) first proposed NAFTA and then sold it not only to North America's leaders but also to his fellow citizens (sweeping away their long-time fears of U.S. dominance over the Mexican economy). The potential advantages for Mexico are enormous: huge new infusions of investment capital—from Europe and Japan (it is hoped) as well as from the United States—will greatly expand job opportunities, accentuate the currently impressive rate of national economic growth, and eventually raise living standards to a level that could vault Mexico into the developed world.

Although this promising scenario is not beyond Mexico's reach, the country still faces a number of serious challenges that it must overcome. One problem is that the Mexican economy, a mere one-twentieth the size of that of the United States, is full of inefficiencies and corruption. Another is that outside investors to date have mainly been lured by Mexico's cheap labor (as the maquiladoras testify); if this perception continues under NAFTA, then the new industrial opportunities that would arrive might not be the correct mix to support a true takeoff of the Mexican economy (see the Focus essay on p. 220). A third dilemma is a heavily geographical one: ongoing development favors only a limited number of places (especially in the north) while leaving untouched vast areas that are home to the approximately half of the Mexican population that remains mired in grinding poverty. Clearly, the country has a long way to go to solve these problems, but its track record over the past decade has demonstrated a remarkable ability to surmount obstacles and keep forging ahead.

1990s (see box this page). With the elimination of existing trade restrictions between the U.S. and Mexico over the next few years, the ties that bind manufacturers to the bor-

In the path of NAFTA: as the agreement takes full effect, Monterrey in the Mexican State of Nuevo León is most advantageously situated.

der zone (not to mention the reason for the existence of maquiladoras in the first place) may disappear.

In preparation for that day, the federal government is vigorously pushing ahead with its new program to upgrade Mexico's infrastructure to developed-world standards—especially in telecommunications, superhighways, and electric-power provision. The northeastern city of Monterrey (2.8 million), which has benefited from all the development trends of recent years, is today often singled out as a model of what a future Mexican growth center should be. Among this city's assets are a highly educated and well-paid labor force; a stable international business community; a thriving high-technology complex of ultramodern industrial facilities that has attracted blue-chip multinational companies; and a new expressway leading toward the Rio Grande that has brought San Antonio, Texas, within a six-hour drive.

Mexico has made impressive gains in industrialization, it is addressing itself with new determination to agrarian reform, and it seeks to integrate all sectors of the population into a truly Mexican nation. After a century of struggle and oppression, this country has lately taken long strides toward overcoming its chronic problems. However, meaningful further progress is threatened by a final challenge: the rapid growth of its huge population. Ultimately, no amount of reform can keep pace with a growth rate (now 2.3 percent yearly) that will, if sustained, produce a Mexican population of twice the size (184 million) of today's 92 million by the year 2025. No political or economic system in the developing world could long withstand the impact of such population growth. Today, there are indications that the annual rate of natural increase may be slowing, but until it is checked and stabilized, Mexico's latest accomplishments remain under a demographic cloud.

THE CENTRAL AMERICAN REPUBLICS

Crowded onto the narrow segment of the Middle American land bridge between Mexico and the South American continent are seven countries collectively known as the Central American republics (Fig. 5-8). Territorially, they are all quite small: only one, Nicaragua, is larger than the Caribbean island of Cuba. Populations range from Guatemala's 10.3 million down to Panama's 2.5 million in the six Hispanic republics, whereas the sole former British territory, Belize (which until 1981 was British Honduras), has only about 225,000 inhabitants.

As elsewhere in Middle America, the ethnic composition of the population is varied, with Amerindian and white minorities and a mestizo majority. The exceptions are Guatemala, where 45 percent of the population remains relatively "pure" Amerindian (Maya and Quiché—see Fig. 5-4) and the remainder is of strongly Amerindian mestizo ancestry; and Belize, where about 50 percent of the population is black or mulatto, a situation reminiscent of the social geography of the Caribbean. Demographics are least complex in Costa Rica, where there is a large white majority of Spanish and other relatively recent European immigrants; in a population of 3.3 million, the black component constitutes under 5 percent of the total and Amerindians less than 1 percent.

The narrowing land bridge on which these republics are situated consists of a highland belt flanked by coastal lowlands on both the Caribbean and Pacific sides. From the earliest times, the people have been concentrated in the upland *tierra templada* (temperate) zone. Here tropical temperatures are moderated by elevation (see the box on p. 292 entitled "Altitudinal Zonation"), and rainfall is adequate for the cultivation of a variety of crops. As noted earlier, the

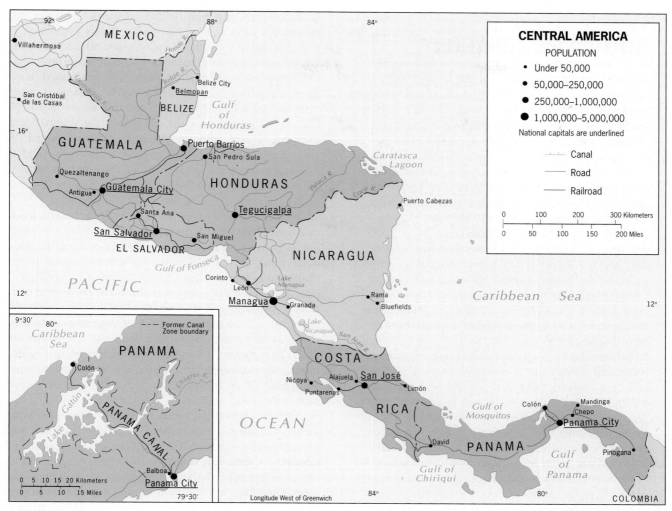

FIGURE 5-8

Middle American highlands are studded with volcanoes, and local areas of fertile volcanic soils are scattered throughout the region. The old Amerindian agglomerations were located in these more fertile parts of the highlands, and this human distribution persisted during the Spanish period.

Today, the capitals of Guatemala (Guatemala City, 946,000), Belize (Belmopan, 4,800), Honduras (Tegucigalpa, 988,000), El Salvador (San Salvador, 1.0 million), Nicaragua (Managua, 1.2 million), and Costa Rica (San José, 879,000), all lie in the interior, most of them at 3,500 feet (1,060 m) or higher in elevation. In all of mainland Middle America, Panama City (951,000) is the only coastal capital (Fig. 5-8). The size of these cities, in countries whose populations average about 5 million, is a reflection of their primacy (as with the dominance of Mexico City over the rest of Mexico). On average, the next-ranking town is only one-fifth as large as the capital city.

The distribution of population within Central America, apart from its concentration in the region's uplands, also exhibits greater densities toward the Pacific than toward the Caribbean coastlands (Fig. I-9). El Salvador, Belize, and (to some degree) Panama are exceptions to the rule that people in mainland Middle America are concentrated in the *templada* zone. Most of El Salvador is *tierra caliente*, and the majority of its 5.9 million people are crowded, hundreds to the square mile (at densities approaching India's), in the intermontane plains lying less than 2,500 feet (750 m) above sea level. In Nicaragua, too, the Pacific-side areas are the most densely populated; the early Amerindian centers lay near Lake Managua, Lake Nicaragua, and in the adjacent highlands. The frequent activity of volcanoes in this Pacific zone is accompanied by the emission of volcanic ash, which settles over the countryside and quickly weathers into fertile soils.

ALTITUDINAL ZONATION

Mainland Middle America and the western margin of South America are areas of high relief and strong local contrasts. People live in clusters in hot tropical lowlands, in temperate intermontane valleys, and even on high plateaus just below the snow line in the Andes. In each of these various zones, distinct local climates, soils, crops, domestic animals, and modes of life prevail. Such **altitudinal zones** are known by specific names as if they were regions with distinguishing properties—as, in reality, they are.

The lowest vertical zone, from sea level to 2,500 feet (about 750 m), is known as the *tierra caliente*, the "hot land" of the coastal plains and low-lying interior basins where tropical agriculture (including banana plantations) predominates. Above this lowest zone lie the tropical highlands containing Middle and South America's largest population clusters, the *tierra templada* of temperate land reaching up to about 6,000 feet (1,850 m). Temperatures here are cooler; prominent among the commercial crops is coffee, and corn (maize) and wheat are the staple grains. Still higher, from about 6,000 feet to nearly 12,000 feet (3,600 m) is the *tierra fría*, the cold country of the higher Andes where hardy crops such as potatoes and barley are the mainstays. Only small parts of the Middle American highlands reach into the *fría zone*, but in South America this environment is much more extensive in the Andes. Above the tree line, which marks the upper limit of the *tierra fría*, lies the *puna* (also known as the *páramos*); this fourth altitudinal zone, extending from about 12,000 to 15,000 feet (3,600 to 4,500 m), is so cold and barren that it can support only the grazing of sheep and other hardy livestock. The highest zone of all is the *tierra helada* or "frozen land," a zone of permanent snow and ice that reaches to the peaks of the loftiest Andean mountains.

These elevation ranges are for highlands lying in the equatorial latitudes. Of course, as one moves poleward of the tropics beyond 15 degrees of latitude, the sequence of five vertical zones is ratcheted downward, with the breaks occurring at progressively lower altitudes.

By contrast, the Caribbean coastal lowlands—hot, wet, and awash in leached soils—support comparatively few people. In the most populous republic, Guatemala, the heart-land also has long been in the southern highlands. Although the large majority of Costa Rica's population is concentrated in the central uplands around San José, the Pacific lowlands have been the scene of major in-migration since banana plantations were first established there. Even in Panama, there is a strong Pacific orientation: more than half of all Panamanians (and this means about 70 percent of the rural population) live in the southwestern lowlands and on adjoining mountain slopes; another 25 percent live and work in the Panama Canal corridor; and of the remainder, a majority (many of them descendants of black immigrants from the Caribbean) live on the Atlantic or Rimland side of the isthmus.

With the possible exception of Costa Rica, Middle America's mainland republics face the same problems as Mexico, only more so; they also share many of the difficulties confronting the Caribbean islands. No present or future challenge is greater than Central America's overpopulation. The region's population explosion began around mid-century, expanding from a base of 9.3 million people in 1950 to a total of 32.4 million in 1994; if current growth rates persist, today's population will double to 65 million by 2020. This amounts to nothing less than an onrushing demographic catastrophe in a region already unable to cope with most of its social, economic, and natural-resource problems.

Although efforts are being made to improve the land-ownership situation, the colonial legacy of hacienda and *peón* hangs heavily over Central America. Generations of people have seen little change in their way of life and have had even less opportunity to bring about any real improvement. Dwellings are still built of mud and straw, sanitary facilities have not yet reached them, schools are terribly overcrowded, and hospital facilities are inadequate. Each year brings a renewed struggle to extract a subsistence livelihood, with little hope that the next year will bring anything better. And today a new threat increasingly overshadows these peoples' fragile existence—the rapid destruction of the tropical forest, which portends environmental disaster (see box pp. 294-295). One of the most jarring contrasts in Middle America is that between the attractive capital city and its immediate surroundings and the desolation of outlying rural areas. Another is that between the splendor, style, and refined culture of the families who own the coffee farms and other enterprises, and the destitute, rag-wearing peasants.

Political Instability

Inequities, repressive government, and the frequent unleashing of armed forces have destabilized Central America for much of its modern history. Today, the region is struggling to emerge from a period of unparalleled turmoil, and the political map (Fig. 5-8) conceals many politico-geographical developments. Nicaragua's Sandinista government, with Soviet and Cuban support, faced a United States-backed force for most of the 1980s that sought to establish its own state within a state. Following the defeat of the Sandinistas in the election of 1990, that rebel military campaign (by the so-called Contras) ceased—but instabilities remain. In El Salvador, a bitter 12-year civil war came to an end in 1992 as leftist rebels and the government resolved many of their differences, and a tenuous peace has settled over the country. In Guatemala, reform after decades of repressive government notwithstanding, a long-entrenched guerrilla insurgency remains active in the western highlands of the country.

The roots of these upheavals are old and deep. Central America is not a large region, but because of its physiography it includes many isolated, comparatively inaccessible locales. Conflicts between Amerindian population clusters and mestizo groups are endemic to the region, and contrasts between the privileged and the poor are especially harsh. Dictatorial rule by local elites followed authoritarian rule by Spanish colonizers. The collision course has been a long one, and this latest episode of violent confrontation was not the first outbreak.

One unprecedented side effect of these recent conflicts has been an enormous flow of refugees. These are not people who, like Haiti's boat people or Mexico's cross-border emigrants, are trying to escape poverty; nor are they comparable to the waves of emigrants who left Cuba for Florida after the communist revolution of 1959 and again during the 1980 Mariel boatlift. Central America's refugees by the tens of thousands have escaped from combat and " and death-squad terrorism. From city as well as countryside they have come, often leaving behind broken families and shattered lives. Some of these refugees have found their way to the United States, but thousands of others cannot leave the region and are forced to exist in the misery of squalid and overcrowded refugee camps.

Guatemala: On the Doorstep

Guatemala, land of the ancient Maya, began its modern existence as an entity of the Mexican Empire following independence from Spain in 1821. Bordered by Mexico to the west and north and by Belize, Honduras, and El Salvador to the east, Guatemala has a lengthy Pacific coastline but a mere window on the Caribbean Sea (Fig. 5-8). Only one of Guatemala's neighbors, El Salvador, felt the full onslaught of rebellion and civil war during the tumultuous 1980s. Geography has shielded the country, bordered by: Mexico, as yet unaffected by turmoil; Belize, a British dependency until a few years ago; and Honduras, among the more stable countries in the region.

Yet Guatemala contains the ingredients that recently led to upheaval elsewhere in Central America. Outside the core area, effective national control dwindles, notably in the west and north. During the past half-century, turbulence has dominated the political scene. In 1944, Guatemala seemed headed for reform after a brutal dictatorship, but 10 years later a military force organized in Honduras captured the capital and took over the government. Ever since, there has been a rebellion against the entrenched rulers in Guatemala City—a rebellion that has reached into the capital itself. Successive military governments (and the pro-military civilian regimes elected since 1985) have tried various means to defeat the rebels, who presently are strongest in the western highlands. Death-squad assassinations of suspected collaborators, kidnapping, and military action have been combined with attempts at economic and social reform in what Guatemalans call the "rifles-and-beans" policy. But in the countryside, the military is feared and the rebels are often viewed as agents of change—where any change is seen as being for the better.

In the meantime, the economy of Guatemala suffers under the burden of huge military expenditures and the effects of continuing strife. Coffee from the highlands and cotton from the drier areas are the main sources of income. Tourism was once a leading industry, but the political situation has destroyed it; foreign investment has all but dried up. Certainly, Guatemala has potential: it produces a wide range of farm crops and has major stands of exploitable timber and substantial nickel deposits. There is evidence that oil lies beneath the northeastern coastal plain in the frontier zone known as the Petén. Yet Guatemala is polarized, and the poles are drifting farther apart. Until the future is clearer, opportunities lie dormant.

Belize: In the Rimland

Strictly speaking, Belize is not a Central American republic in the same tradition as the other six. This country, a wedge of land between northern Guatemala, Mexico's Yucatán Peninsula, and the Caribbean, was a British dependency (called British Honduras) until 1981; thus English, not Spanish, is the *lingua franca* here. Slightly larger than Massachusetts and with a population of only 220,000 (many of African descent), Belize is more reminiscent of a Caribbean island than of a mainland Middle American state.

But in other ways, Central American patterns do prevail. The largest city (and former capital), Belize City (64,000), is the only substantial urban center. Sugar is the principal export, and other commercial crops also generate external income; lobsters and fish are sold on foreign markets as well. Another revenue earner is eco-tourism, based on the natural attractions of the country's still pristine environment.

Belize has not escaped political problems. A treaty between Britain and Guatemala that defined the territory's boundaries, signed in 1859, has been disputed by Guatemala's military rulers. From time to time, the Guatemalans have asserted that the 1859 treaty is void and that Belize be-

TROPICAL DEFORESTATION

As the world vegetation map (Fig. I-8) indicates, two-thirds of Central America was covered by tropical rainforests. We deliberately use the word *was* because destruction of this precious woodland resource is proceeding so swiftly that a typical landscape scene today shows not a rainforest but a scraggly, tree-cleared wasteland ripe for environmental disaster (see photo).

Tropical deforestation in mainland Middle America, as we saw, began with the Spanish colonial era in the sixteenth century. The pace has accelerated incredibly in recent decades; since 1950, over 80 percent of Central America's forests have been decimated. Today, about 3.5 million acres of woodland in Central America and Mexico disappear *each year*; this is equivalent to an area half the size of Belgium, and only West Africa exhibits a faster annual depletion rate. El Salvador has already lost 97 percent of its forests, and the six other republics will reach that stage around the turn of the century. Costa Rica's present situation is typical: the 80 percent of the country that was covered by forests in 1940 had shrunk to 17 percent in 1988 and will be reduced to zero by 2002 if the current tree-removal rate is sustained.

The several causes of tropical deforestation are related to the persistent economic and demographic problems of the developing world. In Central America, the leading cause has been the need to clear rural lands for cattle pasture as many countries (notably Costa Rica) attempt to become meat producers and exporters. Although some gains have been recorded, the price of en-

This scene in Costa Rica shows how badly the land can be scarred in the wake of deforestation. Without roots to bind the soil, tropical rains swiftly erode the unprotected topsoil.

vironmental degradation has been enormous. Because tropical soils are so nutrient-poor, newly deforested areas can function as pastures only for a few years at most.

---

longs to Guatemala. Even when Belize attained independence over a decade ago, that dispute was not finally resolved. Belizeans sometimes express the fear that their country might one day become the Falkland Islands of Central America (see box on p. 331), and they worry that Guatemala's conflict may spill over into their country. Central America's turbulence leaves no country untouched.

Honduras: Mired in Poverty

Honduras occupies a critical place in the political geography of Central America, flanked as it is by unstable Nicaragua, civil-war-exhausted El Salvador, and tense Guatemala. As the map shows, Honduras, in direct contrast to Guatemala, has a lengthy Caribbean coastline and a small window on the Pacific. But that window on the Pacific's Gulf of Fonseca

separates Nicaragua from El Salvador and, in the 1980s, put Honduran territory and waters between the alleged donor and recipients of communist weapons.

Honduras has a democratically elected government, but the military wields considerable power. With 5.8 million inhabitants, about 90 percent mestizo, Honduras is the region's poorest, least developed country. Agriculture, livestock, forestry, and some mining (lead and zinc) form the mainstays of the economy. The familiar Central American products—coffee, bananas, lumber, sugar—earn most of the external income. During the 1960s and 1970s, there was some promising development of light industry around San Pedro Sula (588,000) near the northwestern coast. However, Honduras fell victim to regional strife, investors were afraid to risk their funds, and tourism declined. In recent years refugees streamed into Honduras and American arms flowed through

These fields are then abandoned for other freshly cut lands and quickly become the devastated landscape seen in the photograph. Without the protection of trees, local soil erosion and flooding immediately become problems, affecting (and reducing the output capacity of) still-productive nearby areas.

A second cause of deforestation is the rapid logging of tropical woodlands as the lumber industry increasingly turns from the exhausted forests of the mid-latitudes to harvest the rich tree resources of the equatorial zones, responding to accelerating global demands for new housing, paper, and furniture.

The third major contributing factor is directly related to the population explosion in developing countries: as more and more peasants are required to extract a subsistence from inferior lands, they have no choice but to begin cutting the forest for both firewood and additional crop-raising space, and their intrusion prevents the trees from regenerating.

Although deforestation is a depressing event, it does not seem to tropical pastoralists, farmers, and timber producers to be life-threatening, and perhaps it even seems to offer some short-term economic advantages. Why, then, should there be such an outcry from the scientific community? And why should the World Resources Institute call this "the world's most pressing land-use problem"? The answer is that unless immediate large-scale action is taken, within 10 years the tropical rainforest will be reduced to two large patches—the western Amazon Basin of South America and the middle Zaïre (Congo) Basin of Equatorial Africa—and these will disappear by 2025.

The tropical forest, therefore, must be a very important part of our natural world—and indeed it is. Biologically, the rainforest is by far the richest, most diversified arena of life on our planet: even though it covers only a shrinking 5 percent of the earth's land area, it contains well over half of all plant and animal species. Its loss would mean not only the extinction of millions of species, but also the *end of birth* because the evolutionary process that produces new species would be terminated. Because the rainforests already yield countless valuable medicinal, food, and industrial products, how many potential disease-combatting drugs or new crop varieties to feed undernourished millions would be irretrievably lost?

Perhaps the most ominous (and as yet unknown) consequence of all would be the impact of rainforest disappearance on the planet's climates. Environmental scientists are quite concerned about this coming crisis, and some forecast a global warming trend as the burning of the remaining forests adds vast quantities of carbon dioxide to the air, forming a thickening layer of the gas that would prevent excess heat from escaping the atmosphere. This trapping of warm air—the so-called *greenhouse effect*, because of the obvious analogy—would be further heightened by the loss of the carbon dioxide-absorbing trees and could lead to higher temperatures at every latitude, polar-icecap melting, and significantly raised sea levels that would imperil the world's crowded coastal zones.

the country to the Nicaraguan border—and prospects have remained bleak. This is especially tragic because social and ethnic divisions, which so strongly mark Honduras's neighbors, are not serious here, and the gap between rich and poor, although evident, is not as wide.

El Salvador: Launching a New Era

El Salvador is Central America's smallest country territorially, smaller even than Belize—yet, with a population 30 times as large (5.9 million), the most densely peopled. Again, like Belize, it is one of only two mainland republics that do not have coastlines on the Atlantic as well as Pacific sides. El Salvador adjoins the Pacific in a chain of volcanic mountains. The country's heartland lies behind those mountains in the central interior, where the capital, San Salvador, is located. North of this core area lies another zone of mountains through which the boundary with Honduras was delimited in the 1820s and 1830s. This mountainous interior also has always contained areas beyond effective governmental control.

Unlike neighboring Guatemala, El Salvador's population is quite homogeneous (90 percent mestizo and just 5 percent Amerindian). Yet ethnic homogeneity did not translate into social or economic equality or even opportunity. Other Central American countries were called "banana republics"; El Salvador was a coffee republic, and the coffee was produced on the huge landholdings of a few landowners and on the backs of a subjugated peasant labor force. During the first half of the twentieth century, an elite group known as the "fourteen families" ran the country almost as a feudal holding. The military supported this system and repeatedly suppressed violent and desperate peasant uprisings.

Were it not for great-power involvement, the civil war that engulfed El Salvador from 1980 to 1992 might be viewed as simply another contest between ruling junta and peasant. But the nationwide outbreak of rebellion that began at the end of the 1970s was magnified by alleged Nicaraguan support (with Soviet weapons) for the rebels and U.S. support for the government, which made belated moves toward democracy. The ensuing war ravaged the country, and at least 75,000 people, mostly civilians, were killed. As El Salvador fragmented, its entire economy began to unravel; those who could, fled—on foot into neighboring Honduras, and by air and sea to countries farther away.

In the mid-1990s, trying to put its disastrous civil war behind it, El Salvador is struggling to forge a less polarized society. Deep differences remain over the explosive land reform issue. The government is committed to settling the land claims of former guerrillas and has provided land for subsistence farming to those who request it; but the far more difficult task of land reallocation is only beginning to be faced. When that redistribution does start, it may well turn out to be a limited opportunity for many new landholders: much of the country's farmland was devastated during the civil war, which produced such widespread soil erosion that the agricultural output of this former coffee factory may never again reach pre-1980 levels.

Nicaragua: Turbulent Heart of the Region

When studying Nicaragua, one does well to look again at the map (Fig. 5-8), which underscores the country's pivotal central position. Flanked by Honduras, poorest of the region's countries, to the north, and by Costa Rica, its richest, to the south, Nicaragua occupies the heart of Central America. The Pacific coast follows a southeasterly direction, but the Caribbean coast is oriented north-south so that Nicaragua forms a triangle of land with its capital, Managua, located on the mountainous, Pacific side. Indeed, the core of Nicaragua always has been in this western zone. The Atlantic side, where the mountains and valleys give way to a coastal plain of (disappearing) rainforest, pine savanna, and swampland, has for centuries been home to Amerindian peoples such as the Miskito who have been relatively remote from the focus of national life.

Nicaragua was the typical Central American republic, ruled by a dictatorial government and exploited by a wealthy landowning minority, its export agriculture dominated by huge foreign-owned plantations (run by such large U.S. banana producers as United Brands and Standard Fruit). It was a situation ripe for insurgency, and in 1979 the Sandinista rebels overthrew the Somoza government in Managua. However, Sandinista rule produced its own excesses, and opposition, although weak, persisted. The Miskito and other indigenous peoples resisted Sandinista repression, and in the

borderlands (chiefly along the boundary with Honduras) the Contras fought the Sandinista forces. With big-power support magnifying the conflict, U.S. backing of the Contras gave an ideological advantage to the Sandinistas, who successfully rallied nationalistic fervor at home against the "giant of the north" and its agents.

The civil war ended in 1990; but despite the election of a more democratic regime, Nicaragua today is struggling to overcome increasingly difficult political and economic problems. The anti-Sandinist coalition that elected the postwar government has collapsed because the Sandinistas continue to wield great influence in the running of the country (especially the armed forces). With the Sandinistas threatening a coup if the government fully dismantles its socialist "reforms," instability was returning in 1993 as a "re- Contra" movement gained strength amid rising talk of a resumption of civil war. A leading casualty of this turmoil is the Nicaraguan economy, which has hardly begun its postwar reconstruction and is mainly kept afloat by foreign aid. Unemployment in the early 1990s hovered around the 50-percent level; coffee and cotton production, the prewar export-industry leaders, remain in disarray. And to make matters even worse, the country's annual population-increase rate in 1994 stood at a disastrous 3.1 percent.

Costa Rica: Durable Democracy

As if to confirm what was said about Middle America's endless variety and diversity, Costa Rica differs in very significant ways from its neighbors—and from the norms of Central America. Bordered by two volatile countries (Nicaragua to the north and Panama to the east), Costa Rica is a nation with an old democratic tradition and, in this cauldron, *no* standing army! The country is, in fact, the oldest democracy in Middle and South America, enjoying a freely elected government (except for two brief periods) since 1889. Although the initial Hispanic imprint here was similar to that found elsewhere on the Mainland, early independence, the good fortune to lie remote from regional strife (which fostered an enduring posture of neutrality), and a leisurely pace of settlement allowed Costa Rica the luxury of concentrating on its economic development. Perhaps most important, internal political stability has prevailed over much of the past 175-odd years; the last brush with conflict, nearly 50 years ago, left the nation resolved to avoid further violence, and the armed forces were abolished in 1948 (along with a military establishment, so often the source of trouble throughout Central America).

Costa Rica, like its neighbors, is also divided into environmental zones that parallel the coasts. The most densely settled is the central highland zone, lying in the cooler *tierra templada*. Volcanic mountains prevail in much of this zone, but the heartland is the Valle Central (Central Valley), a fertile 40-by-50-mile (65-by-80-km) basin that contains the leading population cluster focused on the capital city and the country's main coffee-growing area. The capital, San José, is atypical of Middle America—a clean and slumless metropolis that is the most cosmopolitan urban center between Mexico City and South America.

To the east of the highlands are the hot and rainy Caribbean lowlands, a sparsely populated segment of Rimland where many plantations have now been abandoned and replaced by subsistence farmers. Between 1930 and 1960, the (U.S.-based) United Fruit Company shifted most of the country's banana plantations from the crop-disease-ridden Caribbean littoral to Costa Rica's third zone—the plains and gentle slopes of the Pacific coastlands. This gave the Pacific zone a major boost in its economic growth, and today it is the scene of diversifying and expanding commercial agriculture (often requiring irrigation) as well as successful colonization schemes in previously undeveloped valleys and basins.

The long-term development of Costa Rica's economy has given it the region's highest standard of living, literacy rate, and life expectancy. Agriculture continues to dominate, with coffee, bananas, beef, and sugar the leading exports. These commodities, however, have been vulnerable to sharp price fluctuations on the world market in recent years. During the 1980s, low prices and heavy borrowing abroad (to cover the costs of expensive domestic social programs and fuel imports) combined to saddle Costa Rica with an enormous foreign debt, giving the country the dubious distinction of leading Central America in that category. Thus lowered national aspirations are now the order of the day. Politically, Costa Rica remains quite stable, despite its proximity to the region's trouble spots and the presence of thousands of refugees from neighboring and nearby countries. Costa Rica itself harbors no rebel activity at all, and the overwhelming majority of its peace-loving people prefer the country to maintain its neutrality as "the Switzerland of Central America."

Panama: Strategic Canal, Volatile Corridor

The Republic of Panama owes its birth to an idea: the construction of an artificial waterway to connect the Atlantic and Pacific oceans and thereby avoid the lengthy circumnavigation of South America. In the 1880s, when Panama was still an extension of neighboring Colombia, a French company tried and failed to build such a canal here; thousands of workers died of yellow fever, malaria, and other tropical diseases, and the company went bankrupt. By the turn of the century, U.S. interest in a Panama canal (which would shorten the sailing distance between the East and West Coasts by 8,000 nautical miles) rose sharply, and the United States in 1903 proposed a treaty that would permit a renewed effort at construction across Colombia's Panamanian isthmus. When the Colombian Senate refused to go along, Panamanians rebelled, and the United States supported this uprising by preventing Colombian forces from intervening. The Pana-

manians, at the behest of the United States, declared their independence from Colombia, and the new republic immediately granted the United States rights to the Canal Zone, averaging about 10 miles (16 km) in width and just over 50 miles (80 km) in length.

Soon canal construction commenced, and this time the project succeeded as American engineering and technology—and medical advances—triumphed over a formidable set of obstacles. The Panama Canal (see inset map, Fig. 5-8) was opened in 1914, a symbol of U.S. power and influence in Middle America. The Canal Zone was held by the United States under a treaty that granted it "all the rights, powers, and authority" in the area "as if it were the sovereign of the territory."

Two medium-sized vessels—a cruise ship and a container ship—in transit through the Panama Canal. The dimensions of the locks are too small to allow today's larger ships to pass through, and discussions on a new canal (to be located in Panama or elsewhere in Central America) arise from time to time.

Such language might suggest that the United States held rights over the Canal Zone in perpetuity, but the treaty nowhere stated specifically that Panama permanently yielded its own sovereignty in that transit corridor. In the 1970s, as the canal was transferring 20,000 ships per year and generating hundreds of millions of dollars in tolls, Panama sought to terminate U.S. control in the Canal Zone. Delicate negotiations began; in 1977, an agreement was reached on a staged withdrawal by the United States from the territory, first from the Canal Zone and then, by 2000, from the Panama Canal itself. This agreement took the form of two treaties, and following the signing by Presidents Carter and Torrijos, they were ratified in spite of stubborn opposition in the U.S. Senate.

Today, Panama shares some but not all the usual geographic features of the Central American republics. Its population of 2.5 million is about 60 percent mestizo but also contains substantial black, white, and Amerindian sectors. Spanish is the official (and majority) language, but English is in widespread use. Ribbon-like and oriented east-west, Panama's topography is mountainous and hilly, with some mountains reaching higher than 10,000 feet (3,000 m). Eastern Panama, especially Darien Province adjoining Colombia, is densely forested, and here is the only remaining gap in the otherwise complete, intercontinental Pan American Highway. Most of the rural population lives in the uplands west of the canal (much of the urban population is concentrated in the vicinity of the waterway, anchored by the cities at each end of the canal, Panama City and Colón). There, Panama produces bananas, rice, sugarcane, and coffee—and from the sea, shrimp and fishmeal.

Panama today finds itself recovering from the excesses of the Noriega regime, which in the late 1980s brought on international economic sanctions and the 1989 U.S. military invasion that ousted the dictator. Although damage to the country's economy and political stability was considerable, the canal—despite its age and inability to accommodate today's largest ships—remains Panama's focus, its lifeline, its future. It has now been augmented by an oil pipeline that crosses the isthmus, facilitating the interoceanic flow of petroleum, and by the Colón Free Zone (at the Caribbean end of the canal), one of the world's largest free trade areas. Economic revival is also being aided by the continuing development of Panama City as an international banking center.

Relative location was Panama's seminal advantage and has been its ally ever since its birth. Separated by Costa Rica from the turmoil of the other Mainland republics and by the still impassable Darien Gap from turbulent Colombia, Panama still has a future potential that few Middle American countries can match.

THE REALM IN TRANSITION

1. Economic and political transformations are underway in Mexico. Northern Mexico's economic geography is changing as a result of major investment by U.S. and other foreign companies.
2. Although insurgencies in several Central American countries have simmered down, the potential for further conflict remains.
3. Completion date of the transfer of the Panama Canal from the United States to Panama approaches, but the political future of Panama remains cloudy.
4. Emigration from Haiti to the United States is producing serious policy dilemmas. Efforts to stem the tide were coupled with attempts to return democratic government to Haiti via an international oil embargo.
5. The end of the Castro regime, and perhaps communist rule, is in sight in Cuba. The effect on surrounding countries is unpredictable.

◆ PRONUNCIATION GUIDE

Antigua   (an-TEE-gwuh)
Antilles   (an-TILL-eeze)
Arawak   (ARRA-wak)
Archipelago   (ark-uh-PELL-uh-go)
Aruba   (uh-ROO-buh)
Augelli   (aw-JELLY)
Baja   (BAH-hah)
Balsas   (BAHL-suss)
Barbados   (bar-BAY-dohss)
Bauxite   (BAWKS-site)
Belize   (beh-LEEZE)
Belmopan   (bell-moh-PAN)
Bering   (BERRING)
Bonaire   (bun-AIR)
Cacao   (kuh-KAY-oh/kuh-KOW)
Campeche   (kahm-PAY-chee)
Caribbean   (kuh-RIB-ee-un/karra-
   BEE-un)
Cay   (KEE)
Chihuahua   (chuh-WAH-wah)
Ciboney   (see-BOH-nay)
Ciudad Juarez   (see-you-DAHD
   WAH-rez)
Ciudades perdidas   (see-you-
   DAH-dayss pair-DEE-duss)
Cocos   (KOH-kuss)
Colón   (kuh-LOAN)
Cortés, Hernán   (kor-TAYSS,
   air-NAHN)
Costa Rica   (koss-tuh-REE-kuh)
Curaçao   (koor-uh-SAU)
Darien   (dar-YEN)
Ejido[s]   (eh-HEE-doh[ss])
El Paso   (ell-PASSO)
Fonseca   (fahn-SAY-kuh)
Fort-de-France   (for-duh-FRAWSS)
Gaillard   (gil-YARD)
Grenada   (gruh-NAY-duh)
Guadalajara   (gwah-duh-luh-HAHR-uh)
Guadeloupe   (GWAH-duh-loop)
Guatemala   (gwut-uh-MAH-lah)

Guerrero   (geh-RARE-roh)
Hacienda   (ah-see-EN-duh)
Hägerstrand, Torsten   (HAYGER-
   strand, TOR-stun)
Haiti   (HATE-ee)
Hegemony   (heh-JEH-muh-nee)
Hispaniola   (iss-pahn-YOH-luh)
Honduras   (hon-DURE-russ)
Isthmus/isthmian   (ISS-muss/ISS-
   mee-un)
Jamaica   (juh-MAKE-uh)
Junta   (HOON-tah)
Littoral   (LIT-oh-rull)
Maize   (MAYZ)
Managua   (mah-NAH-gwuh)
Maquiladora   (mah-kee-luh-DORR-uh)
Martinique   (mahr-tih-NEEK)
Maya[n]   (MY-uh[un])
Mesoamerica   (MEZZOH-america)
Mestizo   (meh-STEE-zoh)
Mexica   (meh-SHEE-kuh)
Miskito   (mih-SKEE-toh)
Monterrey   (mahnt-uh-RAY)
Mulatto   (moo-LAH-toh)
Nahuatl   (nah-WATTLE)
Nassau   (NASS-saw)
Nicaragua   (nick-uh-RAH-gwuh)
Noriega   (naw-ree-AY-guh)
Nuevo León   (noo-AY-voh lay-OAN)
Ocho Rios   (oh-choe REE-ohss)
Páramos   (PAH-ruh-mohss)
Paseo de la Reforma   (puh-SAY-oh
   day-luh ray-FOR-muh)
Peón   (pay-OAN)
Peones   (pay-OH-nayss)
Petén   (peh-TEN)
Placer[ing]   (PLASS-uh[ring])
Plebiscite   (PLEBBA-site)
Port-au-Prince   (por-toh-PRANSS)
Puerto Plata   (pwair-toh PLAH-tuh)
Puerto Rico   (pwair-toh REE-koh)

Puna   (POONA)
Quiché   (kee-CHAY)
Saba   (SAY-buh/SAH-buh)
Sabinas   (sah-BEE-nuss)
Salinas   (sah-LEE-nuss)
Sandinista   (sahn dee-NEE-stuh)
San José   (sahn hoe-ZAY)
San Juan   (sahn HWAHN)
San Pedro Sula   (sahn pay-droh
   SOO-luh)
Santander   (sahn-tahn-DARE)
Sierra Madre   (see-ERRA MAH-dray)
   Occidental   (oak-see-den-TAHL)
   Oriental   (aw-ree-en-TAHL)
St. Eustatius   (saint yoo-STAY-shuss)
St. Lucia   (saint LOO-shuh)
St. Maarten   (sint MAHRT-un)
Tampico   (tam-PEEK-oh)
Tarascans   (tuh-ruh-SKAHNZ)
Taxco   (TAHSS-koh)
Tegucigalpa   (tuh-goose-ih-GAHL-
   puh)
Tehuantepec   (tuh-WHAHN-tuh-pek)
Tenochtitlán   (tay-noh-chit-LAHN)
Tierra caliente   (tee-ERRA kahl-
   YEN-tay)
Tierra fría   (tee-ERRA FREE-uh)
Tierra helada   (tee-ERRA
   ay-LAH-dah)
Tierra templada   (tee-ERRA
   tem-PLAH-dah)
Tijuana   (tee-WHAHN-uh)
Tobago   (tuh-BAY-goh)
Torrijos   (tor-REE-hohss)
Tula   (TOO-lah)
Valle Central   (VAH-yay sen-TRAHL)
Veracruz   (verra-CROOZE)
Villahermosa   (vee-yuh-air-MOH-suh)
Willemstad   (VILL-um-staht)
Yucatán   (yoo-kuh-TAHN)
Zaïre   (zah-EAR)

HAVANA CUBA Tropic of Cancer

PEN. DE
YUCATÁN HISPANIOLA San Juan
JAMAICA PUERTO RICO
(U.S.A.) GUADELOUPE
(Fr.)

CENTRAL CARIBBEAN SEA MARTINIQUE
(Fr.)
BARBADOS

PUNTA DE GALLINAS TRINIDAD AND TOBAGO
AMERICA Barranquilla Maracaibo Port of Spain
Cartagena Valencia CARACAS
Panamá Ciudad Georgetown
Bolívar GUYANA Paramaribo
Medellín VENEZUELA Cayenne
Nevado del Tolima SANTA FE SURINAME FR.
17 110 DE BOGOTÁ Boa Vista do GUIANA
COLOMBIA Rio Branco GUIANA HIGHLANDS

Quito ILHA DE MARAJÓ Equator
ECUADOR Cotopaxi Belém São Luís
19 347 Manaus (Pará) (Maranhão)
Guayaquil Chimborazo (Manáos)
20 561 Iquitos Leticia Rio Solimões (Amazonas)

Chiclayo Fortaleza
Trujillo Pôrto (Ceará)
Nevs. Huascarán Velho Teresina
22 205 Rio Branco Natal
PERU João Pessoa (Paraíba)
LIMA B R A Z I L RECIFE (Pernambuco)
Callao Cuzco CHAPADA DE Maceió
MATO GROSSO
Volcán Misti La Paz Brasília Salvador
19 098 Nev. Illimani Cuiabá (Bahia)
Arequipa BOLIVIA Diamantina
Mollendo Sucre Belo Horizonte
Potosí Pico da Bandeira Vitória
Iquique CHACO PARAGUAY
SÃO PAULO
Antofagasta Cabo Frio
Salta Asunción Santos RIO DE JANEIRO
Tucumán
Copiapó Corrientes Florianópolis
Coquimbo
Santa Fe Pôrto Alegre
Córdoba Salto
Valparaíso Rosario URUGUAY Rio Grande
SANTIAGO Mendoza BUENOS AIRES
Concepción La Plata MONTEVIDEO
PAMPAS
Bahía Blanca
Valdivia
Puerto Montt Viedma

Comodoro Rivadavia
Monte Golfo San Jorge
Valentín
3 314 FALKLAND IS.
(ISLAS MALVINAS)
(Br.)
Río Gallegos
Stanley
Punta Arenas TIERRA DEL FUEGO
Mt. Sarmiento
8100 ISLA DE LOS ESTADOS
CABO DE HORNOS
(CAPE HORN)

ATLANTIC
OCEAN

PACIFIC
OCEAN

Drake Passage

Relief

Meters		Feet
3050		10 000
1525		5000
610		2000
305		1000
Sea Level		0
152.5		500
1525		5000
3050		10 000
6100		20 000

0 200 400 600 800 1000 Miles
0 400 800 1200 1600 Kilometers

copyright © Rand McNally, 1993

Scale 1:40 000 000; one inch to 630 miles. Lambert's Azimuthal, Equal Area Projection
Elevations and depressions are given in feet

Longitude West of Greenwich

CHAPTER 6

SOUTH AMERICA: CONTINENT OF CONTRASTS

Of all the continents South America has the most familiar shape—that giant triangle connected by mainland Middle America's tenuous land bridge to its sister continent in the north. What we realize less often about South America is that it lies not only south but also mostly east of its northern counterpart as well. Lima, the capital of Peru—one of the continent's westernmost cities—lies farther east than Miami, Florida. Thus South America juts out much more prominently into the Atlantic Ocean than does North America, and South American coasts are much closer to Africa and even to southern Europe than are the coasts of Middle and North America. Lying so far eastward, South America's western flank faces a much wider Pacific Ocean than does North America; the distance from its west coast to Australia is nearly twice that from California to Japan.

As if to reaffirm South America's northward and eastward orientation, the western margins of the continent are rimmed by one of the world's longest and highest mountain ranges, the Andes, a gigantic wall that extends from Tierra del Fuego near the southern tip of the triangle to Venezuela in the far north (Fig. 6-1). Every map of world physical geography clearly reflects the existence of this mountain chain—in the alignment of isohyets (lines connecting places of equal precipitation totals; see Fig. I-6), in the elongated zone of highland climate (Fig. I-7), and in the regional distribution of vegetation (Fig. I-8). Moreover, as Fig. I-9 reveals, South America's biggest population clusters are located along the eastern and northern coasts, overshadowing those of the Andean west.

◆ THE HUMAN SEQUENCE

Although modern South America's largest populations are situated in the east and north, there was a time during the height of the Inca Empire when the Andes Mountains contained the most densely peopled and best organized state on the continent. Although the origins of Inca civilization are

IDEAS & CONCEPTS

Economic geography
Agricultural systems
Land alienation
Isolation
South American culture spheres

Urbanization (2)
Rural-to-urban migration
The "Latin" American city
Insurgent state
Complementarity (3)

REGIONS

The Caribbean North
The Andean West
The Mid-Latitude South

still shrouded in mystery, it has become generally accepted that the Incas were descendants of ancient peoples who came to South America via the Middle American land bridge (possibly following earlier migrations from Asia to North America via the Bering land bridge). But even this is not totally beyond doubt; some scholars maintain that the first settlers in this part of the Western Hemisphere may have reached the Chilean and Peruvian coasts directly by sea from distant Pacific islands. In any case, for thousands of years before the Europeans arrived in the sixteenth century, indigenous Amerindian communities and societies had been developing in South America.

About a thousand years ago, a number of regional cultures thrived in Andean valleys and basins and at places along the Pacific coast. The llama had been domesticated as a beast of burden, a source of meat, and a producer of wool. Religions flourished and stimulated architecture as well as the construction of temples and shrines. Sculpture, painting, and other art forms were practiced. Over these cultures the Incas extended their authority from their headquarters in

TEN MAJOR GEOGRAPHIC QUALITIES OF SOUTH AMERICA

1. South America's physiography is dominated by the Andes Mountains in the west and the Amazon Basin in the central north. Much of the remainder is plateau country.
2. Half of the realm's area and half of its population are concentrated in one country—Brazil.
3. South America's population remains concentrated in peripheral zones. Most of the interior is sparsely peopled.
4. The leading population trend is cityward migration. No realm has urbanized more rapidly since 1980.
5. Regional economic contrasts and disparities, both in the realm as a whole and within individual countries, are strong. In general, the south is the most developed, the northeast the least.
6. Interconnections among the states of the realm, although improving, remain comparatively weak.
7. Strong cultural pluralism exists in the majority of the realm's countries, and this pluralism is often expressed regionally.
8. With the exception of three small countries in the north, the realm's modern cultural sources lie in a single subregion of Europe, the Iberian Peninsula. Spanish is the *lingua franca* except in Portuguese-speaking Brazil.
9. Lingering politico-geographical problems beset the realm. Boundary disputes and territorial conflicts persist.
10. The Catholic church is still a major presence throughout the realm and constitutes one of its unifying elements.

the Cuzco Basin of the Peruvian Andes, beginning late in the twelfth century, to forge the greatest empire in the Americas prior to the coming of the Europeans.

Nothing to compare with the cultural achievements of this central Andean zone existed anywhere else in South America. Beyond the Andean civilizations, anthropologists recognize three groupings of Amerindian peoples: (1) those of the Caribbean fringe, (2) those of the tropical forest of the Amazon Basin and other lowlands, and (3) those called "marginal," whose habitat lay in the Brazilian Highlands, the headwaters of the Amazon River, and most of southern

South America (Fig. 6-3, p. 307). It has been estimated that these Caribbean, forest, and marginal peoples together constituted only about one-quarter of the continent's total native population.

THE INCA EMPIRE

When the Inca civilization is compared to that of ancient Mesopotamia, Egypt, the old Asian civilizations, and the Aztecs' Mexica Empire, it quickly becomes clear that this civilization was an unusual achievement. Everywhere else, rivers and waterways provided avenues for interaction and the circulation of goods and ideas. Here, however, an empire was forged from a series of elongated basins (called *altiplanos*) in the high Andes, created when mountain valleys between parallel and converging ranges filled with erosional materials from surrounding uplands. These *altiplanos* are often separated from one another by some of the world's most rugged terrain, with high snowcapped mountains alternating with precipitous canyons. Individual *altiplanos* accommodated regional cultures; the Incas themselves were first established in the intermontane basin of Cuzco (Fig. 6-3). From that hearth, they conquered and extended their authority over the peoples of coastal Peru and other *altiplanos*. Their first thrust, apparently late in the fourteenth century, was southward.

More impressive than the Incas' military victories was their subsequent capacity to integrate the peoples and regions of the Andean domain into a stable and smoothly functioning state. The odds would seem to have been against them, because as they progressed as far south as central Chile, their domain became ever more elongated, making effective control much more difficult. The Incas, however, were expert road and bridge builders, colonizers, and administrators, and in an incredibly short time they consolidated these new territories. And just before the Spanish arrival (in 1531), they even conquered territories to the north, including Ecuador and part of southern Colombia.

The early sixteenth century was a critical period in the Inca Empire because the conquest of Ecuador and areas to its north for the first time placed stress on the existing administrative framework. Until that time the dominant center of the state had been Cuzco, but now it was decided that the empire should be divided into two units: a southern one ruled from Cuzco and a northern sector focused on Quito. This decision was related to the problem of control over the rebellious north and the possibilities for expansion deeper into Colombia. Thus the empire was now beset by a number of difficulties, notably the uncertain northern frontier and rising tensions between Cuzco and Quito. And just as the Aztec Empire had been ripe for internal revolt when Cortés and his party entered Mexico, so the Spanish arrival in western South

FIGURE 6-1

America happened to coincide with a period of stress within the Inca Empire.

At its zenith, the Inca Empire may have counted more than 20 million subjects. Of course, the Incas themselves were always in a minority in this huge state, and their position be-

came one of a ruling elite in a rigidly class-structured society. The Incas, representative of the emperor in Cuzco, formed a caste of administrative officials who implemented the decisions of their monarch by organizing all aspects of life in the conquered territories. They saw to it that harvests were

◆ FOCUS ON A SYSTEMATIC FIELD

ECONOMIC GEOGRAPHY

In the preceding chapters, we have discussed several concepts, principles, and examples of **economic geography**. In the Introduction this subdiscipline was defined as being concerned with the diverse ways in which people earn a living, and how the goods and services they produce are expressed and organized spatially. In Chapter 3, we identified four sets of productive activities as the major components of the *spatial economy*: (1) primary activities (agriculture, mining, and other extractive industries); (2) secondary activities (manufacturing); (3) tertiary activities (services); and (4) quaternary activities (information and decision-making). Because most of the world's secondary, tertiary, and quaternary activities are located within the developed realms, we focus here on *agriculture*, the dominant livelihood of the hundreds of millions of workers who inhabit the remaining realms of the developing world.

The global distribution of **agricultural systems** is displayed in Fig. 6-2.

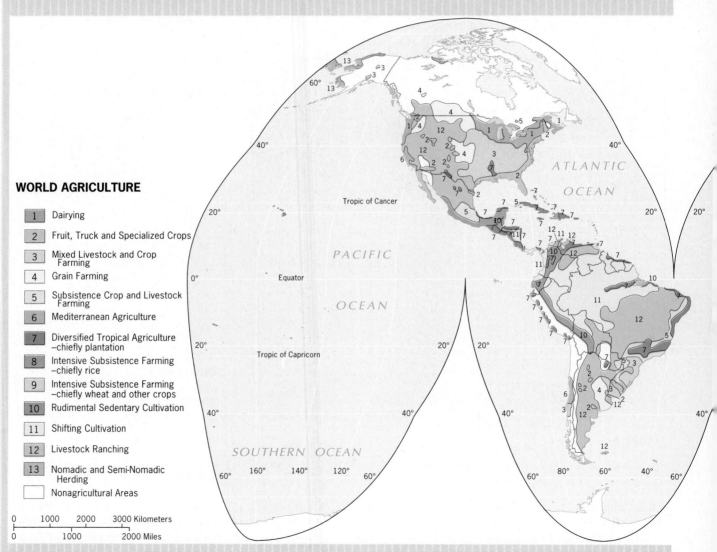

WORLD AGRICULTURE

1	Dairying
2	Fruit, Truck and Specialized Crops
3	Mixed Livestock and Crop Farming
4	Grain Farming
5	Subsistence Crop and Livestock Farming
6	Mediterranean Agriculture
7	Diversified Tropical Agriculture –chiefly plantation
8	Intensive Subsistence Farming –chiefly rice
9	Intensive Subsistence Farming –chiefly wheat and other crops
10	Rudimental Sedentary Cultivation
11	Shifting Cultivation
12	Livestock Ranching
13	Nomadic and Semi-Nomadic Herding
	Nonagricultural Areas

FIGURE 6-2

The spatial organization of agriculture in the advanced commercial economies of Europe and the United States has already been explained in the context of the Von Thünen model, expanded (thanks to modern transportation technology) to the continental scale. In fact, the macro-Thünian framework can even be extended to the world as a whole. Its anchoring "global city" would be the European and North American edges of the North Atlantic Basin, and many of the colonially generated farming systems would fit a sequence of concentric and increasingly distant agricultural zones (such as Middle American fruit and sugar, Argentine beef, Australian wheat, and New Zealand wool). Although this worldwide Von Thünen structuring has eroded with the demise of colonialism, many of that era's features are still apparent (for example, plantation farming), reminding us that most nonsubsistence Third World agriculture remains firmly oriented to the markets of the Western powers that

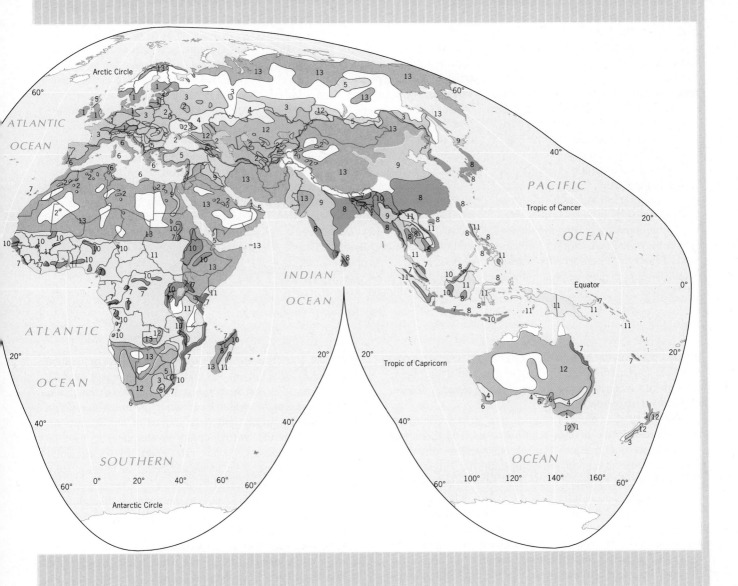

first reorganized primary production in countries of the developing realms. The spatial pattern of those commercial farming systems can be observed in Fig. 6-2 in the distributions of Categories 1 through 4 and 6, 7, and 12.

The six remaining categories are all associated with the various types of noncommercial subsistence agriculture. Shifting cultivation and nomadic/semi-nomadic herding (Categories 11 and 13) are the least intensive of these activities, relegated to marginal environments that—together with vast (unnumbered/white) nonagricultural areas—occupy more than two-thirds of the planet's land surface. Categories 5, 8, 9, and 10 entail more intensive forms of subsistence agriculture. As we noted in the Introduction, the intensive rice farming of Category 8 supports the earth's largest and densest nonurban population clusters on the alluvial soils of Asia's great river valleys and coastal plains; the intensive subsistence cultivation of other crops (Category 9), especially wheat, can also support huge population con-

centrations (northern China and northwestern India are two leading examples). Categories 5 and 10 are occasionally associated with sizeable populations as well, but these farming systems are far less productive and often represent a transition between intensive and extensive agriculture (the latter embodied in Categories 11 and 13).

Looking at the South American portion of Fig. 6-2, commercial and subsistence farming exist side by side here to a greater degree than in any other realm (where one of the two always geographically dominates the other). This, of course, does not represent a planned "balance" between the two; rather, it reflects the continent's deep internal cultural and economic divisions, on which this chapter elaborates.

The commercial agricultural side of South America is expressed in (1) a huge cattle-ranching zone near the coasts (Category 12) that stretches southwest from northeastern Brazil to Patagonia; (2) Argentina's wheat-raising Pampa (Category 4), which is

comparable to the U.S. Great Plains; (3) a Corn Belt-type crop and livestock zone (Category 3) in northeastern Argentina, Uruguay, southern Brazil, and south-central Chile; (4) a number of seaboard tropical plantation strips (Category 7) located in Brazil, the Guianas, Venezuela, Colombia, and Peru; and (5) a Mediterranean-type agricultural zone (Category 6) in Middle Chile.

In stark contrast to these commercial systems, subsistence farming covers the rest of the realm's arable land. Primitive shifting cultivation (Category 11) occurs in the rainforested Amazon Basin and its hilly perimeter; rudimentary sedentary cultivation (Category 10) dominates the Andean plateau country from Colombia in the north to the Bolivian *Altiplano* in the south; and, finally, a ribbon of mixed subsistence farming (Category 5) courses through most of eastern Brazil between the coastal plantation and interior grazing zones.

divided between the church, the community, and individual families; they maintained the public granaries; they made investments to construct and maintain roads, terrace hillsides, and expand irrigation works.

The life of the empire's subjects was strictly controlled by this bureaucracy of Inca administrators, and there was very little personal freedom. Farm quotas were set, and there was no true market economy; the crops, like the soil on which they were grown, belonged to the state. Marriages were officially arranged, and families could live only where the Inca supervisors would permit. Indeed, the family (as a productive entity within the community), not the individual, was considered to be the basic unit of administration. Inca rule was thoroughly effective, and obedience was the only course for its subjects. So highly centralized was the state and so com-

plete the subservience of its tightly controlled population that a takeover at the top was enough to gain power over the entire empire—as the Spaniards quickly proved in the 1530s.

The Inca Empire, which had risen to greatness so rapidly, disintegrated abruptly under the impact of the Spanish invaders. Perhaps the swiftness of its development contributed to its fatal weakness. At any rate, apart from spectacular ruins such as those at Peru's Machu Picchu (see photo p. 308), the empire left behind social values that have remained a part of Amerindian life in the Andes to this day and still contribute to fundamental divisions between the Hispanic and Amerindian populations in this part of South America. For example, the Inca state language, Quechua, was so firmly rooted that it is still spoken by millions of Amerindians living in the highlands of Peru, Ecuador, and Bolivia.

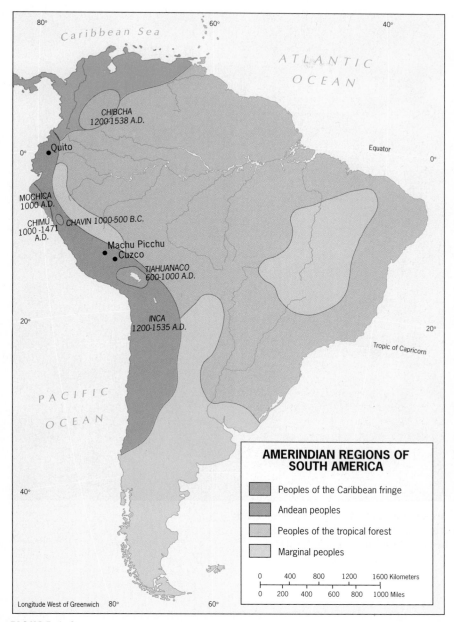

FIGURE 6-3

THE IBERIAN INVADERS

In South America as in Middle America, the location of indigenous peoples determined to a considerable extent the direction of the thrusts of European invasion. The Incas, like Mexico's Maya and Aztec peoples, had accumulated gold and silver in their headquarters, possessed productive farmlands, and constituted a ready labor force. Not long after the 1521 defeat of the Mexica Empire of the Aztecs, the Spanish conquerors crossed the Panamanian isthmus and sailed southward along the continent's northwestern coast. On his first journey, Francisco Pizarro heard of the existence of the Inca Empire. After a landfall in 1527 at Tumbes, on the northern coastal extremity of Peru, he returned to Spain to organize the penetration of the Incan domain.

Pizarro returned to Tumbes with 183 men and two dozen horses in 1531, a time when the Incas were preoccupied with problems of royal succession and strife in the northern provinces. The events that followed are well known, and less than three years later the party rode victorious into Cuzco. Initially, the Spaniards kept intact the Incan imperial structure by permitting the crowning of an emperor who was in

This unusual perspective of the ruins of the ancient Inca city of Machu Picchu (center of photo) reveals the dimensions of the structure and the high Andean relief of the area where it was built. Not until well into the twentieth century did an explorer "discover" this incomparable site.

fact under their control; but soon the land- and gold-hungry invaders were fighting among themselves, and the breakdown of the old order began.

The new order that eventually emerged in western South America placed the indigenous peoples in serfdom to the Spaniards. Great haciendas were formed by **land alienation** (the takeover of former Amerindian lands), taxes were instituted, and a forced-labor system was introduced to maximize the profits of exploitation. As in Middle America, most of the Spanish invaders had little status in Spain's feudal society, but they brought with them the values that prevailed in Iberia: land meant power and prestige, gold and silver meant wealth.

Lima, the west-coast headquarters of the Spanish conquerors, was founded by Pizarro in 1535, about 375 miles (600 km) northwest of the Andean center of Cuzco. Before

long Lima was one of the richest cities in the world, reflecting the enormity of the wealth yielded by the ravaged Inca Empire. Lima soon became the capital of the viceroyalty of Peru, as the authorities in Spain integrated the new possession into their colonial empire (Fig. 6-4). Subsequently, when Colombia and Venezuela came under Spanish control and, later on, when Spanish settlement began to expand in the coastlands of the Rio de la Plata estuary in what is now Argentina and Uruguay, two additional viceroyalties were added to the map: New Granada in the north and La Plata in the south.

Meanwhile, another vanguard of the Iberian invasion was penetrating the east-central part of the continent, the coastlands of present-day Brazil. This area had become a Portuguese sphere of influence, because Spain and Portugal had

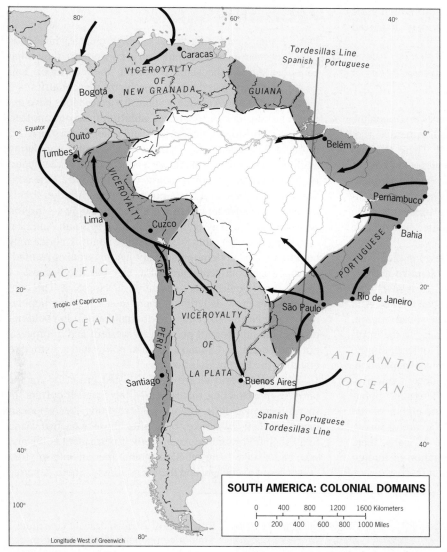

FIGURE 6-4

agreed in the Treaty of Tordesillas (1494) to recognize a north-south line (drawn by Pope Alexander VI) 370 leagues west of the Cape Verde Islands as the boundary between their New World spheres of influence. This border ran approximately along the meridian of 50°W longitude, thereby cutting off a sizeable triangle of eastern South America for Portugal's exploitation (Fig. 6-4).

A brief look at the political map of South America (Fig. 6-5), however, shows that the 1494 treaty did not succeed in limiting Portuguese colonial territory to the east of the agreed-upon 50th meridian. True, the boundaries between Brazil and its northern and southern coastal neighbors (French Guiana and Uruguay) both reach the ocean near 50°W; but then the Brazilian boundaries bend far inland to include al-

most the entire Amazon Basin as well as a good part of the Paraná-Paraguay Basin.

Thus Brazil came to be only slightly smaller than all the other South American countries combined. In population, too, Brazil today accounts for just about half the total of the entire continent. The successful, enormous westward thrust was the work of many Brazilian elements—missionaries in search of converts, explorers in search of quick wealth—but no group did more to achieve this penetration than the so-called *Paulistas*, the settlers of São Paulo. From early in its colonial history São Paulo had been a thriving settlement, with highly profitable plantations and an ever-growing need for labor. The *Paulistas* organized huge expeditions into the interior, seeking Amerindian slaves, gold, and precious stones; at the same time, they were intent on reducing the influence of Jesuit missionaries over the widely dispersed native population.

THE AFRICANS

As Fig. 6-4 shows, the Spaniards initially got very much the better of the territorial partitioning of South America—not just quantitatively but qualitatively as well. There were no rich Amerindian states to be conquered and looted east of the Andes, and no productive agricultural land was under cultivation. The comparatively few eastern Amerindians constituted no usable labor force; it has been estimated that the entire area of present-day Brazil was inhabited by no more than 1 million aboriginal people.

When the Portuguese finally began to develop their New World territory, they turned to the same lucrative activity that their Spanish rivals had pursued in the Caribbean—the plantation cultivation of sugar for the European market. And they, too, found their labor force in the same source region, as millions of Africans were brought in slavery to the tropical Brazilian coast north of Rio de Janeiro. Not surprisingly, Brazil now has South America's largest black population, which is still heavily concentrated in the country's poverty-stricken northeastern states. Today, with the overall population of Brazil at 156.5 million, about one-eighth of the people are black and another 40 percent or so are of mixed African, white, and Amerindian ancestry. Africans, then, definitely constitute the third major immigration of foreign peoples into South America (see Figs. 4-13 and 8-6).

PERSISTENT ISOLATION

Despite their common cultural heritage (at least insofar as their European-mestizo population is concerned), their adjacent location on the same continent, their common language, and their shared national problems, the countries that arose out of South America's Spanish viceroyalties have existed in a considerable degree of **isolation** from one another. Distance, physiographic barriers, and other factors have reinforced this separation. To this day, the major population agglomerations of South America adhere to the coast, mainly the eastern and northern coasts (Fig. I-9). Of all the continents, only Australia has a population distribution that is as markedly *peripheral*, but there are only some 18 million people in Australia as against 312 million in South America.

Compared with other world realms, South America may be described as underpopulated not just in terms of its modest total for a continental area of its size but also because of the resources available for or awaiting development. This continent never drew as large an immigrant European population as did North America. The Iberian Peninsula could not provide the numbers of people that western and northwestern Europe did, and Spanish colonial policy had a restrictive effect on the European inflow.

The New World viceroyalties existed primarily for the purpose of extracting riches and filling Spanish coffers. In Iberia there was little interest in developing the American lands for their own sake; only after those who had made Spanish and Portuguese America their permanent home and who had a stake there rebelled against Iberian authority did things begin to change, and then very slowly. South America

Brazilians of African descent constitute a major component of the population along the country's tropical northeastern coast. This is a busy street in the center of Belém (1.1 million), the large city located just south of the equator and the mouth of the Amazon River.

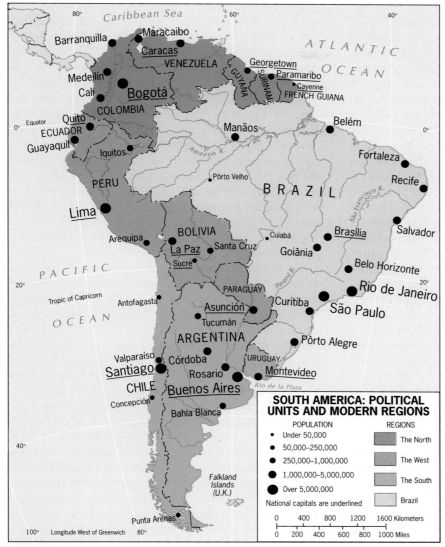

FIGURE 6-5

was saddled with the values, economic outlook, and social attitudes of seventeenth- and eighteenth-century Iberia—not the best tradition from which to begin the task of forging modern nation-states.

INDEPENDENCE

Certain isolating factors had their effect even during the wars for independence. Spanish military strength was always concentrated at Lima, and those territories that lay farthest from this center of power—Argentina and Chile—were the first to establish their independence from Spain (in 1816 and 1818, respectively). In the north, Simón Bolívar led the burgeoning independence movement, and in 1824 two decisive military defeats there spelled the end of Spanish

power in South America. Thus, in little more than a decade, the Hispanic countries had fought themselves free. But this joint struggle did *not* produce unity: no fewer than nine countries emerged from the three former viceroyalties. One, Bolivia, hitherto known as Upper Peru, was named after Bolívar when it was declared independent in 1825. Bolívar's confederacy, the Republic of Gran Colombia, which had achieved independence in 1819, broke up in 1829 and 1830 into Venezuela, Ecuador, and New Granada (the last was renamed Colombia in 1861). Uruguay was temporarily attached to Brazil, but it too attained separate political identity by 1828. Paraguay, once a part of Argentina, also appeared on the map as a sovereign state before 1830 (Fig. 6-5).

It is not difficult to understand why this fragmentation took place. With the Andes intervening between Argentina and Chile and the Atacama Desert between Chile and Peru, overland distances seem even greater than they really are, and these obstacles to contact have proved quite effective. Hence the countries of South America began to grow apart— a separation process sometimes heightened by uneasy frontiers. Friction and even wars have been frequent, and a number of boundary disputes remain unresolved to this day. For example, Bolivia at one time had a direct outlet to the sea through northern Chile but lost this access corridor in a series of conflicts involving Chile, Peru, and (indirectly) Argentina. Chile and Argentina themselves were locked in a longstanding dispute over their Andean boundary. Peru and Ecuador both laid claim to the upper Amazon Basin and the key Peruvian town of Iquitos.

Brazil attained independence from Portugal at about the same time that the Spanish possessions in South America were struggling to end overseas domination, but the sequence of events was quite different. In Brazil, too, there had been revolts against Portuguese control. But the early 1800s, instead of witnessing a decline in Portuguese authority, actually brought the Portuguese government (headed by Prince Regent Dom João) from Lisbon to Rio de Janeiro! Incredibly, Brazil in 1808 was suddenly elevated from colonial status to the seat of empire; it owed its new position to Napoleon's threat to overrun Portugal, which was then allied with the British. Although many expected that a new era of Brazilian progress and development would follow, things soon fell apart. By 1821, with the Napoleonic threat removed, Portugal decided to demote Brazil to its former colonial status. But Dom Pedro, Dom João's son and the new regent, defied his father and led a successful struggle for Brazilian independence. With overwhelming popular support, he was crowned emperor of Brazil in 1822, and loyalist Portuguese forces still in the country were ignominiously deported to Lisbon.

The post-independence relationships between Brazil and its Spanish-influenced neighbors have been similar to the limited interactions among the individual Hispanic republics themselves. Distance, physical barriers, and cultural contrasts served to inhibit positive contact and interaction. Thus Brazil's orientation toward Europe, like that of the other republics, remained stronger than its involvement with the countries on its own continent. (Only in the 1990s are the countries of the South American realm finally coming to recognize the mutual advantages of increasing interaction, led by a pronounced rise in intracontinental trade that could pave the way for greater cooperation in other spheres as well.)

◆ CULTURE AREAS

When we speak of the "orientation" or "interaction" of South American countries, it is important to keep in mind just who does the orienting and interacting, for there is a tendency to generalize the complexities of these countries away. The fragmentation of colonial South America into 10 individual republics and the nature of their subsequent relationships was the work of a small minority of the people in each country. People of African descent in Brazil, at the time of independence, had little or no voice in the course of events; the Amerindians in Peru, numerically a vast majority, could only watch as their European conquerors struggled with each other for supremacy. It would not even be true to say that the European minorities *in toto* governed and made policy: the wealthy, landholding, upper-class elite determined the posture of the state. These were—and in some cases still are— the people who made the quarrels with their neighbors, who turned their backs on wider American unity, and who kept strong the ties with Madrid and Paris (Paris had long been a cultural focus for Middle and South America's well-to-do, and their children were usually sent to French rather than Spanish schools).

So complex and heterogeneous are the societies and cultures of Middle and South America that practically every generalization has to be qualified. Take the one so frequently used—the term *Latin* America. Apart from the obvious exceptions that can be read from the map, such as Jamaica, Guyana, and Suriname, which are clearly not "Latin" countries, it may be improper even to identify some of the Spanish-influenced republics as "Latin" in their cultural milieu. Certainly the white, wealthy upper classes are of Latin European stock, and they have the most influence at home and are most visible abroad; they are the politicians and the businesspeople, the writers and the artists. Their cultural environment is made up of the Spanish (and Portuguese) language, the Roman Catholic church, and the picturesque Mediterranean-style architecture of Middle and South America's cities and towns. These things provide them with a common bond and strong ties to Iberian Europe. However, in the mountains and villages of Ecuador, Peru, and Bolivia are millions of people to whom the Spanish language is still alien, to whom the white people's religion is another unpopular element of acculturation, and to whom decorous Spanish styles of architecture are meaningless when a decent roof and a solid floor are still unattainable luxuries.

South America, then, is a continent of plural societies, where Amerindians of different cultures, Europeans from Iberia and elsewhere, blacks from western tropical Africa, and Asians from India, Japan, and Indonesia have produced a cultural and economic kaleidoscope of almost endless variety. Certainly, calling this human spatial mosaic "Latin" America is not very useful. Is there a more meaningful approach to a regional generalization that would better represent and differentiate the continent's cultural and economic spheres? John Augelli, who also developed the Rimland-Mainland concept for Middle America, made such an attempt. His map (Fig. 6-6) shows that five **South American culture spheres**—internal cultural regions—blanket the realm. This scheme is quite useful if one keeps in mind that South America is undergoing economic change in certain

FIGURE 6-6

areas today and that these culture spheres are generalized and subject to further modification as well.

TROPICAL-PLANTATION REGION

The first culture sphere, the *tropical-plantation* region, in many ways resembles the Middle American Rimland. It consists of several separated areas, of which the largest lies along the northeastern Brazilian coast, with four others along the Atlantic and Caribbean coastlands of northern South America. Location, soils, and tropical climates favored plantation crops, especially sugar. The fact that the aboriginal population was small led to the introduction of millions of African slave laborers, whose descendants today continue to domi-

nate the racial makeup and strongly influence the cultural expression of these areas. The plantation economy later failed, soils became exhausted, slavery was abolished, and the people were largely reduced to poverty and subsistence—socioeconomic conditions that now dominate much of the region mapped as tropical-plantation.

EUROPEAN-COMMERCIAL REGION

The second region on Augelli's map, identified as *European-commercial*, is perhaps the most truly "Latin" part of South America. Argentina and Uruguay, each with a population that is more than 85 percent "pure" European and with a strong Hispanic cultural imprint, constitute the bulk of the

European-commercial region. Two other areas also lie within it: most of Brazil's core area and the central core of Chile. Southern Brazil shares the temperate grasslands of the Pampa and Uruguay (see Fig. I-8), and this area is important as a zone of livestock raising as well as corn production. Middle Chile is an old Spanish settlement zone, home to the approximately one-fifth of the Chilean population who claim pure Spanish ancestry; here, in an area of Mediterranean climate (Fig. I-7), cattle and sheep pastoralism plus mixed farming are practiced. In general, then, the European-commercial region is economically more advanced than the rest of the continent. A commercial economy rather than a subsistence way of life prevails, living standards are better, literacy rates are higher, transportation networks are superior, and (as Augelli has pointed out) the overall development of this region surpasses that of several parts of Europe itself.

AMERIND-SUBSISTENCE REGION

The third region is identified as *Amerind-subsistence*, and it forms an elongated zone along the length of the central Andes from southern Colombia to northern Chile/northwestern Argentina, closely approximating the area occupied by the old Inca Empire. The feudal socioeconomic structure that was established here by the Spanish conquerors still survives. The Amerindian population forms a large, landless peonage, living by subsistence or by working on haciendas far removed from the Spanish culture that forms the primary force in the national life of their country. This region includes some of South America's poorest areas, and what commercial activity there is tends to be in the hands of whites or mestizos. The Amerindian heirs to the Inca Empire live, often precariously, at high elevations (as much as 12,500 feet, or 3800 m) in the Andes. Poor soils, uncertain water supplies, high winds, and bitter cold make farming a constantly difficult proposition; infrequent exchange takes place at remote upland Amerindian markets.

MESTIZO-TRANSITIONAL REGION

The fourth region, *mestizo-transitional*, surrounds the Amerind-subsistence region, covering coastal and interior Peru and Ecuador, much of Colombia and Venezuela, most of Paraguay, and large parts of Argentina, Chile, and Brazil (including the valleys of the Amazon and certain of its tributary rivers). This is the zone of mixture between European and Amerindian—or African in Brazil, Venezuela, and Colombia. The map thus reminds us that countries such as Bolivia, Peru, and Ecuador are dominantly Amerindian and mestizo. In Ecuador, for example, these two groups make up about 90 percent of the total population, and a mere 8 percent can be classified as white. The term *transitional* has an economic connotation, too, because (as Augelli puts it) this region "tends

to be less commercial than the European sphere but less subsistent in orientation than dominantly [Amerindian] areas."

UNDIFFERENTIATED REGION

The fifth region on the map is marked as *undifferentiated* because its characteristics are hard to classify. Some of the Amerindian peoples in the interior of the Amazon Basin had remained almost completely isolated from the momentous changes in South America since the days of Columbus. Although isolation and lack of change are still two notable aspects of this subregion, the ongoing development of Amazonia is reversing that situation. The most remote Amazon backlands plus the Chilean and Argentinean southwest are also sparsely populated and exhibit only very limited economic development; poor transportation and difficult location continue to contribute to the unchanging nature of these areas.

The framework of the five culture spheres just described is necessarily a generalization of a rather complex geographic reality. Nonetheless, even in its simplicity it underscores the diversity of South America's peoples, cultures, and economies.

◆ URBANIZATION

As in other parts of the developing world, people in South America today are leaving the land and moving to the cities. This **urbanization** process intensified sharply after 1950, and it persists so strongly that South America's urban-population percentage is now much more typical of the developed world. In 1925, about one-third of South America's peoples lived in cities and towns, and as recently as 1950 the percentage was just over 40. But by 1975 the continent-wide figure had surpassed 60 percent, and today no less than three out of four South Americans live in urban areas. Of course, these percentages mask the actual numbers, which are even more dramatic. Between 1925 and 1950, the realm's towns and cities grew by about 40 million residents as the urbanized percentage rose from 33 to 42. Then between 1950 and 1975 more than 125 million people crowded into the teeming metropolitan areas—more than *three times* the total for the previous quarter-century—and an additional 100 million since 1975 have swelled the continental total to its current 232 million.

South America's population of 312 million (20 percent larger than that of the United States) has a high growth rate, but nowhere are the numbers increasing faster than in the towns and cities. We usually assume that the populations of rural areas grow more rapidly than urban areas because farm families traditionally have more children than city dwellers. Yet, overall, the urban population of South America has grown

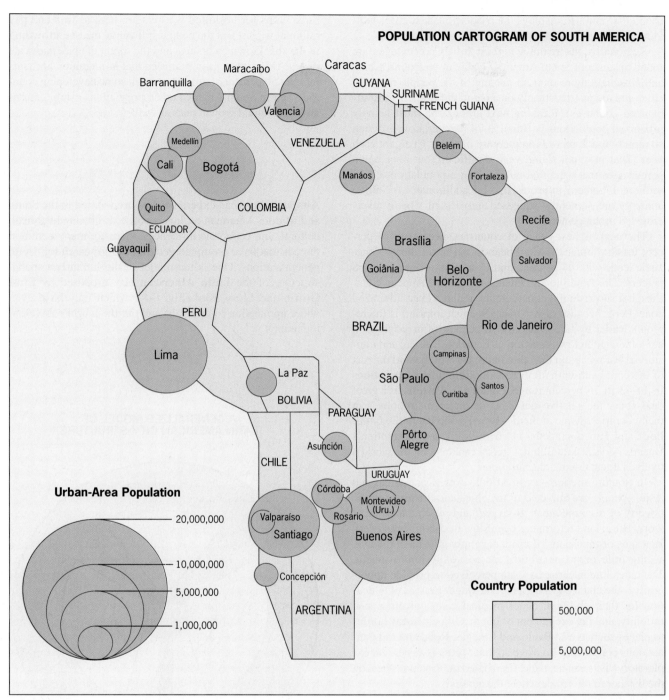

POPULATION CARTOGRAM OF SOUTH AMERICA

Urban-Area Population

- 20,000,000
- 10,000,000
- 5,000,000
- 1,000,000

Country Population

500,000

5,000,000

FIGURE 6-7

annually by nearly 5 percent since 1950, while the rural areas increased yearly by only about 1.5 percent. These figures reveal the dimensions of the **rural-to-urban migration** from the countryside to the cities—still another migration process that affects modernizing societies throughout the developing realms.

The generalized spatial pattern of South America's urban transformation is seen in Fig. 6-7, which shows a *cartogram* of the continent's population (of the type presented for the world in Fig. I-10 on pp. 24-25). But this cartogram goes further than the one in Fig. I-10: it not only shows the 13 South American countries in "population-space" relative to each other but also displays the proportionate sizes of individual large cities within their total national populations.

Regionally, the realm's highest urban percentages are found in southern South America. Today in Argentina, Chile, and Uruguay, more than 85 percent of the people reside in cities and towns (statistically on a par with Europe's most urbanized countries). Ranking next among the most heavily urbanized populations is Brazil at 74 percent, now growing so rapidly that it records an increase of about 1 percent each year. That may not sound very significant, but once again percentages mask the real numbers; in a population of 157 million, 1 percent indicates that an additional 1.6 million annually are somehow squeezed into Brazil's badly overcrowded urban centers.

The next highest group of countries, averaging 73 percent urban, border the Caribbean in the north; here, Venezuela leads with 84 percent, and Colombia follows with 68 percent. The Amerind-subsistence-dominated Andean countries, not surprisingly, constitute the realm's least urbanized zone. Peru, because of its strong Spanish imprint, is the region's leader by far, with fully 70 percent of its population agglomerated in towns and cities. Ecuador, Bolivia, and transitional Paraguay, on the other hand, lag well behind the rest of the continent, with all three exhibiting urban proportions in the 43-to-55-percent range. Figure 6-7 also tells us a great deal about the relative positions of major metropolises in their countries. Three of them—Brazil's São Paulo (20.6 million) and Rio de Janeiro (11.6 million), and Argentina's Buenos Aires (12.2 million)—today rank among the world's twelve largest urban concentrations.

In South America—as in Middle America, Africa, and Asia—people are attracted to the cities and driven from the poverty of the rural areas. Both *pull* and *push factors* are at work. Rural land reform has been slow in coming, and every year tens of thousands of farmers simply give up and leave, seeing little or no possibility for economic improvement. The cities lure because they are perceived to provide opportunity—the chance to earn a regular wage. Visions of education for their children, better medical care, upward social mobility, and the excitement of life in a big city draw hordes to places such as São Paulo and Caracas. Road and rail connections continue to improve so that access is easier and exploratory visits can be made. City-based radio stations beckon the listener to the locale where the action is.

But the actual move can be traumatic. As we saw in the box on Mexico City in Chapter 5, cities of the developing world are surrounded and sometimes invaded by squalid slums, and this is where the uncertain urban immigrant most often finds a first (and frequently permanent) abode in a makeshift shack without even the most basic amenities and sanitary facilities. Many move in with relatives who have already made the transition but whose dwelling can hardly absorb yet another family. And unemployment is persistently high, often exceeding 25 percent of the available labor force. Jobs for unskilled workers are hard to find and pay minimal wages. But the people still come, the overcrowding in the shacktowns worsens, and the threat of epidemic-scale disease rises. It has been estimated that in-migration accounts for over 50 percent of urban growth in some developing countries, and in South America urban populations exhibit unusually high natural growth rates as well.

THE "LATIN" AMERICAN CITY

Although the urban experience has been varied in the South and Middle American realms because of diverse historical, cultural, and economic influences, there are many common threads that have prompted geographers to search for useful generalizations. One is the model of the intraurban spatial structure of the **Latin American city** proposed by Ernst Griffin and Larry Ford (Fig. 6-8), which may have even wider application to cities throughout the world's developing realms.

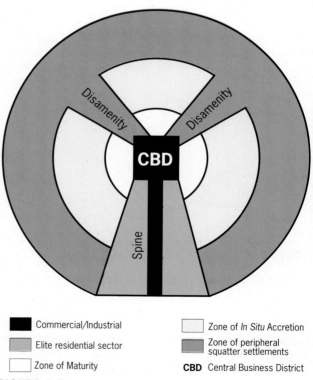

A GENERALIZED MODEL OF LATIN AMERICAN CITY STRUCTURE

■ Commercial/Industrial

▨ Elite residential sector

□ Zone of Maturity

□ Zone of *In Situ* Accretion

▨ Zone of peripheral squatter settlements

CBD Central Business District

FIGURE 6-8

The central plaza, dominated by the cathedral, is a hallmark of the "Latin" American city. Shown here is the *Plaza de Armas* of Cuzco in the mountainous interior of Peru, the city that was the headquarters of the old Inca Empire. Here, as elsewhere in South America, the Iberian urban form was superimposed on the indigenous settlement landscape during the colonial era.

The basic spatial framework of city structure, which blends traditional elements of South and Middle American culture with modernization forces now reshaping the urban scene, is a composite of radial sectors and concentric zones. Anchoring the model is the thriving CBD, which, like its European counterpart, remains the primary business, employment, and entertainment focus of the surrounding metropolitan agglomeration. The landscape of the CBD contains many modern high-rise buildings but also mirrors its colonial beginnings. As we saw on page 269, when the Spanish colonizers laid out their New World cities, they created a central square or *plaza* dominated by a church and flanked by imposing government buildings (see photo above). Lima's *Plaza de Armas*, Bogotá's *Plaza Bolívar*, Montevideo's *Plaza de la Constitución*, and Buenos Aires's *Plaza de Mayo* are classic examples. Early in the South American city's development, the plaza formed the hub and focus of the city, surrounded by shopping streets and arcades. Eventually the city outgrew its old center and new commercial districts formed elsewhere within the CBD, leaving the plaza to serve as a largely ceremonial link with the past.

Emanating outward from the urban core along the city's most prestigious axis is the commercial *spine*, which is surrounded by the *elite residential sector* (shown in green). This widening corridor is essentially an extension of the CBD, featuring offices, shopping, high-quality housing for the upper and upper-middle classes, restaurants, theaters, and such amenities as parks, zoos, and golf courses that give way to wealthy adjoining suburbs, which carry the elite sector beyond the city limits.

The three remaining concentric zones are home to the less fortunate residents of the city (who comprise the great majority of the urban population), with income levels and housing quality decreasing markedly as distance from the city center increases. The *zone of maturity* in the inner city contains the best housing outside the spine sector, attracting middle-class urbanites who invest sufficiently to keep their solidly built but aging dwellings from deteriorating. The adjacent *zone of in situ accretion* is one of much more modest housing interspersed with unkempt areas, representing a transition from inner-ring affluence to outer-ring poverty. The residential density of this zone is usually quite high, reflecting the uneven assimilation of its occupants into the social and economic fabric of the city.

The outermost *zone of peripheral squatter settlements* is home to the impoverished and unskilled hordes that have recently migrated to the city from rural areas. Although housing in this ring mainly consists of teeming, high-density shantytowns, residents here are surprisingly optimistic about finding work and eventually bettering their living conditions—a realistic aspiration documented by researchers who confirm a process of gradual upgrading as squatter communities mature.

A final structural element of many South American cities is the *disamenity sector* that contains relatively unchanging slums, known as *barrios* or *favelas*. The worst of these poverty-stricken areas often include sizeable numbers of people who are so poor that they are forced literally to live in the streets. Thus the realm's cities present almost inconceivable contrasts between poverty and affluence, squalor and comfort; this harsh juxtaposition is frequently observed in the cityscape, as the photo on p. 319 underscores.

As with all geographic models, this particular construct can be criticized for certain shortcomings. Among the criticisms directed at this generalization are the need for a sharper sectoring pattern and a stronger time element involving a multiple-stage approach. Others view this work as a successful abstraction not only of the "Latin" American city but of the Third World city in general. Later, in Chapter 11 (p. 577), we will explore this topic further when we consider a similar model developed for the Southeast Asian city, which incorporates the lasting spatial imprints of colonialism on contemporary urban structure.

◆ SOUTH AMERICA'S REGIONAL GEOGRAPHY

We turn now to the countries of South America and the principal characteristics of their regional geography. In general terms it is possible to group the realm's countries into regional units (Fig. 6-5) because several have qualities in common. Accordingly, the northernmost Caribbean countries,

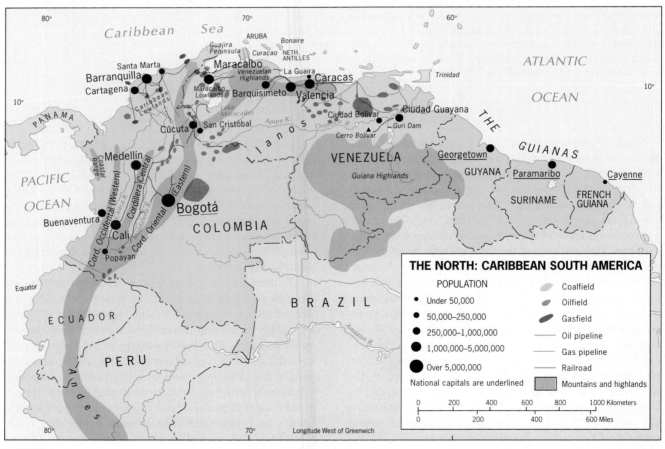

FIGURE 6-9

Venezuela and Colombia, form a regional unit that might also include neighboring Guyana, Suriname, and French Guiana. On the basis of their Amerindian cultural heritage, Andean physiography, and modern populations, the western republics of Ecuador, Peru, and Bolivia constitute a regional entity, to which Paraguay may be added on grounds to be discussed later. In the south, Argentina, Uruguay, and Chile have a common regional identity as the realm's mid-latitude, most strongly Europeanized states. And Brazil by itself constitutes a geographic region in South America because it contains half the continent's land and people; in fact, we will treat that huge country separately in a vignette following the end of this chapter.

THE NORTH:
CARIBBEAN SOUTH AMERICA

As another look at Fig. 6-6 confirms, the countries of the northern tier have something in common besides their coastal location: each has a tropical plantation area signifying early European plantation development, the in-migration of black laborers, and the absorption of this African element into the population matrix. Not only did myriad black workers arrive; many thousands of Asians (from India) also came to South America's northern shores as contract laborers. The pattern is familiar: in the absence of large local labor sources, the colonists turned to slavery and indentured workers to serve their lucrative plantations. Between Spanish Venezuela and Colombia on the one hand and the non-Hispanic "three Guianas" on the other, there is this difference: in the former, the population center of gravity soon moved into the interior and the plantation phase was followed by a totally different economy. In the Guianas, on the other hand, coastal settlement and the plantation economy still predominate (see the box on p. 320).

Venezuela and Colombia have what the Guianas lack (Fig. 6-9): their territories and populations are much larger, their natural environments more varied, their economic opportunities greater. Each also has a share of the Andes Mountains (Colombia's being larger), and each produces oil from reserves that rank among the world's major deposits.

Venezuela

Much of what is important in Venezuela is concentrated in the northern and western parts of the country, where the Venezuelan Highlands form the eastern spur of the north end of the Andes system. Most of Venezuela's 19.8 million people are concentrated in these uplands, which include the capital of Caracas (3.0 million), its early rival Valencia (1.3 million), the commercial and industrial center of Barquisimeto (737,000), and San Cristóbal (286,000) near the Colombian border.

The Venezuelan Highlands are flanked by the Maracaibo Lowlands and Lake Maracaibo to the northwest and by a

Extreme social contrasts, such as this scene near the edge of Caracas (Venezuela's capital), often mark South America's teeming cityscapes where the elite residential sector adjoins lower-income zones.

THE GUIANAS

On the north coast of South America, three small countries lie east of Venezuela and next to northernmost Brazil (Fig. 6-9). In "Latin" South America, these territories are anomalies of a sort: their colonial heritage is Western European, and formerly they were known as British, Dutch, and French Guiana—and called the three *Guianas*. But two of them are independent now: Guyana, Venezuela's neighbor, and Suriname, the Dutch-influenced country in the middle. The easternmost territory, French Guiana, continues under colonial rule.

None of the three countries has a population over 1 million. Guyana is the largest with 840,000, Suriname has 485,000 inhabitants, and French Guiana a mere 110,000. Culturally and spatially, patterns here are Caribbean: peoples of South Asian and African descent are in the majority, with whites forming a small minority. In Guyana, Asians make up just over half the population, blacks and mixed Africans 43 percent, and the others, including Europeans, are tiny minorities. In Suriname, the ethnic picture is even more complicated, for the colonists brought not only Indians from South Asia (now 38 percent of the population) and blacks (31 percent) but also Indonesians (15 percent) to the territory in servitude; the remaining 10 percent of the Surinamese population consists of a separate category of black communities, peopled by descendants of African slaves (known as Bush Negroes) who escaped from the coastal plantations and fled into the forests of the interior. French Guiana is the most European of the three countries; about three-quarters of its small population speak French, and the ethnic majority (68 percent) is of mixed African, Asian, and European ancestry—the *creoles*.

French Guiana, an overseas *département* of France, is the least developed of the three Guianas. There is a small shrimping industry, and some lumber is exported; however, food must be brought in from abroad. By contrast, Suriname has progressed more rapidly despite continuing political instabilities. The plantation economy has given way to production on smaller farms where enough rice (the country's staple crop) is grown to make Suriname self-sufficient. Bananas and citrus fruits are exported, and a variety of fruits and vegetables for the home market are raised on farms in the coastal zone where once sugar plantations prevailed. But Suriname's major income earner is bauxite, mined in a zone across the middle of the country. British-influenced Guyana became independent in 1966 amid internal conflict that basically pitted people of African origins against people of Asian descent. Such problems have continued to plague this country in which the great majority of the people live in small villages near the coast. Bauxite and plantation products earn much of the country's annual revenues.

Unsettled boundary problems affect all three countries. Venezuela has laid claim to all of mineral-rich western Guyana—the Essequibo region, purportedly "stolen" by the British in 1899; this lingering territorial dispute shows no sign of resolution. Guyana, in turn, claims a corner of southwestern Suriname. And part of the border between Suriname and French Guiana also is under contention.

---

large region of savanna country called the *llanos* in the Orinoco Basin to the south and east. The Maracaibo Lowland, once a disease-infested, sparsely peopled coastland, is today one of the world's leading oil-producing areas; much of the oil is drawn from reserves that lie beneath the shallow waters of the lake itself. Actually, Lake Maracaibo is a misnomer, for the "lake" is open to the ocean and is in fact a gulf with a very narrow entry. Venezuela's second city, Maracaibo (1.6 million), is the focus of the oil industry that transformed the Venezuelan economy in the 1970s. After 1980, however, the country suffered the consequences of the global oil depression; like Mexico, Venezuela borrowed heavily against its future oil revenues and still faces the economic burdens of paying off its foreign debt.

The *llanos* on the southern side of the Venezuelan Highlands and the Guiana Highlands in the country's southeast are two of those areas that contribute to South America's image as "underpopulated" and "awaiting development." Whereas the *llanos* are beginning to share in Venezuela's oil production (large reserves have been discovered here), the superior commercial agricultural potential of these savannas and of the *templada* areas (see the box entitled "Altitudinal Zonation" on p. 292) of the Guiana Highlands has yet to be realized. Economic integration of this interior zone with the rest of Venezuela has been encouraged by the discovery of rich iron ores on the northern flanks of the Guiana Highlands, chiefly near Cerro Bolívar. Local railroads connect with the Orinoco River, and from there ores are shipped directly to steel plants in the United States. Ciudad Guayana, barely four decades old, has nearly 750,000 inhabitants; neighboring Ciudad Bolívar today is home to about 280,000 residents. The nearby Guri Dam has been put into service and now supplies about half of Venezuela's electricity. These clusters of development notwithstanding, economic growth has hardly begun to spread outward to surrounding rural areas.

Colombia

Colombia also has a vast area of *llanos* covering about 60 percent of the country. As in Venezuela, this is comparatively empty land far from the national core and much less productive than it could be. Eastern Colombia consists of the upper basins of major tributaries of the Orinoco and Amazon rivers; it lies partly under savanna and partly under rainforest vegetation. Although the Colombian government is promoting settlement east of the Andes, it will be a long time before any part of eastern Colombia matches the Andean zone of the country or even the Caribbean Lowlands of the north. These two regions contain the vast majority of Colombians, and here lie the major cities and productive areas.

Western Colombia is dominated by mountains, but there is a regularity to this terrain. For the most part there are four parallel ranges, generally aligned north-south and separated by wide valleys. The westernmost of these ranges is a coastal belt, less continuous and lower than the other three, which constitute the true Andean mountain chain of Colombia. In this country the Andes separate into three ranges: the Eastern (Oriental), Central, and Western (Occidental) Cordilleras (Fig. 6-9). The valleys between these Cordilleras open northward onto the Caribbean Lowland, where the two important Colombian ports of Barranquilla (1.1 million) and Cartagena (655,000) are located.

Colombia's population of 35.6 million consists of more than a dozen separate clusters, some of them in the Caribbean Lowland, others in the inter-Cordilleran valleys, and still others in the intermontane basins within the Cordilleras themselves. This insular distribution existed before the Spanish colonizers arrived, with the Chibcha civilization (Fig. 6-3) concentrated in the intermontane basins of the Eastern Cordillera. The capital city, Bogotá (corrupted from the Chibchan word *Bacatá*), was founded in one of the major basins in this Andean range at an elevation of 8,500 feet (2,700 m). For centuries the Magdalena valley between the Cordillera Oriental and the Cordillera Central was a crucial link in the key cross-continental route that began in Argentina and ended at the port of Cartagena, and Bogotá benefited from its position along this artery. Today, the Magdalena valley is still Colombia's leading transport corridor, but Bogotá's connections with much of the rest of Colombia still remain quite tenuous. Nonetheless, the capital centers a major metropolis that is home to 5.6 million.

Colombia's physiographic variety is matched by its demographic diversity. In the Andean south, it has a major cluster of Amerindian inhabitants, and here the country begins to resemble its southern neighbors. In the northern coastlands, traces remain of the plantation period and the African population it brought. Bogotá is a great Hispanic cultural headquarters, whose influence extends beyond the country's borders. In the Cauca valley between the Cordillera Central and Cordillera Occidental, the city of Cali (1.8 million) is the commercial focus for a hacienda district where sugar and tobacco are grown. Farther north, the Cauca River flows through a region comprising the provinces of Antioquia and Caldas, whose urban focus is the textile manufacturing city of Medellín (1.7 million) but whose greater importance to the Colombian economy entails the production of coffee. With its extensive *templada* areas along the Andean slopes, Colombia today is the world's second-largest producer of coffee (neighboring Brazil ranks first). In Antioquia-Caldas, coffee is grown on small farms by a remarkably unmixed European population cluster; elsewhere, it is produced on the haciendas so common in this geographic realm.

Following coffee, oil and coal are Colombia's other two leading exports. Petroleum deposits have been worked for decades in the north, adjacent to the (still-disputed) Venezuelan border where the Maracaibo oilfields extend into Colombia. During the mid-1980s, significant discoveries were made in the northeastern province of Arauca (about 200 miles [320 km] south of Lake Maracaibo), and development began immediately; a 500-mile pipeline was quickly completed across the northern Andes to the Caribbean coast, and Colombia was turned into a net oil exporter overnight. In 1991, an even larger oil deposit was discovered in the Cusiana region on the western fringes of the *llanos* about 100 miles (160 km) northeast of Bogotá. Unfortunately, the Arauca and Cusiana oilfields (which also contain large supplies of natural gas) are threatened by local guerrilla insurgencies that have long operated in Colombia's *llanos*; pipelines and other facilities have been attacked, and the future of energy production in this area remains uncertain.

Colombia has had more success exploiting its other recently discovered fossil fuel—coal, a resource that is virtually absent from the South American continent. These rich coal deposits are fortuitously located on the Guajira Peninsula, which juts out into the Caribbean between Lake Maracaibo and the coastal strip that contains Colombia's leading ports; production here has expanded quite rapidly, and the operation at Cerrejón has become one of the world's largest bituminous coal mines. Colombia, of course, is also very well known as a major source of illicit drugs, particularly co-

THE GEOGRAPHY OF COCAINE

It is impossible to discuss northwestern South America today without reference to one of its most widespread activities: the production of illegal narcotics. Of the enormous flow of illicit drugs into the United States each year, much of the heroin and marijuana originate in mainland Middle America; but the most widely used of these illegal substances undoubtedly is cocaine, all of which comes from South America—mainly Bolivia, Peru, and Colombia (the sources of more than 75 percent of the total world supply). Within these three countries, cocaine annually brings in billions of (U.S.) dollars and "employs" thousands of workers, constituting an industry that functions as a powerful economic force. Moreover, the notorious Medellín and Cali cartels, together with other wealthy drug barons who operate the industry, have accumulated considerable power through bribery of politicians, threats of heightened terrorism, and even alliances with guerrilla movements in outlying areas beyond governmental control. The cocaine industry itself is structured within a tightly organized network of territories that encompass the various stages of this narcotic's production.

The first stage of cocaine production is the extraction of coca paste from the coca plant, a raw-material-oriented activity that is located near the areas where the plant is grown. The coca plant was domesticated in the Andes by the Incas centuries ago for use as a stimulant (and in certain rituals), and it is still cultivated and used widely today by the Incas' descendants. Millions of upland peasants chew coca leaves for stimulation and to ward off altitude sickness; they also brew the leaves into coca tea (the area's leading beverage), which they use as a general-purpose medicinal.

The leading zone of coca-plant cultivation is along the eastern slopes of the Andes and adjacent tropical lowlands in Bolivia and Peru. A lesser variety of the plant is now grown in the same physiographic setting in Ecuador and southern Colombia—and increasingly in the nearby upper Amazon Basin of Brazil, which continues to emerge as a major source area. Today, three main areas dominate in the growing of coca leaves for narcotic production (Fig. 6-10): Bolivia's Chaparé district (in the marginal Amazon lowlands northeast of the city of Cochabamba), the Yungas highlands (north of the Bolivian capital, La Paz), and Peru's upper Huallaga valley around the town of Tingo Maria (about 200 miles north of Lima). These areas thrive because they combine many favorable conditions: local environments are conducive to high leaf yields that produce four crops per year (instead of the usual two), and plants here have developed near-total immunity to both disease and insect ravages. The success of these areas has enabled their Amerindian populations to turn coca plants into a cash crop, and operations often reach the scale of plantations, thereby inducing subsistence farmers from other areas to migrate into these now-specialized coca-raising regions. In fact, profitable coca cultivation has led thousands of peasants to abandon the far less lucrative production of food crops, further reducing the capability of nutrition-poor Bolivia to feed itself. Another undesirable side-effect is environmental damage in Peru and Bolivia—notably deforestation and soil erosion—as large areas of wilderness are swiftly transformed into coca-plant "farms." The coca leaves harvested in these eastern Andean regions make their way to nearby centers where coca paste is extracted and prepared; these centers are located at the convergence of rivers and trails, including the Beni rainforests of northernmost Bolivia and particularly the central Bolivian city of Santa Cruz (Fig. 6-10).

The second stage of production involves refining the coca paste (about 40 percent pure cocaine) into cocaine hydrochloride (more than 90 percent pure), a lethal concentrate that is diluted with substances such as sugar or flour before being sold on the streets to consumers. Cocaine refining requires sophisticated chemicals, carefully controlled processes, and a labor force skilled in their supervision, and here Colombia predominates. Much of this activity takes place within its borders—sometimes in the cities, but mostly in the ultramodern forest complexes in the southeastern lowlands where alliances between drug barons and guerrillas help protect against the intrusion of the Bogotá government. Colombia also has other geographic advantages: proximity to the remote upper Amazon rainforests across the Brazilian border (where the cocaine industry is rapidly expanding), and smuggling access to the United States, a short hop by air or sea across the Caribbean.

The final stage of production entails the distribution of cocaine to the U.S. marketplace, which depends on an efficient (but clandestine) transportation network. One favorite avenue for smugglers is to use commercial air and sea links between South America and the United States, concealing the product in countless imaginative ways. More likely, however, is the use of private planes that operate directly out of remote airstrips near the refineries in places beyond effective government control. United States drug interdiction records show that about 75 percent of all cocaine coming into the country travels by noncommercial plane or boat, with Colombia most often the origin but with Brazil steadily increasing its role as a refining and distribution center for outbound "coke." (The same is now happening in Ecuador and Venezuela as new opportunities multiply in the rapidly growing European cocaine market.)

The transport of cocaine to the United States is often a two-step process, making use of such intermediate Mid-dle American transshipment points as the Bahamas, Jamaica, Cuba, and many countries on the mainland. Although South Florida is believed to be a leading port of illegal entry, other coastal points in the U.S. Southeast as well as cities along the Mexican border in the Southwest have recorded steady increases in cocaine smuggling over the past few years. Judging by recent drug seizures in Southern California, the overland route through mainland Middle America has experienced a major upsurge in trafficking during the 1990s.

caine, which undoubtedly ranks among its leading exports as well (see box at left).

Venezuela and Colombia both exhibit a pronounced clustering of often-isolated populations, share a relatively empty interior, and depend on a small number of products for the bulk of their export revenues. The majority of the people in these countries practice subsistence agriculture and labor under the social and economic inequalities common to most of Iberian America.

THE WEST: ANDEAN SOUTH AMERICA

The second regional grouping of South American states encompasses Peru, Ecuador, Bolivia, and Paraguay (Fig. 6-10), four contiguous countries that include South America's only landlocked republics (the latter two). The map of culture spheres (Fig. 6-6) shows the Amerind-subsistence region extending along the Andes Mountains, indicating that these countries have large Amerindian components in their populations. Just over half the people of Peru (population: 23.5 million) are of Amerindian stock, and in Ecuador and Bolivia the figure is also around 50 percent; in Paraguay it is about 95 percent. However, all these percentages are only approximate because it is often impossible to distinguish between Amerindian and "mixed" people of strong Amerindian character. There are, however, other similarities among these countries: their incomes are low; they are comparatively unproductive; and, unhappily, they exemplify the grinding poverty of the landless peonage—a problem that looms large in the future of Ibero-America. Moreover, as noted earlier, these are South America's least urbanized countries; only Lima (7.5 million), the capital of Peru, ranks with Bogotá and Rio de Janeiro as a major-scale metropolis.

Peru

In terms of territory as well as population, Peru is the largest of the four republics. Its half-million square miles (1.3 million sq km) divide both physiographically and culturally into three subregions: (1) the desert coast, the European-mestizo region; (2) the Andean highlands or Sierra, the Amerindian region; and (3) the eastern slopes and adjoining *montaña*, the sparsely populated Amerindian-mestizo interior (Fig. 6-10). It is symptomatic of the cultural division still prevailing in Peru that Lima is not located centrally in a populous basin of the Andes but in the coastal zone; in 1986, Peru's president did propose shifting the capital—to the lush eastern Andean slopes—but no legislative action followed.

At Lima in the heart of the desert coastal strip, the Spanish avoided the greatest of the Amerindian empires, choosing a site some 8 miles (13 km) inland from a suitable anchorage that became the modern outport of Callao. From an economic point of view, the Spanish choice of a headquarters on the Pacific coast proved to be sound, for the coastal region has become commercially the most productive part of the country. A thriving fishing industry based on the cool productive waters of the Peru (Humboldt) Current offshore contributes significantly to the export trade. Irrigated agriculture in some 40 oases distributed all along the arid coast produces cotton, sugar, rice, vegetables, fruits, and wheat; the cotton and sugar are important export products, and the other crops are grown mostly for the domestic market.

The Andean (Sierra) region occupies about one-third of the country and contains the majority of Peru's Amerindian peoples, most of them Quechua-speaking. However, despite the size of its territory and population (nearly half of all Peruvians reside here), the political influence of this region is slight, as is its economic contribution (except for the mines).

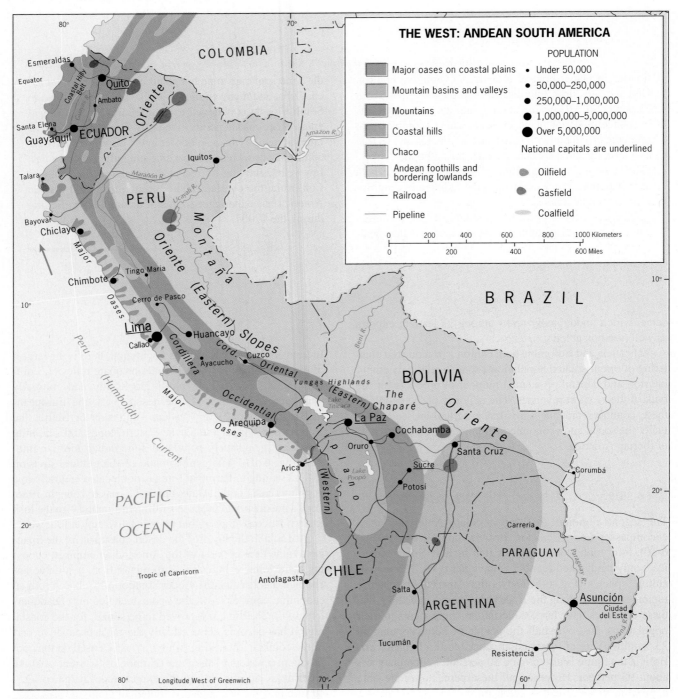

FIGURE 6-10

In the high valleys and intermontane basins, the Amerindian population is clustered either in isolated villages, around which people practice a precarious subsistence agriculture, or in the more favorably located and fertile areas where they are tenants, peons on white- or mestizo-owned haciendas. Most of these people never receive an adequate daily caloric intake or balanced diet of any sort; the wheat produced around Huancayo, for instance, is sent to Lima's European market and would be too expensive for the Amerindians themselves to buy. Potatoes (which can be grown at altitudes of up to 14,000 feet or 4,250 m), barley, and corn are among the subsistence crops here in this *fría* zone, and in the *puna* of the higher basins the Amerindians graze their llamas, alpacas, cattle, and sheep (these vertical zones are discussed in the

INSURGENT STATES

Enormous contrasts exist between Peru's productive, European-mestizo-dominated coastal region and the desolation of the Amerindian-dominated Andean highlands. Such inequities—heightened by rigid social-class systems and repressive government—can lead to insurgency, and this has happened a number of times in the recent history of South and Middle America. Peru has experienced such an upheaval since the early 1980s, but events follow a similar pattern whatever the country. First there are reports of armed rebellions in remote places where the government has little presence. Soon these localized uprisings begin to attain permanence, and the rebels manage to control parts of the country on a more or less continuous basis. Full-scale civil war follows, and the government may fall. Political scientists identify these three stages as: (1) *contention*, the period of initial rebellion; (2) *equilibrium*, when critical areas of the country are in rebel hands; and (3) *counteroffensive*, the stage of ground and air war that will decide the future.

These stages were translated into spatial terms by Robert McColl, who suggested that the equilibrium stage actually marks the emergence of an **insurgent state**, a state within a state (Fig. 6-11). The rebel-controlled insurgent state may not be contiguous, that is, it may be composed of several separate cores; but it undeniably has a headquarters, boundaries, governmental structure and systems of administration, hospitals, schools, and even direct communications with (possibly sympathetic) neighboring countries.

This was the sequence of events in Cuba more than 30 years ago, when then-rebel Fidel Castro managed to establish a stable and expanding insurgent state on the island. In Nicaragua during the 1970s, a rebel group known as the Sandinistas challenged a dictatorial government—first in remote locales, then in a substantial insurgent state, and finally in a full-scale conflict that led to a Sandinista takeover of the country.

In Peru during the 1980s, the Sendero Luminoso rebels created an insurgent state focused on the town of Ayacucho in the west-central Andes. By 1993, that state within a state had expanded to encompass three key sections of the country's Andean zone, with guerrilla operations (in a much wider area of contention) even reaching the outskirts of Lima itself (Fig. 6-11, right map). The rebel drive toward the counteroffensive stage was dealt a setback in 1992 when the government captured Sendero's leader, Abimael Guzmán. But the insurgent state remains well entrenched: it may not appear on the official map of Peru, but its presence is etched in the unusually limited interregional circulation of people and goods.

THE INSURGENT STATE: MODEL AND REALITY

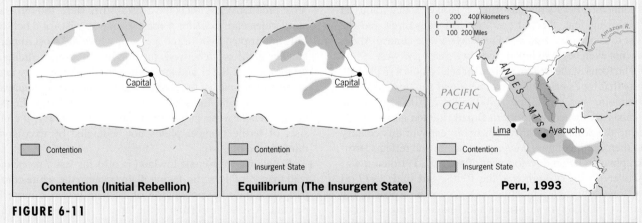

Contention (Initial Rebellion) — Contention

Equilibrium (The Insurgent State) — Contention / Insurgent State

Peru, 1993 — Contention / Insurgent State

PACIFIC OCEAN · Lima · Ayacucho · ANDES MTS. · Amazon R.

0 200 400 Kilometers
0 100 200 Miles

FIGURE 6-11

box on p. 292). The major mineral products from the Sierra are copper, silver, lead, and several other metallic minerals, with the largest mining complex centered on Cerro de Pasco.

The Peruvian Andes are also the setting for South America's most ominous insurgency, the radical-communist *Sen-*

dero Luminoso (Shining Path). This guerrilla movement, which built its support by exploiting the desperate poverty of the Amerindians, established a territorial base in the southern and central Andes (see the box above entitled "Insurgent States"). Today, the Sendero insurgency has spilled out of its

In the Inca heartland: market day in the village of Pisac. Study this interesting photograph for its many geographic qualities: the barrenness of the mountains, the buildings and their graffiti, the modes of dress, the produce from near and far. The stone floor probably is a pre-European surface. The place has electricity. Its fate today is to lie in the heart of the area afflicted by the murderous Sendero Luminoso insurgency.

mountain stronghold to threaten Lima, other coastal cities, and much of the rest of Peru (see Fig. 6-11, right map).

Of Peru's three subregions, the *Oriente* or East—the inland slopes of the Andes and the Amazon-drained, rainforest-covered *montaña*—is the most isolated. A look at the map of permanent (as opposed to seasonal) routes shows how disconnected Peru's regions still are. However marvelous an engineering feat, the railroad that connects Lima and the coast to Cerro de Pasco and Huancayo in the central Andes does not even begin to link the country's east and west.

The focus of the eastern region, in fact, is Iquitos (311,000), a city that looks east rather than west and can be reached by oceangoing vessels sailing 2,300 miles (3,700 km) up the Amazon River across northern Brazil. Iquitos grew rapidly during the Amazon wild-rubber boom early in this century and then declined; now it is growing again and reflects Peruvian plans to begin development of the east. Petroleum was discovered west of Iquitos in the 1970s, and since 1977 oil has flowed through a pipeline built across the Andes to the Pacific port of Bayovar. Meanwhile, traders plying the navigable Peruvian rivers above Iquitos continue to collect such products as chicle, nuts, rubber, herbs, and special cabinet woods.

Ecuador

On the map Ecuador, smallest of the four republics, appears to be just a corner of Peru. But that would be a misrepresentation because Ecuador possesses a full range of regional contrasts. It has a coastal belt; an Andean zone that may be narrow (under 150 miles or 250 km) but by no means of lower elevation than elsewhere; and an *Oriente*—an eastern region that is just as empty and as undeveloped as that of Peru. As in Peru, the majority of the people of Ecuador are concentrated in the Andean intermontane basins and valleys, and the most productive region is the coastal strip. Here, though, the similarities end. Ecuador's coastal zone consists of a belt of hills interrupted by lowland areas, of which the most important lies in the south between the hills and the Andes, drained by the Guayas River and its tributaries. The largest city and commercial center of the country (but not the capital), Guayaquil (2.0 million), forms the focus for this subregion. Ecuador's lowland west, moreover, is not desert country; it consists of fertile tropical plains not bedeviled by excessive rainfall (see Fig. I-6).

Ecuador's west-coast lowland is also far less Europeanized than Peru's Pacific-facing plain because the white component of the total population of 10.5 million is a mere 8 percent. A sizeable portion of the latter is engaged in administration and hacienda ownership in the interior, where most of the 45 percent of the Ecuadorians who are Amerindian also reside. Of the remaining national population, 7 percent is black and mulatto and 40 percent is mestizo (many with a stronger Amerindian ancestry).

The products of this coastal region differ as well from those of Peru. Ecuador is one of the world's leading banana exporters; small farms owned by black and mulatto Ecuadorians and located in the north near Esmeraldas contribute to this total, as do farms on the eastern and northern margins of the Guayas Lowland. Cacao is another important lowland crop, and lucrative coffee is grown on the coastal hillsides as well as in the Andean *templada* zones; cotton and rice are also cultivated, and cattle can be raised. In recent years, the production of petroleum in the jungles of the *Oriente* region has reached substantial proportions, and a trans-Andean pipeline was opened in the 1970s to connect the rich interior oilfields to the port of Esmeraldas. As a result, Ecuador joined the Organization of Petroleum Exporting Countries (OPEC) (which it recently left) and became South America's second-largest exporter of crude oil.

Ecuador is not a poor country, and the coastal strip has undergone vigorous development in recent years. But the Andean interior, where the white and mestizo administrators and hacienda owners are outnumbered by the Amerindians by about four or five to one, is a different story—or rather, a story similar to that of Andean regions in the other countries of western South America. Quito, the capital city (1.8 million), lies in one of the several highland basins in which the Andean population is clustered. Its functions remain primarily administrative, because productivity in the Andean region is too low to stimulate commercial and industrial development. The Ecuadorian Andes do differ from those of Peru in that they apparently lack major mineral deposits. Despite the completion of a railroad linking Quito to Guayaquil on the Pacific coast, the interior of Ecuador remains isolated—and, economically, comparatively inert.

Bolivia

From Ecuador southward through Peru, the Andes broaden until in Bolivia they reach a width of some 450 miles (720 km). In both the Cordillera Oriental and the Cordillera Occidental, peak elevations in excess of 20,000 feet (6,000 m) are recorded; between these two great ranges lies the *Altiplano* proper (this term was defined on p. 302). On the boundary between Peru and Bolivia, freshwater Lake Titicaca—the highest large lake on earth—lies at 12,507 feet (3,700 m) above sea level. Here, in its west, lies the heart of modern Bolivia; here, too, lay one of the centers of Inca civilization—and indeed of pre-Inca cultures. Bolivia's capital, La Paz (1.2 million), is also located on the *Altiplano* at 11,700 feet (3,570 m), one of the highest cities in the world.

Lake Titicaca is what helps make the *Altiplano* livable, for this large body of water ameliorates the coldness in its vicinity, where the snow line lies just above the plateau surface. On the surrounding cultivable land, grains have been raised for centuries in the Titicaca Basin to the extraordinary elevation of 12,800 feet (3,850 m), and to this day this area of Peru and Bolivia supports a major cluster of Amerindian subsistence farmers.

Modern Bolivia is the product of the European impact, an influence that has bypassed some of the Amerindian population clusters. Of course, these indigenous Bolivians no more escaped the loss of their land than did their Peruvian or Ecuadorian counterparts, especially east of the *Altiplano*. What made the richest Europeans in Bolivia wealthy, however, was not land but minerals. The town of Potosí in the Cordillera Oriental became a legend for the immense deposits of silver in its vicinity; copper, zinc, and several alloys were also discovered there. Over most of the past century, Bolivia's tin deposits (which ranked among the world's richest) yielded much of the country's annual export income; but since 1980, declining tin reserves and falling world prices reached the point where Bolivia's outdated mining methods have forced nearly all of the industry to shut down. However, oil and particularly natural gas are now accounting for a growing share of foreign revenues. Bolivia exports gas to Argentina and Brazil; in return, Brazil is committed to assisting the development of the corridor between Santa Cruz (606,000) and Corumbá (Brazil) in the southeastern lowlands, where commercial agriculture—notably soybean production—is on the rise.

Bolivia has had a turbulent history. Apart from endless internal struggles for power, the country first lost its window on the Pacific coast in a disastrous conflict with Chile, then lost its northern territory of Acre to Brazil in a dispute involving the rubber boom in the Amazon Basin, and finally lost 55,000 square miles (140,000 sq km) of southeastern Gran Chaco territory to Paraguay. The most critical, by far, was the loss of its outlet to the sea over a century ago—which may yet be regained from Chile if longstanding negotiations can be concluded successfully.

Although Bolivia has rail connections to the Chilean ports of Arica and Antofagasta, it remains severely disadvantaged by its landlocked situation. Because the Cordillera Occidental and the *Altiplano* form the country's inhospitable western margins, one might suppose that Bolivia would look eastward and that its *Oriente* might be somewhat better developed than that of Peru or Ecuador, but such is not the case (the booming cocaine industry notwithstanding). The densest settlement clusters within its dispersed population of 8.2 million are in the valleys and basins of the Cordillera Oriental, where the mestizo sector is also stronger than elsewhere in the country. Cochabamba (485,000) Bolivia's third-largest city, lies in a basin that forms the country's largest concentration of population; Sucre, the legal capital (La Paz is the *de facto* capital), is located in another. Here, of course, lie major agricultural districts between the barren *Altiplano* to the west and the recently opened tropical savannas to the east.

Paraguay

Paraguay is the only non-Andean country in this region, but ethnically it is no less Amerindian. Of its 4.8 million people, more than 95 percent are mestizo, but with so pervasive an Amerindian influence that any white ancestry is almost totally submerged. Although Spanish is Paraguay's official language, Amerindian Guaraní is more commonly spoken (in fact, the country is probably the world's most completely bilingual). By any measure, Paraguay is the poorest of the four countries of western South America, although it does have opportunities for pastoral and agricultural industries that have thus far gone unrealized. One of the reasons for this must be isolation—the country's landlocked position.

Paraguay's exports, in their modest quantities, have to be shipped through Argentina's Buenos Aires, a long haul via the Paraguay-Paraná River from the capital of Asunción (1.2 million). Cotton, soybeans, meat (dried and canned), and timber reach foreign markets. Grazing in the northern Chaco region is the most important commercial activity, but here cattle generally do not compare favorably to those of Argentina. Hoped-for development in the fertile Paraguayan east around Itaipu Dam on the Paraná border with Brazil (see p. 341) has not yet materialized. Nonetheless, Paraguay's prospects are improving in the 1990s: decades of dictatorship have given way to more democratic government, foreign investment is being attracted, and the country has joined Brazil, Argentina, and Uruguay to form the new Southern Cone Common Market (*Mercosur*) that will begin operating in 1995.

THE SOUTH: MID-LATITUDE SOUTH AMERICA

Argentina

South America's three southern countries—Argentina, Chile, and Uruguay—are grouped into one region. By far the largest in terms of both population and territory is Argentina, whose 1.1 million square miles (2.8 million sq km) and 33.9 million people rank second only to Brazil in this geographic realm. Argentina exhibits a great deal of physical-environmental variety within its boundaries, and the vast majority of the population is concentrated in the physiographic subregion known as the Pampa. Figure I-9 indicates the degree of clustering of Argentina's inhabitants on the land and in the cities of the Pampa (the word means "plain"). It also shows the relative emptiness of the other six subregions (mapped in Fig. 6-12)—the scrub-forest Chaco in the northwest, the mountainous Andes in the west (along whose crestline lies the boundary with Chile), the arid plateaus of Patagonia south of the Rio Colorado, and the undulating transitional terrain of intermediate Cuyo, Entre Rios (also known as "Mesopotamia"), and the North.

The Argentine Pampa is the product of the past 125 years. During the second half of the nineteenth century, when the great grasslands of the world were being opened up (including those of the United States, Russia, and Australia), the economy of the long-dormant Pampa began to emerge. The food needs of industrializing Europe grew by leaps and bounds, and the advances of the Industrial Revolution—railroads, more efficient ocean transport, refrigerated ships, and agricultural machinery—helped make large-scale commercial meat and grain production in the Pampa not only feasible but highly profitable. Large haciendas were laid out and farmed by tenant workers who would prepare the virgin soil and plant it with wheat and alfalfa, harvesting the wheat and leaving the alfalfa as pasture for livestock. Railroads radiated ever farther outward from the capital of Buenos Aires and brought the entire Pampa into production. Today, this Argentine subregion has South America's densest railroad network, and once-dormant Buenos Aires has become the world's tenth-largest metropolitan complex (12.2 million). Yet the Pampa itself has hardly begun to fulfill its productive potential, which could easily double with more efficient and intensive agricultural practices.

Over the decades, several specialized agricultural areas appeared on the Pampa. As we would expect, a zone of vegetable and fruit production became established near the huge Buenos Aires conurbation located beside the estuary of the Rio de la Plata. To the southeast is the predominantly pastoral district, where beef cattle and sheep are raised. In the drier west, northwest, and southwest, wheat becomes the important commercial grain crop, but half the land remains devoted to grazing. Among the exports, cereals usually lead by value, followed by animal feed, vegetable oils, meat, hides and skins, and wool.

Argentina's wealth is reflected in its fast-growing cities, which epitomize the European-commercial cultural character of the realm's southernmost countries. No less than 86 percent of the Argentine population may be classified as urbanized, an exceptionally high figure for a country in a developing realm. An astonishing 35 percent of all Argentineans now reside in the Greater Buenos Aires conurbation, which also contains most of the industries, many of them managed by Italians, Spaniards, and other immigrants. This dominance over national life, however, has disillusioned many Argentines; a plan to relocate the capital southward to the Patagonian frontier was approved in the mid-1980s but has since been abandoned because its costs are unacceptably high. Metropolitan Córdoba (1.3 million) is also a focus of industrial growth. Much of the manufacturing in the major cities is associated with the processing of Pampa products and the production of consumer goods for the domestic market. One out of every four wage earners in the country is engaged in manufacturing—another indication of Argentina's somewhat advanced economic standing.

Argentina's population shows a high degree of clustering and a decidedly peripheral distribution. The Pampa subregion covers only a little more than 20 percent of Argentina's territory, but with more than three-fourths of the people concentrated here the rest of the country cannot be densely populated. Outside the Pampa, pastoralism is an almost uni-

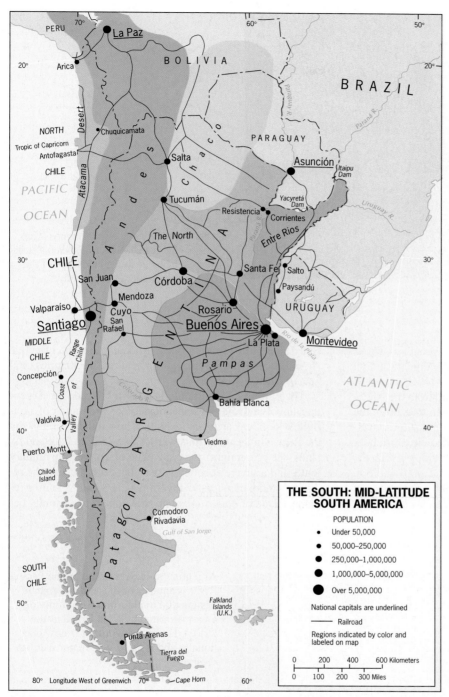

FIGURE 6-12

versal pursuit, but the quality of the cattle is much lower than in the Pampa; in semiarid Patagonia, sheep are raised. Some of the more distant areas are of actual and potential significance to the country: in Patagonia an oilfield is in production near coastal Comodoro Rivadavia; in the far northwest Chaco, Argentina may share with Paraguay significant oil reserves

that are yet to be exploited. Yerba maté, a local tea, is produced in the North and Entre Rios subregions; and the quebracho-tree extract (for tanning leather) that both Paraguay and Argentina export comes from the Paraguay-Paraná valley. Northeastern Argentina is also an area targeted for major industrial development, sparked by newly-completed Yacy-

The street scene in central Buenos Aires, here on the Avenida Florída, bears a striking resemblance to the landscape of the CBD of any major European city.

retá Dam on the Paraná River border with Paraguay (whose hydroelectric potential surpasses that of the gigantic Itaipu Dam farther upstream).

In addition, the streams that flow eastward from the Andes provide opportunities for irrigation. Tucumán (628,000), the focus of Argentina's major sugar-producing district, developed in response to a unique set of physical circumstances and a rapidly growing market in the Pampa cities, which the railroad made less remote. At Mendoza (735,000) and San Juan to the north and San Rafael to the south, vineyards and fruit orchards reflect the economy of the Cuyo subregion.

But despite these sizeable near-Andean outposts, effective Argentina remains the area within a radius of 350 miles (560 km) of Buenos Aires.

The 1980s was a particularly stressful decade for Argentina, involving the "disappearances" of thousands of non-rightist political activists, a humiliating military defeat by the British in the disastrous war for the offshore Falkland Islands (see box at right), the unraveling of a repressive military junta, and a severe economic crisis that accompanied the return to a more democratic government. Today, Argentina remains a land of opportunity and challenge—opportunity be-

Argentina's Yacyretá Dam on the Paraná River is another of those megaprojects designed to transform the economy of a vast surrounding area. Billed as the largest dam in the world, it is now operational—and is flooding a major wildlife reserve.

FALKLANDS OR MALVINAS?

Off the southern tip of Argentina, about 300 miles (500 km) from the mainland, lies a small archipelago that in 1982 suddenly became the scene of a bitter war between the Argentines and the islands' owners, the British. Judging from the map (Fig. 6-12), these Falkland Islands hardly seem likely to occasion a full-scale war. The archipelago consists of two main islands and about 100 smaller ones, whose tiny population of about 1,800 (mainly of English ancestry) ekes out an existence farming sheep in a cold, rainy, treeless environment.

The 1982 conflict over the Islas Malvinas—as the Falkland Islands are called in Argentina—was a reminder of the continuing potential for strife in the aftermath of the colonial era. England first took control of the Falklands in the 1760s but was soon ousted by Spain. When Spanish power in southern South America weakened and the new republic of Argentina emerged, the Spanish abandoned the area, leaving it to the Argentines. In 1820, Argentina claimed the Malvinas and followed this up by establishing a base on one of the main islands. But in 1833, when British seapower was at its zenith, London sent a naval force and expelled the Argentines. In Argentina that action has never been forgotten, and British rule has never been acknowledged.

Negotiations over the future of the islands had been continuing without progress in 1982 when an Argentine force suddenly invaded the Falklands, overpowered the small British garrison, and hoisted Argentina's flag over Stanley, the capital. An outraged British government launched a counterattack by sea and air and, in a war that cost about 200 British and more than 700 Argentinean lives, defeated the invaders a few weeks later.

But the British victory did not resolve the fundamental politico-geographical problem. The Argentines continue to insist that the British takeover of the Malvinas in 1833 remains an unresolved act of war and annexation and that the islands are now legally theirs. The archipelago's nearness to Argentina and its potential importance in maritime boundary-making (see Chapter 11) also figure in Argentina's claim. The British respond that if all forcibly acquired territories in the world were given back to their "original" owners, there would be global chaos; they also point to the close ties of the Falklands' population with Britain and the expressed desires of these settlers to remain under the British flag.

Further negotiations are obviously needed, but the chances for a peaceful resolution of this lingering dispute are not good. The British, after their victory, considerably strengthened the defensive capabilities of the islands and claimed a 150-mile-wide (240-km-wide) "exclusive economic zone" around the islands to protect the Falklands' fisheries. The latter created overlapping zones with Argentinean coastal waters and thereby heightened the risk of incidents between the Argentine navy and foreign fishing vessels as a result of differing interpretations of who controls various marine areas.

---

cause its territory is large, its environments varied, and its resources rich and plentiful; challenge because those advantages have been overshadowed by political instabilities, chronic economic problems, and especially a paralyzing inability to get anything done. Practically every enterprise in Argentina has been bedeviled by these adversities, and the country will have to overcome its disorganization if it is ever to realize anything approaching its full potential.

Uruguay

Uruguay, unlike Argentina or Chile, is compact, small, and fairly densely populated. This buffer state of old became a fairly prosperous agricultural country, in effect a smaller-scale Pampa (though possessing less favorable soils and topography); Figs. I-4 and I-7 show the similarity of physical conditions on the two sides of the Plata estuary. Montevideo, the coastal capital with 1.3 million residents, contains 40 percent of the country's population of 3.2 million; from here, railroads and roads radiate outward into the productive agricultural interior. In the immediate vicinity of Montevideo lies Uruguay's major farming area, which produces vegetables and fruits for the city as well as wheat and fodder crops. Most of the rest of the country is used for grazing sheep and cattle, with wool, hides, and meat dominating the export trade.

Uruguay, of course, is a small country, and its area of 68,000 square miles (176,000 sq km)—less territory even than Guyana—does not leave much room for population clustering. Nevertheless, a special quality of the land area of Uruguay is that it is rather evenly peopled right up to the boundaries with Brazil and Argentina. And of all the countries in South America, Uruguay is the most truly European, notably lacking the racial minorities that mark even Chile and Argentina, but with a sizeable non-Spanish European component in its population.

Chile

For 2,500 miles (4,000 km) between the crestline of the Andes and the coastline of the Pacific lies the narrow strip of land that is the Republic of Chile. On the average just 90 miles (150 km) wide (and only rarely over 150 miles or 250 km in width), Chile is the textbook case of what political geographers call an *elongated state*, one whose shape tends to contribute to external political, internal administrative, and general economic problems. In the case of Chile, the Andes Mountains do form a barrier to encroachment from the east and the sea constitutes an avenue of north-south communication. History has shown the country to be quite capable of coping with its northern rivals, Bolivia and Peru.

As Figs. I-7 and 6-12 indicate, Chile is a three-subregion country. About 90 percent of Chile's 14.1 million people are concentrated in what is called Middle Chile, where Santiago (5.4 million), the capital and largest city (one of the world's smoggiest), and Valparaíso (388,000), the chief port, are located. North of Middle Chile lies the Atacama Desert, which is wider, drier, and colder than the coastal desert of Peru. South of Middle Chile, the coast is broken by a plethora of fjords and islands, the topography is mountainous, and the climate—wet and cool near the Pacific—soon turns drier and colder against the Andean interior. South of the latitude of Chiloé Island there are no permanent overland routes, and there is hardly any settlement; one of the few places in far southern Chile that is now experiencing some development is Punta Arenas on the Strait of Magellan (Fig. 6-12), a growing fishing port, energy producer, and the Western Hemisphere gateway to Antarctica (see box).

These three subregions are also clearly apparent on the map of culture spheres (Fig. 6-6), which displays a mestizo north, a European-commercial zone in Middle Chile, and an undifferentiated south. In addition, a small Amerind-subsistence zone in northern Chile's Andes is shared with Argentina and Bolivia; the Amerindian component (7 percent) in the Chilean population largely originated from the million or so indigenous peoples who lived in Middle Chile.

Some intraregional differences exist between northern and southern Middle Chile, the country's core area. Northern Middle Chile, the land of the hacienda and of Mediterranean climate with its dry summer season, is an area of (usually irrigated) crops that include wheat, corn, vegetables, grapes, and other Mediterranean products; livestock-raising and fodder crops also take up much of the productive land but are giving way to the more efficient and profitable cultivation of fruits for export. Southern Middle Chile, into which immigrants from both the north and from Europe (especially Germany) have pushed, is a better-watered area where raising cattle has predominated; but here, too, more lucrative fruit, vegetable, grain, and other expanding food crops are changing the area's agricultural specializations.

Prior to the 1990s, the arid Atacama region in the north accounted for most of Chile's foreign revenues. The Atacama contains the world's largest exploitable deposits of nitrates, and at first these provided the country's economic mainstay. But this mining industry soon declined after the discovery of methods of synthetic nitrate production early in the twentieth century. Subsequently, copper became the chief export (Chile possesses the world's largest reserves); it is found in several places, but the main concentration lies on the eastern margin of the Atacama Desert near the town of Chuquicamata, not far from the port of Antofagasta.

Chile today is in the midst of a growth boom that is transforming its economic geography. Since the ouster of the re-

Chile is becoming ever more actively involved in the Pacific Basin trading scene. Its leading trade partner today is not neighboring Argentina or Peru, or even the United States, but Japan across the Pacific Ocean. Chilean raw materials (minerals and lumber), fruits, wines, and other goods cross the ocean; but most importantly, manufactured goods (textiles, electronics) are figuring in the export picture. Here, fruits and vegetables are loaded for shipment to Asia.

THE SOUTHERN REALM: ANTARCTICA AND SURROUNDINGS

The Southern realm is the least populous and most remote of the world's geographic regions. It consists of the Antarctic continent and the waters surrounding it—the Southern Ocean. The Antarctic landmass is almost totally covered by an enormous and thick icesheet; the Southern Ocean is a giant swirl of frigid water, moving in an easterly (clockwise) direction around Antarctica. This is not an attractive picture, but the Southern realm has long attracted pioneers and explorers. Antarctica's coasts have been visited by navigators from various countries, by whale and seal hunters, and by explorers who established temporary stations on the margins of the landmass and planted the flags of their nations there. Between 1895 and 1914, the journey to the South Pole became an international obsession; Roald Amundsen, the Norwegian, reached it first in 1911. All this led to the creation of national claims during the ensuing interwar period. The geographic effect was the partitioning of Antarctica into pie-shaped sectors centered on the South Pole (Fig. 6-13). One of the areas of contention between states was the Antarctic Peninsula (facing South America), where British, Argentinean, and Chilean claims overlapped—a situation that remains unresolved. Only a single Antarctic sector is free of such claims: Marie Byrd Land (shown in white on Fig. 6-13).

Why should states be interested in territorial claims in so remote and difficult an area? Both land and sea contain resources that may some day become crucial: protein in the waters, and fossil fuels and minerals beneath the land surface. Antarctica (5.5 million square miles/14.2 million sq km) is almost twice as large as Australia, and the Southern Ocean is nearly as large as the Atlantic. However distant actual exploitation may be, several countries want to keep their stakes in the Southern realm. But the claimant states (those with territorial claims) have recognized the need for cooperation and the potential for conflict. The 1950s witnessed a major international program of geophysical research, the so-called International Geophysical Year. The spirit of cooperation that made this program possible extended to the political sphere and led to the 1961 signing of the Antarctic Treaty (to which 39 countries now subscribe). This agreement ensures continued scientific cooperation, prohibits military activities, safeguards the environment, and holds national claims in abeyance. But the Antarctic Treaty was reconstrued as the Wellington Agreement in 1991, and there has been a growing concern that it might not be strong enough. In particular, it does not settle the question of resource exploitation.

In an age of growing national self-interest and increasing resource needs, the possibility exists that international rivalry in the Southern realm will intensify and produce a confrontation that the treaty cannot prevent. The Southern realm is truly the globe's last frontier, and the partition of its lands and waters is a process fraught with dangers.

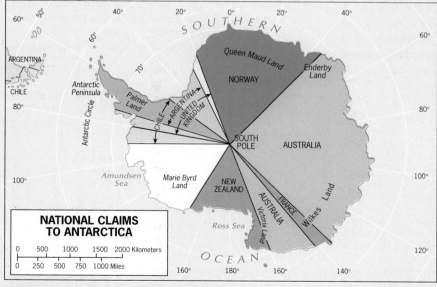

NATIONAL CLAIMS TO ANTARCTICA

FIGURE 6-13

pressive military government in 1989, Chile's economy has become the most robust in all of South and Middle America; between 1990 and 1993, the unemployment rate was cut from 25 to 4 percent, the number of Chileans living in poverty was reduced by one-fifth, and massive foreign investment was attracted. The last is of particular significance because these new international connections enable the export-led Chilean economy to diversify and develop in important new ways. Copper remains the single leading export, but many other mining ventures (including gold) have been launched. In the agricultural sphere, fruit and vegetable production for export has soared because Chile's harvests coincide with the winter farming lull in the affluent countries of the Northern Hemisphere; other primary economic activities that are increasing their foreign earnings are seafood and wood products, especially paper. Industries are benefiting as well, as new factories manufacture an array of goods ranging from basic chemicals to the most sophisticated computer software.

One of the most remarkable aspects of Chile's rapidly internationalizing economy is the new role the country is forging for itself on the prestigious Pacific Basin trading scene. Its greatest breakthrough has been the development of major ties with the Japanese, a relationship that has blossomed so dramatically since the late 1980s that by 1992 Japan had supplanted the United States as Chile's leading trading partner! Despite the enormous transoceanic distance that separates them, the two countries say they see mutual advantages of **complementarity** in their intensifying spatial interaction. The Chileans want to become the gateway to South and Middle America for Japan and other Austrasian countries; the Japanese view Chile as a bountiful source of many critical commodities lacking in their domestic resource base—metallic minerals, timber, fish, and a variety of farm produce. (More suspicious observers are concerned that Japan might try to use its Chilean linkages as a back door to the emerging North American Free Trade community established by the United States, Canada, and Mexico.) Whatever the outcome, these ongoing developments demonstrate that the new eco-nomic-geographic dynamism described in Chapter 4 is not confined to the coastlands of Austrasia: clearly, Chile is now in the process of becoming the first economic tiger on the South American shoreline of the sprawling Pacific Rim.

THE REALM IN TRANSITION

1. Political landscapes are changing in Brazil and in the Southern Cone (Argentina, Chile, and Uruguay) as democratic institutions strengthen.

2. An arc of countries from Colombia to Bolivia is under increasing stress from drug-related problems, ranging from illicit cultivation to aggressive trade cartels. Foreign drug markets and limited economic alternatives sustain this illegal and damaging economy.

3. The discovery of a major oil reserve in interior Colombia during the early 1990s indicates that the full potential of the Caribbean North region is not yet known. The United States remains the region's major customer.

4. South America, of all the realms facing the Pacific Ocean, is the least affected thus far by Pacific Rim developments. In the mid-1990s the major exception was Chile, called by some geographers the first economic tiger on South America's Pacific coast.

5. The penetration, exploitation, and destruction of Brazil's Amazonian interior is the environmental—and social—issue of the century in South America. By the year 2000, one-third of this vast region is likely to be affected.

◆ PRONUNCIATION GUIDE

Acre (AH-kray)
Altiplano (ahl-tee-PLAH-noh)
Amerind (AMMER-rind)
Amundsen, Roald (AH-moon-sun, ROH-ahl)
Antioquia (ahn-tee-OH-kee-ah)
Antofagasta (untoh-fah-GAHSS-tah)
Arauca (ah-RAU-kah)
Arica (ah-REE-kah)
Asunción (ah-soohn-see-OAN)
Atacama (ah-tah-KAH-mah)
Augelli (aw-JELLY)
Ayacucho (eye-uh-KOO-choh)
Barquisimeto (bar-key-suh-MAY-toh)
Barranquilla (bah-rahn-KEE-yah)
Barrio[s] (BAHR-ree-oh[ss])
Bayovar (bye-YOH-vahr)
Belém (bay-LEM)
Beni (BAY-nee)
Bogotá (boh-goh-TAH)
Bolívar, Simón (boh-LEE-vahr, see-MOAN)
Buenos Aires (BWAY-nohss EYE-race)
Cacao (kuh-KAY-oh/kuh-KAU)
Caldas (KAHL-dahss)
Cali (KAH-lee)
Callao (kah-YAH-oh)
Caracas (kah-RAH-kuss)
Cartagena (kar-tah-HAY-nah)
Caste (CAST)
Cauca (KOW-kah)
Central (sen-TRAHL)
Cerrejón (serray-HONE)
Cerro Bolívar (serro boh-LEE-vahr)
Cerro de Pasco (serro day PAH-skoh)
Chaco (CHAH-koh)
Chaparé (chah-pah-RAY)
Chibcha (CHIB-chuh)
Chile (CHILLI/CHEE-lay)
Chiloé (chee-luh-WAY)
Chuquicamata (choo-kee-kah-MAH-tah)
Ciudad Guayana (see-you-DAHD gwuh-YAHNA)
Cochabamba (koh-chah-BUM-bah)
Comodoro Rivadavia (comma-DORE-oh ree-vah-DAH-vee-ah)
Cordillera (kor-dee-YERRA)
Córdoba (KORD-oh-bah)

Corumbá (kor-room-BAH)
Creole[s] (KREE-oal[z])
Cusiana (koo-see-AHNA)
Cuyo (KOO-yoh)
Cuzco (KOO-skoh)
Départment (day-part-MAW)
Ecuador (ECK-wah-dor)
Entre Rios (en-truh-REE-ohss)
Esmeraldas (ezz-may-RAHL-dahss)
Essequibo (essa-KWEE-boh)
Falkland[s] (FAWK-lund[z])
Favela[s] (fah-VAY-lah[ss])
Fría (FREE-uh)
Guajira (gwah-HEAR-ah)
Guaraní (gwah-rah-NEE)
Guayaquil (gwye-ah-KEEL)
Guayas (GWYE-ahss)
Guiana[s] (ghee-AH-nah[z])
Guri (GOOR-ree)
Guyana (guy-AHNA)
Guzmán, Abimael (gooze-MAHN, uh-bee-my-ELL)
Hacienda (ah-see-EN-duh)
Huallaga (wah-YAH-gah)
Huancayo (wahn-KYE-oh)
Iberia (eye-BEERY-uh)
Iquitos (ih-KEE-tohss)
Isohyet (EYE-so-hyatt)
Itaipu (ee-TIE-pooh)
João, Dom (ZHWOW, dom)
Junta (HOON-tah)
La Paz (lah-PAHZ)
Lima (LEE-mah)
Llama (LAH-muh)
Llano[s] (YAH-noh[ss])
Machu Picchu (MAH-choo PEEK-choo)
Magdalena (mahg-dah-LAY-nah)
Magellan (muh-JELLEN)
Malvinas, Islas (mahl-VEE-nahss, EECE-lahss)
Maracaibo (mah-rah-KYE-boh)
Medellín (meh-deh-YEEN)
Mercosur (mair-koh-SOOR)
Mestizo (meh-STEE-zoh)
Mexica (meh-SHEE-kuh)
Montaña (mon-TAHN-yah)
Montevideo (moan-tay-vee-DAY-oh)
Occidental (oak-see-den-TAHL)
Oriental (orry-en-TAHL)
Oriente (orry-EN-tay)

Orinoco (orry-NOH-koh)
Pampa (PAHM-pah)
Paraguay (PAHRA-gwye)
Paraná (pah-rah-NAH)
Patagonia (patta-GOH-nee-ah)
Paulista[s] (pow-LEASH-tah[ss])
Pisac (pea-SAHK)
Pizarro, Francisco (pea-SAHRO, frahn-SEECE-koh)
Plata, Rio de la (PLAH-tah, REE-oh day lah)
Plaza de Armas (PLAH-sah day AR-mahss)
Plaza de la Constitución (PLAH-sah day luh con-stee-too-see-YOAN)
Plaza de Mayo (PLAH-sah day MYE-oh)
Potosí (poh-toh-SEE)
Puna (POONA)
Punta Arenas (POON-tah ah-RAY-nahss)
Quebracho (kay-BROTCH-oh)
Quechua (KAYTCH-wah)
Quito (KEE-toh)
Rio de Janeiro (REE-oh day zhah-NAIR-roh)
San Cristóbal (sahn kree-STOH-bahl)
San Juan (sahn HWAHN)
San Rafael (sahn rah-fye-ELL)
Santa Cruz (sahnta KROOZ)
Santiago (sahn-tee-AH-goh)
São Paulo (sau PAU-loh)
Sendero Luminoso (sen-DARE-oh loo-mee-NOH-soh)
Sucre (SOO-kray)
Suriname (soor-uh-NAHM-uh)
Templada (tem-PLAH-dah)
Tierra del Fuego (tee-ERRA dale FWAY-goh)
Titicaca (tiddy-KAH-kuh)
Tordesillas (tor-day-SEE-yahss)
Tucumán (too-koo-MAHN)
Tumbes (TOOM-bayss)
Uruguay (OO-rah-gwye)
Valparaíso (vahl-pah-rah-EE-so)
Venezuela (veh-neh-SWAY-lah)
Von Thünen (fon-TOO-nun)
Yacyretá (yah-see-ray-TAH)
Yerba maté (YAIR-bah mah-TAY)
Yungas (YOONG-gahss)

EMERGING BRAZIL:
Potentials and Problems

By any measure, Brazil is South America's giant. The country's economy is now the world's ninth-largest, and it has the capacity to grow at rates that far exceed those of developing nations. The population, before 1980 one of the world's fastest-growing, has now slowed its increase, but the next 15 years will still see the addition of about 45 million more Brazilians (a gain of 29 percent). Symbolized by its modern forward capital of Brasília, the vast Amazon-Basin-dominated interior has been opened up by new roads that annually lure at least a quarter of a million new settlers. Brazil's cities are mushrooming; its modern industrial base is the eighth largest on earth. Despite having been

Its potentials notwithstanding, Brazilian society is marked by a gap between its wealthy and poor populations that ranks among the widest on earth. In the landscape, that schism is an all-too-common sight. This squalid shacktown overlooks the capital, Brasília, a showcase planned city when it was first occupied a third of a century ago.

staggered by severe economic problems since the early 1980s, resource-rich Brazil retains the potential to take off toward a future in the developing world.

With its 3.3 million square miles (8.5 million sq km) of equatorial and subtropical South America, Brazil is a giant on the global stage as well: it is the only country to contain both the equator and a tropic—the Tropic of Capricorn (23½°S latitude). Territorially, it is exceeded in size only by Russia, Canada, China, and the United States. In population size, Brazil (at 156.5 million) ranks fifth among the world's countries. Economically, it is rising in the international ranks, and Brazil seems likely to become a world force in the twenty- first century.

Brazil is so large that it has common boundaries with all the other South American countries except Ecuador and Chile (Fig. 6-5). Its environments range from the tropical rainforest of the Amazon River basin (nearly all of which lies within Brazil's borders) to the temperate conditions of the Argentine Pampa (Fig. I-7). Its mineral resources are known to include enormous iron and aluminum ore (bauxite) reserves, extensive tin and manganese deposits, and major (recently discovered) oil- and gasfields. Other significant energy developments involve massive hydroelectric plants, including newly completed Itaipu Dam, one of the world's biggest. Brazilians have also successfully substituted sugarcane-based alcohol (gasohol) for gasoline, and so many of its cars (nearly 40 percent) now use this fuel that costly petroleum imports have been sharply reduced since the 1970s. Besides these natural endowments, Brazil's soils sustain an agricultural output that has burgeoned in recent years, today ranking the country as a global leader in the production and export of soybeans, coffee, and orange-juice concentrate.

◆ REGIONS

Brazil is a very large country, but its landscapes are not as spectacularly diverse as those of several much smaller South American republics. Brazil has no Andes Mountains, and the countryside consists mainly of plateau surfaces and low hills (Fig. 6-1). Even the lower-lying Amazon Basin, which covers almost 60 percent of the country, is not entirely a plain: between the tributaries of the great river lie low but extensive tablelands. To the southeast, the plateau surface of the Brazilian Highlands rises slowly eastward, but its highest segments fail to reach 10,000 feet (3,000 m). Along the coastline, there is a steep escarpment leading from plateau surface to sea level that leaves almost no living space along its base. Thus, although Brazil has some 4,500 miles (7,200 km) of Atlantic frontage, there is relatively little coastal plainland, and cities such as Rio de Janeiro are squeezed between mountain slopes and the shore. Under these physio-

TEN MAJOR GEOGRAPHIC QUALITIES OF BRAZIL

1. Territorially, Brazil ranks fifth largest in size among the world's countries. Its area covers nearly half of the continent, and Brazil borders every South American country except Chile and Ecuador.

2. Brazil is large, but its physiographic diversity does not match its size. The great Andes chain lies outside Brazil. Landscapes consist mainly of plateaus, low hills, and the vast, undulating Amazon Basin.

3. Brazil's population is larger than the total of all the other South American countries combined.

4. Brazil's regional development proceeded most rapidly along its Atlantic margins and was slowest in the Amazonian interior. Today, however, Amazonia is one of the world's most active settlement frontiers.

5. Brazilian governments have strived to focus national attention on the opportunities of the interior. Brasília, the forward capital completed in 1960, is a leading manifestation of that effort.

6. Brazil exhibits a strong national culture: a single language and the domination of one religious faith have constituted powerful unifying forces.

7. Although still ranked as a developing nation, Brazil generates economic indicators that point to "takeoff" conditions. Rapid urbanization and growing industrial strength also characterize the country.

8. In recent decades, Brazil has had one of the world's highest rates of population growth. These rates have abated since the mid-1980s, but the country's doubling time of 37 years still ranks below the global average of 41 years.

9. Brazil is a federation in which the eastern States of its modern core area dominate national affairs. Strong central authority and military involvement in government have recently prevailed, but civilian rule returned in 1985.

10. Although it is the giant of South America, Brazil's relationships with its neighbors remain comparatively distant.

graphic circumstances, Brazil is most fortunate to possess several very good natural harbors.

Brazil is a federal republic consisting of 26 States and the federal district of the capital, Brasília (Fig. B-1). As in

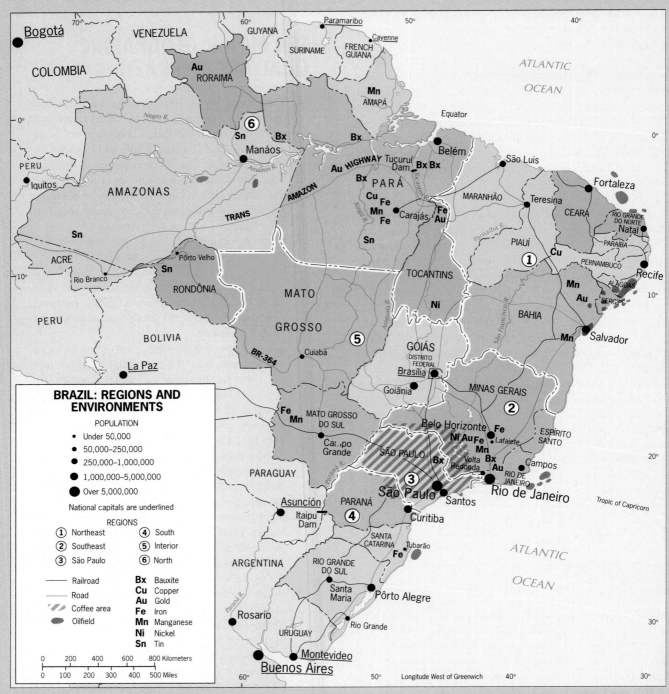

FIGURE B-1

the United States, and for similar reasons, the smallest States lie in the northeast and the larger ones farther west. The State of Amazonas is the biggest, twice the size of Texas with over 600,000 square miles (1.5 million sq km), but its huge area contains only about 2.3 million people. At the other extreme, Rio de Janeiro State (17,000 square miles/

44,000 sq km) on the southeastern coast has a population of more than 15 million. But the State with the largest population by far is São Paulo, now exceeding 35 million and still growing prodigiously.

Although Brazil is about as large as the 48 contiguous United States, it does not possess the clear physiographic regionalism familiar to us. Apart from the distinctive escarpment marking the Atlantic edge of the Brazilian Highlands, there are no mountain barriers, no well-defined coastal plains, and no naturally demarcated desert regions. Therefore, the six regions discussed below have no absolute or even generally accepted boundaries. In Fig. B-1, the regional boundaries have been drawn to coincide with the borders of States, making identifications easier.

THE NORTHEAST

The Northeast was Brazil's source area, its culture hearth. The plantation economy took root here at an early date, attracting Portuguese planters who soon imported the coun-

try's largest group of African slaves to work in the sugar fields. But the ample and dependable rainfall that occurs along the coast soon gives way to lower and more variable patterns in the interior savanna (see Fig. I-7), and today most of the Northeast is poverty-stricken, hunger-afflicted, overpopulated, and subject to devastating droughts. In fact, extended dry periods occur so regularly that this region (home to almost 50 million people) has become known as the *Polygon of Drought*. One of the worst of these episodes, lasting four years, occurred in the early 1980s. An even more severe drought began in 1990 and continued through 1993, triggering an outbreak of cholera as well as rural violence to demand a relief effort by the federal government. Ironically, there is ample groundwater, but the region's huge population of peasant farmers cannot afford

The old town of Salvador is a legacy of the city's heyday, a wealth drawn from whales and crops and built on African slave labor. Dozens of magnificent churches stand among ornate mansions and splendid, spacious townhouses, each an individual architectural accomplishment. Barrel-tiled roofs, intricate wrought iron, meticulously inlaid sidewalks, balconies, arched windows, and elaborate doors and gates attest to the riches once available here. This part of old Salvador is called Pelourinho, and has been designated an "historic treasure" by UNESCO (an agency of the United Nations). Today the district is undergoing gentrification, a revitalization effort that brings in affluent new residents as well as expensive shops. At the same time, however, this reconstruction is displacing most of the neighborhood's poor black population—and a rich African cultural heritage that dates back almost 300 years.

to drill the necessary wells. Without such irrigation, human and animal overpopulation are combining to deplete the natural vegetation and thereby hasten the encroachment of aridity.

Sugar still remains the chief crop along the moister coast, and livestock herding prevails in the drier inland back-country—the *sertão*—with beef cattle in the better grazing zones and goats elsewhere. The comparatively small areas of successful commercial agriculture—cotton in Rio Grande do Norte, sisal in Paraíba, and sugar cane in Pernambuco—stand in sharp contrast to the myriad patches of shifting subsistence agriculture located nearby.

The Northeast today is Brazil's great contradiction. In the cities of Recife (2.6 million) and Salvador (2.8 million), the architecture still bears the imprint of an earlier age of wealth, but thousands of peasants without hope—driven from the land by deteriorating conditions—constantly arrive to expand the usual surrounding shantytowns. As yet, few of the generalizations about emerging Brazil apply here, in South America's largest, most poverty-stricken corner. The federal government, aware of the Northeast's plight, directed investments to the region in the 1980s to help diversify its economic geography. The most visible effort was a huge petrochemical complex built near Salvador, which created thousands of jobs, attracted further (including foreign) investment, and boosted the region's industrial base in general. But even this project barely made a dent in alleviating the economic problems of the impoverished Northeast—where 55 percent of the people still live in the countryside, where subsistence farming on tiny plots remains the rule rather than the exception, and where 30 percent of Brazil's population generates only 15 percent of the country's gross domestic product.

THE SOUTHEAST

In the State of Bahia, a transition occurs toward the south. The coastal escarpment becomes more prominent, the plateau higher, and the terrain more varied; annual rainfall increases and is seasonally more dependable. The Southeast has been modern Brazil's core area, with its major cities and leading population clusters. Gold first drew many thousands of settlers, and other mineral finds also contributed to the influx (Rio de Janeiro itself served as the terminus of the "Gold Trail"), but ultimately the region's agricultural possibilities ensured its stabilization and growth. The mining towns needed food, and prices for foodstuffs were high; farming was stimulated, and many farmers came (with their slave workers) to Minas Gerais State to till the soil. Eventually, a pastoral industry came to predominate, with large herds of beef cattle grazing on planted pastures.

The post-World War II era brought another mineral age to the region, based not on gold or diamonds but on the iron ores around Lafaiete and the manganese and limestone carried to the steelmaking complex at Volta Redonda (Fig. B-1). Iron mining has now become one of Brazil's leading economic activities; in 1990 (the latest year for which data were available at press time), Brazil was second only to the then Soviet Union in total production and was the world's largest exporter of iron ore. In 1965 Brazil produced only 23 million metric tons of iron ore (versus 98 million in the United States that year), but by 1990 the output was 150 million metric tons (against only 59 million in the United States). Moreover, industrial diversification in the Southeast region has proceeded apace. Belo Horizonte is the rapidly growing metallurgical center of Brazil, with a population of 4.3 million—12 times what it was in 1950. Rio de Janeiro continues in second place nationally (after São Paulo) among the manufacturing cities and is now home to 11.6 million.

SÃO PAULO

The State of São Paulo, long a part of the Southeast region, today claims its own regional identity because it is the focus of ongoing Brazilian development. Currently, no less than 22 percent of the national population reside here; moreover, São Paulo State accounts for over 40 percent of Brazil's gross domestic product and for about 60 percent of its manufacturing activity. Led by its vast coffee plantations (known as *fazendas*), this State is also the leading foreign-exchange earner of the republic, and it leads Brazil in the production of numerous other crops as well. Brazilian farmers have enormously increased their soybean output nationally (soybeans now exceed coffee in acreage and value), surpassing China in the 1970s and regularly taking first or second place among world producers since 1980. Another leading agricultural pursuit is the production of orange-juice concentrate; São Paulo State now produces more than double the annual output of Florida, thanks to a climate free of winter freezes, ultramodern processing plants, and a fleet of specially equipped tankers to ship the concentrate to markets in the United States—and more than 50 other countries ranging from Japan to Russia.

Matching this prodigious agricultural output is the State's industrial strength. The São Paulo metropolis is now the country's leading manufacturing complex, with Brazil's foremost industry (automobiles) based in the city of São Paulo's southern suburbs. The State does not have a mineral base to compare to that of Minas Gerais; nonetheless, it has become the leading industrial region not only in Brazil but in all of South America. The revenues derived from the coffee plantations provided the necessary investment capital, hydroelectric power from the slopes of the coastal escarpment produced the needed energy, and immigration from Portugal, Italy, Japan, and elsewhere con-

Panoramic view of São Paulo, the dynamic pulse of late-twentieth-century Brazil. South America's largest metropolis (and the world's third biggest) continues to grow rapidly and has just surpassed the 20-million mark.

tributed the labor force. São Paulo State lay juxtaposed between raw-material-producing Minas Gerais and the States of the South (Fig. B-1). Communications were improved, notably with the major outport of Santos (1.4 million) but also with the interior hinterland. As the capacity of the domestic market grew, the advantages of central location and agglomeration secured São Paulo's primacy. Metropolitan São Paulo today is truly the pulse of Brazil; its burgeoning population of 20.6 million (which grows by at least 500,000 per year) also makes it the realm's biggest urban agglomeration—and the third largest in the world, after Tokyo and Mexico City.

THE SOUTH

Three States make up the southernmost Brazilian region: Paraná, Santa Catarina, and Rio Grande do Sul. The contribution of recent European immigrants to the agricultural development of southern Brazil, as in neighboring Uruguay and Argentina, has been considerable. Many came not to the coffee country of São Paulo State but to the available lands farther south. Here the newcomers occupied fairly discrete areas. Portuguese rice farmers clustered in the valleys of the major rivers of Rio Grande do Sul, and that State now produces one-quarter of Brazil's annual rice crop. The Germans occupied the somewhat higher areas to the north and in Santa Catarina, where they were able to carry on the type of mixed farming with which they were familiar: corn, rye, potatoes, and hogs as well as dairying. The Italians selected the highest slopes and established thriving vineyards. The markets for this produce, of course, are the large urban areas to the north. Paraná, in contrast, exports its coffee harvest to overseas markets.

Unlike the Northeast and Southeast, the South has never been a boom area. It does, however, have a stable, progressive, and modern agricultural economy; farming methods here are the most advanced in Brazil. The diversity of its European heritage is still reflected in the regional towns, where German and other European languages are preserved. More than 25 million people live in the three southern States, and the region's importance is steadily increasing. Coal from Santa Catarina and Rio Grande do Sul, shipped north to the steel plants of Minas Gerais, was a crucial element in Brazil's industrial emergence. Local industry is growing as well, especially in Pôrto Alegre (3.7 million) and Tubarão, where South America's single largest steel-making facility opened in 1983.

The Brazilian South is also the scene of one of this century's greatest construction projects, Itaipu Dam (completed in 1991; see photo p. 342). Located astride the Paraná River on the Brazil-Paraguay border in the southwest corner of Paraná State (Fig. B-1), Itaipu's dimensions are truly awesome: at 600 feet (180 m) in height and 5 miles (8 km) in length, it is six times bigger than Egypt's Aswan High Dam—so big that in 1992 (its first year of operation at full capacity) it supplied *40 percent* of Brazil's total electric power! Much of this energy was consumed in the rapidly expanding São Paulo metropolis, 500 miles (800 km) to the east. It was hoped that the Itaipu area itself would also become a growth center and enjoy considerable benefits as well; in anticipation, hundreds of thousands migrated there, but opportunities have fallen well below expectations, particularly on the Paraguayan side of the border where the Itaipu work force was disbanded in the late 1980s.

THE INTERIOR

Interior Brazil is often referred to as the Central-West, or *Centro-Oeste*. This is the region that Brazil's developers hope to make a part of the country's productive heartland, and in the 1950s the new capital of Brasília was deliber-

Brazil, too, has dammed the Paraná River—here at Itaipu, upstream from the Argentinean Yacyretá project seen on p. 330. Before the latter was completed, *this* was "the world's largest."

ately situated on its margins (Fig. B-1). By locating the new capital city in the untapped wilderness 400 miles (650 km) inland from its predecessor, Rio de Janeiro, the country's leaders dramatically signaled the opening of Brazil's age of development and a new thrust toward the west. As we saw earlier, in Japan's shifting of its capital from Kyoto to Tokyo when it entered its modernization stage, governments have used the relocation of their own headquarters to emphasize the beginning of new eras and new geographic linkages. Moreover, numerous countries have recently shifted their capitals (among them Subsaharan Africa's Tanzania and Nigeria); in South America, both Argentina and Peru have lately considered such moves.

Brasília is also noteworthy in another regard: it represents what political geographers call a **forward capital**. There are times when a state will relocate its capital to a sensitive area, perhaps near a zone under dispute with an unfriendly neighbor, in part to confirm its determination to sustain its position in that contested zone. An example was the decision by Pakistan in the 1960s to move its capital from coastal Karachi to northern Islamabad near the disputed territory of Kashmir. These are called "forward" capitals because of their position in an area that would be first to be engulfed by conflict in case of strife with a neighbor. Brasília, of course, does not lie in or near a contested zone, but Brazil's interior has been an internal frontier, one to

be conquered by a developing nation; in that drive, the new capital occupies a decidedly forward position.

Despite the growth of Brasília since 1960 (from zero to 3.4 million inhabitants today), the economic integration of the Interior into the rest of the country will take considerable time. This is a vast upland area, largely a plateau at an elevation of over 3,000 feet (900 m), covered with savanna vegetation (tropical grasslands containing widely spaced trees). Like Minas Gerais, the Interior was the scene of gold rushes and some discoveries; but unlike its eastern neighbor, it did not present agricultural alternatives when those mineral searches petered out. Today, the three States of the Interior—Goiás, Mato Grosso, and Mato Grosso do Sul—have a combined population of about 12 million (another million or so were detached from northern Goiás in 1990 to create the new State of Tocantins), so that the average density is under 25 per square mile (10 per sq km). Nevertheless, this represents a noteworthy increase, for in 1960 the Centro-Oeste contained just 2.9 million people. Thus this Brazilian region's population has quadrupled over the past third of a century—a pace well ahead of the national growth rate.

Despite these gains, the Centro-Oeste continues to face the problems common to tropical savanna regions everywhere: soils are not especially fertile, and the vegetation is susceptible to damage by overgrazing. In the absence of major mineral finds, pastoralism remains the chief economic activity. But there is always the fear that what happened to the Northeast could happen here, and that Brazil's determination to open the Interior might reproduce the serious environmental problems of Amazonia (discussed in the next section). At the moment, however, that day seems far off: one can fly for hours above this huge area and see only an occasional small settlement and a few widely spaced roads (communications are best in the south). What the Interior needs is investment—in the clearing and opening of the alluvial soils in the region's river valleys (where soybean production is now expanding), in experimental farms, in the provision of electric power, and in further mineral exploration. The first step was the relocation of the national capital—an enormously expensive venture, but a mere beginning in view of what the Interior really requires.

THE AMAZONIAN NORTH

The largest and most rapidly developing region—whose population has nearly tripled to 15 million since 1980—is also the most remote from the core of Brazilian settlement: the seven States of the Amazon Basin (Fig. B-1). This was the scene of the great rubber boom at the turn of the century, when the wild rubber trees in the *selvas* (tropical rainforests) produced huge profits and the central Amazon city of Manáos (1994 population: 1.7 million) enjoyed a brief period of wealth and splendor. The rubber boom ended in 1910, however, when plantations elsewhere (notably in Southeast Asia) began to produce rubber more cheaply, efficiently, and accessibly. For most of the seven decades that followed, Amazonia was a stagnant hinterland—but all that changed dramatically during the 1980s. New development has been spawned throughout this awakening region, which has become the scene of the world's largest migration into virgin territory as over 250,000 new settlers arrive every year. Most of this activity is occurring south of the Amazon River, in the tablelands between the major waterways and along the Basin's wide rim.

Two ongoing development schemes are especially worth noting because they are quintessential expressions of what is occurring here. The first is the *Grande Carajás Project* in eastern Pará State, a huge, multifaceted scheme centered on one of the world's largest-known deposits of iron ore in the Serra dos Carajás hills (Fig. B-1). In addition to a vast mining complex, other new construction here includes the Tucuruí Dam on the nearby Tocantins River and a 535-mile (850-km) railroad to the Atlantic port of São Luis (975,000). This ambitious development project also emphasizes further mineral exploitation (including bauxite, manganese, and copper), cattle raising, crop farming, and forestry. If expansion plans are implemented, Grande Carajás will one day cover one-sixth of all Amazonia.

Understandably, tens of thousands of settlers have descended on this area in the past few years. Those seeking business opportunities have been in the vanguard, but they have been followed by masses of lower-income laborers and peasant farmers in search of jobs and landownership. Surveying the initial stage of this colossal enterprise in his recent book, *Passage Through El Dorado,* Jonathan Kandell compared it to the westward surge of nineteenth-century pioneers in the United States and used the words "energy, hope, greed, and savagery" to describe the turmoil of the competition to succeed in this remote and often hostile environment.

The second leading development scheme, known as the *Polonoroeste Plan,* is located about 500 miles (800 km) to the southwest in the 1,500-mile-long Highway BR-364 corridor that parallels the Bolivian border and connects the western Brazilian towns of Cuiabá, Pôrto Velho, and Rio Branco (Fig. B-1). Although the government had planned for the penetration of the North to proceed via the east-west Trans-Amazon Highway, the migrants of the 1980s and 1990s have preferred to follow BR-364 and settle within the Basin's southwestern rim zone, mostly in Rondônia State. Agriculture has been the dominant activity here, attracting affluent growers and ranchers from the South, plantation workers from São Paulo and Paraná states displaced by agrarian mechanization, and subsistence farmers from all over the country. The common denominator has been the quest for land, and bitter conflicts continue to break out between peasants and landholders as the Brazilian government cautiously pursues the persistent and volatile issue of land reform.

The usual pattern of Amazonian settlement is something like this. As main and branch highways are cut through the wilderness, settlers (enticed by cheap land) follow and move out laterally to clear spaces for farming. Crops, usually maize (corn) or upland rice are planted, but within three years the heavy equatorial-zone rains leach out soil nutrients and accelerate surface erosion. As soil fertility declines, pasture grasses are then planted, and the plot of land is soon sold to cattle ranchers (most are associated with big agribusiness corporations based in Rio or São Paulo). The peasant farmers then move on to newly opened areas, clear more land for planting, and the cycle repeats itself. As long as open spaces remain, this is a profitable pursuit for all parties, but it assures the widespread establishment of a low-grade land use that will ultimately concentrate most of the earnings in the hands of large landowners.

Perhaps worst of all is the environmental impact of this extensive grazing system that involves so little attachment between farmers and their lands: it requires clearing enormous stands of tropical woodland. In the early 1990s, between 20,000 and 30,000 square miles (52,000 and 78,000 sq km) of rainforest were disappearing *annually* in Amazonia—an area larger than the state of West Virginia.

Here, just one day ago, stood magnificent trees of the Amazonian rainforest, their canopies hundreds of feet above the ground, testimony to hundreds of millions of years of evolution. Macaws squawked, jaguars prowled, primates nested here. And then this world of natural wonders fell in moments to the saws and fires of human making, and it is silent now, save for the crackling of the last cinders.

Moreover, the deforestation crisis here in the Brazilian North (see photo above) is exacerbated by the scattered development; hence, the remaining selvas are dotted with a myriad of expanding clearings that are constantly coalescing. And because this deforestation also accounts for over half the total worldwide tropical deforestation, Amazon woodland destruction has much wider implications; as the box entitled "Tropical Deforestation" on pp. 294-295 describes, this is an environmental crisis of global proportions, with future consequences that could negatively affect every form of terrestrial life.

◆ POPULATION PATTERNS

Brazil's population of 156.5 million is as diverse as that of the United States. In a pattern that is familiar in the Americas, the indigenous inhabitants of the country were deci-

mated following the European invasion; estimates of the number of Amerindians who survive today in small communities deep in the Amazonian interior vary, but their total is not likely to exceed 200,000—well under 10 percent of the number thought to have been there when whites first arrived. Africans came in great numbers, too, and today there are about 17 million blacks in Brazil. Significantly, however, there was also much racial mixing, and nearly 70 million Brazilians (about 45 percent of the national population) have combined European, African, and minor Amerindian ancestries. The remaining 70 million or so, now no longer in the majority, are of European descent.

Until Brazil became independent in 1822, the Portuguese were virtually the only Europeans to settle in this country. But after independence other European settlers were encouraged to come, and many Italians, Germans, and Eastern Europeans arrived to work on the coffee plantations, farm in the south, or try their luck in business. Immigration reached a peak during the 1890s, a decade in which nearly

1.5 million newcomers reached Brazilian shores. The complexion of the population was further diversified by the later arrival of Lebanese and Syrians, many of whom opened small shops.

Even the Japanese have had an impact on Brazilian social geography, and today they constitute a bustling community of more than 1.2 million (the largest anywhere outside Japan) that grew from the nucleus of 781 pioneers who arrived in 1908. At first they performed menial jobs on the coffee *fazendas,* but now the fully integrated ethnic Japanese of São Paulo State have risen to the topmost ranks in Brazil's easy-going immigrant society as highly successful farmers, urban professionals, business leaders—and traders with Japan.

Brazilian society, to a greater degree than is true elsewhere in the Americas, has made progress in dealing with its racial divisions. To be sure, blacks are still the least advantaged among the country's population groups except for the Amerindians. But ethnic mixing in Brazil is so pervasive that hardly any group is unaffected, and official census statistics about "blacks" and "Europeans" are meaningless. What the Brazilians do have is a true national culture, expressed in an adherence to the Catholic faith (though weakening, this is still the world's largest Roman Catholic country), in the universal use of a modified form of Portuguese as the common language, and in a set of lifestyles in which vivid colors, distinctive music, and a growing national consciousness and pride are fundamental ingredients.

This is not to suggest an absence of regional variety in Brazilian socio-geographical patterns. The black component of the population in the Northeast, for instance, remains far stronger than it is in the Southeast. After their initial concentration in the Northeast, black workers were taken southward to Bahia and Minas Gerais as the economic heartland moved in that direction. Today the black population remains strongest in these areas and in Rio de Janeiro; it is weak in the States of the South, more recently settled and more completely taken over by Europeans. Nor has Brazil escaped problems of racial prejudice: lately, black leaders and others in the country have publicly voiced their objections against discrimination. Yet, although Brazil may not be the multiracial society it is sometimes portrayed to be, it also does not have the history of overt and legally sanctioned racism that has characterized a number of other plural societies.

Another population problem that has beset Brazil—one that could still resurface and jeopardize the overall development potential—was its persistently high birth rate. As recently as the late 1970s, the country grew at the very rapid pace of 2.8 percent annually. But the 1980s saw a turnaround and then a steady drop, and by the opening of the 1990s this rate was down to (a still high) 2.0 percent. Although a reversal is possible at any time, demographers expect the decline to continue. As of the mid-1990s, the rate of increase had leveled off at 1.9 percent—but that translates into a doubling time of only 37 years and a surpassing of the 200-million level in 2010. One of the most noteworthy post-1980 population trends has been a substantial plunge in the fertility rate, from 4.4 children per woman in 1980 to less than 3.0 today. Surprisingly, this slowdown in population increase took place in the absence of an active birth control policy by the Brazilian government (and also in the face of disapproval by the decreasingly influential Catholic church). Demographers believe that Brazil's self-induced decline is the result of three factors: a rapid spread in contraceptive usage, the negative influence on family formation of a decade-long period of economic stagnation, and newly widespread access to television (which seems to be reshaping the attitudes and aspirations of the Brazilian people).

A final glimpse at the population geography of Brazil underscores the country's unevenness—a concluding theme in the section on economic development that follows. Figure B-2 is a map of the population density of the Brazilian States, with the density of each State shown as a flat platform "pushed up" from a base plane. Its height is determined by the magnitude of its density, measured on the scale shown next to the population surface. The "peaks" and "troughs" of this three-dimensional surface etch sharply the density contrasts of a nation in which 90 percent of the people are still crowded into 10 percent of the available space, yielding a high average density on that tenth of land of 410 people per square mile (157 per sq km).

The map in Fig. B-2 also demonstrates some recent technological advances in cartography. The computer-drawn perspective we see here is that of an astronaut looking down on Brazil from a point in space approximately above Santiago, Chile. The total surface appears as a series of steps that are joined to form a number of analogous mountains, ridges, depressions, and plains that represent variations in population density across Brazil. The *SYMVU* computer program used to generate the map permits us to view this surface from any direction and angle and to make any number of manipulations; software packages of this kind are constantly increasing in sophistication and today are also available for personal computers.

◆ DEVELOPMENT PROBLEMS

The country's great potential notwithstanding, Brazil's overall development during the past two decades has been unspectacular. Prior to the mid-1970s, particularly between 1968 and 1973, Brazil embarked on a period of spectacular growth—its economy expanding at an annual rate of 9 percent, its exports doubling in value every two years, and its new investments making real headway in improving health

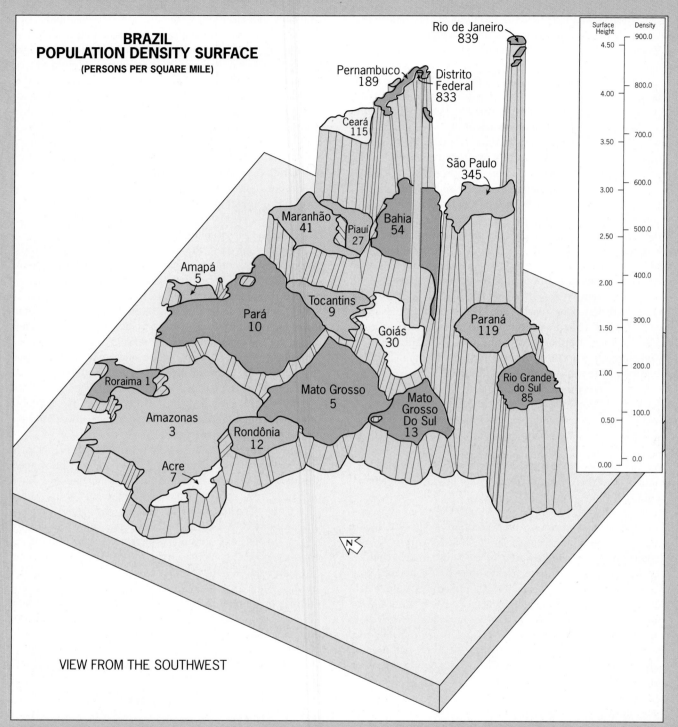

BRAZIL
POPULATION DENSITY SURFACE
(PERSONS PER SQUARE MILE)

Rio de Janeiro
839

Pernambuco
189

Distrito
Federal
833

Ceará
115

São Paulo
345

Maranhão
41

Piauí
27

Bahia
54

Amapá
5

Tocantins
9

Pará
10

Goiás
30

Paraná
119

Roraima 1

Mato Grosso
5

Mato
Grosso
Do Sul
13

Rio Grande
do Sul
85

Amazonas
3

Rondônia
12

Acre
7

Surface Height	Density
4.50	900.0
4.00	800.0
3.50	700.0
3.00	600.0
2.50	500.0
2.00	400.0
1.50	300.0
1.00	200.0
0.50	100.0
0.00	0.0

VIEW FROM THE SOUTHWEST

FIGURE B-2

standards, agriculture, and living conditions. But the global energy crisis of 1974 quickly wrote an end to those years of the Brazilian "economic miracle" as the nation's huge new imported-fuel bills thwarted its forward drive.

After marking time in the economic doldrums that endured through the rest of the 1970s, Brazil's military rulers in 1980 vowed to resume the developmental push by borrowing heavily abroad, thereby breaking a longstanding guideline to keep away from foreign financial involvements so as to preserve full control over the country's economy. All too soon, however, Brazilians found themselves seriously mired in the same economic morass as Mexico, Argentina, and a number of other South American countries, having accumulated a staggering debt in foreign loans that surpassed U.S. $90 billion by 1984.

What had gone wrong? Basically, the answer lay with the worldwide economic recession of the early 1980s: falling commodity prices on the international market suddenly and sharply reduced Brazil's income and caused a sizeable decline in exports. The government, deeply committed to several development programs and unwilling to trim them back, redoubled its borrowing abroad (at ever-higher interest rates) to make up the difference. As runaway deficits mounted up at home from continued massive governmental spending, inflation soared from 80 to nearly 150 percent annually between 1980 and 1983.

Increasingly, Brazil found itself unable to keep up payments on its rapidly climbing overseas debt. To stave off bankruptcy, the International Monetary Fund agreed to supply emergency funds so that Brazil could begin to repay its loans, but only if a severe domestic austerity program was initiated at once. Stiff economic and energy conservation measures were swiftly imposed, and the government stuck to them despite sporadic outbreaks of social unrest among the poor of the Northeast and Rio de Janeiro State. A brief turnaround occurred in the mid-1980s, and the Brazilian economy's performance improved dramatically. But the restoration of civilian rule in 1985 was marred by ineffective leadership; inflation swiftly resumed its upward spiral, and by 1987 the foreign debt had soared to U.S. $110 billion.

This enormous foreign debt has grown even larger since the late 1980s (reaching U.S. $125 billion in 1993) because Brazil became mired in another deep recession (again of global proportions) that began in 1989 and has persisted into the mid-1990s. By late 1993, after four years of zero economic growth, Brazil was experiencing more than a 10 percent loss in gross domestic product per capita, unemployment to the extent that one-third of its manufacturing industries lay idle, and a surge in the population living below the poverty line. Government policies in the early 1990s seemed only to aggravate these difficulties; however, the impeachment of Brazil's president in 1993 (on unrelated corruption charges) provided a new regime with another chance at putting the nation's economic house in order. Even as Brazil is waiting for an economic recovery, however, the immediate outlook is not all that bleak.

Agriculture presents Brazil with great opportunities, as we have seen. Problems in this sector continue as well, but these have not prevented the country from becoming the second-biggest farm producer in the Western world. Apart from the ample revenues derived from coffee sales overseas, Brazil has also been selling ever-greater quantities of soy-beans, sugar, and orange-juice concentrate on the international market. The range of crops Brazilian farmers can grow is considerable; however, much land that could be cultivated is not, agricultural methods still need improvement, and land reform in the all-important coffee *fazendas* has come too slowly. Millions of peasants in Brazil still practice shifting cultivation when environmental conditions do not really demand it, so that returns from the soil too often are minimal. It has been estimated that productivity per farm worker in the United States is at least 50 times higher than in Brazil; mechanization has barely begun in some Brazilian regions. However, despite the rush to the cities, the labor force engaged in agriculture has grown in recent years, reflecting Brazil's still underdeveloped condition.

In this context, we should keep in mind just how far Brazil has yet to go from its present **takeoff stage of development**. The total area of Brazil is 44 percent bigger than that of newly expanded Europe, but the total market value of its goods and services produced in 1989 was only 58 percent higher than that of the Netherlands alone (which has a population one-tenth the size of Brazil's); calculated on a per capita basis, the Brazilian figure is a mere 16 percent of the Dutch level of output. Brazil's awakening is sometimes compared to the first stage of Japan's modernization, but such analogies may still be premature.

GROWTH POLES

When the military regime took control of Brazil in 1964, it embarked on a development program that is still referred to as the *Brazilian Model*. Essentially, this program involved the government in shared participation with private enterprise so that public interests, private concerns, and foreign investors would not operate at mutual disadvantage—and to the disadvantage of Brazil as a whole. Agriculture and industry were supported and promoted but under certain priorities and guidelines.

Among the problems always facing national development planners is the remote region where opportunities lie but where the investments needed to exploit them are very high. If an area is already developing and attracting immigrants, should money be spent to improve existing facilities or should a new, more distant location be stimulated? In such decisions the **growth-pole concept** becomes relevant. The term is almost self-explanatory: a growth pole is a location where a set of industries, given a start, will expand and spawn ripples of development in the surrounding area. Certain conditions must exist, of course. There would be little point in selecting Rio Branco as a growth pole in the southwestern corner of the Amazonian North region (Fig. B-1), where there is no real prospect for development in the immediate hinterland and where there are still relatively few people and only minor agricultural or industrial activities to stimulate. But growth-pole theory did help to shape the decision to build Brasília in the 1950s, for in surrounding, underpopulated Goiás State, a new market of 700,000 people constituted a stimulus for many activities (close to 7 million people reside in the area today).

MULTINATIONAL ECONOMIC INFLUENCES

As we noted above, the Brazilian Model involves governmental intervention in the economy, including various forms of control over foreign investors. This was a significant dimension of Brazilian development policy, for large corporations played a major role in Brazil's economic surge between the late 1960s and 1980. In fact, the power of multinational corporations in underdeveloped countries has recently become a matter for concern among the leaders of those countries because these global corporations, backed by enormous financial resources, can influence the economics as well as the politics of entire states.

Brazil's leaders had long welcomed foreign investment, but as economic progress accelerated in the 1960s and 70s, they perceived the risks involved in foreign control over Brazilian firms. Multinational corporations can introduce and spread technological advances, and they can provide capital and increase employment—but in the process, they gain control over the industrial and agricultural export sectors of the economy, and through their efficiency they can throttle local competition and damage business oriented toward local markets. Aware of such impacts, the government of Brazil for a time prohibited foreign commercial banks from entering the Brazilian economy and imposed regulations requiring multinational corporations active in Brazil to keep more of their huge profits inside the country. But the severe economic recession of the early 1980s forced the government to abandon its strictures against foreign borrowing, which subsequently accelerated so rapidly that since 1985 Brazil has consistently ranked among the world's leading debtor nations.

STAGGERING INTERNAL INEQUITIES

Our survey of South America's largest country has underscored its highly variable character and the sharp divisions between modern and traditional Brazil. This is truly a nation of profound contrasts, with perhaps the widest gap separating the rich and the poor to be found anywhere on earth. Astonishingly, the wealthiest 2 percent of Brazil's citizens control 70 percent of the land and annually earn as much as the poorest *two-thirds* of the population. Even though it now ranks among the world's leading food exporters, well over half of all Brazilians suffer from chronic malnutrition. And, as we have observed, even though considerable regional development is taking place, the benefits are overwhelmingly channeled toward already affluent individuals and powerful corporations—thereby further widening that enormous gulf between the haves and the have-nots.

Significantly, 35 percent of the Brazilian population is under 15 years of age. A great majority of this demographic cohort lives in or near poverty, especially in the drug-infested *favelas* (inner-city slums) and shantytown rings that are mushrooming in Brazil's teeming urban areas. Therefore, unless the country's resources and access to its opportunities are redistributed in a more equitable manner, the future will continue to hold potentially dangerous social problems and the threat of civil disorder (so long held in check). Civilian governments since 1985 have clearly demonstrated their reluctance to move boldly against these challenges, caving in to the political pressures of the powerful landowners to move very slowly on the crucial issue of land reform, and refraining from launching meaningful new social programs. Thus promising Brazil may now possess the ninth-largest national economy, but progress toward membership in the developed world remains clouded by the reality that most of its people have so far been shut out from participating in this historic crusade.

◆ PRONUNCIATION GUIDE

Amazonas   (ahma-ZOAN-ahss)

Bahia   (bah-EE-yah)

Belo Horizonte   (BAY-loh haw-ruh-ZONN-tee)

Brasília   (bruh-ZEAL-yuh)

Bauxite   (BAWKS-site)

Centro-Oeste   (SENTRO oh-ESS-tee)

Cuiabá   (koo-yuh-BAH)

Favela[*s*]   (fah-VAY-lah[ss])

Fazenda   (fah-ZENN-duh)

Goiás   (goy-AHSS)

Grande Carajás   (GRUNN-dee kuh-ruh-ZHUSS)

Islamabad   (iss-LAHM-uh-bahd)

Itaipu   (ee-TYE-pooh)

Karachi   (kuh-RAH-chee)

Kashmir   (KASH-meer)

Kyoto   (kee-YOH-toh)

Lafaiete   (lah-fuh-YAY-tuh)

Manáos   (muh-NAUSS)

Mato Grosso   (mutt-uh-GROH-soh)

Mato Grosso do Sul   (mutt-uh-GROH-soh duh-SOOL)

Minas Gerais   (MEE-nuss zhuh-RICE)

Nigeria   (nye-JEERY-uh)

Pará   (puh-RAH)

Paraguay   (PAHRA-gwye)

Paraíba   (pah-rah-EE-buh)

Paraná   (pah-rah-NAH)

Pelourinho   (peh-loo-REEN-yoh)

Pernambuco   (pair-nahm-BOO-koh)

Polonoroeste   (POLLOH-nuh-roh-ESS-tee)

Pôrto Alegre   (POR-too uh-LEG-ruh)

Pôrto Velho   (POR-too VELL-yoo)

Recife   (ruh-SEE-fuh)

Rio Branco   (REE-oh BRUNG-koh)

Rio Grande do Norte   (REE-oh GRUN-dee duh NORTAH)

Rio Grande do Sul   (REE-oh GRUN-dee duh SOOL)

Rio de Janeiro   (REE-oh day zhah-NAIR-roh)

Rondônia   (roh-DOAN-yuh)

Salvador   (SULL-vuh-dor)

Santa Catarina   (SUN-tuh kuh-tuh-REE-nuh)

Santiago   (sahn-tee-AH-goh)

Santos   (SUNT-uss)

São Luis   (sau loo-EECE)

São Paulo   (sau PAU-loh)

Serra dos Carajás   (SERRA doo kuh-ruh-ZHUSS)

Sertão   (sair-TOWNG)

Sisal   (SYE-sull)

SYMVU   (SIM-view)

Tanzania   (tan-zuh-NEE-uh)

Tocantins   (toke-un-TEENS)

Tubarão   (too-buh-RAUNG)

Tucuruí   (too-koo-roo-EE)

Uruguay   (OO-rah-gwye)

Volta Redonda   (vahl-tuh rih-DONN-duh)

CHAPTER 7

NORTH AFRICA/SOUTHWEST ASIA: THE CHALLENGE OF ISLAM

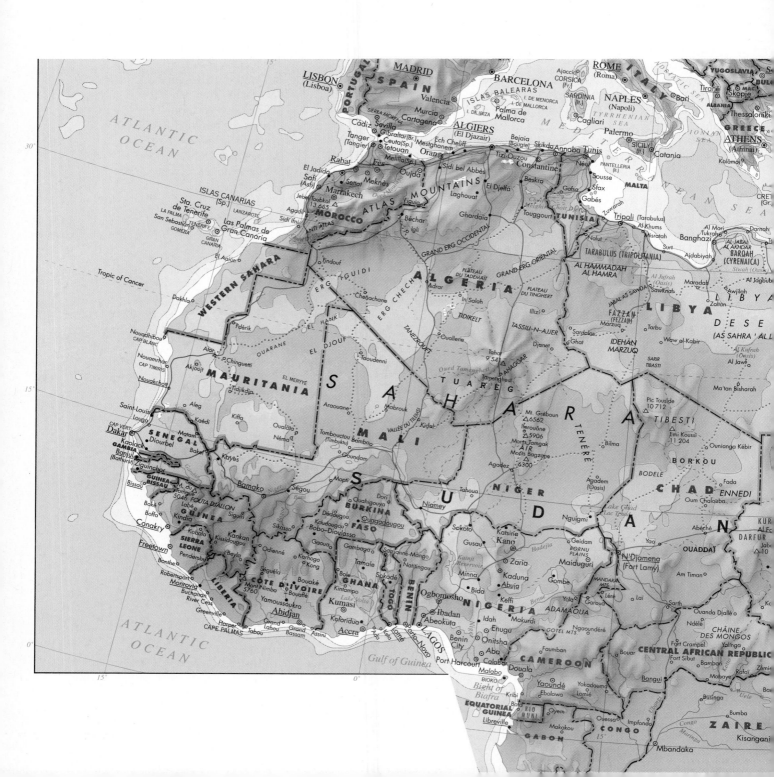

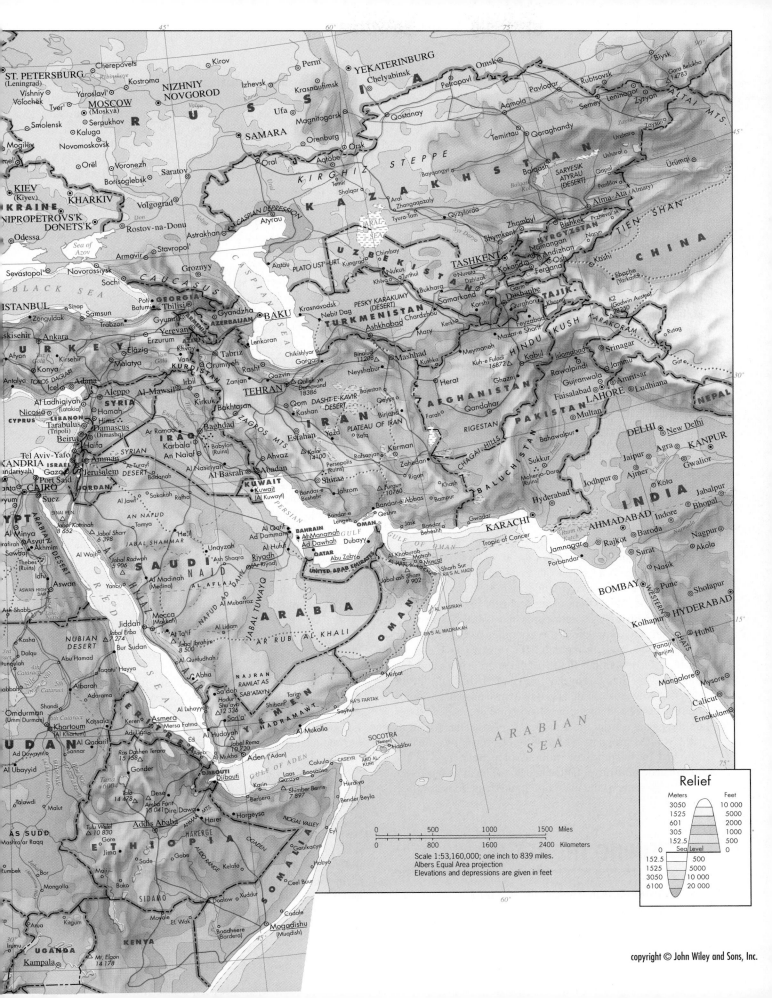

IDEAS & CONCEPTS

Cultural geography
Culture hearth (3)
Fertile Crescent
Hydraulic civilization
Climate change
Diffusion
    Expansion diffusion
    Relocation diffusion
Boundaries (2)
State expansion
Nomadism
Buffer state

REGIONS

Egypt and the Lower Nile Basin   Arabian Peninsula
The Maghreb and Libya   Empire States
African Transition Zone   Turkestan
Middle East

From Morocco on the shores of the Atlantic to the mountains of Afghanistan, and from the Horn of Africa to the steppes of inner Asia, lies a vast geographic realm of enormous cultural complexity. It lies at the crossroads where Europe, Asia, and Africa meet, and it is part of all three; throughout history, its influences have radiated to these continents and to practically every other part of the world as well. This is one of humankind's primary source areas. On the Mesopotamian Plain between the Tigris and Euphrates rivers (in modern-day Iraq) and on the banks of the Egyptian Nile arose civilizations that must have been among the very earliest. In the soils, plants were domesticated that are now grown from the Americas to Australia. Along its paths walked prophets whose religious teachings are still followed by hundreds of millions of people. And in the final decade of the twentieth century, the heart of this realm is beset by some of the most bitter and dangerous conflicts on earth.

◆ DEFINING THE REALM

It is tempting to characterize this geographic realm in a few words, to stress one or more of its dominant features.

It is, for instance, often called the "Dry World," containing as it does the vast Sahara and Arabian deserts. But most of the realm's people live where there is water—in the Nile delta, along the Mediterranean coastal strip (or *tell*) of northwesternmost Africa, along the Asian eastern and northeastern shores of the Mediterranean Sea, in the Tigris-Euphrates Basin, in far-flung desert oases, and along the lower mountain slopes of Iran south of the Caspian Sea as well as those of Turkestan to the northeast. True, we know this world region as one where water is almost always at a premium, where peasants often struggle to make soil and moisture yield a small harvest, where nomadic peoples and their animals circulate across dust-blown flatlands, where oases are islands of sedentary farming and trade in a sea of aridity. But it also is the land of the Nile, the lifeline of Egypt, and the crop-covered *tell* of the northern coast of Morocco, Algeria, and Tunisia.

Before we investigate this realm further, it is useful to look again at Fig. I-7, the map of world climates. Note the considerable degree of coincidence between the extent of the *B* (desert and steppe) climates and the limits of both the North African and Southwest Asian segments. In the case of North Africa, the border between this realm and Subsaharan Africa comes close to matching the southern limits of the *BSh* zone. In Southwest Asia, deserts, steppes, and mountain climates dominate. Except for the Mediterranean coasts, this is a realm of very low and highly variable annual precipitation, of searing daytime heat and chilling nighttime cold, of strong winds and dust-laden air. Soils are thin; mountain slopes carry little vegetation. To all this, water brings exception—not only along coasts and rivers but also in oases and *qanats*, tunnels dug into water-bearing rock strata at an angle, so that the water drains to the surface. The population map (Fig. I-9) reminds us how scattered and isolated the population clusters are—as a matter of necessity.

AN ARAB REALM

North Africa/Southwest Asia is also often referred to as the "Arab World." Once again, this implies a uniformity that does not actually exist. In the first place, the name *Arab* is applied loosely to the peoples of this area who speak Arabic and related languages, but ethnologists normally restrict it to certain occupants of the Arabian Peninsula—the Arab "source." In any case, the Turks are not Arabs, and neither (for the most part) are the Iranians or the Israelis. Moreover, although it is true that Arabic is spoken over a wide region that extends from Mauritania in the west across all of North Africa to the Arabian Peninsula, Syria, and Iraq in the east, there are many areas within the realm where it is not used by most of the people.

In Turkey, for example, Turkish is the major language, and it has Ural-Altaic rather than Semitic or Hamitic roots. In Iran, the Iranian language belongs to the Indo-European linguistic family. In Ethiopia, Amharic is spoken by the ruling plateau people; although it is more closely related to Arabic than is Iranian, it nonetheless remains a distinct language as well. The same is true of Hebrew, which is spoken in Israel. Other "Arab World" languages that have separate ethnological identities are spoken by the Tuareg people of the Sahara, the Berbers of northwestern Africa, and the peoples of the transition zone between North Africa and Subsaharan Africa to the south.

AN ISLAMIC REALM

Another name given to this realm is the "World of Islam." The prophet Muhammad (Mohammed) was born in Arabia in A.D. 571, and in the centuries that followed his death in 632, Islam spread into Africa, Asia, and Europe. This was the age of Arab conquest and expansion; their armies penetrated southern Europe, their caravans crossed the deserts, and their ships plied the coasts of Asia and Africa. Along these routes they carried the Muslim (Islamic) faith, converting the ruling classes of the states of the West African savanna, threatening the Christian stronghold in the highlands of Ethiopia, penetrating the deserts of inner Asia, and pushing into India and even the island extremities of Southeast Asia. Islam was the religion of the marketplace, the bazaar, the caravan. Where necessary, it was imposed by the sword, and its protagonists aimed directly at the political leadership of the communities they entered. Today, the Islamic faith with its more than 1 billion followers extends well beyond the limits of the realm discussed here (Fig. 7-2): it is the major religion in northern Nigeria, in Pakistan, and in Indonesia; it is influential in East Africa; and it still prevails in its old strongholds of Albania, Kosovo, and Bosnia in Eastern Europe. Moreover, in South Asia more than 100 million Muslims live in Hindu-dominated India, the world's largest religious minority.

On the other hand, the "World of Islam" is not entirely Muslim either. In Israel, Judaism is the prevailing faith; Christianity remains strong in Lebanon; and ancient Coptic Christian churches still exist in Egypt. Thus the connotation of "World of Islam" when applied to North Africa/Southwest Asia is far from satisfactory: the religion prevails far beyond these areas, and within the realm there are a number of countries in which Islam is not the dominant faith.

"MIDDLE EAST"

Finally, this realm is frequently called the "Middle East." That must sound quite odd to someone in, say, India, who might think of a Middle West rather than a Middle East! The name, of course, reflects the biases of its source: the "Western" world, which saw a "Near East" in Turkey, a

TEN MAJOR GEOGRAPHIC QUALITIES OF NORTH AFRICA/ SOUTHWEST ASIA

1. This realm contains several of the world's great ancient culture hearths and some of its most durable civilizations.
2. The North Africa/Southwest Asia realm is the source of several world religions, including Islam, Christianity, and Judaism.
3. This realm is predominantly but not exclusively Islamic (Muslim). That faith pervades cultures from Morocco in the west to Afghanistan in the east.
4. North Africa/Southwest Asia is the "Arab World," but significant population groups in this realm are not of Arab ancestry.
5. The population of North Africa/Southwest Asia is widely dispersed in discontinuous clusters.
6. Natural environments in this realm are dominated by drought and unreliable precipitation. Population concentrations occur where the water supply is adequate to marginal.
7. The realm contains a pivotal area in the "Middle East," where Arabian, North African, and Asian regions intersect.
8. North Africa/Southwest Asia is a realm of intense discord and bitter conflict, reflected by frequent territorial disputes and boundary frictions.
9. The end of Soviet rule and the revival of Islam in Turkestan have the effect of expanding this realm into inner Asia.
10. Enormous reserves of petroleum lie beneath certain portions of the realm, bringing wealth to these favored places. But overall, oil revenues have raised the living standards of only a small minority of the total population.

"Middle" East in Egypt, Arabia, and Iran, and a "Far" East in China and Japan. Still, the term has taken hold, and it can be seen and heard in everyday usage by journalists as well as members of the United Nations. In view of the complexity of this realm, its transitional margins, and its far-flung areal components, the name Middle East need be faulted only for being imprecise—it does not make a single-factor region of North Africa/Southwest Asia, as do the terms Dry World, Arab World, and World of Islam.

CULTURAL GEOGRAPHY

Cultural geography involves the application of the concept of culture to geographic problems. This multifaceted concept was discussed in the Introduction, and subsequent chapters have shown it to be a key to the systematic understanding of differences and similarities among human societies. In practice, cultural geography is a wide-ranging and comprehensive field whose concerns we can subdivide into five major components: (1) cultural landscapes, (2) culture hearths, (3) cultural diffusion, (4) cultural ecology, and (5) culture regions.

As we noted on p. 20, the term *cultural landscape* refers to the composite of human imprints on the earth's surface. In addition to studying the artifacts of material (tangible) culture, many geographers now also study the landscapes of popular culture. This represents a broadening of the field to include vernacular culture—culture in the context of its sustainers, the people who maintain and nurture it. *Popular culture* is the ever-changing "mass" culture of urbanized and industrialized society that is less tradition-bound, more open, individualistic, and class-structured than strongly traditional *folk culture*, the durable way of life found in the comparatively isolated rural areas. North Africa/Southwest Asia, like most developing realms, is dominated by the latter

(the 1979 overthrow of the shah's regime in Iran was an object lesson in the limitations to bucking this status quo). Among the topics recently researched by vernacular-culture geographers are music and art styles, sports, and food and drink preferences.

Culture hearths are the crucibles of cultural growth and achievement, the source areas from which radiated ideas, innovations, and ideologies that changed the world beyond. Southwest Asia contains the hearths of three major religions—Islam, Christianity, and Judaism; the first developed in western Arabia, the second and third near the eastern shore of the Mediterranean. Even more importantly, this geographic realm was the source of entire civilizations—in Mesopotamia and the Nile valley—as the world map indicates (see Fig. 7-4).

Cultural diffusion is the process of dissemination, the outward spreading of an idea or innovation from its hearth to other places (and frequently to other cultures). Today, most of the world's cultures are the products of innumerable innovations that arrived in an endless, centuries-long stream. Often it is possible to trace the route and timing of the adoption of a particular innovation, so the phenomenon of diffusion is a valuable element in the study of cultural geography (the dynamics of spatial

diffusion are treated in the box entitled "Diffusion Processes" on pp. 368-369). Figure 7-1 shows the result of the diffusion of certain food taboos throughout the North Africa/Southwest Asia realm and in certain neighbors to the southeast. Adherents of Islam and Judaism avoid pork, and those of Hinduism do not eat beef; as the map demonstrates, these food taboos accompanied the spread of their associated religions across northern Africa and much of southern Eurasia.

Cultural ecology is concerned with the multiple relationships between human cultures and their physical environments. However, as we saw in Chapter 5, this tradition of geography in the past attracted proponents of *environmental determinism* and similar viewpoints, and the idea of a mutual and balanced human-habitat interaction did not gain currency until the 1930s. Environmentalism in American geography was countered in the 1920s from two directions. From Western Europe came the short-lived doctrine of *possibilism*, which argued against climatic and physiographic determinism by claiming that people, through their cultures, are free to choose from a number of environmental "possibilities." The second and far more persuasive attack came from within, through the scholarship of Carl Sauer, this century's most

◆ REGIONS IN THE REALM

Identifying and delimiting regions in this vast geographic realm is a considerable challenge. Not only are population clusters widely scattered, but cultural transitions—internal as well as external—make it difficult to discern a regional framework.

As we have noted on several earlier occasions, the world's regional geographic framework is subject to change. When Columbus sailed for the New World, the entire Balkan region of Eastern Europe was under the sway of the Muslim Ottoman Empire, and Islam cast its shadow over Vienna and Venice. In those times, this was not just a North African/Southwest Asian realm but an Eastern European one as well. Little more than a century ago, after the Austrian and

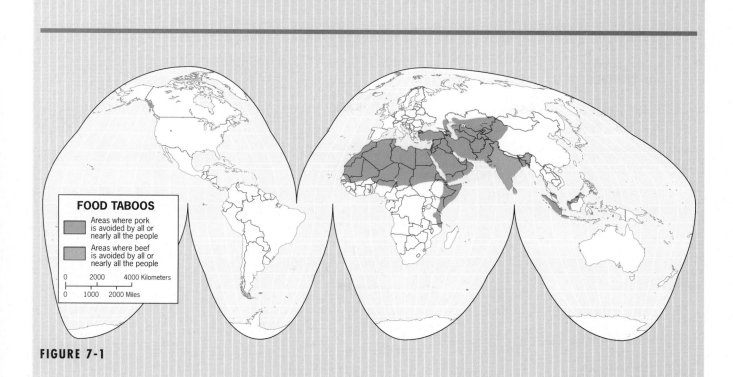

FIGURE 7-1

distinguished cultural geographer. His position, all but unanimously adopted today, regards humans and the natural environment as co-equal partners in an interacting unity, with humans the agents of a stage-by-stage modification of the physical landscape.

The study of *culture regions* involves the identification and mapping of phenomena to delimit territories occupied by human communities that share a particular culture. These can range from local culture complexes to continental-scale assemblages (culture realms) that are even used by some geographers to organize their approach to world regional geography. Here, too, cultural geographers are especially interested in the changing distribution of phenomena over time as innovations emanate from culture hearths and disperse across great distances. They study not only the agents and pathways of transmission but cultural barriers as well. For example, the spatial expansion of Islam was far more rapid than the spreading of the Arabic language because recipient cultures were relatively amenable to religious change but resisted attempts to modify their linguistic traditions.

This concern with hearths and diffusion, of course, underscores the fact that the five elements of cultural geography discussed here are not mutually exclusive, but constantly interact with one another as scholars seek deeper understanding of humanity's rich and fascinating culture systems.

Austro-Hungarian empires had wrested the upper Danube Basin from the Turks, the Muslims still ruled over much of Romania, Bulgaria, Serbia, Bosnia, Albania, and Greece. Not until the second decade of the twentieth century did the Ottomans lose the last of their European holdings, which finally erased the European Muslim region from the map. Ever since, Christian Greece and Muslim Turkey have been antagonistic neighbors.

In Asia, too, Islam penetrated far northward, only to be subjugated. The Muslims spread northward from Persia into the Transcaucasian corridor between the Black and the Caspian seas, but east of the Caspian lay a much larger Muslim region: Turkestan. Here developed a jigsaw of Islamic societies based on the steppes and oases of inner Asia. The Russian tsars, however, had other plans for this region (see Chapter 2). During the nineteenth century, even as the

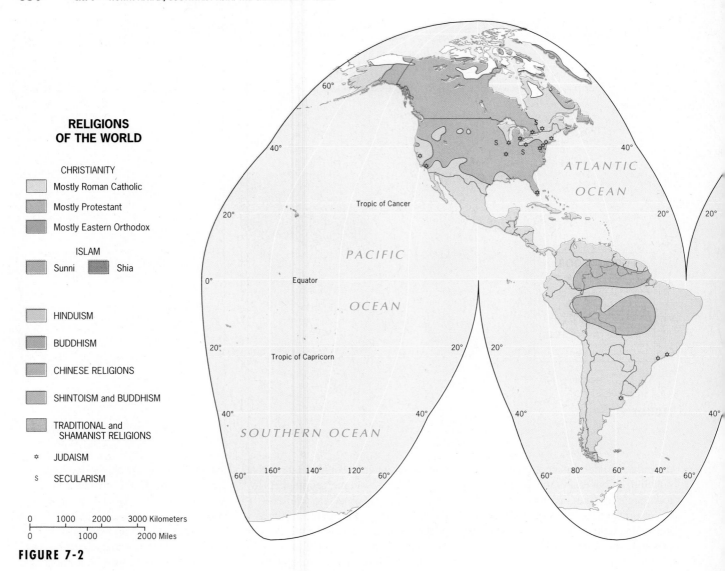

RELIGIONS OF THE WORLD

CHRISTIANITY
- Mostly Roman Catholic
- Mostly Protestant
- Mostly Eastern Orthodox

ISLAM
- Sunni
- Shia

- HINDUISM
- BUDDHISM
- CHINESE RELIGIONS
- SHINTOISM and BUDDHISM
- TRADITIONAL and SHAMANIST RELIGIONS
- ✡ JUDAISM
- s SECULARISM

0   1000   2000   3000 Kilometers
0   1000   2000 Miles

FIGURE 7-2

Ottomans were losing Eastern Europe, the Muslim states of Turkestan proved no match for the Russian armies. Later, when the Soviet communists inherited the Russian Empire, they laid out a boundary framework for their Muslim colonies (which they called Soviet Central Asia) and proceeded to try to extinguish Islam in favor of Moscow's official atheism.

In Eastern Europe, old Christian traditions soon expunged Islam's imprints except in Albania, Kosovo, and Bosnia, where remnants of it survived. But in the five Soviet republics of Turkestan, Islam proved more durable. After the Soviet Empire collapsed in 1991, Islam immediately reasserted itself, and a wave of Muslim fervor swept over the region. As we shall see at the end of this chapter, today Turkey, Iran, and even Pakistan all seek influence in this newly reopened Islamic frontier.

The realm's regional framework, therefore, is changing once again. We must now recognize the inner Asian region as part of it. Change also is affecting Transcaucasia, where the freed former Soviet republic of Azerbaijan displays intense Muslim militancy. The borders of this realm have always been volatile; the mid-1990s are no exception.

THE REGIONAL MATRIX

The vast North Africa/Southwest Asia realm is divided into seven regions (Fig. 7-3). As both Figs. I-14 and 7-3 suggest, it is appropriate to recognize two sets of regions: those to the south of a line connecting the northeastern corner of the Mediterranean Sea and the head of the Persian Gulf, and those to the north. This is a line of contrast, conflict,

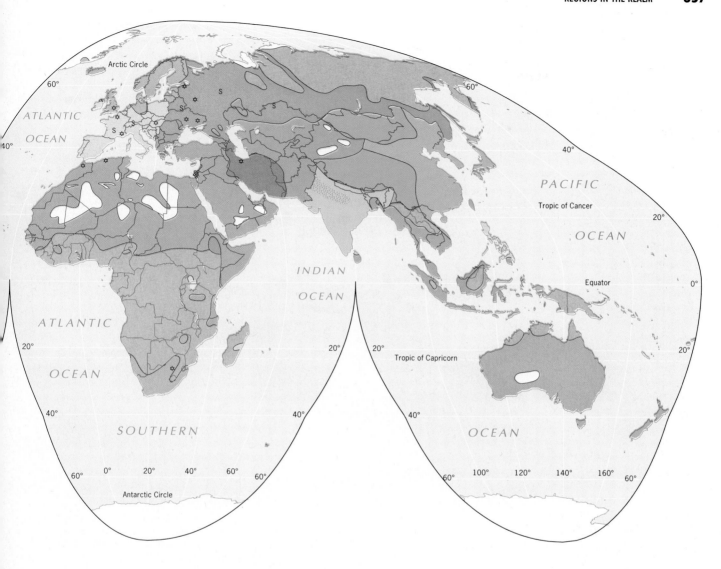

and transition that not only separates Arab from non-Arab states but also divides different Islamic sects. Astride this line lies a nation without a state: the Kurds, victims of intraregional rivalries.

The following are the regional components of this geographic realm:

1. *Egypt and the Lower Nile Basin*. This region in many ways constitutes the heart of the realm as a whole. Egypt (together with Iran and Turkey) is one of the realm's three most populous countries. It is the historic focus of this part of the world and a major political and cultural force. It shares with its southern neighbor, Sudan, the waters of the lower Nile River.

2. *The Maghreb and Its Neighbors*. Western North Africa (the *Maghreb*) and the areas that border it also form a region, consisting of Algeria, Tunisia, and Morocco at the center and Libya, Chad, Niger, Mali, and Mauritania along the broad periphery. The last four of these countries also lie astride or adjacent to the broad transition zone where the Arab-Islamic realm of northern Africa merges into Subsaharan Africa.

3. *The African Transition Zone*. From southern Mauritania in the west to Somalia in the east, across the entire African landmass at its widest extent, the realm dominated by Islamic culture interdigitates with that of Subsaharan Africa. No sharp dividing line can be drawn here: people of African ethnic stock have adopted the Muslim faith and Arabic language and traditions. As a result, this is less a region than a zone of transition.

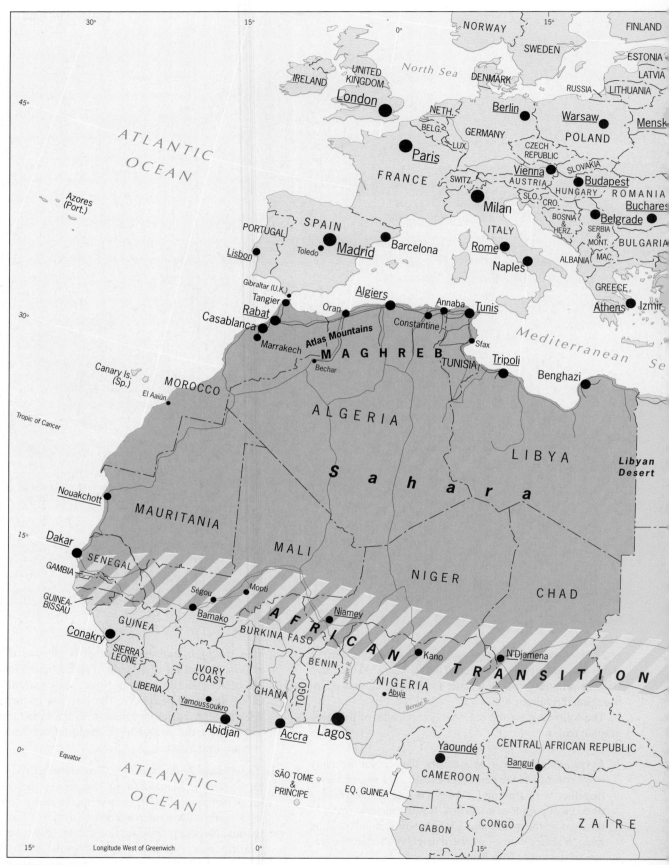

FIGURE 7-3

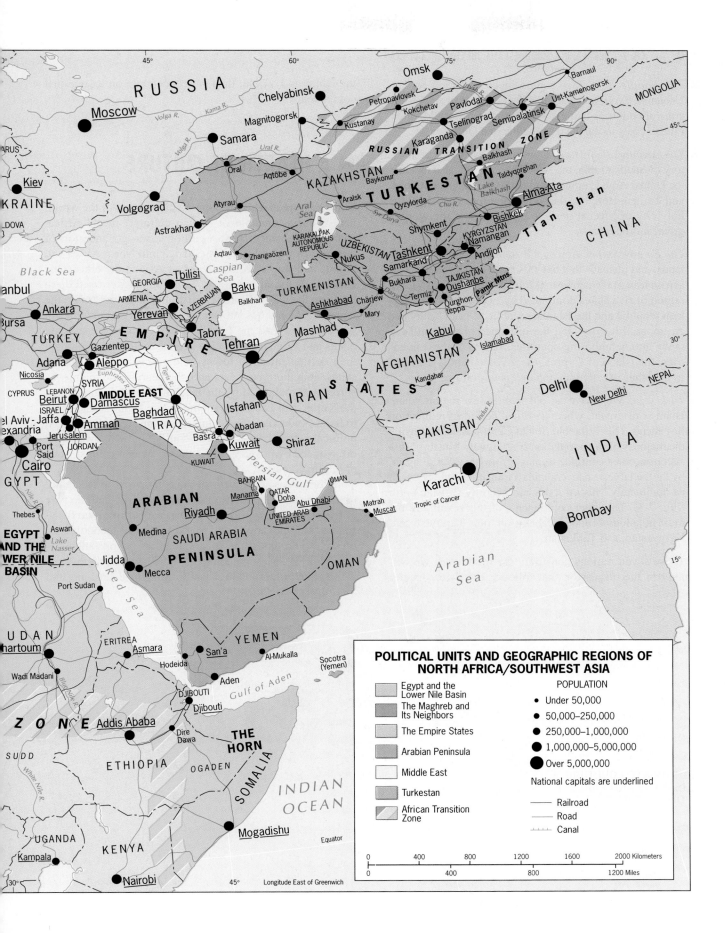

POLITICAL UNITS AND GEOGRAPHIC REGIONS OF NORTH AFRICA/SOUTHWEST ASIA

Egypt and the Lower Nile Basin

The Maghreb and Its Neighbors

The Empire States

Arabian Peninsula

Middle East

Turkestan

African Transition Zone

POPULATION

· Under 50,000

● 50,000–250,000

● 250,000–1,000,000

● 1,000,000–5,000,000

● Over 5,000,000

National capitals are underlined

Railroad

Road

Canal

0   400   800   1200   1600   2000 Kilometers

0   400   800   1200 Miles

4. *The Middle East*. This region includes Israel, Jordan, Lebanon, Syria, and Iraq. In effect, it is the crescent-like zone of countries that extends from the eastern Mediterranean coast to the head of the Persian Gulf.

5. *The Arabian Peninsula*. Dominated by the large territory of Saudi Arabia, the Arabian Peninsula also includes the United Arab Emirates, Kuwait, Bahrain, Qatar, Oman, and Yemen. Here lies the source and focus of Islam, the holy city of Mecca; here, too, lie many of the world's greatest oil deposits.

6. *The Empire States*. Across the zone of mountains, highlands, and plateaus that extends from Turkey in the west across Iran in the center to Afghanistan in the east, lies a region of five states dominated by two: the secular Turkish republic and the Islamic republic of Iran. Both have an imperial history, and the influence of each still extends into neighboring countries. In one of these countries, the island of Cyprus, the involvement is Turkish; in the other, Azerbaijan, the link is with Iran.

7. *Turkestan*. We revive an old name for a region where Islam is resurgent, where a critical period of redirection lies ahead. This is the old Soviet Central Asia, and while the majority of the population is nominally Muslim, Russian minorities continue to play a role here. Kazakhstan, by far the largest state, contains a transition zone between its dominantly Muslim south and majority-Russian north (Fig. 7-3). The other components of this region are the most populous country, Uzbekistan, plus Turkmenistan, Kyrgyzstan, and Tajikistan.

These seven regions constitute the framework we will employ in the discussion that follows. First we consider the realm as a whole; then we concentrate on its regional components.

◆ A HEARTH OF CULTURE

This geographic realm occupies a pivotal part of the world: here Eurasia, crucible of human cultures, meets Africa, source of humanity itself. A million years ago, ancestors of our species walked from East Africa into North Africa and Arabia and spread from one end of Asia to the other. One hundred thousand years ago, *Homo sapiens* crossed these lands on the way to Europe, Australia, and, eventually, the Americas. Ten thousand years ago, human communities in what we now call the Middle East began to domesticate plants and animals, learned to irrigate their fields, enlarged their settlements into towns, and formed the earliest of states. One thousand years ago, the heart of the realm was stirred and mobilized by the teachings of Muhammad and the Koran, and Islam was on the march from North Africa to India. And today, this realm is a cauldron of religious and political activity, weakened by conflict but empowered by oil, plagued by poverty but fired by a fundamentalist wave of religious revival.

In the basins and valleys of the great rivers of this realm (the Tigris and Euphrates of modern-day Iraq and the Nile of Egypt) lay two of the world's earliest **culture hearths**— sources of innovations and ideas, traits and technologies, lifestyles and landscapes that spread far beyond their nuclei (Fig. 7-4). Mesopotamia, the "land between the rivers," was a zone of fertile alluvial soils, abundant sunshine, and

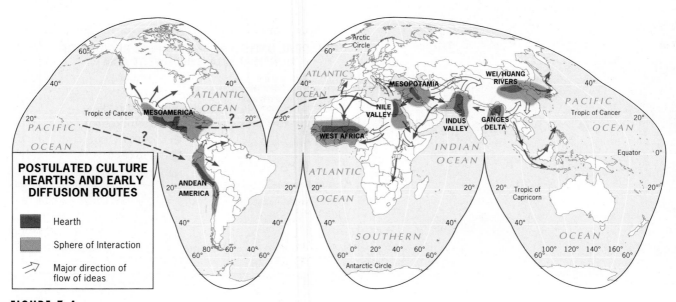

FIGURE 7-4

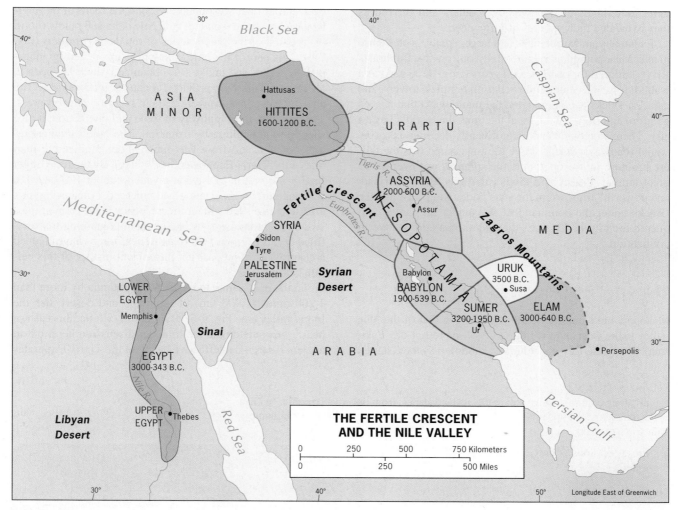

FIGURE 7-5

ample water, and here, in the Tigris-Euphrates lowland between the head of the Persian Gulf and the uplands of present-day Turkey, arose one of humanity's first civilizations. Mesopotamia's agricultural know-how, which involved planned sowing and harvesting and the distribution of surplus grain for storage and later consumption, radiated to villages far away. Eventually the wider region was anchored by a **Fertile Crescent**, a zone of advanced farming that extended from the Mediterranean coast to very near the Persian Gulf (Fig. 7-5).

MESOPOTAMIA

Before the agricultural innovations from Mesopotamia spread far and wide, this region (like others in the realm) was a patchwork of small farm villages containing up to 200 inhabitants. Now farm surpluses made life more secure, and some of the villagers could engage in nonfarm work (for example, toolmaking, record-keeping, or policing). Now certain villages, advantageously located or otherwise favored, grew into towns. Urbanization had begun along with social class formation; military forces were established to protect the towns; competition among cities grew. Some cities acquired large hinterlands, and the region became a mosaic of early states. Some of this is recorded in the world's earliest maps: the ancient Mesopotamians sketched the layout of towns and fields in fine-grained clay, baked by the sun. A few of these old maps have survived.

Why did certain ancient cities thrive while others declined? Various theories have been advanced to explain the success of some and the failure of others. One of these proposes that irrigation was the key to progress and power. Knowledge of irrigation not only produced higher yields of food crops but was also a weapon: the city government could deny water to farmers, villages, and towns situated in less fortuitous locations. The **hydraulic civilization** theory holds that cities able to control irrigated farming over large hinterlands held power over others and thrived as they expanded their spheres of influence.

Eventually, powerful cities and large, durable states arose in the Mesopotamian region. Babylon, on the Euphrates River, endured for nearly 2,000 years (from 4100 B.C.) as a walled, fortified center endowed with temples, towers, and palaces. Docks accommodated the many boats that carried goods to and from what was, at the time, the world's largest city. There were wide and magnificently decorated processional roadways; rising above the townscape was the tallest structure in Mesopotamia, the *ziggurat* or tower of the great temple. Priests and kings ruled city and hinterland, and powerful armies kept the state under control. When the Greeks under the command of Alexander the Great conquered the city in 331 B.C., they claimed a cornerstone of civilization.

THE NILE AND INDUS VALLEYS

Mesopotamia's culture hearth lay between that of the Nile valley to the west and the Indus valley (in present-day Pakistan) to the east. Egypt's cultural evolution may have started

even earlier than Mesopotamia's, and the focus of this civilization lay above (south of) the Nile delta and below (north of) the first of the Nile's series of rapids or *cataracts* (Fig. 7-5). This part of the Nile valley lies surrounded by inhospitable desert, and unlike Mesopotamia (which lay open to all comers), the Nile provided a natural fortress here. The ancient Egyptians converted their security into progress. The Nile was their highway of trade and interaction; it also supported agriculture by irrigation. The Nile's cyclical regime of ebb and flow was much more predictable than that of the Tigris-Euphrates river system. By the time Egypt finally fell victim to outside invaders (about 1700 B.C.), a full-scale urban civilization had emerged. Ancient Egypt's artist-engineers left a magnificent legacy in the form of massive stone monuments, some of them containing treasure-filled crypts of prominent rulers. These tombs have enabled archeologists to reconstruct the ancient history of this culture hearth.

To the east, separated from Mesopotamia by more than 1,200 miles (1,900 km) of mountain and desert, lies the Indus valley (see Fig. 9-6). By modern criteria, this eastern hearth lies outside the realm under discussion here; but in ancient times it had effective ties with the Tigris-Euphrates

Egypt's long cultural history still marks its landscapes, from the great pyramids at Giza to the temples of the upper Nile valley. These lions flank the approach to a Pharaonic palace at Luxor.

region (Fig. 7-4). Mesopotamian innovations reached the Indus region early, and eventually the cities of the Indus became power centers of a civilization that reached far into present-day northern India.

LASTING LEGACIES

Today, the world continues to benefit from the accomplishments of the ancient Mesopotamians and Egyptians. They domesticated cereals (wheat, rye, barley), vegetables (peas, beans), and fruits (grapes, apples, peaches); they also domesticated many animals (horses, pigs, sheep). They advanced not only irrigation and agriculture but also calendrics, mathematics, astronomy, government, engineering, metallurgy, and a host of other skills and technologies. As time went on, many of their innovations were adopted and then modified by other cultures in the Old World and eventually in the New World as well. Europe was the greatest beneficiary of these legacies of Mesopotamia and ancient Egypt, whose achievements constituted the very foundations of "Western" civilization.

◆ STAGE FOR ISLAM

Today, many of the great cities of this realm's culture hearths are archeological curiosities. In some instances, new cities have been built on the sites of the old, but the great ancient cultural traditions of North Africa/Southwest Asia went into a deep decline after many centuries of continuity. It cannot escape our attention that a large number of the ruins of these ancient urban centers are located in what is now desert. Presuming that they were not built in the middle of these drylands, it is tempting to conclude that **climate change**, associated with shifting environmental zones in the wake of the last Pleistocene glacial retreat, destroyed the old civilizations. Indeed, some geographers have suggested that the momentous innovations in agricultural planning and irrigation technology may have been made in response to changing environmental conditions as the riverine communities tried to survive.

The scenario is not difficult to imagine: as outlying areas began to fall dry and farmlands were destroyed, people congregated in the already crowded river valleys—and every effort was made to increase the productivity of lands that could still be watered. Eventually overpopulation, destruction of the watershed, and perhaps reduced rainfall in the rivers' headwater areas combined to deal the final blow. Towns were abandoned to the encroaching desert; irrigation canals filled with drifting sand; remaining croplands dried up. Those who could migrated to areas reputed still to be pro-

ductive. Others stayed, their numbers dwindling, increasingly reduced to subsistence.

As old societies disintegrated, new power emerged elsewhere. First the Persians, then the Greeks, and later the Romans imposed their imperial designs on the tenuous lands and disconnected peoples of the Middle East. Roman technicians converted North Africa's farmlands into irrigated plantations whose products went by the boatload to Roman Mediterranean shores. Thousands of people were carried in bondage to the cities of the new conquerors. Egypt was quickly colonized, as was the area we now define as the Middle East. Significantly, the Arab settlements on the Arabian Peninsula were more distant and therefore more remote from these invasions—and hence somewhat more secure in their isolation.

RELIGIOUS ROOTS

The ancient cities of the Middle East were centers of religious authority as well as military power. There were numerous local gods, represented by priests whose authority was confirmed by their impressive temples. Yet not all of the region's peoples were sedentary (settled) farmers: always there were those who trekked with their livestock in perpetual search of water and pasture. The British geographer William Fisher, in his book *The Middle East*, notes that these pastoral peoples tended to combine their *nomadic* lifeways with their own religious view: the notion of a single god with whom a close individual relationship, lasting a lifetime and beyond, can exist. "Practically all of the great monotheistic religions of the world," states Fisher, "seem to have arisen on the margins of the great deserts of Eurasia, where pastoral peoples come into contact with the polytheism of settled areas." And indeed, long before the faith that was to galvanize this realm arose, other religions emerged: Zoroastrianism in what is today Iran, and Judaism and Christianity in modern-day Israel. But the teachings of Zoroaster remained confined to Persia, Judaism was devastated by the Babylonians and again, later, by the Romans, and Christianity became an alien faith as its center moved northwestward to Rome. Until the early seventh century A.D., none of the religions that had their origins in this realm became dominant within it.

MUHAMMAD THE PROPHET

In a remote place on the Arabian Peninsula, where Arab communities had been little affected by the foreign invasions of the Middle East, an event occurred early in the seventh century that was to change the course of history and affect the destinies of people in many parts of the world.

Islam's impact has spread far beyond its Arabian hearth, a diffusion etched on cultural landscapes from Morocco to Indonesia. The Selangor State Mosque (in the Southeast Asian country of Malaysia) combines traditional Islamic design with highly modern construction: the characteristic blue dome is made of aluminum, not tile.

In a town called Mecca, about 45 miles (70 km) from the Red Sea coast in the Jabal Mountains, a man named Muhammad in the year A.D. 613 began to receive a series of revelations from Allah (God). Muhammad (571–632) already was in his early forties, and he had barely 20 years to live. Convinced after some initial self-doubt that he was indeed chosen to be a prophet, Muhammad committed his life to the fulfillment of the divine commands he believed he had received. Arab society was in social and cultural disarray, but Muhammad forcefully taught Allah's lessons and began the transformation of his culture. His personal power soon attracted enemies, and he was forced to flee Mecca for the safer haven of Medina, where he continued his work. Mecca, of course, later became Islam's holiest place.

The precepts of Islam in many ways constituted a revision and embellishment of Judaic and Christian beliefs and traditions. There is but one god, who occasionally communicates through prophets; Islam acknowledges that Moses and Jesus were such prophets. What is earthly and wordly

is profane; only Allah is pure. Allah's will is absolute; Allah is omnipotent and omniscient. All humans live in a world created for their use, but only to await a final judgment day.

Islam brought to the Arab world not only the unifying religious faith it had lacked but also a new set of values, a new way of life, a new individual and collective dignity. Islam dictated observance of the *Five Pillars:* (1) repeated expressions of the basic creed, (2) the daily prayer, (3) a month of daytime fasting (*Ramadan*), (4) the giving of alms, and (5) at least one pilgrimage to Mecca. And the faith prescribed and proscribed in other spheres of life as well. Alcohol, smoking, and gambling were forbidden. Polygamy was tolerated, although the virtues of monogamy were acknowledged. Mosques appeared in Arab settlements, not only for the (Friday) sabbath prayer, but also as social gathering places to bring communities closer together. Mecca became the spiritual center for a divided, widely dispersed people for whom a collective focus was something new.

THE FLOWERING OF ISLAMIC CULTURE

The conversion of a vast realm to Islam was not only a religious conquest: the rise of Islam was accompanied by a glorious explosion of Arab culture. In science, in the arts, in architecture, and in other fields, Arab society far outshone Europe. While remnants of the Roman Empire languished, Arab energies soared.

When the wave of Islamic diffusion reached the Maghreb, the Arabs saw, on the other side of the narrow Strait of Gibraltar, an Iberia ripe for conquest and ready for renewal. An Arab-Berber alliance, the *Moors*, invaded Spain in 711, and before the end of the eighth century all but northern Castile and Catalonia were under Arab control.

It took seven centuries for Spain's Catholic armies to recapture all of the Arabs' Iberian holdings, but by then the Muslims had made an indelible imprint on the Spanish cultural landscape. The Arabs brought unity and imposed the rule of Baghdad, and their works soon overshadowed what the Romans had wrought. *Al-Andalus*, as this westernmost outpost of Baghdad was called, was endowed with thousands of magnificent mosques, castles, schools, gardens, and public buildings (see photo at right). The ultimately victorious Christians destroyed most of the less durable art (pottery, textiles, furniture, sculpture) and burned the contents of libraries, but the great Islamic structures survived, including the Alhambra in Granada, the Giralda in Seville, and the Great Mosque of Córdoba, three of the world's greatest architectural achievements. While Spanish culture became Hispanic-Islamic culture, the Muslims were transforming their cities from Turkestan to the Maghreb in the image of Baghdad. The lost greatness of a past era still graces those townscapes today.

Islam's penetration of Iberia permanently transformed the cultural landscapes of many Spanish cities. The Mesquita Mosque in Córdoba is only one of numerous Islamic legacies in Spain's southern province of Andalucía.

THE ARAB EMPIRE

The stimulus, spiritual as well as political, that Muhammad provided was so great that the Arab world was mobilized almost overnight. The prophet died in 632, but his faith and fame continued to spread like wildfire. Arab armies formed, invaded, and conquered, and Islam was carried throughout North Africa. By the early ninth century, the Muslim world included emirates or kingdoms extending from Egypt to Morocco, a caliphate occupying most of Spain and Portugal, and a unified region encompassing Arabia, the Middle East, Iran, and much of Pakistan. Muslim forces had attacked France and Italy and penetrated deeply into inner Asia. Ultimately, the Arab Empire extended from Morocco to tropical Asia and from Turkestan to the southern margins of the Sahara (Fig. 7-6). The original capital was at Medina in Arabia, but in response to these strategic successes it was moved, first to Damascus and then to Baghdad. In the fields of architecture, mathematics, and science, the Arabs far overshadowed their European contemporaries (see the box entitled "The Flowering of Islamic Culture"); they also established institutions of higher learning in several cities,

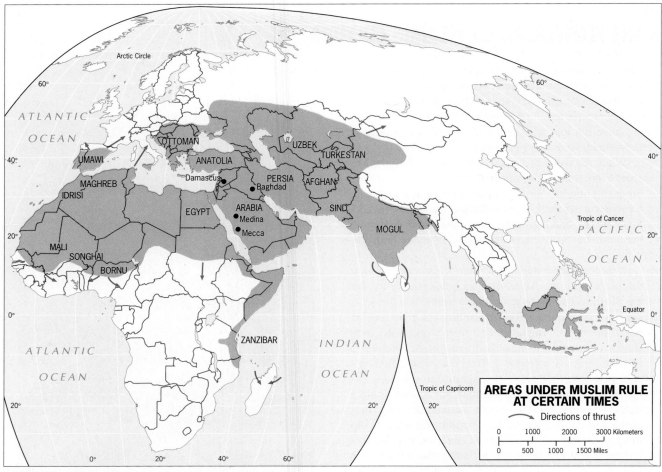

FIGURE 7-6

including Baghdad, Cairo, and Toledo in Spain. The faith had spawned a culture and is still at the heart of that culture today.

The spread of the Islamic faith throughout North Africa/Southwest Asia (and beyond) occurred in waves radiating from Medina and the nearby holy city of Mecca. Islam went by camel caravan and by victorious army; it was carried by pilgrim and sailor, scholar and sultan. Its dissemination through so vast and discontinuous a realm is an example of the process of *spatial diffusion* (see the box on pp. 368-369 entitled "Diffusion Processes").

The worldwide diffusion of Islam continues today. In the United States, it is emerging in the religious movement commonly called the Black Muslims but officially known as the American Muslim Mission. There are Islamic communities in South Africa, the Philippines, and other far-flung places. This is not the first time that the Middle East has generated an idea that affected much of the world. Agricultural methods, metallurgical techniques, architectural styles, and countless other innovations that developed

in this realm have, throughout the course of recorded history, been adopted by societies across the globe.

◆ DIVISIVE FORCES

Islam brought purpose, energy, and regulation to a fractious part of the world. It ushered in an age of empire, a time of seemingly limitless expansion and conversion of heathens. The realm of Islam grew ever larger—into southern Europe, western Africa, southwestern Asia, India, and beyond (see Fig. 7-6). Achieving expansion was one thing; attaining lasting unity was quite another. Even as Islam's domain spread outward, the seeds of division were sown.

THE SUNNI-SHIA DIVISION

Muhammad received truths and revelations from Allah, but *not* among these was any guidance as to Islam's religious leadership after the prophet's death. Should Islam, after Muhammad, be led by a prominent person in the community where the prophet lived, or should the new leader be a blood relative of Muhammad himself?

In Mecca and Medina, several candidates had large numbers of supporters. The first chosen successor was the father of Muhammad's wife (and thus not a blood relative). The next two leaders were men who had been devout followers of the prophet. None of this, however, satisfied the supporters of Muhammad's cousin. When he, a man named Ali, was made *caliph* (successor), his followers proclaimed that the first true heir of the prophet had been installed. By the time Ali died, the Muslims were divided into two sects: the *Sunnis*, who believed that the first appointed successors of Muhammad were as legitimate as any, and the *Shi'ites*, who held that only Ali was the prophet's sanctioned heir. From the beginning of this disagreement, the numbers of Muslims who took the Sunni side far exceeded those who regarded themselves as Shia (followers) of Ali. The great expansion of Islam was largely propelled by Sunnis; the Shi'ites survived as small minorities scattered throughout the realm. Today, about 85 percent of all Muslims are Sunnis.

But the Shi'ites were masters of political manipulation who vigorously promoted their version of the faith. In the early sixteenth century their work paid off: the royal house of Persia (modern-day Iran) made Shi'ism the only legal religion throughout its vast empire. That domain extended from Persia into lower Mesopotamia (modern Iraq), into Azerbaijan, and into western Afghanistan and Pakistan. As the map of religions (Fig. 7-2) shows, this created for Shi'ism a large culture region and gave the faith unprecedented strength. Iran remains the bastion of Shi'ism in the realm today, and the appeal of Shi'ism continues to radiate into neighboring countries and even farther afield.

The differences between Sunni and Shia versions of Islam have had profound consequences for the realm. Sunni Muslims believe in the effectiveness of family and community in the solution of life's problems, whereas Shi'ites follow their infallible *imams* (mosque officials who lead worshippers in prayer), the sole source of true knowledge. Sunnis tend to be comparatively reserved; Shi'ites often are passionate and emotional. For example, the death of Ali's son, Husayn, who would have been caliph, is commemorated annually with intense processions during which the marchers beat themselves with chains and cut themselves with sharp metal instruments. To many Sunnis, this is an unseemly demonstration of excess.

During the last decades of the twentieth century, Shi'ism gained unprecedented influence in the realm. In its heartland, Iran, a *shah* (king) tried to secularize the country and to limit the power of the imams; he provoked a revolution that cost him the throne and made Iran an Islamic republic—in fact, a Shi'ite republic. Before long, Iran was at war with Sunni-dominated Iraq, and Shi'ite parties and commu-

nities elsewhere were invigorated by the newfound power of Shi'ism. From Arabia to the Maghreb, Sunni-ruled countries warily watched their Shi'ite minorities, newly imbued with religious fervor. Mecca, the holy place for both Sunnis and Shi'ites, became a battleground during the week of the annual pilgrimage, and for a time the (Sunni) Saudi Arabian government denied entry to Iranian Shi'ite pilgrims. The split between Sunni and Shia is not the only fracture in Islam, but it is geographically the most consequential.

FUNDAMENTALISM

Another cause of intra-Islamic conflict lies in the resurgence of *fundamentalism*. In Shi'ite Iran, the imams wanted to reverse the shah's moves toward liberalization and secularization: they wanted to (and did) recast society in traditional, fundamental Islamic molds. An *ayatollah* (leader under Allah) replaced the shah; Islamic rules and punishments were instituted. Women, considerably liberated and educated during the shah's regime, resumed roles more in keeping with traditional Islam. The vestiges of Westernization, encouraged by the shah, disappeared. The war against Iraq began as a conflict over territory but became a holy war that cost hundreds of thousands of lives.

Islamic fundamentalism did not rise in Iran alone, nor was it confined to Shi'ite communities. Many Muslims— Sunnis as well as Shi'ites—in all parts of the realm disapproved of the erosion of traditional Islamic values, the corruption of society by European colonialists and later by Western modernizers, and the declining power of the faith in the secular state. While economic times were good, such dissatisfaction stayed beneath the surface. But when jobs were lost and incomes declined, the appeal of a return to fundamental Islamic ways increased.

This set Muslim against Muslim in all the regions of the realm. Fundamentalists fired the faith with a new militancy, challenging the status quo in countries ranging from Pakistan to Algeria. The militants forced their governments to ban "blasphemous" books, to resegregate the sexes in schools, to enforce traditional dress codes, to legitimize religious-political parties, and to heed the wishes of the *mullahs* (teachers of Islamic ways). Militant Muslims proclaimed that democracy inherited from colonialists and adopted by Arab nationalists was incompatible with the rules of the Koran.

In Algeria, the growing power of the fundamentalists led to a crisis. In late 1991, during democratic elections, the so-called Islamic Salvation Front (ISF) showed great strength. It was clear that the ISF would defeat the government in the second round of voting scheduled some weeks later; and leaders of the ISF stated that they would transform Algeria into an Islamic republic. The government preempted the militants by resigning, leaving the door open for the armed forces to take control. Algeria was plunged into a long and

DIFFUSION PROCESSES

Diffusion is defined as the spread or dissemination of a phenomenon across space and through time. Phenomena subject to diffusion are many: culture traits, technological innovations, diseases (the diffusion of AIDS is a recent example), political ideas, and religious practices are among examples.

Diffusion is a spatial process and is therefore of geographic interest. Understanding it allows us to reconstruct the dispersal of cultural and technological ideas in the past; it also helps us predict the impact of future epidemics or the likelihood that a certain invention (say, a television set with higher-resolution images) will find a market to justify the cost of production.

The study of spatial diffusion, because of its time dimension, is rooted in historical geography. The American cultural geographer Carl Sauer studied the diffusion of ancient agriculture. The Swedish geographer Torsten Hägerstrand in 1952 published a fundamental work on this topic entitled *The Propagation of Innovation Waves*. This carried the study of diffusion processes from the descriptive and historical approach to the theoretical and predictive.

Diffusion takes place when an idea, invention, virus, or other phenomenon spreads through a population. Sometimes an innovation spreads throughout a population and is adopted by virtually everyone. At other times, an innovation wave (or outbreak of a virus) is weak and fades away before it reaches the great majority of the people. Islam's propagation wave was enormously strong, and the faith "saturated" the vast majority of those in its path. Hägerstrand's work identified four stages in the diffusion process: a *primary* stage, when an innovation appears at its source and begins to be adopted there (for Islam, Muhammad's teachings in and around Medina in the early decades of the seventh century); a *diffusion* stage, the full-scale spreading of the innovation as it radiates outward, creating new centers of dispersal as it does; the *condensing* stage, when isolated and outlying areas are

finally reached by the innovation wave; and the *saturation* stage, when the process slows down and ends.

Geographers recognize two broad categories of diffusion process: expansion diffusion and relocation diffusion. Each category includes several ways in which diffusion can take place.

The spread of Islam (Fig. 7-7) is an example of **expansion diffusion**, although in later stages Islam dispersed by relocation diffusion. Expansion diffusion proceeds as the term implies: an innovation develops in a source area and stays strong there while spreading across an ever larger population and territory. Adjacent individuals in a locationally fixed population are thus affected as the innovation wave passes by. This kind of expansion diffusion, in which local proximity is important and virtually everyone is affected, is called *contagious diffusion*. The term, appropriately, comes from medical geography, and viruses spread by contagious diffusion—as we know all too well from our exposure to cold and influenza viruses. An innovation can also spread by leapfrogging over a wider area than the local dispersion pattern associated with contagious diffusion. At the national and continental scales, the urban hierarchy becomes the main channel of diffusion. This process of *hierarchical diffusion* starts in the primate or largest city; the innovation subsequently courses downward through progressively smaller cities, towns, and villages until the saturation stage is reached (by transformation to contagious diffusion in the hinterlands of the smaller urban centers). Nationwide flu outbreaks commonly follow such paths; similarly, new products and fashions first appear in the largest metropolitan areas and then trickle down the national urban system.

In the **relocation diffusion** process, phenomena are spread not within a fixed population but by adopting individuals who move and carry innovations (or infections) with them. As Fig. 7-2 shows, there is no contiguity between the major arena of Islam (extending from Maurita-

deep crisis of confrontation and violence; in 1992, the head of state was assassinated by fundamentalists.

Already, Sudan, Pakistan, and of course Iran are officially Islamic republics. To many orthodox Sunnis who prefer to keep mosque and state separate, what has happened in Sudan is frightening. In 1989, a democratically elected government there was overthrown in a military coup. The army allowed the leaders of the National Islamic Front to institute Islamic laws; "nonbelievers" were purged, and the *sharia* criminal code was introduced. (The sharia laws pre-

scribe corporal punishment, amputations, stoning, and lashing for major and minor offenses.) Algeria's fundamentalists often referred to Sudan as their model as the elections approached, because Sudan proved that an Islamic state could indeed be established in a Sunni-dominated country. The specter of a Sudan-style regime in Algiers helped precipitate the crisis of 1991- .

The rift between more liberal Muslims and fundamentalists poses a major challenge for the future. Militant Muslims confront the governments of Egypt, Jordan, Tunisia,

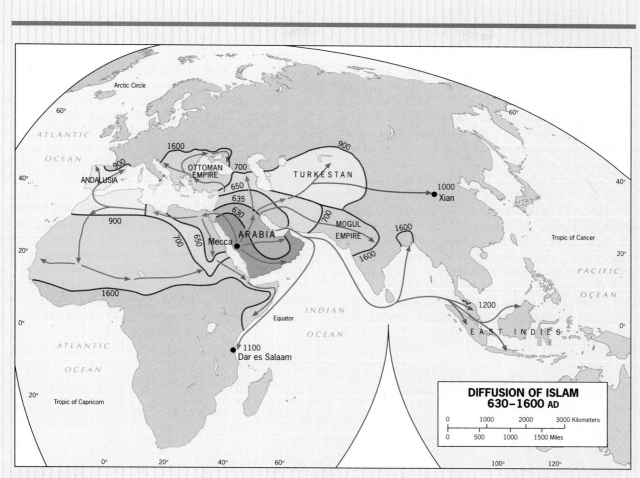

FIGURE 7-7

nia to Pakistan) and Indonesia, where the Muslim faith also took hold. Islam was brought to Indonesia by Arab and Persian seafarers, who created centers of dispersal on those distant islands. Relocation diffusion also carried

Islam to the southern tip of Africa and across the Atlantic to the Americas.

The diffusion of Islam continues today. In its condensing stage, this durable religion still wins converts.

and even Turkey; these governments have reacted in various ways (in Egypt, a combination of appeasement and containment; in Tunisia and Algeria, repression; in Jordan, co-option). Other countries, such as Morocco and Saudi Arabia, clearly are vulnerable. Hanging in the balance is the future of the realm.

We should not be led to believe that fundamentalism and religious militancy are exclusively Islamic phenomena. The fundamentalist notion also challenges Christian, Judaic, Hindu, and Buddhist communities and societies. No-

where, however, does the fundamentalist drive match the intensity and vigor it displays in the Muslim realm.

OTHER RELIGIONS

Still another source of division in the North Africa/Southwest Asia realm lies not within Islam but between Islam and other religions. Christianity had arisen even earlier than Islam, not far from the eastern shores of the Mediterranean Sea. When Christianity emerged, the Romans were in control of much of the *Levant* (as this area, extending from eastern Greece to northern Egypt along the Mediterranean coast, is sometimes called). Christianity won acceptance in the Roman Empire, and the faith diffused westward and northward to tribes living beyond the Romans' domain. By the time Islam arose, Christianity was established from Spain to Turkey. Like Islam, Christianity was dividing into sects: Roman Catholicism, the Eastern Orthodox churches, and later the Protestant denominations. In Europe as in North Africa and Southwest Asia, religion often was a cause for war; strife between Catholics and Protestants in Northern Ireland continues to this day.

When the wave of Islamic diffusion reached into Iberia from North Africa, however, Christians in Europe mobilized not only to meet the threat but also to resurrect Christianity in the region of its source. The Arabs had colonized the whole of southern and central Spain and Portugal, had occupied Sicily, and threatened Rome itself; but by the eleventh century the Islamic diffusion wave had weakened and the Muslims were repulsed. Now began the Crusades, a series of military expeditions whose aim was to reestablish Christianity in the heart of the Muslim domain. The

campaign at first succeeded in creating Christian footholds in several locales, including Jerusalem, but the Muslim forces fought back vigorously. In 1187, Jerusalem fell to Muslim armies, and by the end of the thirteenth century the Christians had been persuaded to halt their attempts to penetrate the realm of Islam.

The aftermath of the Crusades, however, still marks the cultural landscape of the realm today. A substantial Christian minority (about one-fourth of the population) remains in Lebanon; small Christian minorities also survive in Jerusalem and in Syria, Egypt, and Jordan. Strained relations between the long-dominant Christian minority in Lebanon and the Muslim majority (more than half of it Shi'ite) contributed to the disastrous armed struggle that engulfed this country during the 1970s and 1980s.

The Jewish religion, Judaism, predated both Christianity and Islam by many centuries; it, too, had its source in the Levant. The rise of Islam overpowered the smaller Jewish communities, but it was the Christians, not the Jews, who waged centuries of holy war against the Muslims. Jewish communities existed in most areas of the realm, from Morocco to Iran and from Turkey to the Horn of Africa. Then, in 1948, Judaism gained territorial expression as the newly

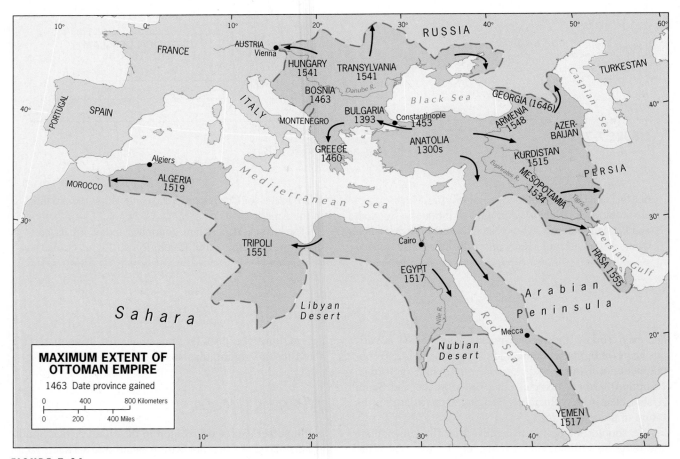

FIGURE 7-8A

constituted state of Israel, and Palestine, the area between the Mediterranean Sea and the Jordan River northeast of the Sinai Peninsula, became a battleground for Arab and Jew. As we will note later, the relationships between individual Arab countries and Israel have created further friction among Muslim states in the realm. Jerusalem—holy city for Judaism, Christianity, and Islam—lies in the crucible of regional conflict.

TERRITORY AND BOUNDARIES

It is a geographic twist of fate that the last great expansion of Islamic power into Europe bequeathed the heartland of the North Africa/Southwest Asia realm with some of its most difficult and divisive problems: issues of territory and boundary. Long after the Muslims had withdrawn from Spain and after the Crusades had run their course, a powerful Islamic empire arose centered on what is today northwestern Turkey, in the area of Anatolia. Here lay the Byzantine remnant of the Roman Empire, and the city of Constantinople (now Istanbul) was the "Rome" of the eastern churches of Christianity. But the Ottomans (named after

Osman I, who started the Islamic drive) conquered Constantinople in 1453, and soon controlled much of Eastern Europe. They were on the doorstep of Vienna; they also captured western Persia and Mesopotamia, Egypt, and a large part of the Maghreb (Fig. 7-8A). The Ottoman Empire under Suleyman the Magnificent, who ruled from 1522 to 1560, was the most powerful state of its time in all of western Eurasia.

Today, Turkey is all that remains of this vast domain. After Suleyman's reign, the Austrians and the Russians pushed the Ottoman boundaries back, the Balkan countries freed themselves, and Egypt threw off the Ottoman yoke. After World War I and following the creation of the modern state of Turkey, the European colonial powers took over the territories that were to become Syria, Iraq, Jordan, Lebanon, and Israel.

The collapse of the Ottoman-Turkish Empire paved the way for the Europeans to achieve what the Crusades could not: to take control over the Islamic societies of North Africa, the Middle East, and Arabia (Fig. 7-8B). Syria and Lebanon came under French rule, as did Algeria and Tunisia and neighboring Morocco. The British controlled Egypt and Sudan, Iraq, and the areas of Jordan and Israel (then Pales-

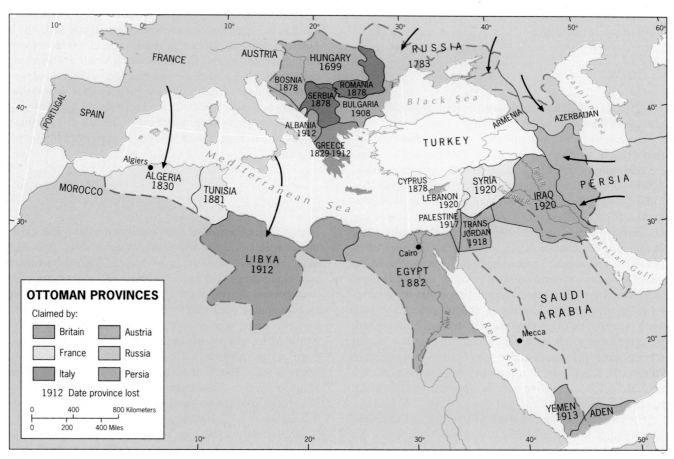

FIGURE 7-8B

ESTABLISHING BOUNDARIES

States have capitals, core areas, administrative divisions, and boundaries. Boundaries are sensitive parts of the anatomy of a state: just as people are territorial about their individual properties, so nations and states are sensitive about their territories and limits. Boundaries, in effect, are contracts between neighboring states. The *definition* of a boundary is contained in an elaborate treaty that verbally describes its precise location. Cartographers then perform the *delimitation* (official mapping) of what the treaty stipulates. Certain boundaries are actually placed on the ground as fences, walls, or other artificial barriers: this represents the *demarcation* of the boundary. In 1993, Kuwait was in the process of demarcating its new, United-Nations-defined desert boundary by digging a deep trench to reinforce it, in the hope of preventing a surprise Iraqi overrun of the kind it experienced in 1990. The world map shows that some boundaries have a sinuous form while others are straight lines. Boundaries can therefore be classified as *geometric* (straight-line or arc), *physiographic* (coinciding with rivers or mountain crests), or *anthropogeographic* (marking breaks or transitions in the cultural landscape).

tine). Even the Italians established themselves in Libya and also along the shores of the Red Sea in Eritrea.

To delimit their colonial holdings and to facilitate administration, the Europeans laid out a framework of political **boundaries**. Those boundaries (with some later modifications) established the politico-geographical map (Fig 7-3) of the realm. As Fig. I-9 reminds us, this realm's population of 448 million is clustered, fragmented, and strung out in river valleys, coastal zones, and crowded oases. Long stretches of boundary lie across uninhabited territory, often in ruler-straight lines. There seemed to be no need to adjust these *geometric* boundaries to physical or cultural features in the landscape. Elsewhere, though, boundaries were created to separate populations or to allocate water and soil. Such boundaries, often inadequately defined or delimited, were destined to produce conflict, and many did. In still other places—for instance, along the frontier between Saudi Arabia and Yemen—no firm boundary definition was accomplished, and the respective nations were left with potential trouble.

The colonial period did much to stabilize and functionally organize the states of this vast realm, but it also left a legacy of unsettled territorial quarrels. When Iraq invaded Kuwait in 1990 (see box p. 386), its regime proclaimed that the boundary between the two countries was colonial, non-Islamic, and therefore null and void. It would not be the last of such actions or claims (see the box entitled "Establishing Boundaries").

◆ THE POWER AND PERIL OF OIL

Travel through the cities and towns of North Africa and Southwest Asia, talk to students and shopkeepers, taxi drivers and travellers, and you will hear the same refrain over and over again: leave us in peace, let us do things our traditional way. Our problems—with each other, with the world—seem always to result from outside interference. The stronger countries of the world exploit our weaknesses and magnify our quarrels. We want to be left alone.

That wish might have come closer to fulfillment were it not for two relatively recent events: the creation of the state of Israel and the discovery of some of the world's largest oil reserves. We defer our discussion of the founding of Israel until the regional section of this chapter. Here the focus is on the realm's most valuable export product: oil.

It is sometimes said that oil is the realm's most valuable *resource*. That, of course, is not accurate. The great majority of the people here continue to be farmers. The realm's most valuable resources are water and tillable soil. Oil is indeed a resource, an energy resource, and it is in demand throughout the world. But there was a Muslim world long before the first barrel of oil was taken from the ground, and there are tens of millions of people whose lives, to this day, are only indirectly affected by oil revenues. Egypt has very little oil (a few wells produce oil in the Sinai Peninsula). Morocco has almost none. Turkey depends heavily on oil imports.

LOCATION OF KNOWN RESERVES

In very general terms, oil (and associated natural gas) exists in this realm in three discontinuous zones (Fig. 7-9). The most productive of these zones extends from the southern and southeastern part of the Arabian Peninsula northwestward around the rim of the Persian Gulf, reaching into Iran and continuing northward into Iraq, Syria, and southeastern Turkey, where it peters out. The second zone lies across North Africa and extends from north-central Algeria eastward across northern Libya to Egypt's Sinai Peninsula, where it ends. The third zone begins on the margins of the realm in eastern Azerbaijan, continues under the Caspian Sea into Turkmenistan and Kazakhstan, and also reaches into Uzbekistan, Tajikistan, and Kyrgyzstan.

The search for oil goes on, not only in this realm but also in many other areas of the world. As a result, estimates of the realm's reserves are subject to change as new

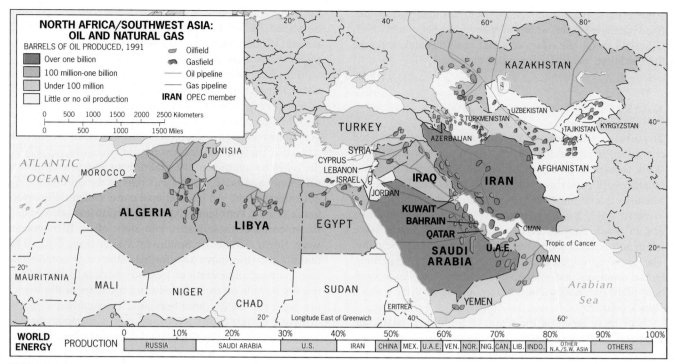

FIGURE 7-9

discoveries are announced. Current assessments suggest that more than 65 percent of the world's known oil reserves lie in the North Africa/Southwest Asia realm.

In terms of production, however, two other realms—North America and Russia—rank as the world's leaders. As we have noted elsewhere, oil production in the former Soviet Union reached as much as 25 percent of the world's annual total in some years, and the United States produces about 15 percent. But some of the former Soviet Union's reserves and refineries now lie outside Russia (see Fig. 2-14), and Russia's production has diminished because of financial and logistical difficulties. Thus the oil production of such countries as Saudi Arabia (largest in the realm), Iran, Kuwait, the United Arab Emirates, and Libya is of great importance to the rest of the world, notably the energy-demanding developed world. Supertankers displacing 300,000 tons, too large to transit the Suez Canal, leave the oil terminals in the Persian Gulf to sail thousands of miles to meet fuel demands in Japan, Europe, the United States (whose own production supplies less than half of what is needed), and other markets.

When the political boundaries of this realm were defined, mostly by the colonial powers, the dimensions of the oil reserves existing beneath the desert sand were not known. Some exploration had taken place, and production had begun in Iran in 1908; a few wells in the Sinai area of Egypt had yielded oil as early as 1913. But the major discoveries came later: the great Kirkuk reserve in Iraq, then

under British rule (1927); the fields in Kuwait, then a British protectorate (1938); the reserves in Saudi Arabia (1936); those in the United Arab Emirates (1958–1960); and the enormous deposits in Libya (1959), after the country had been made independent following Italy's occupation and World War II. By the time the oil was found, therefore, the boundaries already had been drawn. Some countries found themselves with riches undreamed of when the Turkish Empire collapsed; others were less fortunate.

The presence of oilfields in the realm thus created still another arena of conflict among the peoples and countries here. Along the eastern coast of the Arabian Peninsula, for example, lay a number of small, strongly traditional, relatively isolated and stagnant emirates and sheikdoms. Arabia's dominant state was Saudi Arabia, unified during the 1920s by its leader, Ibn Saud. Slavery, long practiced here, was not officially abolished until Ibn Saud's son was deposed in 1964. The small coastal entities, from Kuwait to Oman, relied on British protection or association to maintain their status as discrete political units. When it became clear that their territories contained major oil reserves, these countries felt threatened by their more powerful neighbors. When Kuwait was recognized as an independent state by Britain in 1961, Iraq claimed it as a "historic" province, and British troops had to prevent an Iraqi invasion (in 1990, Kuwait was not so fortunate). When Bahrain attained independence in 1971, it established a special protective relationship (against Saudi Arabia) with Iran, then still ruled by the shah. In 1992, Saudi

Arabia was at odds with Yemen over their inadequately defined common border, where oil reserves are believed to exist. Oil could enrich the rulers of small countries and empower them at home, but it also imperiled the peace.

A FOREIGN INVASION

The oil-rich countries of the realm (Fig. 7-9) had a coveted energy resource, but they did not possess the equipment or skills to exploit it. These had to be introduced from the outside world, and this entailed what many tradition-bound Muslims feared most: the penetration of the vulgarities of Western ways. In *The Middle East*, William Fisher states that in Saudi Arabia the two major cultural forces were Islam and Aramco, the joint Arab-American oil company. The growing network of oil and natural gas pipelines, tanker terminals, and storage and refinery complexes symbolized the international interactions arising from the realm's "black gold."

Thus there arose, in the realm of traditional Islam, the trappings of modernization. In Arab countries with comparatively small populations and large oil production, the impact of huge oil revenues transformed the cultural landscape: gleaming glass-encased office buildings towered over mosques, superhighways crossed ancient camel paths, modern airports outdid oases as desert hubs of activity. Desalinization plants allowed larger clusters of people to live in water-deficient places. First Kuwait, then the United Arab Emirates achieved the highest per capita income in the world. Arabs invested their newfound money in enterprises all over the world. Ruling elites brought in hundreds of thousands of workers from other Arab countries and from as far away as India and Sri Lanka to do the menial work they disdained. When Iraq invaded Kuwait in 1990, only 42 percent of the country's population were Kuwaiti; according to World Bank data, another 42 percent were temporary workers from other Arab countries (mostly Palestinians) and about 15 percent were workers from Asian states. The fears of traditionalist Muslims had come true.

While oil prices were high and incomes soared, national economies did well and political tensions (with the notable exception of Iraq and Iran) were contained. In the 1970s, the oil-rich countries of the realm were able to manipulate oil supply and price, creating a crisis in the oil-consuming developed world. An international cartel called OPEC (Organization of Petroleum Exporting Countries) coordinated the interests of these states, and eight of its 13 members were North African or Southwest Asian (Fig. 7-9). (The realm's countries have also forged some other noteworthy supranational organizations [see box at right].) But OPEC did not include *all* the oil-producing countries of the world, nor were the OPEC members able to establish a really effective cartel. In the 1980s, oil prices dropped sharply because an oil glut developed, and the economies of the oil-exporting countries were adversely affected.

What has been the effect of oil revenues in those countries endowed with large reserves? In a very general way, oil's impact has been inversely proportional to the size of those countries' populations. The small countries on the Arabian Peninsula, all with fewer than 3 million people, have been transformed; the "oil sheikdoms" are the envy of the realm. Saudi Arabia (17.2 million), a vast and more populous country, has huge oil reserves and a large income, but the result has been to create strong regional contrasts. On the Persian Gulf coast, opposite peninsular Qatar

Oil wealth has brought modernization to Arabian cities on the Persian Gulf. This is the townscape of central Dubai in the United Arab Emirates, which reflects the area's oil-derived investments.

SUPRANATIONALISM IN THE REALM OF ISLAM

The countries of North Africa and Southwest Asia often have been in conflict with one another, but they also have sought to establish international organizations. States of this realm, for instance, form a crucial bloc of OPEC (see Fig. 7-9), and the countries of the Maghreb formed the Arab Maghreb Union (UMA) to further their joint interests.

The largest supranational organization in the realm is the League of Arab States (LAS), often called the Arab League, founded in 1945 "to strengthen relationships between members and to promote Arab aspirations." The League has often been severely divided over major issues, especially over Israel (Egypt, a charter member, was suspended from 1979 to 1990 because it had signed a peace treaty with the Jewish state) and, more recently, Iraq and its aggression toward Kuwait. In 1993, membership in the LAS was as follows:

Algeria	Morocco
Bahrain	Oman
Djibouti	Qatar
Egypt	Saudi Arabia
Iraq	Somalia
Jordan	Sudan
Kuwait	Syria
Lebanon	Tunisia
Libya	United Arab Emirates
Mauritania	Yemen

The Palestine Liberation Organization (PLO) also had the rank of full member of the LAS, although the Palestinians did not have a territorial state.

The five countries of Turkestan have begun the process of forging a multi-state organization, but in 1993 these former Soviet republics still were tied to Russia in the ephemeral Commonwealth of Independent States (CIS).

An important sign of wider regional cooperation is the Islamic Development Bank (IDB), whose members include not only many participants in the Arab League but also such countries as Afghanistan, Bangladesh, Brunei, Cameroon, Malaysia, Pakistan, and Turkey. The IDB, whose purpose is to provide financial support for Islamic enterprises, is the first supranational organization to cross the Arab/non-Arab divide.

and island Bahrain, Saudi Arabia looks ultramodern, with high-technology cities, factories, and facilities that reflect the oil boom—a reflection that carries over to the capital, Riyadh, 250 miles (400 km) inland. But take a detour out of this core area, toward the Red Sea coast (say, to Medina or Jidda) or toward the border with Yemen, and Saudi Arabia becomes the land of desert, oasis, and camel, of vast distance and slow change. Here the oil boom still seems remote, and it is Islam that matters most.

Economic decisions and political stability also have had much to do with the fortunes of the oil-rich countries. Saudi Arabia invested heavily in the development of industries that will outlast the oil reserves and brought in technological know-how from its trading partners, especially the Japanese. Kuwait, with little to develop in its own small territory, made major investments abroad, so that in the years before the Iraqi invasion of 1990, its income from foreign investments actually exceeded that derived from its oil exports. But Iraq (19.6 million), in urgent need of more efficient agriculture, has wasted enormous amounts of oil money on military equipment and expensive wars. Most of the decade of the 1980s was spent in a costly struggle with neighboring Iran; when this conflict abated, Iraq invaded Kuwait, with disastrous consequences when a U.S.-dominated United Nations force in 1991 not only evicted the Iraqis but severely damaged Iraq's cities, towns, roads, dams, and other infrastructure.

Iran, too, suffered from political misfortune. Under the shah Iran acquired the largest and most powerful military establishment in the realm. Much money was lost through corruption (and to illegal diversion to foreign accounts of the elite), but still there was enough left over to fund some major development programs, including industrialization in the major cities, reforms in agriculture, and improvements in education. Again the oil centers on Iran's Persian Gulf coast, notably Abadan, and the capital, Tehran, displayed the effects of the oil boom in their townscapes; however, as in Saudi Arabia, the rural areas saw much less change. When the shah was ousted by the fundamentalist revolution in 1979, only about 3,000 of Iran's more than 60,000 villages had running water available. Those who hoped that the religious regime of the Ayatollah Khomeini would bring not only peace but also a fairer distribution of the oil wealth were disappointed. In 1982, after nearly two years of sporadic skirmishes with its neighbor, Shi'ite Iran launched a major offensive against Sunni-dominated Iraq, embarking on a war that would cost hundreds of thousands of young Iranians their lives. Again, oil fueled destruction, not development.

In North Africa, the country with the smallest population, Libya (4.8 million), also has the largest oil production and has felt the impact of this oil bonanza far more strongly than neighboring Algeria or Tunisia. The Libyan regime has used its oil revenues not only to foster modernization (while excluding Western influences) but also to gain a measure of external political influence in the realm and beyond, far in excess of its modest demographic dimensions.

Perhaps the best reflection of reality is the map. Figure 7-9 shows a system of oil and gas pipelines and export terminals that strongly resembles the exploitive interior-to-coast railroad lines in a mineral-rich colony of the past. Such a pattern spells disadvantage for the exporter, whether colony or independent country. As the failure of the OPEC cartel proved once again, markets, not raw material exporters, dominate international trade. Oil brought this realm into contact with the outside world in ways unforeseen just a century ago. Oil has strengthened and empowered some of the realm's peoples; it has dislocated and imperiled others. It has been a double-edged sword.

◆ REGIONS AND STATES

Earlier in this chapter we identified the seven regions of the vast and multicultural North Africa/Southwest Asia realm (Fig. 7-3). Despite its historic and environmental unifying properties, this is a realm of diversity and variety. Turkey seeks entry into the European Community; Sudan penetrates Subsaharan Africa; Kazakhstan straddles the border with Russia. We turn now to these regional components of the realm.

EGYPT AND THE LOWER NILE BASIN

Egypt, anchor of the lower Nile Basin, occupies a pivotal location in the heart of a realm that extends for over 6,000 miles (9,600 km) longitudinally and some 4,000 miles (6,400 km) latitudinally. At the northern end of the Nile and of the Red Sea, at the eastern end of the Mediterranean Sea, in the northeastern corner of Africa across from Turkey to the north and Saudi Arabia to the east, adjacent to Israel, to Islamic Sudan, and to militant Libya, Egypt lies in the crucible of this realm. Because it owns the Sinai Peninsula (recently lost to and regained from Israel), Egypt, alone among states on the African continent, has a foothold in Asia, a foothold that gives it a coast overlooking the strategic Gulf of Aqaba. Egypt also controls the Suez Canal, vital link between the Indian and Atlantic oceans and lifeline of Europe. It is hardly necessary to further justify Egypt's designation (together with northern Sudan) as a discrete region.

The Greek scholar Herodotus described Egypt as the gift of the Nile, but ancient Egypt also was a product of natural protection. The lower Nile valley—the heart of Egypt—lay encircled by inhospitable desert, open only to the sea in the north. The pharaohs needed to trade for wood and metal (of which there was little in Egypt), but this trade was carried on by Phoenicians and Greeks who

brought their goods to Egypt's northern outposts. The Egyptians had a natural fortress, and in their isolation they converted security into stability and progress. Modern Egypt's cultural landscape still carries the marks of antiquity's accomplishments in the form of great stone sculptures and pyramids that bear witness to the rise of a culture hearth more than 5,000 years ago.

The Nile River

Egypt's Nile is the aggregate of two great branches upstream: the White Nile, which originates in the streams that fill Lake Victoria in East Africa, and the Blue Nile, whose source lies in Lake Tana in Ethiopia. The two Niles converge at Khartoum (2.5 million), now Sudan's capital. Fly northward from Khartoum along the Nile toward Egypt, and you will realize how small and vulnerable this ribbon of

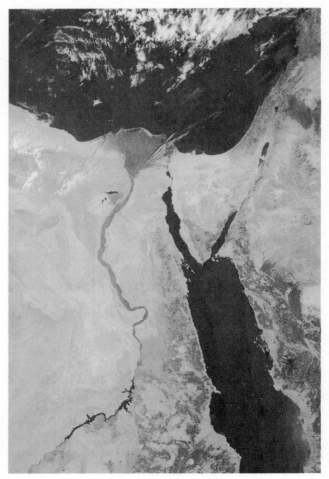

Satellite view of the area of the northern Red Sea and the eastern Mediterranean. Note the thin ribbon of life traversing the desert west of the Red Sea: the valley of the Nile River. Nearly 60 million Egyptians depend on this slender thread of water.

water looks in the vast wasteland of the Sahara (see photo p. 376). And on this ribbon depend the lives of tens of millions of people.

About 95 percent of Egypt's 58.4 million people live within a dozen miles (20 km) of the river's banks, or in its delta (Fig. 7-10). It has always been this way; the Nile rises and falls seasonally, watering and replenishing crops and fields on its banks. In April and May, the river usually is at its lowest level, a trickle through the desert. Then during the summer months it rises, to reach flood stage at Cairo (9.7 million), the capital, in October. The Nile may then be as much as 20 feet (6 m) above its lowest stage. In Novem-

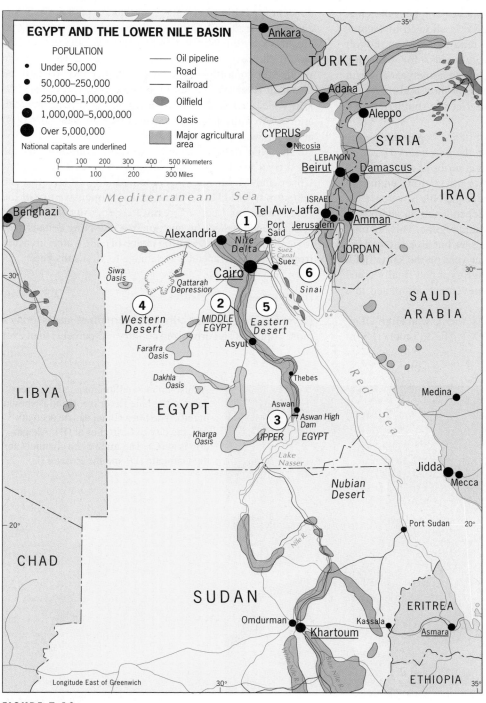

FIGURE 7-10

ber and December it falls rapidly, and then more slowly toward its springtime minimum.

In ancient times this regime made possible the invention of *basin irrigation*. The Egyptians built earthen ridges to cordon off fields along the low banks of the river, creating numerous artificial basins. The silt-laden river waters poured into these basins during the Nile's flood stage; while the water was still high, the exits were closed. This trapped the water and caused the deposition of its fertile load of alluvium. Later, when the river's level dropped, the exits were opened again and the water was allowed to escape. Left behind was rejuvenated soil, ready for sowing. For all its disadvantages (variable flood levels, for example, would leave many fields unirrigated in some years), basin irrigation continued to be practiced for thousands of years. Some basin irrigation was still practiced in southern Egypt as recently as the 1970s!

The construction of permanent dams, begun during the nineteenth century, made possible the *perennial irrigation* of Egypt's farmlands. By building a series of artificial barriers across the river (with locks for navigation), engineers were able to control the floods, raise the Nile's water level, and free the farmers from their dependence on sharp seasonal fluctuations of the river's natural regime. This not only expanded the country's cultivable area but also allowed the growing of more than one crop per year on the same field. By the mid-1980s, all farmland in Egypt had finally come under perennial irrigation, an achievement completed in a single century.

The greatest of all Nile projects, the Aswan High Dam, was begun in 1958 and opened in 1971. The 365-foot (110-m) high dam creates Lake Nasser, one of the world's largest artificial lakes (Fig. 7-10). As the map shows, the inundated area extends into Sudan, where 50,000 Nile valley residents had to be resettled. The effect of the Aswan High Dam was to increase Egypt's irrigable land by nearly 50 percent and to provide oil-poor Egypt with about 40 percent of its electricity.

Egypt has often been described as one elongated oasis, and Fig. 7-10 confirms this: virtually all of the population is concentrated in a riverine ribbon that covers barely 5 percent of the land area. Drive southward from Cairo, and you are struck by the narrowness of "green Egypt" in so many places; the oasis ranges from 3 to 15 miles (5 to 25 km) in width. But go northward from the capital, and you will see the great river fan out across a delta over 150 miles (250 km) wide, anchored in the west by the great city of Alexandria (3.6 million) and in the east by Port Said (505,000) at the gateway to the Suez Canal.

The Nile delta's channels are now controlled as the great river itself is, and they irrigate twice as much land as Upper and Middle Egypt combined (Fig. 7-10). But the delta is a troubled area. The ever more intensive use of the Nile's water and silt upstream are depriving the delta of much-needed replenishment. And the low-lying delta is geologically subsiding, creating fears of salt-water invasion from the Mediterranean and damage to vital soils here.

Egypt's Regional Prominence

Egypt has changed markedly in modern times and is today a more populous and urbanized country (45 percent) than ever

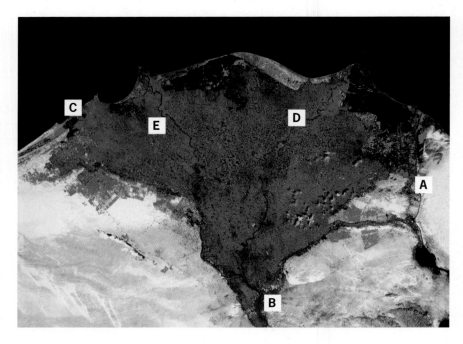

A false-color satellite image of the great delta of the Nile River (cropland is shown in red). The Suez Canal can be seen at (A), the city of Cairo lies at (B), Alexandria is at (C). The main delta channels are the Damietta (D) and the Rosetta (E).

before. However, Egypt's subsistence farmers, the *fella-heen*, still struggle to make their living off the land, as did the peasants in the Egypt of 5,000 years ago. In a society that at one time was the source of countless innovations, the tools of too many of these farmers often are as old as any still in use—the hand hoe, the wooden buffalo-drawn plow, and the sickle. Water continues to be drawn from wells by the wheel and bucket, and the hand-turned Archimedean screw (to move water over riverbank ridges) can still be seen in service. The peasant still lives in a small mud dwelling, which would look very familiar to the farmer of many centuries ago; neither would that distant ancestor be surprised at the poverty, recurrent disease, high infant mortality rates, and lack of tangible change in the countryside. Despite the Nile dams and irrigation projects, Egypt's available farmland per capita has steadily declined during the past two centuries as population growth keeps nullifying gains in agricultural productivity.

Yet by many measures, Egypt remains this realm's most important and influential country: Egypt lies spatially, culturally, and ideologically at the heart of the Arab world. Long under the heel of foreigners, Egypt reasserted itself during the 1950s, when its monarchy was overthrown and the Suez Canal was nationalized. When the Egyptians, in 1956, successfully withstood a last attempt by the British and French (with Israeli help) to recapture the Canal corridor, Egyptian stature in the realm soared. Later, two wars with Israel (in 1967 and 1973) confirmed Egypt's regional prominence, but the country's freedom of action was eroded by its need for foreign aid. In 1978, when Egypt signed a peace accord with Israel, other Arab countries suspended its membership in the Arab League. When Egypt joined the United Nations force that expelled Iraq from Kuwait following the 1990 invasion, some of the realm's states accused Cairo of pandering to American wishes.

Today, Egypt remains the cornerstone of the Arab world. Its population size rivals that of Turkey and Iran in the Islamic realm, but Egypt is the dominant *Arab* country. A secular state with a democratically elected government, Egypt in the mid-1990s is buffeted by the forces of militant Islam in the form of the Muslim Brotherhood. Since the 1987 elections, the Muslim Brotherhood has been the main opposition party, and it has at times resorted to violence and intimidation to promote its goal—an Islamic republic.

Cairo at the Core

As Fig. 7-10 indicates, Egypt contains six subregions: ① the Nile delta, or Lower Egypt; ② the Nile valley from Cairo to Thebes, or Middle Egypt; ③ the Nile valley from Thebes to the Sudan border, or Upper Egypt; ④ the Western Desert, containing several large oases and the Qattarah Depression; ⑤ the Eastern Desert and the Red Sea coast, and ⑥ the Sinai Peninsula. The great majority of Egyptians live and work in Lower and Middle Egypt, the country's core area. Alexandria, Egypt's leading port and second industrial center, is the urban focus of Lower Egypt, the Nile delta. Upper Egypt contains the Aswan High Dam and Lake Nasser. The Sinai Peninsula produces enough oil to meet Egypt's demand, and Egypt today exports a modest amount of this fuel. But the hub of Egypt is Middle Egypt, and at its focus lies Cairo, largest city in the entire realm.

With almost 10 million people, metropolitan Cairo is home to one-sixth of the entire Egyptian population. Not only is Cairo the realm's largest city, it is also the largest urban center on the whole African continent. The Muslim Arabs who chose the site in A.D. 969 as the center of their new empire knew their geography well: here the Nile valley opens into the delta. Cairo became and remained Egypt's primate city, older Alexandria notwithstanding.

Cairo now is the world's twentieth-largest metropolis, and it shares with other cities of the underdeveloped world the staggering problems of crowding, poverty, inadequate sanitation, poor health conditions, and unemployment. Even among such cities, Cairo is noteworthy for its stunning social contrasts. Along the Nile waterfront, elegant skyscraper hotels and apartment buildings rise above carefully manicured, Parisian-looking surroundings. Ask to be allowed on the roof of one of these tall buildings, and you will see the pyramids shimmering in the distance—but in another direction, the urban landscape extends, gray, dusty, and almost featureless, as far as the eye can see. Not far away, more than a million squatters live in the sprawling cemetery known as the City of the Dead (see photo p. 380). On the outskirts, millions reside in overcrowded shantytowns of mud huts and hovels.

And yet Cairo is the dominant city not only for Egypt but for a wider sphere, the cultural capital of the Arab world, with centers of higher learning, magnificent museums, a symphony orchestra, national theater, and opera company. Although Cairo always has been primarily a center of government, administration, and religion, it also is a river port and an industrial complex (textiles, food processing, and iron and steel production rank among its manufactures). Countless thousands of small handicraft indus- tries are scattered throughout the neighborhoods, and the grand *bazaar* (traditional market square) throbs daily with the trade in small items.

Below Cairo, in the delta to the north, lie vast fields of the cotton that constitutes Egypt's main cash crop (exported in raw form as well as finished textiles). Above the capital, food crops dominate in the Nile-hugging strip of farmlands that is Middle Egypt. Despite the expansion of its irrigated croplands, Egypt in the mid-1990s imported more than 60 percent of the food it needs, and population growth continued at about 2.5 percent annually. Egypt's planners know that reducing this rather high growth rate would improve the economic situation, but fundamentalist Mus-

The City of the Dead, located at the eastern edge of the old city of Cairo, bakes under the desert sun. Today, this graveyard is overrun by hordes of impoverished squatters who occupy its stifling tombs.

lims object to any programs that promote family planning. The government is wary of militant Islam; thus development gains are wiped out by demographics. This does not bode well for the future, and Egypt's future is crucial to that of the entire realm.

THE MAGHREB AND LIBYA

The countries of northwestern Africa are collectively called the *Maghreb*, but the Arab name for them is more elaborate than that: *Djezira-al-Maghreb*, or "Isle of the West," in recognition of the great Atlas mountain range rising like a vast island from the waters of the Mediterranean Sea to the north and the sandy flatlands of the Sahara to the south.

The countries of the Maghreb (sometimes spelled *Maghrib*) are Morocco, last of the North African kingdoms; Algeria, a secular republic beset by the religious-political problems we noted earlier; and Tunisia, smallest and most Westernized of the three (Fig. 7-11). Libya, facing the Mediterranean between the Maghreb and Egypt, is unlike any other North African country: an oil-rich desert state whose population is almost entirely clustered in settlements along the coast.

Whereas Egypt is the gift of the Nile, the Atlas Mountains form the nucleus of the settled Maghreb. These high ranges wrest from the rising air enough orographic rainfall (see the box on p. 179) to sustain life in the intervening valleys, where good soils support productive farming. From the vicinity of Algiers eastward along the coast into Tunisia, annual rainfall averages more than 30 inches (75 cm), a total more than three times as high as that recorded for Alexandria in Egypt's delta. Even 150 miles (240 km) inland, the slopes of the Atlas still receive over 10 inches (25 cm) of rainfall. The effect of the topography can be read on the world map of precipitation (Fig. I-6): where the highlands of the Atlas terminate, desert conditions immediately begin.

The Atlas Mountains are structurally an extension of the Alpine system that forms the orogenic backbone of Europe, of which Switzerland's Alps and Italy's Appennines are also parts. In northwestern Africa, these mountains trend southwest-northeast and commence in Morocco as the High Atlas, with elevations close to 13,000 feet (4,000 m). Eastward, two major ranges appear that dominate the landscapes of Algeria proper: the Tell Atlas to the north, facing the Mediterranean, and the Saharan Atlas to the south, overlooking the great desert. Between these two mountain chains, each consisting of several parallel ranges and foothills, lies a series of intermontane basins (analogous to South America's Andean altiplanos but at lower elevations), markedly drier than the northward-facing slopes of the Tell Atlas. In these valleys, the rain shadow effect of the Tell Atlas is reflected not only in the steppe-like natural vegetation but also in land-use patterns: pastoralism replaces cultivation and stands of short grass and bushes blanket the countryside.

Berbers and Colonists

The countries of the Maghreb are sometimes referred to as the Barbary States, in recognition of the region's oldest inhabitants, the Berbers. The Berbers' livelihoods (nomadic pastoralism, hunting, some farming) changed as foreign invaders, first the Phoenicians and then the Romans, entered their territory. The latter built towns and roads, laid out farm fields and irrigation canals, and introduced new methods of cultivation. Then came the Arabs, conquerors of a different sort. They demanded the Berbers' allegiance and their conver-

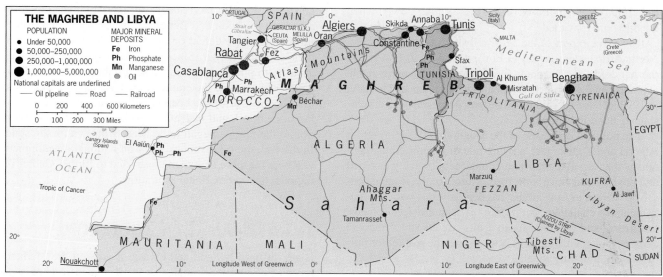

FIGURE 7-11

sion to Islam, radically changed the political system, and organized an Arab-Berber alliance (the *Moors*) that pushed across the Straits of Gibraltar into Spain and colonized a large part of southwestern Europe. After the Moors' power declined, the Ottoman Turks established a sphere of influence along North Africa's coasts. But the most pervasive foreign intervention came during the nineteenth and twentieth centuries when the European colonial powers—chiefly France but also Spain—established control and (in the now-familiar sequence of events) delimited the region's political boundaries.

During the colonial era, well over a million Europeans came to settle in North Africa—most of them French, and a large majority to Algeria—and these immigrants soon dominated commercial life. They stimulated the renewed growth of the region's towns, and Casablanca (3.3 million today), Algiers (3.7 million), and Tunis (2.1 million) rose to become the urban foci of the colonized territories. Although the Europeans dominated trade and commerce and integrated the North African countries with France and the European Mediterranean world, they did not confine themselves to the cities and towns. They recognized the agricultural possibilities of the favored parts of the *tell* (the lower Tell Atlas slopes and narrow coastal plains that face the Mediterranean) and established thriving farms. Agriculture here, not surprisingly, is of the Mediterranean variety: Algeria soon became known for its vineyards and wines, citrus groves, and dates; Tunisia has long been one of the world's leading exporters of olive oil; Moroccan oranges went to many European markets. Staples such as wheat and barley were also among the exports.

Despite the proximity of the Maghreb to Mediterranean Europe and the tight integration of the region's territories within the French political framework, nationalism emerged as a powerful force here. Morocco and Tunisia secured independence mainly through negotiation, but in Algeria a costly revolutionary war was fought between 1954 and 1962. It was not difficult for the nationalists to recruit followers for their campaign; the justification for it was etched in the very landscape of the country, in the splendid shining residences of the landlords and the miserable huts of the peasants. But the revolution's success brought new troubles. Hundreds of thousands of French people left Algeria, and the country's agricultural economy fell apart, making an orderly transition impossible. Productive farms went to ruin, exports declined badly, and needed income was lost.

Oil and Emigrants

The fortunes of the Maghreb states since independence have varied. Morocco became embroiled in a territorial conflict involving Western Sahara to its south, a former Spanish possession. A small but resilient group of Western Saharan residents resisted Morocco's absorption of this desert area, and Algeria and Mauritania gave them some support. This strained relations between Morocco and its neighbors, but in Morocco itself the campaign had a unifying effect at a time when the king faced growing anti-monarchist sentiment. In Algeria, oil revenues replaced farm produce as the major export and revenue earner; reserves not only of oil but also of natural gas are linked by pipelines to Algerian as well as Tunisian ports for shipment to Europe (Fig. 7-11). For more than a decade, Algeria's income rose with world oil prices, but then a combination of dropping prices and diminishing reserves reversed the upward trend. Tunisia, too, went through an oil boom, but its supplies dwindled even faster, and other sources of income were needed. Fortunately these were available: Tunisia had not abandoned its agricultural economy (citrus fruits, olives and olive oil, textiles, leather goods) when the oil boom prevailed.

The other export of the Maghreb states has been people. North Africans emigrated to Europe, mainly to France, by the hundreds of thousands in search of work unavailable at home. Smaller emigration streams led to Spain and to Italy; in 1993, at least 1.5 million Algerians, Moroccans, and Tunisians resided in Europe.

The political landscape of the Maghreb is changing. Both Algeria and Tunisia face a rising tide of Islamic fundamentalism that strengthened as national economies weakened and poverty and frustration rose. Morocco, the leading conservative force in Arab affairs, has dealt severely with the fundamentalist challenge, but its non-oil economy (phosphates, fertilizers, and Mediterranean fruits are its chief income earners, along with tourism) is weak, and the market for militant Islam will thus widen.

The Desert Coast

Between the eastern terminus of the Atlas range (in Tunisia) and Egypt's Nile delta, the Sahara reaches the very shores of the Mediterranean Sea. Here lies Libya, small in population (4.8 million), large in area (nearly three times bigger than Texas), and rich in oil.

Almost rectangular in shape, Libya is a country whose four corners matter most (Fig. 7-11). What limited agricultural possibilities exist lie in the northwest in Tripolitania, centered on the capital, Tripoli (3.3 million), and in the northeast in Cyrenaica, where Benghazi (1.1 million) is the urban focus. Between these two coastal clusters, which are home to 90 percent of the Libyan people, lies the Gulf of Sidra, a deep Mediterranean bay. Libya has claimed this gulf as its territorial sea, but this claim has not been accepted by other countries. Libya's two southern corners are the desert Fezzan, a mountainous area on the southwestern border with Algeria and Niger, and the sparsely populated Kufra Oasis in the southeast. Despite its huge size and tiny population, Libya has claimed a sector (the Aozou Strip) of its southern neighbor, Chad, and has pursued this claim with armed forces.

But the (economically) most important part of Libya lies in the northeastern interior south of Benghazi. Major oil reserves here have made Libya one of the world's leading oil exporters. Libya's longtime ruler, Muammar Gadhafi, used the country's oil revenues not only to improve Libya's infrastructure (roads, ports, water supplies, electricity provision) but also to strengthen the military and to support revolutionary Arab causes elsewhere in the realm and beyond. For instance, Libya strongly supported Sudan's reconstitution as an Islamic republic, and Libya often is accused by Egypt of covertly supporting fundamentalists opposed to the Cairo government.

Libya's interventions attracted responses that have ranged from disapproval in the Arab world to retaliation from abroad. In 1986, United States warplanes bombed targets in and around Tripoli and Benghazi; in 1989, Chadian forces inflicted a major defeat on the Libyan army in the area of their disputed border. Since the end of the 1980s, Libya's capacity to pursue its stated causes (among them the elimination of Israel and the eradication of Western influences in the Muslim world) has been greatly reduced by the declining price of oil, and in 1992 the voices of political opposition to the regime began to be heard from within the state. In the end, the oil revenues spent on national development will bear far more fruit than those used to further Gadhafi's regional and international aspirations.

THE AFRICAN TRANSITION ZONE

Islam diffused by land and by sea. It crossed the Mediterranean Sea and the Indian Ocean; it gained footholds on many coasts from East Africa to Indonesia. In North Africa, Islam also crossed the desert to the south. It came up the Nile by boat and across the Sahara by caravan, and it reached the peoples of the interior steppes and savannas on the other side. There, along a wide stretch of Africa known today as the *Sahel* (an Arabic word for "border" or "margin"), Islam's proselytizers converted millions to the Muslim faith. As we will discover in the next chapter, the zone between the desert and the forest was one of Africa's culture hearths, and here lay large and durable states. Their rulers and subjects adopted Islam, and soon huge throngs of West Africans marched along the savanna corridor eastward on their annual pilgrimages to Mecca.

Then came Europe's colonial powers, and the modern political map of Africa took shape. From Senegal to Sudan, Muslims and non-Muslims were thrown together in countries not of their making. Thus superimposed on the cultural landscape and on the political mosaic, a zone of transition marks the map (see Fig. 7-3).

A Corridor of Instability

The African Transition Zone affects primarily Senegal, Mauritania, Mali, Burkina Faso, Niger, Nigeria, Chad, Sudan, and the countries of the "Horn" of Africa. As the map shows, from Sudan the zone sweeps eastward into Ethiopia and then curves southward to the border area between Kenya and Somalia. This is why Africa's Horn, a geographic term used to identify the general area of Ethiopia, Eritrea, Djibouti, and Somalia, is not a geographic region: it straddles two geographic realms.

The African Transition Zone, as Fig. 7-3 indicates, is a wide corridor across which Islam yields to traditional African (animist) and Christian religions, Islamic local laws yield to Western legal systems, and the realm of the Muslim North gives way to the realities of Subsaharan Africa. In the west, French colonialism grafted a European blueprint of development onto traditional Islam: Senegal and its large port, Dakar (now with a population of 2.0 million), was its focus, with roads radiating inland toward neighboring Mali and beyond. Eastward, the Sahel countries of Niger and Chad were administrative divisions of the French colonial empire that achieved independence with little economic opportunity. Chad, with a population of 5.5 million, is about 45 percent Muslim (that is, the northern part of the country), 35 percent Christian, and 20 percent animist (the Christians and animists are concentrated in the African south).

Conflict across the transition zone has afflicted Chad time and again, but in neighboring Sudan the strife between Muslims of the north and non-Muslims of the south has been even costlier. When, in 1989, the Khartoum government proclaimed an Islamic republic in which the *sharia* laws would prevail, the last chance for a unified and stable future may have been lost. Unlike Chad, Muslims in Sudan's 28.2 million population are in the majority (70 per-

cent), but the Islamic, Arabized north has been unable to conquer and subjugate the African south. War began almost immediately after independence in 1956, and the first conflict (lasting from 1956 to 1972) cost an estimated half-million lives. After just one decade of accommodation, the war was renewed in 1983 and has taken even more lives than the first. The African Transition Zone is an area of chronic instability.

This also is true of the African Horn (Fig. 7-12). Here the dominant state, Ethiopia, centered on the high-plateau capital of Addis Ababa (2.2 million) and long ruled by the Amhara, a Christian minority, held sway over large Muslim populations in coastal Eritrea and in the eastern lowlands adjoining Muslim Somalia. In Ethiopia, the transition zone becomes a spatially narrow but ideologically deep cultural chasm. Ethiopia in the 1980s became the "Balkans" of Africa, creating more than 2 million refugees in a struggle that pitted Muslim Eritreans against non-Muslim Ethiopi-

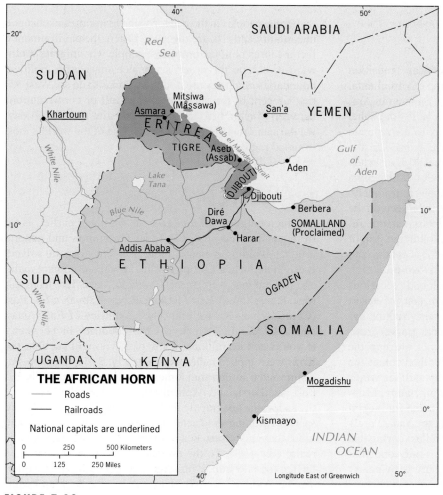

FIGURE 7-12

ans. In 1990, other peoples also rose up against the regime in Addis Ababa, and in 1991 the dictatorship collapsed. But that did not solve the politico-geographical problems in the Horn. Eritrea, the coastal territory, wanted independence: this was achieved in 1993—thereby rendering Ethiopia a landlocked country (Fig. 7-12). Moreover, the Oromo, a cluster of peoples in southern Ethiopia who constitute 42 percent of the country's population of 54.5 million, continue to demand a larger voice in the country's future government.

Eastern Ethiopia is part of the African Transition Zone, and here the country merges into the Islamic world. The Muslims of the Ogaden (the area in the semiarid lowlands of the east) are Somali, related to the peoples of Somalia but long under the rule of the Christian Amhara. Their future in the Ethiopia of the twenty-first century is uncertain because Somalia, too, has disintegrated in civil war (requiring the humanitarian intervention of U.S. forces in 1992)—not Muslim against non-Muslim, but Muslim clan against clan in a fierce struggle for power. Even tiny Djibouti at the mouth of the Red Sea, site of a major French military base, has not escaped conflict as the two ethnic groups dominant in the population, the Afars and the Issas, have gone to war against each other.

Environmental Problems

As if all this strife were not enough, the African Transition Zone is one of the world's most problematic environmental areas. This east-west zone borders the great Sahara (an Arabic word for "desert"), and in that part of it called the Sahel (running from west of the Horn all the way to the Atlantic coast [see Fig. 8-10]), droughts occur frequently. In the 1970s, the world became aware of the Sahel's plight when a multi-year dry spell killed more than 300,000 people and 5 million livestock. The Sahel is a steppe environment, and the *cultural* African Transition Zone can be seen on the map to coincide substantially with the *climatic* steppe region (see Fig. I-7). In the 1980s, Ethiopia fell in the grip of a devastating drought that caused countless deaths and hastened the collapse of the regime in Addis Ababa.

Life on the edge of the desert is difficult and precarious (see photo above). Geographers now know that the desert is not static but expands and contracts in fairly regular cycles. When it contracts, its margins turn green, grasses grow, and the bush fills out. People and their livestock move in, the herds expand, and settlements flourish. But then comes the inevitable expansion of the Sahara: the rains fail, the winds blow sand and dust southward, the grazing lands disappear, people must move, and livestock succumb for lack of food and water. A recent study showed that the Sahara grew about 15 percent (roughly 500,000 square miles) during just the four years between 1980 and 1984; yet only one year later, after a moist summer season, it had shrunk by nearly 300,000 square miles.

Desert sand meets the shrinking pastureland in the heart of the Sahel in southern Niger. This is one of the most revealing photographs ever taken of desertification caused by overgrazing.

For the peoples in the path of such environmental changes, uncertainty and risk are the rules. Often, the environmental factor causes political problems: people who migrate southward to escape the desert's advance must cross political boundaries, and they are not always free to do so. West Africa's recent historical geography is full of confrontations between Muslim herders and non-Muslim farmers. Today, the African Transition Zone remains one of the world's most troubled regions.

THE MIDDLE EAST

The regional term *Middle East*, as we noted earlier, is far from satisfactory—but it is in such common and general use that avoiding it creates more problems than are solved. It originated when Europe was the world's dominant realm and when things were "near," "middle," and "far" from Europe: hence a *Near* East (Turkey, Egypt, Libya), a *Far* East (China, Japan, Korea, and other countries of East Asia), and a *Middle* East (many of the countries in between). If you check definitions of the past, you will see that the terms were applied rather inconsistently: Syria, Lebanon, Palestine, even Jordan sometimes were included in the "Near" East, and Persia and Afghanistan in the "Middle" East.

Today, the geographic designation *Middle East* has more specific meaning. And at least half of it has merit: this region, more than any other, lies at the middle of the vast Islamic realm (Fig. 7-3). To the north and east of it, respectively, lie Turkey and Iran, with Afghanistan and Muslim Turkestan beyond the latter. To the south lies the Arabian Penin-

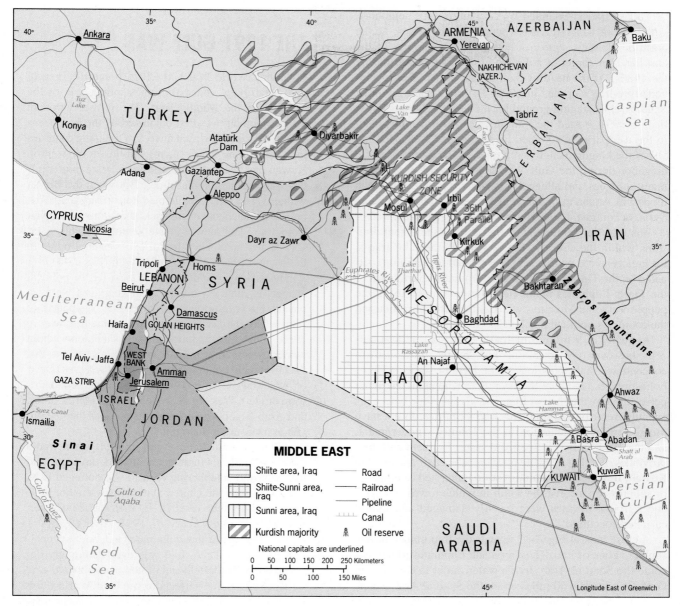

FIGURE 7-13

sula. And to the west lie the Mediterranean Sea and Egypt, and the rest of North Africa. This, then, is the pivotal region of the realm, the very heart of it.

Five countries form the Middle East (Fig. 7-13): Iraq, largest in population (19.6 million) and territorial size, facing the Persian Gulf; Syria, next in both categories and fronting the Mediterranean; Jordan, linked by the narrow Gulf of Aqaba to the Red Sea; Lebanon, whose survival as a unified state has come into question; and Israel, Jewish nation in the crucible of the Muslim world.

Iraq

California-sized Iraq comprises nearly 60 percent of the total area of the Middle East and has 40 percent of the region's population. With major oil reserves and large areas of irrigated farmland, Iraq also is best endowed with natural resources. Iraq is heir to the early Mesopotamian states and empires that emerged in the basin of the Tigris and Euphrates rivers, and the country is studded with significant archeological sites.

Today, Iraq is bounded by as many as six neighbors, with most of which it has adversarial relationships. To the north lies Turkey, source of both of its vital rivers. To the east lies Iran, target of a decade of destructive war during the 1980s. At the head of the Persian Gulf is Kuwait, invaded by Iraq's armies in 1990. To Iraq's south lies vast Saudi Arabia, staging ground for the United Nations forces that, in 1991, ousted the Iraqi invaders from Kuwait. And to the west, Iraq is adjoined by Jordan and Syria. Jordan, last of the kingdoms in this region, was Iraq's most helpful neighbor during the Gulf War and afterward; it accepted many refugees and allegedly transferred cargo to Iraq via its port at Aqaba in defiance of United Nations sanctions. Syria, in contrast, joined the campaign against Iraq.

The regional geography of Iraq reveals the potent divisions within the country (Fig. 7-13). The heart of Iraq is the area centered on the capital, Baghdad (4.5 million), situated on the Tigris River amid the productive farmlands of the Tigris-Euphrates plain. Here the great majority of the people are Sunni Muslims, who dominate the core area and the country's political machine.

As many as 9 million of Iraq's nearly 20 million citizens, however, are not Sunnis but Shi'ites, most of them concentrated in the populous south. Here lie some of Shia's holiest places, and here the Ayatollah Khomeini, escaping persecution by the shah of Iran, was given refuge before his eventual return to Tehran. But Iraq's Shi'ites do not have proportional representation in the country's government, and the 1991 Gulf War spelled disaster for them and their provinces. Fearing a Shi'ite uprising after their ouster from Kuwait, Baghdad's remaining forces unleashed a campaign of suppression against the south in which towns, villages, and religious shrines were heavily damaged and untold casualties occurred.

A close look at Fig. 7-13 reveals the importance of this southern zone of Iraq. The Tigris and Euphrates join to become the Shatt al Arab, Iraq's water outlet to the Persian Gulf. Over its last 50 miles or so, the Shatt al Arab waterway also becomes the boundary between Iraq and Iran. Iraq's territorial claim to land on the Iranian side of the waterway precipitated the decade of war between the two countries beginning in 1980.

Also note the situation of Iraq's most important southern city, Basra (728,000), and the position of the boundary with Kuwait. In this area lie several major oilfields; one, the Rumailah oilfield, extends from beneath Iraqi soil into Kuwait. In 1990, Iraq claimed that Kuwait was draining oil from this field by drilling slanted wells under Iraqi territory; that the boundary between the two countries was not agreed upon; and that Kuwait was failing to adhere to OPEC rules on oil production and pricing. The Shatt al Arab waterway still was filled with the wreckage of the Iran-Iraq War, so that the annexation of Kuwait would also give Iraq a new outlet to the Persian Gulf. This combination of justifications and potential returns persuaded Iraq

THE 1991 GULF WAR

Following World War I, the British and the French, with Arab help, ended the longstanding rule of the Turks over the Middle East. The former Ottoman Turkish domain was first colonized by the Europeans and later divided into a patchwork of fledgling countries in which ruling royal (Arab) families took control. Kuwait had been administered from Basra in present-day Iraq during colonial times. But when the British withdrew in 1961, they defined a boundary that separated the independent, oil-rich sheikdom of Kuwait from Iraq. This newly created mini-state, however, very nearly landlocked Iraq from its natural outlet to the sea at the head of the Persian Gulf (Fig. 7-13).

The arrival of independence also led to political change as some of the old Arab kingdoms were replaced by revolutionary regimes—as happened in Iraq, where the Baathist extremists came to power in 1968. Today historic kingdoms and sheikdoms adjoin modern, sometimes militaristic republics. Add to this the uneven distribution (Fig. 7-9) of oil reserves (many of which were undiscovered when the boundaries were drawn), Western dependence on those resources, and Arab-Israeli tension, and it is obvious that the potential for conflict is strong.

All these forces converged in 1990 when Iraq's dictator, Saddam Hussein, threatened, invaded, and then annexed neighboring Kuwait. An outraged U.N. coalition led by the United States began a major military build-up that resulted in their attack (Desert Storm), begun on January 16, 1991. After 45 days of a one-sided air war that culminated in a brief but decisive ground invasion, Iraq was overwhelmingly defeated on the battlefield and ousted from Kuwait.

Most of the after-effects of the Gulf War are felt inside Iraq, including bomb damage and an international trade embargo; millions continue to live in misery. Saddam Hussein, however, remains in power. Most of his energies have been directed toward the brutal repression of two opposition groups—the Kurds of northern Iraq and the Shi'ites of the south—who rebelled at the end of the Gulf War, believing (incorrectly) that they had allied support.

to embark on its disastrous invasion of Kuwait in 1990 (see box above).

The core area, centered on Baghdad, and the Shi'ite south are two of Iraq's major subregions; a third subregion lies in the north. Here most of the people are Sunni Muslims, but

A FUTURE KURDISTAN?

Political maps of the region do not show it, but where Turkey, Iraq, and Iran meet, the cultural landscape is not Turkish, Iraqi, or Iranian. Here live the Kurds, a fractious and fragmented nation of about 24 million (no certainty exists about their exact numbers). More Kurds live in Turkey than in any other country (perhaps as many as 10 million); possibly as many as 7 million live in Iran, 3 million in Iraq, and smaller numbers in Syria, Armenia, and even Azerbaijan (see Fig. 7-13).

The Kurds have lived in this isolated, mountainous frontier zone for more than 2,000 years. Their ancestry is not Turkish, Persian, or Arab: their origin is unknown, but when you visit their domain, you are struck by the number of people with fair hair and blue eyes. Historical geographers believe that the Kurds have a northern origin. Their languages, differentiated by locale and by dialect, are distinct from Turkish, Persian, and Arabic. The Kurds are a nation, their nomadic ways now gone in favor of farm and town.

The Kurds may be a nation, but they have no state; nor do they have the international exposure that people of other stateless nations (such as the Palestinians) receive. Baghdad's brutal repression of the Iraqi Kurds created more than 2 million refugees in 1991 and made world headlines—but soon the world forgot again.

The Kurds have been pawns in an endless series of geopolitical struggles. The shah of Iran encouraged their rebellion against Iraq during a quarrel in the 1970s, then abandoned them to retribution when the conflict ended. The Turks as recently as the 1980s forbade the public use of Kurdish speech in the Kurdish area of southeastern Turkey. And the Kurds have often engaged in warfare among themselves.

Many Kurds dream of a day when their fractured homeland will become a nation-state. Most would agree that the city of Diyarbakir, now in Turkey, would become the capital; it is the closest any Kurdish town comes to a dominant city. After the 1991 Gulf War, the United Nations established a "security zone" for Kurds in Iraq, extending from the 36th parallel northward to the borders with Turkey and Iran (Fig. 7-13). This was done to encourage the refugees to return and to protect them against further Iraqi mistreatment. But this also may be the closest the Kurds will come, in the foreseeable future, to their dream of a free Kurdistan. The stakes—territory, oil, farmland, river sources—are too great for the states in this immediate area to agree to move their boundaries.

they are not Arabs: this is the land of the Kurds. Fewer than 4 million Kurds live in the mainly mountainous areas of northern Iraq, and they constitute perhaps 15 percent of the total Kurdish population in the realm. The Kurds are minorities in all the countries they inhabit, and governments like to undercount their minorities. Geographers estimate that there may be as many as 24 million Kurds, with the largest number in Turkey, the next largest minority in Iran, then in Iraq, Syria, and small clusters even in Armenia and Azerbaijan (Fig. 7-13).

As the map shows, Iraq's Kurds occupy a sensitive area of the country: the huge oil reserves on which the city of Kirkuk (353,000) is situated lie on the margins of the Kurdish domain. During the Iran-Iraq War and again during the Gulf War, the Kurds rose against their Iraqi rulers and briefly succeeded in taking control over parts of their domain, but Iraq's response always ended such efforts, often with great loss of life. In the late 1980s, Baghdad even resorted to the use of cyanide and mustard gas against Kurdish villagers. At the end of the Gulf War, the United Nations established a security zone between the 36th parallel and Iraq's northern border to encourage refugees who had crossed into Turkey and Iran to return. However, as Fig. 7-13 shows, a large part of Iraq's Kurdish domain lies outside this safety zone. Once again, the Kurds are what they have always been: victims of more powerful neighbors (see box at left).

Iraq's infrastructure and economy were shattered during the Gulf War, but in fact Iraq had wasted much of its potential on the earlier conflict with Iran—and on mismanagement, corruption, and inefficiency. With its good agricultural land and its enormous oil income, Iraq should be one of the economic success stories of the entire realm. Instead, it is one of its tragedies.

Syria

Of all the countries in the Middle East, Iraq is the only one without a joint border with Israel. Syria, as Fig. 7-13 indicates, has a short boundary with the Jewish state—and there it lost a piece of territory to the Israelis, the Golan Heights, in 1967.

Like Lebanon and Israel, Syria has a Mediterranean coastline where unirrigated agriculture is possible. Behind this densely populated northwestern coastal belt, Syria has a much larger interior than its neighbors, but the areas of productive capacity are quite dispersed. Damascus, the capital (2.1 million), was built on an oasis and is considered to be the world's oldest continuously inhabited city; although it lies in the dry rain shadow of the coastal mountains, it is surrounded by a district of irrigated agriculture. Damascus lies in the southwestern corner of the country (Fig. 7-13), in close proximity to the Golan Heights and Israel. In the far north, near the Turkish border, lies another old caravan-route center, Aleppo (1.9 million), at the northern end of Syria's important cotton-growing area; here the Orontes River, whose headwaters flow through northern Lebanon's strategic Bekaa Valley, is the chief source of irrigation water. Syria's wheat belt, stretching east along the northern border, also focuses on Aleppo. In the eastern part of the country, the Euphrates valley and the far northeast (the Jazirah) are being developed for large-scale mechanized wheat and cotton farming with the aid of pump irrigation systems. Although future water supplies are uncertain (upstream, Turkey has built dams on the Euphrates that affect the river's flow in both Syria and Iraq), production of these crops is rising, and more than half the Syrian harvest now comes from this area. Recent discoveries of oil also add greatly to the importance of this long-neglected part of the country.

Southward and southeastward, Syria turns into desert, and the familiar sheep, camel, and goat herders move endlessly across the parched countryside. There may be as many as a half million of them, a noticeable proportion of Syria's 14.8 million people. In contrast to Israel and, to a lesser degree, to Lebanon as well, Syria is very much a country of farmers and peasants; only 50 percent of the people live in cities and towns of any size. But again, Syria produces adequate harvests of wheat and barley and normally does not need staple imports. It also exports these two grains, but its biggest source of external revenue after oil remains cotton. In addition, Syria is a country where opportunities for the expansion of agriculture still exist; their realization would improve the cohesion of the state and bring its separate subregions into a tighter spatial framework.

Jordan

None of this can be said for Jordan (3.8 million), the desert kingdom that lies east of Israel and south of Syria. It, too, was a product of the Ottoman collapse, but it suffered heavily when Israel was created, more than any other Arab state. In the first place, Jordan's trade used to flow through Haifa, now an Israeli port, so that Jordan has to depend on destabilized Lebanon's harbors or the tedious route via Aqaba in the far south. Second, Jordan's final independence in 1946 was achieved with a total population of only about 400,000, including nomads, peasants, villagers, and a few urban dwellers. Then, with the partition of Palestine and the creation of Israel in 1948, Jordan received more than half a million Arab refugees; it soon also found itself responsible for another half million Palestinians who, although living on the western side of the Jordan River, were incorporated into the state. Thus refugees outnumbered residents by more than two to one, and internal political problems were added to external ones—not to mention the economic difficulties of beginning national life as a very poor country.

Nonetheless, Jordan has survived with U.S., British, and other aid, but its problems have hardly lessened. Many Jor-

danian residents still have only minimal commitment to the country, do not consider themselves its citizens, and give little support to the hard-pressed monarchy. Dissatisfied groups constantly threaten to drag the country into another conflict with Israel; the 1967 war was disastrous for Jordan, which lost the West Bank (its claim was formally surrendered to the Palestine Liberation Organization in 1988) as well as its sector of Jerusalem (the kingdom's second-largest city). Where hope for progress might lie—for example, in the development of the Jordan River valley—political conflicts intrude. The capital city, Amman (1.2 million), reflects the limitations and poverty of the country. Without oil, without much farmland, without unity or strength, and overwhelmed with refugees, Jordan presents one of the bleaker pictures in the Middle East.

Lebanon

Lebanon, Israel's northern coastal neighbor on the Mediterranean Sea, is one of the exceptions to the rule that the Middle East is the world of Islam: one-fourth of the population of 3.6 million adheres to the Christian rather than Muslim faith. Less than one-eighth the territorial size of Jordan (and only half the size of Israel), Lebanon has a long history of trade and commerce, beginning with the Phoenicians of old who were based here. Lebanon must import much of its staple food, wheat; the coastal belt below the mountains, although intensively cultivated, normally cannot produce enough grain to feed the entire population.

But normality has not prevailed in Lebanon during recent times. The country fell apart in 1975 when a civil war broke out between Muslims and Christians. This was a conflict with many causes. Lebanon for several decades had functioned with a political system that divided power between these two leading communities, but the basis for that system had become outdated. In the 1930s, Muslims and Christians in Lebanon were at approximate parity; but over the ensuing decades, the Muslims increased their numbers at a much faster rate than the more urbanized and generally wealthier Christians (many of whom have emigrated from Lebanon in recent years). The Muslims' displeasure with an outdated political arrangement (developed during the French occupation) was expressed during several outbreaks of rebellion prior to full-scale civil war. By then, Lebanon had also become a base for over 300,000 Palestinian refugees; these people, many of them living in squalid camps, were never satisfied with Lebanon's moderate posture toward Israel. When the first fighting between Muslims and Christians broke out in the northern coastal city of Tripoli, the Palestinians joined the conflict on the Muslim side. In the process, Lebanon was wrecked.

Beirut, the capital, once a city of great architectural beauty and often described as the Paris of the Middle East, was heavily damaged as Christian, Sunni Muslim, Shi'ite

Muslim, Druze, and Palestinian militias—as well as the Lebanese and Syrian armies—fought for control. By the end of the l980s, Beirut approached total destruction, and only the poorest 150,000 war-ravaged residents remained of the 1.5 million who had lived there as recently as the late 1970s. Since 1990, the conflict has abated; Beirut today is embarked on a long rebuilding process, and its population (now 1.8 million) is rising once more.

As the Muslims' strength intensified, the Christians concentrated in an area along the coast between Beirut and Tripoli (Fig. 7-13). In the meantime, Israel had taken control over a security zone in the south, and various competing Muslim factions occupied and controlled parts of the country. In 1976, the Syrians involved themselves in the situation, trying to pacify Lebanon through a military presence. Eventually the Syrians, too, fell victim to Lebanon's fractious character; by backing various client factions, Syria helped reduce the violence outside Beirut but then found itself inextricably enmeshed in the country's troubles.

In the mid-1990s, an exhausted Lebanon finally seemed to be moving toward some form of internal accommodation, but old enmities still threatened the fragile peace. Whether Lebanon will survive as a political entity remains uncertain.

Israel

Israel lies at the very heart of the Arab world (Fig. 7-3). Its neighbors are Lebanon and Syria to the north and northeast, Jordan to the east, and Egypt to the southwest—all in some measure still resentful of the creation of the Jewish state in their midst. Since 1948, when Israel was created as a homeland for the Jewish people on the recommendation of a United Nations commission, the Arab-Israeli conflict has overshadowed all else in the Middle East.

Indirectly, Israel was the product of the collapse of the Ottoman Empire. Britain gained control over the Mandate of Palestine, and British policy supported the aspirations of European Jews for a homeland in the Middle East, embodied in the concept of Zionism. In 1946, the British granted independence to the territory lying east of the Jordan River, and "Transjordan" (now the state of Jordan) came into being. Shortly afterward, the territory west of the Jordan River was partitioned by the United Nations, and the Jewish people got slightly more than half of it—including, of course, some land that had been occupied by Arabs. Jews actually owned only about 8 percent of Palestine's land, but they constituted more than one-third of the area's population.

As soon as the Jewish people declared the independent state of Israel on May 14, 1948, the new country was attacked by its Arab neighbors, who rejected the scheme. In the ensuing battle, Israel not only held its own but gained some crucial territory in its central and northern areas as well as in the Negev Desert to the south (Fig. 7-14). At the end of this first Arab-Israeli war, in 1949, the Jewish population controlled 80 percent of what had been Palestine west of the Jordan River. As Fig. 7-14 shows, Israel now owned not only the territory allocated to it by the United Nations but additional areas facing Egypt, Lebanon, and the West Bank (the area west of the Jordan River shown on the map). During the conflict, troops from Transjordan (the independent Arab kingdom east of the Jordan River) invaded this West Bank area, and in 1950 the king formally annexed it to his country, which he renamed Jordan.

This early conflict proved to be only the first in a series of wars between Israel and its Arab neighbors. In 1967, a week-long military conflict resulted in a major Israeli victory: Israel took the Golan Heights from Syria, the West Bank from Jordan, and the Sinai Peninsula (up to the Suez Canal itself) from Egypt. In 1973, another brief war led to Israel's withdrawal from the Suez Canal to truce lines in the Sinai Peninsula, and later all of the conquered Sinai Peninsula was returned to Egypt.

Since then, a fragile peace has been sustained, punctuated by terrorist incursions from Lebanon and Jordan, Israeli retaliation, and Israeli encroachment on Lebanon in a security zone extending along their joint border (but on the Lebanese side). In the meantime, the greatest challenge facing Israel has been internal: the achievement of an accommodation between the Israeli government and the Arabs (Palestinians) under its control. One prominent issue has been the displacement of Palestinians and their subsequent quest for a homeland: when Israel was created, more than 600,000 Palestinian Arabs were forced to leave the area to seek refuge in neighboring countries. Today, almost 4 million Arabs who call themselves Palestinians live in Jordan (where they constitute about 55 percent of the total population) and other Arab countries (see the box on p. 391 entitled "The Palestinian Dilemma"). When Iraq invaded Kuwait in 1990, some 400,000 Palestinians were working and living in that oil-rich skeikdom; when most of them chose to side with Iraq, they faced expulsion after the war, and many left for Jordan.

Palestinians inside and outside Israel and its occupied territories have been hoping for a homeland—if not an independent state, then a territory with substantial autonomy. The West Bank might have become such a territory, but Israel has built numerous Jewish settlements there, making future concessions to the Palestinians difficult. Despite Palstinian (and repeated U.S.) objections, Jewish immigration into the West Bank increased over the years: in 1977 only 5,000 Jews lived in the West Bank, but by 1993 about 125,000 had settled there. This movement into the West Bank was spurred in the late 1980s and early 1990s by the arrival in Israel of hundreds of thousands of Jewish immigrants from the (then) Soviet Union. Although the construction of new settlements was halted in 1992, Jews now

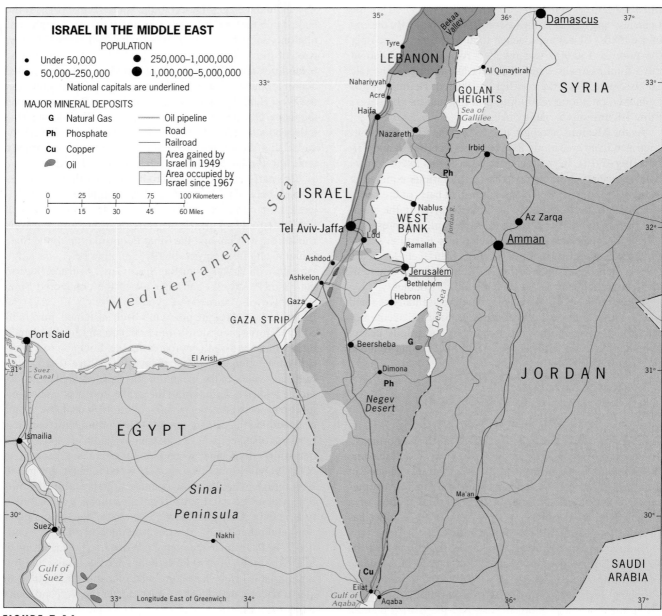

FIGURE 7-14

make up about 15 percent of the area's population, and they are dispersed across all of it (Fig. 7-15).

Another problem has been the city of Jerusalem (544,000), holy place for Jews, Muslims, and Christians alike. In the original U.N. blueprint for Palestine, the city was to have become internationalized; instead, Jerusalem became divided between the Jewish state and Jordan, an arrangement that satisfied neither side and caused continuous friction. Then, during the 1967 war, the holy city fell into Israeli hands when Jewish forces captured the entire West Bank

territory. Israel regards Jerusalem as its capital, but not all countries recognize the city as such; the United States has maintained its embassy in Tel Aviv.

Comparative Prosperity A major irritant, too, has been Israel's rapid rise to strength and prosperity amid the underdevelopment so common to the Middle East. In terms of physical geography, there is nothing particularly productive about most of Israel's land; neither is its territory that large (8,000 square miles/20,000 sq km —smaller than

THE PALESTINIAN DILEMMA

Ever since the creation of Israel in 1948 in what had been the British Mandate of Palestine, Arabs who called Palestine their homeland have lived as refugees in neighboring countries. A large number have been assimilated into the societies of Israel's neighbors, but many still live in refugee camps. The Palestinians call themselves a nation without a state (much as the Jews were before Israel was founded), and they demand that their grievances be heard. Until Palestinian hopes for a national territory in the Middle East are considered by the powers that have influence in the region, a full settlement between Israel and the Arab countries may remain unattainable. Current estimates of Palestinian populations in the realm are as follows:

Israel and Occupied Territories		2,650,000
Israel	900,000	
West Bank	1,000,000	
Gaza Strip	750,000	
Jordan		2,100,000
Lebanon		400,000
Syria		350,000
Saudi Arabia		250,000
Iraq		70,000
Egypt		60,000
Kuwait		30,000
Libya		25,000
Other Arab states		515,000
TOTAL		6,450,000

In 1987, Palestinians in Gaza and on the West Bank launched an uprising called the *intifada*, which claimed thousands of casualties. But behind the scenes, U.S. Secretary of State James Baker and the Bush Administration worked to bring the Israelis and the Palestinians together. Following the further mediation of Norwegian diplomats, an historic event took place on the White House lawn in Washington on September 13, 1993: Israeli Prime Minister Yitzak Rabin and Palestinian leader Yassir Arafat shook hands to seal an agreement that will turn over responsibility for the Gaza Strip (Fig. 7-14) and for the Arab city of Jericho in the West Bank (Fig. 7-15) to the Palestinians. This, Palestinians believe, is the first critical step toward the formation of a Palestinian state adjacent to Israel. But other crucial issues remain to be resolved: the boundaries of the Jericho zone, Jewish settlements within Gaza, access to Jericho, and much more. Still, the breakthrough has occurred.

Massachusetts). Yet Israel has been transformed by the energies of its settlers and, importantly, by heavy investments and contributions made by Jews and Jewish organizations elsewhere in the world, especially the United States. It is often said that in Israel the desert has been converted into farmland, and this is increasingly so. The irrigated acreage has been enlarged to many times its 1948 extent, and water is now carried even into the Negev Desert itself, where new food-production technologies are achieving some astonishing results in agricultural output. Particularly important are advances in computer-controlled *fertigation*, whereby brackish desert water mixed with fertilizers is piped directly to the roots of each plant (see photo). Fruit, vegetable, and grain yields have been spectacular under the harshest extremes of temperature and aridity, and Israeli technicians are now training agronomists in dozens of countries in the use of automated desert agricultural technology.

Without an appreciable resource base, industrialization also presents quite a challenge to Israel. Evaporation of waters in the zone bordering the Dead Sea has left deposits of potash, magnesium, and salt, and there is rock phosphate in the Negev. But there is very little fuel available within Israel; no coal deposits are known, although some oil has been found in the Negev. To circumvent the Suez Canal, an oil pipeline leads from the port of Eilat on the Gulf of Aqaba to the refinery at Haifa; a second, larger pipeline has been constructed to link Eilat and Ashkelon. Imported oil formerly came from Iran, but since the Iranian revolution in 1979, Israel has been forced to turn elsewhere for its petroleum.

One source is Egypt. When Israel ceded a section of the occupied Sinai Peninsula where oilfields had been found

The barren Negev Desert is the country's least-favored land resource, but the Israelis are making it bloom by pioneering ultrasophisticated fertigation techniques. Clearly, this is a food-production breakthrough of the utmost importance for water-deficient farming areas the world over.

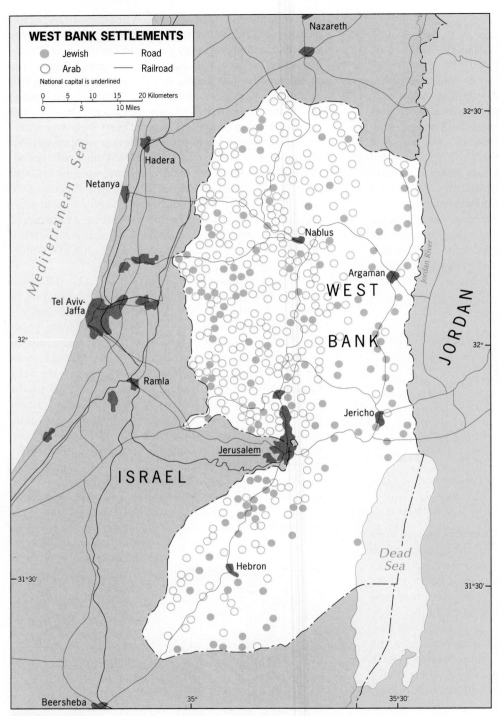

FIGURE 7-15

and developed, Egypt, having agreed to normalize its relations with Israel, also agreed to sell oil to Israel in return. The small steel plant at Tel Aviv uses imported coal and iron ore but was built for strategic reasons; it is not an eco-

nomic proposition. Thus the only manufacturing industry for which Israel has any domestic raw materials is the chemical industry, which has seen substantial growth. For the rest, however, Israel must depend on the considerable skills of its labor force; the specialized field of diamond-cutting, in which the Israelis have a global reputation, is an outstanding example. Many technicians and highly skilled craftspeople have been among the hundreds of thousands of Jewish immigrants who came to Israel since its creation, and the nation's best course was to make maximum use of this personnel.

In effect, then, Israel is a Western-type, developed country within the Middle East. Its population of 5.4 million is highly urbanized, with just over 90 percent of the people living in towns and cities. Israel's core includes the two major metropolises, Tel Aviv-Jaffa (2.3 million) and Haifa (282,000), and the coastal strip between them (Fig. 7-14). In total, this core area incorporates over three-quarters of the country's population, although the proportion is less if the territories conquered in the 1967 war are added (including East Jerusalem's Arab sector, with over 200,000 residents).

A Race Against Time Despite the persistence of antagonisms, time may be slowly eroding some barriers between Israel and the Arabs. U.S.-sponsored negotiations between Israel and its Arab neighbors, including the Palestinians, began in 1992, and Egyptian President Mubarak visited Israel to further the "peace process." For a growing number of Islamic countries the question no longer is whether Israel has a right to exist but rather what boundaries will be acceptable to all, and how the status of the occupied territories will be determined. Among the negotiating points, despite Jewish immigration into the West Bank, is a complicated "land for peace" program that would redefine the layout of this area.

As some external problems ease, however, divisions within Israel loom ever more threateningly. Today, the country's 4.5 million Jews are divided by a social gap based on ethnicity, income level, and cultural heritage. On the one hand, there are the majority Sephardic Jews of Asian-North African origin who largely occupy the lower echelons of Israeli society; on the other, there are the minority Ashkenazim Jews of European and American background who dominate the professions and higher-income groups. Constant frictions affect the relationships between these groups, reflecting a clash between the Islamic and Western cultural values of their source areas; residential and educational segregation mark the social geography of the two populations. Thus the younger generations of the Sephardim and Ashkenazim are not encouraged to integrate and commingle, perpetuating a schism that could one day threaten national unity. Moreover, Orthodox Jews, who advocate the dominance of religious law over civil Israeli law, are often at odds with nonreligious Jews who prefer to live in a secular society—a division that has sometimes resulted in violence.

In the volatile Middle East and in the realm at large, Israel is in a race against time. Only a satisfactory settlement with its Palestinian minority, plus normalized relations with moderate and secular Arab states, can guarantee it a stable future in a realm in transition.

THE ARABIAN PENINSULA

About the regional identity of the Arabian Peninsula there can be no doubt: south of Jordan and Iraq, the entire peninsula is encircled by water. This is a region of old-style emirates and sheikdoms made wealthy by oil; here, too, lies the source of Islam.

As a region, the Arabian Peninsula is environmentally dominated by a desert habitat and politically dominated by the Kingdom of Saudi Arabia (Fig. 7-16). With its huge territory of 830,000 square miles (2,150,000 sq km), Saudi Arabia is the realm's third biggest state; only Sudan and Algeria are somewhat larger. On the peninsula, Saudi Arabia's neighbors (moving clockwise from the head of the Persian Gulf) are Kuwait, Bahrain, Qatar, the United Arab Emirates, the Sultanate of Oman, and the new Republic of Yemen (created in 1990 through the unification of former North Yemen and South Yemen). Together, these countries on the eastern fringes of the peninsula contain about 18 million inhabitants; the largest by far is Yemen, with 11 million. The interior boundaries of these states, however, are still inadequately defined, here in one of the world's last remaining frontier-dominated zones.

Saudi Arabia

Saudi Arabia itself has only 17.2 million inhabitants in its vast territory, but the kingdom's importance is reflected in Fig. 7-9: the Arabian Peninsula contains the earth's largest concentration of known petroleum reserves. Saudi Arabia occupies most of this area, and by some estimates may possess as much as one-quarter of all the world's remaining oil. These reserves lie in the eastern part of the country, particularly along the Persian Gulf coast and in the Rub al Khali (Empty Quarter) to the south.

The national state that is Saudi Arabia was only consolidated as recently as the 1920s through the organizational abilities of King Ibn Saud. At the time, it was a mere shadow of its former greatness as the source of Islam and the heart of the Arab world. Apart from some permanent settlements along the coasts and in scattered oases, there was little to stabilize the country; most of it is desert, with annual rainfall almost everywhere under 4 inches (10 cm). The land surface rises generally from east to west, so that the Red Sea is fringed by mountains that reach nearly 10,000 feet (3,000 m). Here the rainfall is slightly higher, and there are some farms (coffee is a cash crop). These mountains also contain known deposits of gold, silver, and other met-

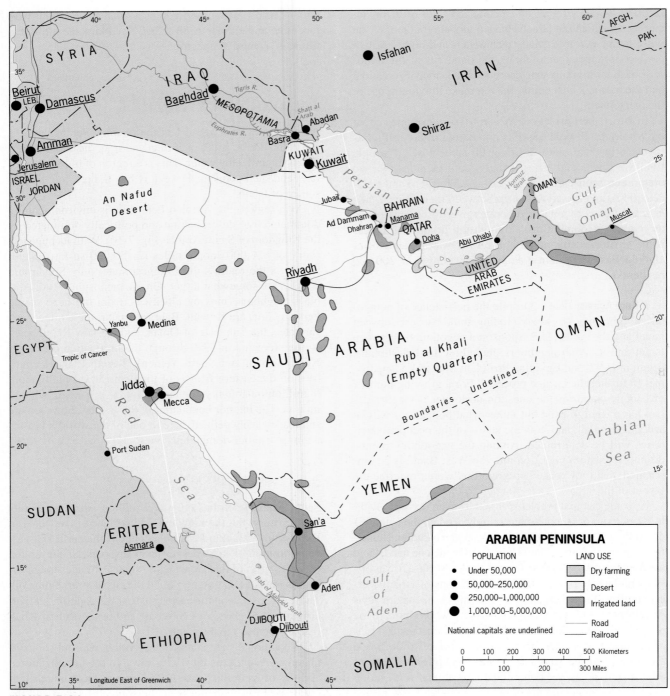

FIGURE 7-16

als, and the Saudis hope to diversify their exports by adding minerals from the west to the oil from the east.

Figure 7-16 reveals that most economic activities in Saudi Arabia are concentrated in a wide belt across the "waist" of the peninsula, from the boom town of Dhahran on the Persian Gulf through the national capital of Riyadh (2.6 million) in the interior to the Mecca-Medina area near the Red Sea. A fully effective internal transportation and communications network, one of the world's most modern, has recently been completed. But in the more remote zones of the interior, Bedouin nomads still ply their ancient caravan routes across the vast deserts. For several decades, Saudi Arabia's aristocratic royal families were virtually the sole beneficiaries of their country's incredible wealth, and there was hardly any impact on the lives of villagers and nomads.

That is now changing, however. Agriculture in particular is receiving major government investments because the Saudis want to prevent the food weapon from being used against them (as they themselves have occasionally wielded the oil weapon). As a result, widespread well drilling has significantly enhanced water supplies to support crops, and surpluses were quickly produced. These achievements, however, have been enormously expensive, for they necessitate the pumping of water from aquifers far below the desert surface as well as the construction of extensive center-pivot irrigation systems. Despite these efforts, there is now a growing realization that the Saudi Arabian goal of attaining self-sufficiency in food production is elusive. To begin with, the underground water supplies are proving to be a one-time-only, nonrenewable resource; desalinization

plants serve the coastal cities and towns and some interior places as well, but even an oil-rich kingdom cannot afford the price of raising its crops with desalinized water. Even more importantly, the Saudi population is growing at a much faster pace (3.5 percent, thereby doubling in just 20 years) than the rate at which the domestic food supply can be increased.

The country's rulers also have instituted reforms in housing, medical care, and education, and since 1970 they have spent hundreds of billions of dollars on national development programs. The collapse of oil prices during the 1980s slowed these programs' progress, but overall living standards have improved significantly. Industrialization, too, has been stimulated. The newly planned city of Jubail, north of Dhahran on the Persian Gulf, has become an industrial center with state-of-the-art petrochemical and metal fabrication plants. Similarly, Yanbu, north of Jidda (1.5 million) on the Red Sea coast, lies at the end of a trans-Arabian oil pipeline, an incipient industrial outpost for the west. Still, as we

Booming Jubail on the shore of the Persian Gulf, one of the world's most modern, state-of-the-art planned cities. This is the focus of Saudi Arabia's budding petrochemical industry.

noted earlier in this chapter, much of Saudi Arabia remains tradition-bound, and in the villages of the interior, the age of oil and development seems remote.

On the Periphery

Five of Saudi Arabia's six neighbors on the Arabian Peninsula face the Persian and Oman gulfs (Fig. 7-16) and are monarchies in the Islamic tradition. All five also derive substantial revenues from oil. Their populations range from 0.5 to 2.7 million; these are not strong or (OPEC aside) influential states.

They do, however, display considerable geographic diversity. *Kuwait*, at the head of the Persian Gulf, comes close to cutting Iraq off from the open sea; no matter how the aftermath of the Gulf War is settled, that issue will continue to exist here. *Bahrain* is an island state, a tiny territory, its oil reserves now dwindling. Bahrain's approximately 630,000 people are 50 percent Shi'ite and only 35 percent Sunni. Neighboring *Qatar* consists of a peninsula jutting out into the Persian Gulf, a featureless, sandy wasteland made habitable by oil (now declining in importance) and natural gas (rising simultaneously). *The United Arab Emirates*, a federation of seven emirates, faces the Persian Gulf between Qatar and Oman. The sheik is an absolute monarch in each of the emirates, and the seven sheiks form the Supreme Council of Rulers. In terms of oil revenues, however, there is no equality: two emirates—Abu Dhabi and Dubai—have most of the reserves. And the eastern corner of the Arabian Peninsula is occupied by the Sultanate of *Oman*, another absolute monarchy centered on the capital, Muscat (430,000). In Fig. 7-16, note that Oman consists of two parts: the large eastern corner of the peninsula and a small but critical cape to the north that protrudes into the Persian Gulf to form a narrow choke point—the Hormuz Strait (Iran lies on the opposite shore). Tankers that leave the other states must negotiate this narrow channel at slow speed, and during politically tense times they have had to be protected by warships.

This leaves what is, in many ways, Saudi Arabia's most substantial and potentially most difficult neighbor: *Yemen*. Our chart (Appendix A) states that Yemen has an area of 203,900 square miles, but that is an estimate: the boundary between Yemen and Saudi Arabia was never properly defined. In the early 1990s, following the unification of the two Yemens, exploration for oil was begun in the promising borderland; Saudi Arabia immediately demanded that it be stopped until the boundary has been defined and delimited. (The Saudis' relationship with the Yemenis, never very friendly, deteriorated after Yemen chose Iraq's side during the 1991 Gulf War, and Saudi Arabia expelled some 1 million Yemenis working in the kingdom.)

The size of Yemen's population also is uncertain and is variously reported between 11 and 13 million. Significantly, nearly half the people are Shi'ites. San'a (611,000), for-

merly the capital of North Yemen, was chosen as the headquarters of the unified country, and Aden (547,000) is the only port of consequence. The new Yemen was born as a multi-party, secular, democratic state, the only one in the region. But its economy is by far the weakest, with very limited oil production.

Yemen occupies a strategic part of the peninsula. Between Yemen and Djibouti in Africa's Horn, the mouth of the Red Sea narrows to the confined Bab el Mandeb Strait, another choke point on Arabia's periphery. The immediate prospects for instability, however, are on land, not at sea. Royal, Sunni, oil-rich, modernizing Saudi Arabia stands in stark contrast to democratic, strongly Shi'ite, agricultural, poor, underdeveloped Yemen. Such contrast, against a backdrop of territorial disagreement, spells trouble.

THE EMPIRE STATES

North of the eastern Mediterranean Sea, the Middle East, and the Persian Gulf lies a tier of states that connects the turbulent present to a powerful, imperial past. Five states constitute this region, where Arab ethnicity gives way but Islamic culture continues. Two of these states, Turkey and Iran, are modern (or rather, current) manifestations of a greater history. Three others, Afghanistan, Azerbaijan, and Cyprus, have seen imperial power descend on them (Fig. 7-17).

So many things change along the border between the Middle East and these Empire States that we seem to be entering a different realm here. The physical geography strengthens this impression: we leave the rocky and sandy expanses of desert and enter the higher relief of mountains and plateaus. Gone is the Arabic language we heard from Morocco to Oman and from Syria to the Horn of Africa. Gone, too, is the Arab nationalism that so strongly influences political geography from the Maghreb to the Middle East. But here to stay is the overarching imprint of Islam, in cities, villages, and countryside, not only in the cultural landscapes but also in ways of life and views of community and world. And when we get beyond the plateaus of Turkey and the mountains of Iran, we find that this part of the realm, too, has its deserts, oases, caravans, and clustered populations. We have crossed a regional boundary, but we have not left the realm.

Turkey

Before we focus on Turkey, center of the once-vast Ottoman Empire that brought Islam to the doorstep of Vienna and to the shores of the Arabian Peninsula, let us consider who the Turks are and why Turkey's influence may again expand far beyond its present borders. Historical geographers report that it is the Chinese who, perhaps as early as the third century, called *Tukiu* the nomadic peoples then living far from modern Turkey, in the steppes and forests

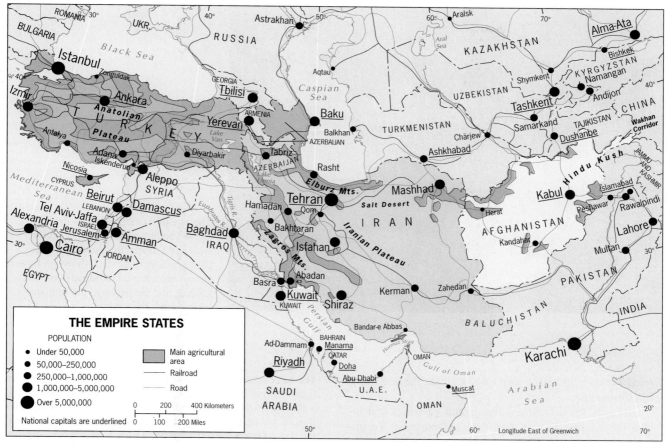

FIGURE 7-17

of Siberia. By the sixth century, these nomadic Turks had created an empire stretching from Mongolia to the Black Sea. Over time they migrated into what is today northern Iran and into the domain of the collapsing Byzantine Empire, Turkey. In the process, they spread their *Turkic* language far and wide; but despite this wide diffusion, basic Turkish remains a *lingua franca* for a vast area extending from western China to southern Bulgaria.

The Turks were thinly spread, and their loosely knit states rose and fell. The Uzbeks, led by Tamerlane, with Samarkand as his capital city, conquered territory extending from the Volga to Pakistan during the second half of the fourteenth century. The Kazakhs and Uygurs also founded large empires, but the Turkish peoples had little cultural continuity—until the arrival of Islam. The Arab-propelled Islamic wave touched the Turkish peoples before the end of the seventh century and diffused throughout their domain over the next several centuries. Its effect was to make the far-flung Turkish peoples part of a multicultural, multinational, but uniform religious civilization. Turkish lands now had linkage, from the Aegean Sea in the west to western China in the east. Their renewed vigor sent them on imperial conquests in various directions, and thus the Islamic

wave penetrated India from Turkestan and Europe from Turkey.

Earlier in this chapter, we noted the rise and decline of the Turkish Ottoman Empire. At the beginning of the twentieth century, the country we now know as Turkey lay at the center of this decaying and corrupt empire, ripe for revolution and renewal. This occurred in the 1920s and thrust into national and international prominence a leader who became known as the father of modern Turkey: Mustafa Kemal Atatürk.

The ancient capital of Turkey was Constantinople (now Istanbul), located on the strategic straits connecting the Black and Mediterranean seas. But the struggle for Turkey's survival had been waged from the heart of the country, the Anatolian Plateau, and it was here that Atatürk decided to place the new seat of government. Ankara (today home to 2.8 million), the new capital, possessed certain advantages: it would remind the Turks that they were (as Atatürk always said) Anatolians; it lay nearer the center of the country than Istanbul; and it could therefore act as a stronger unifier. Istanbul (7.8 million), located on the western side of the Bosporus waterway, lies on the threshold of Europe, with the minarets and mosques of this largest and most

The city of Istanbul, Turkey straddles a natural boundary between Southwest Asia and Europe: the Bosporus. This photo, looking eastward, shows an Eastern European cultural landscape in the foreground, and the mosques and minarets of Islam across the strait.

varied Turkish city rising above a somewhat Eastern European townscape.

Although Atatürk moved the capital eastward and inward, his orientation was westward and outward. To implement his plans for Turkey's modernization, he initiated reforms in almost every sphere of life within the country. Islam, formerly the state religion, lost its official status. The state took over most of the religious schools that had controlled education. The Roman alphabet replaced the Arabic. Islamic law was replaced by a modified Western code. Symbols of old—wearing beards, wearing the fez—were prohibited. Monogamy was made law, and the emancipation of women began. In Muslim society, women have generally been denied access to education, freedom of movement, or social contact; in a few countries, they still must cover their faces in public. The new government took pains to stress Turkey's separateness from the Arab world, and ever since it has remained aloof from the affairs that engage the other Islamic states.

What Atatürk wanted for Turkey, of course, could not be achieved in a short time. Turkey is largely an agricultural country, and with three-quarters of its inhabitants subsistence or near-subsistence farmers, most living in small villages in often isolated rural areas, some opposition to change, especially rapid change, had to be expected. In fact, the government had to yield to the devoutly Muslim peasants on some issues. Still, Atatürk's directives continue to be pursued seven decades after he came to power.

Turkey is a mountainous country of generally moderate relief. The highest mountains lie near its eastern border with Armenia and Iran. Here elevations reach 6,500–10,000 feet (2,000–3,000 m), but westward the altitudes decline. The

The huge reservoir behind Atatürk Dam on the upper Euphrates is now filling up, changing dramatically the hydrography of southeastern Turkey. The downstream countries of Syria and Iraq, however, cast a wary eye toward projects that could affect the flow of water through their territories.

THE PROBLEM OF CYPRUS

In the northeastern corner of the Mediterranean Sea (see Fig. 7-17), much farther from Greece than from Turkey or Syria, lies Cyprus (population 740,000). Since ancient times this island has been predominantly Greek in its population, but in 1571 it was conquered by the Turks, under whose control it remained until 1878. Following the decay of the Ottoman state, the British soon took control of the island. By the time Britain was prepared to offer independence to much of its empire in the years following World War II, it had a problem in Cyprus because the Greek majority among the 600,000 population preferred *enosis*—union with Greece. It is not difficult to understand that the 20 percent of the Cypriot population that was Turkish wanted no such union.

By 1955, the dispute had reached the stage of violence because differences between Greeks and Turks are deep, bitter, and intense. It had become impossible to find a solution to a problem in which the residents of this island country think of themselves as Greeks or Turks first rather than Cypriots. Nonetheless, in 1960 Cyprus was granted independence under a complicated constitution designed to permit majority rule but with a guarantee of minority rights.

The fragile order finally broke down in 1974. As a civil war engulfed the island, Turkish armed forces intervened (about 30,000 soldiers were still present in the early 1990s), and a major redistribution of population occurred.

The northern 40 percent of Cyprus became the stronghold of more than 100,000 Turkish Cypriots; only a few thousand Greeks remained there. The rest of the island—south of the U.N.-patrolled "Green Line" that was demarcated across the country to divide the two ethnic communities—became the domain of the Greek majority, now including nearly 200,000 in refugee camps. As in Lebanon, partition looms as a very serious prospect; the fundamental differences between Greek and Turkish Cypriots are still unresolved, hatreds have deepened, and a form of de facto apartheid now exists on the island.

In 1983, again bringing the stalemated conflict to the attention of the world, the Turkish community (today comprising one-fifth of the population) seceded by declaring itself the new independent Turkish Republic of Northern Cyprus. The situation is especially difficult because of the many factions involved. The Turkish Cypriots, many of whom have expressed a desire for reintegration, are discrete from the Turkish nation and express a strong attachment to their island homeland. The Greek Cypriot majority is ideologically divided, and the crisis of 1974 began not as a Greek-Turkish conflict but as a result of a coup by Greek Cypriot soldiers against the Greek-dominated government of the island. Internal factionalism and external involvement have made Cyprus a pawn of foreign powers—and a casualty of history.

Anatolian Plateau in the heart of the country is dry enough to be classified as a steppe (see Fig. I-7, the area marked *BSk*). Here the people live in small villages, grow subsistence cereals, and raise livestock. But the best farmlands lie in the coastal zones, small as they are. There is little coastal lowland along the Mediterranean: the areas around Antalya, Adana, and Iskenderun are the only places where the mountains retreat somewhat from the sea. The northern Black Sea coast is also quite narrow, but it is comparatively moist; it gets winter as well as summer rainfall (though with a cool-season maximum) totaling 30 to 105 inches (75 to 260 cm) yearly—uncommonly high for this water-deficient realm. Thus as Fig. I-9 reflects, these coastal areas, especially the Aegean-Marmara lowlands, are the most densely populated parts of Turkey and the most productive ones as well.

In population size, Turkey is one of the three largest countries in the realm, with 61.9 million inhabitants (both Egypt and Iran exhibit similar totals). As Fig. 7-17 suggests, Turkey has a superior surface communications network; it is also the realm's most industrialized country,

despite the dominance of agriculture in its export economy. The production of cotton has stimulated a textile industry, and a small steel industry has been established based on a coalfield near Zonguldak, near the Black Sea, and an iron-ore deposit several hundred miles away in east-central Turkey. In the southeast, Turkey has found some oil, and it may share in the zone of oilfields of which the famed Kirkuk reserve in nearby Iraq is a part. With its complex geology, Turkey possesses a variety of mineral deposits that provide ample opportunity for future development. Spatially, this development seems destined to reinforce the country's westward orientation.

In the mid-1990s, Turkey faces several competing challenges. Abroad, these involve: (1) the government's long-range hope to join the European Community; (2) the unsettled problem of Cyprus (see box above), hence relationships with Greece; (3) relationships with neighboring Iraq, whose major rivers have their sources in Turkey and whose major oil pipeline lies across Turkey (see Fig. 7-9); and (4) the opening of Turkestan to foreign influences, among which Turkey hopes to be dominant.

At home, the challenges are: (1) a still-weak economy, whose per capita gross domestic product still is only slightly more than one-quarter of that of none-too-wealthy Greece; (2) the rise of Islamic fundamentalism, often in opposition to current government plans and programs; (3) the Kurdish question, involving the relationships between the large Kurdish minority in southeastern Turkey (see Fig. 7-13) and the government in Ankara; and (4) related to this, the issue of human rights, a key condition for entry into any European union.

Turkey's great opportunity lies in its relative location. It could become the bridge between the European and Islamic realms, a vital link in the changing geography of western Eurasia.

Iran

Iran is the second state with a history of empire in this northern region of the North Africa/Southwest Asia realm. In 1971, the shah and his family celebrated the 2,500th anniversary of Persia's first monarchy with unmatched scenes of royal splendor. But before the end of that decade, revolution engulfed Iran and Shi'ite Islamic fundamentalism drove the shah from power. The monarchy was replaced by an Islamic republic.

As Fig. 7-17 shows, Iran occupies a critical area in this turbulent realm. It controls the entire corridor between the Caspian Sea and Persian Gulf. To the west it adjoins Iraq and Turkey, both historic adversaries. To the north (west of the Caspian Sea), Iran borders Azerbaijan and Armenia, where Muslims and Christians are at war. To the east Iran meets Pakistan and Afghanistan, and east of the Caspian Sea Iran faces volatile Turkmenistan. In many places, Iran seems to spill over into its neighbors: the Kurdish domain connects it to Iraq and Turkey; Azerbaijan is both an Iranian province and an adjacent independent state; Baluchistan is divided between Iran and Pakistan, and even in the Persian Gulf Iran reaches outward. In 1993, Iran asserted sovereignty over several small islands located in the Gulf's choke point, the Hormuz Strait.

Today, after 15 years of retribution, devastating war, and economic decline, Iran appears to be seeking a new course. Shi'ite Islam continues to dominate in the affairs of state (in 1989 the regime imposed the death penalty against author Salman Rushdie, who had written what was regarded as a blasphemous novel), but practical issues are causing significant changes in Iranian policy. Once again, international interconnections are responsible: at a time when Turkey, Pakistan, and even India are vying for Iran's hinterlands, the Iranians cannot afford to continue their postrevolutionary isolation.

Iran, as Fig. 7-17 demonstrates, is a country of mountains and deserts. The heart of the country is an upland, the Plateau of Iran, that lies surrounded by even higher mountains including the Zagros Mountains in the west, the Elburz Mountains along the southern shores of the Caspian

Sea, and the mountains of the Khurasan region to the northeast. The Iranian Plateau is therefore actually a highland basin marked by salt flats and wide expanses of stone and sand. On the hillsides, where the topography wrests a modicum of moisture from the air, lie some fertile soils. Elsewhere, only oases break the arid monotony—oases that for countless centuries have been stops on the area's caravan routes.

Iran has little arable land. The moist zone along the Caspian Sea coast supports a ribbon of settlement, and the area north of the head of the Persian Gulf, centered on Abadan (525,000), also contains fertile soils. Still, most people depend directly on farming or livestock herding. Despite the limitations of environment and livelihoods, Iran's population is 63.7 million—and growing rapidly at an annual rate of 3.3 percent.

In ancient times, Persepolis in southern Iran was the focus of the powerful Persian kingdom. Then, as now, people clustered in and around the oases or depended on *qanats* (underground tunnels leading from the mountains) for their water supply, as Persepolis did. The urban focus of modern Iran, Tehran (7.5 million), lies far to the north on the southern slopes of the Elburz Mountains. This mushrooming metropolis, which also remains partially dependent on the *qanat* system for its water, rose from a caravan station to become the capital of a modernizing state.

Iran indeed modernized during the regime of Shah Muhammad Reza Pahlavi (1941–1979), although its myriad villages and nomadic communities (see the box entitled "People on the Move") are reminders that change does not come quickly or easily to a country as large and tradition-bound as Iran. Thus the modernization process operated to intensify local and regional contrasts, facilitating the revolution that was to come—an upheaval that had its roots in the fundamental Islamic traditions of the great majority of the people. Like Atatürk and his successors in Turkey, the shah sought to bring major reforms to Iran (a task made possible by the country's enormous oil income). Apart from the social changes brought by this "White Revolution," industrialization was introduced, some agriculture was modernized, and a kind of domestic peace corps was established to improve medical services and literacy in rural areas. But the shah's reign was sustained by the intervention of foreign interests, and his policies ran counter to the Islamic traditions of the majority of his people. Muslim leaders opposed him, among them the exiled Ayatollah Khomeini. In time, this fanatical religious leader became the symbol of the Islamic revolution that exploded around his return in 1979 and continued to run its course beyond his death 10 years later.

Among the circumstances that brought Khomeini to power were the vast material inequities that the modernization effort had engendered. The contrasts between the advantaged elite in Tehran and the poverty of urban slum dwellers, the haves in cities like Isfahan (1.8 million) and Shiraz (1.5 million) and the have-nots nearly everywhere

PEOPLE ON THE MOVE

Camp has been broken, the camel caravan formed, and the purposeful journey of these nomads resumes—here in the Ahaggar Mountains of southernmost Algeria in the very heart of the Sahara.

Countless thousands of people in North Africa/Southwest Asia are on the move; movement is a permanent part of their lives. They travel with their camels, goats, and other livestock along routes that are almost as old as human history in this realm. Most of the time, they do so in a regular seasonal pattern, visiting the same pastures year after year, stopping at the same oases, pitching their elaborate tents near the same stream. This is a form of cyclical migration—**nomadism**.

Nomadic movement, then, is not simply an aimless wandering across boundless dry plains. Nomadic peoples know their domain intimately; they know when the rains have regenerated the pastures they will make their temporary locale, and they know when it is necessary to move on. Some nomadic peoples remain in the same location for several months every year, and their portable settlements take on characteristics of permanence—until, on a given morning, an incredible burst of activity accompanies the breaking of camp. Amid ear-piercing bleating of camels, the clanking of livestock's bells, and the shouting of orders, the entire community is loaded onto the backs of the animals and the journey resumes (see photo). The leaders of the groups know where they are going; they have been making this circuit all their lives.

Among the members of the community there is a division of labor; some are skilled at crafts and make leather and metal objects for sale or trade when contact is made with the next permanent settlement. A part of the herd of livestock traveling with the caravan may belong to townspeople, who pay for their care; nomads are not free from the tentacles of the cities.

Although nomadic pastoralism remains a significant way of life on the Iranian Plateau, in the Arabian Peninsula, and elsewhere in the sparsely populated areas of the realm, the number of people involved is steadily declining today. A major cause is the building of new roads, which increasingly criss-cross the deserts of North Africa and Southwest Asia and enable camels to be replaced by faster trucks that can carry much bigger loads. At the same time, severe environmental problems, such as drought and *desertification*—desert expansion resulting from human degradation of adjacent, fragile semiarid zones—are convincing many to give up the rigors of nomadism. And important, too, are rising tensions over international boundaries lying in deserts, which prompt governments to fortify and restrict movements across them, thereby severely impeding circulation along a number of routes traditionally followed by nomads.

else, who remained mired in a web of debt and dependency—all this produced wide opposition to the shah's policies. Ruthless political repression, torture, and a lack of avenues for the expression of alternative ideas also contributed to the breakdown of the order the shah's authoritarian rule had wrought.

The wealth generated by petroleum could not transform Iran in ways that might have staved off the revolution. Modernization remained but a veneer: in the villages away from Tehran's polluted air, the holy men continued to dominate the lives of ordinary Iranians. As elsewhere in the Muslim world, urbanites, villagers, and nomads remained enmeshed in a web of production and profiteering, serfdom, and indebtedness that has always characterized traditional society here. The revolution swept this system away, but in the end it did not improve the lot of Iran's millions. A devastating war with Iraq (1980–1990), into which Iran ruthlessly poured hundreds of thousands of its young men, sapped the coffers as well as the energies of the state. When it was over, Iran was left poorer, weaker, and aimless, its revolution spent on unproductive pursuits.

In the mid-1990s, Iran's government is reaching outward in new ways. Still avowedly fundamentalist and devotedly Shi'ite, the modern Iranians realize that their country's future is being built outside its borders—in Transcaucasia, in Turkestan, and in Afghanistan. Iran's old imperial rival, secular Turkey, is everywhere in Iran's hinterlands, promoting its brand of Islam and its views on statehood—and extending bank credits and loans. And so Iran must compete: to strengthen its ties with Shi'ites outside its borders, to promote its model of an Islamic republic, to organize mosques, and to rally fundamentalists. Even as its representatives disperse into Iran's neighboring lands, Iran is purchasing weapons from former Soviet republics and is resurrecting its old ties with Moscow. The revolution may have ended a monarchy, but it has not extinguished all ties to an imperial past.

Azerbaijan*

As Fig. 7-17 shows, the boundaries of Iran over most of their length lie along coastlines or across deserts. But in the west and northwest, the situation is different. In the west, in the border area with Iraq and Turkey, lies Iran's part of "Kurdistan" (Fig. 7-13). And in the northwest lies the Iranian province of Azerbaijan. This also is the name of a new country across the border to the north: former Soviet Azerbaijan.

The Azeris (short for Azerbaijanis) on both sides of the border are of the same ancestry: they are a Turkish people divided between the (then) Russian and Persian empires by a treaty signed in 1828. Since then, Iran's Azeris have been assimilated into the Persian (later Iranian) state, but Soviet

Azerbaijan became an independent republic in 1991 as a result of the collapse of the U.S.S.R. When the Soviet Azeris mounted their campaign to oust the Soviets, thousands of Iranian Azeris came to the border to shout support and to smuggle arms across. The two peoples had not forgotten their common heritage.

Independent Azerbaijan has a population of 7.4 million. Estimates of the Azeri population in Iran vary but are as high as 15 million (with about 4 million living near the Azerbaijan border), making this a substantial nation. Obviously, the potential for irredentism exists: Azerbaijan, like Iran, is dominantly Shi'ite, its religion steadily reviving after the Soviet period. Iran may face an unfamiliar challenge here: an independent, greater Azerbaijan may appeal to Azeris as much as an autonomous Kurdistan interests the nearby Kurds.

Afghanistan

The easternmost country in this region is Afghanistan, a landlocked, mountainous, divided, embattled product of foreign rivalries. Afghanistan exists because the British and the Russians, competing in this area during the nineteenth century, agreed to create it as a cushion or **buffer state** between them. This is how Afghanistan acquired the narrow corridor of land leading from the main territory eastward, the Wakhan Corridor (Fig. 7-17).

The Wakhan Corridor has the effect of making China one of Afghanistan's neighbors. The country's more important adjoining states are Pakistan to the east and south, Iran to the west, and, following the collapse of the Soviet Union, Turkmenistan, Uzbekistan, and Tajikistan to the north. With so many neighbors, Afghanistan is fragmented ethnically and culturally. Its population of about 17.7 million has no majority; the largest minority are the Pushtuns (or Pathans), whose homeland lies in the east and southeast along the border with Pakistan. In the north, the Tajiks have a territorial base contiguous to neighboring Tajikistan, and the Uzbeks are concentrated farther westward. Turkmen, Hazara, and smaller minority groups create a complicated ethnic mosaic in a country that is also divided by physical geography. Texas-sized Afghanistan is a desert plateau in the west and south, and the center and northeast are dominated by the soaring ranges of the Hindu Kush.

There is little to unite the peoples of Afghanistan, and much to divide them. Sunni Islam prevails here, but the faith has not been an intercultural bond. During the 1970s, the Soviet Union began to see an opportunity in the country's weakness, and in 1979 the Soviets invaded Afghanistan to prop up a puppet government. This *did* create a common cause, and while millions of refugees streamed into Pakistan and Iran, anti-Soviet rebels, with Western weapons and other assistance, mounted a fierce resistance. In 1989 the Soviets conceded defeat and withdrew, and in 1992 the government in Kabul (2.1 million) collapsed and was replaced by a revo-

---

*Azerbaijan is treated in greater detail as part of the Transcaucasia vignette on pp. 165-169.

lutionary council representing some (but not all) the factions. The situation began to resemble that of the pre-Soviet past: a feudal pattern with a weak and ineffectual government in the capital.

Afghanistan thus remains one of the realm's weakest and poorest countries. Urbanization still is below 20 percent, circulation is minimal, agricultural and pastoral subsistence remain the dominant livelihoods, and there is little national integration. Fruits, grown in the fertile lowlands adjacent to the Hindu Kush in the north, and carpets are the main exports. A small oil and natural gas industry was emerging in the north, but the Soviets, as they departed, capped the wells.

Afghanistan does have one coveted asset, one it cannot sell—its relative location. The country lies at the juncture of powerful Islamic states and near others. Its weakness may attract intervention by neighbors that already have stakes in the country, ethnic bridgeheads the Soviets could not claim. Pakistan may have an interest in redefining its boundary with Afghanistan; Tajikistan may have similar objectives. Will Afghanistan become the Lebanon of this region?

TURKESTAN

The old regional term *Turkestan* has reappeared on the map of the Islamic world. In the five former Soviet republics of Central Asia—now independent countries in which Islam is reviving even as politics and economics are changing—a momentous transformation is taking place. The configuration of this region is not yet final, but its general contents are clear: it will comprise most or all of five states. As shown in Fig. 7-18, these are: (1) *Kazakhstan*, territorially larger than all four others combined but situated astride an ethnic transition zone between Russia and Turkestan; (2) *Uzbekistan*, the most populous state, located at the heart of the region; (3) *Turkmenistan,* large but sparsely peopled and fronting the Caspian Sea; (4) *Tajikistan*, its population spilling over into adjoining Afghanistan and into China; and (5) *Kyrgyzstan*, land of the great Tian Shan range.

This is Turkestan today, but the region so defined does not incorporate all of what once was the Turkic domain. Turkic peoples inhabit all of northwestern China (turn to Fig. 10-12 on p 536 for the evidence), and the resurrection of an Islamic Turkestan will challenge the Chinese on their western frontier.

In Turkestan, Islam was long suppressed by the Soviet communist regime, but the faith was not extinguished. It is now reviving (as is Christianity in Russia itself), and a new wave of Islamic diffusion is spreading over the region. In previous editions of this book, politico-geographical realities forced us to classify this region as part of the former Soviet Union; but today Turkestan is being reconnected to the realm of Islam.

Islam returns to Turkestan. During the Soviet period, Islam was discouraged in various ways; but it survived, especially among older citizens. Now the mosques are functioning again, and the oldsters can transfer their faith to the young. These Muslims in Tashkent, Uzbekistan are leaving their mosque after Friday prayers.

Kazakhstan

In one respect, Kazakhstan resembles Chad and Sudan in the African Transition Zone: part of it lies in the realm of Islam, and part of it lies in a neighboring realm. As we noted in Chapter 2, the neighboring realm is Russia. Northern and northeastern Kazakhstan have been Russified; the south remains Turkestan.

In southern Kazakhstan, the physical and cultural geography resume their dry-world, Islamic-world characteristics, two of the defining qualities of the North Africa/ Southwest Asia realm. From the Caspian Sea in the west to the Chinese border in the east, Kazakhstan is a land of vast desert and steppe; scattered, water-dependent population clusters; few permanent surface communications; and no-

FIGURE 7-18

madic herding. The largest settled population concentrations lie in the better-watered east, where the capital, Alma-Ata (1.3 million), also is located.

The Kazakhs are united by their common descent, by their Turkic language, and by Sunni Islam. But southern Kazakhstan is by no means a culturally homogeneous area. The Kazakhs themselves constitute a minority of an estimated 40 percent of the republic's total population (Russians, concentrated in the north, make up about 38 percent). In southern Kazakhstan, Kazakhs are in the majority, but they share their land with numerous minorities, large and small. The final Soviet census of 1989 indicates that nearly 100 minorities can be identified in all of Kazakhstan, many of them in the Kazakh-dominated south.

In Chapter 2 we mentioned the grand Soviet planning projects in Turkestan, notably those centered on the Aral Sea area. As Fig. 7-18 shows, the Aral Sea lies on the border between Kazakhstan and Uzbekistan, and here the Soviets brought millions of acres of desert land into production through massive irrigation. However, while the cotton fields expanded, natural draininage into the Aral Sea was disrupted, pesticides polluted the groundwater, and an ecological and human disaster occurred. The Aral Sea has been destroyed, deprived of replenishment; illnesses caused by chemical contamination have devastated the people of the area.

The Soviets also plundered Kazakhstan's considerable mineral and fuel resources. The city of Karaganda (687,000) was founded to facilitate the mining of a huge coalfield; today, that city looks to Moscow, not to Mecca, for its future. In the west, Kazakhstan possesses major oil supplies, and some estimates suggest that the reserves in the Tengiz Basin at the northeastern end of the Caspian Sea (now being developed under contract with a major U.S. oil company) may be among the largest in the world; there are also large refineries and, near Guryev (Atyrau), a nuclear power station. All this gives Kazakhstan bargaining power during the transition now underway, but the Kazakhs are a loosely knit nation, and national unity will be both necessary and difficult to achieve.

Kazakhstan's political and cultural mixes remain explosive, touching off both anti-Russian and inter-ethnic strife. And, in another familiar pattern, there are latent territorial disputes: Kazakhstan's new government has made known its dissatisfaction with parts of its boundaries with Uzbekistan and Turkmenistan.

Uzbekistan

The former Soviet republic of Uzbekistan lies at the heart of Turkestan, a populous, California-sized country with 22.5 million inhabitants, nearly 75 percent Uzbeks. To a greater degree even than Kazakhstan, Uzbekistan was blighted by Soviet economic policy. To make Uzbekistan the world's third-largest cotton producer, vast farmlands in the arid Aral Sea area were opened up for irrigation. Now locally grown foods are unsafe, drinking water is polluted, infant mortality has soared, and cancer rates are high.

At the other (eastern) end of Uzbekistan lies the Fergana valley, densely populated and poverty-ridden. Here, too, lies the capital, Tashkent (2.2 million), on the very doorstep of Kazakhstan. Between these extremities, Uzbekistan typifies Turkestan: expanses of dry, open desert and steppe broken only by oases and ribbons of water. Uzbekistan is therefore a disjointed country, but the Uzbeks are the largest national group in Turkestan. Converted many centuries ago to Sunni Islam, they ruled Central Asia until their khanates of Khiva and Bukhara were absorbed into the Soviet Union in 1924. Russians never came here in great numbers, at least not compared to Kazakhstan: they constitute about 7 percent of the population today, a proportion that is declining.

As Fig. 7-18 shows, the far northwestern part of Uzbekistan, the zone bordering the Aral Sea, was designated the Karakalpak Autonomous Soviet Socialist Republic during the Soviet period. The justification for this action was the ethnic distinctiveness of the Karakalpak people who inhabit this area; they are much more closely related to the Kazakhs than to the Uzbeks. Between 1925 and 1936, the Karakalpaks' domain actually was an administrative part of the Russian S.F.S.R., but in 1936 it was assigned to the Uzbek S.S.R. Its status in present Uzbekistan is uncertain; Uzbeks form a minority here. The Uzbeks also form minorities in neighboring republics: nearly 25 percent in Tajikistan, 13 percent in Kyrgyzstan, and about 10 percent in Turkmenistan. The press

in Tashkent has editorialized about a "greater Uzbekistan" that should perhaps be renamed to reflect the Uzbeks' regional (not just national) strength.

Uzbekistan is an important, populous, centrally located state in the region. Its continuing stability is crucial to the region's future. Unfortunately, the threats to that stability are serious: relations with neighbors are tense; irredentism toward Uzbek minorities in other republics is occurring; friction with internal non-Uzbek minorities is frequent; and Muslim fundamentalism is growing. The future will be full of challenges.

Turkmenistan

In the southwest of Turkestan lies the desert republic of Turkmenistan, inhabited by just 4.1 million people. Extending eastward from the shores of the Caspian Sea to the boundary with Afghanistan, and bordering Iran for over 700 miles (1,100 km), this area was part of the old Muslim Turkestan before it became a Soviet republic in 1925. As large as Nevada and Utah combined, Turkmenistan was the frontier domain of many nomadic peoples when Soviet efforts to modernize the area began. Although 72 percent of the republic's inhabitants are Turkmen, many ethnic divisions still fragment the predominantly Muslim population. In addition, there are Russians (9 percent, mostly in the towns), Uzbeks (another 9 percent), Kazakhs, Ukrainians, Tatars, and Armenians.

Soviet efforts to stabilize the population and to make it more sedentary centered on a massive project: the Kara Kum Canal, begun in the 1950s to bring water from mountains to the east into the heart of the desert. In 1993, the canal, which eventually will reach the Caspian Sea, was more than 700 miles (1,100 km) long, and about 3 million acres had been brought under cultivation. Cotton, corn, fruits, and vegetables are now farmed. But many Turkmen people still herd sheep, and Astrakhan fur remains a valuable export.

During the disintegration of the U.S.S.R., the Soviet regime accused Iran of irredentism toward Turkmenistan's Muslims. But the people here are mostly Sunni Muslims, and in 1989 they gave asylum to thousands of Christian Armenians who fled across the Caspian Sea in their escape from Azeri persecution in Baku. Ashkhabad (459,000), the capital, lies on the only major east-west road, on the railroad between the Caspian port of Krasnovodsk and the east, and on the Kara Kum Canal. To the west lie oil and natural gas reserves that may, in time, help lift Turkmenistan out of its economic dormancy.

Tajikistan

In the rugged terrain of southeastern Turkestan, the political geography is as complex as it is in Transcaucasia (see pp. 165-169). Tajikistan, Kyrgyzstan, and eastern Uzbekistan look like pieces of a jigsaw puzzle, their territories interlocked and, in Tajikistan's case, functionally fragmented (Fig. 7-18).

Iowa-sized Tajikistan borders Afghanistan and China. Its eastern reaches are dominated by the gigantic Pamirs, where high glacier-sustaining mountains feed rivers such as the Amu-Darya, source of irrigation water for the neighboring desert republics of Uzbekistan and Turkmenistan. The tallest peaks in this range, both over 23,000 feet (7,000 m), tower above some of the world's most spectacular highland terrain.

The Tajiks are a separate people in the region; they are of Persian (Iranian) origin, not Turkic, and speak a Persian (and thus Indo-European) language. They constitute 62 percent of the population of 5.9 million, but a larger part of the Tajik nation inhabits neighboring Afghanistan (there also is a Tajik minority across the border to the east in China [see Fig. 10-12 on p.536]). In their republic the Tajiks share their land with Uzbeks, concentrated in the west and northwest (24 percent), Russians (a shrinking 7 percent), Armenians, and others, including the small Ismaili Muslim community for whom the Gorno-Badakhshan Autonomous Region was created by the Soviets in several Pamir valleys. Most Tajiks, despite their Persian affinities, are Sunni Muslims, not Shi'ites.

The capital of Tajikistan, Dushanbe (634,000), lies in the west, a forward capital in a sense because it lies near areas in Uzbekistan to which Tajikistan has laid claim. When the Tadzhik Soviet Socialist Republic (as the Russian communists spelled it) was established in 1929, this was an area of semi-nomadic herding and handicraft industries. Soviet rule and economic planning brought large industrial enterprises, mining, and irrigated agriculture; but most Tajiks remained farmers, herders, and producers of subtropical fruits (in the warmer valleys) and grains.

Along with the other four Central Asian states, Tajikistan declared its independence as the Soviet Union began to unravel in 1991. But the transition almost immediately deteriorated into rioting, bloodshed, and civil war. As Moscow's grip relaxed, fundamentalist Muslims rebelled against the entrenched communist leadership and seized the upper hand through much of 1992. But Tajikistan's newly independent neighbors and Russia, alarmed by the prospect of the forging of a Tajik Islamic republic with strengthening ties to nearby Iran and adjacent Afghanistan, gave support to the old rulers to counterattack the rebels. This effort soon succeeded—destroying much of Tajikistan's economy in the process—and by early 1993, the former communists had regained control of the country. Tajikistan's relative location as well as its volatile political situation (the fundamentalists survive as a guerrilla force) make for a particularly challenging future.

Kyrgyzstan

The great mountain ranges of the Tian Shan straddle the center of Kyrgyzstan, a republic whose outline on the map seems to relate to neither topography nor culture. Once part of the Turkic Empire, this territory fell to Russian expansionism during the nineteenth century when it was a remote frontier land. Once under Soviet control, it acquired its present boundaries and gained identity as a republic.

Kyrgyzstan's boundaries were defined on the basis of nationality as interpreted by Soviet planners in the 1920s, but today the indigenous inhabitants, the Kyrgyz (formerly spelled Kirghiz), represent just over 50 percent of the population of 4.7 million. Russians constitute more than 20 percent, Uzbeks about 13 percent. Ukrainians, Tatars, and other minorities also inhabit pockets of mountain-divided, Nebraska-sized Kyrgyzstan. Like their neighbors to the west and south, most Kyrgyz are Sunni Muslims.

The capital, Bishkek (687,000), lies on the north slope, very near the border with Kazakhstan (Fig. 7-18). Surface communications between northern Kyrgyzstan and neighboring Kazakhstan are better than those within Kyrgyzstan itself; the Tian Shan remains a formidable barrier, although a tunnel is under construction. In the meantime, towns in southwestern Kyrgyzstan are linked by road (and some by rail) to Uzbekistan, from where many immigrants have come to work in factories and to farm.

Pastoralism is the republic's mainstay. In addition to sheep and cattle, the Kyrgyz raise yaks for meat and milk; the yak can survive on high-altitude pastures unsuitable for other livestock. Irrigated lowlands adjacent to the mountains yield wheat, fruits, and vegetables. Most industries are processing and packaging plants for locally produced textiles and foods and turn out consumer goods. The economy, in general terms, remains typical of a Third World country. What potential there is centers on oil and natural gas, textile manufacturing, and (possibly in the future) tourism. Remote and disjointed, Kyrgyzstan has little national identity. In the aftermath of the Soviet Empire, it remains an internally fragmented, isolated, poverty-afflicted entity in a region with an uncertain destiny.

The Future: The Realignment of Turkestan

For more than a century, Russian tsars and Soviet dictators kept Central Asia's Muslim peoples under tight control. The Soviets laid out a new boundary framework in what was Turkestan; they revolutionized local economies; they collectivized domestic farming and vastly expanded it in the fragile ecology of the desert of Kazakhstan and Uzbekistan. In their final decade, they sought to expand their sphere of influence in the region by intervening in Afghanistan, a military and political adventure that ended in disaster and made refugees out of millions of Afghans.

At the end of 1991, the five republics of former Soviet Central Asia joined their former Russian masters in the so-called Commonwealth of Independent States (see p. 120). But even as relationships with Moscow were being revived, signs of a new era appeared. The Soviets had widely discouraged Islamic teaching and worship: now the calls to Friday prayer are ringing out again from the mosques of Alma-

Ata, Bishkek, Tashkent, Dushanbe, and Ashkhabad as well as those of the smaller cities and towns. The Soviets had suppressed intraregional rivalries and boundary disputes, but now the old ethnic conflicts are erupting once again. The Soviet system was represented by millions of Russian colonists who held the power and took much land; the disengagement will not be easy.

For what may lie ahead, let us also look beyond the region's borders. The word "Turk" appears here in many forms: in the name of the region (Turkestan), in the name of one of the republics (Turkmenistan), in the Turkic languages spoken here. In the days of the Ottoman Empire's vast wealth and power, Turkey was the source and the beacon for this sprawling domain. Today, Turkish interest in Central Asia is reviving. Turkey is also Southwest Asia's beacon of another kind: it is a secular state in a Muslim-dominated realm, a state that adheres to the principle that religion and government—mosque and state—must be separate. In Central Asia, Soviet domination suppressed but did not extinguish Islam. The kind of Islamic revival that will take place concerns Turkey greatly. If it takes a course toward Turkey's model, then Turkey's leadership may again be felt in an area from which it has been long excluded. At the same time, Turkey wants to join the European Community; by establishing itself in Turkestan, it could become a bridge between two often adversarial realms.

But the Turks are not the only potential contestants for such involvement. Geographically better situated are the Iranians, who gave strong irredentist support to the Azerbaijanis when they rose against Soviet domination in 1990. The Iranians hold the card of Islamic fundamentalism, and in Turkestan fundamentalism is on the rise. True, the Iranians are Shi'ite Muslims and most of Turkestan's Muslims are Sunnis, but the next decade of Islamic revival will present a window of opportunity for conversion. The former shah of Iran had dreams of a greater Iranian political empire; the current Islamic leaders of Iran, now an Islamic state by law, have visions of a greater religious empire. Turkestan to the north—and Transcaucasia—present an opportunity the shah never had.

And there is a third potential competitor: Pakistan. The Islamic state of Pakistan was severely affected by the Soviet intervention in Afghanistan: several million refugees, many of them Tajiks, came across the border into Pakistan and changed the political geography of its northwestern border zone (see p 479). Now Pakistan adjoins a weakened, divided Afghanistan whose large Tajik population has affinities with the Tajiks of Turkestan. Should the region become destabilized again, Pakistan—whose military power and perhaps even nuclear capability are making it a major regional force—may find itself encouraged, if not compelled, to assert itself in a region that has long buffeted it.

Still another country has a major stake in the course of events in Turkestan: India. With a Muslim minority of more than 100 million within its borders, India has a vital interest in the region's future. India faces Islam's upsurge with trepidation but not with inaction. Visit the cities of Turkestan, and Indian entrepreneurs are everywhere, establishing linkages that will, in time to come, bear political as well as economic fruit. Said one of these businesspeople to this chapter's author: "We Indians do not want to become the Israelis of South Asia. We see Central Asia's Islamic revival as a challenge, sure, but also as an opportunity to improve India's future."

And so events in Turkestan confirm again the interconnectedness of our world. Russians try to redirect their influence on the region; Islamic proselytizers from Iran, bankers from Turkey, politicians from Germany, economists from the United States, merchants from Pakistan, and managers from India are now engaged in a region long closed to the outside world. So we are witnessing the beginnings of a major realignment, the further expansion and integration of an already strengthening Muslim world that will stretch not only from Morocco to Indonesia but also from Kazakhstan to Somalia. It is a vital lesson in regional geography: realms and regions are anything but static.

THE REALM IN TRANSITION

1. Countries in all regions of this geographic realm are currently affected by Islamic fundamentalism and militancy.
2. The decline of atheistic communism is giving scope to the revival of the Islamic faith, notably in Turkestan but also in Transcaucasia. Here this realm is expanding.
3. The margins of this realm are in ferment. Afghanistan shows signs of fragmentation. Ethiopia has fractured, yielding the Islamic state of Eritrea. Morocco's quest to integrate former Western Sahara is unfulfilled.
4. The role of Turkey is expanding in several critical spheres. Secular Turkish governments seek to counter Islamic fundamentalism with nonreligious alternatives.
5. The search for an accommodation between Israel and the Palestinians has entered a new and sensitive phase. The rewards of continuing success will be enormous, the price of failure incalculable.

◆ PRONUNCIATION GUIDE

Abadan   (ahba-DAHN)

Abu Dhabi   (ah-boo-DAH-bee)

Adana   (AHD-uh-nuh)

Addis Ababa   (adda-SAB-uh-buh)

Aden   (AID-nn)

Aegean-Marmara   (uh-JEE-un MAH-muh-ruh)

Afars   (AH-farz)

Ahaggar   (uh-HAG-er)

Al Andalus   (ahl ahnda-LOOSE)

Aleppo   (uh-LEP-poh)

Algeria   (al-JEERY-uh)

Algiers   (al-JEERZ)

Ali   (ah-LEE)

Allah   (AHL-ah)

Alma-Ata   (ahl-muh-uh-TAH)

Amhara   (am-HAH-ruh)

Amharic   (am-HAH-rick)

Amman   (uh-MAHN)

Amu-Darya   (uh-moo-DAHR-yuh)

Anatolia   (anna-TOH-lee-uh)

Andalucía   (ahn-duh-loo-SEE-uh)

Ankara   (ANG-kuh-ruh)

Antalya   (ant-ull-YAH)

Aozou   (OO-zoo)

Aqaba   (AH-kuh-buh)

Aral   (ARREL)

Aramco   (uh-RAM-koh)

Archimedean   (arka-MEEDY-un)

Armenia   (ar-MEENY-uh)

Ashkelon   (OSH-kuh-lahn)

Ashkenazim   (osh-kuh-NAH-zim)

Ashkhabad   (ASH-kuh-bahd)

Aswan   (as-SWAHN)

Atatürk   (ATTA-tyoork)

Ayatollah   (eye-uh-TOH-luh)

Azeri   (ah-ZERRY)

Azerbaijan   (ah-zer-bye-JAHN)

Baathist   (BATH-ist)

Bab el Mandeb   (bab ull MAN-dub)

Babylon   (BABBA-lon)

Bahrain   (bah-RAIN)

Baku   (bah-KOO)

Baluchistan   (buh-loo-chih-STAHN)

Basra   (BAHZ-ruh)

Bedouin   (BEH-doo-in)

Beirut   (bay-ROOT)

Bekaa   (buh-KAH)

Benghazi   (ben-GAH-zee)

Berber   (BERR-berr)

Bishkek   (BISH-kek)

Bosnia   (BOZZ-nee-uh)

Bosporus   (BAHSS-puh-russ)

Brunei   (broo-NYE)

Bukhara   (boo-KAHR-ruh)

Burkina Faso   (ber-keena FAHSSO)

Byzantine   (BIZZ-un-teen)

Cairo   (KYE-roh)

Caliph   (KAY-liff)

Caliphate   (KALLA-fate)

Castile   (kass-STEEL)

Constantinople   (kon-stant-uh-NOPLE)

Córdoba   (KORR-duh-buh)

Cyprus   (SYE-pruss)

Cyrenaica   (sear-uh-NAY-icka)

Damascus   (duh-MASK-uss)

Damietta   (dam-ee-ETTA)

Dhahran   (dah-RAHN)

Diyarbakir   (dih-yahr-buh-KEER)

Djezira-al-Maghreb   (juh-ZEER-uh ahl-mahg-GRAHB)

Djibouti   (juh-BOODY)

Druze   (DROOZE)

Dubai   (doo-BYE)

Dushanbe   (doo-SHAHM-buh)

Eilat   (AY-laht)

Elam   (EE-lum)

Elburz   (el-BOORZ)

Emirate   (EMMA-rate)

Enosis   (ee-NOH-sis)

Eritrea   (erra-TRAY-uh)

Ethiopia   (eeth-ee-OH-pea-uh)

Euphrates   (yoo-FRATE-eeze)

Fatah   (fuh-TAH)

Fellaheen   (fella-HEEN)

Fergana   (fare-GAHN-uh)

Fezzan   (fuh-ZANN)

Gadhafi, Muammar   (guh-DAHFI, MOO-uh-mar)

Ganges   (GAN-jeeze)

Gaza   (GAH-zuh)

Gezira   (juh-ZEER-uh)

Gibraltar   (jih-BRAWL-tuh)

Giralda   (hih-RAHL-duh)

Giza   (GHEE-zuh)

Golan   (goh-LAHN)

Gorno-Badakhshan   (gore-noh-bah-dahk-SHAHN)

Guryev   (GHOOR-yeff)

Hägerstrand, Torsten   (HAYGER-strand, TOR-stun)

Haifa   (HYE-fah)

Hamitic   (ham-MITTICK)

Hazara   (huh-ZAHR-ah)

Herodotus   (heh-RODDA-tuss)

Hierarchical   (hire-ARK-uh-kull)

Hindu Kush   (hin-doo KOOSH)*

Hormuz   (hoar-MOOZE)

Husayn   (hoo-SINE)

Hussein, Saddam   (hoo-SAIN, suh-DAHM)

Ibn Saud   (ib'n sah-OOD)

Imam   (ih-MAHM)

Interdigitate   (intuh-DID-juh-tate)

Intifada   (intee-FAH-duh)

Iran   (ih-RAN/ih-RAHN)

Iranian   (ih-RAIN-ee-un)

Iraq   (ih-RAK/ih-RAHK)

Iraqi   (ih-RAKKY)

Irredentism   (irra-DENT-tism)

Isfahan   (iz-fuh-HAHN)

Iskenderun   (iz-ken-duh-ROON)

Islam   (iss-LAHM)

Ismaili   (izz-MYE-lee)

Israel   (IZ-rail)

Issas   (ee-SAHZ)

Istanbul   (iss-tum-BOOL)

Jabal   (JAB-ull)

Jazirah   (juh-ZEER-uh)

Jerusalem   (juh-ROO-suh-lum)

Jubail   (joo-BILE)

Judaic   (joo-DAY-ick)

Judaism   (JOODY-ism)

Kabul   (KAH-bull)

Kara Kum   (kahr-ruh KOOM)

Karaganda   (karra-gun-DAH)

Karakalpak   (karra-kal-PAK)

Kazakh   (KUZZ-uck)

Kazakhstan   (kuzz-uck-STAHN/KUZZ-uck-stahn)

Khartoum   (kar-TOOM)

Khiva   (KEE-vuh)

Khomeini   (hoh-MAY-nee)

Khurasan   (koor-uh-SAHN)

Kibbutz   (kih-BOOTS)*

Kirkuk   (keer-KOOK)*

Koran   (kaw-RAHN)

Kosovo   (KOSS-oh-voh)

Krasnovodsk   (kruzz-noh-VAUGHTSK)

*Double "o" pronounced as in *book*.

Kufra (KOO-fruh)
Kurd (KERD)
Kurdistan (KERD-duh-stahn)
Kuwait (koo-WAIT)
Kyrgyz (KEER-geeze)
Kyrgyzstan (KEER-geeze-stahn)
Levant (luh-VAHNT)
Lingua franca (LEEN-gwuh
    FRUNK-uh)
Luxor (LUK-soar)
Maghreb (mahg-GRAHB)
Malaysia (muh-LAY-zhuh)
Mali (MAH-lee)
Mauritania (maw-ruh-TANEY-uh)
Medina (muh-DEENA)
Mesopotamia (messo-puh-TAY-
    mee-uh)
Mesquita (meh-SKEE-tuh)
Minaret (MINNA-ret)
Monogamy (muh-NOG-ah-mee)
Mosque (MOSK)
Mubarak, Hosni (moo-BAH-rahk,
    HOZZ-nee)
Muhammad (moo-HAH-mid)
Mullah (MOOL-ah)*
Muscat (MUH-skaht)
Muslim (MUZZ-lim)
Nasser, Gamal Abdel (NASS-er,
    guh-MAHL AB-dul)
Negev (NEH-ghev)
Niger (nee-ZHAIR)
Nigeria (nye-JEERY-uh)
Ogaden (oh-gah-DEN)
Oman (oh-MAHN)
Oromo (AW-ruh-moh)
Orontes (aw-RAHN-teeze)
Pahlavi, Shah Muhammad Reza
    (puh-LAH-vee, shah mooh-HAH-
    mid RAY-zuh)
Pakhtuns (puck-TOONZ)

Pakistan (PAH-kih-stahn)
Palestine (PAL-uh-stine)
Pamir (pah-MEER)
Pathan (puh-TAHN)
Persepolis (per-SEPP-uh-luss)
Persia (PER-zhuh)
Pharaonic (fair-ray-ONNICK)
Phoenicians (fuh-NEE-shunz)
Polygamy (puh-LIG-ah-mee)
Polytheism (polly-THEE-izm)
Port Said (port-sah-EED)
Pushtun (PAH-shtoon)
Qanat (KAH-naht)
Qatar (KOTTER)
Qattarah (kuh-TAR-ruh)
Ramadan (rahma-DAHN)
Riyadh (ree-AHD)
Rub al Khali (rube ahl KAH-lee)
Rumailah (roo-MYE-luh)
Sahara (suh-HARRA)
Sahel (suh-HELL)
Samarkand (sahm-ahr-KAHND)
San'a (suh-NAH)
Saudi Arabia (SAU-dee uh-RAY-
    bee-uh)
Sedentary (SEDDEN-terry)
Selangor (suh-LANG-gher)
Semitic (seh-MITTICK)
Senegal (sen-ih-GAWL)
Sephardic (suh-FAHRD-ick)
Sharia (SHAH-ree-uh)
Shatt al Arab (shot ahl uh-RAHB)
Sheikdom (SHAKE-dum)
Shia (SHEE-uh)
Shi'ite (SHEE-ite)
Shiraz (shih-RAHZ)
Sinai (SYE-nye)
Somalia (suh-MAHL-yuh)
Suez (SOO-ez)
Suleyman (SOO-lay-mahn)

Sumer (SOO-mer)
Sunni (SOO-nee)
Syr-Darya (seer-DAHR-yuh)
Syria (SEARY-uh)
Tajik (TUDGE-ick)
Tajikistan (tah-JEEK-ih-stahn)
Tana (TAHNA)
Tashkent (tahsh-KENT)
Tatar (TAHT-uh)
Tehran (tay-uh-RAHN)
Tel Aviv-Jaffa (tella-VEEVE-
    JOFF-uh)
Tengiz (TEN-ghiz)
Thebes (THEEBZ)
Tian Shan (tyahn-SHAHN)
Tigris (TYE-gruss)
Toledo (toh-LAY-doh)
Transcaucasia (tranz-kaw-
    KAY-zhuh)
Tripoli (TRIPPA-lee)
Tripolitania (trip-olla-TANEY-yuh)
Tuareg (TWAH-regg)
Tukiu (too-KYOO)
Tunis (TOO-niss)
Tunisia (too-NEE-zhuh)
Turkestan (TERK-uh-stahn)
Turkmenistan (terk-MEN-uh-stahn)
Ural-Altaic (YOOR-ullu-al-TAY-ick)
Uygur (WEE-ghoor)
Uzbek (OOZE-beck)
Uzbekistan (ooze-BECK-ih-stahn)
Wakhan (wah-KAHN)
Yanbu (YAN-boo)
Yemen (YEMMON)
Zagros (ZAH-gruss)
Ziggurat (ZIG-goo-raht)
Zionism (ZYE-un-ism)
Zonguldak (zawngle-DAHK)
Zoroaster (zorro-AST-tuh)
Zoroastrian (zorro-ASTREE-un)

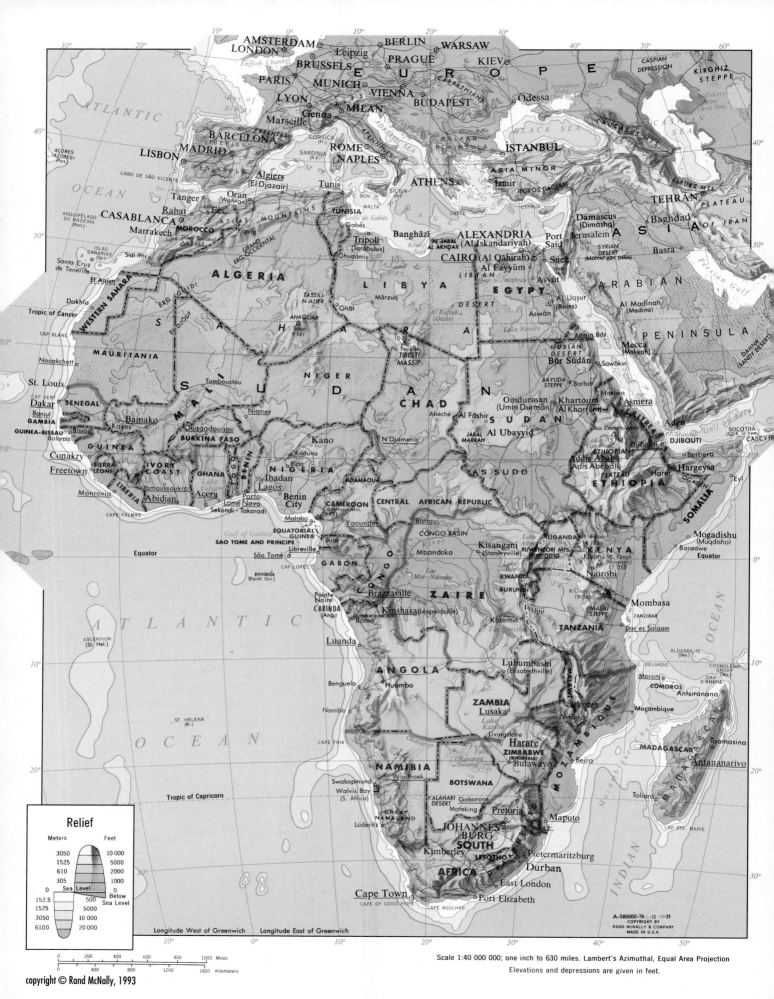

CHAPTER 8

SUBSAHARAN AFRICA: REALM OF REVERSALS

The African continent lies at the heart of the earth's land-masses. Situated astride the equator, Africa looks westward to the Americas, northward to Europe and Asia, eastward to Australia, and southward to Antarctica. Flanked by the Atlantic and Indian oceans, Africa stretches from the Mediterranean Sea in the north to the Southern Ocean in the south. Africa contains the world's greatest desert and some of its mightiest rivers, snowcapped mountains and lake-filled valleys, parched steppes and dripping rainforests. It is a land-mass of escarpments and plateaus, where altitude moderates the tropical heat. Africa is the source of humanity, the scene of the emergence of the first communities. Africa is also a realm of great problems. Ravaging diseases afflict its peoples. Environmental deterioration is widespread. In some areas, population growth is the highest in the world, and the food supply cannot keep up. After slave-raiding and colonial exploitation, Africa now suffers from numerous political conflicts that have made millions homeless.

The African geographic realm extends from near the southern edge of the Sahara southward to the Cape of Good Hope. Of course, as we know, the Saharan boundary is not sharp; languages, religions, and ways of life change across a wide transition zone. But the Africa of Algeria and Egypt stands a world apart from the Africa of Nigeria and Zaïre. Our focus in this chapter is on the latter—Subsaharan Africa.

◆ REALMS AND BOUNDARIES

We have seen that geographic realms and regions are parts of a global system of spatial classification and are defined to help us understand our complex world. Few of the realm boundaries shown in Fig. I-1 are as difficult to establish as is this northern limit of the African realm. And yet, when we examine such maps as those of world religions, African languages, and human racial groups, the line shown on Fig. I-1 appears on every one. Islam, as we observed earlier, penetrated Africa by various routes, but in Subsaharan Africa it loses the dominance it enjoys in North African

IDEAS & CONCEPTS

Medical geography	Green Revolution
Geography of languages	Regional complementarity (4)
Relative location (3)	Colonial spatial organization
Rift valleys	Periodic markets
Continental drift	Sequent occupance
Endemic diseases	Landlocked state

REGIONS

West Africa
East Africa
Equatorial Africa
Southern Africa

countries (Fig. 7-2). Note how the realm boundary crosses Sudan, the country immediately south of Egypt: northern Sudan is overwhelmingly Islamic, but the south is almost entirely non-Islamic. On the language map (Fig. 8-1), note that northern Sudan speaks Arabic, whereas the south speaks Sudanic languages. Similarly, we could observe the transition from Caucasian to African on a map of human racial groups. Add to this the contrasting ways of life between north and south, and it is clear that this segment of the African realm boundary is beyond dispute.

AFRICAN LANGUAGES

The linguistic evidence is the strongest of all, not only in Sudan but across all of Africa. It certainly has validity west of Sudan where, as Fig. 8-1 shows, the Afro-Asiatic languages (such as widely used Arabic) give way to the Niger-Congo languages spoken throughout most of the heart of Africa. Crossing this linguistic frontier is a vivid geo-

MEDICAL GEOGRAPHY*

Medical geography is that branch of the discipline concerned with people's health. Practitioners of medical geography study the spatial aspects of health and illness. Where are diseases found? How do they spread? Are there specific kinds of environments in which certain illnesses are located? How are disease and environment related? How do the changes that societies undergo affect the health of their populations? Where do people go when they seek health care? How does the location of a health-care facility affect the opportunity to obtain care? These and many other related questions are of interest to medical geographers.

Insights into disease occurrence or disease origin can sometimes be gained simply by mapping the distribution of that disease. In a famous early case, Dr. John Snow was able to relate the deadly effects of cholera in 1854 to a contaminated water source by mapping the locations of deaths in a portion of London (see the map in Appendix B); the spatial pattern clearly showed that many deaths occurred within a few blocks of a pump used as a major source of drinking water. Although we now know much more about this disease and its causes, the medical community in the mid-nineteenth century was still arguing whether such invisible things as "germs" actually existed! It took a map to provide evidence that, in this case at least, something about an apparently healthful water supply was fatal to many who drank from it.

This example illustrates a major concern among medical geographers with *disease ecology*. Disease ecology studies the manner and consequences of interaction between the environment and the causes of morbidity and mortality. *Morbidity* is the condition of illness, whereas *mortality* is the occurrence of death. By "environment" we mean the varied aspects of nature and society that bear on people's lives; geographers have a long tradition of dealing with the full array of physical and human environmental factors. Medical geographers study in an integrated manner the complex interaction of a disease *agent* (the pathogen or illness-causing organism itself), a possible disease *vector* (the intermediate transmitter of the pathogen), the physical and social environments, and even the cultural behavior patterns of individuals potentially at risk.

Geographers map the distributions of disease (as did Dr. Snow) in order to identify the spatial context in which the illness occurs and to monitor changes in these patterns. For instance, even when a primary agent is known and can be treated, as in the case of rubella (German measles), outbreaks continue to occur. Medical geographers studied the location of recurrent appearances of rubella and suggested that one of the secondary causes might well be the lack of information by the population at risk about the availability of vaccines. More difficult is the study of *degenerative* diseases (such as heart disease, cancer, and stroke) because of the multiplicity of factors that lie behind these illnesses. The contributions of medical geographers—with their ecological, integrated, and descriptive approaches—are especially useful

for suggesting hypotheses to other medical researchers who deal directly with the causes of these diseases and their effects on individuals.

Analysis of the diffusion of illnesses has also been given a great deal of attention by medical geographers. The notion of *contagious diffusion* (see p. 368) is drawn directly from our experience with diseases. In Africa, for example, the gradual spread of cerebro-spinal meningitis, from its early appearance in Ethiopia in 1927 across the Sahel to Senegal in 1941, was associated with the main east-west transport routes traversing the continent. Just as important, however, was the more rapid spread that occurred during the dry seasons, when contact at water sources was at its highest.

In a separate study of the diffusion of cholera in Africa during the 1970s, Robert Stock identified at least four distinct route patterns in the complex sequence of cholera diffusion that affected much of West and East Africa. Each of the four—coastal, riverine, urban-hierarchical, and radial contact diffusion—was identified by a careful mapping of the direction and timing of the disease's spread. Work such as Stock's is especially useful for those attempting to prevent disease diffusion; policies designed to interrupt one type of disease spread may be unsuccessful if applied in areas where other route patterns are more important.

Not all studies of disease ecology by medical geographers suggest natural environmental causes for disease occurrence. When studying cancer of the esophagus in Malawi and Zambia during the 1960s and 1970s, Neil McGlashan noted a higher rate among people living in the eastern part of the

*Contributed by Stephen S. Birdsall

region. After mapping a number of physical and cultural features and considering a variety of explanations, he argued that the distribution of this form of cancer was matched by the distribution of a particular form of locally distilled spirit. Therefore, the argument went, the cancer was related to either the distillation process or the materials used in this particular beverage. The actual cause of the cancer remains unknown, but McGlashan's early association of several mapped patterns provided a focus for subsequent medical research.

There are times, too, when studies focus on secondary patterns of activity as the source of the illness. During the 1970s, when most gasoline produced in the United States contained a lead additive to boost octane, a number of researchers examined the possible relationship between this fuel additive and patterns of lead poisoning among children. One study, for example, found a much higher level of lead in the blood of children who lived within 300 feet (90 m) of a main road than among those who lived at greater distances. Another study found that lead levels in the blood of youthful residents of a northern U.S. city varied seasonally: levels were significantly higher during the summer months, when children spent more time outdoors.

If studies of disease ecology are one major concern for medical geographers, a second involves the *delivery of health care*. At a very basic level, equitable health-care delivery is made difficult by the uneven spatial distribution of those who seek care and those who provide it. If people live close to a hospital or clinic, they are more likely to use those facilities when illness

strikes—or so one would expect. But how far will an individual travel to obtain care? The answer, studies show, depends on the individual's physical characteristics (age, sex), social context (marital status, proximity to family or neighbors), economic resources (ability to pay for travel or service), attitude toward "illness" (is a doctor necessary?), barriers to treatment (racial or religious discrimination), and much more. An answer to the question "where does one go to obtain health care?" is not obvious. A great deal of research has been done to identify the importance of each variable in terms of its effect on the health care that is sought or obtained.

In some countries, such as the United States, doctors exercise considerable personal choice in deciding where they will practice medicine. Some researchers have pointed to the heavy concentration of doctors in urban areas—specifically in the more affluent sections of metropolitan areas—as an example of inequality in health care: patients least able to afford it must bear high travel costs or do without the care. In other countries, the government may regulate the distribution of health care to some degree, but many parts of the world have too few medical professionals for the population's need. In the late 1980s, there were more than 6,500 people per physician in Nigeria, and about 23,000 people per physician in Zaïre. These figures compared with 404 people per physician in the United States.

Related to the study of health-care delivery, medical geographers are also involved at the planning stage. One of the major research efforts by geographers has been to develop methods of

determining where activities *should* be located under ideal conditions. Planners seeking to identify the optimum locations for hospitals, clinics, medical offices, and even sites for emergency medical service centers have used such techniques of medical geography. These so-called "location allocation" models have been developed primarily for the developed world; elsewhere, questions about the number of practitioners and the available alternatives to "Western" medicine may be more important than their spatial distribution.

Different types of questions are pressing in low-income countries. How does economic development affect the health of a population? Although the ability to pay for health care may improve on the average, new diseases may also be introduced or spread by development projects. Irrigation schemes, for example, have been found to provide new habitats for disease vectors and spread the incidence of malaria and schistosomiasis. The labor migration encouraged by urban and industrial growth has been shown to spread contagious diseases such as hepatitis when workers return home after contracting the illness, or when they bring it to the urban center as they arrive looking for work.

Clearly, there are no easy or obvious paths to the resolution of health-care questions, however vital that resolution may be. Medical geography makes its contribution to what must be a truly multidisciplinary approach through the identification and analysis of spatial patterns, locational associations, and the full environmental contexts within which morbidity and mortality occur.

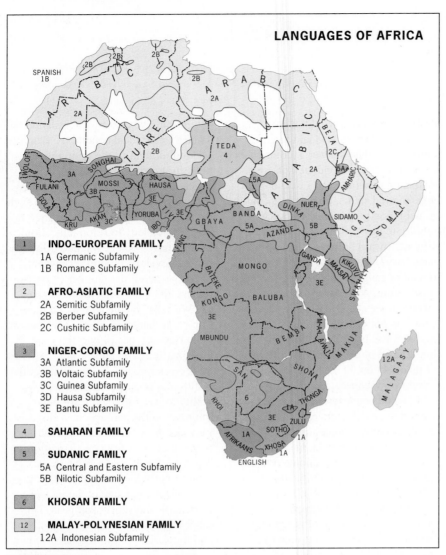

LANGUAGES OF AFRICA

SPANISH 1B

1 **INDO-EUROPEAN FAMILY**
   1A Germanic Subfamily
   1B Romance Subfamily

2 **AFRO-ASIATIC FAMILY**
   2A Semitic Subfamily
   2B Berber Subfamily
   2C Cushitic Subfamily

3 **NIGER-CONGO FAMILY**
   3A Atlantic Subfamily
   3B Voltaic Subfamily
   3C Guinea Subfamily
   3D Hausa Subfamily
   3E Bantu Subfamily

4 **SAHARAN FAMILY**

5 **SUDANIC FAMILY**
   5A Central and Eastern Subfamily
   5B Nilotic Subfamily

6 **KHOISAN FAMILY**

12 **MALAY-POLYNESIAN FAMILY**
   12A Indonesian Subfamily

FIGURE 8-1

graphic experience. Quite suddenly, the *lingua franca* of Arabic ceases to be useful; a desert-margin trade language, Hausa, replaces it. But where Hausa fades out, the intricate linguistic mosaic of Africa takes over—dozens of languages in a single country, hundreds in one region, as many as a thousand in the realm as a whole. East of Sudan, the linguistic basis for the realm's limit is less evident. Amharic and Galla, spoken in Ethiopia, are quite closely related to Arabic and are part of the realm where Afro-Asiatic languages prevail. But the peoples of the Ethiopian Highlands are not Muslims; Islam dominates only north and east of this historic Coptic Christian area (Fig. 7-2). Based on the religious criterion and on traditions and ways of life, the realm's boundary bends southward to include the heart of Ethiopia and then turns eastward to separate Somalia from Kenya.

One of the most interesting aspects of the Subsaharan Africa realm is its **geography of languages**. Here, in an

area about one-seventh that of the inhabited world, is spoken one-third of all languages by a population constituting less than one-tenth of all humankind. This is one of Africa's most distinct regional properties, part of the richness of its mosaic of cultures—and one of its problems in modern times. In the classification of languages, we use terms employed in biology so that languages thought to have a shared but possibly distant origin are grouped into a family; when their relationship is closer, they belong together in a subfamily. The Niger-Congo family of languages (Category 3 on the map), therefore, contains five subfamilies, and the Bantu subfamily (3E) is the most extensive of these (Fig. 8-1). Scholars studying African languages do not yet agree on the regional distribution of African language clusters. It is agreed that the Khoisan family (6) represents the oldest surviving African languages, spoken over a far larger area of the continent before the Niger-Congo

TEN MAJOR GEOGRAPHIC QUALITIES OF AFRICA

1. The physical geography of Africa is dominated by the continent's plateau character, variable rainfall, soils of low fertility, and persistent environmental problems in farming.
2. The majority of Africa's peoples remain dependent on farming for their livelihood. Urbanization is accelerating, but most countries' populations remain below 40 percent urban.
3. The people of Africa continue to face a high incidence of disease, including malaria, sleeping sickness, and river blindness.
4. Most of Africa's political boundaries were drawn during the colonial period without regard for the human and physical geography of the areas they divided. This has caused numerous problems.
5. Considerable economic development has occurred in many scattered areas of Africa, but much of the realm's population continues to have little access to the goods and services of the world economy.
6. The realm is rich in raw materials vital to industrialized countries.
7. Patterns of raw-material exploitation and export routes set up during the colonial period still prevail in most of Subsaharan Africa. Interregional connections are poor.
8. Africa has increasingly been drawn into the competition and conflicts among the world's major powers. The continent contains about one-third of the world's refugee population.
9. Africa's population growth rate is by far the highest of any continent's in spite of a difficult agricultural environment, numerous hazards and diseases, and periodic food shortages. Some of the best land is used to produce such cash crops as coffee, tea, cocoa, and cotton for sale overseas.
10. Even though post-independence dislocations, civil wars, and massive losses of life have plagued some parts of Africa, other areas have shown relative stability, cohesion, and economic growth.

Eurafricans living in South Africa, the realm's southernmost state.

RACIAL PATTERNS

The concept of race remains under constant debate among anthropologists and other scholars. Attempts to define race on purely genetic bases have not proven satisfactory. Race, like culture, is a complex and multifaceted concept. In addition, "race" is frequently misused in everyday thinking as a substitute for preconceptions about social potential. Categories of individual physical characteristics are erroneously assumed to represent visible indicators of behavior. A great deal of evidence to the contrary notwithstanding, this socialized approach to the concept of race continues to generate a great deal of disagreement. Despite all these difficulties, it is useful to treat human races as broad categories of genetically defined features. In the case of Subsaharan Africa, there is geographic coincidence between the African racial group and the realm as defined by other criteria.

Separated by the Sahara from lands to the north and by vast oceans from lands to the west and east, Africa is an almost insular continent. External influences reached Africa more by sea than over land; even Islam came to East Africa by Persian and Arab boats. Europe's colonial powers came first to West Africa's coasts, then rounded the southern Cape and advanced along the eastern shores in search of adequate harbors. No Tatar hordes invaded from afar; no Mongol armies swept across the countryside. Until Islam penetrated from the north, and until European colonizers pushed inland from their coastal stations, Subsaharan Africa lay isolated. Only its northernmost peoples interacted regularly with North African societies, and even these contacts were limited by the capacities of desert caravans.

◆ ENVIRONMENTAL BASES

Africa has an unusual location. No other landmass is so squarely positioned astride the equator, reaching almost as far to the north as to the south; this location has much to do with Africa's vegetation, soils, agricultural potential, and population distribution. Africa also is a very large landmass, containing about one-fifth of all the earth's land surface. It is 4,800 miles (7,700 km) from the north coast of Tunisia to the southern coast of South Africa, and 4,500 miles (7,200 km) from coastal Senegal in the extreme western *Bulge* of Africa to the tip of the *Horn* in easternmost Somalia. These distances have environmental as well as human implications. Much of Africa is far from marine sources of moisture. Many of Africa's peoples live remote from routes to the outside world.

diffusion. It is also evident that Madagascar's languages belong to a non-African, Malay-Polynesian family (12), revealing the Southeast Asian origins of a large sector of that island's population. Afrikaans (1A) is an Indo-European language, a derivative of Dutch spoken by Europeans and

PHYSIOGRAPHY

Not only is Africa's **relative location** unusual, but its physical geography is unique. Take, for example, the distribution of the world's linear mountain ranges. Every major landmass has at least one mountainous backbone: South America's Andes, North America's Rocky Mountains, Europe's Alps, and Asia's Himalayas. Yet Africa, covering one-fifth of the land surface of the earth, has nothing comparable. The Atlas Mountains of the far north occupy a mere corner of the landmass, and the Cape Ranges of the far south are not of continental dimensions. And where Africa does have high mountains, as in Ethiopia and South Africa, these are really deeply eroded plateaus—or, as in East Africa, high, snowcapped volcanoes. Missing in Africa are those elongated, parallel ranges of the Andes or Alps.

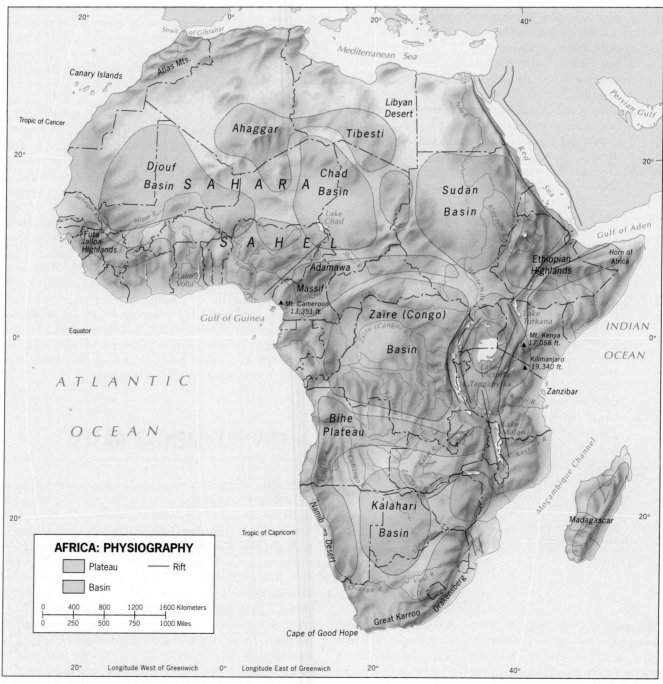

FIGURE 8-2

Lakes and Rift Valleys

This discovery should stimulate us to look more closely at the physiographic map of Africa (Fig. 8-2). What else is unusual about Africa's landscapes? As the map shows, Africa has a set of great lakes, concentrated in the east-central portion of the continent.

With the single exception of Lake Victoria, these lakes are remarkably elongated, from Lake Malawi in the south to Lake Turkana in the north. What causes this elongation and the persistent north-south alignment that can be observed in these lakes? The lakes occupy portions of deep trenches cutting through the East African plateau, trenches that can be seen to extend well beyond the lakes themselves. Northeast of Lake Turkana, such a trench cuts the Ethiopian Highlands into two sections, and the entire Red Sea looks much like a northward continuation of it. On both sides of Lake Victoria, smaller lakes lie in similar trenches, of which the western one runs into Lake Tanganyika and the eastern one extends completely across Kenya, Tanzania, and Malawi (Fig. 8-2).

The technical term for these trenches is **rift valleys**. As the name implies, they are formed when huge parallel cracks or faults appear in the earth's crust and the strips of crust between them sink or are pushed down to form great linear valleys. Altogether, these rift valleys stretch more than 6,000 miles (9,600 km) from the north end of the Red Sea to Swaziland in Southern Africa. In general, the rifts from Lake Turkana southward are between 20 and 60 miles (30 and 90 km) wide, and the walls, sometimes sheer and sometimes step-like, are well defined (see photo below).

River Courses

Next, our attention is drawn to Africa's unusual river systems (Fig. 8-2). Africa has several great rivers, with the Nile and Zaïre (or Congo) ranking among the most noteworthy in the world. The *Niger* rises in the far west of Africa, on the slopes of the Futa Jallon Highlands, but first flows inland toward the Sahara. Then, after forming an interior delta, it suddenly elbows southeastward, leaves the desert area, plunges over falls as it cuts through the plateau area of Nigeria, and creates another large delta at its mouth. The *Zaïre (Congo)* River begins as the Lualaba River on the Zaïre-Zambia boundary; for some distance, it actually flows northeast before turning north, then west and southwest, finally cutting through the Crystal Mountains to reach the ocean. Note that the upper courses of these first two

As the East African rift valleys opened, elongated strips of the earth's crust collapsed into them along their sides. The step-like terrain that resulted is clearly visible in this view of central Kenya's rift taken from atop the valley wall.

rivers appear quite unrelated to the continent's coasts where they eventually exit. In the case of the *Zambezi* River, whose headwaters lie in Angola and northwestern Zambia, the situation is the same; the river first flows south, toward the inland delta known as the Okovango Swamp, and then turns northeast and southeast, eventually to reach its delta south of Lake Malawi. Finally, there is the famed erratic course of the *Nile* River, which braids into numerous channels in the Sudd area of southern Sudan and, in its middle course, actually reverses direction and flows southward before resuming its flow toward the Mediterranean delta in Egypt. With so many peculiarities among Africa's river courses, could it be that all have been affected by the same event at some time in the continent's history? Perhaps—but first let us look further at the map.

Plateaus and Escarpments

All continents have low-lying areas—witness the Gulf-Atlantic Coastal Plain of North America or the riverine lowlands of Eurasia and Australia. But as the map shows, coastal lowlands are few and of limited extent in Africa. In fact, it is reasonable to call Africa a *plateau continent*; except for coastal Moçambique and Somalia and along the northern and western coasts, almost the entire continent lies above 1,000 feet (300 m) in elevation and fully half of it is over 2,500 feet (800 m) high. Even the Zaïre (Congo) Basin, Equatorial Africa's huge tropical lowland, lies well over 1,000 feet above sea level in contrast to the much lower-lying Amazon Basin across the Atlantic.

Although Africa is mostly plateau, this does not mean that the surface is completely flat and unbroken. In the first place, the rivers have been eroding the surface for millions of years and have made some fairly good cuts in it. For example, Victoria Falls on the Zambezi is 1 mile (1,600 m) wide and over 300 feet (90 m) high. Volcanoes and other types of mountains, some of them erosional leftovers, stand well above the landscape in many areas; the Sahara, where the Ahaggar and Tibesti mountains both reach about 10,000 feet (3,000 m) in elevation, is no exception. In several places the plateau has sagged under the weight of accumulating sediments. In the Zaïre Basin, for example, rivers have transported sand and sediment downstream for tens of millions of years and, for some reason, dropped their erosional loads into what was once a giant lake the size of an interior sea. Today the lake is gone, but the thick sediments that press this portion of the African surface into a giant basin are proof that it was there. And this was not the only inland sea. To the south, the Kalahari Basin was filling with sediments that now constitute that desert's sand; and far to the north, in the Sahara, three similar basins lie centered on Sudan, Chad, and what today is Mali (the Djouf Basin).

The margins of Africa's plateau are of significance, too. Much of the continent, because it is a plateau, is surrounded by an escarpment. In Southern Africa, where this feature is especially pronounced, the *Great Escarpment* (as it is called there) marks the plateau's edge along many hundreds of miles; here the land drops precipitously from more than 5,000 feet (1,500 m) in elevation to a narrow, hilly coastal belt. From Zaïre to Swaziland and intermittently on or near most of the African coastline, a scarp bounds the interior upland. Such escarpments are found in other parts of the world, too: in Brazil at the eastern margins of the Brazilian Highlands, and in India at the western edge of its Deccan Plateau. But Africa, even for its size, has a disproportionately large share of this topographic phenomenon.

CONTINENTAL DRIFT AND AFRICA

Africa's remarkable and unusual physiography was one of the pieces of evidence that the geographer Alfred Wegener used, early in this century, to construct his hypothesis of **continental drift**. According to this idea (first introduced on p. 9), all the landmasses on earth were assembled into one giant continent named *Pangaea*; the southern continents constituted *Gondwana*, the southern part of this supercontinent (Fig. 8-3). After a long (geologic) period of unity, this huge landmass began to break up more than 100 million years ago. Africa, which lay at the heart of Gondwana, attained the approximate shape we see today when North and South America, Antarctica, Australia, and India drifted radially away. Later, geophysicists gave the name *plate tectonics* to this process, and it is now understood that the breakup of Pangaea was only the most recent in a series of continental collisions and separations that spans much of the earth's history (see Fig. I-3).

Africa's situation at the heart of the supercontinent explains much of what we see in its landscapes today. The Great Escarpment is a relic of the giant faults (fractures) that formed when the neighboring landmasses split off. The rift valleys are only the most recent evidence of the pulling forces that affect the African plate; the Red Sea is an advanced stage of such rifts, and we may expect East Africa to separate from the rest of Africa much as Arabia did earlier (and Madagascar before that). Africa's rivers once filled lakes within the continent, but they did not reach the distant ocean; today, the lakes are drained by rivers that eroded inland when the oceans began to wash Africa's new shorelines.

And why does Africa not have those mountain chains that led us to look at the map more closely? The answer may lie in the direction and distance of plate motion. South America moved far westward, its plate colliding with the plate under the waters of the Pacific Ocean; the Andes crumpled up, accordion-like, in this collision. India moved northeastward, wedging into the Asian landmass; similarly, the Himalayas rose upward. But Africa moved comparatively little, being more affected by pulling tensional forces that create rifts than by the pressures of plate collision that generate the uplifting of mountains. Decidedly, the physio-

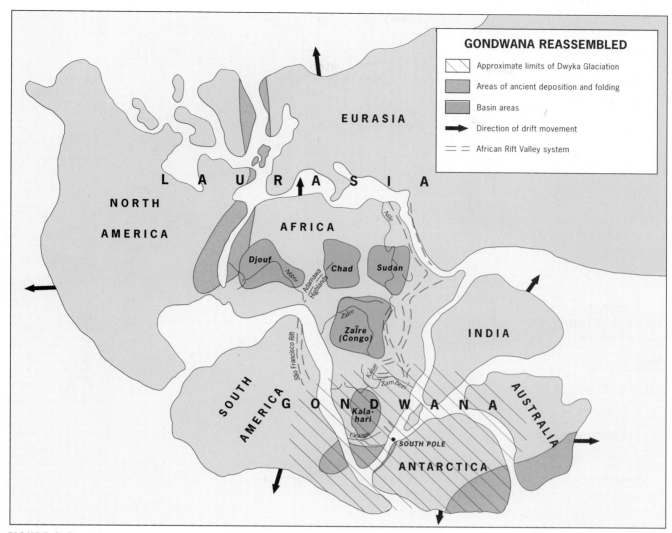

FIGURE 8-3

graphic map reveals more than just the location of rivers, plains, and lakes.

CLIMATE AND VEGETATION

Africa's climatic and natural vegetation distributions are almost symmetrical about the equator (Figs. I-7 and I-8). With the equator virtually bisecting the continent, the atmospheric conditions that affect the African surface tend to be similar in both halves. The hot, rainy climate of the Zaïre (Congo) Basin merges gradually, both north and south, into climates with distinct winter dry seasons; outside the basin, therefore, trees are less majestic and then less numerous as one moves away from the equator. Eventually, the humid equatorial region is left behind entirely as annual rainfall becomes less abundant and less reliable. Now the tree-studded grasslands of the savanna dominate the landscape, a wide zone of sometimes park-like vegetation that

has long been the principal home of the continent's great herds of wildlife. Moving poleward, the savanna gives way to the drier tropical steppe, the sparse grasslands that are so easily overgrazed by livestock. Beyond the steppe lies the desert: the Sahara in the north, the Kalahari in the south. A close look at Fig. I-7 reveals that even the dry-summer subtropical climate, known as the Mediterranean climate, occurs on both the north and the south coasts of Africa, although the zone at the southern tip is very small.

It is interesting to compare this pattern to that of annual rainfall (Fig. I-6). Note how much larger the region of heavy precipitation is in the equatorial zone of South America, in contrast to that of Africa. Only the western zone of Equatorial Africa and a strip along the southern West African coast receive substantial rainfall. Much of eastern and southern Africa receive modest rainfall amounts, especially when we remember the considerable evaporation that takes place. Africa's topography—its high eastern plateau, escarpments, and generally elevated interior—coupled with its bulk cre-

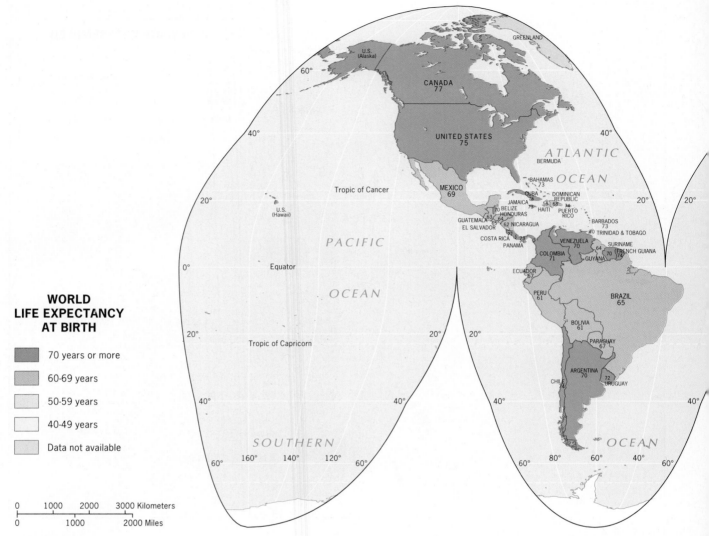

**WORLD
LIFE EXPECTANCY
AT BIRTH**

- 70 years or more
- 60-69 years
- 50-59 years
- 40-49 years
- Data not available

0   1000   2000   3000 Kilometers

0   1000   2000 Miles

FIGURE 8-4

The Maasai are among East Africa's cattle herders, and they must move frequently in search of fresh pastures. Note the sparse vegetation in the steppelands across which these cattle are moving, and the incipient gulleys already forming (right foreground).

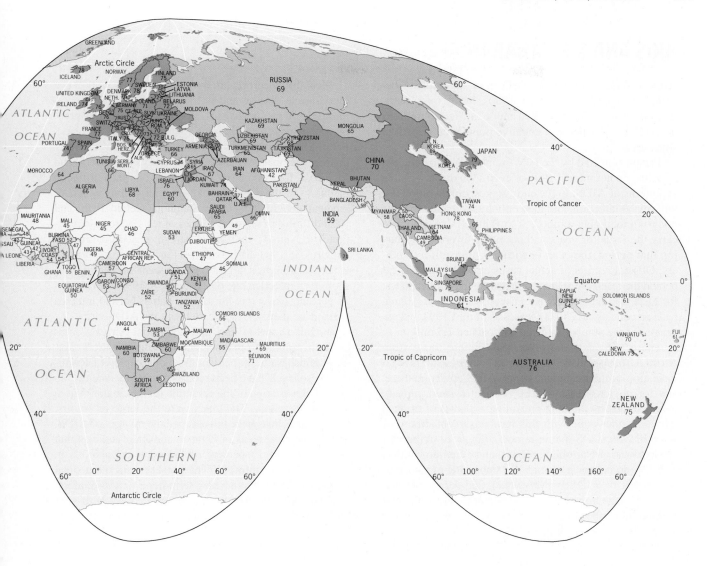

ates large areas with insufficient moisture. Here the soils are poor, the natural vegetation is fragile, and the environment endangered by burgeoning human demands.

◆ ENVIRONMENT, PEOPLE, AND HEALTH

In average terms, Africa south of the Sahara is not a densely peopled realm. Less than one-tenth of the world's population—about 530 million people—inhabit this subcontinent of difficult environments, and a majority of them remain farmers. The situation is reflected in Fig. I-9: note that much of the realm is moderately populated and that large clusters are few. The largest human concentrations occur in Nigeria, around Lake Victoria in East Africa, and in several smaller areas of Southern Africa.

What Fig. I-9 does not reveal, however, is the overall health of the African population. The tragic truth is that Africa suffers from local and regional famines, that diets are generally not well balanced, and that life expectancies are lower here than in any other world geographic realm (Fig. 8-4). Why is so much of Africa so afflicted? The realm's physical geography has much to do with it.

Tropical areas are breeding grounds for organisms that carry disease, such as flies, mosquitoes, fleas, and even snails. Many people depend on river and well water that may carry these vectors (disease transmitters) or may be contaminated. Inadequate or imbalanced diets weaken the body and make it susceptible to illness; many Africans are thus at risk because food availability is limited. Africa's soils and climates limit the range of foods that can be produced locally, and people generally do not have the money to spend on meats or other expensive supplements. As a result, millions of Africans spend their entire lives in a state of poor health.

AIDS AND SUBSAHARAN AFRICA

AIDS—Acquired Immune Deficiency Syndrome—became a pandemic in the early 1990s. In the mid-1990s, its impact on Africa was devastating, afflicting millions and creating prospects of negative population growth in some areas.

Persons infected by the HIV (human immunodeficiency virus) do not immediately or even soon display visible symptoms of AIDS. In the early stages, only a blood test will reveal infection, and then only by indicating that the body is mobilizing antibodies to fight HIV. People can carry the virus for years without being aware of it; during that period, they can unwittingly transmit it to others. Official reports of actual cases of AIDS thus lag far behind the reservoir of those infected. In the United States, for instance, the Centers for Disease Control in Atlanta had recorded about 250,000 cases as of mid-1993, but estimates of the number infected approached 1.5 million.

These U.S. data, however, pale before those from Africa. Official statistics for African countries still give no indication of the magnitude of the AIDS epidemic there, and for obvious reasons. The medical system, already overwhelmed by the long-prevailing maladies of tropical Africa, cannot cope with this new onslaught. Many of those ill with AIDS live in remote villages or in the vast shantytowns of Nairobi, Kinshasa, and other large cities, and they do not see a doctor. Yet staggering evidence of the impact of AIDS is everywhere, and surveys by the World Health Organization have begun to unveil the real magnitude of the disease in Subsaharan Africa: over 6 million HIV cases as of late 1992.

Most disconcerting are the trends. Because AIDS is a blood-borne disease and is transmitted most efficiently through (often undetected) blood contact, sexual practices in a society can presage routes of spatial diffusion. Patronage of prostitutes along the East African highway linking Kampala, Uganda, and Kenya's Nairobi and Mombasa, for example, is known to be one such diffusion pathway. In their 1991 book, *The Geography of AIDS*, medical geographers Gary Shannon, Gerald Pyle, and Rashid Bashshur reported a series of tests on female prostitutes in Nairobi: in 1981, 4 percent tested positive for the virus; in 1984, 61 percent; and in 1986, 85 percent. In Blantyre, Malawi's largest city, pregnant women in 1984 showed an infection rate of 2 percent; in 1990, the rate was 22 percent. About one-third of the babies born to infected women are themselves infected with the AIDS virus.

Infection rates in East and Equatorial Africa's cities are high, ranging from 8 percent of adults in Zaïre's capital of Kinshasa to 30 percent in Kigali, capital of Rwanda. These levels have been reached relatively recently, so the real devastation of the epidemic in Subsaharan Africa has not yet begun. The cities were the large reservoirs, but the rural areas have not escaped the ravages of AIDS. Surveys in rural areas of Uganda and Zaïre indicate that between 8 and 12 percent of the adult populations in those places are infected. The present state of medical knowledge holds out no long-term hope for those infected. Thus AIDS, before the end of the century, will alter the population-growth rates of much of tropical Africa.

Children who do not get enough protein in their diets, for instance, develop such food-deficiency disorders as kwashiorkor and marasmus; they may survive, but their resistance to diseases of older age will be reduced.

Diseases strike populations in different ways. A local or regional outbreak, affecting many people is known as an *epidemic*. When a disease spreads worldwide, as some forms of influenza have done in recent years, the phenomenon is called a *pandemic*. AIDS (discussed in the box above) may have begun as an epidemic in a part of Equatorial Africa in the 1970s; by the early 1990s, it had assumed pandemic proportions, killing victims from Kenya to Romania to the United States. A disease can affect a population in still another way. Some illnesses do not come in a violent attack but invade and inhabit the body, establishing a kind of equilibrium with it. These diseases sap energies and shorten lifetimes, but they are not usually the cause of eventual

death. Such **endemic diseases** afflict tens of millions of Africans.

Undoubtedly, *malaria*, endemic to most of Africa and a killer of up to 1 million children per year, is the worst. In recent years malaria has again been on the rise, and in rural areas it shows growing resistance to drugs that once kept it under control. The malarial mosquito is the vector of the parasite, and the mosquito prevails in almost all of inhabited Africa. Africans who survive childhood are likely to suffer from malaria to some degree, with a debilitating effect.

African sleeping sickness is transmitted by the tsetse fly and now affects most of tropical Africa (Fig. 8-5A). This fly infects not only people but also their livestock. Its impact on Africa's population has been incalculable. The disease appears to have originated in a West African source area about A.D. 1400 (Fig. 8-5B); since then, it has inhibited the

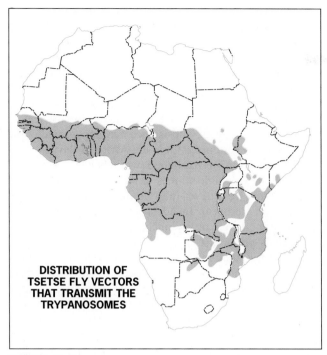

FIGURE 8-5A

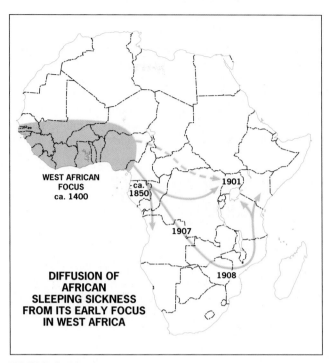

FIGURE 8-5B

development of livestock herds where meat would have provided a crucial balance to seriously protein-deficient diets. It channeled the migrations of cattle herders through fly-free corridors in East Africa (Fig. 8-5A), destroying herds that moved into infested zones. Most of all, it ravaged the human population, depriving it not only of potential livelihoods but also of its health.

Still another serious and widespread disease is *yellow fever*, also transmitted by mosquitoes. Yellow fever is endemic in the wetter tropical zones of Africa, but it sometimes appears in other areas in epidemic form, as in 1965 in Senegal when 20,000 cases were reported (there were undoubtedly thousands more). This is another disease that strikes children, who, if they survive, acquire a certain level of immunity.

To this depressing list must be added *schistosomiasis* (also called bilharzia), which is transmitted by snails. The parasites enter via body openings when people swim or wash in slow-moving or standing water infested by the snails. Many development projects in Africa involving irrigation have inadvertently introduced schistosomiasis into populations previously not exposed to these parasites. Internal bleeding, loss of energy, and pain result, although schistosomiasis is not itself fatal. It is endemic today to more than 200 million people worldwide, most of them residing in Africa.

These are the major and more or less continentwide diseases; there are many others of regional and local distribution. The dreaded river blindness, caused by a parasitic worm transmitted by a small fly, is endemic in the savanna belt

A young child acts as the "eyes" of a victim of river blindness.

south of the Sahara from Senegal east to Kenya; in northern Ghana alone, it blinds a large percentage of the adult villagers. And animals and plants as well are attacked by Africa's ravages. Besides sleeping sickness, livestock herds are also afflicted by rinderpest. Crops and pastures are stripped periodically by swarms of locusts (the late 1980s were particularly bad years), which number an average of 60 million insects each, travel thousands of miles, and devour the plant life of entire countrysides.

◆ THE PREDOMINANCE OF AGRICULTURE

The great majority of Africans live today as their predecessors lived—by subsistence farming, herding, or both. Not that the African environment is particularly easy for the cultivator who tries to grow food crops or raise cattle: tropical soils are notoriously unproductive, and much of Africa suffers from excessive drought.

Most African families still depend on subsistence farming for their living. Along the coast (especially the west coast) and the major rivers, some communities depend primarily on fishing. Otherwise, the principal mode of life is farming—with grain crops dominant in the drier areas and root crops in the more humid zones. In the methods of farming, the sharing of the work between men and women, the value and prestige attached to herd animals, and in other cultural aspects, subsistence farming provides the opportunity to gain insight into the Africa of the past. The subsistence form of livelihood was changed only very indirectly by colonialism; tens of thousands of villages all across Africa were never fully brought into the economic orbit of the European invaders, and life in these settlements went on more or less unchanged.

Africa's pastoralists more often than not mix farming with their livestock-raising pursuits, and very few of them are exclusively herders. In Africa south of the Sahara, there are two belts of herding activity: one extending laterally along the West African savanna-steppe transition zone and connecting with the East African area where the Maasai

African women, some carrying their babies, work in a field in Uganda. In rural areas, women perform a multitude of tasks ranging from water fetching and cooking to farming and housekeeping. The workload is often unevenly divided between men and women, the women doing far more of the work in addition to childbearing and child raising.

A GREEN REVOLUTION FOR AFRICA?

The gap between world population and food production has been narrowed through the **Green Revolution**, the development of more productive, higher-yielding types of grains. Where people depend mainly on rice and wheat for their staples, the Green Revolution pushed back the specter of hunger. The Green Revolution has had less impact in Africa, however. In part, this relates to the realm's high rate of population growth (which is substantially higher than that of India or China). Other reasons have to do with Africa's staples. Rice and wheat support only a small part of Africa's population. Corn (maize) supports many more, along with sorghum, millet, and other grains; in moister areas, root crops such as the yam and cassava as well as the plantain (similar to the banana) supply the bulk of calories. These crops were not priorities in Green Revolution research.

Lately there have been a few signs of hope. Even as terrible famines struck several parts of the continent and people went hungry in places as widely scattered as Moçambique and Mali, scientists worked to accomplish two goals: first, to develop strains of corn and other crops that would be more resistant to Africa's virulent crop diseases, and second, to increase the productivity of those strains. But these efforts faced serious problems: an average African acre of corn, for example, yields only about 0.5 tons of corn, whereas the world average is 1.3 tons.

Now a virus-resistant corn variety has been developed that also yields better than the old, and it is being introduced in Nigeria and other countries. And hardier types of yams and cassava also are evolving, again raising yields significantly.

Africa needs more than this if the cycle of food deficiency is to be reversed. It has been estimated that since 1970, food production has fallen by 1 percent every year. Inefficient farming methods, inadequate equipment, soil exhaustion, apathy, fatalism, and male dominance also must be overcome if Subsaharan Africa is to feed the 635 million mouths expected to live in the realm by 2000. Military governments (of which Africa has many) have been especially neglectful in their agricultural policies, which require quiet persuasion and effective extension systems rather than glamorous, inspiring campaigns. That there is an alternative, a way to success, has been proven in recent years by Zimbabwe, where the farming sector of the economy thrives and a surplus of staples is produced. But Zimbabwe has a comparatively small population (11 million), inherited a sound agrarian sector from colonial times, and since 1980 has largely been spared the instabilities that have bedeviled other parts of Africa. Today, there are some bright spots, but the battle for food sufficiency in Africa is far from being won.

---

drive their herds, the other centering on the plateau of South Africa. Especially in East and Southern Africa, cattle are less important as a source of food than they are as a measure of their owners' wealth and prestige in the community. Hence African cattle-owners in these areas have always been more interested in the size of their herds than in the quality of the animals. The staple foods here are the grain crops of corn, millet, and sorghum (together with some concentrations of rice), which overlap with the herding areas. Probably a majority of Africa's cattle-owners are sedentary farmers, although some pastoralists—such as the Maasai—engage in a more or less systematic cycle of movement, following the rains and seeking pastures for their livestock.

Cattle, of course, are not the only livestock in Africa. Chickens are nearly ubiquitous, and there are millions of goats everywhere—in the forest, in the savanna, in the steppe, even in the desert; the goats always seem to survive, and no African village would be complete without them. Where conditions are favorable, goats multiply swiftly, and then they denude the countryside and promote soil erosion. Goats, thus, constitute both an asset to their individual owners and a serious liability to the state.

A gradual change in the involvement of Africans in the cash economy has taken place during the past three decades. Excluded during the colonial period from most activities that could produce income for themselves and their families, many Africans have chosen, since independence, to make changes that promise to bring in money. Farmers have introduced cash crops onto their small plots, at times replacing the subsistence crops altogether. These developments notwithstanding, Africa continues to fall further behind in meeting its food needs. Across the continent, agricultural output is declining. Per capita food production in Subsaharan Africa during the mid-1990s was more than 10 percent below what it was a decade earlier. Even though total production is higher today, with 104 million more mouths to feed in 1994 than in 1984, all food increases are more than offset by the rapid rate of population growth (see the box entitled "A Green Revolution for Africa?").

◆ AFRICA'S PAST

Africa is the cradle of humanity. Research in Tanzania, Kenya, and Ethiopia has steadily pushed back the date of the earliest origin of human prototypes by hundreds of thousands, even millions, of years. It is therefore something of an irony that comparatively little is known about Subsaharan Africa from 5,000 to 500 years ago—that is, prior to the onset of European colonialism. This is only partly due to the colonial period itself, during which African history was neglected, many African traditions and artifacts were destroyed, and many misconceptions about African cultures and institutions arose and became entrenched. It is also a result of the absence of a written history over most of Africa south of the Sahara until the sixteenth century— and over a large part of it until much later than that. The best records are those of the savanna belt immediately south of the Sahara, where contact with North African peoples was greatest and where Islam achieved a major penetration.

The absence of a written record does not mean, as some scholars have suggested, that Africa does not have a history as such prior to the coming of Islam and Christianity. Nor does it mean that there were no rules of social behavior, no codes of law, no organized economies. Modern historians, encouraged by the intense interest shown by Africans generally, are now trying to reconstruct the African past, not only from the meager written record but also from folklore, poetry, art objects, buildings, and other such sources. Much has been lost forever, though. Almost nothing is known of the farming peoples who built well-laid terraces on the hillsides of northeastern Nigeria and East Africa or of the communities that laid irrigation canals and constructed stone-lined wells in Kenya; and very little is known about the people who, perhaps a thousand years ago, built the great walls of Zimbabwe (see photo below). Porcelain and coins from China, beads from India, and other goods from distant sources have been found in Zimbabwe and other points in East and Southern Africa, but the trade routes within Africa itself—let alone the products that circulated on them and the people who handled them—still remain the subject of guesswork.

AFRICAN GENESIS

Africa on the eve of the colonial period was in many ways a continent in transition. For several centuries, the habitat in and near one of the continent's most culturally and economically productive areas—West Africa—had been changing. For 2,000 years, probably more, Africa had been innovating as well as adopting ideas. In West Africa, cities were developing on an impressive scale; in Central and Southern Africa, peoples were moving, readjusting, sometimes struggling with each other for territorial supremacy. The Romans had penetrated to southern Sudan, North African peoples were trading with West Africans, and Arab *dhows* were sailing the waters along the eastern coasts, bringing Asian goods in exchange for gold, copper, and a comparatively small number of slaves.

Consider the environmental situation in West Africa as it relates to the past. As Figs. I-6 through I-8 indicate, the environmental regions in this part of the continent exhibit a decidedly east-west orientation. The isohyets (lines of equal rainfall totals) run parallel to the southern coast (Fig. I-6); the climatic regions, now positioned somewhat differently from where they were two millennia ago, still trend strongly east-west (Fig. I-7); the vegetation map, although generalized, also reflects this situation (Fig. I-8), with a coastal forest belt yielding to savanna (tall grass with scattered trees in the south; shorter grass in the north), which gives way, in turn, to steppe and desert.

These stone ruins in Zimbabwe mark the rise of an early African city, but the functions of its many buildings remain a mystery. They have been interpreted as fortifications, as religious structures, and as political symbols. There is no doubt that when this center arose, whatever its role, African society in this area changed dramatically.

Early Trade

Essentially, then, the situation in West Africa was such that over a north-south span of a few hundred miles, there was an enormous contrast in environments, economic opportunities, modes of life, and products. The peoples of the tropical forest produced and needed goods that were quite different from the products and requirements of the peoples of the dry, distant north. As an example, salt is a prized commodity in the forest, where the humidity precludes its formation, but salt is in plentiful supply in the desert and steppe. Thus the desert peoples could sell salt to the forest peoples, but what could be offered in exchange? Ivory and spices could be sent north, along with dried foods. Thus there was a degree of **regional complementarity** between the peoples of the forests and the peoples of the drylands. And the savanna peoples—those located in between—were beneficiaries of this situation, for they found themselves in a position to channel and handle the trade (an activity that is always economically profitable).

The markets in which these goods were exchanged prospered and grew, and there arose a number of true cities in the savanna belt of West Africa. One of these old cities, now an epitome of isolation, was once a thriving center of commerce and learning and one of the leading urban places in the world—Timbuktu. Others, predecessors as well as successors of Timbuktu, have declined, some of them into oblivion. Still other savanna cities continue to have considerable importance, such as Kano in the northern part of Nigeria.

Early States

States of impressive strength and amazing durability arose in the West African culture hearth (see Fig. 7-4). The oldest state about which anything is known is Ghana. Ancient Ghana was located to the northwest of the coastal country that has taken its name in the post-colonial period. It covered parts of present-day Mali and Mauritania along with some adjacent territory. Ghana lay astride the upper Niger River and included gold-rich streams flowing off the Futa Jallon Highlands, where the Niger has its origins. For a thousand years, perhaps longer, old Ghana managed to weld various groups of people into a stable state. The country had a large capital city complete with markets, suburbs for foreign merchants, religious shrines, and, some miles from the city center, a fortified royal retreat. There were systems of tax collection for the citizens and extraction of tribute from subjugated peoples on the periphery of the territory; tolls were levied on goods entering the Ghanaian domain; and an army maintained control. Muslims from the northern drylands invaded Ghana about A.D. 1062, when the state may already have been in decline. Even so, Ghana continued to show its strength: the capital was protected for no less than 14 years. However, the invaders had ruined the farmlands, and the trade links with the north were de-

stroyed. Ghana could not survive, and it finally broke apart into a number of smaller units.

In the centuries that followed, the focus of politico-territorial organization in the West African culture hearth shifted almost continuously eastward—first to ancient Ghana's successor state of Mali, which was centered on Timbuktu and the middle Niger River valley, and then to the state of Songhai, whose focus was Gao, also a city on the Niger and one that still exists today. One possible explanation for this eastward movement may lie in the increasing influence of Islam; Ghana had been a pagan state, but Mali and its sucessors were Muslim and sent huge, rich pilgrimages to Mecca along the savanna corridor south of the desert. Indeed, hundreds of thousands of citizens of the modern Republic of Sudan trace their ancestry to the lands now within northern Nigeria, their ancestors having settled there while journeying to or from Mecca.

Unmistakably, the West African savanna region was the scene of momentous cultural, political, and economic developments for many centuries, but it was not alone in its progress in Africa. In what is today southwestern Nigeria, a number of urban farming communities became established, the farmers having concentrated in these walled and fortified places for reasons of protection and defense; surrounding each "city of farmers" were intensely cultivated lands that could sustain thousands of people clustered in the towns. In the arts, too, Nigeria produced some great achievements, and the bronzes of Benin are true masterworks. In the region of the Zaïre River mouth, a large state named Kongo existed for centuries. In East Africa, trade on a large scale with China, India, Indonesia, and the Arab domain brought crops, customs, and merchandise from these distant regions. In Ethiopia and Uganda, populous kingdoms emerged. Nevertheless, much of what Africa was in those earlier centuries has yet to be reconstructed. Yet with all this external contact, it was clearly not isolated.

THE COLONIAL TRANSFORMATION

The period of European involvement in Subsaharan Africa began in the fifteenth century. This period was to interrupt the path of indigenous African development and irreversibly alter the entire cultural, economic, political, and social makeup of the continent. It started quietly enough, with Portuguese ships groping their way along the west coast and rounding the Cape of Good Hope not long before the opening of the sixteenth century. Their goal was to find a sea route to the spices and riches of the Orient. Soon, other European countries were sending their vessels to African waters, and a string of coastal stations and forts sprang up. In West Africa, the nearest part of the continent to European spheres in Middle and South America, the initial impact was strongest. At their coastal control points, the Europeans traded with African middlemen for the slaves who were wanted on New World plantations, for the gold that had been flowing northward across the desert, and for ivory and spices.

Suddenly, the centers of activity lay not with the cities of the savanna but in the foreign stations on the Atlantic coast. As the interior declined, the coastal peoples thrived. Small forest states rose to power and gained unprecedented wealth, transferring and selling slaves captured in the interior to the European traders on the coast. Dahomey (now

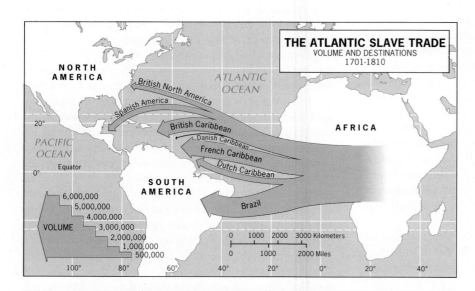

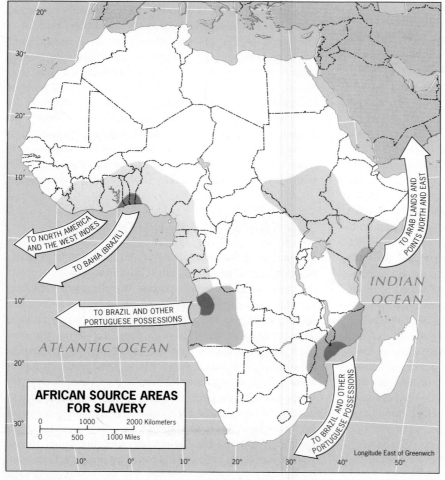

FIGURE 8-6

called Benin) and Benin (now part of neighboring Nigeria) were states built on the slave trade; when the practice of slavery eventually came under attack in Europe, abolition was vigorously opposed in both continents by those who had inherited the power and riches it had brought.

Although it is true that slavery was not new to West Africa, the kind of slave trading introduced by the Europeans certainly was. In the savanna states, African families who had slaves usually treated them comparatively well, permitting marriage, affording adequate quarters, and absorbing them into the family. The number of slaves held in this way was small; probably the largest number of persons in slavery in pre-colonial Africa were in the service of kings and chiefs. In East Africa, however, the Arabs had introduced (long before the Europeans) the sort of slave trading that was first brought to West Africa by whites: African middlemen from the coast raided the interior for slaves and marched them in chains to the Arab *dhows* that plied the Indian Ocean. There, packed by the hundreds in specially built vessels, they were carried off to Arabia, Persia, and India. It is sad but true that Europeans, Arabs, and Afri-

cans combined to ravage the Black Continent, forcing perhaps as many as 30 million persons away from their homelands in bondage (Fig. 8-6). Families were destroyed, as were whole villages and cultures; and those affected suffered a degree of human misery for which there is no measure.

The European presence on the West African coast brought about a complete reorientation of trade routes, for it initiated the decline of the interior savanna states and strengthened the coastal forest states. Moreover, it ravaged the population of the interior through its insatiable demand for slaves; but it did not lead to any major European thrust toward the interior, nor did it produce colonies overnight. The African middlemen were well organized and strong, and they managed to maintain a standoff with their European competitors, not just for a few decades but for centuries. Although the European interests made their initial appearance in the fifteenth century, West Africa was not carved up among them until nearly 400 years later, in many areas not until after the beginning of the twentieth century.

As fate would have it, European interest was to grow strongest—and ultimately most successful—where African

Cape Town, white South Africa's "mother city," looking southward across Table Mountain, which dominates this magnificent scene. It was here, very close to Africa's southern tip, that the first Dutch settlers landed in 1652. Today, this port— historically and potentially one of the most strategic in the world—is one of Africa's best-equipped.

organization was weakest. In the middle of the seventeenth century, the Dutch chose the shores of South Africa's Table Bay, where Cape Town lies today, as the site for permanent settlement. Their initial purpose was not colonization but rather establishment of a resupply station for the months-long voyage to and from Southeast Asia and their East Indies colonies; the southern tip of Africa was the obvious halfway point. There was no intent to colonize here because Southern Africa was not known as a productive area, and more worthwhile East Africa lay in the spheres of the Portuguese and the Arabs. Probably the Hollanders would have elected to build their station at the foot of Table Bay whatever the indigenous population of the interior, but they happened to choose a location about as far away from the centers of Bantu settlement as they could have found. Only the San (Bushmen) and their rivals, the Khoikhoi (Hottentots), occupied Cape Town's hinterland.

When conflicts developed between Amsterdam and Cape Town and some of the settlement's residents decided to move into the hinterland, they initially faced only the harassment of small groups of these two peoples rather than the massive resistance that the well-organized Bantu states would probably have offered. To be sure, a confrontation eventually did develop between the advancing Europeans and the similarly mobile Bantu Africans, but it began decades after Cape Town was founded and hundreds of miles from it. Unlike some of the West African way-stations, Cape Town was never threatened by African power, and it became a European gateway into Southern Africa.

Elsewhere in the realm, the European presence remained confined almost entirely to the coastal trading stations, whose economic influence was very strong. No real frontiers of penetration developed; individual travelers, missionaries, explorers, and traders went into the interior, but nowhere else in Africa south of the Sahara was there an invasion of white settlers comparable to Southern Africa's.

Penetration

After more than four centuries of contact, Europe finally laid claim to all of Africa during the second half of the nineteenth century. Parts of the continent had been "explored," but now representatives of European governments sought to expand or create African spheres of influence for their homelands. Cecil Rhodes for Britain, Karl Peters for Germany, Pierre de Brazza for France, and Henry Stanley for the king of Belgium were some of the leading figures who helped shape the colonial map of the continent. In some areas, such as along the lower Zaïre River and in the vicinity of Lake Victoria, the competition among the European powers was especially intense. Spheres of influence began to crowd each other; sometimes they even overlapped. So, in late 1884, a conference was convened in Berlin to sort things out. At this conference, the ground-

work was laid for the now-familiar political boundaries of Africa (see box at right).

As the twentieth century opened, Europe's colonial powers were busily organizing and exploiting their African dependencies. The British, having defeated the Boers of Dutch heritage in South Africa, came very close to achieving their Cape-to-Cairo axis: only German East Africa interrupted a vast empire that stretched southward from Egypt and Sudan through Uganda, Kenya, Nyasaland (now Malawi), and the Rhodesias (now Zambia and Zimbabwe) to South Africa (Fig. 8-7, 1910 map). The French took charge of a vast realm that reached from Algiers in the north and Dakar in the west to the Zaïre River in Equatorial Africa. King Leopold II of Belgium held personal control over his Belgian Congo (now Zaïre). Germany had colonies scattered in all sections of the continent except the north. The Portuguese controlled two huge territories, Angola and Moçambique, along the flanks of Southern Africa, and a small entity in West Africa known as Portuguese Guinea (now Guinea-Bissau). Italy's possessions in tropical Africa were confined to the Horn, and even Spain got into the act with a small dependency consisting of the island of Fernando Póo (now called Bioko) and the mainland area called Rio Muni (Equatorial Guinea). The only places where the Europeans did not overwhelm African desires to remain independent were Ethiopia, which fought some heroic battles against Italian forces, and Liberia, where African-Americans retained control.

The two world wars also had some effect on this colonial map of Africa. In World War I, Germany's defeat resulted in the complete loss of its colonial possessions. The territories in Africa were placed under the administration of other colonial powers by the League of Nations' mandate system. In World War II, fascist Italy launched a briefly successful campaign against Ethiopia, but the ancient empire was restored to independence when the allied forces won the war. Otherwise, the situation in colonial Africa in the late 1940s—after a half-century of colonial control—remained quite similar to the one that arose out of the Berlin Conference (Fig. 8-7, 1950 map).

Policies and Repercussions

Geographers—especially political geographers—are interested in the ways in which the philosophies and policies of the colonial powers were reflected in the spatial organization of the African dependencies. These colonial policies can be expressed in just a few words. For example, Britain's administration in many parts of its vast empire was referred to as *indirect rule*, since indigenous power structures were sometimes left intact and local rulers were made representatives of the crown. Belgian colonial policy was called *paternalism* in that it treated Africans as though they were children who needed to be tutored in Western

THE BERLIN CONFERENCE

In November 1884, the imperial chancellor and architect of the German Empire, Otto von Bismarck, convened a conference of 14 powerful states (including the United States) to settle the political partitioning of Africa. Bismarck not only wanted to expand German spheres of influence in Africa but also sought to play off Germany's colonial rivals against one another to the Germans' advantage. The major colonial contestants in Africa were the British, who held beachheads along the West, South, and East African coasts; the French, whose main sphere of activity was in the area of the Senegal River and north of the Zaïre (Congo) Basin; the Portuguese, who now desired to extend their coastal stations in Angola and Moçambique deep into the interior; King Leopold II of Belgium, who was amassing a personal domain in the Congo; and Germany itself, active in areas where the designs of other colonial powers might be obstructed, as in Togo (between British holdings), Cameroon (a wedge into French spheres), South West Africa (taken from under British noses in a swift strategic move), and East Africa (where the effect was to break the British design for a Cape-to-Cairo axis).

When the conference convened in Berlin, more than 80 percent of Africa was still under traditional African rule.

Nonetheless, the colonial powers' representatives drew their boundary lines across the entire map. These lines were drawn through known as well as unknown regions, pieces of territory were haggled over, boundaries were erased and redrawn, and sections of African real estate were exchanged in response to urgings from European governments. In the process, African peoples were divided, unified regions were ripped apart, hostile societies were thrown together, hinterlands were disrupted, and migration routes were closed off. All of this was not felt immediately, of course, but these were some of the effects when the colonial powers began to consolidate their holdings and the boundaries on paper soon became barriers on the African landscape (Fig. 8-7).

The Berlin Conference was Africa's undoing in more ways than one. The colonial powers superimposed their domains on the African continent; when independence returned to Africa after 1950, the realm had by then acquired a legacy of political fragmentation that could neither be eliminated nor made to operate satisfactorily. The African politico-geographical map, is thus a permanent liability that resulted from three months of ignorant, greedy acquisitiveness during the period of Europe's insatiable search for minerals and markets.

---

ways. Although the Belgians made no real efforts to make their African subjects culturally Belgian, the French very much wanted to create an "Overseas France" in their African dependencies. French colonialism has been identified as a process of *assimilation*—the acculturation of Africans to French ways of life—and France made a stronger cultural imprint in the various parts of its huge colonial empire. Portuguese colonial policy had objectives similar to those of the French, and the African dependencies of Portugal were officially regarded as "Overseas Provinces" of the state. If you sought a one-word definition of Portuguese colonial policy, however, the term *exploitation* would emerge most strongly: few colonies made a greater contribution (in proportion to their known productive capacities) to the economies of their colonial masters than did Moçambique and Angola.

Colonial policies have geographic expressions as well, and the **colonial spatial organization** of the ruling powers has become the political infrastructure of contemporary independent Africa. As the map shows (Fig. 8-7), Britain possessed the most far-flung colonial empire in Africa.

British colonial policy tended to adjust to individual situations. In colonies, white-settler minorities had substantial autonomy, as in Kenya and Southern Rhodesia (now Zimbabwe); in protectorates, the rights of African peoples were guarded more effectively; in mandate (later trust) territories, the British undertook to uphold League of Nations' (later United Nations') administrative rules; and in one case, the British shared administration with another government (Egypt) in Sudan's *condominium*. Britain's colonial map of Africa was a patchwork of these different systems, and the independent countries that emerged reflect the differences. Nigeria, which had been a colony in its south and an indirectly ruled protectorate in its north, became a federal state accommodating major internal differences. Kenya, the former colony, became a highly centralized unitary state after the Africans wrested control of the productive core area from the whites.

In contrast to the British, the French placed a cloak of uniformity over their colonial territories in Subsaharan Africa. Contiguous and vast, although not very populous, France's colonial empire extended from Senegal eastward

COLONIZATION AND LIBERATION

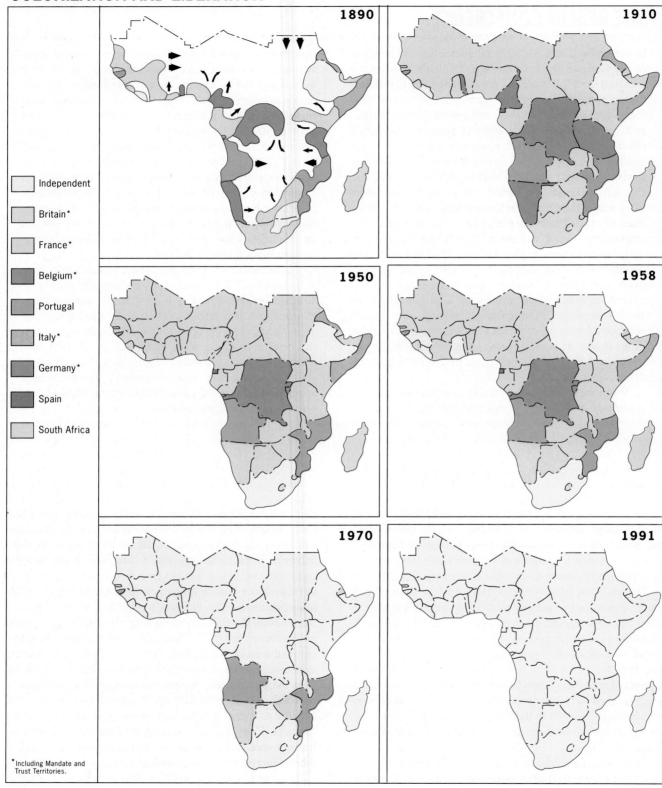

Independent

Britain*

France*

Belgium*

Portugal

Italy*

Germany*

Spain

South Africa

*Including Mandate and
Trust Territories.

FIGURE 8-7

to Chad and southward to the former French Congo (now the Congo Republic). This huge area was divided into two units, French West Africa (focused on Dakar) and French Equatorial Africa (whose headquarters was in Brazzaville).

France itself, as we know, is the classic example of the centralized unitary state whose capital is the cultural, political, and economic focus of the nation, overshadowing all else. The French brought their concept of centralization to Africa as well. In France, all roads lead to Paris; in Africa, all roads were to lead to France, to French culture, to French institutions. For the purposes of assimilation and acculturation, French West Africa, half the size of the entire United States of America (although with a population of only some 30 million in 1960), was divided into administrative units, each centered on the largest town and all oriented toward the governor's headquarters at Dakar. As the map shows, great lengths of these boundaries were straight lines delimited across the West African landscape; history tells us to what extent they were drawn, not on the basis of African realities but for France's administrative convenience. The present-day state of Burkina Faso, for instance, existed as an entity until 1932 when, because of administrative problems, it was divided among Ivory Coast, Soudan (now Mali), and Dahomey (now Benin). Then in 1947, the territory was suddenly recreated and named Upper Volta. Little did the boundary-makers expect that what they were doing would one day affect the national life of an independent country.

Belgian administration in the Congo (now Zaïre) provides another insight into the results, in terms of spatial organization, of a particular set of policies of colonial government. Unlike the French, the Belgians made no effort to acculturate their African subjects. The policies of paternalism actually consisted of rule in the Congo by three sometimes competing interest groups: the Belgian government, the managements of huge mining corporations, and the Roman Catholic Church. Each of these groups had major regional spheres of activity in this vast country.

As the map shows (p. 444), Zaïre has a corridor to the ocean along the Zaïre River between Angola to the south and the Congo Republic (former French Congo) to the north. The capital, Kinshasa (formerly Leopoldville), lies at the eastern end of this corridor; not far from the ocean lies the country's major port, Matadi. It was in this corridor that the administrative and transport core area of Zaïre developed, and this was the place from which the decisions made in Brussels were promulgated by a governor-general. The economic core of the Congo (and now of Zaïre) lay in the country's southeast, in the province then known as Katanga, which was administered from the copper-mining headquarters of Lubumbashi (formerly Elisabethville). This is a portion of the northernmost extension of Southern Africa's great mining belt, the balance of which continues into Zambia as that country's Copperbelt. From hundreds of miles around, workers streamed toward the mines (Fig. 8-8), a pattern of regional labor migration that has persisted for decades and is still going strong.

Between the former Belgian Congo's administrative and economic core areas lay the Congo (Zaïre) Basin, once a profitable source of wild rubber and ivory. The colony's six administrative subdivisions were laid out in such a way that each incorporated part of the area's highland rim and part of the forested basin. As the 1960 date of independence approached, it seemed briefly that each of these Congolese provinces, centered on its own administrative capital, might break away and become an independent African country (as each French African dependency did in West and Equatorial Africa). In the end, however, the country held together, and Leopoldville became Kinshasa, capital of Zaïre, Subsaharan Africa's largest state in territorial size.

Portugal's rule in Angola and Moçambique was designed to exploit four assets: (1) labor supply for the interior mines, especially in Moçambique; (2) transit functions and port facilities—Moçambique from South Africa and Southern Rhodesia, Angola from the Copperbelt and Katanga; (3) agricultural production, particularly cotton from northern Moçambique and coffee from Angola; and (4) minerals, mainly from diamond- and oil-rich Angola. In this effort, the Portuguese created a system of rigid control that involved forced labor and the compulsory farming of certain crops. In Moçambique particularly, the country was divided into a large number of small districts that were tightly controlled. Movement and communication, even within a single African ethnic area, were kept to a minimum. Accordingly, Portuguese colonial rule was often described as the harshest of all the European systems; for a long time, it seemed unlikely that an independence movement could be mounted. But when independence came to Angola's and Moçambique's neighbors (Zaïre and Tanzania), Portugal's days were numbered. As elsewhere in recently colonial Africa, however, the imprint of colonialism in former Portuguese Africa remains a strong, pervasive element in the regional geography of the continent.

Besides the geographic consequences that flowed from the various forms of colonial policy, the current (and future) map of the realm was affected in many other ways as well. In spite of the differences among individual colonial policies, all were paternalistic, assimilative, and exploitive to some degree. However difficult it was to unify territories within colonial spheres, it has proven virtually impossible to resolve the problems of redefining truly African regional interests without regard to the colonial heritage. With few exceptions, the primary development of capital cities within each colonial territory has been strengthened even further since independence. And much of the transport network at independence reflected the colonial approach to development: railways (if they existed) and major roads (however defined) almost always facilitated the movement of goods between the interior

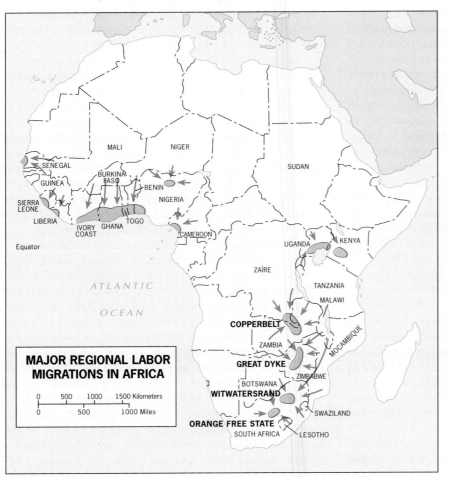

FIGURE 8-8

and coastal outlets but offered little basis for interior circulation. The geographic patterns established during the colonial period are more often a hindrance to development than an aid.

◆ AFRICA'S REGIONS TODAY

On the face of it, Africa would seem to be so massive, so compact, and so unbroken that any attempt to justify a contemporary regional breakdown is doomed to failure. No deeply penetrating bays or seas create peninsular fragments as in Europe. No major islands (other than Madagascar) provide the broad regional contrasts we see in Middle America. Nor does Africa really taper southward to the peninsular proportions of South America. And Africa is not cut by an Andean or a Himalayan mountain barrier. Given Africa's colonial fragmentation and cultural mosaic, is regionalization possible? Indeed it is.

Maps of environmental distributions, population concentrations, ethnic patterns, historic culture hearths, and colonial frameworks yield the four-region structure represented in Fig. 8-9. *West Africa* includes the countries of the western coast and Sahara margin from Senegal and Mauritania in the west to Nigeria and Niger in the east. *Equatorial Africa* is delineated from Nigeria by physiographic as well as cultural breaks: the Adamawa Highlands (the volcanic range containing dangerous, gas-emitting lakes) coincide with the border between British-influenced Nigeria and French-acculturated Cameroon. But Chad, located northeast of Cameroon, presents a problem. It possesses both West African and Equatorial African properties; the Sahara-influenced north of Chad is a world apart from the savanna-forested south. So our map shows Chad to be an area of transition between the two regions. Equatorial Africa consists of Zaïre and its neighbors to the north, including the southern part of Sudan. *East Africa* as a geographic region is centered on three countries: Kenya, Tanzania, and Uganda. Two territorially smaller states, Rwanda and Burundi, also lie within East Africa, and the highland portion of Ethiopia is part of this region as well. As the map shows, both Equatorial and East Africa lie astride the

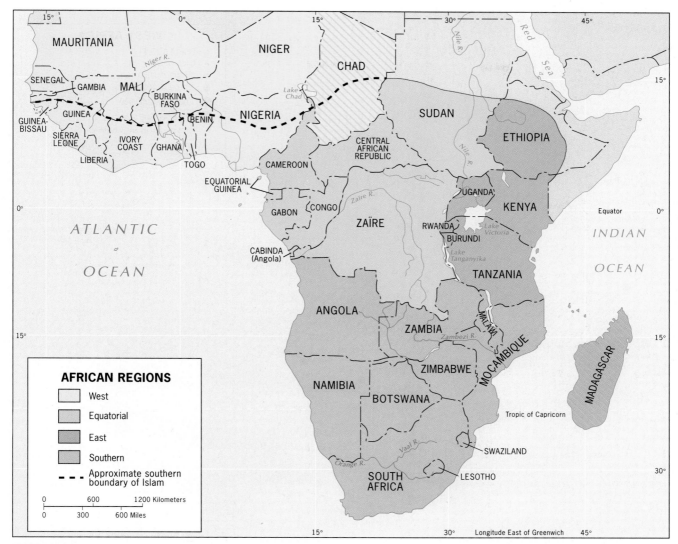

FIGURE 8-9

equator; but whereas Equatorial Africa is mainly lowland country, with large areas of rainforest and equatorial environments, East Africa is mainly highlands, where cooler and generally drier conditions prevail. *Southern Africa* is the region extending from the southern borders of Zaïre and Tanzania to the continent's southern tip. Madagascar, despite its nearness to Southern Africa, cannot be included in this geographic region without qualification, as we will see.

One additional and familiar regional frame of reference is Africa's Horn, the triangular zone expressed by the coastal margins of Somalia (see also pp. 382-384). The Horn, however, is not regionally comparable to the four units mapped in Fig. 8-9. It lies astride the African Transition Zone between (Islamic) North Africa and Subsaharan Africa, and is a mosaic of black African and non-African cultures and traditions. The Horn is an area, but it is not a geographic region.

WEST AFRICA

West Africa occupies most of Africa's Bulge, extending south from the margins of the Sahara to the Gulf of Guinea coast, and from Lake Chad west to Senegal (Fig. 8-10). Politically, the broadest definition of this region includes all those states that lie to the south of Morocco, Algeria, and Libya and to the west of Chad (itself sometimes included) and Cameroon. Within West Africa, a rough division is sometimes made between the very large, mostly steppe and desert states that extend across the southern Sahara (Chad could also be included here), and the smaller, better-watered coastal states.

Apart from once-Portuguese Guinea-Bissau and long-independent Liberia, West Africa comprises four former British and nine former French dependencies. The British-influenced countries (Nigeria, Ghana, Sierra Leone, and

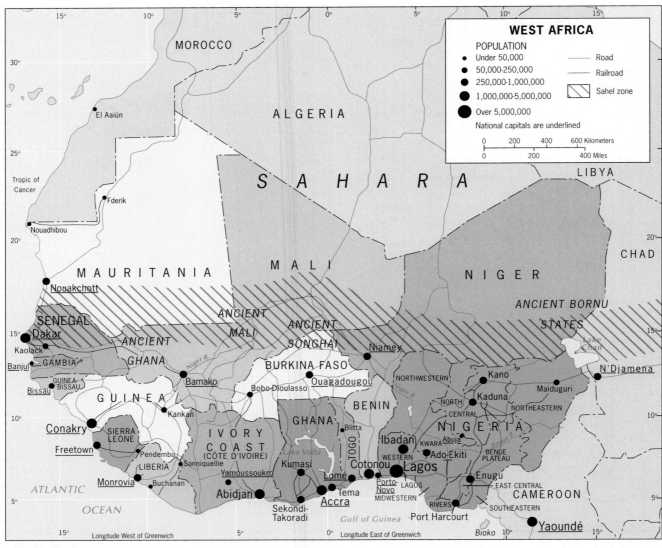

FIGURE 8-10

Gambia) lie separated from one another, whereas Francophone West Africa is contiguous. As Fig. 8-10 shows, political boundaries extend from the coast into the interior so that from Mauritania to Nigeria, the West African habitat is parceled out among parallel, coast-oriented states. Across these boundaries, especially across those between former British and former French territories, there is very little interaction. For example, in terms of value, Nigeria's trade with Britain is about 100 times as great as its trade with nearby Ghana. The countries of West Africa are not interdependent economically, and their incomes are to a large extent derived from the sale of their products on the international market. But the African countries do not control the prices that their goods can command on these world markets, and when prices are low, they face serious problems.

Given these cross-currents of subdivision within West Africa, what are the justifications for the concept of a single West African region? First, there is the remarkable cultural and historical momentum of this part of the realm. The colonial interlude failed to extinguish West African vitality, expressed not only by the old states and empires of the savanna and the cities of the forest but also by the vigor and entrepreneurship, the achievements in sculpture, music, and dance, of peoples from Senegal to Nigeria's southeastern Iboland. Second, West Africa contains a set of parallel east-west ecological belts, clearly reflected in Figs. I-6 to I-8, whose role in the development of the region is pervasive. As the transport-route pattern on the map of West Africa indicates, overland connections within each of these belts, from country to country, are quite poor; no coastal or interior railroad was ever built to connect this tier of countries. Yet, spatial interaction is stronger across these belts, and some north-south economic exchange does take place, notably in the coastal consumption of meat from cattle raised in the northern savannas. And third, West Africa received an early and crucial imprint from European colonialism, which—with its maritime commerce and the slave trade—transformed the region from one end to the other. This impact was felt all the way to the heart of the Sahara, and it set the stage for the reorientation of the whole area, from which emerged the present patchwork of states.

The effects of the slave trade notwithstanding, West Africa today is Subsaharan Africa's most populous region (Fig. I-9). In these terms, Nigeria (whose census results are in doubt, but with perhaps 100 million people) is Africa's largest state; Ghana (17.0 million) ranks high as well. As Fig. I-9 shows, West Africa also claims regional identity in that it constitutes one of Africa's major population clusters. The southern half of the region, understandably, is home to the majority of the people. Mauritania, Mali, and Niger include too much of the unproductive Sahel's steppe and the arid Sahara to sustain populations comparable to those of Nigeria, Ghana, or Ivory Coast. However, this is not to say that only the coastal areas of West Africa are densely populated; the interior savannalands also contain sizeable clusters.

The peoples along the coast reflect the modern era introduced by the colonial powers: they prospered in their newfound roles as middlemen in the coastward trade. Later, they were in a position to experience the changes of the colonial period; in education, religion, urbanization, agriculture, politics, health, and many other endeavors, they adopted new ways. The peoples of the interior, in contrast, retained their ties with a very different era in African history. Distant and aloof from the main theater of European colonial activity and often drawn into the Islamic orbit, they experienced significantly less change. But the map reminds us that Africa's boundaries were not drawn to accommodate such differences. Both Nigeria and Ghana possess population clusters representing the interior as well as the coastal peoples, and in both countries the wide gap between north and south has produced political problems.

Nigeria: West Africa's Cornerstone

When Nigeria achieved full independence in 1960, it was endowed with a federal political structure that consisted of three regions based on the three major population clusters within its borders—two in the south and one in the north. Around the Yoruba core in the southwest lay the Western Region. The Yoruba are a people with a long history of urbanization, but they are also farmers; in the old days, they protected themselves in walled cities around which they practiced intensive agriculture. The colonial period brought coastal trade, increased urbanization, cash crops (the main-

stay, cocoa, was introduced from Fernando Póo [now Bioko] in the 1870s), and, eventually, a measure of security against encroachment from the north. Lagos, the country's first federal capital and now a teeming conurbation of 10.3 million people, grew up around port facilities on the region's south coast. A new, more centrally located capital is today being completed at Abuja (Fig. 8-10). Ibadan (1.5 million), also one of Subsaharan Africa's largest cities, evolved from a Yoruba settlement founded in the late-eighteenth century. At independence, Nigeria's Western Region, more than any other part of Nigeria, had been transformed by the colonial experience.

East of the Niger River and south of the Benue River, the Ibo population formed the core of the Eastern Region. Iboland, although coastal, lay less directly in the path of colonial change, and its history and traditions also differed sharply from those of Western Nigeria. Little urbanization had taken place here, and even today, although over one-third of the Western Region's people live in cities and towns, less than 25 percent of the Eastern Region's population is urbanized. With more than 20 million people, the rural areas of Eastern Nigeria are densely peopled. Over the years, many Ibo have left their crowded habitat to seek work elsewhere—in the west, in the far north, in Cameroon, and even on the island of Bioko.

The third federal region at independence was at once the largest and the most populous: the Northern Region. It extended across the full width of the country from east to west and from the northern border southward beyond the Niger and Benue rivers. This is Nigeria's Muslim North, centered on the Hausa-Fulani population cluster, where the legacy of a feudal social system, conservative traditionalism, and resistance to change hung heavily over the country.

Nigeria's three original regions, then, lay separated not only by sheer distance but also by tradition and history and by the nature of colonial rule. In Nigeria, even physiography and biogeography conspire to divide south from north: across the heart of the country (and across much of West Africa at about the same latitude) stretches the so-called *Middle Belt*—poor, unproductive, disease- and tsetse-ridden country that forms a relatively empty barrier between the northern and southern regions. Not surprisingly, Nigeria's three-region federal system failed. Interregional rivalries and inter-ethnic suspicions led to civil war between 1967 and 1971, when the Eastern Region tried to secede as a separate political entity called Biafra. The original three regions were subdivided, rearranged, and subdivided once again in an attempt to devise a system that would prevent such disasters in the future.

Nigeria today is a cornerstone of the new Africa, a country whose plans for economic growth are based on substantial oil reserves in the area of the Niger delta. By the early 1980s, over 90 percent of Nigeria's export revenues were derived from the sale of petroleum and petroleum products, and cities from Port Harcourt (535,000) to Lagos reflected the oil boom of the south. This heavy dependence

The survival of Nigeria as a unified state is an African success story. The Nigerians have overcome strong centrifugal forces in a multi-ethnic country that is dominantly Muslim in the north, Christian in the south. Here, a mosque in Kano, a major city of the north, attracts a multitude of the faithful for Friday (sabbath) prayers. But now, some radical Muslim cleries are calling for an Islamic Republic in Nigeria. Can this country avoid the fate of Sudan?

on petroleum, however, made the country's ambitious development plans hostage to world oil prices—and lower prices in recent years have thrown the Nigerian economy into disarray.

Coastal States

Westward from Nigeria, West Africa's coastline presents a complex and fascinating cultural and political mosaic. Benin, Nigeria's neighbor, takes its name from one of the region's powerful kingdoms of the past. Many West Africans were carried as slaves from this area to Brazil, and today there is a growing cultural and emotional link between the people of Benin and those of African ancestry in and near Salvador in Bahia State (see Fig. 8-6) on the opposite side of the Atlantic. Next to Togo, Benin's western neighbor (briefly a German colony and later administered by France), lies Ghana. This country, once known as the Gold Coast, also took its name from the ancient West African state. Ghana was the first West African state to achieve independence (in 1957), but like so many other African countries, it has suffered from ineffective government, economic mistakes, corruption, and instability. Optimistic predictions heralding the success of the British "Westminster Model" in Africa were erased by the harsh realities of tribalism and military coups. Ghana's economy, once prosperous from the sale of cocoa, collapsed under Moscow-style planning programs; money was siphoned off to foreign accounts, and thousands of the most capable Ghanaians left the country. Two of the projects promoted by the first president of Ghana, Kwame Nkrumah, are prominent on the country's map: the great Volta River Dam and the port of Tema near the capital city of Accra (1.7 million). Neither

fulfilled its expectations, and Nkrumah died in dishonor and exile.

Compared to Ghana's fate, Ivory Coast fared much better, in large part because the country enjoyed several decades of stability. *Côte d'Ivoire* (the country's official name) was one of France's two most important West African possessions (the other was Senegal), and the president who negotiated independence in 1960 was still at the helm in the 1990s. During its first 30 years, Ivory Coast's capital, Abidjan (2.8 million), grew into a major urban and industrial center; the cocoa- and coffee-based economy was diversified, and living standards rose. French involvement in the country remained quite strong. But problems also emerged. The Sahelian drought drove cattle herders into northern Ivory Coast, and friction with local farmers whose crops were trampled created a challenge for the government. Worse, President Félix Houphouët-Boigny engineered the movement of the capital to Yamoussoukro (his home town) and embarked on a multimillion-dollar project to build there a Roman Catholic basilica to rival that of St. Peter's in Rome; resentment over these costs was among the causes behind unprecedented street disorders as the country's fourth decade of independence opened. As economic growth slowed, the stability of Ivory Coast was threatened.

France's other major coastal colony, Senegal (whose capital is Dakar, with 2.0 million people), was the anchor of its West African empire and remains a leading regional state today. As Fig. 8-10 shows, Senegal lay in the path of the Sahel's environmental onslaught, and the country's north suffered especially severely from drought and dislocation. Senegal's economic geography remains fragile: the country depends on the export of peanuts and phosphates and on its fishing industry. Droughts, reduced market

This astonishing structure is the Roman Catholic basilica, Notre Dame de la Paix (Our Lady of Peace), built near the capital of Ivory Coast. It was funded by the country's president; Pope John Paul II, after some hesitation, attended the consecration. Will archeologists of the future conclude that this part of West Africa was a vast region of Catholicism?

prices for its exports, and a high rate of population growth have inhibited development—the story of many an African country.

Between Ivory Coast and Senegal, no fewer than four countries face the Atlantic Ocean: Liberia, the one corner of West Africa that was never colonized and to which many freed slaves from America settled upon their return to Africa; Sierra Leone, a former British dependency that also was the destination of returning slaves; Guinea, one of coastal West Africa's least-developed countries from the former French Empire; and Guinea-Bissau, a small Portuguese-influenced country. All four suffer severely from the economic, demographic, medical, and political problems that afflict the region (and the realm) as a whole, but Liberia has ailed the most. In 1990, Liberia's political order collapsed in the chaos of a ruinous civil war that pitted ethnic groups against each other. More than half of all Liberians became refugees, and over 1 million crossed, destitute, into neighboring countries.

Periodic Markets

The great majority of the people of West Africa are not involved in the production of exports for world markets but subsist on what they can grow and raise—and trade. Their local transactions take place at small markets in villages. These village markets are not open every day but operate at regular intervals. In this way, several villages in an area get their turn to attract the day's trade and exchange, and each benefits from its participation in the wider network of interactions. People come to these **periodic markets** on foot, by bicycle, on the backs of their animals, or by whatever other means available. Periodic markets are not exclusively a West African phenomenon. They also occur in interior Southeast Asia, in China, and in Middle and South America as well as in other parts of Africa. The intervals between market days vary. In much of West Africa, village markets tend to be held on every fourth day, although some areas have two-day, three-day, or eight-day cycles.

The bustle and excitement of a periodic market in West Africa: a village in Mali is the center of action for a day. Piles of firewood (background), baskets, pots, textiles, and countless foods are for sale.

Periodic markets, then, form an interlocking network of exchange places that serves rural areas where there are few or no roads. As each market in the network gets its turn, it will be near enough to one portion of the area so that the people who live in the vicinity can walk to it, carrying what they wish to sell or trade. In this way, small amounts of produce filter through the market chain to a larger regional market, where shipments are collected for interregional or perhaps even international trade.

What is traded, of course, depends on where the market is located. A visit to a market in West Africa's forest zone will produce very different impressions from a similar visit in the savanna zone. In the savanna, sorghum, millet, and shea-butter (an edible oil drawn from the shea-nut) predominate, and you will see some Islamic influences on the local scene. In the southern forest zone, such products as yams, cassava, corn, and palm oil change hands; here, too, one is more likely to see some imported manufactured goods passing through the market chain, especially near the relatively prosperous areas of cocoa, coffee, rubber, and palm oil production. In general, however, the quantities of trade are very small—a bowl of sorghum, a bundle of firewood—and their value is low; these markets serve people who mostly live at or near the subsistence level.

EAST AFRICA

East Africa is highland, plateau Africa, mainly savanna country that turns into steppe toward the drier northeast. Great volcanic mountains rise above a plateau that is cut by the giant rift valleys. The pivotal physical feature is Lake Victoria, where the three major countries' boundaries come together (Fig. 8-11) and on whose shores lie the primary core area of Uganda and the secondary cores of both Kenya and Tanzania. With limited mineral resources (the chief ones are diamonds in Tanzania south of Lake Victoria and copper in Uganda to the west of the lake), most people in this region depend on the land—and on the water that allows crops to grow and livestock to graze. In much of East Africa, rainfall amounts are marginal or insufficient. The

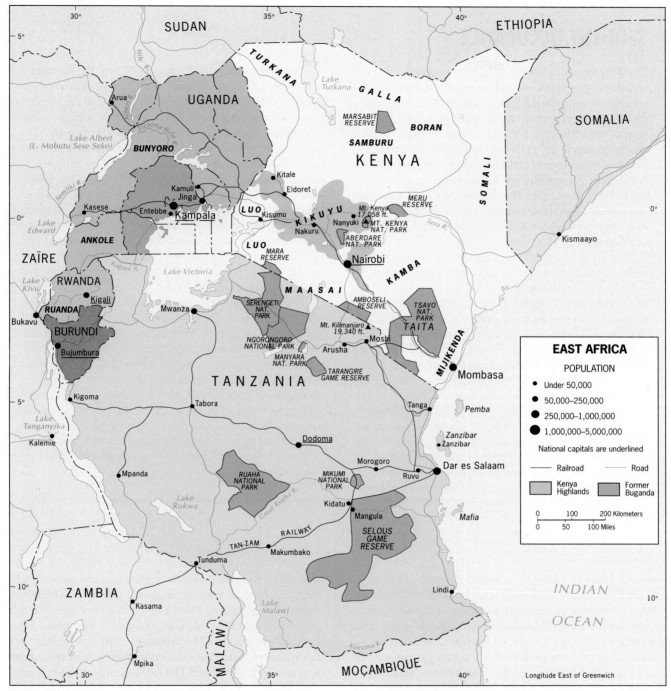

FIGURE 8-11

heart of Tanzania is dry and tsetse- and malaria-ridden, and it still suffers occasional food shortages. Eastern and northern Kenya consist largely of dry steppe country, land that is subject to frequently recurring drought. As Fig. I-6 shows, the wettest areas lie spread around Lake Victoria, and Uganda receives more rainfall than its neighbors.

Tanzania

Tanzania (a name derived from *Tan*ganyika plus *Zan*zibar) is the largest East African country. Its area exceeds that of the four other countries combined, although its population of 29.4 million is only slightly larger than Kenya's (28.1

SEQUENT OCCUPANCE

The African realm affords numerous opportunities to illustrate the geographic concept of **sequent occupance**. This concept involves the study of an area that has been inhabited—and transformed—by a succession of residents, each of whom has left a lasting cultural imprint. A place and its resources are perceived differently by peoples of different technological and other cultural traditions. These contrasting perceptions are reflected in the contents of the cultural landscape. That landscape today, therefore, is a composite of these contributions, and the challenge is to reconstruct the contributions made by each successive community.

The idea of sequent occupance is applicable in rural as well as urban areas. The ancient San peoples used the hillsides and valleys of Swaziland, for instance, to hunt and to gather roots, berries, and other edibles. Then the cattle-herding Bantu found the same slopes to be good for grazing and, in the valleys, planted corn and other food crops. Next came the Europeans, who laid out sugar plantations in the lowlands; but after using the higher slopes for grazing, they planted extensive forests, and lumbering became the major upland industry.

The Tanzanian coastal city of Dar es Salaam (1.8 million) provides a striking urban example of sequent occupance. Its Indian Ocean site was first chosen for settlement by Arabs from Zanzibar to serve as a mainland retreat. Next, it was selected by the German colonizers as a capital for their East African domain, and it was given a German layout and architectural imprint. When the Germans were ousted following their defeat in World War I, a British administration took over in "Dar," and the city entered still another period of transformation; a large Asian population soon left it with a zone of three- and four-story apartment houses that seem transplanted directly from India. Then, in the early 1960s, Dar es Salaam became the capital of newly independent Tanzania, under African control for the first time. Thus, in less than one century, Dar es Salaam experienced four quite distinct stages of cultural dominance, and each stage of the sequence remains etched into the cultural landscape. Indeed, a fifth stage has now begun, since national government functions were recently moved from Dar es Salaam to the new interior capital of Dodoma.

Because of Africa's cultural complexity and historical staging, numerous examples of this kind exist. The Dutch, British, Union, and Republican periods can be observed in the architecture and spatial structure of Cape Town and other urban places in South Africa's Cape Province; Mombasa in Kenya carries imprints from early Persian, Portuguese, Arab, British, and several African communities. In the countryside, African traditional farmlands taken over by Europeans and fenced and farmed are now being reorganized by new African proprietors, and contrasting attitudes toward soil and slope are clearly imprinted in the landscape.

---

million). Uganda, with 18.8 million, ranks next; however, the small countries of Rwanda and Burundi together total almost 15 million, so East Africa contains some very densely populated areas. Although all these states lie in the same region, they display some strong internal differences.

Tanzania has been described as a country without a primary core area because its zones of productive capacity and its population (Fig. I-9) lie dispersed—mostly near the country's margins on the east coast, near the shores of Lake Victoria in the northwest, near Lake Tanganyika in the far west, and near Lake Malawi in the interior south. (Kenya, however, has several good-quality agricultural areas and a strongly concentrated core area centered on its capital of Nairobi.) Moreover, Tanzania is a country of many ethnic groups, none with a numerical or locational advantage to enable domination of the state.

During its period of independence, Tanzania's government embarked on an experiment of "self-help" socialism that involved great dislocation: thousands of new villages were established without adequate planning. There was opposition, of course, but the new order prevailed. All this occurred while Tanzania was a haven for insurgents fighting the Portuguese in Moçambique, a difficult merger with the island of Zanzibar was accomplished, and the capital was moved from coastal Dar es Salaam (see the box entitled "Sequent Occupance") to Dodoma in the interior. Eventually, the socialist program, heralded as a model for the entire Third World, failed. In the 1990s, Tanzania is trying a new system that combines socialism with market principles.

Kenya

Kenya's different path of development is reflected by the tall buildings of Nairobi (2.1 million), the capital and primate city, and the productive farms of the nearby highlands. But Kenya's development is concentrated, not spa-

Nairobi, Kenya is East Africa's largest city. As elsewhere in Africa (and in much of the world), the downtown skyscrapers do not represent the townscape with which most inhabitants are familiar. But they do represent the new elites, the successors of the colonists who founded the city in the first place.

tially dispersed like Tanzania's, so the evidence of post-independence development is strong in the core area but very limited in the sparsely peopled interior. As Fig. 8-11 shows, both Tanzania and Kenya have a single-line railroad that traverses the entire country from the major port in the east (Mombasa [640,000] in the case of Kenya) to the far west, and feeder lines come from north and south to meet this central transport route. In Kenya, that central railroad and its branches in the highlands really represent the essence of the country, but in Tanzania the railroad lies in the "empty heart" of the country, and it is the branch lines that lead to the productive and populated peripheral zones. The Chinese-built Tan-Zam Railway was to provide Tanzania with a link to neighboring landlocked Zambia as well as access to a poorly developed domestic area with relatively good agricultural potential in the southern highlands. Unfortunately, problems of operating costs and maintenance practices sidetracked the project, and the Tan-Zam railroad has been a failure.

Kenya does not possess major known mineral deposits. Besides its coffee and tea exports, Kenya earns substantial revenues from a tourist industry that grew rapidly during the 1960s and 1970s until East African political difficulties reduced the flow of visitors. In the mid-1970s, the total number of foreign visitors to Kenya was approaching half a million, and the industry became the largest single earner of foreign exchange. Most of Kenya's famous wildlife reserves lie in the drier and more remote parts of the country, but population pressure is a growing problem even in those areas. In the mid-1990s, Kenya also suffered from political problems involving a difficult transition from one-party rule to democracy. Poaching threatened animal species, tourism dwindled, and Kenya's stability was at risk. Simultane-

ously, Kenya's rate of population growth remained among the world's highest (see Appendix A, columns 5 and 6).

Uganda

Uganda contained the most important African political entity in the region when the British entered the scene during the second half of the nineteenth century. This was the Kingdom of Buganda (shown in dark brown in Fig. 8-11), which faced the north shore of Lake Victoria, had an impressive capital at Kampala, and was stable—as well as ideally suited for indirect rule over a large hinterland. The British established their headquarters at nearby Entebbe on the lake (thus adding to the status of the kingdom) and proceeded to organize their Uganda Protectorate in accordance with the principles of indirect rule. The Baganda (the people of Buganda) became the dominant people in Uganda, and when the British departed, they bequeathed to Uganda a complicated federal system that was designed to perpetuate Baganda supremacy.

Although a **landlocked state** dependent on Kenya for an outlet to the ocean, Uganda at independence (1962) had better economic prospects than many other African countries. It was the largest producer of coffee in the British Commonwealth. It had major cotton exports and also exported tea, sugar, and other farm products. Copper was mined in the southwest; and an Asian immigrant population of about 75,000 played a leading role in the country's commerce. Nevertheless, political disaster overtook the country. Resentment at Baganda overlordship fueled revolutionary change, and a brutal dictator, Idi Amin, took control in 1971. He ousted the Asians, exterminated his opposition, and destroyed the economy. Eventually, a military campaign supported by neighboring Tanzania drove Amin from power, but by then (1979) Uganda lay in ruins. Recovery has been slow, complicated by the high cost of the AIDS epidemic, which struck Uganda with particular severity. The departure of the British colonizers set in motion a sequence of calamities from which the country may never fully recover.

EQUATORIAL AFRICA

This region, like East Africa, lies astride the equator, but in Equatorial Africa elevations are lower than in the highland east. As a geographic region (Fig. 8-12), Equatorial Africa consists of Zaïre, Congo, the Central African Republic, Gabon, Cameroon, Equatorial Guinea, and the southern reach-

FIGURE 8-12

es of Chad. (The transitional character of Chad was noted previously; see Fig. 8-9.)

Zaïre

The giant of Equatorial Africa is Zaïre, the former Belgian Congo. As Fig. 8-12 shows, Zaïre is large territorially; it contains a wide range of resources, the region's largest cities, and its greatest potential for development. But the country has been hampered by transport and communication problems, by a lack of national cohesion, by falling prices for its exports, and by autocratic and corrupt government.

For so large a country, Zaïre's population of 40 million seems modest. The country's core area lies around the capital, Kinshasa (4.2 million), and along its narrow corridor to the port of Matadi (270,000) near the Atlantic Ocean. But its economic focus has long been in the distant southeast, around the city of Lubumbashi (760,000), where most of the valuable mineral resources (chiefly copper) lie. The great Zaïre River system would seem to create a natural transport network, but a series of rapids interrupts this route, and several transshipments are necessary. Under normal circumstances, it is easier to send exports through Angola than through Zaïre itself. Internal circulation in Zaïre is plagued by breakdowns and delays because the huge forested basin that forms the center of the country remains a formidable obstacle even to modern equipment. During the independence period, Zaïre's infrastructure has virtually disintegrated. Trains no longer run, bridges have collapsed, and roads are impassable.

In the mid-1990s, Zaïre's social order also broke down. A dictatorial president confronted a hostile populace and a restive military, and riots rocked the cities. Of all of Africa's lost opportunities, Zaïre's may turn out to be the most costly—the failure of a giant at the very heart of the realm.

Lowland Neighbors

Only one of the five remaining countries of the Equatorial region is landlocked: the Central African Republic, one of Africa's least developed states. Congo, lying to the west across the river from Zaïre, has few natural resources but does enjoy some locational advantages. Brazzaville (825,000), its capital, was France's Equatorial African headquarters, and Pointe-Noire (475,000) was its major port. This gave Congo a transit function that brought in revenues, but unfortunately the country itself has little to trade. Its western neighbor, Gabon, has brighter prospects: its diversifying economic geography includes oil reserves, manganese, uranium, and iron ores as well as a productive (but destructive) logging industry. But the agricultural sector is weak, and large investments in transport systems were necessary (even though Gabon is only one-ninth the size of Zaïre).

Cameroon, north of Gabon, contains tropical forests in the south and open savannas in the north, but the real con-

A familiar scene in West and northern Equatorial Africa: longhorn cattle (here in Cameroon) being driven southward from northern pastures to southern markets.

trasts are from west to east. By Equatorial African standards, western Cameroon has progressed considerably because of some oil reserves, a diversified agricultural output, exports of forest products, and modest industrial growth. Here lie the capital, Yaoundé (1.1 million), and the port of Douala (1.3 million). By comparison, the interior east lags in isolation and remoteness.

By any ranking, Equatorial Africa as a whole remains the least developed of Subsaharan Africa's regions. Subsistence is the main activity, and economic transactions involve mostly the export of raw materials.

SOUTHERN AFRICA

Southern Africa, as a geographic region, consists of all the countries and territories lying south of Equatorial Africa's Zaïre and East Africa's Tanzania (Fig. 8-13). Thus defined, the region extends from Angola and Moçambique (on the Atlantic and Indian ocean coasts, respectively) to the turbulent Republic of South Africa, and includes a half-dozen landlocked states. Also marking the northern limit of the

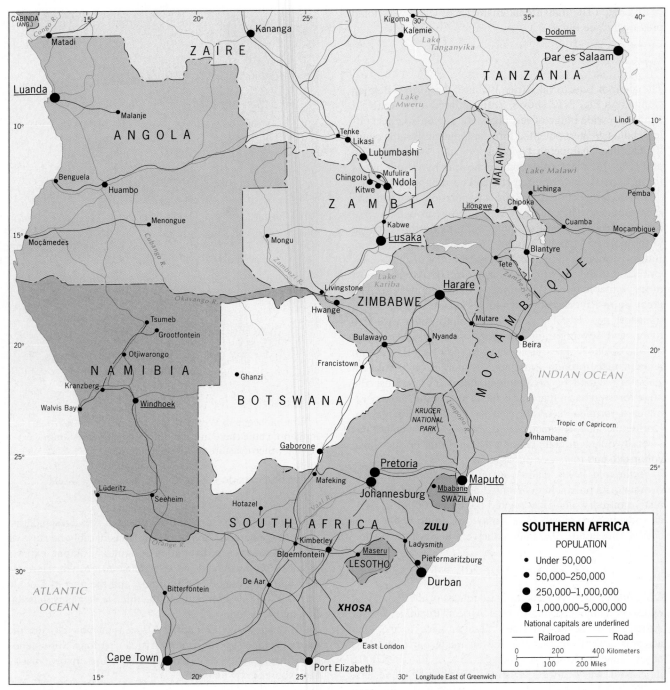

FIGURE 8-13

region are Zambia and Malawi. Zambia is nearly cut in half by a long land extension from Zaïre, and Malawi penetrates deeply into Moçambique. The colonial boundary framework, here as elsewhere, produced many liabilities.

Southern Africa constitutes a geographic region in physiographic as well as human terms. Its northern zone marks

the southern limit of the Zaïre Basin in a broad upland that stretches across Angola and into Zambia (the yellow corridor extending eastward from the Bihe Plateau in Fig. 8-2). Lake Malawi is the southernmost of the East African rift valley lakes; Southern Africa has none of East Africa's volcanic and earthquake activity. Much of the region is plateau

MADAGASCAR

Maps showing the geographic regions of the Subsaharan Africa realm often leave Madagascar unassigned. And for good reason: Madagascar differs strongly from Southern Africa, the region to which it is nearest—just 250 miles (400 km) away. It also differs from East Africa, from which it has received some of its cultural infusions. Madagascar is the world's fourth-largest island, a huge block of Africa that separated from the main landmass 80 million years ago. About 2,000 years ago, the first settlers arrived—not from Africa (although perhaps via Africa) but from Southeast Asia. Malay communities strengthened in the interior highlands of the island, which resembles Africa in a prominent eastern escarpment and a central plateau. Here was formed a powerful kingdom, the empire of the Merina. Its language, Malagasy, of Malay-Polynesian origin, became the indigenous tongue of the entire island (Fig. 8-1).

The Malay and Indonesian immigrants brought Africans to the island as wives and slaves, and from this forced immigration evolved the African component in Madagascar's population of 12.7 million. In all, nearly 20 discrete ethnic groups coexist, among which the Merina (3.7 million) and Betsimisaraka (1.9 million) are the most numerous. Like mainland Africa, Madagascar experienced colonial invasion and competition. Portuguese, British, and French colonists appeared after 1500, but the Merina were well organized and long resisted the colonial conquest. Eventually, Madagascar became part of France's colonial empire, and French became the *lingua franca* of the educated elites.

As would be expected, the staple food of Madagascar is rice, not corn. Export crops include coffee, vanilla, and cloves. There is mining of chromite and graphite, and Madagascar has potential oil reserves in coastal and offshore areas. But the period since independence in 1960 has been difficult. Once a rice exporter, the island now requires imports of that staple as population growth increased and agricultural efficiency declined. The flow of exports also shrank as political turmoil engulfed Madagascar. For some years named the Malagasy Republic, Madagascar witnessed authoritarian government, repression, rebellion, and instability. In 1975, following a military coup, a new government was elected and a new constitution written. The country regained its official name: Democratic Republic of Madagascar. During the 1960s and early 1970s, Madagascar was an isolated country, almost completely closed to outsiders. But the 1980s introduced a new period of international ties and assistance, research, and reconstruction, and 1993 witnessed a peaceful presidential election.

Madagascar's capital, Antananarivo (955,000), is the country's primate city, its architecture and atmosphere combining traces of Asia and Africa. The countryside, too, is reminiscent of Southeast Asian villages and terraced farms on an African physical landscape. And in what remains of Madagascar's forests are the remnants of one of the world's most unusual fauna, now threatened with extinction by the rising tide of humanity.

country, and the Great Escarpment is much in evidence here. There are two pivotal river systems: the Zambezi (which forms the border between Zambia and Zimbabwe) and the Orange-Vaal (South African rivers that combine to demarcate southern Namibia from South Africa).

The social, economic, and political geographies of Southern Africa also confirm the regional definition. Landlocked Zambia, whose economic core area is the Copperbelt northwest of Ndola, always looked southward for its outlets to the sea, its electric power, and its fuels. Malawi's core area lies in the south of the country, and its outlets, too, are southward. Moçambique, with its lengthy Indian Ocean coastline, has served as an exit not only for Malawi but for Zimbabwe (through the port of Beira) and South Africa (through Maputo) as well. Offshore Madagascar, however, remains a separate entity (see box above).

Southern Africa is the continent's richest region in material terms. A great zone of mineral deposits extends through the heart of the region from Zambia's Copperbelt through Zimbabwe's Great Dyke and South Africa's Bushveld Basin and Witwatersrand, to the goldfields and diamond mines of the Orange Free State and northern Cape Province in the heart of South Africa. Ever since colonial exploitation of these minerals began, large numbers of migrant laborers have come to work in the mines (Fig. 8-8). The range and volume of minerals mined in this belt are enormous, from the copper of Zambia and the chrome and asbestos of Zimbabwe to the gold, chromium, diamonds, platinum, coal, and iron ore of South Africa. However, not all of Southern Africa's mineral deposits lie in this central backbone. There is coal in western Zimbabwe at Hwange and in central Moçambique near Tete. In Angola, petroleum

An explosion rocks the walls of an open-pit copper mine at Chingola in Zambia's Copperbelt. The decline in world prices of Zambia's mineral exports produced an economic crisis from which the country has not recovered.

from oilfields along the north coast heads the export list; diamonds are mined in the northeast, manganese and iron on the central plateau. Namibia produces copper, lead, and zinc from a major mining complex anchored to the town of Tsumeb in the north, and diamond deposits are worked along the beaches facing the Atlantic Ocean in the south.

This is a mere summary of Southern Africa's mineral wealth, and it is matched by the variety of crops cultivated in the region. Vineyards drape the slopes of the Cape Ranges in South Africa; apple orchards, citrus groves, banana plantations, and fields of sugarcane, pineapples, and cotton reflect South Africa's diverse natural environments. In Zimbabwe, tobacco has long been the leading commercial crop, but cotton also grows on the plateau, and tea thrives along the eastern escarpment slopes. Angola is a major coffee producer. The staple crop for the majority of Southern Africa's farmers is corn (maize), but wheat and other cereals are also grown. Even the pastoral industry is marked by variety: not only are dairying and beef cattle herds large, but South Africa also exports sizeable quantities of wool.

Despite this considerable wealth and productivity, the countries of Southern Africa are not well off. Only the most advantaged parts of South Africa can be described as developed; but severely underdeveloped areas exist even within that country. As elsewhere in Africa, political crises and associated armed conflicts have devastated entire states.

The Northern Tier

Within the four countries that extend across the northern part of the region—Angola, Zambia, Malawi, and Moçambique—problems abound. Angola (population 9.4 million), formerly a Portuguese dependency, had a thriving economy at independence but was engulfed by civil strife that was worsened by outside involvement. Its government initially chose a Marxist-communist course, and when a rebel movement emerged in the south, the South African government supported it financially and militarily. The Angolan leaders called on Moscow for help, and thousands of Cuban troops arrived in the capital of Luanda (2.2 million) to help combat the southern insurgents. In addition, political conflict in neighboring Namibia spilled over into Angola as opponents of the South African administration of Namibia sought refuge there.

The impact of all this on the Angolan economy was devastating. Farms were abandoned, railroads mined, port facilities damaged, and funds diverted to military use. As Fig. 8-13 shows, a small part of Angola lies separated from the main territory along the coast to the north. This small exclave, Cabinda, contains oil reserves that continued to produce revenues for the Luanda government through the post-independence crisis—or Angola would have been totally bankrupted by the war. In 1993, hopes that a U.N.-brokered peace and supervised elections would end Angola's crisis were dashed when fighting resumed.

On the opposite coast, Moçambique (17.5 million) was even less fortunate. Also a former Portuguese domain that chose a Marxist course for development, Moçambique had far fewer resources than Angola. Before independence, its chief sources of income came from its cashew and coconut plantations and from its relative location. As the map shows, Moçambique's port of Beira (370,000) is ideally situated to handle the external trade of Zimbabwe and southern Malawi, and its capital, Maputo (2.2 million), is the closest port to South Africa's Witwatersrand mining and industrial complex. Goods streamed through these ports, and when hydroelectric facilities on the Zambezi River (the Cabora Bassa Dam) neared completion, prospects seemed fair. Today, however, Moçambique is often identified as one of the world's most severely underdeveloped countries, subject to famines, social dislocation, political disorder, and economic chaos. Again, a rebel movement, supported for a time by South Africa, contributed to this crisis. The hydroelectric project was damaged; the port facilities at Beira ceased functioning; Maputo's transit role all but ceased. The plantations lay untended, and the tourist industry, once a promising sector, was destroyed. Moçambique by the early 1990s had changed its political course, and an accommodation with the South Africans was reached. But it will take generations for the country to climb from the depths of impoverishment to which it has been condemned.

Between Angola and Moçambique lie three landlocked states that formed part of the British colonial empire: Zambia, Malawi, and Zimbabwe. Zambia (9.0 million) and Malawi (9.3 million) are less developed than Zimbabwe, although Zambia contains the mineral-rich Copperbelt. The decline of world prices for its minerals, plus the problems and costs associated with their long-distance transportation, have hurt the economy of Zambia. Malawi's economic geography is almost completely agricultural, with a variety of crops that include tea, cotton, tobacco, and peanuts; this has helped cushion the economy against market swings and has also involved a larger percentage of the labor force in productive work. Still, more workers labor in the mines of other countries than are gainfully employed within Malawi itself.

Southern States

Six countries constitute Africa's southernmost tier of states and form a distinct subregion within Southern Africa: Zimbabwe, Namibia, Botswana, Swaziland, Lesotho, and the Republic of South Africa. As Fig. 8-13 shows, four of the six are landlocked. Botswana occupies the heart of the Kalahari Desert and surrounding steppe; only 1.4 million people inhabit this Texas-sized country. Lesotho (2.0 million) is encircled by South African territory, and Swaziland (900,000), ancestral home of the Swazi nation, is largely enclosed by the Republic as well. All three of these countries depend heavily on the income earned by workers who are employed by South African mines, factories, and farms.

Zimbabwe Zimbabwe, located directly north of South Africa, is an important African state of 11 million people. When the British moved this country toward sovereignty, the local white minority declared independence from London in 1965, hoping to retain control. A bitter civil war ensued, which resulted in a settlement and black majority rule in 1980. After independence, friction between the Shona majority and the Ndebele minority clouded the future, but the majority's goal of a one-party state was achieved.

Zimbabwe is a well-defined country. To the north lie the Zambezi River and the giant Kariba Dam. To the east the plateau descends to Moçambique's coastal plain; here, local soils and climates vary, and a range of crops is grown. To the west lie Botswana and the Kalahari's arid plains. And to the south, beyond the Limpopo River, lies South Africa, now facing the struggle from which Zimbabwe is emerging.

The core area is defined by the mineral-rich Great Dyke and its environs, extending southwest from the vicinity of the capital, Harare (1.1 million), to the country's second city, Bulawayo (590,000). Copper, asbestos, and chromium are among the major mineral exports, but Zimbabwe is not just an ore-exporting country. Farms produce tobacco, cotton, tea, sugar, and other crops (corn is the local staple), and many whites stayed on their farms after independence, encouraged by the government to do so. Coal is mined in the far west of the country and hauled by rail to an industrial complex located about midway between Harare and Bulawayo.

Although Zimbabwe still confronts serious environmental, social, economic, and political problems, in virtually every respect it is ahead of its neighbors in the region (except South Africa). The political compromise, the mixed economy, the retention of a productive segment of the white minority, and an avoidance of involvement in neighbors' conflicts have served Zimbabwe well amid the turbulence of a changing region. It was Zimbabwe's great misfortune to find itself, during the decade from 1983 to 1993, at the heart of a zone struck by the worst drought in its recorded history. Livestock died, farms lay abandoned, and, inevitably, social and political tensions rose. Yet Zimbabwe survived without the regional conflicts that tormented other countries in the grip of nature's wrath.

Namibia Namibia became an independent African state in 1990, after more than 70 years of South African administration and two decades of active black opposition. As Fig. 8-13 shows, Namibia is bounded to the south by the Orange River and to the north and east by mainly geometric (straight-line) boundaries. A finger-like corridor of Namibian territory extends in the northeast between Angola and Botswana to Zambia.

Namibia is named after a desert, and appropriately so. The Kalahari and Namib deserts dominate the physiography; only in the far north do conditions moderate somewhat. This huge country (about as large as Texas and Oklahoma combined) is inhabited by only 1.5 million people, the majority of them concentrated in the north against the Angolan border. The centrally located capital, Windhoek (185,000), lies far from this population concentration. The major port, Walvis Bay, actually lay in an exclave of South African territory that did not become part of Namibia until 1994. Much of Namibia consists of vast ranches owned by white farmers; the enormous expanses of sheep pastures in the south stand in stark contrast to the village-settled, small-farm-dominated north. Between these areas lies Namibia's mineral belt, southwest of Tsumeb. From mines in this zone, the country exports copper, lead, zinc, and smaller amounts of other minerals. The most valuable minerals lie elsewhere, however—along the south coast, where diamonds abound.

Given these assets, Namibia's future would appear bright. But much of the country's productive capacity remains in foreign hands, and many mines would close if wages paid to workers from the north were to be raised substantially. Despite its small population, Namibia needs food imports to supplement what is produced locally. The country's economy, at independence, was tightly linked to South Africa's, and this situation cannot be changed by political action except at the risk of disaster. For Namibia, the long haul toward true sovereignty has just begun.

The giant of Southern Africa is the Republic of South Africa, the dominant force in the region, now facing the challenge of change. From Namibia to Angola to Moçambique, the republic has influenced and sometimes controlled the sequence of events—even after independence came to these neighbors. Now, however, South Africa itself confronts the prospect of a new order, and its energies will go toward channeling that transition rather than toward foreign adventures. We discuss this unique African state in the separate vignette that follows.

THE REALM IN TRANSITION

1. Although parts of Europe are engulfed by tribal warfare, many African countries—economically poorer and culturally more diverse—remain stable, an accomplishment to be noted.
2. The spread of AIDS continues to afflict tropical African countries, with devastating social and economic consequences.
3. The difficult and strife-ridden transformation of South Africa has the potential to affect race relations in many other parts of Africa and the world, including the United States.
4. After decades of comparative stability in postcolonial Africa's boundary framework, significant changes now loom. Eritrea's independence and Somaliland's attempted secession are cases in point.
5. Although global attention focuses on the destruction of large tracts of the Amazonian rainforest, Africa's remaining forests, too, are retreating under the logger's saw. Many species of animals and plants are being lost.

◆ PRONUNCIATION GUIDE

Abidjan   (abb-ih-JAHN)
Abuja   (uh-BOO-juh)
Accra   (uh-KRAH)
Adamawa   (add-uh-MAH-wuh)
Afrikaans   (uff-rih-KAHNZ)
Ahaggar   (uh-HAH-gahr)
Amin, Idi   (uh-MEEN, iddy)
Antananarivo   (anta-nana-REE-voh)
Baganda   (bah-GAHN-duh)
Bantu   (ban-TOO)
Beira   (BAY-ruh)
Benin   (beh-NEEN)
Benue   (BANE-way)
Betsimisaraka   (bet-simmy-sah-
   RAH-kah)
Biafra   (bee-AH-fruh)
Bihe   (bee-HAY)
Bilharzia   (bill-HARZEE-uh)
Bioko   (bee-OH-koh)
Boer   (BOOR)
Botswana   (bah-TSWAHN-uh)
Brazzaville   (BRAHZ-uh-veel)
Buganda   (boo-GAHN-duh)
Bulawayo   (boo-luh-WAY-oh)
Burkina Faso   (burr-keena FAHSSO)
Burundi   (buh-ROON-dee)
Cabinda   (kuh-BIN-duh)
Cabora Bassa   (kuh-boar-rah
   BAHSSA)
Chingola   (ching-GOH-luh)
Dahomey   (dah-HOH-mee)
Dakar   (duh-KAHR)
Dar es Salaam   (dahr ess suh-
   LAHM)
de Brazza, Pierre   (duh-BRAH-zah,
   PYAIR)
Deccan   (DECKEN)
Dhow   (DOW)
Djibouti   (juh-BOODY)
Djouf   (JOOF)
Dodoma   (DOH-duh-mah)
Douala   (doo-AHLA)
Drakensberg   (DRAHK-unz-berg)
Entebbe   (en-TEBBA)
Ethiopia   (eeth-ee-OH-pea-uh)
Fernando Póo   (fer-nahn-doh POH)
Fulani   (foo-LAH-nee)
Futa Jallon   (food-uh juh-LOAN)
Gabon   (gah-BAW)
Gao   (GHAU)
Ghana   (GAH-nuh)

Ghanaian   (gah-NAY-un)
Gondwana   (gond-WON-uh)
Guinea   (GHINNY)
Guinea-Bissau   (ghinny-bih-SAU)
Harare   (huh-RAH-ray)
Hausa   (HOW-sah)
Houphouët-Boigny, Félix   (ooh-
   FWAY bwah-NYEE, fay-LEEKS)
Hwange   (WAHNG-ghee)
Ibadan   (ee-BAHD'N)
Ibo [land]   (EE-boh [land])
Kalahari   (kallah-HAH-ree)
Kampala   (kahm-PAH-luh)
Kano   (KAH-noh)
Kariba   (kuh-REE-buh)
Katanga   (kuh-TAHNG-guh)
Kenya   (KEN-yuh)
Khoikhoi   (KHOY-khoy)
Khoisan   (khoy-SAHN)
Kibo   (KEE-boh)
Kigali   (kih-GAH-lee)
Kilimanjaro   (kil-uh-mun-JAH-roh)
Kinshasa   (kin-SHAH-suh)
Kwashiorkor   (kwuh-shee-OAR-koar)
Lagos   (LAY-gohss)
Lesotho   (leh-SOO-too)
Liberia   (lye-BEERY-uh)
Limpopo   (lim-POH-poh)
Lingua franca   (LEAN-gwuh
   FRUNK-uh)
Lualaba   (loo-uh-LAH-buh)
Luanda   (loo-AN-duh)
Lubumbashi   (loo-boom-BAH-shee)
Maasai   (muh-SYE)
Madagascar   (madda-GAS-kuh)
Maize   (MAYZ)
Malagasy   (malla-GASSY)
Malawi   (muh-LAH-wee)
Malay   (muh-LAY)
Mali   (MAH-lee)
Maputo   (mah-POOH-toh)
Marasmus   (muh-RAZZ-muss)
Matadi   (muh-TAH-dee)
Mauritania   (maw-ruh-TAY-nee-uh)
Merina   (meh-REE-nuh)
Moçambique   (moh-sum-BEAK)
Mombasa   (mahm-BAHSSA)
Nairobi   (nye-ROH-bee)
Namib   (nah-MEEB)
Namibia   (nuh-MIBBY-uh)
Ndebele   (en-duh-BEH-leh)

Ndola   (en-DOH-luh)
Nkrumah, Kwame   (en-KROO-muh,
   KWAH-mee)
Niger [Country]   (nee-ZHAIR)
Niger [River]   (NYE-jer)
Nigeria   (nye-JEERY-uh)
Nyasaland   (nye-ASSA-land)
Olduvai   (OLE-duh-way)
Okovango   (oh-kuh-VAHNG-goh)
Pangaea   (pan-GAY-uh)
Pointe-Noire   (pwahnt-nuh-WAHR)
Polygamy   (puh-LIG-ahmee)
Rhodesia   (roh-DEE-zhuh)
Rinderpest   (RIN-duh-pest)
Rio Muni   (ree-oh-MOOH-nee)
Rwanda   (roo-AHN-duh)
Sahara   (suh-HARRA)
Sahel   (suh-HELL)
San   (SAHN)
Schistosomiasis   (shistoh-soh-MYE-
   uh-siss)
Senegal   (sen-ih-GAWL)
Shari   (SHAH-ree)
Shona   (SHOH-nuh)
Sisal   (SYE-sull)
Songhai   (SAWNG-hye)
Soudan   (soo-DAH)
Sudd   (SOOD)*
Swazi [land]   (SWAH-zee [land])
Tanganyika   (tan-gun-YEEKA)
Tanzania   (tan-zuh-NEE-uh)
Tema   (TAY-muh)
Tete   (TATE-uh)
Timbuktu   (tim-buck-TOO)
Tsetse   (TSETT-see)
Tsumeb   (SOO-meb)
Turkana   (ter-KANNA)
Uganda   (yoo-GAHN-duh/yoo-
   GANDA)
Vaal   (VAHL)
Wegener   (VAY-ghenner)
Windhoek   (VINT-hook)
Witwatersrand   (WITT-waw-terz-rand)
Yamoussoukro   (yahm-uh-SOO-kroh)
Yaoundé   (yown-DAY)
Yoruba   (YAH-rooba)
Zaïre   (zah-EAR)
Zambezi   (zam-BEEZY)
Zimbabwe   (zim-BAHB-way)

*Double "o" pronounced as in "book."

SOUTH AFRICA:
Crossing the Rubicon

IDEAS & CONCEPTS

Plural society (2)
Core-periphery relationships
Apartheid
Separate development

Infrastructure (2)
Domestic colonialism
Regional polarization

The Republic of South Africa occupies the southernmost portion of the African continent, but it lies at the center of world attention. Long in the grip of one of the world's most notorious racial policies (*apartheid* and its derivative, "separate development"), South Africa today is a country in transition. The framework of apartheid, the racial separation of South Africa's peoples, is being demolished, and the grand design of "separate development"—the creation of a mosaic of ethnic states within the republic—is being abandoned. Although the dimensions of change may not be as great as those affecting the former Soviet Union and Eastern Europe, its significance is no less. And the challenge facing South Africa may be the greatest of all.

South Africa stretches from the warm subtropics in the north to Antarctic-chilled waters in the south. With a land area in excess of 470,000 square miles (1.2 million sq km) and a heterogeneous population of 43.9 million, South Africa

Central Johannesburg lies at the heart of South Africa's second-largest metropolis, which, in turn, forms the core of the Witwatersrand conurbation. Mineral wealth built much of the CBD, and Johannesburg remains one of the Southern Hemisphere's chief financial centers.

TEN MAJOR GEOGRAPHIC QUALITIES OF SOUTH AFRICA

1. South Africa's relative location at the southern tip of the African continent assigns the republic considerable strategic importance.
2. South Africa is Africa's only truly temperate-zone country. Latitude and elevation combine to produce a range of natural environments unmatched on the continent.
3. South Africa contains a wide range of minerals, some of them strategically important. The country is rich in coal, but it has no significant petroleum reserves.
4. The historical geography of South Africa involves the in-migration of populations from other parts of Africa, from Europe, and from Asia.
5. South Africa is Africa's sole remaining state in which minority rule is exercised by people of European descent. The country's plural society consists of four major cultural components, each itself divided.
6. South Africa's cultural pluralism is strengthened by its regional expression. A mosaic of "traditional" regions marks the republic.
7. South Africa's white population is larger than the white-settler populations of all other Subsaharan African countries combined, even during the height of the colonial era.
8. South Africa's political geography has been dominated by apartheid and its regional expression, "separate development." Although the white-minority government now seeks to retreat from these premises, the system is deeply entrenched.
9. Political and politico-geographical practices by the South African government have caused many countries to cut their economic ties with the republic, isolating South Africa from much of the world.
10. Although South Africa is often identified as Africa's only developed country, the republic consists of juxtaposed, sharply contrasting developed and underdeveloped regions.

swana work in South Africa's mines, factories, and fields. South African armed forces have intervened in external civil wars even as South African grain went to feed its neighbors.

With such power and influence, South Africa would seem to be the cornerstone of what we have defined as the Subsaharan Africa realm. But this is not the case. South African strength has been built through a system of domination by the white minority over the black majority. This system has concentrated most wealth and virtually all power in the hands of the white minority. Thus South Africa's policies—political, economic, social—have not been policies of consensus. They have been, as apartheid was, policies of the minority. This is now changing, and a new map of South Africa is taking shape.

◆ HISTORICAL GEOGRAPHY

As Fig. 8-7 reminds us, South Africa lay shielded from advancing decolonization by a *buffer zone* (see the 1970 map, bottom left) until the 1970s. Once African independence reached South Africa's borders in Moçambique and Zimbabwe, and when even Namibia (long ruled by South Africa) became independent, South Africa's geopolitical situation had changed drastically. The republic's comparatively comfortable isolation had come to an end.

South Africa is a **plural society** with the greatest ethnic complexity of all the African countries. Its peoples came not only from other parts of the continent but from Europe and Asia as well. The oldest inhabitants, the San, who speak the "click" language shown as Khoisan on Fig. 8-1, were being overpowered and enslaved by the majority peoples, the Bantu, even before the Europeans arrived to colonize the area. As European settlement developed at the Cape and Cape Town grew into a substantial port town, a major African empire—one of the most powerful ever to emerge in the realm—arose with its core and headquarters in the present-day province of Natal (Fig. SA-1). This empire, the kingdom of the Zulu (see Fig. 8-13), long battled its European enemies but eventually was defeated. The Zulu nation, however, remains a strong force in South Africa, and Zulu leaders know that but for the white invasion, they might well be South Africa's rulers today.

To complicate matters further, the first white settlers in South Africa, the Dutch, who founded Cape Town in 1652, were followed shortly after 1800 by the British. Eventually, the two European communities went to war against each other, and the Boer War (1899–1902) was won by the British. But the *Boers*, descendants of the Dutch settlers, negotiated a favorable settlement, out of which came the Union of South Africa (1910). In the decades that followed, the Boers (who came to be known as *Afrikaners*) steadily gained numerical and political strength. In 1948, they were able to defeat the British in national (white) elections, and in 1961 they proclaimed the Republic of South Africa (still

is the dominant state in Southern Africa. It contains the bulk of the region's minerals, the majority of its good farmlands, its largest cities, best ports, most productive factories, and most developed transport networks. Mineral exports from Zambia and Zimbabwe move through South African ports. Workers from as far away as Malawi and as nearby as Bot-

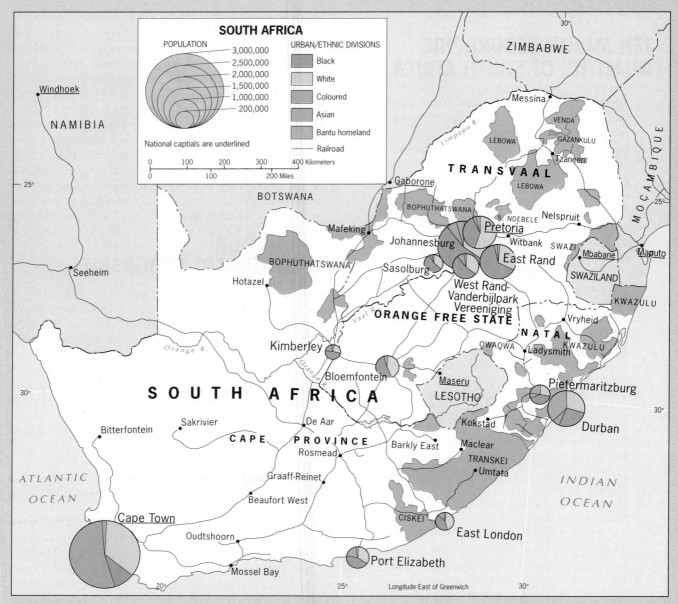

FIGURE SA-1

the country's official name today) and left the British Commonwealth.

Afrikaners, British, Zulu, and other black peoples are not the only players in the social geography of South Africa's drama. At the Cape, from the earliest years of white settlement there was much intermarriage, and a population cluster of mixed ancestry evolved. Today, these people form the majority population in Cape Town and in a large portion of the city's hinterland. They are known in South Africa as the *Coloured* people, and today they number about 3.5 million (8 percent of the country's population). A majority of these people speak the Afrikaner language and attend Afrikaner churches.

The British, following their penetration of South Africa, also changed the demographic mosaic, but not by inter-

marriage. Needing labor for their sugar plantations in Natal, where the defeated Zulu refused to work, the British during the nineteenth century began to bring to South Africa indentured laborers from a distant colony. South Asians (or *Indians*, as they are called in South Africa) came by the thousands, and most stayed after their contracts were up. Today there are more than 1 million Indians in the republic, heavily concentrated in the city of Durban (which is about one-third Indian) and Natal Province.

The ethnic diversity of South Africa is illustrated by Fig. SA-1. The largest cluster of urbanized population lies in the southern Transvaal, heart of the country's core area. All the cities in this cluster, including the country's second largest, Johannesburg (1.9 million), and the administrative capital, Pretoria (1.1 million), have black African majorities and large white minorities, with very small Coloured and Asian sectors. Compare this to Durban (1.2 million) on the east coast of Natal, where African, Asian, and white sectors each form about one-third of the population, with a very small Coloured minority. Next, look at the pie chart for the country's biggest city, Cape Town (2.7 million): here the Coloured people are in the majority, the white sector is quite large, and the black African minority is rather small. Now compare the proportion of black Africans in Cape Town with that of Port Elizabeth (744,000), due east along the coast; in Port Elizabeth, the black Africans form the largest population sector, and the Coloured minority is smaller even than the white sector. South Africa's metropolises, depending on their location, differ enormously.

South Africa, therefore, is Africa's most pluralistic and heterogeneous society. Of its 44 million people, approximately 34 million are black Africans, representing not only the great Zulu nation but also the Xhosa, Sotho, and other major peoples. About 5.3 million are white, of European ancestry; a majority of them are Afrikaners, and these are the people who hold political and economic power. Even the Coloured (3.5 million) and Asian (1.1 million) communities are fragmented by ethnic background, language, and religion. Geography and history seem to have conspired to divide, not unite, here at the southern tip of the continent.

◆ PHYSICAL GEOGRAPHY

To gain some understanding of the complexities of South Africa's problems, the physical geography is important. In South African historical geography, landform and landscape have played significant roles. The cradle of the modern state lies where two oceans meet, in the shadow of Table Mountain (see the photo on p. 429). White settlement expanded into the interior, channeled by the valleys that separate the Cape Ranges, of which Table Mountain is a part. Here, the seventeenth-century Europeans first contested the land with the San peoples. The southward migration of the Bantu had not yet carried the San's other enemies into this part of what is now Cape Province. Confrontation between white and Bantu was years away.

The Cape Ranges dominate the southern extremity of Africa, but the heart of South Africa is not mountainous; it is a plateau, an elevated flatland with some hilly topograpy. Much of this plateau is covered by tall natural grass, giving it a prairie-like appearance; in South Africa this is referred to as the *veld*, a Dutch (also Afrikaans) word meaning "field." The plateau is rimmed by the sometimes sheer, sometimes step-like Great Escarpment (often thousands of feet high), which separates a fairly narrow coastal strip from the extensive veld above. The Great Escarpment, too, has played its role in the country's historical geography. The Zulu Empire emerged in the hilly land between scarp and sea, where the Zulu parlayed security into

organization and power. The Dutch (Boers), when they embarked on their penetration of the interior, had much trouble ascending the rocky walls to the grasslands above. Even in modern South Africa, a map of surface routes reveals where the escarpment lies. It was and is a barrier, and it separates contrasting environments.

In the interior, the plateau veld slopes and rolls in giant waves of landscape. Much of the surface lies above 5,000 feet (1,500 m), and this is the *highveld* with its cold winters and cool, quite untropical summers. In lower areas lies the *middleveld*, as in the basins of major streams; below the escarpment is the *lowveld*, where South Africa takes on more tropical-African qualities. Onto the highveld and middleveld swept the Zulu armies more than 150 years ago, defeating weaker enemies and absorbing lands and peoples. In those days the whites were not yet a factor, and a Zulu era seemed about to unify the region. But the Boers, seeking escape from their British competitors, entered the arena, and the inevitable war ensued. By the late 1830s, Zulu power was broken. The Boers controlled the highveld, and the British held the coast inland to the Great Escarpment.

South Africa's great rivers also figured in the human drama, as Fig. SA-1 confirms. In the high Drakensberg (Dragons Mountains) rises the Orange River, tinted ochre by the local soils. The Orange crosses the country in its westward flow to the Atlantic, forming the boundary between two provinces (Cape and Orange Free State) and, in its lower reaches, the international boundary with Namibia. It is joined by the Vaal (Gray) River, across which lies— not surprisingly—the province of Transvaal. That province, in turn, is bounded by the Limpopo River, the boundary between South Africa and Botswana as well as Zimbabwe. These rivers and their tributaries did more than supply water in an area of chronic water shortage. From the volcanic basalts of the Drakensberg, the Orange carried fragments of rock downstream, including diamonds. Where the Orange and Vaal unite lies one of the world's great diamond accumulations, and nearby arose a city whose name, a century ago, was synonymous with gemstones—Kimberley.

◆ ECONOMIC GEOGRAPHY

The discovery of diamonds in the late 1860s opened a new age in the country's development. At that time, the British held sway over the Cape (below the Great Escarpment) and Natal (also below the plateau wall). The Boers had established two independent republics on the highveld. In certain areas, such as in present-day Lesotho and Swaziland, African peoples still held their lands.

Certainly, minerals had been found prior to the diamonds near the Orange-Vaal confluence; in fact, the Bantu mined copper and other ores centuries earlier. But the diamond finds set into motion a new economic geography. Fortune hunters, capitalists, and black workers came to Kimberley by the thousands. Rail lines were laid from the coast to this growing new market. Equipment was needed; land was surveyed; goods never before seen in South Africa passed through the port of Cape Town. The diamonds also brought new frictions. Until the 1860s, both the colonists and the colonized had been rural people—farmers, pastoralists. Settlements (even Cape Town and Durban) had remained comparatively modest. But now began an age of mining and industry, of machinery and labor. For the Boers, it was the beginning of the end of their republics, and the end of their isolation.

Just 25 years later, what remained of Boer isolation and aloofness ended finally and totally. In what is today Johannesburg, diggers struck what was to be, for quite some time, the world's greatest goldfield. A new and even larger stream of foreigners and their money came, now to the very heart of the Boers' South African republic. Just 30 miles (50 km) from Pretoria, the Boer capital, lay the mushrooming city of Johannesburg. The Boers tried to keep control—after all, this was their country—but the invaders paid little attention. Conflict was inevitable, and before century's end, Boer and Briton were at war.

It was a war that soon ended in Boer defeat and the exile of their president, Paul Kruger. It also was a war of which much of the world disapproved. The British Empire was seen as a powerful bully, imposing its will on the underdog Boers. The Boers knew, however, that the battle was theirs to fight alone, as had been the case earlier against the Zulu. They had no European "homeland" to turn to; they had no allies; they were not colonizers who could abandon the conflict and withdraw from Africa. Their Afrikaner successors today feel much the same way.

Through this era of economic growth and accompanying conflict, South Africa's potential became clear. Its varied climates and soils yielded a wide range of grains, vegetables, and fruits. In addition to diamonds and gold, there were numerous other minerals; from the central Transvaal to the central Orange Free State lay a series of great deposits. Cheap labor was plentiful. When the British and Boers negotiated their Boer War settlement, neither side took the interests of the Africans into account. Both saw the economic potentials.

During the twentieth century, South Africa proved to be even richer than had been foreseen. Additional goldfields were discovered in the Orange Free State. Coal was found to be present in abundance, and so was iron ore; this gave rise to a major iron and steel industry. Other metallic minerals—chromium and platinum among them—yielded large revenues on world markets. Asbestos, manganese, copper, nickel, antimony, and tin were mined and sold; a thriving metallurgical industry developed in South Africa itself. Capital flowed into the country, white immigration continued, cities and industries mushroomed, and the Union prospered.

In agriculture, too, South Africa's capacities were unmatched elsewhere in Subsaharan Africa. Tropical fruits became Natal's specialties; other fruits ranging from apples to grapes stood in large orchards at the Cape. Great fields of wheat draped the Cape elsewhere; on the highveld there developed the Maize Triangle, South Africa's Corn Belt. In drier areas of the eastern Cape, thousands of sheep made South Africa one of the world's leading wool producers. Locally produced raw materials from cotton to plastics formed the basis of an expanding industrial complex. South Africa developed Africa's only true networks of internal transportation (as opposed to export routes). Its ports even handled goods for neighbors and others as far away as Zaïre and Malawi. When World War II cut off the country from its overseas trading partners, there were economic problems, but local industries thrived, exempt from outside competition and benefiting from a large and comparatively wealthy domestic market.

◆ POLITICAL GEOGRAPHY

Economic success carried the seeds of political problems. South Africa by now had developed **core-periphery relationships**. The great cities, bustling industries, mechanized farms, and huge ranches had begun to resemble a developed country, much like Western Europe or the United States. But beyond the skyscrapers and white-owned corridors lay quite another South Africa, where conditions were more like those in rural Tanzania or Zambia. From the villages in this periphery came the workers, the cheap labor that kept mines, factories, and commercial farms going—and kept profits high. Many of those workers settled on the edges of the cities where the industries and mines were located, so that South African cities had their downtowns, their white suburbs, and their black "townships" and squatter settlements. (One of these townships outside Johannesburg is well-known *Soweto*, short for *South Western Township*.) Thus there was racial segregation in the Union long before the 1940s—as was the case in Africa's other colonies as well.

But there also was flexibility. In and around Cape Town, the Coloured community had some representation in local (and, through white representatives, even in State) government. In Durban there had been some progress toward racial accommodation between Asians and other residents. Then in 1948 came a momentous political change, one that was to lead to the politico-geographical transformation of the country. In that year the Afrikaners defeated the English-speaking people and their party in the (whites-only) election of the national government. Elected as prime minister (like the U.S. president, the most powerful position) was an Afrikaner named Daniel François Malan. His platform had been a one-word solution to the problems of multiracial South Africa—**apartheid**.

Apartheid (an Afrikaans word meaning, literally, apartness or separation) was not just an abstract theory that would

For millions of black South Africans, this is the scene of daily life: barren terrain, ramshackle housing, inadequate (if any) services. Many such settlements have sprung up near the urban centers, and are now growing rapidly.

be softened by the checks and balances of government. Now that the Afrikaners, nearly 50 years after their Boer War defeat, had control of their country again, they set about transforming it. Just when the rest of the world was examining the fairness of social doctrines, South Africa embarked on a path of opposite direction. "Europeans (Whites) Only" signs appeared where they had not been before. Buses and trains, long informally and habitually segregated, now were segregated as a matter of law. Social services, education, and other institutions were more rigorously segregated than before. Small residential areas near inner cities, where some African, Asian, or Coloured housing remained, were evacuated and the land sold to white investors.

Apartheid, in theory, was not simply the separating of black from white. It was also the separating of Coloured from white, Asian from Zulu. Its chief objective, pursued through a set of strict new laws, was to minimize contact among the many ethnic groups and cultural communities in South Africa. It applied even within the white community. So-called "parallel-medium" schools, where classes were taught in both English and Afrikaans, were divided according to the same principle. White children who spoke Afrikaans would henceforth go to an Afrikaans-language-only school; white children who spoke English went to their own school—elsewhere. Behind apartheid of this kind lay a wider vision, a grand design of a South Africa that would consist of a mosaic of racially based states, where each ethnic or cultural community would have its own "homeland." Again, this was not just theory. During the 1950s, the government set this plan into motion, and the map of South Africa was transformed as these "homelands" began to appear on it (the bluish areas in Fig. SA-1). Apartheid had now become **separate development**.

Opposition to the *Bantustan* program, as separate development was called, mounted inside as well as outside South Africa. Apartheid itself had already elicited passive as well as violent opposition, and as early as the 1960s, death, imprisonment, and exile were imposed on anti-apartheid activists. (One of those imprisoned was Nelson Mandela, a prominent black opponent of the system.) Still, the program went ahead. The underlying principle was geopolitical: in multiracial South Africa, a Zulu, for instance, would have his or her own national homeland and would be a citizen there. Outside this homeland, in the white part of South Africa (or in another black homeland, or in a future Coloured or Asian homeland), that Zulu would be a foreigner just as any other noncitizen would be. Conversely, a white person would not have any citizen's rights in the Zulu homeland or in any of the other nine black homelands to be created.

In 1976, the first of the homelands, the Transkei (Fig. SA-1), homeland of the Xhosa nation, was declared independent. Wedged between the Great Escarpment and the Indian Ocean and between Natal and eastern Cape Province, Belgium-sized Transkei was to be the model state-within-a-state. Next came Bophuthatswana (1978), with its infamous capital, Sun City; it was followed by tiny Venda in 1979 and the impoverished Ciskei in 1982.

This politico-geographical reorganization of South Africa was no mere reservation-style warehousing. Enormous investments were made in the creation of national **infrastructures**—in the capital cities, roadways, agriculture, officialdom, and other trappings of nationhood. But the map reveals an undisputable reality: the total area of the black homelands was to be a mere fraction of all of South Africa. Only about 14 percent of the country was to be set aside as homelands for the 70 percent of the people who are black

Africans. Moreover, much of the land in the homelands is of low productivity; most of the known minerals lie in the white areas. The conclusion is inescapable: to survive, workers from the homelands would have to apply for temporary work permits in South Africa's mines, in its factories, and on its farms. Many scholars have described the separate development program as an instance of **domestic colonialism**; some likened it to the reign of the (now-defunct) Soviet Empire over its Muslim republics and postwar Eastern Europe.

During the 1980s, a number of circumstances turned the world's attention to South Africa. First, African opposition, long overshadowed by decolonization struggles elsewhere, now united to face one final challenge. An exiled organization, the African National Congress (ANC), could infiltrate the republic from neighboring states; with international support, the ANC was able to raise the consciousness of many people and their governments. The name of Nelson Mandela, not known to many outsiders in the 1960s and 1970s, now became a symbol of black resistance. Second, the schedule of homeland creation ran into trouble inside South Africa. The fifth independent homeland was to be Kwazulu, home of the most powerful African nation. But the Zulu were in a position to refuse their designation—and did so. Through their leader, Chief Mangosuthu Buthelezi, the Zulu rejected homeland status, and the program was thrown into disarray. Third, the South African economy after 1980 was not nearly as strong as it had been when the separate development program was launched. We have seen the problems of oil-rich as well as oil-poor economies in previous chapters, and South Africa lacks this critical energy resource. When rising petroleum costs coincided with declines in the price of gold and other exports, South Africa did not have the money to carry out its plans as it had before. And fourth, many countries and companies stopped doing business with South Africa. In the United States, "disinvestment" became a rallying point for those opposed to South Africa's social policies; several Western European governments severely cut back their trade connections with South Africa.

The impact of these events in South Africa could be felt in various ways. Within the white power structure, a debate arose over the future of apartheid and separate development. Many changes were made as the "petty" apartheid of the 1950s began to disappear. For those who could afford it, the major hotels in the large cities were "internationalized"—that is, they could accommodate people of all races. The "Whites Only" signs on park benches and at bus stops also began to vanish. More important, the plan whereby citizens of the Transkei (and other homelands) would be foreigners in South Africa was dropped. All residents of South Africa would again have South African citizenship; those with homeland credentials would have dual citizenship.

But the "grand" apartheid design of separate development could not be swept away overnight. The governing party, the Afrikaner-based National Party, began to face opposition from conservative members, who argued that reforms were coming too quickly. A splinter party (the Conservative party) soon formed to opposed any further concessions, whatever the pressures from inside or outside South Africa. This conservative opposition continues to grow today, and it has a spatial dimension. Much of it is concentrated in the small towns and farms of the *platteland*, the rural zones of the Transvaal and the Orange Free State. These are the areas to which Afrikaners moved when they sought to escape British domination; here the Boer War was most vigorously fought; and here the vestiges of Afrikaner orthodoxy are strongest. Thus **regional polarization** within the white constituency has a significant and potentially troubling regional expression.

◆ WILL THE STATE SURVIVE?

If you were to read the South African press on a daily basis, you would be constantly reminded of a growing concern in the republic: that this ethnically fractured country might go the way of Yugoslavia. Leaders of virtually all the more than 20 parties, groups, and factions that have participated in conferences convened to design South Africa's future frequently remind their followers that failure might mean the "Yugoslavization" of their country.

Certainly, the potential for such a collapse exists. Already, a virtual state of civil war exists between the Zulu organization, Inkatha, and militants of the African National Congress; by late 1993, more than 7,000 people had been killed in this conflict, most of them in the province of Natal. A crime wave, part of it racially generated, is sweeping the country. Radicals at both ends of the political spectrum, bent on destroying any progress toward peaceful resolution, have staged assassinations to polarize the various sides (and for the first time, reminiscent of Kenya in the 1950s and Zimbabwe in the 1970s, whites are dying in this violence as well as others). Splinter groups, seeking a greater share of power than their numbers warrant, have accused Mandela and the Afrikaner Nationalist leader, F.W. de Klerk, of collusion to rule post-apartheid South Africa.

Nonetheless, CODESA (Conference Toward a Democratic South Africa) has survived and continued, albeit intermittently and despite walkouts, angry withdrawals, stoppages, and boycotts. Even parties that earlier said they would never participate, such as the white Conservatives (who disapprove of de Klerk's cooperation with the ANC), have come to the table. In mid-1993, agreement was reached on democratic elections and a five-year transition to majority rule—an achievement that, if carried out successfully, would be a beacon of hope for factionalized countries all over the world.

South Africa's fitful progress must be seen, however, against the dimensions—especially the geographic dimensions—of the obstacles ahead. International sanctions and

South African police in armored vehicles attempt to control ethnic violence in Alexandra, Johannesburg. Zulu Inkatha members broke through police barriers, and a furious gun battle ensued—an oft-repeated occurrence in politically changing South Africa.

the ANC's own slogan of "liberation before education" have combined not only to devastate the economy but also to yield a veritable army of unemployed and unemployable young Africans, most of them living in the squalid slums on the cities' outskirts. Talk of secession is rife in Natal, where the Inkatha movement fires Zulu resentment over the Mandela–de Klerk (Nobel-Peace-Prize-winning) alliance. And reincorporating the "homelands" is not simply a matter of removing border posts. The police of Ciskei opened fire on an ANC march that crossed "national" borders. The ruler of Bophuthatswana in March 1993 proclaimed that only a national referendum among his "country's" citizens could undo the independence they were granted by Pretoria years ago.

Still, barring another disaster of historic proportions, South Africa may indeed "cross the Rubicon," as former President P. W. Botha (who, in late 1980s, began the negotiations that led to Nelson Mandela's release from prison) described the process ahead. In the mid-1990s apartheid is being swept away; the immediate concern is whether Republic of South Africa (or *Azania*, as black South Africans want to rename their country) can survive the transition period between autocracy and democracy. The larger question in the minds of many South Africans is whether democracy in South Africa will fare better than it has in other countries on this troubled continent.

◆ PRONUNCIATION GUIDE

Afrikaans (uff-rih-KAHNZ)
Afrikaner (uff-rih-KAHN-nuh)
Apartheid (APART-hate)
Autocracy (aw-TOCK-ruh-see)
Azania (uh-ZAY-nee-uh)
Bantu (ban-TOO)
Bantustan (BAN-too-stahn)
Boer (BOOR)
Bophuthatswana (boh-pooh-taht-SWAHNA)
Botha (BAW-tuh)
Botswana (bah-TSWAHN-uh)
Buthelezi, Mangosuthu (boo-teh-LAY-zee, mungo-SOO-too)
Ciskei (SISS-skye)

Drakensberg (DRAHK-unz-berg)
Durban (DER-bun)
Inkatha (in-KAH-tah)
Johannesburg (joh-HANNIS-berg)
Khoisan (khoy-SAHN)
Kruger (KROO-guh)
Kwazulu (kwah-ZOO-loo)
Lesotho (leh-SOO-too)
Limpopo (lim-POH-poh)
Malan, Daniel François (muh-LAHN, DAN-yell frawn-SWAH)
Mandela (man-DELLA)
Moçambique (moh-sum-BEEK)
Namibia (nuh-MIBBY-uh)
Natal (nuh-TAHL)

Ochre (OH-ker)
Pretoria (prih-TOR-ree-uh)
San (SAHN)
Sotho (SOO-too)
Soweto (suh-WETTO)
Swaziland (SWAH-zee-land)
Transkei (TRUN-skye)
Transvaal (TRUNZ-vahl)
Vaal (VAHL)
Veld (VELT)
Voortrekker (FOR-trecker)
Xhosa (SHAW-suh)
Zimbabwe (zim-BAHB-way)
Zulu (ZOO-loo)

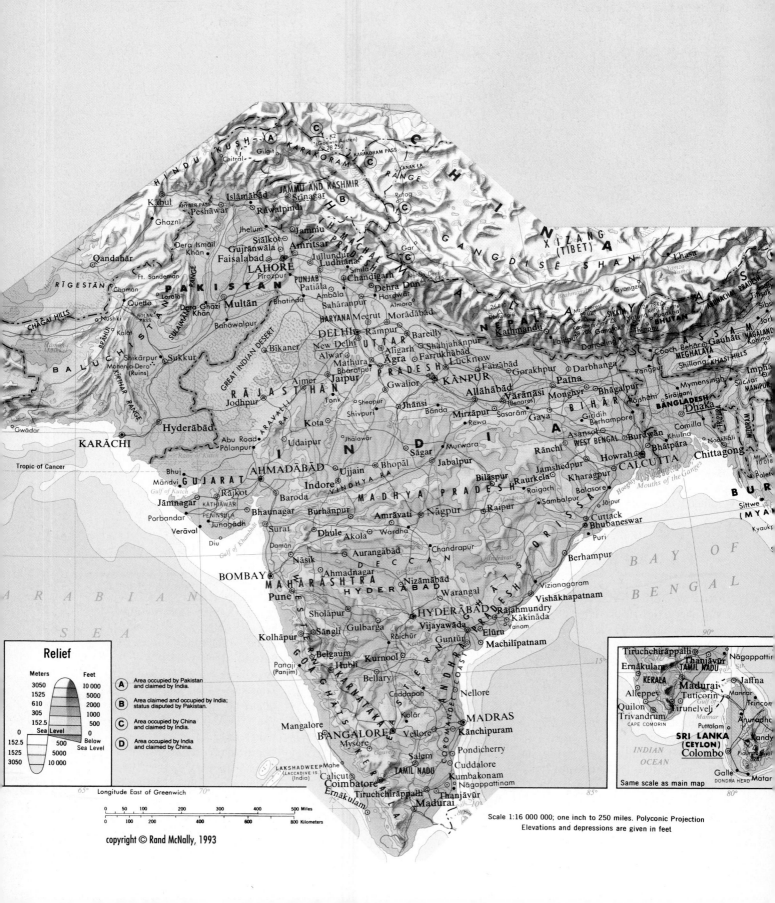

HINDU KUSH–YA
KARAKORAM
K2
(Godwin Austen)
28 250
KARAKORAM PASS
RANGE
YANAK LA
Chitral
Gilgit
C
JAMMU AND KASHMIR
Rutog
C
CHINA
Kābul
Ghaznī
Islāmābād
Srīnagar
Rāwalpindi
KHYBER PASS
Peshāwar
B
C
Gar
Khasa
Jhelum
Jammu
Yamzo
XIZANG
(TIBET)
GANGDISE SHAN
Qandahār
Dera Ismāil
Khān
Siālkot
Gujrānwāla
Amritsar
Jullundur
Ludhiāna
Simla
C
Nanda Devi
25 645
Gyangzê
Faisalābad
LAHORE
Chandīgarh
Fīrozpur
Dehra Dūn
RĪGESTĀN
Chaman
PUNJAB
Patiāla
Hardwār
Almora
26 810
Mt. Everest
29 028
24 784
SIKKIM
ARUNACHAL PRADESH
PAKISTAN
Quetta
Loralāi
Multān
Bhatinda
Ambāla
Sahāranpur
Dhaulagiri
Kanchenjunga
28 208
Gangtok
Darjeeling
Thimphu
BOLAN
PASS
Bahāwalpur
HARYANA
Merut
Morādābād
NEPAL
Kāthmāndu
Lālitpur
BHUTAN
CHĀGAI HILLS
Nushki
Kalāt
Dera Ghāzi
Khān
DELHI
Rāmpur
Bareilly
Cooch Behār
Rangpur
Gauhāti
ASSAM
MEGHALAYA
Kohīma
NAGALAND
Shikārpur
Sukkur
New Delhi
UTTAR
Alīgarh
Shāhjahānpur
Lucknow
Faizābād
Gorakhpur
Darbhanga
Shillong
KHĀSI HILLS
Imphāl
MANIPUR
Shillong
BALUCHISTAN
Bīkaner
Alwar
Mathura
Āgra
Farrukhābād
Rājshāhi
Silchar
KIRTHAR
RANGE
BRĀHUI
RANGE
SULAIMĀN
RANGE
Hyderābād
Mohenjo-Daro
(Ruins)
Jaipur
Bharatpur
PRADESH
KĀNPUR
Allāhābād
Vārānasi
(Benares)
Patna
Monghyr
BIHĀR
Bhāgalpur
Mymensingh
BANGLADESH
Dhaka
MIZORAM
Gwādar
RĀJASTHĀN
Jodhpur
Tonk
Gwalior
Jhānsi
Bānda
Mirzāpur
Sasarām
Gaya
Girīdih
Berhampore
Rājshāhi
Comilla
Noākhāli
Chittagong
BUR
(MYA
Tropic of Cancer
KARĀCHI
Ajmer
Sheopur
Shivpuri
Rewa
Murwāra
Ranchi
Asansol
WEST BENGAL
Burdwān
Howrah
CALCUTTA
Bhātpāra
Mt. Vic.
10 018
Poleti
ARAVALLI
RA.
Abu Road
Pālanpur
Udaipur
Jhālawār
Kota
INDIA
Sāgar
Jabalpur
Bilāspur
Raurkela
Kharagpur
Balāsore
Jājpur
Hooghly
Mouths of the Ganges
Bhuj
Māndvi
GUJARAT
Rājkot
AHMADĀBĀD
Ujjain
Bhopāl
MADHYA
PRADESH
Raigarh
Sāmbalpur
Cuttack
Sittwe
(MYA)
Kyaukpyu
Gulf of Kutch
Jāmnagar
Porbandar
KĀTHIĀWĀR
PENINSULA
Bhaunagar
Junāgadh
Diu
Indore
Baroda
Burhānpur
VINDHYA RA.
Narmada
Amrāvati
Nāgpur
Raipur
Chandrapur
ORISSA
Jājpur
Bhubaneswar
Puri
Berhampur
Verāval
Gulf of Khambhāt
Surat
Dhule
Akola
Wardha
Daman
Nāsik
Aurangābād
DECCAN
Godāvari
BAY OF
BENGAL
BOMBAY
Ahmadnagar
MAHĀRĀSHTRA
Nizāmābād
HYDERĀBĀD
Warangal
Vizianagaram
Vishākhapatnam
Pune
Sholāpur
HYDERĀBĀD
Rajahmundry
Kākināda
Yanam
ARABIAN
SEA
Sāngli
Gulbarga
Vijayawāda
Elūru
Machilīpatnam
Kolhāpur
Belgaum
Raichūr
Guntūr
Panaji
(Panjim)
Hubli
Kurnool
Bellary
KARNĀTAKA
EASTERN GHĀTS
WESTERN GHĀTS
ANDHRA PRADESH
COROMANDEL COAST
Nellore
15°
Mangalore
Cuddapah
Kolār
BANGALORE
Mysore
Vellore
MADRAS
Kānchipuram
Pondicherry
LAKSHADWEEP
(LACCADIVE IS.)
(India)
Mahe
Salem
Cuddalore
Kumbakonam
Nāgappattinam
Calicut
Coimbatore
KERALA
TAMIL NADU
Tiruchchirāppalli
Thanjāvūr
Madurai

Relief

Meters	Feet
3050	10 000
1525	5000
610	2000
305	1000
152.5	500
Sea Level	Sea Level
0	0
152.5	500
1525	5000
3050	10 000
	Below Sea Level

A Area occupied by Pakistan and claimed by India.

B Area claimed and occupied by India; status disputed by Pakistan.

C Area occupied by China and claimed by India.

D Area occupied by India and claimed by China.

Longitude East of Greenwich

65°   70°

0   50 100   200   300   400   500 Miles
0   100   200   400   600   800 Kilometers

copyright © Rand McNally, 1993

Scale 1:16 000 000; one inch to 250 miles. Polyconic Projection
Elevations and depressions are given in feet

Inset:
Tiruchchirāppalli
Thanjāvūr
Nāgappattinam
Ernākulam
TAMIL NADU
KERALA
Madurai
Jaffna
Alleppey
Tuticorin
Tirunelveli
Quilon
Trivandrum
CAPE COMORIN
Gulf of Mannar
Mannar
Anurādha
SRI LANKA
(CEYLON)
Colombo
Kandy
INDIAN
OCEAN
Galle
DONDRA HEAD
Matar
Trinco
Puttalam

Same scale as main map

85°   80°

CHAPTER 9

SOUTH ASIA: RESURGENT REGIONALISM

From Iberia to Arabia and from Malaysia to Korea, Eurasia is a landmass ringed by peninsulas. Among these, however, none can compare to the great triangle of India that juts deep into the Indian Ocean—a vast, varied, volatile part of the world.

India, the world's second most populous country (after China), lies at the heart of the South Asian geographic realm. With a population that will exceed 1 billion before the end of the twentieth century, India dominates—but this is a realm of giants. To India's east lies Bangladesh, with nearly 120 million inhabitants; to its west lies Pakistan, home to another 130 million. Only Nepal, in the mountainous north, and Sri Lanka, in the insular south, have modest populations of roughly 20 million each.

This huge population cluster lies in a comparatively well-defined physical realm. To the north, South Asia is separated from the rest of the continent by the earth's greatest mountain range, the Himalayas, and its extensions east and westward. To the east, mountains and dense forests separate South Asia from neighboring Southeast Asia. Only the west affords avenues of entry, none of them easy. From there, across deserts and through mountain passes, came influences that repeatedly heightened the complexity of South Asian society. Today, South Asia is a Babel of languages, a Jerusalem of religions, a Lebanon of politics—and yet its conflicts have remained localized. Surprisingly, only one major political change has occurred over the past half-century of postcolonial independence. But how long will this record stand?

◆ REALM AND REGIONS

To the north, east, and south, the South Asian realm is demarcated by mountains, forests, and coastlines, and as such constitutes one of the world's best-defined **physiographic realms**. As maps of religions and languages (see Figs. 7-2 and 9-7) underscore, South Asia is also clearly defined in terms of cultural criteria. The main problem of regional definition lies in the northwest, where physical and cultural boundaries are less clear.

IDEAS & CONCEPTS

Population geography	Social stratification
Population distribution	Irredentism (2)
Population density (2)	Forward capital (3)
Doubling time	Federal state
Population explosion	Centrifugal forces (3)
Demographic transition	Centripetal forces (2)
Physiographic realm	Caste system
Wet monsoon	

REGIONS

Pakistan	Sri Lanka
India	The North
Bangladesh	(Nepal & Bhutan)

As here defined, South Asia consists of five regions: (1) India; (2) the northern mountain territories from Kashmir to Nepal and Bhutan; (3) the southern island areas of Sri Lanka and the Maldives; (4) the east, centered on Bangladesh; and (5) the west, Pakistan. The largest region, India, divides into several subregions to be discussed later.

The inclusion of Pakistan in the South Asian (rather than the Islamic North Africa/Southwest Asia) realm can be debated. The political boundary between India and Pakistan forms a strong cultural divide as well: Pakistan is an Islamic state, whereas India is a secular state in which the Hindu religion and its ways of life dominate. On this basis it could be argued that Pakistan is not part of the South Asian realm, but consider the following: ethnic continuity links Pakistan more strongly to India than to Iran or Afghanistan. Historic migrations and diffusions into India have advanced across Pakistan. During the period of British colonial rule, Pakistan was part of Britain's South Asian empire, and English remains the universal language in both countries. And of India's 918.6 million people, more than 100 million are Muslims. Another

POPULATION GEOGRAPHY

Population geography is a field of human geography that focuses on the *distribution* and *growth* of populations, on the *composition* of populations, and on their *movement*. As always in geography, these topics are studied in a spatial context. We want to know how people have adjusted to and exploited the opportunities of the geographic areas they occupy.

Population distribution summarizes the way people have organized themselves, collectively and individually, in their overall environment. Worldwide, as we saw in the Introduction (Fig. I-9), humanity has formed three primary concentrations (all on the Eurasian landmass) and several secondary ones. This chapter deals with one of these primary clusters.

Maps can depict the distribution of population in various ways. Figure 9-1 is a *dot map* in which each dot represents a certain number of people. At a glance, this map reflects how South Asia's population, over more than 40 centuries, has responded to the opportunities and limitations of terrain and climate. Millennia ago and even today, the great rivers of South Asia (the Ganges, Brahmaputra, and Indus) form the focus of life for hundreds of millions. But people also have moved to the cities—to Bombay and Calcutta and Delhi. The discovery and exploitation of mineral and energy resources have also affected the pattern and density of population in South Asia.

A *pattern*, as we have noted earlier, is the geometric arrangement of a distribution. For example, the concentration of people along the fertile floodplains of rivers creates a *linear* pattern of population distribution.

When people move from the countryside into towns and cities, the pattern becomes *clustered*. The term *dispersion* describes the areal spread of a population, its dispersal or scattering over an area. Obviously, there are many other geographic adjectives with which to define patterns; sometimes the cultural landscape does seem to display a square, rectangular, concentric, or some other geometric configuration.

Population density is another prominent topic in this field. How many people are crowded together on a certain unit of land area? How many more before this area's carrying capacity is reached or exceeded? Parts of South Asia are among the world's most densely populated areas, but other zones of the realm are sparsely peopled. Geographers distinguish between *arithmetic density*, the number of people per unit area regardless of its environment, and *physiologic density*, the number of people per unit area of arable land. India's arithmetic density in 1994 was 743 per square mile (the U.S. figure is just 71). India's physiologic density, however, was 1,456. This higher figure is more relevant because no country has a population that is evenly spread over its total area, and in developing countries a large percentage of the people still depend directly on the productivity of their land.

Population growth is a key topic of population geography. Appendix A provides information not only on the total populations of countries but also on their rate of natural increase, **doubling time**, and life expectancy (in addition to arithmetic density and percentage of urban dwellers). For the world as a whole, population growth is simply the overall rate of natural increase—that is, the excess of births over deaths during a given period (for

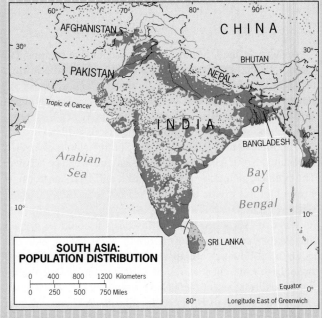

**SOUTH ASIA:
POPULATION DISTRIBUTION**

0   400   800   1200 Kilometers

0   250   500   750 Miles

FIGURE 9-1

statistical purposes, normally one calendar year). But for an individual country, province, county, or municipality, population growth involves not only births and deaths but also *immigration* and *emigration*. Any time we look at a map of population growth, we should take into account these additional factors of movement (the migration process is discussed in the box on 252-253).

The human population of the world as a whole has grown explosively over the past two centuries and especially during the twentieth century. In 1820, the population of the entire world was about 1 billion (India alone will surpass this number during the late 1990s). Even as recently as 1930, world population was 2 billion. But by then, the **population explosion** was well underway, and in 1975 it reached 4 billion. It will surpass 8 billion some time after 2010.

Will this cycle go on indefinitely? It had better not—and here geography comes into play. The bulk of the world's increasing numbers of people are born in economically still-developing countries. But in developed states, population growth has leveled off. These developed countries (which were identified in the Introduction and discussed in Part I) have gone through a so-called **demographic transition**. As their economic development and associated industrialization and urbanization proceeded, high birth and high death rates gave way to a period of high birth and lower death rates; eventually, the birth rates came down too (Fig. 9-2). The result: low growth, as in much of Western Europe, Japan, and elsewhere in the developed world. So the hope is that progress toward economic well-being will also bring about a full demographic transition for India, Pakistan,

Egypt, and other countries where population growth is still in the high-rate stage (Stage 1, Fig. 9-2). As we noted in Chapter 4, the economic tigers on the Pacific Rim seem now to be experiencing just this final phase of the demographic transition.

The specter of hunger and famine still haunts hundreds of millions of people who depend on locally grown crops for their survival, but the fact is that the world can produce adequate food for all—adequate food if not balanced nutrition. Again, the fate of millions often is a matter of geography: of the remoteness and isolation of people in need, of dislocation and refuge for others trapped in cycles of political violence. Today, Africa is most severely afflicted by these circumstances, but South Asia has seen its share.

Population geographers also are interested in the *composition* of populations, in the proportions of the sexes, the lengths of people's lives, the ages of each population sector—all, of course, in spatial context. Fast-growing populations produce high numbers of youngsters who must be nourished, educated, and, later, employed. As long as population growth continues at a high rate, national economies struggle to provide for these youngsters; for many people, things get worse, not better. But what of those countries whose population growth has leveled off or has even begun to decline? There, the looming problem is the growing population of "oldsters," long-lived retirees who need state pensions, medical services, and other social amenities. The burden to provide these amenities falls on a shrinking work force in the middle years and is not sufficiently replenished by young people—because birth rates are now so low.

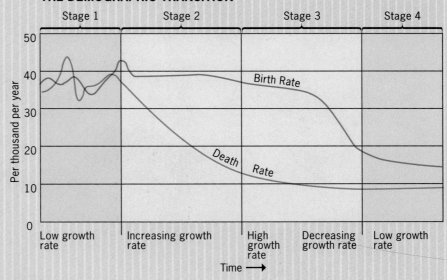

THE DEMOGRAPHIC TRANSITION

FIGURE 9-2

Population composition can be represented on the map, but it can also be displayed in population profiles that reveal, at a glance, a population's *age-sex structure*. Figure 9-3 compares such a structure for India with that of Japan. India's pyramid-like profile has the familiar lower-age bulk of developing countries; Japan's constricted profile represents the aging population typical of a developed economy and society. India's hope is to reduce the preponderance of people in the youngest age groups. Japan's hope is to sustain its vitality and prosperity even as its youngsters dwindle in number.

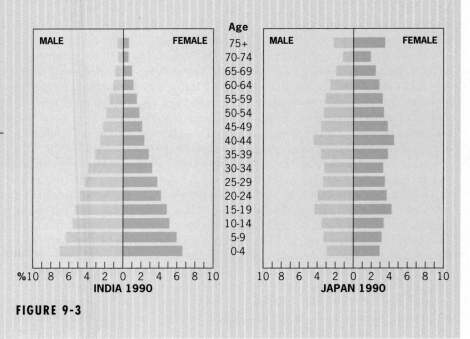

FIGURE 9-3

The Himalayas form a towering wall that separates South Asia from the Chinese Realm. The cloud-lined mountain to the left is Everest (elevation 29,028 feet/ 8848 m).

100 million Muslims live in Bangladesh, so the Islamic population of the rest of South Asia far exceeds that of Pakistan. The cultural divide between the west and the rest, therefore, is less sharp than it might appear. Hence Pakistan forms the western flank of South Asia and belongs in this geographic realm.

PHYSIOGRAPHIC REGIONS

Before we look into the complex and fascinating cultural geography of South Asia, it is useful to gain some insight into the physical stage of this crowded realm. South Asia is a realm of immense physiographic variety, of snow-capped peaks and forest-clad slopes, of vast deserts and wide river basins, of great plateaus and spectacular shores. It is the land of tropical cyclones and monsoons, of devastating droughts and destructive floods, of irrigated basins and of unproductive uplands. South Asia, in short, is a world unto itself.

In very general terms, it is possible to recognize three rather clearly defined physiographic zones in South Asia: the northern mountains, the southern peninsular plateaus, and, between these two, a belt of river lowlands. The *northern mountains* extend from the Hindu Kush and Karakoram range in the west to the Himalayas (Mt. Everest, the world's tallest peak, lies in Nepal) in the center and the ranges of Bhutan in the east. The *southern plateaus* are dominated by the massive Deccan, a plateau built of basalt that poured out when India separated from Africa during the breakup of Gondwana (see Fig. 8-3). And the belt of *river lowlands* extends eastward in a vast crescent from Pakistan's lower Indus valley (the area known as Sind) through the wide plain of the Ganges valley of India and on across the great double delta of the Ganges and Brahmaputra in Bangladesh (Fig. 9-4).

Northern Mountains

This broad three-zone division of South Asia's physiography merely introduces the diversity of South Asian landscapes. Each zone contains much variety: the mountains of the west and north, for example, are dry and barren in Baluchistan and in the Afghan borderlands; their valleys become green and tree-studded in Kashmir; the relief reaches a maximum in Nepal; and then, to the east, the ranges become densely forested in the eastern Assam-Myanmar (Burma) borderland zone. Moreover, these northern mountains do not simply rise out of the river valleys below: there is a continuous belt of transitional foothills between the lofty ranges and the low-lying river basins.

River Lowlands

The belt of alluvial lowlands that extends from the Indus plain to the Brahmaputra valley also is anything but uniform. This physiographic region is often called the North

TEN MAJOR GEOGRAPHIC QUALITIES OF SOUTH ASIA

1. South Asia is well-defined physiographically, extending from the southern slopes of the Himalayas to offshore Sri Lanka and the Maldive Islands.
2. Two river systems, the Ganges-Brahmaputra and the Indus, form crucial lifelines for hundreds of millions of people in this realm. The annual wet monsoons are a critical environmental element.
3. India lies at the heart of the world's second-largest population cluster, which by 2010 will be the first.
4. No part of the world faces demographic problems with dimensions and urgency comparable to those of South Asia.
5. All the states of South Asia suffer from underdevelopment. Food shortages occur; nutritional imbalance prevails.
6. Agriculture in South Asia in general is comparatively inefficient and less productive than in other parts of Asia.
7. The great majority of South Asia's peoples live in villages and subsist directly on the land.
8. Strong cultural regionalism marks South Asia. The Hindu religion dominates life in India; Pakistan is an Islamic state; Buddhism thrives in Sri Lanka.
9. The South Asian realm's politico-geographical framework results from the European colonial period, but important modifications took place after the European withdrawal.
10. India constitutes the world's largest and most complex federal state.

Indian Plain, and its internal environmental contrasts are clearly reflected in Figs. I-6 to I-8. The Indus, which rises in Tibet and penetrates the Himalayas to the west in its course to the Arabian Sea, receives its major tributaries from the *Punjab* ("land of five rivers"); this physiographic subregion—situated between the Indus and North Indian plains—extends into Pakistan as well as India. The subregion of the lower Indus is characterized by its minimal precipitation, its desert soils, and its irrigation-based clusters of settlement. This is the heart of Pakistan, and the farmers here grow wheat for food and cotton for sale.

Hindustan, the central subregion extending from eastern Punjab near the historic city of Delhi to the Ganges-Brahmaputra delta at the head of the Bay of Bengal, is wet

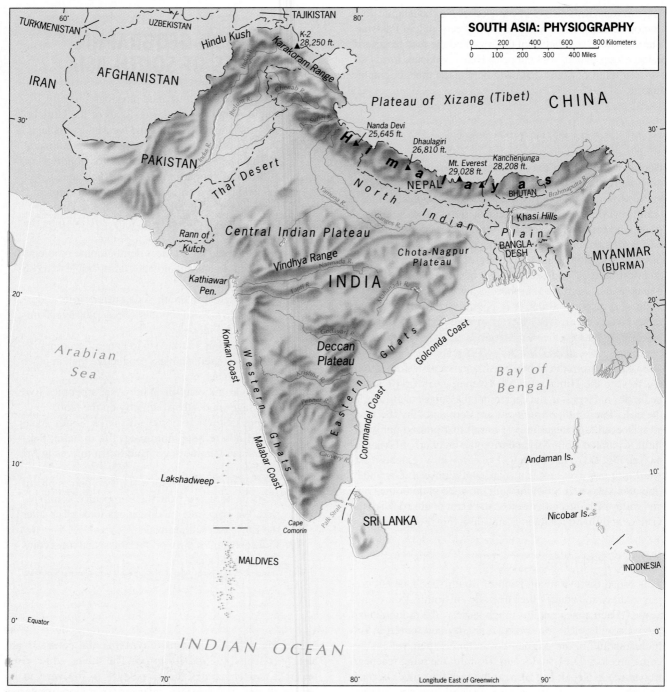

FIGURE 9-4

country. The precipitation here exceeds 100 inches (250 cm) in sizeable areas, and 40 inches (100 cm) almost everywhere (Fig. I-6). Deep alluvial soils cover much of this subregion, and the combination of rainfall, river flow, good soils, and a long growing season make this India's most productive area. As Fig. 9-1 confirms, this is also India's largest and most concentrated population core. Rural densities in places exceed 1,000 people per square mile (385 per sq km); moreover, lying between Delhi and delta-side Calcutta is a chain of the country's leading urban centers, connected by the densest portion of the Indian transportation network. In the moister east, rice is the chief food crop; jute is the main commercial product, notably in Bangladesh and the rest of the delta. To the west, near Punjab and around the drier margins of the Hindustan subregion, wheat and such drought-resistant cereals as millet and sorghum are cultivated.

The easternmost extension of the North Indian Plain comprises the lower Brahmaputra valley in Assam. This valley is much narrower than the Gangetic (Ganges) plain and very moist, and it suffers from frequent flooding; the river is of limited use for navigation. Up on the higher slopes, tea plantations have been developed; in the lower elevations, rice is grown. Assam is just barely connected with the main body of India through a corridor only a few miles wide that runs between the southeastern corner of Nepal and the northwestern tip of Bangladesh; it remains one of South Asia's frontier areas.

Southern Plateaus

Turning now to the peninsular component of the subcontinent, also referred to as plateau India, we find once again that there is more variety than first appearances suggest. There are physiographic bases for dividing the plateau into a northern zone (the Central Indian plateau would be an appropriate term) and a southern sector, already well-known as the Deccan plateau. The dividing line can be drawn on the basis of roughness of terrain (along the Vindhya range) or rock type (much of the Deccan is lava covered); and the Tapti and Godavari rivers—in the west and east, respectively—form a clear lowland corridor between the two regions (Fig. 9-4). The Deccan (meaning "south") has been tilted to the east, so that the major drainage is toward the Bay of Bengal. This plateau is also marked along much of its margin by a mountainous escarpment called the Ghats (the word means "hills"), which descends to the fairly narrow coastal plains below. Parallel to these surrounding coasts are the Eastern and Western Ghats, which meet near the southern tip of the subcontinent.

Peninsular India possesses a coastal lowland zone of varying width. Along the southwestern littoral lies the famous Malabar Coast, and north of it the Konkan Coast. Along the southeastern shore is the Coromandel Coast and, to its northeast, the Golconda Coast leading toward the Ganges delta. These physiographic subregions lie wedged between the interior plateau and the Indian Ocean; the Malabar-Konkan Coast is the more clearly defined, as the Western Ghats are more prominent and higher in elevation than the Eastern Ghats. Thickly forested and steep, the Malabar-Konkan escarpment dominates the Arabian Sea coastland and limits its width to an average of less than 50 miles (80 km); the Coromandel-Golconda coastal plain is wider, and its interior margins are less pronounced.

Although not very large in terms of total area, these two coastal subregions have had and continue to have much importance in the Indian state. Figure I-6 indicates how

INDIA

Our use of the name *India* for the heart of the South Asian realm derives from the Sanskrit word *sindhu*, used to identify the ancient civilization in the Indus valley. This word became *sinthos* in Greek descriptions of the area and then *sindus* in Latin. Corrupted to *indus*, which means "river," it was first applied to the region that now forms the heart of Pakistan. Subsequently, it was again modified to *India* to refer generally to the land of river basins and clustered peoples from the Indus in the west to the lower Brahmaputra in the east.

well-watered the Malabar-Konkan Coast is (the triggering onshore airflow, or *wet monsoon*, is diagrammed and discussed in the box on pp. 468–469). This rainfall supply and the balmy tropical temperatures have combined with the coast's fertile soils to create one of India's most productive farming areas. On the lowland plain, rice is grown; on the adjacent slopes, spices and tea are cultivated. Of course, this combination of favorable circumstances for intensive agriculture led to the emergence of southern India's major coastal population concentration (Fig. 9-1). Along these coasts the Europeans, beginning with the Greeks and Romans, made contact with India. Later the coasts became spheres of British influence, and it was during this period that two of India's greatest cities, Bombay and Madras, began their growth.

◆ THE HUMAN SEQUENCE

The Indian subcontinent is a land of great river basins. Between the mountains of the north and the uplands of the peninsula in the south lie the broad valleys of the Ganges, the Brahmaputra, and the Indus. In one of these, the Indus (see the box above entitled "*India*"), lies evidence of the realm's oldest civilization, contemporary to and interacting with ancient Mesopotamia. Unfortunately, much of the earliest record of this civilization lies buried beneath the present water table in the Indus valley, but those archaeological sites that have yielded evidence indicate that here was a quite sophisticated culture with large, well-organized cities. As in Mesopotamia and the Nile valley, considerable advances were made in the technology of irrigation, and the civilization was based on the productivity of the Indus lowlands' irrigated soils (Fig. 9-6).

DYNAMICS OF THE WET MONSOON

Land and water surfaces heat and cool at different rates. Although the oceans with their various water layers warm and cool rather slowly, solid landmasses heat and cool rapidly in response to the temperature of the air with which they come in contact. In winter, the cold land surface often develops an overlying high-pressure cell from which winds blow toward lower-pressure areas associated with adjacent warmer seas; in summer, the process reverses as the land heats up and generates a low-pressure cell that sucks in moisture-laden air lying above the now higher-pressure (and relatively cooler) ocean surface. The bigger the land area, the more pronounced these effects of "continentality" become—a phenomenon that reaches its grandest scale in the central and eastern portions of the Eurasian landmass.

In southern and eastern Asia, particularly, this mechanism shapes a massive seasonal reversal of onshore and offshore windflows known as *monsoons* (an old Arabic

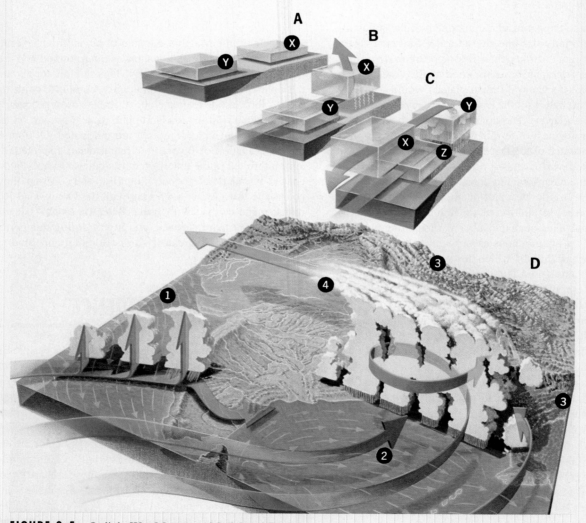

FIGURE 9-5   India's Wet Monsoon: Model and Reality.

word meaning "season"). During the cool winter months, a high-pressure cell over the land generates an offshore airflow—the *dry monsoon*. But in the hot summer season, a strong onshore flow of rain-bearing winds predominates as low-pressure cells form over the heated continent. It is this **wet monsoon** that is the key to India's agricultural possibilities and seasonal life rhythms, providing the subcontinent with sufficient moisture to support its huge population. Monsoonal windflows, which help balance temperatures between the tropics and mid-latitudes, actually occur over a 10,000-mile-long (16,000-km-long) broad crescent of central Africa and southern Asia, stretching east from the West African coast to India and then northeast as far as Japan. The monsoons function as an enormous heat engine, transferring vast masses of air put into motion by temperature and pressure differences above continents and nearby oceans.

During summer, the solid land surface heats up far more rapidly than the sea (with its many layers of cooler water that constantly mix). An air parcel in contact with the ground—labeled **X** in diagram **A** of Fig. 9-5—quickly becomes warmer than its offshore counterpart (**Y**). In diagram **B**, parcel **X** begins to rise, cool, and shed its moisture, making room for air parcel **Y** to flow inland to equalize the now-lowered surface air pressure. Parcel **Y** contains even more moisture than **X**, and soon that water content is precipitated as rain as **Y** itself is heated, rises, and cools (remember that as air becomes cooler, it can hold less moisture). These movements are illustrated in diagram **C**, which also shows parcel **Z** moving in to re-place **Y**; original parcel **X** has cooled and dried at high altitude as it moved out over the ocean.

In diagram **D**, this model is applied to India. The subcontinent is influenced by two branches of the wet summer monsoon, both originating in the Indian Ocean to the southwest, with each moisture-laden airstream shaped by regional topography. The Arabian Sea branch ① saturates the Malabar-Konkan coastal strip, surges over the Western Ghats (where orographic uplift removes much of the remaining moisture), and becomes a relatively dry eastward flow across the interior Deccan plateau. The Bay of Bengal branch ② comes ashore over the Ganges-Brahmaputra delta but is quickly blocked by the Himalayan wall to the north ③. As diagram **D** illustrates, surface winds now rise in a spiral over Bangladesh and are steered westward by the mountain barrier onto the Gangetic plain. Massive rainfall occurs throughout this zone, supplying the water necessary to support its intensive agriculture. This wet-monsoon-generated precipitation gradually tapers off across Hindustan ④ as the distance from the Bay of Bengal increases.

These critical monsoonal circulation patterns continue for as long as the land surface remains significantly warmer than the surrounding sea. Although this weather system does provide sufficient rainfall in most years, it can vary without warning—producing droughts (such as occurred in 1987) that constitute a serious environmental hazard for the hundreds of millions who must rely on the wet monsoon.

DEVELOPMENT IN THE RIVER LOWLANDS

The Indus valley civilization was to India what the lower Nile culture hearth was to Africa: ideas and innovations diffused from there eastward and southward through the many different communities and societies that coexisted in the peninsula (see Fig. 7-4). But far more than the ancient Egyptians did in Africa, the Indus culture brought coherence to India, and over an even longer period of diffusion. By the time the Indus civilization, anchored by the cities of Harappa and Mohenjo-Daro, began to decline about 4,000 years ago, many Indian cultures shared the legacy of the Harappans.

Now a new force entered the scene: the *Indo-Europeans* (or *Aryans*), who, from about 3500 B.C. onward, invaded the Indus valley from the direction of Iran, adopted many of the innovations of the Indus civilization, and pushed their frontier of settlement eastward beyond the valley into the Ganges lowland, where they founded an urban culture of their own. Pushing southward into the peninsula, they conquered and absorbed the tribes they found there. Their language, *Sanskrit*, began to differentiate into the linguistic complex of modern-day India (Fig. 9-7).

In the centuries following this invasion, Indian culture went through a period of growth and development. From a formless collection of isolated tribes and their villages, regional organization emerged. Towns developed, arts and

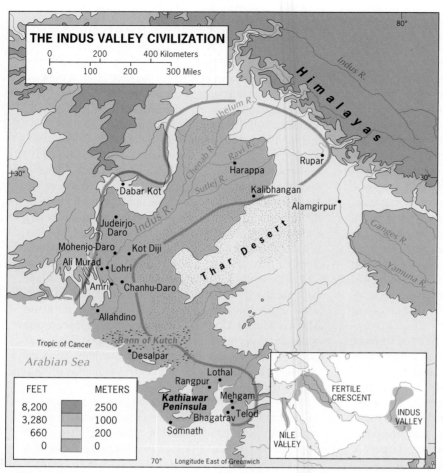

THE INDUS VALLEY CIVILIZATION

FEET	METERS	
8,200	2500	
3,280	1000	
660	200	
0	0	

FIGURE 9-6

crafts blossomed, and trade with Southwest Asia increased. Most important of all, *Hinduism* emerged from the beliefs and practices brought to India by the Indo-Europeans, and a whole new way of life, based on this faith, evolved. A multi-layered **social stratification** developed, controlled and administered by powerful priests. Ruling Brahmans stood at the head of a complex, bureaucratic hierarchy—a *caste system*—in which soldiers, artists, merchants, peasants, and all others had their place. But all was not peaceful in the India of, say, 3,000 years ago. Aggressive and expansionist kingdoms arose, competing and struggling with each other for greater power and influence.

It was in one of these kingdoms, located in northeastern India, that Prince Siddhartha, better known as Buddha, was born. His birth in the sixth century B.C. (more than 2,500 years ago) was unremarkable, but his actions were unique: he voluntarily gave up his princely position to seek salvation and enlightenment through religious meditation. His teachings demanded the rejection of earthly desires and prescribed a reverence for all forms of life. He walked the

length and breadth of India and attracted a substantial following, but his teachings did not have a major impact on Hindu-dominated society during his lifetime. That impact came later, when the ruler of a powerful Indian state decided to make Buddhism the state religion.

Before this occurred, however, South Asia was buffeted by yet another series of outside penetrations from the west and northwest. First the Persians pushed into the Indus basin, and next the Greeks, under Alexander the Great, invaded not only the Indus valley but also the very heartland of India, the Gangetic plain, late in the fourth century B.C.

The Southern Peninsula

While all this was happening in northern India, the peninsular south lay comparatively isolated, protected by distance from the arena of cultural innovation, infusion, and conflict. Southern India had been occupied by ancient peoples long before the Indus and Ganges civilizations arose—peoples whose historic linkages are not clear. Their physical appear-

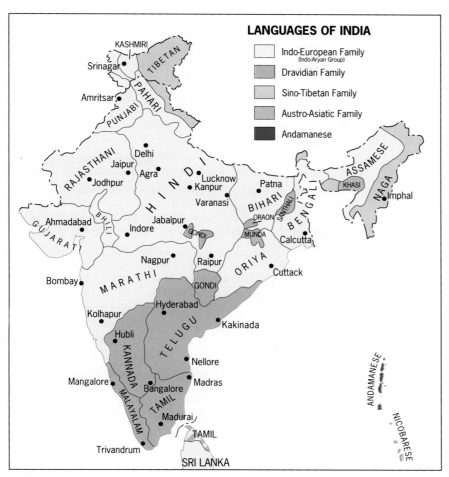

FIGURE 9-7

ance suggests connections to Africans and indigenous Australians; their languages, too, are distinctive and not related to those of the Indo-European north. As Fig. 9-7 indicates, the south is a distinct subregion of India: both the peoples and their languages are known as *Dravidian*. There has, of course, been considerable intermixture with the north, but southern India's regional discreteness still prevails. The four major Dravidian languages—Telugu, Tamil, Kanarese (Kannada), and Malayalam—all have long literary histories. Today, Telugu and Tamil are spoken by nearly one-fifth of India's more than 900 million citizens.

AŚOKA'S MAURYAN EMPIRE

When the Greeks withdrew from the Ganges basin and the Hindu heartland was once again free, a powerful empire arose there—the first true empire in the realm. This, the Mauryan Empire, extended its influence over India as far west as the Indus valley (thus incorporating the populous Punjab)

and as far east as Bengal (the double delta of the Ganges and Brahmaputra); it reached as far south as the modern city of Bangalore.

The Mauryan Empire was led by a series of capable rulers who achieved considerable stability over a vast domain. Undoubtedly the greatest of these leaders was Aśoka, who reigned for nearly 40 years during the middle part of the third century B.C. Aśoka was a believer in Buddhism, and it was he who elevated this religion from obscurity to regional and ultimately global importance.

In accordance with Buddha's teachings, Aśoka reordered his government's priorities from conquest and expansion to a Buddhist-inspired search for stability and peace. He sent missionaries to the outside world to carry Buddha's teachings to distant peoples, thereby also contributing to the diffusion of Indian culture. As a result, Buddhism became permanently established as the dominant religion in Sri Lanka (formerly Ceylon), and it established footholds as far afield as Southeast Asia and Mediterranean Europe. Ironically, Buddhism thrived in these remote places even as it declined in India itself. With Aśoka's death, the faith lost its strongest supporter.

The Mauryan Empire represented the greatest of political and cultural achievements of India in its day, and when it collapsed, late in the second century A.D., India fragmented into a jigsaw of states. Once again, India lay open to infusions from the west and northwest, and across present-day Pakistan they came: Persians, Afghans, Turks, and others driven from their homelands or attracted by the lands of the Ganges.

THE POWER OF ISLAM

In the late tenth century, the wave of Islam came rolling like a giant tide across the subcontinent, spreading from Persia in the west and Afghanistan in the northwest. Of course, the Indus valley lay directly in the path of this Islamic advance, and virtually everyone was converted. Next the Muslims penetrated the Punjab, the subregion that lies astride the present Pakistan-India border, and there perhaps as many as two-thirds of the inhabitants became converts. Then Islam crossed the bottleneck where Delhi is situated and diffused eastward and southeast into the Gangetic plain and the subregion known as Hindustan—India's evolving core area. Here Islam's proselytizers had less success, persuading perhaps one in eight Indians to become Muslims. In the meantime, Islam arrived at the Ganges delta by boat, and present-day Bangladesh became overwhelmingly Islamic. (To the south of the Ganges heartland, however, Islam's diffusion wave lost its energy: Dravidian India never came under Muslim influence.)

Islam's vigorous, often violent onslaught brought far-reaching changes to Indian society. As in West Africa, Islam often was superimposed by political control: the rulers of states were converted, and their subjects followed. By the early fourteenth century, a sultanate centered at Delhi controlled more of the subcontinent than even the Mauryan Empire had earlier. Later, the Islamic Mogul Empire (the similarity to the word *Mongol* is no coincidence) constituted the largest political entity ever to unify the realm in precolonial times. To many Hindus of lower caste, Islam formed a welcome alternative to the rigid socio-religious hierarchy in which they were trapped at the bottom. Thus Islam was the faith of the ruling elites and of the disadvantaged, a powerful cultural force in the heartland of Hinduism.

Just as Islam weakened in southern Europe, so its force became spent in vast and populous India. For all the Muslims' power, they never managed to convert even a majority of South Asians to their faith. They dominated the northwest corner of the realm (present-day Pakistan), where Lahore became one of Islam's greatest cities. But in all of what is today India, less than 15 percent of the population became and remained Muslim. And throughout the period of Islamic intervention, the struggle for cultural supremacy went on. Placid Hinduism and aggressive Islam did not easily coexist.

EUROPEAN INTRUSION

Into this turbulent complexion of religious, political, and linguistic disunity still another element soon intruded: European powers in search of raw materials, markets, and political influence. The Europeans profited from the Hindu-Muslim contest, and they were able to exploit local rivalries, jealousies, and animosities. British merchants gained control over the trade with Europe in spices, cotton, and silk goods, ousting their competitors—the French, Dutch, and Portuguese. The British East India Company's ships also took over the intra-Asian sea trade between India and Southeast Asia, long in the hands of Arab, Indonesian, Chinese, and Indian merchants. In effect, the East India Company became India's colonial administration.

As time went on, however, the East India Company faced problems it could not solve. Its commercial activities remained profitable, but it became entangled in a widening effort to maintain political control over an expanding Indian domain. The Company proved to be an ineffective governing agent at a time when the increasing Westernization of India brought to the fore new and intense frictions. Christian missionaries were challenging Hindu beliefs, and many Hindus believed that the British were out to destroy the caste system. Changes also came in public education, and the role and status of women began to improve. Aristocra-

The Taj Mahal tomb, located in Agra, about 100 miles south of Delhi. This architectural jewel is the most famous reminder of the Muslim contribution to the evolution of Hindustan.

cies saw their positions threatened as Indian landowners had their estates expropriated. Finally, in 1857, the three-month Sepoy Rebellion broke out and changed the entire situation. It took a major military effort to put it down, and from that time the East India Company ceased to function as the government of India. Administration was turned over to the British government, the Company was abolished, and India formally became a British colony—a status it held for the next 90 years until this *raj* (rule) ended in 1947.

COLONIAL TRANSFORMATION

Four centuries of European intervention in South Asia greatly changed the realm's cultural, economic, and political directions. Certainly, the British made positive contributions to Indian life, but colonialism also brought serious negative consequences. In this respect there are important differences between the South Asian case and that of Subsaharan Africa. When the Europeans came to India, they found a considerable amount of industry, especially in metal goods and textiles, and an active trade with both Southwest and Southeast Asia in which Indian merchants played a leading role. The British intercepted this trade, changing the whole pattern of Indian commerce.

India now ceased to be South Asia's manufacturer, and soon the country was exporting raw materials and importing manufactured goods—from Europe, of course. India's handicraft industries declined; after the first stimulus, the export trade in agricultural raw materials also suffered as other parts of the world were colonized and linked in trade to Europe. Thus the majority of India's people (who were farmers then as now) suffered an economic setback as a result of the manipulations of colonialism. Although in total volume of trade the colonial period brought considerable increases, the composition of the trade India now supported by no means brought a better life for its people.

Neither did the British manage to accomplish what the Mauryans and the Moguls had tried to do: unify the subcontinent and minimize its internal cultural and political divisions. When the crown took over from the East India Company in 1857, about 750,000 square miles (nearly 2 million sq km) of Indian territory were still outside the British sphere of influence. Slowly, the British extended their control over this huge unconsolidated area, including several pockets of territory already surrounded but never integrated into the previous corporate administration. Moreover, the British government found itself obligated to support a long list of treaties that had been made by the company's administrators with numerous Indian princes, regional governors, and feudal rulers.

These treaties guaranteed various degrees of autonomy for literally hundreds of political entities in India, ranging in size from a few acres to Hyderabad's more than 80,000 square miles (200,000 sq km). The British Crown saw no alternative but to honor these guarantees, and India was carved up into an administrative framework under which there were more than 600 "sovereign" territories in the subcontinent. These "Native States" had British advisors; the large British provinces such as Punjab, Bengal, and Assam had British governors or commissioners who reported to the viceroy of India, who in turn reported to Parliament and the monarch in London. In all, this near-chaotic amalgam of modern colonial control and traditional feudalism reflected and in some ways deepened the regional and local

Colonial remnant: the Queen Victoria Memorial (silver-domed building, right) that overlooks a tributary of the Hooghly River in Calcutta.

disunities of the Indian subcontinent. Although certain parts of India quickly adopted and promoted the positive contributions of the colonial era, other areas rejected and repelled them, thereby adding yet another element of division to an increasingly complicated human spatial mosaic.

Colonialism did produce assets for India. The country was bequeathed one of the best railroad and highway transport networks of the colonial domain. British engineers laid out irrigation canals through which millions of acres of land were brought into cultivation. Settlements that had been founded by Britain developed into major cities and bustling ports, as did Bombay (15.1 million inhabitants today), Calcutta (11.7 million), and Madras (5.9 million). The latter are still three of India's largest urban centers, and their cityscapes bear the unmistakable imprint of colonialism. Modern industrialization, too, was brought to India by the British on a limited scale. In education, an effort was made to combine English and Indian traditions; the Westernization of India's elite was supported through the education of numerous Indians in Britain. Modern practices of medicine were also introduced. Moreover, the British administration tried to eliminate features of Indian culture that were deemed undesirable by any standards—such as the burning alive of widows on the funeral pyres of their husbands, female infanticide, child mar-

riage, and the caste system. Obviously, the task was far too great to be achieved successfully in barely three generations of rule, but post-independence India itself has continued these efforts where necessary.

PARTITION

Even before the British government decided to yield to Indian demands for independence, it was clear that British India would not survive the coming of self-rule as a single political entity. As early as the 1930s, the idea of a separate Pakistan was being promoted by Muslim activists, who circulated pamphlets arguing that British India's Muslims were a nation distinct from the Hindus and that a separate state consisting of Sind, Punjab, Baluchistan, Kashmir, and a portion of Afghanistan should be created from the British South Asian empire in this area. The first formal demand for such partitioning was made in 1940, and, as later elections proved, the idea had almost universal support among the realm's Muslims.

As the colony moved toward independence, a political crisis developed: India's majority Congress Party would not even consider partition, and the minority Muslims re-

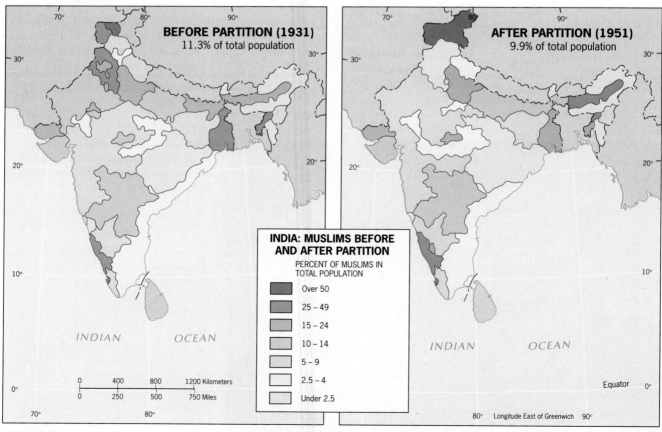

FIGURE 9-8

Flight was one response to the 1947 partition of what had been British India, resulting in one of the greatest mass population transfers in human history. Here, two trainloads of eastbound Hindu refugees fleeing (then) West Pakistan arrive at the station in Amritsar, the first city inside India.

fused to participate in any future unitary government. But partition would be no simple matter. True, Muslims were in the majority in the western and eastern sectors of British India, but there were many Islamic clusters scattered throughout the realm (Fig. 9-8). Any new boundaries between Hindus and Muslims to create an Islamic Pakistan and a Hindu India would have to be drawn right through areas where both sides coexisted. People by the millions would be displaced.

Nor were Hindus and Muslims the only people affected by partition. The Punjab area, for example, was home to millions of Sikhs, whose leaders were fiercely anti-Muslim. But a Hindu-Muslim border based only on those two groups would leave the Sikhs in Pakistan. Even before independence day, August 15, 1947, Sikh leaders talked of revolt, and there were some riots. But no one could have foreseen the dreadful killings and mass migrations that followed the creation of the boundary and the formation of independent Pakistan and India. Just how many people felt compelled to participate in the ensuing migrations will never be known; 15 million is the most common estimate. It was human suffering on an incomprehensible scale.

Even that huge number of refugees, however, hardly began to "purify" either India of Muslims or Pakistan of Hindus. After the initial mass exchanges, there still were tens of millions of Muslims in India (right map, Fig. 9-8); on the Indian side of the border with Pakistan, there still was a large Muslim cluster. Today, the Muslim minority in Hindu-dominated India is regarded as the world's largest minority, far more than a mere remnant of the days when Islam ruled the realm.

◆ PAKISTAN: THE NEW CHALLENGE

If Egypt is the gift of the Nile, then Pakistan is the gift of the Indus. The Indus River and its major tributary, the Sutlej, sustain the ribbons of life that form the heart of this populous country (Fig. 9-9).

Territorially, Pakistan is not very large by Asian standards; its area is about the same as that of Texas plus Louisiana. But Pakistan's population of 129.2 million makes this one of the world's 10 most populous states. Among Muslim countries (officially, it is known as the Islamic Republic of Pakistan) only Indonesia is larger, but Indonesia's Islam is much less pervasive than Pakistan's. Pakistan is an active participant in the worldwide resurgence of Islamic fervor; Indonesia is not—at least not yet.

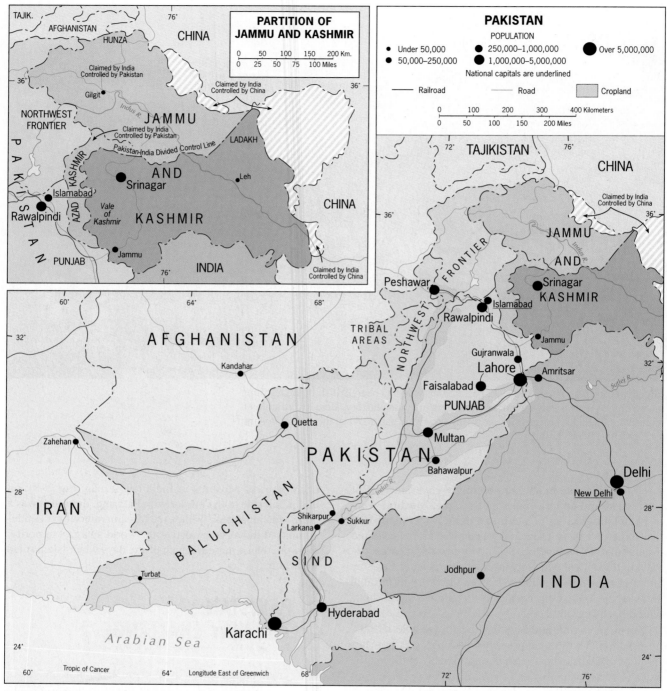

FIGURE 9-9

Pakistan lies like a giant wedge between Iran and Afghanistan to the west and India to the east. This wedge extends from the shores of the Arabian Sea in the south, where it is widest, to the snowcapped mountains of the north, where it reaches a point. Here in the far north of Pakistan, the boundary framework is complex and jurisdictions are uncertain. India, China, Pakistan, and, until recently, the Soviet Union have all vied for control over parts of this mountainous northern frontier. The Soviets, by virtue of their temporary rule over Afghanistan, controlled the Wakhan Corridor, shown on both the main map and the inset of Fig. 9-9. The Chinese have claimed areas to the east (inset map). But the major territorial conflict has been between Pakistan and India over the areas of Jammu and Kashmir. This is an issue that has plagued relations between these two countries for decades (see the box entitled "The Problem of Kashmir").

THE PROBLEM OF KASHMIR

Kashmir is a territory of high mountains surrounded by Pakistan, India, China, and, along several miles in the far north, Afghanistan (Fig. 9-9). Although known simply as Kashmir, the area actually consists of several political divisions, including the State properly referred to as *Jammu and Kashmir* (one of the 562 Indian States at the time of independence) and the administrative areas of Gilgit in the northwest and Ladakh (including Baltistan) in the east. The main conflict between India and Pakistan over the final disposition of this territory has focused on the southwest, where Jammu and Kashmir are located.

When partition took place in 1947, the existing States of British India were asked to decide whether they would go with India or Pakistan. In most of the States, this issue was settled by the local authority, but in Kashmir there was an unusual situation. There were about 5 million inhabitants in the territory at that time, nearly half of them concentrated in the basin known as the Vale of Kashmir (where the capital, Srinagar, is located). Another 45 percent of the people were concentrated in Jammu, which leads down the foothill slopes of the Himalayas to the edge of the Punjab. The small remainder of the population was scattered through the mountains, including Pakhtuns in Gilgit and other parts of the northwest. Of these population groups, the people of the mountain-encircled Vale of Kashmir are almost all Muslims, whereas the majority of Jammu's population is Hindu.

But the important feature of the State of Jammu and Kashmir was that its rulers were Hindu, not Muslim, although the overall population was more than 75 percent Muslim. Thus the rulers were faced with a difficult decision in 1947—to go with Pakistan and thereby exclude themselves from Hindu India, or to go with India and thereby incur the wrath of the majority of the people. Hence, the maharajah of Kashmir sought to remain outside both Pakistan and India and to retain the status of an autonomous unit. This decision was followed, after the partitioning of India and Pakistan, by a Muslim uprising against Hindu rule in Kashmir. The maharajah asked for the help of India, and Pakistan's forces came to the aid of the Muslims. After more than a year's fighting and through the intervention of the United Nations, a cease-fire line was established that left Srinagar, the Vale, and most of Jammu and Kashmir (including nearly four-fifths of the territory's population) in Indian hands (Fig. 9-9, inset map). In due course, this line began to appear on maps as the final boundary settlement, and Indian governments have proposed that it be so recognized.

Why should two countries, whose interests would be served by peaceful cooperation, allow a distant mountain-land to trouble their relationship to the point of war? There is no single answer to this question, but there are several areas of concern for both sides. In the first place, Pakistan is wary of any situation whereby India would control vital irrigation waters needed in Pakistan. As the map shows, the Indus River, the country's lifeline, crosses Kashmir. Moreover, other tributary streams of the Indus originate in Kashmir, and it was in the Punjab that Pakistan learned the lessons of dealing with India for water supplies. Second, the situation in Kashmir is analogous to the one that led to the partition of the whole subcontinent: Muslims are under Hindu domination. The majority of Kashmir's people are Muslims, so Pakistan argues that free choice would deliver Kashmir to the Islamic Republic. A free plebiscite is what the Pakistanis have sought—and the Indians have thwarted. Furthermore, Kashmir's connections with Pakistan prior to partition were much stronger than those between Kashmir and India, although India has invested heavily in improving its links to Jammu and Kashmir since the military stalemate. In recent years, it did seem more likely that the cease-fire line (now called the Divided Control Line) would indeed become a stable boundary between India and Pakistan. (The incorporation of the State of Jammu and Kashmir into the Indian federal union was accomplished as far back as 1975, when India was able to reach an agreement with the State's chief minister and political leader.)

The drift toward stability, however, was sharply reversed in the late 1980s as a new crisis engulfed Kashmir. Extremist Muslim groups demanding independence escalated a long-running insurgency into a new phase of violence in 1988; this brought on a swift crackdown by the Indian military, whose harsh tactics prompted the citizenry to far more strongly support the separatists. By the early 1990s, Kashmir was under tight curfew, the Pakistanis were charging the Indians with widespread human-rights violations, and the Indians were accusing the Pakistanis of inciting secession and supplying arms. With India prepared to go to any length to prevent the loss of Kashmir—whose populace now clearly aspires to break free of Indian control—the specter of the third war with Pakistan over this territory since 1947 loomed ominously.

When Pakistan became an independent state following the partition of British India (1947), its capital was Karachi on the south coast, near the western end of the Indus delta. As the map shows, however, the present capital is Islamabad (525,000), near the larger city of Rawalpindi (1.3 million) in the north, not far from Kashmir. By moving the capital from the "safe" coast to the embattled interior and by placing it on the doorstep of contested territory, Pakistan made clear to its neighbors its intent to stake a claim to its northern frontiers. And by naming the city Islamabad, Pakistan proclaimed its Muslim foundation, here in the face of the Hindu challenge. This politico-geographical use of a national capital can be very effective, and Islamabad exemplifies the principle of the **forward capital**.

SUBREGIONS

Pakistan evolved as the western flank of South Asia, separated by desert expanse and mountain slope from the vast geographic realm that lies to the west, the Muslim-dominated North Africa/Southwest Asia realm. The area of present-day Pakistan shared its Muslim heritage with this western domain, but for colonial control and guidance it had to look eastward to Delhi. When independence approached, Pakistan's Muslim leaders insisted on the creation of an Islamic state, and in 1947 the fateful boundary was drawn.

Sind

On physiographic as well as cultural-geographic grounds (as we have shown), Pakistan was and is a discrete region of the South Asian realm, but the country contains four well-defined subregions (Fig. 9-9). In the lower Indus valley in the southeast lies *Sind*, centered on the cities of Karachi

(9.8 million) on the coast and Hyderabad (1.1 million) upriver. The Arabs invaded Sind in the eighth century and imposed Islam on what had been a Buddhist area; from here, they disseminated the Muslim faith in all directions. In modern times, after the partition of the British Indian empire, millions of refugees flooded into Sind, nearly doubling the population and ballooning the sizes of cities and towns. But agriculture is Sind's mainstay, and the expansion of irrigated croplands in the Indus lowland, improvement of farming methods, and introduction of more productive seed strains made Sind a Pakistani breadbasket based on wheat and rice. Commercially, cotton is king here, giving rise to a major textile industry in the cities and towns.

Baluchistan

To the west of Sind lies *Baluchistan*, a vast desert area of flat plains and spectacular mountains, sparsely peopled and little developed. Physiographically, Baluchistan extends into southeastern Iran, where it is also called Baluchistan. This is desert country, remote and often forbidding, where camel caravans still move along ancient routes. It is also an area rich in mineral potential, and some oil and natural gas have been found here.

Punjab

To the north of Sind, where the Indus is joined by the Sutlej River, lies the heart of modern Pakistan, *Punjab*. Actually, the boundary between Pakistan and India was superimposed right across the greater Punjab, so that both countries now include subregions named Punjab (sometimes spelled *Panjab* in India). Pakistan's Punjab is home to nearly 60 percent of the country's population, so this truly is its core area. Three cities anchor this core: Lahore, the outstanding

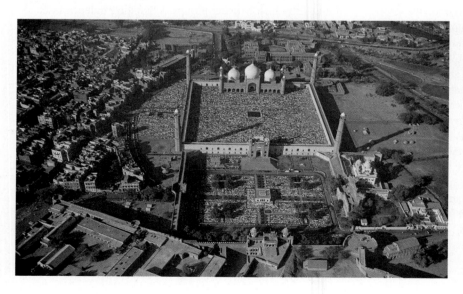

Lahore is the religious, cultural, and historic focus of Pakistan. It is a city of architectural grandeur and regional import, the region being Pakistan's Punjab. Here, nearly half a million worshippers fill the vast courtyard of the Badshahi Mosque.

center of Islamic culture in the realm, Faisalabad (1.9 million), and Multan (1.2 million).

Lahore, now home to 5.0 million people, lies close to the India-Pakistan border. The city grew rapidly following partition. Founded about two thousand years ago, Lahore became established as a great Muslim center during the Mogul period. As a place of royal residence, Lahore was adorned with numerous magnificent buildings, including a great fort, several palaces, and many mosques, which to this day remain monuments of history displaying superb stonework and tile and marble embellishments. The site of an old university and magnificent gardens, Lahore was the focus of a huge hinterland in pre-partition times. In those days its reach extended south to the port city of Karachi in Sind, north to Peshawar, and east to Delhi and Hindustan. Although its Indian hinterland was partitioned off in 1947, Lahore retained its importance as a cultural, historic, and industrial headquarters of the nation.

Northwest Frontier

The fourth subregion of Pakistan is appropriately called the *Northwest Frontier*. This is the area that faces turbulent and disintegrating Afghanistan, and it has been the scene of massive refugee influxes, local separatist movements, and social dislocation. The Northwest Frontier is dominated by mountain ranges, mountain-encircled basins, and strategic passes. The most famous pass of all is the Khyber Pass to Afghanistan. The Turks invaded the upper Indus valley by this route; later the Moguls streamed through it on their way to India; and in recent decades millions of war-weary emigrants have walked this road to safety in Pakistan.

The largest city in the Northwest Frontier province is Peshawar (1.7 million), but after Baluchistan this is Pakistan's least urbanized area. Peshawar lies in an alluvium-filled, fertile valley where fields of wheat and corn cover the countryside. But with its large refugee population (which, during the 1980s, reached in excess of 4 million), the Northwest Frontier needs help from the outside. Both the Pakistan government and the United Nations have aided the province with funds and relief supplies.

The ethnic and cultural backgrounds of the population in the Northwest Frontier vary. Even before the refugee influx from Afghanistan, a large part of the population consisted of Pathans (also called Pakhtuns, Pashtuns, or Pushtuns), who were closely related to the Pathans of central Afghanistan. At times, the Pathans of Afghanistan urged their kinspeople in Pakistan to demand greater autonomy, if not outright independence—a practice we know as **irredentism**. When the Soviets invaded Afghanistan in 1979 and war broke out, several million Pathans moved from Afghanistan to the Northwest Frontier. Pakistan not only managed to accommodate this refugee population but also countered the secession movement by improving the integration of the Northwest Frontier province with the rest of the country through road building, economic aid programs, and enhanced educational opportunities.

In the mid-1990s, Pakistan's Northwest Frontier remained a sensitive zone. It is only 170 miles (270 km) from Peshawar to Afghanistan's capital, Kabul, and conditions in Afghanistan have always influenced the Northwest Frontier. When the Soviets withdrew from Afghanistan during the late 1980s, Afghan refugees began to return home—but soon, Afghanistan's diverse ethnic-cultural groups (see p. 402) were in conflict again, this time over control of the capital and the government. A new outflow of refugees started, and the prospect of a collapsing Afghanistan, right on Pakistan's border, arose. Today, Pakistan's relationship with, and role in, a future Afghanistan is a crucial question for the entire region.

LIVELIHOODS

For all its size and growing regional influence, Pakistan remains a severely underdeveloped country. In the mid-1990s, urbanization was just approaching 30 percent. Population growth was high at 3.1 percent, implying a doubling time of only 23 years. Subsistence farming of food crops still occupied the great majority of the people. Life expectancy for a child born today is just 56 years. Illiteracy still afflicts 70 percent of the population.

Nevertheless, Pakistan during its nearly half-century of independence has made significant economic progress. Irrigation has expanded enormously; land reform has progressed; and the cotton-based textile industry has generated important revenues. Despite a very limited mineral resource base, some manufacturing growth has taken place, including a steel mill at Port Qasim near Karachi.

To the rest of the world, Pakistan sells cotton textiles, carpets and tapestries, and leather goods—and rice. Despite its large and growing population, Pakistan in the early 1990s still was able to export rice, which is a measure of the success of the agricultural expansion program. As Appendix A shows, Pakistan has the highest per-capita gross national product (GNP) of all the states of mainland South Asia. Certainly, Pakistan has exceeded the expectations that experts had for it at independence.

EMERGING REGIONAL POWER

A country's underdevelopment is not necessarily a barrier against its emergence as a power to be reckoned with in international affairs. When Pakistan became independent in 1947, the country was weak, disorganized, and divided. Just 50 years later, Pakistan is a major military force and, probably, a nuclear power.

To say that Pakistan in 1947 was a divided country is no exaggeration. In fact, upon independence, present-day

Pakistan was united with present-day Bangladesh, and the two countries were called West Pakistan and East Pakistan, respectively. The basis for this scheme was Islam: in Bangladesh, too, Islam is the state religion. Between the two Islamic wings of Pakistan lay Hindu India. But there was little else to unify the easterners and westerners, and their union lasted less than 25 years. In 1971, a costly war of secession led to independence for East Pakistan, which then took the name of *Bangladesh*; at the same time, West Pakistan became *Pakistan*.

To Pakistan, the loss of Bangladesh was no disaster: in virtually every respect, Bangladesh was (and remains) even more severely underdeveloped than Pakistan (see appropriate data in Appendix A). Pakistan's challenges lay closer to home—in Kashmir, in the Northwest Frontier, and, most important, to the east in India. India had supported Bangladesh in its campaign for independence, and the two countries did not resume diplomatic relations until five years later. Ever since, the political relationship between India and Pakistan has been tense. Apart from the conflict in Kashmir, their other differences are numerous. India remained a democracy while Pakistan became a military dictatorship; India's treatment of its Muslim minorities frequently riled Pakistan; India developed a close relationship with the same Soviet Union seen as a threat by Islamabad. During the period of the Cold War, India tilted toward Moscow while Pakistan, despite frequent disputes, was favored by Washington. In 1979, the United States cut off aid to Pakistan because Washington believed that the Pakistanis were secretly preparing to build nuclear weapons; but when the Soviet invasion of Afghanistan later that year put Moscow's armed forces on the doorstep of northwest Pakistan, the United States sold F-16 fighter planes to the Pakistani regime.

Mutual suspicion raised the military stakes in both Pakistan and India, and it is likely that both countries now possess nuclear arms. In the meantime, Pakistan now finds itself not only near disintegrating Afghanistan but also close to the scene of one of the great geopolitical, economic, and ideological transformations of the twentieth century: Turkestan (see pp. 403-407). No longer merely a newly decolonized, severely underdeveloped country trying to survive, Pakistan today is a powerful bulwark of traditional Sunni Islam, militant in its regional aspirations and situated in a critical part of the changing world.

◆ INDIA: FROM DEMOCRACY TO THEOCRACY?

Nearly three-quarters of the great land triangle of South Asia is occupied by a single country—India, the world's most populous democracy and, in terms of human numbers, the world's largest federation. Consider this: India has nearly

as many inhabitants as live in all the countries of Subsaharan Africa plus North Africa/Southwest Asia *combined*—74 of them. At present rates of population growth, India not only will outnumber these two realms at some time during the next century but will overtake even China. If India holds together as a single state, it is likely to become the world's most populous country, bar none.

That India has endured as a unified country is one of the politico-geographical miracles of the twentieth century. India is a cultural mosaic of immense ethnic, religious, linguistic, and economic diversity and contrast; it is a state of many nations. The period of British colonialism gave India the underpinnings of unity: a single capital, an interregional transport network, a *lingua franca*, a civil service. And upon independence in 1947, India adopted a **federal** system of government, giving regions and peoples some autonomy and identity, and allowing others to aspire to such status. Unlike Africa, where federal systems failed and where military dictatorships replaced them (notably in Nigeria), India, for nearly a half century now, has remained essentially democratic and has retained its federal framework.

This does not mean that India has not suffered from *centrifugal forces* (see pp. 484-485). Border conflicts, frontier wars, civil strife, and international clashes (most recently involving nearby Sri Lanka) have buffeted and at times threatened India's stability. But India has prevailed where others in the postcolonial world have failed.

POLITICAL SPATIAL ORGANIZATION

The politico-geographical map of India shows a federation divided into 25 States and 7 Union Territories (UTs)(Fig. 9-10). Delhi/New Delhi, the Union Territory that contains the capital, is the only Union Territory that has substantially more than 1 million inhabitants; the other six UTs are small territorially as well as demographically.

Several Indian States have territories and populations that are themselves larger than many countries and nations of the world. As Fig. 9-10 shows, the (areally) largest States lie on the great southward-pointing peninsula. Uttar Pradesh (139.1 million inhabitants)* and Bihar (86.4) constitute much of the Ganges River basin and are the core area of modern India (see the box on p. 482 entitled "Solace and Sickness from the Holy Ganges"). Maharashtra (78.9 million), anchored by the great coastal city of Bombay, also has a population larger than that of most of the world's countries. West Bengal, the State that adjoins Bangladesh, has 68.1 million residents, 11.7 million of whom live in its urban focus, Calcutta.

These are staggering numbers, and they do not decline much toward the south. Southern India consists of four States

---

*All State population data are for 1991, the year of India's latest census.

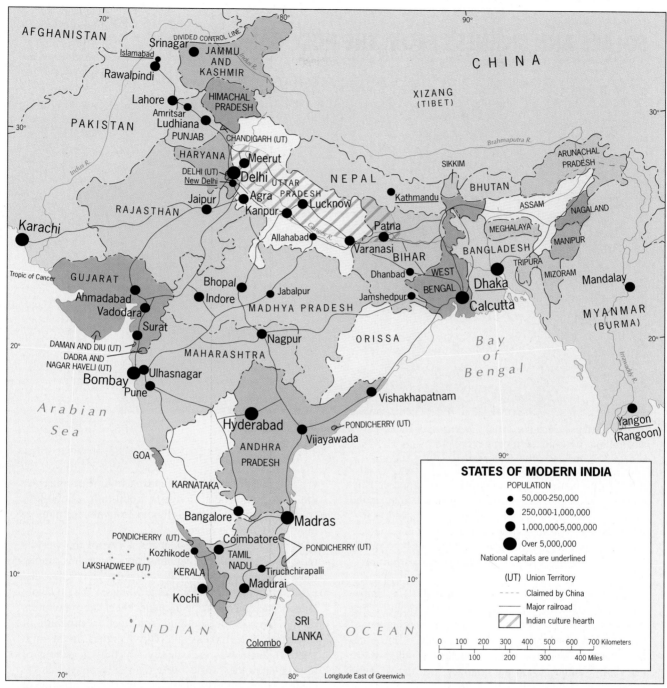

FIGURE 9-10

linked by a discrete history and by their distinct Dravidian languages. Facing the Bay of Bengal are Andhra Pradesh (66.5 million) and Tamil Nadu (55.9 million), both part of the hinterland of the great city of Madras, located on the coast very near their joint border. Facing the Arabian Sea are Karnataka (45.0 million) and Kerala (29.1 million). Kerala, often at odds with the federal government in New Del-

hi, long has had the highest literacy rate in India and one of the lowest rates of population growth. Strong local government and strictly enforced policies are credited with these achievements here. "It's a matter of geography," explained a teacher in the Kerala city of Cochin (1.4 million). "We are here about as far away as you can get from the capital, and we make our own rules."

SOLACE AND SICKNESS FROM THE HOLY GANGES

The stone steps beside the Ganges in Varanasi, India. This is Hinduism's holiest city, and millions descend these steps every year.

Stand on the banks of the Ganges River in Varanasi, Hinduism's holiest city, and you will see people bathing in the holy water, drinking it, and praying as they stand in it—while the city's sewage flows into it nearby and the partially cremated corpses of people and animals float past. It is one of the world's most compelling—and disturbing—sights.

The Ganges (*Ganga*, as the Indians call it) is Hinduism's holy river. Its ceaseless flow and spiritual healing power are, in the eyes of Hindus, earthly manifestations of the almighty. Therefore, tradition has it, the river's water is immaculate, and no amount of human (or other) waste can really pollute it. On the contrary: just touching the water can wash away a believer's sins.

At Varanasi (1.1 million), Allahabad (968,000), and other cities and towns along the Ganges, the river banks are lined with Hindu temples, decaying ornate palaces, and dozens of wide stone staircases called *ghats*. These stepped platforms lead down to the water, enabling thousands of bathers to enter the river (see photo above). They come from the city and from afar, many of them pilgrims in need of the healing and spiritual powers of the water. It is estimated that more than a million people enter the river somewhere along its 1,600-mile (2,800-km) course every day. During religious festivals, the number may be 10 times as large.

By any standards, the Ganges is one of the world's most severely polluted streams, and thousands among those who enter it become ill with diarrhea or other diseases; many die. In 1986, Prime Minister Rajiv Gandhi launched a major scheme to reduce the level of pollution in the river, a decade-long construction program of sewage treatment plants and other facilities. In the mid-1980s, Gandhi was told, nearly 400 million gallons of sewage and other wastes were being disgorged into the Ganges every day. The plan called for the construction of nearly 40 sewage treatment plants in riverfront cities and towns.

Many Hindus, however, did not support this costly program to clean up the Ganges. To them, the holy river's spiritual purity is all that matters. Getting physically ill is merely incidental to the spiritual healing power that a drop of the *Ganga's* water contains.

As Fig. 9-10 shows, India's smaller States lie mainly in the northeast, on the far side of Bangladesh, and in the northwest, toward Jammu and Kashmir. North of Delhi, India lies flanked by China and Pakistan, and physical as well as cultural landscapes change from the flatlands of the Ganges to the hills and mountains of spurs of the Himalayas. In the State of Himachal Pradesh, forests cover the hillslopes and living space is reduced by relief; only 5.2 million people live here, many in small, comparatively isolated clusters. Before independence and political consolidation, the colonial government called this area the "Hill States."

But the map becomes even more complex in the distant northeast, beyond the narrow corridor between Bhutan and Bangladesh. The dominant State here is Assam, famed for its tea plantations and important to India because of its oil and gas production, amounting to more than 40 percent of the domestic total. Assam attained full statehood in 1972, just after Bangladesh became independent from Pakistan. The great majority of its 22.4 million residents live in the lowland of the Brahmaputra River, upstream from Bangladesh; illegal migration from Bangladesh into Assam has at times strained relations between India and its Muslim neighbor.

In the Brahmaputra valley, Assam resembles the India of the Ganges. But in almost all directions from Assam, things change. To the north, in sparsely populated Arunachal Pradesh (865,000), we are in the Himalayan offshoots again. To the east, in Nagaland (1.2 million), Manipur (1.8 million), and Mizoram (690,000), lie the forested and terraced hillslopes that separate India from Burma (Myanmar). This is an area of numerous ethnic groups (more than a dozen in Nagaland alone) and of frequent rebellion against Delhi's government. And to the south, the States of Meghalaya (1.8 million) and Tripura (2.8 million), hilly and still wooded, border the teeming floodplains of Bangladesh. Here in the country's northeast, where peoples are always restive and where population growth still soars, India faces one of its strongest regional challenges.

India's Changing Map

The present map of India's States and UTs (Fig. 9-10) is not the one with which India was born as a sovereign state in 1947, and it is sure to change again in the future. After independence, the government first had to contend with several hundred "princely states," fiefdoms whose rights had been protected during the colonial period. These were absorbed into the States, and the privileged "princely orders" were phased out by 1972.

Next, the Indian government reorganized the country on the basis of its major regional languages (see Fig. 9-7). Hindi, spoken by more than one-third of the population, was designated the country's official language, but 13 other major languages also were given national status by the Indian constitution, including the four Dravidian languages of the south. English, it was anticipated, would become India's common language, its *lingua franca* at government, administrative, and business levels. And indeed, English not only remained the language of national administration but also became the chief medium of commerce in growing urban India. English was the key to better jobs, financial success, and personal advancement, and the language constituted a common ground in higher education.

The newly devised framework based on the major regional languages, however, proved to be unsatisfactory to many communities in India. In the first place, many more languages are in use than the 14 that had been officially recognized. Demands for the establishment of additional States soon arose. In 1960, the State of Bombay was divided into two language-based States, Gujarat and Maharashtra. Other changes were made to accommodate linguistic pressures.

A second problem involved the smaller ethnic groups of the northeast. The Naga, a group of peoples whose domain had been incorporated into Assam State, rebelled soon after India's independence. A protracted war brought federal troops into the area; following a truce and a decade of negotiations, Nagaland was proclaimed a State in 1961. This led the way for other politico-geographical changes in India's problematic northeastern wing.

Still another dilemma involves India's Sikh population. The Sikhs (the word means "disciples") adhere to a religion that was created about five centuries ago to unite warring Hindus and Muslims into a single faith. This faith's principles rejected negative aspects of Hinduism *and* Islam, and it gained millions of followers in the Punjab and adjacent areas. During the colonial period, many Sikhs supported British administration of India, and by doing so they won the respect and trust of the British, who employed tens of thousands of Sikhs as soldiers and policemen. By 1947, there was a large Sikh middle class in the Punjab. When independence came, many left their rural homes and moved to the cities to enter urban professions. Today, they still exert a strong influence over Indian affairs, far in excess of the 2 percent of the population they constitute.

After independence, the Sikhs demanded that the original Indian State of Panjab (Punjab) be divided into a Sikh-dominated west and a Hindu-majority east. The government agreed, so that Punjab as now constituted (Fig. 9-10) is India's Sikh stronghold, whereas neighboring Haryana State is mainly Hindu. Unfortunately, this redelimitation did not defuse all the pressures in the area. A militant Sikh minority demanded that an even more autonomous Sikh State be created in the Punjab, to be called *Khalistan*. Gradually this minority gained strength, and when the radical Sikh leaders held the State in a grip of violence and intimidation, the federal government sent troops to contain them. This led to a disastrous confrontation in 1984 at the Golden Temple in the capital, Amritsar, Sikhism's holiest shrine. In its aftermath, two of then-Prime Minister Indira Gandhi's Sikh bodyguards assassinated her, and the crisis deepened. To this

Cultural strife in the world's largest democracy: Hindus stage their 1992 attack on the mosque at Ayodhya. The mosque stands on a spot regarded by Hindus as a holy site; the conflict involving this shrine cost numerous lives and inflamed Hindu-Muslim animosities throughout the country.

day, relationships between Sikhs and non-Sikhs are tense, cycles of violence persist, and otherwise prosperous Punjab is in disarray.

These ethnic, cultural, and regional problems are but a sample of the stresses on India's federal framework. There is no Muslim State in India, but India has more than 100 million Muslims within its borders—the largest cultural minority in the world. As Fig. 9-8 shows, the percentage of Muslims is highest in remote Jammu and Kashmir, but it also is substantial in such widely dispersed States as Karnataka, Gujarat, and West Bengal. It should also be noted that the Muslim population in the 1990s (roughly 12 percent) constitutes a larger percentage than it did after partition (9.9 percent). Moreover, as a sector of India's population, the Muslim minority today is among the most rapidly growing.

Centrifugal Forces

In Chapter 1 we introduced the concept of **centrifugal** and **centripetal forces** (see box p. 60), respectively the dividing and unifying forces that continuously affect all states. No country in the world contains greater cultural diversity than India, and variety in India comes on a scale unmatched anywhere else on earth. Such diversity spells strong centrifu-

gal forces, although, as we will see, India also has powerful consolidating bonds.

Among the centrifugal forces, Hinduism's stratification of society into castes remains a pervasive reality. Under Hindu dogma, castes are fixed layers in society whose ranks are based on ancestries, family ties, and occupations. The **caste system** may have its origins in the early social divisions into priests and warriors, merchants and farmers, craftspeople and servants; it may also have a racial basis, as the Sanskrit term for caste is color. Over the centuries, its complexity grew until India came to possess several thousand castes, some with a few hundred members, others containing millions. Thus, in city as well as village, communities were segregated according to caste, ranging from the highest (priests, princes) to the lowest (the untouchables).

A person was born into a caste based on his or her actions in a previous existence. Hence, it would not be appropriate to counter such ordained caste assignments by permitting movement (or even contact) from a lower caste to a higher one. Persons of a particular caste could perform only certain jobs, wear only certain clothes, worship only in prescribed ways at particular places. They or their children could not eat with, play with, or even walk with people of a higher social status. The untouchables occupying the lowest tier were the most debased, wretched members of this rigidly structured social system. Although the British ended the worst excesses of the caste system, and postcolonial Indian leaders—including Mohandas Gandhi (the great spiritual leader who sparked the independence movement) and Jawaharlal Nehru (the first prime minister)—have worked to modify it, centuries of class consciousness are not wiped out in a few decades. In traditional India, caste provided stability and continuity; in modernizing India, it constitutes an often painful and difficult legacy.

Today, it is possible to discern a geography of caste—a degree of spatial variation in its severity. Cultural geographers estimate that about 15 percent of all Indians are of lower caste, about 40 percent of backward caste (one important rank above the lower caste), and some 18 percent of upper caste (at the top of which are the *Brahmans*, men in the priesthood). (The caste system does not extend to the Muslims, Sikhs, and other non-Hindus in India, which is why the percentages above do not total 100.) Not only the colonial government but also successive Indian governments have tried to help the lowest castes. This has had more effect in the urban areas than in the rural parts of India. In the isolated villages of the countryside, the untouchables often are made to sit on the floor of their classroom (if they go to school at all); they are not allowed to draw water from the village well because they might pollute it; and they must take off their shoes, if they wear any, when they pass higher-caste houses. But in the cities, untouchables have reserved places in the schools, a fixed percentage of government jobs at State as well as federal levels, and a quota of seats in national and State legislatures. Much of this was the result of efforts by Mohandas (Mahatma) Gandhi, who took a special interest in the fate of the untouchables (*harijans*) in Indian society.

The caste system remains a powerful centrifugal force, not only because it fragments society but also because efforts to weaken it often result in further division (see the box entitled "Jharkand: A New State in the Making?"). Gandhi himself was killed, only a few months after independence, by a Hindu fanatic who opposed his work for the least fortunate in Indian society. Today, India is being swept by a wave of Hindu fundamentalism that is caused, at least in part, by continuing efforts to help the poorest. Higher castes see themselves as disadvantaged, and they take refuge in a "return" to fundamental Hindu values. A small political party, the Hindu-nationalist Bharatiya Janata Party, is now gaining widespread support and is poised to challenge the older, established parties that have long governed India. If Indian politics fragments along religious lines, the miracle of Indian unity may come to an end.

Centripetal Forces

In the face of all these divisive forces, what bonds have kept India unified for so long? Without question, the dominant binding force in India is the cultural strength of Hinduism, its sacred writings, holy rivers, and general influence over Indian life. For the great majority, Hinduism is a way of life as much as it is a faith, and its diffusion over virtually the entire country (Muslim, Sikh, and Christian minorities notwithstanding) brings with it a national coherence that constitutes a powerful antidote to regional divisiveness.

Further, communications in much of this populous country are better than they are in many other developing countries, and the continuous circulation of people, ideas, and goods helps bind the disparate state together. Before independence, opposition to British rule was a shared philosophy, a strong centripetal force. After independence, the preservation of the union was a common objective, and national planning made this possible.

India's capacity for accommodating major changes and its flexibility in the face of regional and local demands have also served as a centripetal force. Boundaries have been shifted; internal political entities have been created, relocated, or otherwise modified; and secessionist demands have been handled with a mixture of federal power and cooperative negotiation. Indians in South Asia have accomplished what Europeans in Yugoslavia could not, and India's history of success is itself a centripetal force.

No discussion of India's binding forces would be complete without mention of the country's strong leadership. Gandhi, Nehru, and their successors did much to unify India by the strength of their compelling personalities. For many years, leadership was a family affair: Nehru's daughter, Mrs. Indira Gandhi, twice took decisive control (in 1966 and 1980) following episodes of governmental weakness, and *her* son, Rajiv Gandhi (who later was also assassinated), served as prime minister during the late 1980s. That era now appears to be over, and the crucial question of India's leadership again hangs in the balance.

THE POPULATION DILEMMA

The human population of the realm's seven countries today totals more than 1.2 billion—more than one-fifth of all humankind. India alone has almost 920 million inhabitants, second only to China among the countries of the world. Such rapid population growth poses a threat to national development. India's federal government and the governments of its States have enacted legislation and implemented programs to reduce the rate of population growth. Although these initiatives have had some effect, India's population in the mid-1990s was still growing at an annual rate of 2.0 percent (the realm as a whole was expanding by 2.2 percent).

To gain an idea of the meaning of these numbers, consider the *doubling time* they imply. When a population grows at an annual rate of 2.0 percent, it doubles in only 34 years. If population growth in India does not slow down signifi-

JHARKAND: A NEW STATE IN THE MAKING?

Where the Indian States of Bihar, West Bengal, Orissa, and Madhya Pradesh meet, in the zone west of Calcutta, lies one of India's poorest areas. This is tribal India, where meager subsistence farming supports millions of families, where schools are few, hospitals virtually unknown, and serfdom a lingering condition. The people here are at the bottom rung of the caste ladder, the poorest of the poor. They have had no say in State or federal affairs; industrialization has passed them by; planners have ignored them; land has been stolen from them.

Until now. In the 1980s, people in this four-State area mapped below (Fig. 9-11) began to organize themselves politically, and in Bihar their party won several seats in the legislature. Soon there were demands for a separate State whose tribal interests would be paramount and whose representatives would press for change at the fed-

eral level. This new State would be called *Jharkand*. In 1992, this intensifying demand was punctuated by road and railroad blockades, strikes, and bombings.

But neither the federal government nor the government of the most-affected State, Bihar, has been prepared to negotiate with the separatists. As Fig. 9-16 shows, southern Bihar is part of India's major eastern industrial region; although the tribespeople here have been little affected by industrialization, the revenues generated by this manufacturing activity are critical to Bihar's budget. Losing the south to a Jharkand State would not only put these industries under a rival State government but would cause a crisis in (remaining) Bihar's finances. Once again, the lines between the establishment and the ethnic upstarts are drawn—and still another part of India's map hangs in the balance.

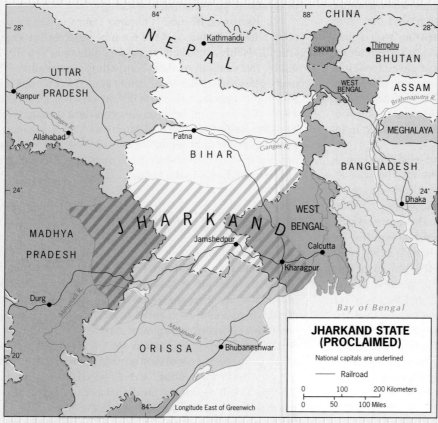

FIGURE 9-11

cantly, India will have 1.84 billion inhabitants by the year 2028. If that is to be the case, there is little chance that India will make any real economic progress. The great majority of its people will remain mired in poverty, and food crises will again occur.

Doubling time is a powerful indicator of a country's (or a region's) prospects. The faster a population doubles, the less likely it is to improve living conditions or opportunities for its people. Examine column 6 of Appendix A, and you will see doubling times ranging from under 20 to over 300 years. South Asia as a whole had a doubling time in 1994 of only 33 years; Europe's was 338 years. The world's rate of natural increase today is 1.7 percent, which means a doubling time of 41 years. South Asia, therefore, is growing faster than the world average. Indeed, as the global map of population growth (Fig. 9-12) indicates, South Asia is growing even faster than East Asia. If present conditions do not change, South Asia will soon become the world's largest population cluster, and India will surpass China to become the world's most populous country.

Not all the countries of South Asia are growing at the same rate. Pakistan, among the three largest countries of the realm, is growing fastest, at 3.1 percent; in the mid-1990s, Pakistan's population was doubling in 23 years. Bangladesh, too, was mushrooming at 2.4 percent, doubling in 29 years. The lowest growth rate was recorded in Sri Lanka, with only 1.5 percent (doubling in 46 years). (Not surprisingly, Sri Lanka also has the realm's highest GNP per capita: economic development and reduced growth rates are closely related.)

India's Demographic Challenge

India is the giant of South Asia, and the country's struggle to contain its population explosion will set the course for the realm. Figure 9-13 (p. 490) shows India's population growth since 1921. During the first decades of the twentieth century, the gap between birth rates and death rates began to widen. Birth rates remained high, but death rates declined as medical services improved, food distribution networks became more effective, agricultural production expanded, and costly wars were suppressed by the colonial regime. As the graph reveals, India's growth rate reached an all-time high of 2.22 percent during the decade of 1971–1981. In terms of actual numbers, however, the decade of 1981–1991 recorded the largest increase: 161 million.

In this as in so many other respects, there is not just one India but several different and distinct Indias. During the decade of its most rapid population growth (1971–1981), the highest growth rates were recorded in the States of the northeast and northwest; only three States had growth rates below 2 percent (Fig. 9-14A). During the period from 1981 to 1991, no fewer than eight States (including tiny Goa) had growth rates below 2 percent (Fig. 9-14B). Comparing these

two maps tells us that little has changed in India's heartland, where the populous states of Uttar Pradesh and Bihar show slight decreases but West Bengal and Madhya Pradesh display equally slight increases in growth rates. Some slight reductions are seen in the States of the northeast, but the important declines are recorded in the northwest, west, and south. Kerala and neighboring Tamil Nadu have growth rates *below* that of Sri Lanka.

Does this mean that the *demographic transition* (see p. 463) has taken hold in these areas of India and that the rest of the country will soon follow? Unfortunately, there is as yet no evidence for this. Vigorous propaganda campaigns, federal and State support for family planning programs, and even compulsory sterilization tactics have been implemented. In Kerala and Tamil Nadu, education rather than rapid economic development has resulted in a widespread acceptance of family planning. But the key to India's future lies in its core area, in the great lowland of the Ganges. There, change is coming perilously slowly.

Urbanization

As we saw in Chapter 6, the swiftly increasing urbanization of developing countries is one of the world's leading population processes in the 1990s. Although India may be known in part for its teeming cities with their hordes of homeless street dwellers (perhaps 400,000 in Calcutta alone), urbanization has proceeded more slowly here and only reached the 26 percent level in 1994. Once again, however, we must take note of India's massive proportions and realize that 26 percent of its population means 239 million people (almost the size of the United States).

The latest findings show an unprecedented upsurge in cityward migration, and urban India today is growing more than twice as fast (about 5 percent yearly) as the country's overall population. Among the reasons for this shift cited by Indian planners are a dramatic loosening of ties between poor peasants and their villages, and the widespread establishment in the cities of villagemen or "caste brothers" who are able to help their relatives and friends make similar moves to the burgeoning urban residential colonies of newcomers (increasingly defined by language and custom). Thus, even by Third World standards, Indian cities are places of staggering social contrasts (see lower photo p. 491). Not surprisingly, as crowding intensifies, social stresses multiply; sporadic rioting, often attributable to the actions of rootless urban youths unable to find employment, has affected numerous cities in recent years.

India's modern urbanization also has its roots in the colonial period, when the British selected Calcutta, Bombay, and Madras as regional trading centers and as coastal focal points for their colony's export and import traffic. All were British military outposts by the late seventeenth century. Madras, where a fort was built in 1640, lay in an area where the Brit-

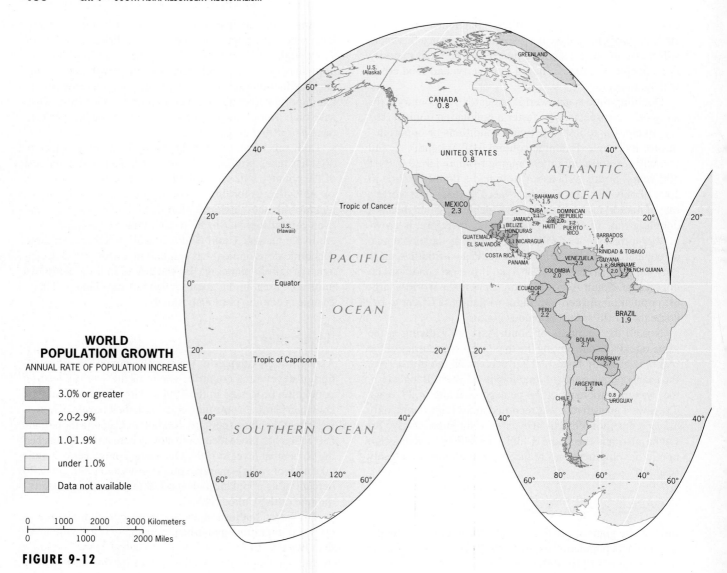

WORLD POPULATION GROWTH

ANNUAL RATE OF POPULATION INCREASE

- 3.0% or greater
- 2.0-2.9%
- 1.0-1.9%
- under 1.0%
- Data not available

FIGURE 9-12

ish faced few challenges; Bombay (1664) had the advantage of being the closest of all Indian ports to Britain; and Calcutta (1690) was positioned on the margin of India's largest population cluster and had the most productive hinterland. Calcutta lies 80 miles (130 km) from the coast on the Hooghly River, and myriad Ganges delta channels connect it to its hinterland. This natural transport network made the city an ideal colonial headquarters. But the British had to contend with Indian rebelliousness in the region; in 1912, they moved the colonial capital from Calcutta to the safer interior city of New Delhi, built adjacent to the old Mogul headquarters of Delhi (today these coalesced urban areas have a total population of 9.9 million).

Indian urbanization reveals several regional patterns. In the northern heartland, the west (the wheat-growing zone) is more urbanized than the east (where rice forms the main staple crop). This undoubtedly relates to the differences between wheat farming and the labor-intensive, small-plot cultivation and multiple cropping of rice. In the west, urbanization is around 40 percent (high by Indian standards); in the east, only about 20 percent of the population resides in urban centers. Furthermore, India's larger (100,000-plus) cities are concentrated in three regions: (1) the northern plains from Punjab to the Ganges delta, (2) the Bombay-Ahmadabad area, and (3) the southern part of the peninsula, which includes Madras and Bangalore (4.7 million). The only interior metropolises with populations over 1.5 million not located within one of these regions are centrally positioned Nagpur (1.8 million) and the capital of Andhra Pradesh, Hyderabad (5.3 million).

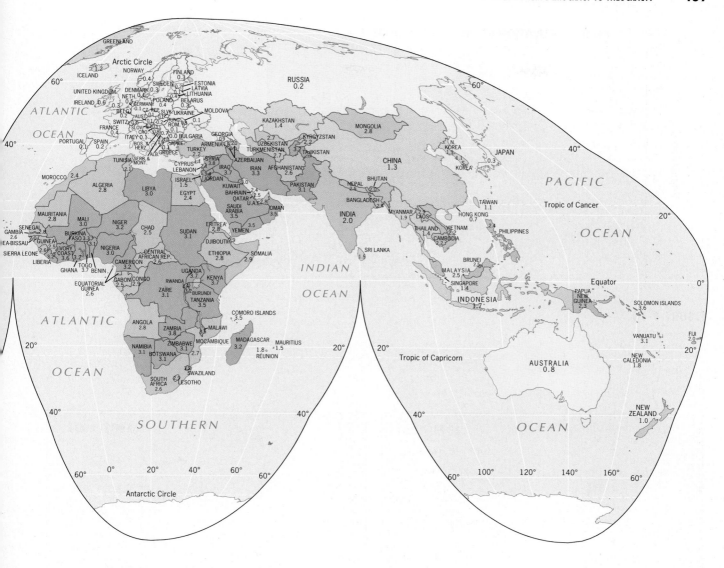

INDIAN DEVELOPMENT

If India has faced problems in its great effort to achieve political stability and national cohesion, these are more than matched by the difficulties that lie in the way of economic growth and development. The large-scale factories and power-driven machinery of the colonial powers wiped out a good part of India's indigenous industrial base. Indian trade routes were taken over. European innovations in health and medicine sent the rate of population growth soaring, without introducing solutions for the many problems this spawned. Surface communications were improved and food distribution systems became more efficient, but local and regional food shortages occurred (and still do) as droughts frequently caused crop failures. Today, nearly half of India's 900-million-plus people live in abject poverty, and the prospects of reducing that high level of human misery anytime soon are not encouraging. (Yet even with its modest annual per-capita income [U.S. $350], the sheer *size* of India's population creates a very big overall economy—the world's sixth-largest according to the latest rankings.)

Agriculture

India's underdevelopment is nowhere more apparent than in its agriculture. Traditional farming methods continue, and yields per acre and per worker remain low for virtually every crop grown under this low-technology system. Moreover, movement of agricultural commodities is hampered by the transportation inefficiencies of the traditional farming sys-

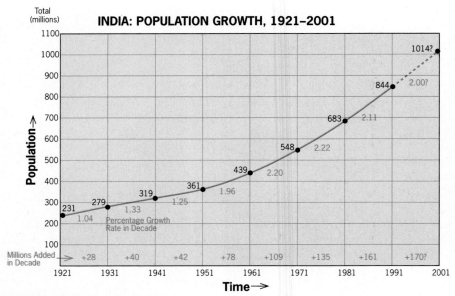

FIGURE 9-13

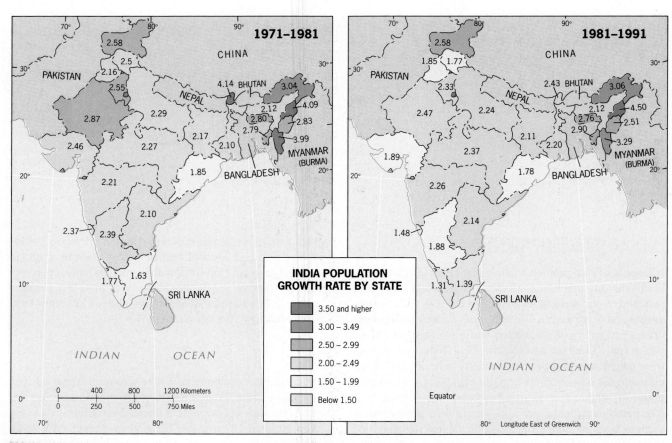

FIGURE 9-14

Agriculture in India, overall, is among the least efficient in all of Asia. Poor organization, inadequate equipment, lack of capital, and nature's constraints combine to keep yields comparatively low.

tem: in 1987, only 36 percent of India's 600,000 villages were accessible by motorable road, and today animal-drawn carts still outnumber motor vehicles nationwide.

As the total population grows, the amount of cultivated land per person declines. Today, this *physiologic density* is 1,456 per square mile (563 per sq km). However, this is nowhere near as high as the physiologic density in neighboring Bangladesh, where the figure is more than twice as great (3,318 and 1,282, respectively). But India's farming is so inefficient (see photo above) that this is a deceptive comparison. More than two-thirds of India's huge working population depends directly on the land for its livelihood, but the

Indian cities reflect sharp social divisions. These modern, middle-to-upper-class apartment buildings rise above squatters dwellings in New Delhi.

great majority of Indian farmers are poor and unable to improve their soils, equipment, or yields. Those areas in which India has made substantial progress toward the modernization of its agriculture (as in Punjab's wheat zone) remain islands in a sea of agrarian stagnation.

This stagnation has persisted in large measure because India failed, after independence, to implement a much-needed nationwide land reform program. In the late 1980s, about one-quarter of India's entire cultivated area was still owned by less than 5 percent of the country's farming families, and little land redistribution was taking place. Perhaps half of all rural families own either as little as an acre—or no land at all. Independent India inherited inequities from the British colonial period, but the individual States of the federation would have had to cooperate in any national land reform program. As always, the large landowners retained considerable political influence, so the program never got off the ground.

To make matters worse, much of India's farmland is badly fragmented as a result of local rules of inheritance, thereby inhibiting cooperative farming, mechanization, shared irrigation, and other opportunities for progress. Not surprisingly, land consolidation efforts have had only limited success except in the States of Punjab, Haryana, and Uttar Pradesh, where modernization has gone farthest. Certainly, official agricultural development policy, at the federal as well as State level, has also contributed to India's agricultural malaise and the uneven distribution of progress. Unclear priorities, poor coordination, inadequate information dissemination, and other failures have been reflected in the country's disappointing output.

It is instructive to compare Fig. 9-15, showing the distribution of crop regions and water supply systems in India, with Fig. I-6, which shows mean annual precipitation in India and the world. In the comparatively dry northwest, notably in the Punjab and neighboring areas of the upper Ganges, wheat is the leading cereal crop; here, India has made major gains in annual production through the introduction of high-yielding varieties developed in Mexico. This innovation was part of the so-called Green Revolution (see the box on p. 425 in Chapter 8) of the 1960s, when strains of wheat and rice were developed that were so much more productive than existing varieties that they were dubbed "miracle" crops. Introducing these new seeds also led to the expansion of cultivated areas, the development of new irrigation systems, and the more intensive use of fertilizer (a mixed blessing, for fertilizers are expensive and the "miracle" crops are more heavily dependent on them).

Toward the moister east, and especially in the wet-monsoon-drenched areas (Fig. 9-5), rice takes over as the dominant staple. About one-fourth of India's total farmland lies under rice cultivation, most of it in the States of Assam, West Bengal, Bihar, Orissa, and eastern Uttar Pradesh and along the Malabar coastal strip facing the Arabian Sea. These areas receive over 40 inches (100 cm) of rainfall annually, and irrigation supplements precipitation where necessary.

India has more land devoted to rice cultivation than any other country, but yields per acre remain among the world's lowest—despite the introduction of "miracle rice." Nevertheless, the gap between demand and supply has narrowed, and in the late 1980s India actually *exported* some grain to Africa as part of a worldwide effort to help refugees there. The situation remains precarious, however. As the population map (Fig. 9-1) shows, there is a considerable degree of geographic covariation between India's rice-producing zones

The same paddyfields before and after the arrival of the monsoon rains in Goa State on the west-central Arabian Sea coast, revealing the stunning contrast in the agricultural landscape of the dry and wet seasons.

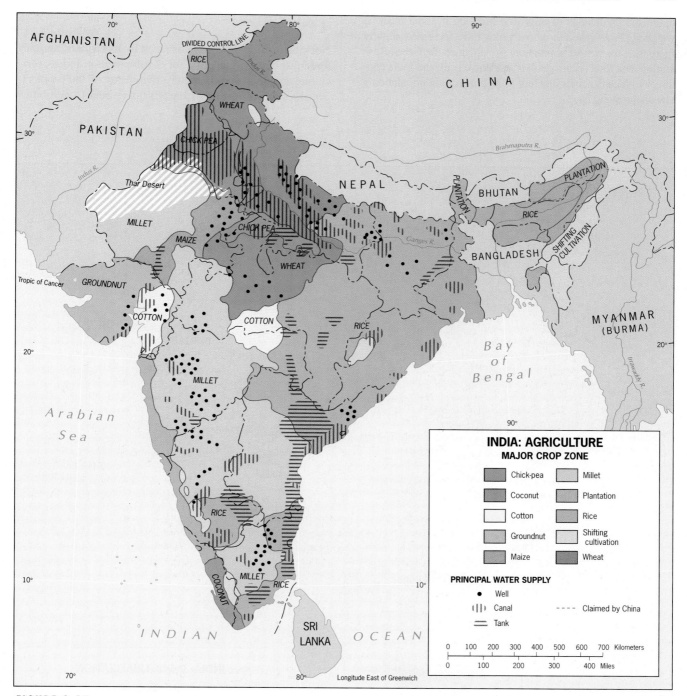

FIGURE 9-15

and its most densely populated areas. India is just one poor-harvest year away from another food crisis.

Of course, rice is not the only grain crop of India. As Fig. 9-15 shows, India's diverse environments permit the cultivation of wheat (especially in the northwest), millet, corn (maize), and other cereals. But subsistence continues to be the

fate of many tens of millions of Indian villagers who cannot afford fertilizers, cannot cultivate the new and more productive strains of rice or wheat, and cannot escape the cycle of poverty. Perhaps as many as 160 million of these people do not even own a plot of land and must live as tenants, always uncertain of their fate. This is the enduring reality against which optimistic predictions of improved nutrition in India must be weighed. True, rice and wheat yields have increased slightly more than the rate of population growth since the Green Revolution. But food security still is an elusive goal, and India continues to face the risks inherent in its burgeoning population's always-growing needs.

Industrialization

Notwithstanding the problems faced by its farmers, agriculture must be the foundation for development in India. Agriculture employs approximately two-thirds of the workers, generates most of the government's tax revenues, contributes

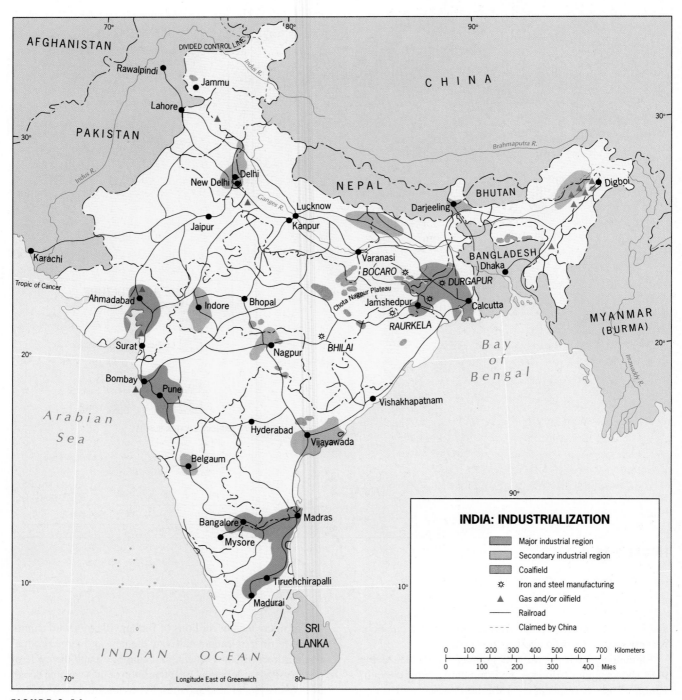

INDIA: INDUSTRIALIZATION

- Major industrial region
- Secondary industrial region
- Coalfield
- ☼ Iron and steel manufacturing
- ▲ Gas and/or oilfield
- — Railroad
- --- Claimed by China

FIGURE 9-16

India now is the world's sixth-largest economy, and has the third-largest industrial labor force. But factory equipment usually is obsolete, and pollution is generated in massive amounts. This metallurgical plant near New Delhi exemplifies these woes.

many of its chief exports by value (cotton textiles, tea, fruits and vegetables, jute products, and leather goods all rank high), and produces most of the money the country can spend in other sectors of the economy. Add to this the compelling need to grow more and more food crops, and India's heavy investment in agriculture is understandable.

In 1947, India inherited the mere rudiments of an industrial framework. After more than a century of British control over the economy, only 2 percent of India's workers were engaged in industry, and manufacturing and mining combined produced only about 6 percent of the national income. Textile and food-processing industries dominated. Although India's first iron-making plant was opened in 1911 and the first steel mill began operating in 1921, the initial major stimulus for heavy industrialization came after the outbreak of World War II. Manufacturing was concentrated in the largest cities: Calcutta led, Bombay was next, and Madras ranked third.

The geography of manufacturing today still reflects those beginnings, and industrialization in India has proceeded slowly, even after independence (Fig. 9-16). Calcutta now anchors India's eastern industrial region—the Bihar-Bengal district—where jute manufactures dominate, but cotton, engineering, and chemical industries also exist. On the nearby Chota Nagpur Plateau to the west, coal-mining and iron and steel manufacturing have developed.

On the opposite side of the subcontinent, two industrial areas dominate the western manufacturing region: one centered on Bombay and the other focused on Ahmadabad (3.7 million). This region, lying in Maharashtra and Gujarat states, specializes in cotton and chemicals, with some engineering and food processing. Cotton textiles have long been an industrial mainstay in India, and this was one of the few industries to derive some benefit from the nineteenth-century economic order imposed by the British. With the local cotton harvest,

the availability of cheap yarn, abundant and inexpensive labor, and the power supply from the Western Ghats' hydroelectric stations, the industry thrived and today outranks Britain itself in the volume of its exports.

Finally, the southern industrial region chiefly consists of a set of linear, city-linking corridors focused on Madras, specializing in textile production and light engineering activities. In the 1990s, all of India's manufacturing regions are increasing their output of ready-to-wear garments—another legacy of the early development of cotton textiles. Today, clothing has become India's second-leading export by value; the production of gems and jewelry, another growing specialization, ranks first.

When India achieved independence, its government immediately set out to develop its own industrial base, both to lessen dependence on imported manufactures and to cease being an exporter of raw materials to the developed world. In the process, emboldened by the early successes of the Green Revolution, Indian planners actually overspent on their industrial programs—but the problem of rapid population growth now bedeviled industry as it did agriculture. Unemployment was to be reduced as industrialization progressed; instead, unemployment rose. Per capita incomes would rise as high-value products rolled off new assembly lines; but income rose very little, and today it is still below U.S. $400 per year. Modern factories would safely and harmoniously blend into the high-density urban environment; unfortunately, the 1984 toxic gas leak at Union Carbide's pesticide plant in Bhopal (1.3 million) constituted the world's worst industrial disaster to date, killing about 3,500 and injuring an astounding 200,000.

Despite some imbalances and inefficiencies, India's industrial resource base is quite well endowed. Limited high-quality coal deposits are exploited in the Chota Nagpur district; in combination with large lower-grade coalfields elsewhere, the country's total output is high enough to rank it among the world's 10 leading coal producers. In the absence of known major petroleum reserves (some oil comes from Assam, Gujarat, Punjab, and offshore from Bombay), India must spend heavily on energy every year. Major investments have been made in hydroelectric plants, especially multipurpose dams that provide electricity, enhance irrigation, and facilitate flood control. India's iron ores in Bihar (northwest of Calcutta) and Karnataka (in the heart of the Deccan) may rank among the largest in the world. Jamshedpur (918,000), located west of Calcutta in the eastern industrial region, has emerged as India's leading steelmaking and metals-fabrication center. Yet India still exports iron ore as a raw material to developed countries (mainly Japan)—in the Third World, entrenched patterns are difficult to break.

Indian industrialization was also assisted by a major infrastructural advantage. In contrast to many other former colonial and now underdeveloped countries, India possesses a well-developed network of railroads with over 60,000 miles (100,000 km) of track. The British colonizers built much of this system, but once again they bequeathed to India a liability as well as an asset. The railways were laid out, of course, to facilitate exploitation of interior hinterlands and to improve India's governability; this effective transport network permitted the British to move their capital from coastal Calcutta to the deep interior at Delhi. But the railroad system never evolved as a unified whole. Different British colonial companies constructed their own networks, and (reminiscent of Australia) no fewer than four separate railway gauges came into use. The two widest gauges now constitute about 90 percent of the whole system, but many transshipments are still necessary. India's government continues to standardize the rail system when tracks must be replaced, but the expense is considerable and progress remains slow. Yet India's railway network connects most parts of the country and all its manufacturing complexes (Fig. 9-16). And in terms of overall length, the system ranks as the world's fourth largest.

◆ BANGLADESH: PERSISTENT POVERTY

On the map of South Asia, Bangladesh looks like another State of India: the country occupies the area of the double delta of India's great Ganges and Brahmaputra rivers, and it lies almost completely surrounded by India on its landward side (Fig. 9-10). But Bangladesh is an independent country, born in 1971 following its brief war for independence against Pakistan, with a territory about the size of Wisconsin. Today it is one of the poorest and least developed countries on earth,

with a population of 116.8 million that is doubling in under 30 years.

NATURAL HAZARDS

In the spring of 1991, Bangladesh was struck by yet another in an endless series of natural disasters: a devastating hurricane (or cyclone, as these tropical storms are called in this part of the world) that killed perhaps as many as 150,000 people (the exact toll will never be known). The storm, on a curving northward path across the Bay of Bengal, pushed a surging wall of water nearly 20 feet high across the islands and flatlands of the delta and swept most of the southeastern port city of Chittagong off the map (Fig. 9-17). The storm surge forced its way well inland along the winding channels of the Ganges-Brahmaputra delta, causing death and destruction even far from the exposed coastlands along the bay. When the waters receded, the bodies of countless people and animals were carried out to sea, later to wash up on the beaches. It was a catastrophe of unimaginable proportions—but it was not, by far, the worst calamity that Bangladesh has suffered. During the twentieth century, 8 of the 10 costliest natural disasters in the entire world have struck this single country. What is it that makes Bangladesh so vulnerable?

Look at Fig. 9-17 again. The land of Bangladesh lies just barely above sea level; the deltaic plain of the Ganges-Brahmaputra is a labyrinth of stream channels (see satellite image on p. 26). Only in the extreme east and southeast do these flatlands yield to hills and mountains. The delta's alluvial soils are extremely fertile, and every available patch of it is under crops: rice and wheat for subsistence, jute and tea for cash. The rivers' annual floods renew the farmlands' fertility by bringing silt; at the seaward margins of the delta, the silt piles up to form new islands. Even as this new land builds up, people move in to farm it. The crush of ever more mouths to feed compels this migration.

Now consider the inverted-funnel shape of the Bay of Bengal (see Fig. 9-4). Cyclones form often in this warm-water, humid-air environment (in contrast to the Arabian Sea, with its drier air and desert coasts, on the western side of India). When cyclones form in the Bay of Bengal, they often move northward along a rightward curve. As they do so, the water that piles up ahead of the storm has no place to go: the bay becomes ever narrower. And so, time and again, storm surges rise across the delta, sweeping people, livestock, and crops from the land. After the storm abates, the returning outrush of water causes added devastation as the delta's normally placid channels become troughs of raging torrents.

Unlike the comparatively wealthy Dutch, the Bangladeshis cannot combat their environmental enemy. Flood and storm warning systems are insufficient; escape plans and routes are inadequate. A program was recently begun to construct

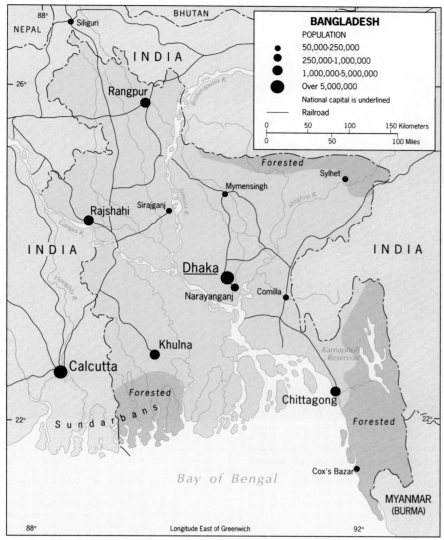

FIGURE 9-17

concrete, storm-proof shelters on pillars, to which trapped villagers might flee. Few were saved by the available shelters when the 1991 cyclone struck. Even this is too costly for Bangladesh to bear.

STABILITY AND SUBSISTENCE

Bangladesh's economic condition is reflected by its GNP per capita (a mere U.S. $200) and by its level of urbanization (only 14 percent in the mid-1990s). This is a land of subsistence farmers, with one of the highest physiologic densities in the world: 3,318 per square mile (1,282 per sq km). But there *are* bright spots: higher-yielding varieties of rice have helped close the gap between supply and demand, and wheat now plays a larger role on the farmlands, a rotation that improves food security. But diets remain unbalanced; overall, nutrition is barely adequate.

The other bright spot is political: despite occasional election-time skirmishes and notwithstanding some latent border disputes with India, the country has been relatively stable over the past decade. This has been crucial to its survival, because countries with fragile subsistence economies suffer disproportionately when political struggles cause dislocation. (Southern Africa's Moçambique, now even poorer than Bangladesh, is a case in point.) Nonetheless, the situation in Bangladesh *is* fragile. The country's infrastructure was badly damaged during the 1971 war for independence, and the communications system has never been repaired. Moreover, corruption

The intensively-farmed, low-lying alluvial plain and double (Ganges-Brahmaputra) delta of Bangladesh is frequently subjected to flooding, particularly during the peak of the wet-monsoon period. This scene shows the 1988 seasonal flood, one of the century's worst.

in government and a recurrent insurgency problem in the forested, mountainous southeast pose additional challenges for the years to come.

Dhaka, the centrally situated capital (8.8 million), and the devastated port of Chittagong (2.9 million) are the only urban centers of consequence in this dominantly rural country. Another measure of Bangladesh's economic misfortune lies in its transport system: there still is not a single road bridge across the Ganges River anywhere in the country and only one such railway crossing. The Brahmaputra River has not been bridged at all. When you travel from Dhaka to any town some distance away, you must be prepared for crawling road traffic and time-consuming ferry transfers; much of the country can be reached only by boat, thousands of which ply Bangladesh's many waterways.

The Indian State of West Bengal adjoins Bangladesh to the west, and this part of India resembles Bangladesh (*Bangla* means Bengal) in physical as well as human ways. West Bengal, too, occupies part of the delta of the Ganges River.

And like most Bangladeshis, the people here are Bengali. It was religion that put the boundary on the map in 1947: more than 80 percent of the people of Bangladesh adhere to Islam. But here in India's east, the cultural divide is much less sharp than it is vis-à-vis Pakistan in the west. More than 15 percent of the Bangladeshis are Hindu, while West Bengal contains a large Muslim minority.

Even more so than in much of India, population growth is Bangladesh's prime challenge. But let us not lose sight of the global context. A child born in Bangladesh will consume, during an equivalent lifetime, only 3 percent of the total consumption of a child born in the United States (by "consumption" we mean food, energy, minerals, and all other natural resources). Put another way, a single American child will consume what about 33 Bangladeshi children do. True, Bangladesh faces a population dilemma. But populous Bangladesh strains the world far less than a developed country of the same dimensions. In Bangladesh, survival is the leading industry; all else is luxury.

◆ SRI LANKA: ISLAND OF THE SOUTH

Sri Lanka (known as Ceylon prior to 1972), the compact, pear-shaped island located just 22 miles (35 km) across the

Palk Strait from the southern tip of the Indian peninsula, is the fourth independent state to have emerged from the British sphere of influence in South Asia (Fig. 9-18). Sovereign since 1948, Sri Lanka has had to cope with political as well as economic problems, some of them quite similar to those facing India and Pakistan, and others quite different.

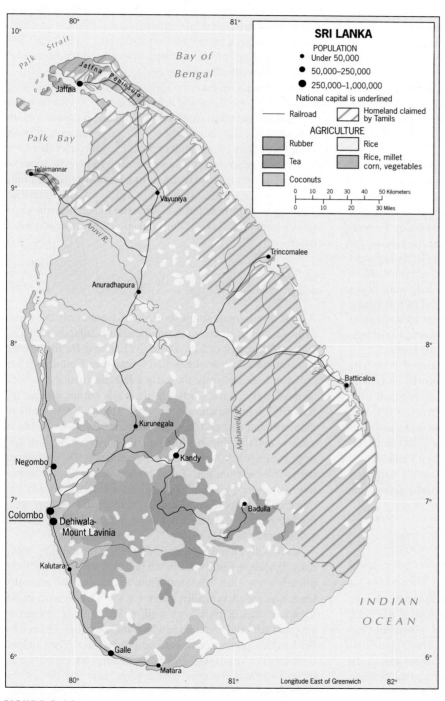

FIGURE 9-18

There were good reasons to create a separate independence for Sri Lanka. This is neither a Hindu nor a Muslim country; the majority—some 70 percent—of its 18.2 million people are Buddhists. Furthermore, unlike India or Pakistan, Sri Lanka is a plantation country (a legacy of the European period), with export agriculture still the mainstay of the external economy.

The majority of Sri Lanka's people are not Dravidian but are of Aryan origin with a historical link to ancient northern India. After the fifth century B.C., their ancestors began to migrate to Ceylon, a relocation that took several centuries to complete and brought to this southern island the advanced culture of the northwestern portion of the subcontinent. Part of that culture was the Buddhist religion; another component was the knowledge of irrigation techniques. Today, the descendants of these early invaders, the *Sinhalese*, speak a language (Sinhala) belonging to the Indo-European linguistic family of northern India.

The darker-skinned Dravidians from southern India never came in sufficient numbers to challenge the Sinhalese. They introduced the Hindu way of life, brought the Tamil language to northern Sri Lanka, and eventually came to constitute a substantial minority (now 18 percent) of the country's population. Their numbers were markedly strengthened during the second half of the nineteenth century when the British brought hundreds of thousands of Tamils from the adjacent mainland to Ceylon to work on the plantations that were being laid out. Sri Lanka has sought the repatriation of this ethnic element in its population, and an agreement to that effect was even signed with India. In 1978, however, Tamil was granted the status of a national language of Sri Lanka.

Sri Lanka is not a large island (about the size of West Virginia), but it displays considerable topographic diversity. The upland core lies in the south, where elevations reach over 8,000 feet (2,500 m) and sizeable areas exceed 5,000 feet (1,500 m). Steep, thickly forested slopes lead down to an encircling lowland, most of which lies below 1,000 feet (300 m). Northern Sri Lanka, capped by the Jaffna Peninsula, is entirely low lying. Rivers, the sources of ricefield irrigation waters, flow radially from the interior highland across this lowland rim.

PRODUCTIVITY

Since the decline of the Sinhalese Empire, focused on centrally located Anuradhapura, the moist southwest has been the leading zone of productive capacity (the plantations are concentrated here) and the population core. Three plantation crops have been successful: coconuts in the hot lowlands, rubber up to about 2,000 feet (600 m), and tea, the product for which Sri Lanka is most famous, in the highlands above. Sri Lanka is one of the world's largest tea exporters, and this commodity accounts for about one-fourth of the country's annual exports by value.

Whereas Sri Lanka's plantation agriculture has always been very productive and quite efficient, the same cannot be said for the island's ricelands. As recently as the 1960s, it was necessary to import half the rice consumed in Sri Lanka, a situation that was detrimental to the general economic situation. Accordingly, the government made it a priority to reconstruct plainland irrigation systems, repopulate the lowlands (until its successful eradication, malaria was an obstacle to settlement there), and intensify rice cultivation. The result was a substantial increase in rice production, and Sri Lanka achieved self-sufficiency.

In a country so heavily agricultural, it is not surprising to find very little industrial development except for factories that process plantation and farm products. Sri Lanka appears to have very little in the way of mineral resources; graphite is the most valuable mineral export. The industries that have developed, other than those processing foodstuffs, depend on Sri Lanka's relatively small local market; they include cement, paper, textiles, shoes, china, glassware, and the like. The majority of these industries cluster in and near the capital, Colombo (677,000), the largest city and the country's major port.

FRAGMENTATION

A look at the map suggests that Sri Lanka, off the coast of India, might share some situational advantages with Taiwan, off the coast of China. But Sri Lanka is far from being an economic tiger on the Indian Ocean rim. Certainly it has opportunities, but since the early 1980s this nation has been tormented by conflict between its major ethnic groups, the majority Sinhalese and the minority Tamils. The Tamils of the north and east had long argued that they were unable to achieve equal rights in education, employment, landownership, and linguistic and political representation. Their demands led to insurrection, and Tamil leaders called for a Cyprus-like partitioning of the island. Since 1984, a civil war has raged in Sri Lanka, with Tamil extremists demanding nothing less than an independent *Eelam* in their domain (see the striped area in Fig. 9-18).

Eventually, India became embroiled in the conflict. More than 50 million Tamils live in the Indian State of Tamil Nadu, directly across the Palk Strait from Sri Lanka, and naturally there was much sympathy there for the Tamil cause in Sri Lanka. It was in India's as well as Sri Lanka's interest to suppress the civil war, and in 1987, Indian troops, invited by Sri Lanka's government, entered the conflict. They helped contain the "Tamil Tigers," as the insurgents called themselves, but at great cost. Three years later the last of the Indian forces withdrew as inconclusive peace negotiations continued. Then, in 1991, Rajiv Gandhi, India's former prime

MINISCULE, MENACED MALDIVES

Imagine a country of more than a thousand tiny islands whose combined area is just 115 square miles (less than 300 sq km), located more than 400 miles (650 km) from the nearest continent, where the highest elevation is barely more than 6 feet (2 m) above sea level in a region where tropical storms prevail. Add a population of about 230,000, and we have described the Republic of the Maldives, southernmost political entity of the South Asian realm (Fig. 9-4).

As you sail toward the capital, the island of Male (50,000), the Maldives look like large lily pads on the ocean. Small boats ply the waters, linking the inhabited islands (about 200 have settlements) to each other and to the capital. Male's townscape is dominated by a large mosque, and a few buildings rise above two stories. Here, buildings have replaced the ubiquitous palm trees that almost completely cover the other islands.

The inhabitants of the Maldives speak a language that is related to an old form of Sinhalese, and it is believed that the first settlers came from Sri Lanka. If they were Buddhists, though, their successors were converted to Islam, and for centuries during the colonial period, when Portuguese and Dutch and British ruled here, the Maldives were a sultanate. Three years after the Maldives became independent, the sultanate was abolished, and the country has been a republic since 1968.

The Maldives' magnificent, palm-fringed beaches lure visitors from colder climes, and tourism has become the major industry, although fishing remains important as well. The people here have a special reason to worry about predictions of global warming and sea-level rise: if these come true, their entire country will disappear.

minister who headed the government when Indian troops fought in Sri Lanka, was assassinated near the Tamil city of Madras while campaigning to return to office.

As the 1990s wore on, the cycle of violence continued, punctuated by acts of terrorism. The casualty was Sri Lanka's future; apart from the enormous cost of the war, industries (such as tourism) suffered and investment dwindled. At independence, Sri Lanka seemed to have South Asia's brightest prospects; today, the country struggles for survival.

About 400 miles (650 km) southwest of Sri Lanka, well off of India's southern tip on the Arabian Sea side of the subcontinent, lie the Maldives. This island chain constitutes the remainder of South Asia's southernmost, insular region, and is profiled in the box to the left.

◆ THE MOUNTAINOUS NORTH

South Asia, as we noted earlier, is one of the world's most clearly defined geographic realms in physical as well as cultural terms. Walls of mountains stand between India and China—mountains that defy penetration even in this age of modern highways. And those mountains are more than barriers: South Asia's great life-giving rivers rise here, their courses fed by melting snow, sustaining tens of millions in the valleys and plains far below. Control over those source areas has caused centuries of conflict, and the results are etched on the political map.

As the map shows, a tier of landlocked countries and territories lies across this mountainous northern zone. From Afghanistan in the west and through Jammu and Kashmir and Nepal to Bhutan in the east, these isolated, remote, vulnerable entities are products of a long and complicated frontier history. Their vulnerability is underscored by the recent misfortunes of one of them, Afghanistan, and the disappearance (as a separate country) of another, Sikkim. The latter, wedged between Nepal and Bhutan, was absorbed by India in 1975 and made one of its 25 States. The kingdoms of Nepal and Bhutan, however, retain their independence.

NEPAL

Nepal, the size of Illinois and containing a population of 20.9 million, lies directly northeast of India's Hindu coreland. It is a country of three geographic zones (Fig. 9-19): a southern, subtropical, fertile lowland called the Terai; a central belt of Himalayan foothills with swiftly flowing streams and deep valleys; and the spectacular high Himalayas themselves (topped by Mt. Everest) in the north. The capital, Kathmandu (385,000), lies in the east-central part of the country in an open valley of the central hill zone.

Nepal is a materially poor but culturally rich country. The Nepalese are a people of many sources, including India, Tibet, and interior Asia; about 90 percent are Hindu, but Nepal's Hinduism is a unique blend of Hindu and Buddhist ideals. Thousands of temples and pagodas ranging from the simple to the ornate grace the cultural landscape, especially in the val-

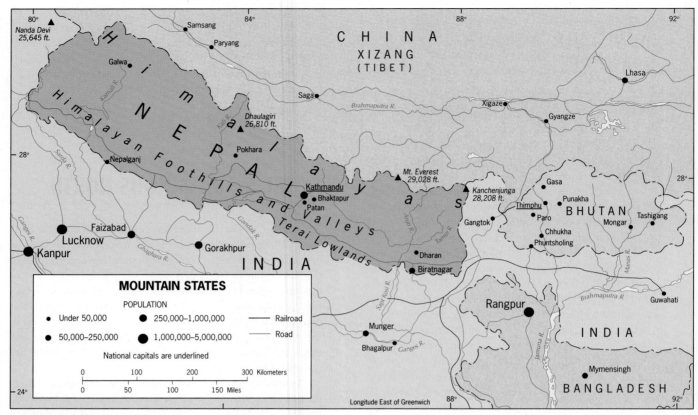

FIGURE 9-19

ley of Kathmandu, the country's core area. Although over a dozen languages are spoken, 90 percent of the people also speak Nepali, a language related to Indian Hindi.

Nepal's problems are those of underdevelopment and centrifugal political forces. Living space is limited, population pressure is high and rising steadily, and environmental degradation is a consequence. Deforestation is particularly severe—over one-third of Nepal's alpine woodlands have been cut over into wastelands since the 1960s—as the growing population of subsistence farmers is forced to expand into higher-altitude wilderness zones for sufficient crop-raising space (on steep terraces) and to obtain the firewood that supplies most of Nepal's energy needs. But soil quality is poor throughout the uplands, new farms are soon abandoned after a few seasons of declining productivity, and the land denudation process intensifies. Moreover, the steep slopes and the awesome power of the wet-monsoon rains accelerate soil erosion in treeless areas, and so much silt is now transported out of the Himalayas that (according to some researchers) river flooding is heightening in the crowded lower Ganges and Brahmaputra basins. With about half its farmland already abandoned to erosion and with 95 percent of its population engaged in subsistence agriculture (rice, corn, wheat, and millet), Nepal today faces a serious ecological crisis.

As the data in Appendix A underscore, Nepal is a very severely underdeveloped country; its per capita GNP (U.S. $170) is the lowest in the entire realm of South Asia, lower even than that of Bangladesh. The country's infrastructure is weak, and regionalism is strong. In terms of political geography, support for the old monarchy in the core area was not enough to forestall a nationwide demand for more democracy that, during the late 1980s, created a period of costly disruption. In 1991, democratic elections ushered in a new era.

Subdivisions

But the end of absolute monarchy did not solve Nepal's economic woes. Nepal needs integration and improved communications; the southern Terai zone, with its tropical lowlands resembling neighboring India, is a world apart from the hills of the central zone. And the peoples of the west have origins and traditions quite different from those in the east. There is always a fear of domination by the giant to the south, but even relations with neighboring Bhutan have been problematic, especially because Nepal now has representative government whereas Bhutan continues to be an absolute monarchy. Landlocked, regionally fragmented, economically deteriorating, and culturally splintered, Nepal faces the future with many

Deforestation is one of Nepal's major environmental threats. Against the spectacular backdrop of the main Himalaya range, the country's lower mountain slopes are denuded as population pressures force farming and fuel-wood extraction to ever higher elevations.

liabilities and few assets. Survival as a coherent state is its greatest challenge.

BHUTAN

East of Nepal lies the Kingdom of Bhutan, territorially only one-third as large and with just 755,000 inhabitants. As Fig. 9-19 shows, an extension of India's territory lies between Nepal and Bhutan (the State of Sikkim, referred to earlier),

but more than one-quarter of Bhutan's population is of Nepalese origin. Following Nepal's overthrow of the monarchy, some Nepalese leaders began to express concern over the fate of Nepalese citizens of Bhutan, and relations between the two countries worsened. In 1993, Nepalese people from Bhutan were moving as refugees to Nepal.

Not only was Bhutan the sole remaining absolute monarchy in the realm, but here, Buddhism is the state religion—although most Nepalese in Bhutan are Hindus. Bhutan's official language is related to Tibetan, and the country is more isolated and northward-looking even than Nepal. The capital, Thimphu, has a mere 32,000 residents and is located in the west-central part of the country.

Bhutan had a special position in South Asia during the period of British colonial control and was allowed to remain a separate entity. After India became independent in 1947, a treaty between the two was drawn up in 1949 that permitted Bhutan to continue to control its internal affairs but that required its rulers to consult India on matters of foreign relations.

This would appear to have been a prescription for eventual annexation by India, but Bhutan has survived—in part because of its protective remoteness and also because it has been self-sufficient in food. Bhutan's is a subsistence economy, but it has sustained the country; the monarchy has treated it like the mountain fortress it is. Certainly there are economic possibilities: forestry, hydroelectric power generation, and tourism have potential; moreover, the country is known to posess deposits of limestone, coal, slate, dolomite, lead, graphite, copper, talc, marble, gypsum, graphite, mica, pyrites, beryl, tufo. But development has been minimal, and for visitors this is a country where time seems to have stood still. The clock, nevertheless, is ticking, and even Bhutan will not escape the interactions that have forged this volatile South Asian geographic realm.

THE REALM IN TRANSITION

1. The rise of Hindu fundamentalism, expressed politically through a militant party, may usher in a new era in Indian politics and is changing India's political landscape.
2. India's enormous Muslim minority, long fragmented and divided geographically as well as doctrinally, is being galvanized and radicalized in reaction to rising Hindu militancy.
3. The conflict between India and Pakistan over Jammu and Kashmir continues to destabilize their joint northern frontier.
4. China continues to lay claim to substantial parts of Indian territory in their joint borderlands, although a 1993 agreement in-principle, signed by the heads of state, may lead to a settlement of the issue.
5. Pakistan's disagreements with India continue, and there are fears of nuclear armaments on both sides. But Pakistan increasingly is turning to the west and north (Afghanistan and Turkestan) to expand its horizons and its influence.

◆ PRONUNCIATION GUIDE

Agra (AHG-ruh)

Ahmadabad (AH-muh-duh-bahd)

Allahabad (ALLA-huh-bahd)

Amritsar (um-RIT-sahr)

Andhra Pradesh (ahn-druh pruh-DESH)

Anuradhapura (unna-rahd-uh-POOR-uh)

Arunachal Pradesh (AHRA-NAHTCH-ull pruh-DESH)

Aryans (AHR-yunz)

Aśoka (uh-SHOH-kuh)

Assam (uh-SAHM)

Ayodhya (uh-YOAD-yuh)

Badshahi [Mosque] (bud-SHAH-hee [MOSK])

Baluchistan (buh-loo-chih-STAHN)

Bangalore (BANG-guh-loar)

Bangladesh[i] (bang-gluh-DESH [ee])

Bengal[i] (beng-GAHL [ee])

Bharatiya Janata (buh-RUH-tee-uh juh-NAH-tuh)

Bhopal (boh-PAHL)

Bhutan (boo-TAHN)

Bihar (bih-HAHR)

Brahmaputra (brahm-uh-POOH-truh)

Buddhism (BOOD-izm)

Caste (CAST)

Ceylon (seh-LONN)

Chittagong (CHITT-uh-gahng)

Chota Nagpur (choat-uh NAHG-poor)

Cochin (koh-CHIN)

Colombo (kuh-LUM-boh)

Coromandel (kor-uh-MANDLE)

Deccan (DECKEN)

Delhi (DELLY)

Deltaic (dell-TAY-ick)

Dhaka (DAHK-uh)

Dravidian (druh-VIDDY-un)

Eelam (EE-lum)

Faisalabad (fye-SAHL-ah-bahd)

Gandhi (GONDY)

Ganga (GUNG-guh)

Ganges/Gangetic (GAN-jeez/ gan-JETTICK)

Ghats (GAHTSS)

Gilgit (GILL-gutt)

Goa (GO-uh)

Godavari (guh-DAH-vuh-ree)

Golconda (goll-KON-duh)

Gujarat (goo-juh-RAHT)

Gurkha (GHOOR-kuh)

Harappa (huh-RAP-uh)

Harijan (hah-ree-JAHN)

Haryana (hah-ree-AHNA)

Himachal Pradesh (huh-MAHTCH-ull pruh-DESH)

Himalayas (him-AHL-yuz/ himma-LAY-uz)

Hindu Kush (hin-doo KOOSH)*

Hindustan (hin-doo-STAHN)

Hooghly (HOO-glee)

Hyderabad (HIDE-uh-ruh-bahd)

Islamabad (iss-LAHM-uh-bahd)

Jaffna (JAHF-nuh)

Jammu (JUH-mooh)

Jamshedpur (JAHM-shed-poor)

Jamuna (JUM-uh-nuh)

Jharkand (JAR-kahnd)

Kabul (KAH-bull)

Kanarese (KAHN-uh-reece)

Karachi (kuh-RAH-chee)

Karakoram (kahra-KOR-rum)

Karnataka (kahr-NAHT-uh-kuh)

Kashmir (KASH-meer)

Kathmandu (kat-man-DOOH)

Kerala (KEH-ruh-luh)

Khalistan (kahl-ee-STAHN)

Khyber (KYE-burr)

Konkan (KAHNG-kun)

Ladakh (luh-DAHK)

Lahore (luh-HOAR)

Littoral (LIT-uh-rull)

Madhya Pradesh (mahd-yuh pruh-DESH)

Madras (muh-DRAHSS)

Maharashtra (mah-huh-RAH-shtra)

Malabar (MAL-uh-bahr)

Malayalam (mal-uh-YAH-lum)

Malaysia (muh-LAY-zhuh)

Maldives (MALL-deeves)

Male (MAH-lee)

Manipur (man-uh-POOR)

Mauryan (MAW-ree-un)

Meghalaya (may-guh-LAY-uh)

Meghna (MAIG-nuh)

Mizoram (mih-ZOR-rum)

Mohenjo Daro (moh-hen-joh-DAHRO)

Multan (mool-TAHN)

Myanmar (mee-ahn-MAH)

Naga[land] (NAHGA-[land])

Nagpur (NAHG-poor)

Nehru, Jawaharlal (NAY-roo juh-WAH-hur-lahl)

Nepal (nuh-PAHL)

Orissa (aw-RISSA)

Pagoda (puh-GOH-duh)

Pakhtuns (puck-TOONZ)

Pakistan (PAH-kih-stahn)

Palk (PAWK)

Pashtuns [see Pushtuns]

Pathans (puh-TAHNZ)

Peshawar (puh-SHAH-wahr)

Punjab (pun-JAHB)

Pushtuns (PAH-shtoonz)

Qasim (kah-SEEM)

Raj (RAHDGE)

Rajasthan (RAH-juh-stahn)

Rajiv (ruh-ZHEEV)

Rawalpindi (rah-wull-PIN-dee)

Sepoy (SEE-poy)

Shi'ite (SHEE-ite)

Siddhartha (sid-DAHR-tuh)

Sikh (SEEK)

Sikhism (SEEK-izm)

Sikkim (SICK-um)

Sinhala (sin-HAHLA)

Sinhalese (sin-hah-LEEZE)

Sri Lanka (sree-LAHNG-kuh)

Srinagar (srih-NUG-arr)

Sunni (SOO-nee)

Sutlej (SUTT-ledge)

Taj Mahal (TAHJ muh-HAHL)

Tamil [Nadu] (TAMMLE [NAH-doo])

Tapti (TAHP-tee)

Telugu (TELLOO-goo)

Terai (teh-RYE)

Thimphu (thim-POOH)

Tripura (TRIP-uh-ruh)

Turkestan (TER-kuh-stahn)

Uttar Pradesh (ootar-pruh-DESH)

Varanasi (vuh-RAHN-uh-see)

Vindhya (VIN-dyuh)

*Second double "o" pronounced as in "book"

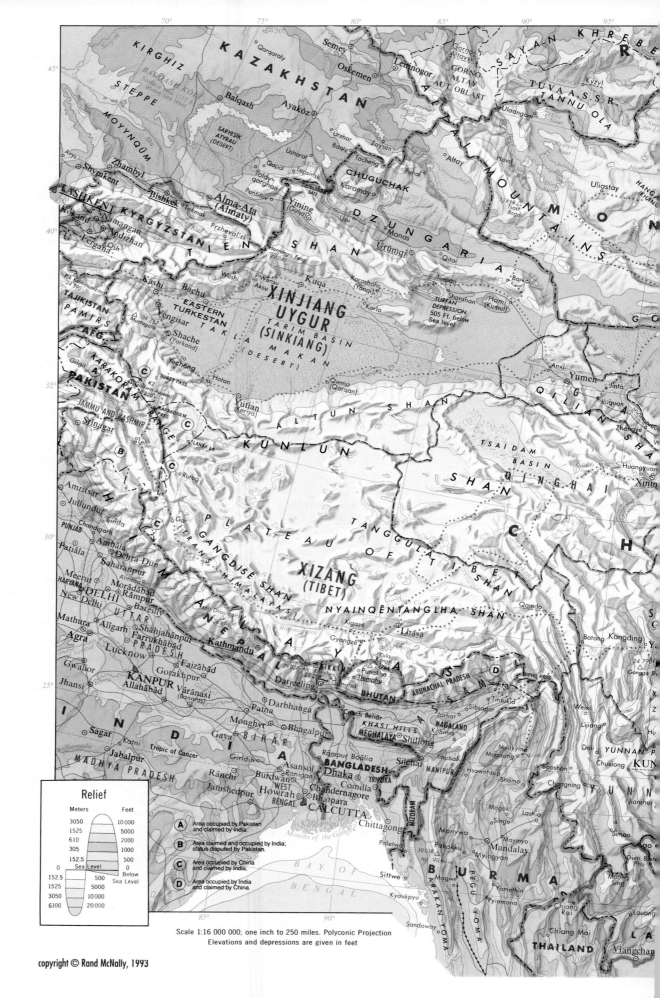

Scale 1:16 000 000; one inch to 250 miles. Polyconic Projection
Elevations and depressions are given in feet

copyright © Rand McNally, 1993

CHAPTER 10

CHINA: THE LAST EMPIRE?

IDEAS & CONCEPTS

Resource conservation
Extraterritoriality
Hegemony
Core area (2)
Buffer state (2)
Collectivization
Communization
Economic tiger (2)
Special Economic Zones (SEZs) (2)
Open Cities
Open Coastal Areas

REGIONS

China Proper
  Eastern Lowlands
  Shaanxi and Sichuan
  The South
  The Northeast
Inner Mongolia
Xinjiang
Xizang (Tibet)

It is often said and written, these days, that the disintegration of the Soviet Union marked the collapse of the world's last remaining great empire. Russian imperialism forged a vast colonial empire inherited by Moscow's communist despots, but now the pieces of that empire are independent states. The age of empires, say some scholars, is over.

From a geographic viewpoint, that verdict is premature. By many measures an empire far more populous than Russia's ever was still dominates the eastern quadrant of Asia. It is a realm without democracy or multi-party elections. It is a power core in control of territories that are colonies in all but name. It is a land of numerous disadvantaged minorities. It is a domain that still lays claim to territories beyond its borders. It is a regime that threatens neighbors. It is China, the last of the twentieth century's devolving empires.

China has not managed to shut out the winds of political and economic change that are sweeping the world. China's east is affected by the momentous changes marking the western Pacific Rim (as discussed in Chapter 4), and China's west lies exposed to the postcolonial transformation of Turkestan (as noted in Chapter 7). And the very heart of China was riven by student-led, labor-supported pro-democracy movements that crested tragically in the center of Beijing at Tiananmen Square in 1989. None of this, however, has impelled China's communist rulers to do what Soviet leaders did. The Communist Party's monopoly over the politics of China remains as strong as ever. The world may have turned its back on communist dogma, but not China. And China, we should remember, contains more than one-fifth of all humankind. Those who now write about a post-communist world are—again—too hasty.

◆ A LAND OF CONTRADICTIONS

China is nonetheless changing in fundamental ways. In the political sphere, authoritarian communism persists, but in the economic arena, China has opened its doors to capitalist enterprise. China's rulers are determined to keep political control, thereby to avoid the turmoil that has plagued devolving empires from the British to the Russian. But they also are determined not to miss the opportunities of market economies and to share in the economic opportunities arising on the Pacific Rim.

In the early 1990s, while North American and European economies stagnated, China's was growing at an annual rate of 9 percent. Geographically, this growth is transforming the country's structure. Zones along China's Pacific coast are booming, their urban landscapes transformed by new skyscrapers and modern factories. But China's vast interior is comparatively unchanged, still awaiting the impact of the new eco-

TEN MAJOR GEOGRAPHIC QUALITIES OF CHINA

1. China's population represents over one-fifth of all humankind. Territorially, China ranks third among the world's countries.
2. China is one of the world's oldest continuous civilizations.
3. China's civilization developed over a long period in considerable isolation, protected by physiographic barriers and by sheer distance from other source areas.
4. The Chinese state and national culture evolved from a core area that emerged in the north, near the present capital of Beijing. China's culture hearth has remained there ever since.
5. Foreign intervention had disastrous impacts on Chinese society, from European colonialism to Japanese imperialism. Intensified regionalism and territorial losses are only two of the many resulting afflictions.
6. China occupies the eastern flank of Eurasia. Its sphere of influence was reduced by Russian expansionism in East Asia.
7. China's enormous population is heavily concentrated in the country's eastern regions. Western zones remain comparatively empty and open, and are also more arid and far less productive.
8. China's communist-designed transformation after 1949 involved unprecedented regimentation and the imposition of effective central authority, with results that are perhaps permanently imprinted on the cultural landscape.
9. China's recent modernizing drive notwithstanding, the country remains a dominantly rural society with limited urbanization and industrialization.
10. Rural China is a land of enduring traditions. Neither the Communist Revolution nor the modernization drive has truly changed the villagers' way of life. Many old values persist, and the teachings of Kongfuzi (Confucius) are still remembered.

nomic policies. In time, China will have to face up to its fundamental political-economic contradiction. When that moment arrives, communist dogma will confront capitalist energy, and China will face the social consequences of unequal incomes and the spatial problems of regional inequalities. It will take more than a military clampdown on protesters in city squares to resolve China's paradox.

NAMES AND PLACES

In 1958, the government of the People's Republic of China adopted the so-called *pinyin* system of standard Chinese, which replaced the Wade-Giles system used since colonial times. This was not done to teach foreigners how to spell and pronounce Chinese names and words but to establish a standard form of the Chinese language throughout China. The pinyin system is based on the pronunciation of Chinese characters in Northern Mandarin, the Chinese spoken in the region of the capital and in the north in general.

The new linguistic standard caught on rather slowly outside China, but today it is in general use. The old name of the capital, Peking, has become Beijing. Canton is now Guangzhou. The Yangtze Kiang (River) is now the Chang Jiang. Tientsin, Beijing's port, is now Tianjin. A few of the old names persist, however. The Chinese call their colony Xizang, but many maps still carry the name Tibet.

Personal names, too, were affected by pinyin usage. China's long-time ruler, Mao Tse-tung, now is called Mao Zedong. His eventual successor, Teng Hsiao-ping, is more simply Deng Xiaoping.

And remember: the Chinese, when they write their names, use the last name first. Xiaoping, therefore, is Mr. Deng's first name. To the Chinese, it is Clinton Bill and Gore Al, not the other way around!

Whatever happens, its impact will be felt globally, and it will be critical for us in the Western world to understand China as best we can. Many years ago, two prominent geographers, Halford Mackinder and Nicholas Spykman, debated the geopolitical prospects of Eurasia's power cores. Mackinder, nearly a century ago, assessed the distribution of resources and populations, environmental conditions and natural advantages, and concluded that the "heartland" power, Russia or its successor, would become the continent's dominant force. Spykman interpreted the data differently. The rim of Asia, he argued, contained the demographic, environmental, locational and resource potentials to produce even greater power. Spykman prophetically used the term "rimland" as early as 1944. Today, Asia's rimlands are emerging, and the former Soviet heartland is in decline.

Where will China's ascent lead? When an economy grows at an annual rate of 9 percent, the total national wealth doubles every eight years; hence, China is gaining rapidly in its quest for development. We should remember how rapidly Japan rose, not only when it modernized after the Meiji

Restoration but also from the ashes of World War II. China now is poised to take what its communist godfather, Mao Zedong, called a "great leap forward." Yet this great leap will occur because China's rulers, in the post-Mao period, have been able to loosen the reins on the economy. If China's new-age communists can hold the country together, if they can resist political reform while guiding economic change, the future will witness China's turn as a challenger for the role of global superpower—if not world domination.

As we noted in Chapter 4, the Pacific is becoming the Atlantic of yesteryear. Across it, and along its margins, the United States and China will trade and compete. The nineteenth century saw the rise of European colonial empires that spanned the world. The twentieth century witnessed Nazi Germany's quest for world domination and the Soviet Union's emergence as a global superpower. The twenty-first century will see the ascent of China—if the last empire lasts.

CHINESE PERSPECTIVES

When we in the Western world chronicle the rise of civilization, we tend to focus on the historical geography of Southwest Asia, the Mediterranean, and Western Europe. Ancient Greece and Rome were the crucibles of culture; Mediterranean and Atlantic waters were the avenues of its diffusion. China lay remote, so we believe, barely connected to this Western realm of achievement and progress. When an Italian adventurer named Marco Polo visited China during the thirteenth century and described the marvels he had seen, his work did little to change European minds. Europe was and would always be the center of civilization.

The Chinese, naturally, take quite a different view. Events on the western edge of the great Eurasian landmass were deemed irrelevant to theirs, the most advanced and refined culture on earth. Roman emperors were rumored to be powerful, and Rome surely was a great city, but nothing could match the omnipotence of China's rulers. Certainly the Chinese city of Xian far eclipsed Rome as a center of sophistication. Chinese civilization existed long before ancient Greece and Rome emerged, and it was still there long after they collapsed. China, the Chinese teach themselves, is eternal. It was, and always will be, the center of the civilized world.

We should remember this when we study China's regional geography, because 4,000 years of Chinese culture and perception will not change overnight—not even in a generation. Time and again, China overcame the invasions and depredations of foreign intruders, and afterward the Chinese would close off their vast country against the outside world. Barely two decades ago, in the early 1970s, there were just a few *dozen* foreigners in the entire country with its (then) nearly 1 billion inhabitants. The institutionalization of communism required this, and, following a quarrel, even the Russian advisers had been thrown out. But then China's rulers

RESOURCE CONSERVATION

Among the many fields of study that make geography what it is, that of **resource conservation** is one of the oldest—and one of the most interesting. Understanding what natural resources are, how they have been (and are being) used, how they may be conserved, and how they enable nations and governments to pursue their economic as well as political goals keeps our researchers and analysts busy.

Even defining the term *resource* is not as easy as it might seem. Economic geographers often classify the resources of a national economy into three groups: land, labor, and capital. Geographers generally tend to focus on the first of these three categories—the land—under which are subsumed soil and vegetation, metallic and nonmetallic minerals, fuels, and all other raw materials that form part of the natural composition of the earth. Hence we concentrate our attention on *natural resources*, those that occur in nature and are put to human use.

Resources are natural substances, but they do not *become* resources until they are utilized. To our distant hunting and gathering ancestors, copper was not a resource; it did not become so until it was deemed to have value as a malleable substance for the creation of various implements; its worth increased when, in combination with tin, it produced bronze. To the ancient Egyptians, petroleum was not a resource; uranium did not become a resource until the twentieth century. Undoubtedly there are substances on and in the earth that will become resources in the future, but that are not resources today. In other words, human development determines and defines the resource: what is or is not a resource is a cultural matter.

Every civilization in the history of humanity has depended on its particular set, or *complex*, of resources. Technological development leads to the use of new resources. Ours is the era of energy resources: coal fired the flames of the Industrial Revolution, and oil, natural gas, and nuclear-energy uranium followed. Modern industries also consume unprecedented amounts and varieties of metals and alloys, out of which most of today's weapons are forged.

A distinction should be made between *nonrenewable* and *renewable* resources, although this concept is more complicated than it seems at first. Some resources *do* renew themselves (if we do not interfere with the process). Leave a patch of soil to rest, and it will regenerate over time. A cut-over forest will regenerate—not to the old-growth quality, but trees will grow again. The hydrologic cycle ensures that rivers will flow indefinitely (barring climate change or human destruction of drainage basins). Other resources are finite; they *can* be used up. Certain scarce ores, energy resources (such as oil) or other natural substances, theoretically at least, could run out.

Economic geographers remind us that some conditions apply to this apparently simple distinction. Renewable resources are *not* inexhaustible: soil, for example, can be overused, resulting in erosion and destruction. And nonrenewable resources such as oil may actually never run out: when the last drop has been discovered, the cost of the final barrels will be so high that no one can afford it. So, theoretically, no nonrenewable resource will ever be completely exhausted.

From these comments you will note that the term resource encompasses

---

decided that an opening to the Western world would be advantageous, and U.S. President Richard Nixon was invited to visit Beijing. That historic occasion, in 1972, ended this latest period of isolation—as always, on China's terms. Since then, China has been open to tourists and businesses, teachers and investors. Tens of thousands of Chinese students were sent to study at American and other Western institutions. Long-suppressed ideas flowed into China, and the pro-democracy movement arose. China's rulers knew that their violent repression of this movement would anger the world, but this did not matter: foreign condemnation was deemed irrelevant. Foreigners in China had done much worse. And Westerners had no business interfering in China's domestic affairs.

It would be wrong to suggest that China's rulers act on the world stage without any concern for the consequences. When, during the 1992 presidential campaign in the United States, then-President George Bush approved the sale of jet fighter aircraft to Taiwan, China objected vehemently but did not take retaliatory economic action that would have damaged its own progress. Earlier, China had sold Silkworm missiles to Southwest Asian countries in full knowledge of American disapproval and stopped doing so only when U.S. officials specified the disadvantages to China of continuing these sales. But China still operates from a position of weakness; early during the next century, it will be far stronger militarily as well as economically. Then the colossus on the Pacific will be a far more formidable power with which to negotiate.

more than minerals and fuels. Not only soil and water but also the natural vegetation, wildlife, fish and other marine life also constitute resources. All these are tangible resources. Geographers like to point out that something as intangible as relative location can also constitute a resource. Would Singapore or Hong Kong have arisen in inner Asia? They have no measurable tangible resources, but they have the intangible spatial advantage: location.

The protection and conservation of natural resources, as noted earlier, is an old geographic pursuit. It is fascinating to compare the topics taught under this rubric a half century ago (when the earth's human population was barely 40 percent of what it is today) with our current concerns. Back then, soil erosion, stream control, dust-bowl concerns, and similar issues were prominent. Take such a college course today, and you hear about pesticide dangers, atmospheric pollution and acid rain, oil-spill dangers, ozone depletion, radiation hazards, and similar worries. Our huge numbers and our unchecked industrialization have made endangered species of some of our

most precious resources, the atmosphere and ocean waters among them.

And even as some countries seek to reduce their consumption of (and dependence on) certain resources, thereby protecting the natural environment, others are rushing headlong into their own industrial age. China's cities today are forests of pollution-belching skyscrapers, its rivers indescribably polluted, its mines scarring the countryside. But China, now that its turn has come, will not accept the urgings of other countries that it limit its production of pollutants or its construction of huge hydroelectric dams (thus drowning vast historic and scenic lands). The conservation and protecton of resources is a luxury that will come later, when China has caught up to is competitors.

And speaking of competition, this, ultimately, is what natural resources have been all about. One definition of political power in this world holds that such power lies in "the capacity of a country to use its tangible and intangible resources in such a way as to influence the behavior of other countries." We could argue over that concept: Japan rose to world-power status without a

major domestic resource base, but it acquired a colonial empire that contained all it needed. This means that a nation's skills and energies, too, constitute a resource. Britain, on the other hand, had the resources to propel it to the center of the world stage, and Germany used its domestic resource base to launch and sustain two world wars.

Which brings us back to China. In the competitive world of the future, China has what its predecessors on the world stage lacked: an enormous, historically regimented population (unlike Britain or Japan); the capacity to feed itself (unlike the higher-latitude former Soviet Union); a vast and varied resource base including oil, coal, metals; and a coastal location open to the Pacific Ocean (unlike the effectively landlocked U.S.S.R.). In resource-laden Eurasia, the challenge for world power has had an eastward-moving focus, from Britain and France to Germany and Russia. That eastward movement may reach its culmination in a challenge to come from China, whose natural resources could sustain such a takeoff throughout the twenty-first century.

◆ ETERNAL CHINA

We call this geographic realm the Chinese World because, as many Chinese see it, China is a world unto itself. But China is not the only country in this East Asian sphere. To the north, flanked on three sides by China, lies the Republic of Mongolia. And to the east, China's world overlaps with countries of the western Pacific Rim, including South Korea, Taiwan, and Hong Kong. North Korea has a lengthy border with China; in the mid-1990s, this still was a communist dictatorship based on the Maoist model. Compared to North Korea, China seems a positively liberal country, politically as well as economically.

Most of China's 1.2 billion people are heirs to what may indeed be the world's oldest continuous national culture and civilization. The present capital, Beijing, lies near the nucleus of ancient China. Near the city lies a cave in which were found fossils of "Peking Man," a representative of *Homo erectus*, possibly an ancestor of modern humans, who lived here half a million years ago. Anthropologists are not yet in agreement on human origins, but one theory holds that modern humans evolved from several regional predecessors, including an East Asian species of *Homo erectus*. If this turns out to be the case, modern Chinese people may be able to trace their ancestries deep into the Pleistocene.

In any case, Chinese culture and civilization have ancient roots, traceable at least 4,000 years. Again, we in the Western world look to Mesopotamia and Southwest Asia generally for clues to the origin of states, but the process may have started even earlier in China. In the Lower Basin of the Huang He (Yellow River), near the confluence of the Huang and Wei rivers, early Chinese societies developed state-like organizations. At times, a single state seems to have dominated; at other times, there was fragmentation and competition. China's early culture hearth expanded and contracted as its environmental and other fortunes fluctuated. But the thread of culture never broke, and over more than 40 centuries the Chinese created a society with strong traditions, values, and philosophies. Unlike Mesopotamia, this was an isolated society, confident in its strength, continuity, and ultimate superiority. It could reject and repel foreign influences or sometimes absorb and assimilate them. Yet through most of its long period of gestation, China lay remote, protected by distance and physiography from outside influences.

RELATIVE LOCATION FACTORS

The Chinese themselves, throughout their nation's history, have contributed to the isolation of their country from foreign influences. As we noted, the most recent episode of exclusion and closure happened just a few decades ago. It is one of China's recurrent traditions, spawned by China's relative location and Asia's physiography. Many a writer has commented on China's "splendid isolation"—an isolation made possible by geography.

A physiographic map of East Asia offers an explanation (Fig. 10-1). To China's north lie the rugged mountain ranges of eastern Siberia and the vast Gobi Desert. To the northwest, beyond Xinjiang, the mountains open—but onto the huge, dry Kirghiz Steppe. To the west and southwest lie the legendary Tian Shan and Pamirs, snowcapped ranges that might as well be thousand-foot-high walls. To the south, as a result

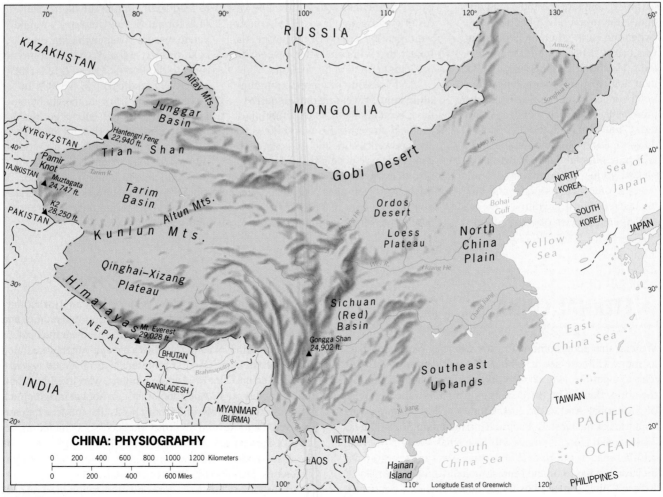

FIGURE 10-1

In the Tian Shan: western China's mountain wall shows the effects of recent glaciation and mass earth movements. This valley is part of the historic Silk Road that linked China to Europe.

of China's colonization of Tibet (Xizang), China's empire is bounded by the incomparable Himalayas, the effective boundary between the Chinese and South Asian realms. And high relief also separates China from much of Southeast Asia. On the physiographic map, China looks like a mountain-, desert-, and forest-encircled fortress.

Equally telling is the factor of distance. Until recently, China always lay far from the modern source areas of industrial innovation and change. True, China—as the Chinese emphasize—was itself such a hearth, but China's contributions to the outside world were very limited. China did interact to some extent with Korea, Japan, Taiwan, and parts of Southeast Asia, and millions of Chinese did emigrate to neighboring areas. But compare this to the impact of the Arabs, who ranged far and wide and who brought their knowledge, religion, and political influence to areas from Mediterranean Europe to Bangladesh and from West Africa to Indonesia. Later, when Europe became the center of change, China found itself farther removed, by land or sea, than almost any other part of the world.

Today, modern communications notwithstanding, China still is distant from anywhere else on earth. Going by rail from Beijing to Moscow, the heart of China's Eurasian neighbor, is a tedious journey of several days. Direct overland connections with India are practically nonexistent. Communications with Southeast Asian countries, though improving, remain tenuous.

The most significant changes in China's relative location and external connections are occurring on the Pacific Rim, although political impediments still exist. China's co-communist alliance with North Korea has weakened as a result of North Korea's lack of enthusiasm for China's economic reforms. In contrast, links with South Korea are strengthening after a long period of isolation. And Japanese business and investment are making huge inroads in China. For the first time in history, China not only lies spatially near a world-class hearth of technological innovation and financial strength; it also has opened its doors to Japan's products and ideas.

Additional changes are in the offing. In 1997, China takes control over Hong Kong (see pp. 242-244). This leaves Taiwan as the unresolved issue: although China has recognized the independence of devolved imperial fragments from Estonia to Tajikistan, it will not acknowledge Taiwan's sovereignty. This is the kind of decision that imperial rulers accustomed to ignoring world opinion feel they can afford to make.

EXTENT AND ENVIRONMENT

China's total area is only very slightly larger than that of the United States, including Alaska: each has about 3.7 million square miles (9.5 million sq km). As Fig. 10-2 reveals, the longitudinal extent of China and the 48 contiguous U.S. States also is quite similar. Latitudinally, though, China is considerably wider. Miami, near the southern limit of the United States, lies halfway between Shanghai and Hong Kong. China's area extends well into the tropics. Southern China thus takes on characteristics of tropical Asia. In the northeast, too, China incorporates much of what would in North America be Quebec and Ontario. Westward, China's land area becomes narrower and similarities increase. Of course, China has no west coast!

Now compare the climate maps of China and the United States in Fig. 10-3 (which are enlargements of the appropriate portions of the world climate map in Fig. I-7). Note that both China and the United States have a large southeastern

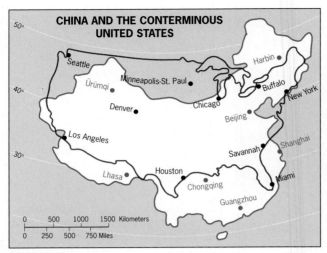

FIGURE 10-2

climatic region marked *Cfa* (that is, humid, temperate, warm-summer), flanked in China by a zone of *Cwa* (where winters become drier). Westward in both countries, the *C* climates yield to colder, drier climes. In the United States, moderate *C* climates develop again along the Pacific coast; China, however, stays dry and cold as well as high in elevation at equivalent longitudes.

Note especially the comparative location of the U.S. and Chinese *Cfa* areas in Fig. 10-3. China's lies much farther to the south. In the United States, the *Cfa* climate extends beyond 40° North latitude, but in China, very cold and generally winter-dry *D* climates take over at the latitude of Virginia. Beijing has a warm summer but a bitterly cold and long winter. Northeastern China, in the general latitudinal range of Canada's lower Quebec and Newfoundland, is much more severe than its North American equivalent. Harsh environments prevail over vast regions of China—but, as we will see later, nature makes up for this in spectacular fashion. From the climatic zone marked *H* (for highlands) in the west come great life-giving rivers whose wide basins contain enormous expanses of fertile soils. Without these, China would not have a population more than four times that of the United States.

If you were to travel in China, environments and distances would at times seem quite familiar. From Shanghai to the capital, Beijing, is not much farther than from Washington, D.C. to Chicago. Flying cross-country would take about the same amount of time, given similar aircraft. But be prepared: China is not yet a country of modern transport facilities. Efficient airports, superhighways, or bullet trains have yet to arrive. Still, return visitors are struck by the continuous improvement in China's communication systems and especially by the growth of its railroad services. Developing, modernizing China is on the move.

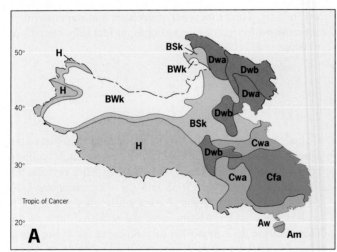

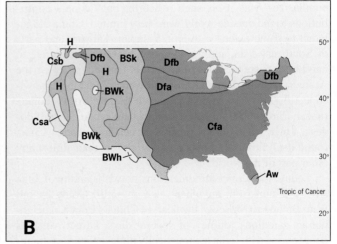

CLIMATES OF CHINA AND THE CONTERMINOUS UNITED STATES
After Köppen-Geiger

A HUMID EQUATORIAL CLIMATE	B DRY CLIMATE	C HUMID TEMPERATE CLIMATE	D HUMID COLD CLIMATE	E COLD POLAR CLIMATES
Am Short dry season	**BS** Semiarid	**Cf** No dry season	**Df** No dry season	**H** Unclassified highlands
Aw Dry winter	**BW** Arid	**Cw** Dry winter	**Dw** Dry winter	
	h=hot **k**=cold	**Cs** Dry summer **a**=hot summer **b**=cool summer		

FIGURE 10-3

◆ PHYSIOGRAPHIC REGIONS

The gigantic stage of China has been the scene of momentous human events: the invention of certain forms of agriculture, the development of ancient urban centers, the formation of many states. China has witnessed migrations, invasions, revolutions. China's empire has gone through expansions and contractions, through times of plenty and of famine.

To understand the human geography of today's (and tomorrow's) China, we must first investigate the country's physical geography. For those of us who live in North America, this is not too difficult to do: in physiographic terms, China again bears some similarities to the United States. When we examined North America's physical geography, we identified about a dozen physiographic regions, each with well-

defined subregions. We can discern seven first-order regions in China (Fig. 10-4).

In a general way, we may say that China, like the United States, has low-lying plains in the east and deserts and mountains in the west. China even has two rivers that rival the Mississippi: the Huang (Yellow) River, which flows into the Bohai Gulf of the Yellow Sea, and the Chang Jiang, which reaches the East China Sea near Shanghai. The fertile alluvial basins and deltas of the lower courses of these rivers support the world's largest agglomeration of rural population. For as long as there has been a China, the North China Plain and the Lower Chang Basin have been its breadbasket—or rather, its rice and wheat bowls.

Let us look at Fig. 10-4 in a general way first. We identify seven regions, four of which face the Pacific coast (①) through ④). These four eastern regions encompass what is

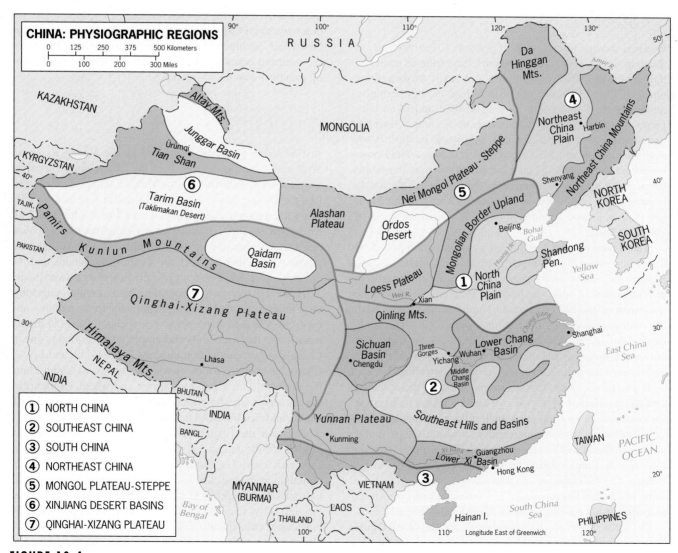

FIGURE 10-4

often called *China Proper*, the "real" China with its vast cultivated plains, huge population clusters, large cities, and long cultural history. Region ⑤ marks the transition zone from China Proper into Mongolia, land of the Gobi Desert. Region ⑥ is a distant, remote land of deserts, steppes, and mountains that faces Russia and the new republics of Turkestan. And region ⑦ consists mainly of annexed Tibet (Xizang), including vast expanses of high-elevation plateaus and spectacular mountains.

To summarize, China divides into three broad environmental zones:

1. The river basins and highlands of eastern China (regions ① through ④)
2. The arid northwest (regions ⑤ and ⑥)
3. The frigid mountains and plateaus of the southwest (region ⑦)

① NORTH CHINA

It seems strange to call region ① *North* China because, as the map shows, there are regions in China situated even farther northward. But convention is not easily broken. Over much of China's history, this was indeed the northernmost region, beyond which lay an uncontrolled and often dangerous frontier. When the Chinese chose Beijing as their national headquarters, the city was a *forward capital*; it lay near the margins of China's undisputed domain. The Lower basin and delta of the Huang became known as the North China Plain, and the regional term "North China" obviously is permanent. Nevertheless, Northeast China and the Mongol Plateau-Steppe (regions ④ and ⑤) lie to the north of "North" China.

The wedge-shaped North China region is dominated by the great Huang River, which rises on the Qinghai-Xizang Plateau and, fed by melting snow, follows a tortuous path to the Bohai Gulf. Today, the Huang's main channel lies to the northwest of the hilly Shandong Peninsula, but in the past, the river swung widely (and wildly) across its deltaic zone, causing devastating and death-dealing floods. A massive levee-building project has stabilized the Huang in its main course, and most of the delta's channels are now controlled.

If you travel across this enormous lowland, there will be virtually no relief as far as your eye can see. Wheat fields cover the countryside, and you pass small villages at regular intervals. The surface drops imperceptibly toward the east, and if you go in that direction, the Shandong Peninsula's wooded hills rise in sharp contrast to this flatness. And the crops change! On the slopes of the Shandong Peninsula's hills lie some of China's finest vineyards; Chinese winegrowers refer to the peninsula optimistically as "China's California."

Go in the other direction, westward, and the North China Plain narrows and then yields to quite a different landscape. Here in the west lies the so-called *Loess Plateau*, a large area

named after the powdery, wind-blown deposit that accumulated here in great thickness. Loess probably has its origin in rocks pulverized by the Late Cenozoic glaciers that advanced into Eurasia as they did into North America. Persistent winds picked up this dusty material and then deposited it here in the Loess Plateau in thicknesses reaching 250 feet (75 m). Loess is very fertile, and its fertility (unlike that of ordinary soil) does not decrease with depth. When compacted, it is easily excavated; people have lived in hollowed-out loess "caves" for thousands of years. These opportunities have attracted millions of settlers to the Loess Plateau, but at a price: earthquakes, when they strike here, cause the collapse of dug-out underground dwellings. Over the centuries, countless hundreds of thousands of people have lost their lives as a result.

The northern sector of the North China region is formed by the Mongolian Border Upland, a hilly, forest-covered area that separates the North China Plain from the dry steppes on the Mongolian desert margin. Visitors to the Great Wall segment north of Beijing can see the change in topography, but you have to go farther north to see the climatic change from broad-leaf deciduous trees to steppe grasses and brush. At the northern boundary of the region, the countryside opens up into the expanses of region ⑤.

② SOUTHEAST CHINA

Just as North China is dominated by the Huang River, so Southeast China is the region of the Chang Jiang, the river long known to Americans as the Yangtze. This "middle" river, like the Huang, rises in region ⑦, and it touches virtually every subregion in Southeast China before reaching the East China Sea. As Fig. 10-4 shows, the Chang traverses the Yunnan Plateau, then the populous Sichuan Basin, follows the physical boundary between the Qinling Mountains and the uplands of the Southeast, and finally enters its Lower Basin.

As we will see when we consider China's human geography, the Chang rivals the Huang in importance to the nation—and even exceeds it in some ways. The Sichuan Basin in the interior, with its fertile soils and ample water, contains one of China's largest population clusters. The Middle and Lower Chang basins may not be as wide as the North China Plain, but they yield huge harvests of rice and wheat. (Remember the climatic map: the climate here is warmer than in North China, and rice becomes the principal grain crop.)

South of the Chang River, Southeast China becomes a jumble of hills and basins, and now the natural vegetation is evergreen forest. The basins are densely peopled, and as the precipitation map (Fig. I-6) shows, water supply from the hills is ample. The physical scene here is varied: you travel by road through hilly, even mountainous countryside, and then the landscape opens up into a slope-encircled basin with villages along the foothills and paddies dividing the flatlands. Upslope, tea farms use the higher relief. Soon you leave this area

These terraces may not look as well-kept as those elsewhere in Asia (such as the Philippines or Indonesia), but they make productive a narrow, steep-sided valley in an area of Southeast China.

and the hills close in again, but shortly the scene is repeated, and another narrow valley opens into a green lowland.

In the south of Southeast China lies an important river we should probably call the "river of many names." Between Guangzhou and the estuary around Hong Kong, this river has the name probably best known to Westerners: the Pearl River. Upstream, it is known as the Xi (West) River. Still farther upstream, it takes on other names; in the Yunnan Plateau, where the river receives several tributaries, it is uncertain just which is the main artery. The map suggests that this Xi River is no Chang, and this impression is soon verified. Not only is the Xi a much shorter stream, but for a great part of its course, it lies in mountainous or hilly terrain. As a result, it is subject to dangerous floods, and only below Wuzhou is it controlled by levees and flood channels.

③ SOUTH CHINA

Using geomorphic, climatic, vegetative, and other data, Chinese geographers recognize a southernmost physiographic region that extends from the margins of the valley of the Xi

River to the borders with Southeast Asia. This region, which includes the island of Hainan, is tropical China, reminiscent in many ways (not just physiographic) of neighboring Vietnam, Laos, and Myanmar (Burma). Chinese geographers often refer to this as "monsoon" China, although, as Fig. 10-3A shows, only Hainan Island and the tip of its neighboring peninsula display true monsoon (*Am*) regimes.

Nevertheless, the vegetation here is tropical forest, average temperatures are the highest in China, and multiple cropping of rice (two and in some places even three harvests a year) is routine. This region lies mostly within the tropics, and the harsh winters of North China are but a dim memory in this part of the country.

④ NORTHEAST CHINA

Northeast China consists of three well-defined subregions. Here, in China's northernmost latitudes, China takes on Siberian qualities, and winter cold is the dominant fact of life.

The heart of Northeast China is formed by still another of China's great lowlands, the Northeast China Plain. No major river dominates here, however, and the plain is an erosional, not a depositional, feature (except in the extreme south below the latitude of Shenyang). No great farm-based population developed here, but that does not mean that the Northeast has not been (or will not be) an important region in China. Manchu invaders used it as a base to challenge Beijing's rulers. Japanese colonizers conquered and exploited it. And today, Northeast China is an integral part of China's industrial heartland.

The centrally-positioned plain is bordered by two major mountain zones: the Da Hinggan Mountains to the northwest and the Northeast China Mountains to the east. Mineral and fuel deposits in these mountains attracted the Japanese, and form the basis for the current industrial activity occurring here. Yet, as we will note later, Northeast China may for some time lag behind the rapidly developing industrial areas on China's Pacific Rim; the region lies nearly landlocked, remote from all this action. Still, its time may come again. When Russia's Far East is opened up, Northeast China will find its relative location much improved.

⑤ PLATEAU-STEPPE OF MONGOLIA

As the climatic map (Fig. 10-3A) reminds us, we are now entering arid China, land of the Gobi Desert margins. This region is dominated by the wind-swept plains shown in Fig. 10-4, but it also includes the Ordos Desert. Here the Huang River enters and leaves the region, much as the Niger River in West Africa elbows into and then out of the Sahara.

Physiographically, the northern part of this region forms the northward-sloping, southern rim of the Gobi. There is somewhat more moisture here than in the heart of the desert, but even so, the natural vegetation is a patchy grassland with some scrub. Droughts are chronic, summers are searingly hot, and winters are bitterly cold. Vicious winds often blow up sand and dust. This would seem to be the domain of nomads and migrants, but as we will see, the Chinese government has invested heavily in stabilizing and settling the area. To make this possible, the Huang River afforded the best prospects. In the Ordos Desert (Mu Us Shamo), the course of the Huang River has become a ribbon of settlement, including several sizeable towns.

Vast areas of interior China are steppe or desert environments, capable of supporting only a sparse human population. This is typical countryside on the margin of the Gobi Desert in Gansu Province. Note the livestock grazing on the thin grasses.

⑥ DESERT BASINS OF XINJIANG

The Alashan Plateau, west of the Ordos Desert, is a physiographic transition zone from the desert and steppe of the Mongolian borderland in the east to the great alternation of mountains and basins of Xinjiang (Sinkiang) to the west. The great divide in this vast western realm is the Kunlun Shan, which extends from China's western extreme all the way across the heart of its territory to the very margins of the Loess Plateau. North of these Kunlun Mountains lie Xinjiang's mountains and basins; southward lies the high plateau of Qinghai and Xizang (Tibet).

The Xinjiang region comprises two huge mountain-enclosed basins (and several smaller basins not shown in Fig. 10-4). The two largest basins are separated by the Tian Shan, a mountain range that stretches from the border with Kyrgyzstan in the west to southern Mongolia in the east. Climatically, these are dry areas; the Tarim Basin is, in fact, a desert (the Taklimakan), whereas the Junggar Basin to its north is steppe country. Both the Tarim and Junggar basins are areas of *endoreic* (internal) drainage: the rivers that rise in the surrounding highlands and flow to the basin floor do not continue to the sea. These streams are the chief source of water flowing onto the region's rough gravels, washed down in thick layers from the mountains. The waters disappear below the surface as they reach these coarse deposits, but they then re-emerge where the gravels thin out (oases have long existed at these sites). Along the southern margin of the Taklimakan Desert lies a string of oases that at one time formed stations on the long westward trade route (the renowned Silk Road) to Europe.

⑦ HIGH PLATEAUS AND MOUNTAINS OF QINGHAI-XIZANG

This enormous region encompasses the world's greatest assemblage of lofty, snow-covered mountain ranges and high-altitude plateaus. It was formed and continues to be modified by the convergent collision of two of the world's great continental tectonic plates, the Eurasian and the Indian-Australian (see Fig. I-3). At the southern margin of this region lies the Himalaya range, and to the north, the Kunlun Shan marks its limit. The *average* elevation here is close to 15,000 feet (4,500 m), with the highest mountains standing more than 10,000 feet *above* this level. The valleys, where most of the people live, descend to about 5,000 feet (1,500 m) above sea level.

The central Qinghai-Xizang Plateau is desolate and barren, cold, windswept, and treeless; here and there some patches of grass sustain a few herd animals. In the northeastern quadrant lies the Qaidam Basin, structurally a *graben* (meaning that the land dropped between geologic faults and was not eroded into a depression). The Qaidam Basin is known for its variety of desert landforms, with shifting sand dunes, wind-sculpted yardangs (low sand ridges), and playas (dry lake beds). A large, thick salt crust has developed, and blocks of salt are used to pave the few roads in the area.

The Qinghai-Xizang region covers nearly one-quarter of China's total area, and it includes all of former Tibet. Even including the subjugated Tibetan people, this quarter of China's domain contains only about 1 percent of the country's inhabitants. Thus China may be the world's largest nation in population size, but it still has vast empty lands.

◆ EVOLUTION OF THE CHINESE STATE

On this huge and varied physical stage took place the drama of China's evolution. It has been suggested that the early Chinese, perhaps as long as 4,000 years ago, received stimuli from the river-based civilizations of Southwest and South Asia, but the evidence is thin. The Chinese are proven innovators, and Chinese cultural individuality was established very long ago.

China's history and historical geography are chronicled as *dynasties*. From very early on, rulers of the same family succeeded each other, often over periods lasting several centuries, until the lineage broke down. Each dynasty left its imprint on Chinese culture and on the map of China's evolving empire. Thus we refer to "Han," "Ming," or "Manchu" China, using dynastic names to identify significant phases in Chinese national evolution.

The oldest dynasty of which much is known is the Shang (sometimes called Yin), which formed when China was taking shape in the area of the confluence of the Huang and Wei rivers (Fig. 10-5). This dynasty lasted from about 1770 B.C. until around 1120 B.C. Walled cities were built during the Shang period, and the Bronze Age commenced during Shang rule. For more than a thousand years after the beginning of the Shang Dynasty, North China was the center of development in this part of Asia. The Zhou Dynasty (roughly 1120–221 B.C.) sustained and consolidated what had begun during the Shang period.

Eventually, agricultural techniques and population numbers combined to press settlement in the obvious direction—southward, where the best opportunities for further expansion lay. During the brief Qin (Chin) Dynasty (221–207 B.C.), the lands of the Chang Jiang were opened up, and settlement spread as far south as the Lower Xi Basin.

HAN CHINA

A pivotal period in the historical geography of China now lay ahead: the Han Dynasty (207 B.C.–A.D. 220). The Han rul-

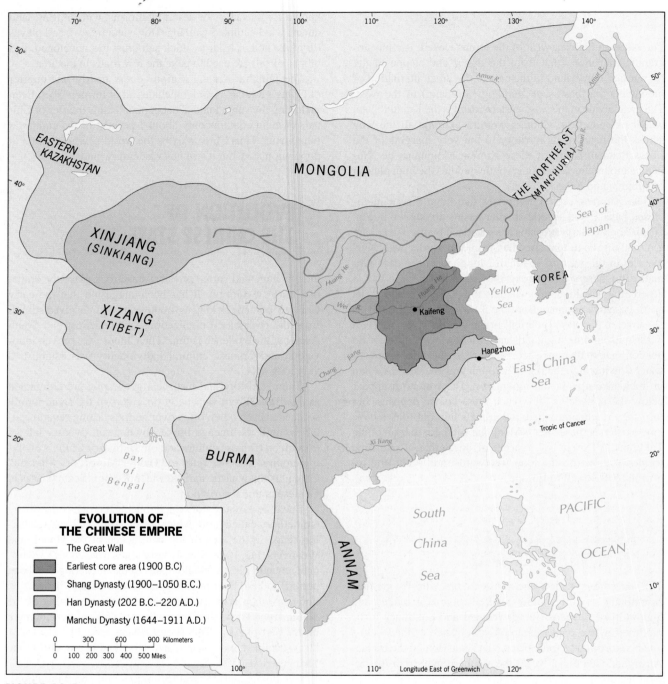

**EVOLUTION OF
THE CHINESE EMPIRE**

—— The Great Wall

Earliest core area (1900 B.C)

Shang Dynasty (1900–1050 B.C.)

Han Dynasty (202 B.C.–220 A.D.)

Manchu Dynasty (1644–1911 A.D.)

| 0 | 300 | 600 | 900 Kilometers |
| 0 | 100 200 300 400 | 500 Miles | |

FIGURE 10-5

ers brought unity and stability to China, and they enlarged the Chinese sphere of influence to include Korea, the Northeast, Mongolia, Xinjiang, and, in Southeast Asia, Annam (located in what is now central Vietnam). Thus the Chinese rulers established control over the bases of China's constant harassers—the nomads of the surrounding steppes, deserts, and mountains—and protected (in Xinjiang) the main overland

avenue of westward contact between China and the rest of Eurasia.

The Han period was a formative one in the evolution of China (Fig. 10-5). China's military power grew stronger than ever before. Change came to the systems of landownership as the old feudal order broke down and private individual property was recognized. The silk trade grew into China's

The Great Wall of China actually is a series of walls, some segments parallel to each other. In the field it looks rather like an elongated building studded with towers at irregular intervals. Completed ca. 200 B.C., the Great Wall was built to separate eastern China's sedentary farmers from the pastoral herders of the Asian interior. Here, autumn colors grace the wall's surroundings north of Beijing.

first external commerce. To this day, most Chinese, recognizing that much of what is China first came about during this period, still call themselves the *People of Han.*

Han China was the Roman Empire of East Asia. Great achievements were made in art, architecture, the sciences, and other spheres. Like the Romans, the Han Chinese had to contend with unintegrated hostile peoples on their empire's margins. In China, these peoples occupied the mountainous country south of the national territory, and the Han rulers built military outposts to keep these frontiers quiet. Again, like the Roman Empire, the China of Han fell into decline and disarray, and more than a dozen successor states arose after A.D. 220 to compete for primacy. Not until the Sui Dynasty (581–618) did consolidation begin again, to be continued during the Tang Dynasty (618–907), another period of national stability and development.

LATER DYNASTIES

Following the Tang period, when China once again was a great national state, the Song Dynasty (960–1279) carried the culture to unprecedented heights despite continuing problems with marauding nomadic peoples. This time the threat came from the north; in 1127, the Song rulers had to abandon their capital at Kaifeng on the Huang River in favor of

Hangzhou in the Southeast. Nevertheless, China during the Song Dynasty was in many ways the world's most advanced state. It had several cities of more than 1 million people, paper money was in use, commerce intensified, literature flourished, schools multiplied in number, the arts thrived as never before, and the philosophies of Kongfuzi (Confucius), for many centuries China's guide, were modernized, printed, and mass-distributed for the first time. Eventually Song China fell to conquerors from the outside, the Mongols led by Kublai Khan. The Mongol authority, known as the Yuan Dynasty (1279–1368), made China part of a vast empire that extended all the way across Asia to Eastern Europe. But it lasted less than a century and had little effect on Chinese culture. Instead, it was the Mongols who adopted Chinese civilization!

In 1368, a Chinese local ruler led a rebellion that ousted the Mongols, signaling the ascent of the great indigenous Ming Dynasty (1368–1644). Under the Ming rulers China's greatness was restored, its territory once again consolidated from the Great Wall in the north to Annam in the south. Finally, in 1644, another northern, foreign nomadic people forced their way to control over China. They came from the far Northeast (see the box entitled "Manchuria: The Northeast"), but unlike the Mongols, they sustained and nurtured Chinese traditions of administration, authority, and national culture. The Manchu (or Qing) Dynasty (1644–1911) extended the Chinese sphere of influence to include Mongolia, Xinjiang,

MANCHURIA: THE NORTHEAST

Three provinces, Liaoning, Jilin, and Heilongjiang, constitute China's Northeast, a region bounded by Russia to the north and east, Mongolia to the west, and North Korea to the southeast (see Fig. 10-13). This was the home of the conquering Manchus and, later, the scene of foreign domination. Russians and Japanese struggled for control over the area; ultimately, Japan created a dependency here in the 1930s that it called Manchukuo. More recently, the region came to be called Manchuria, but this is not an accepted regional appellation in China. Chinese geographers refer to the region encompassed by these three northeastern provinces as simply the Northeast—a practice followed in this chapter.

Xizang, Burma (now Myanmar), Indochina, eastern Kazakhstan, Korea, and a large part of what is today the Russian Far East (Fig. 10-5). But the Manchus also had to contend with the rising power of Europe and the West, and the encroachment of Russia; large portions of the Chinese spheres in interior and Southeast Asia were lost to China's voracious competitors.

◆ A CENTURY OF CONVULSION

China long withstood the European advent in East Asia with a self-assured superiority based on the strength of its culture and the continuity of the state. There was no market for the British East India Company's rough textiles in a country long used to finely fabricated silks and cottons. There was little interest in the toys and trinkets the Europeans produced in the hope of bartering for Chinese tea and pottery. And even when Europe's sailing ships made way for steam-driven vessels and newer and better factory-made textiles were offered in trade for China's tea and silk, China continued to reject the European imports that were still, initially at least, too expensive and too inferior to compete with China's handmade goods. Long after India had fallen into the grip of mercantilism and economic imperialism, China was able to maintain its established order. This was no surprise to the Chinese; after all, they had held a position of undisputed superiority among the countries of eastern Asia as long as could be remembered, and they had dealt with overland and seaborne invaders before.

A (LOST) WAR ON DRUGS

The nineteenth century finally shattered the self-assured isolationism of China as it proved the superiority of the new Europe. On two fronts, economic and political, the European powers destroyed China's invincibility. In the economic sphere, they succeeded in lowering the cost and improving the quality of manufactured goods, especially textiles, and the handicraft industry of China began to collapse in the face of unbeatable competition. In the political sphere, the demands of the British merchants and the growing English presence in China led to conflicts. In the early part of the century, the central issue was the importation into China of opium, a dangerous and addictive intoxicant. Opium was destroying the very fabric of Chinese culture, weakening the society, and rendering China an easy prey for colonial profiteers. As the Manchu government moved to stamp out the opium trade in 1839, armed hostilities broke out, and soon the Chinese sustained their first defeats. Between 1839 and 1842, the Chinese fared badly, and the First Opium War signaled the end of Chinese sovereignty.

British forces penetrated up the Chang Jiang and controlled several areas south of that river (Fig. 10-6); Beijing hurriedly sought a peace treaty. As a result, leases and concessions were granted to foreign merchants. Hong Kong Island was ceded to the British, and five ports, including Guangzhou (Canton) and Shanghai, were opened to foreign commerce. No longer did the British have to accept a status that was inferior to the Chinese in order to do business; henceforth, negotiations would be pursued on equal terms. Opium now flooded into China, and its impact on Chinese society became even more devastating. Fifteen years after the First Opium War, the Chinese tried again to stem the disastrous narcotic tide, and again they were bested by the foreigners who had attached themselves to their country. Now the cultivation of the opium poppy in China itself was legalized. Chinese society was disintegrating, and the scourge of this drug abuse was not defeated until the revival of Chinese power early in the twentieth century.

But before China could reassert itself, much of what remained of China's sovereignty was steadily eroded away (see the box entitled "Extraterritoriality"). The Germans obtained a lease on Qingdao on the Shandong Peninsula in 1898, and in the same year the French acquired a sphere of influence in the far south at Zhanjiang (Fig. 10-6). The Portuguese took Macau (Macao); the Russians obtained a lease on Liaodong in the Northeast as well as railway concessions there; even Japan got into the act by annexing the Ryukyu Islands and, more importantly, Formosa (Taiwan) in 1895.

After millennia of cultural integrity, economic security, and political continuity, the Chinese world lay open to the aggressions of foreigners whose innovative capacities China had denied to the end. But now ships flying European flags lay in the ports of China's coasts and rivers; the smokestacks

FIGURE 10-6

of foreign factories rose above the landscapes of its great cities. The Japanese were in Korea, which had nominally been a Chinese vassal; the Russians had entered China's Northeast. The foreign invaders even took to fighting among themselves, as did Japan and Russia in "Manchuria" (as the foreigners now called the Northeast) in 1904.

RISE OF A NEW CHINA

In the meantime, organized opposition to the foreign presence in China was gathering strength, and this century opened with a large-scale revolt against all outside elements. Bands of revolutionaries roamed the cities as well as the country-

EXTRATERRITORIALITY

Sha Mian Island in Guangzhou (Canton), a colonial-era enclave where foreigners resided beyond the jurisdiction of the Chinese.

A sign of China's weakening during the second half of the nineteenth century was the application in its cities of a European doctrine of international law—**extraterritoriality**. This principle denotes a situation in which foreign states or international organizations and their representatives are immune from the jurisdiction of the country in which they are present. This, of course, constitutes an erosion of the sovereignty of the state hosting these foreign elements, especially when the practice goes beyond the customary immunity of embassies and persons in diplomatic service.

In China, extraterritoriality reached unprecedented proportions. The best residential suburbs of the large cities,

for example, were declared to be "extraterritorial" parts of foreign countries and were made inaccessible to Chinese citizens. Sha Mian Island in the Pearl River in Guangzhou was a favorite extraterritorial enclave of that city (see photo above). A sign at the only bridge to the island stated "No Dogs or Chinese." In this way, the Chinese found themselves unable to enter their own public parks and many buildings without permission from foreigners. Christian missionaries fanned out into China, their bases fortified with extraterritorial security. To the Chinese, this involved a loss of face that contributed to a bitter opposition to the presence of all foreigners—a resentment that finally exploded in the Boxer Rebellion of 1900.

side, attacking not only the hated foreigners but also Chinese citizens who had adopted Western cultural traits. Known as the Boxer Rebellion (after a loose translation of the Chinese name for these revolutionary groups), the 1900 upris-

ing was put down with much bloodshed. Simultaneously, another revolutionary movement was gaining support, aimed against the Manchu leadership itself. In 1911, the emperor's garrisons were attacked all over China, and in a few months the 267-year-old dynasty was overthrown. Indirectly, it too was yet another casualty of the foreign intrusion, and it left China divided and disorganized.

The end of the Manchu era and the proclamation of a republican government in China did little to improve the country's overall position. The Japanese captured Germany's holdings on the Shandong Peninsula, including the city of Qingdao, during World War I; when the victorious European powers met at Versailles in 1919 to divide the territorial spoils, they affirmed Japan's rights in the area. This led to another significant demonstration of Chinese reassertion as nationwide protests and boycotts of Japanese goods were organized in what became known as the May Fourth Movement. One participant in these demonstrations was a charismatic young man named Mao Zedong.

Nonetheless, China after World War I remained a badly divided country. By the early 1920s, there were two governments—one in Beijing and another in the southern city of Guangzhou (Canton), where the famous Chinese revolutionary, Sun Yat-sen, was the central figure. Neither government could pretend to control much of China. The Northeast was in complete chaos, petty states were emerging all over the central part of the country, and the Guangzhou "parliament" controlled only a part of Guangdong Province in the Southeast. Yet it was just at this time that the power groups that were ultimately to struggle for supremacy in China were formed. While Sun Yat-sen was trying to form a viable Nationalist government in Guangzhou, the Chinese Communist Party was formed by a group of intellectuals in Shanghai. Several of these intellectuals had been leaders in the May Fourth Movement, and in the early 1920s they received help from the Communist Party of the Soviet Union. Mao Zedong was already a prominent figure in these events.

Initially, there was cooperation between the new Communist Party and the Nationalists led by Sun Yat-sen. The Nationalists were stronger and better organized, and they hoped to use the communists in their anti-foreign (especially anti-British) campaigns. By 1927, the foreigners were on the run; the Nationalist forces entered cities and looted and robbed at will while aliens were evacuated or, failing that, sometimes killed. As the Nationalists continued their drive northward and success was clearly in the offing, internal dissension arose. Soon, the Nationalists were as busy purging the communists as they were pursuing foreigners. The central figure to emerge in this period was Chiang Kai-shek. Sun Yat-sen died in 1925, and when the Nationalists established their capital at Nanjing (Nanking) in 1928, Chiang was the country's leader.

THREE-WAY STRUGGLE

The post-Manchu period of strife and division in China was quite similar to other times when, following a lengthy period of comparative stability under dynastic rule, the country fragmented into rival factions. In the first years of the Nanjing government's **hegemony**, the campaign against the communists intensified and many thousands were killed. Chiang's armies drove them ever deeper into the interior (Mao himself escaped the purges only because he was in a remote rural area at the time); for a while, it seemed that Nanjing's armies would break the back of the communist movement in China.

The Long March

A core area of communist peasant forces survived in the zone where the provinces of Jiangxi and Hunan adjoin in southeastern China, and these forces defied Chiang's attempts to destroy them. Their situation grew steadily worse, however, and in 1933 the Nationalist armies were on the verge of encircling this last eastern communist stronghold. The communists decided to avoid inevitable strangulation by leaving. Nearly 100,000 people—armed soldiers, peasants, local leaders—gathered near Ruijin and started to walk westward in 1934. This was a momentous event in modern China, and among the leaders of the column were Mao Zedong and Zhou Enlai. The Nationalists rained attack after attack on the marchers, but they never succeeded in wiping them out completely; as the communists marched, they were joined by new sympathizers.

The Long March (see the route in Fig. 10-6), as this drama has come to be called, first took the communists to Yunnan Province, where they turned north to enter western Sichuan. They then traversed Gansu Province and eventually reached their goal, the mountainous interior near Yanan in Shaanxi Province. The Long March covered nearly 6,000 miles (10,000 km) of China's more difficult terrain, and the Nationalists' continuous attacks killed an estimated 75,000 of the original participants. Only about 20,000 survived the epic migration, but among them were Mao and Zhou, who were convinced that a new China would arise from the peasantry of the rural interior to overcome the urban easterners whose armies could not eliminate them.

The Japanese

While the Nanjing government was pursuing the communists, foreign interests made use of the situation to further their own objectives in China. The Soviet Union held a sphere of influence in Mongolia and was on the verge of annexing a piece of Xinjiang. Japan was dominant in the Northeast, where it had control over ports and railroads. The Nanjing government tried to resist the expansion of Japan's sphere of influence. The effort failed, and the Japanese set up a puppet state in the region; they appointed a ruler and called their new possession Manchukuo.

The inevitable full-scale war between the Chinese and the Japanese broke out in 1937. For a while, the Chinese communists and the Nationalists stopped fighting each other in order to concentrate on the war against Japan, but soon their factional war erupted again. Now Chinese communists were fighting Chinese Nationalists while both fought the Japanese. In the process, China broke up into three regions: the Japanese sphere in the north and east, the Nationalists' domain centered on their capital of Chongqing in the Sichuan Basin, and the communist zone in the interior west. The Japanese, by pursuing and engaging Chiang's Nationalist forces, gave the communists an opportunity to build their strength and foster their reputation and prestige in China's western areas.

The Japanese campaign in China was accompanied by unspeakable atrocities. Millions of Chinese citizens were shot, burned, drowned, subjected to gruesome chemical and biological experiments, and otherwise victimized. When China's economic reforms of the 1980s and 1990s led to a renewed Japanese presence in China, the Chinese public and its leaders called for Japanese acknowledgment and apology for these wartime abuses. In Japan, this pitted apologists against strident nationalists, causing a political crisis. In 1992, Emperor Akihito visited China and referred to the war but stopped short of a formal apology. On this issue, the book is not yet closed.

Communist China Arises

After Japan was defeated by the U.S.-led Western powers in 1945, the civil war in China quickly resumed. The United States, hoping for a stable and friendly government in China, sought to mediate the conflict but did so while recognizing the Nationalist faction as the legitimate government. Furthermore, the United States aided the Nationalists militarily, destroying any chance of genuine and impartial mediation. By 1948, it was clear that Mao Zedong's well-organized militias would defeat Chiang Kai-shek. Chiang kept moving his capital—back to Guangzhou, seat of Sun Yat-sen's first Nationalist government, then back to Chongqing. Late in 1949, following a series of disastrous defeats in which hundreds of thousands of Nationalist forces were killed, the remnants of Chiang's faction gathered Chinese treasures and valuables and fled to the island of Taiwan. There, they took control of the government and proclaimed their own Republic of China.

Meanwhile, in Beijing, standing in front of the assembled masses at the Gate of Heavenly Peace on Tiananmen Square, Mao Zedong, on October 1, 1949, proclaimed the birth of the People's Republic of China.

◆ CHINA'S HUMAN GEOGRAPHY

After nearly a half century of communist rule, China is a society transformed. It has been said that the year 1949 actually marked the beginning of a new dynasty not so different from the old, an autocratic system that dictated from the top.

Mao Zedong, in that view, simply bore the mantle of his dynastic predecessors. Only the family lineage had fallen away; now communist "comrades" would succeed each other.

And certainly some of China's old traditions continued during the communist era, but in many other ways China became a totally overhauled society. Benevolent or otherwise, the dynastic rulers of old China headed a country in which—for all its splendor, strength, and cultural richness—the fate of landless people and of serfs often was undescribably miserable; in which floods, famines, and diseases could decimate the populations of entire regions without any help from the state; in which local lords could (and often did) repress the people with impunity; in which children were sold and brides were bought. The European intrusion made things even worse, bringing slums, starvation, and deprivation to millions who had moved to the cities. The communist regime, dictatorial though it was, attacked China's weaknesses on many fronts, mobilizing virtually every able-bodied citizen in the process. Land was taken from the wealthy. Farms were collectivized. Dams and levees were built by the hands of thousands. The threat of hunger for millions receded. Health conditions improved. Child labor was reduced. Mao's long tenure (1949-1976) and apparent omnipotence may remind us of the dynastic rulers' frequent longevity and absolutism, but the new China he left behind is vastly different from the old.

In several areas, however, Mao's strict adherence to Marxist dogma had problematic consequences for China. One is familiar to anyone who has studied communist planned economies anywhere: the control by government over all productive capacity, not only agriculture but also industry. China's industrialization program bequeathed the country with thousands of inefficient, costly, uncompetitive, state-owned manufacturing plants. Another of Mao's dictums had to do with population. Like the Soviets (and influenced by a horde of Soviet communist advisers and planners), Mao refused to impose or even recommend any form of population policy. As a result, China's population grew explosively during his rule. After Mao's death in 1976, reform-minded communists imposed a one-child policy on China's families and used tough tactics, including forced abortions, to sustain that policy. And, as we noted at the beginning of this chapter, limited market activity was introduced to counter the effects of state-controlled manufacturing.

THE POLITICAL MAP

Before we proceed to investigate the emerging human geography of China, we should acquaint ourselves with the country's political and administrative framework (Fig. 10-7). For administrative purposes, China is divided into 3 central-government-controlled municipalities, 5 autonomous regions, and 21 provinces.

The three central-government-administered municipalities are the capital, Beijing (12.4 million) and its nearby port

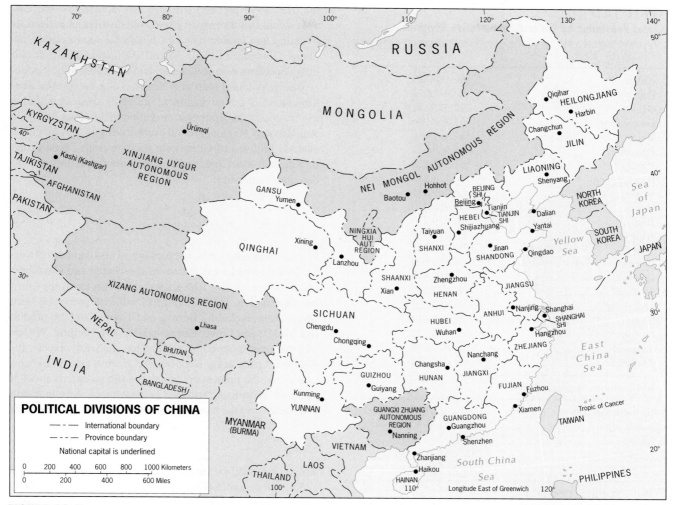

FIGURE 10-7

city, Tianjin (10.7 million), and China's largest metropolis, Shanghai (15.2 million). These are, respectively, the foci of the core areas of China's two most populous and important regions, and direct control over them provides the regime with added security.

The five autonomous regions were established in recognition of the non-Han minorities living there. Some laws that apply to Han Chinese do not apply to certain minorities. As we saw in the case of the former Soviet Union, however, demographic changes and population movements affect such regions, and policies of the 1940s may not work in the 1990s. Han Chinese immigrants now outnumber several minorities in their own regions. The five autonomous regions (A.R.s) are: (1) Nei Mongol A.R. (Inner Mongolia); (2) Ningxia Hui A.R. (adjacent to Inner Mongolia); (3) Xinjiang Uygur A.R. (northwest corner); (4) Guangxi Zhuang A.R. (far south, bordering Vietnam); and (5) Xizang A.R. (Tibet).

China's 21 provinces, like U.S. States, tend to be smallest in the east and larger toward the west. Territorially smallest are the three easternmost provinces on China's coastal bulge:

Zhejiang, Jiangsu, and Fujian. The two largest are Qinghai, flanked by Tibet, and Sichuan, China's Midwest.

As with all large countries, some provinces are more important in the national picture than others. The province of Hebei nearly surrounds Beijing and occupies much of the core of the country. The province of Shaanxi is centered on the great ancient city of Xian. In the southeast, momentous economic developments are taking place in the province of Guangdong, whose urban focus is Guangzhou (4.1 million). When, in the pages that follow, we refer to a particular province or region, Fig. 10-7 is a useful locational guide.

When it comes to population, many Chinese provinces have more inhabitants than most of the world's countries. The latest findings from the country's 1990 census are that China's most populous province, Sichuan, has more than 107 million residents; Shandong and Henan provinces, covering much of the North China Plain, each contain close to 85 million. By contrast, the Xizang Autonomous Region (Tibet) has fewer than two-and-a-quarter million inhabitants (Table 10-1, p. 528).

TABLE 10-1
China: Population by Major Administrative Divisions

Provinces (1990)	
Anhui	56,180,813
Fujian	30,048,224
Gansu	22,371,141
Guangdong	62,829,236
Guizhou	32,391,066
Hainan	6,557,482
Hebei	61,082,439
Heilongjiang	35,214,873
Henan	85,509,535
Hubei	53,969,210
Hunan	60,659,754
Jiangsu	67,056,519
Jiangxi	37,710,281
Liaoning	39,459,697
Qinghai	4,456,946
Shaanxi	32,882,403
Shandong	84,392,827
Shanxi	28,759,014
Sichuan	107,218,173
Yunnan	36,972,610
Zhejiang	41,445,930
Autonomous Regions (1990)	
Guangxi Zhuang	42,245,765
Nei Mongol	21,456,798
Ningxia Hui	4,655,451
Xinjiang Uygur	15,155,778
Xizang	2,196,010
Municipalities (1995 est.)	
Beijing	12,430,000
Shanghai	15,165,000
Tianjin	10,746,000

Sources: 1990 Census of China; United Nations, Population Division, 1991.

◆ REGIONS OF THE CHINESE REALM

A map of China's population distribution (Fig. 10-8) reveals the continuing relationship between the physical stage and its human occupants. In technologically advanced countries, we have noted, people shake off their dependence on what the land can provide; they cluster in cities and in other areas of economic opportunity. This has the effect of depopulating rural areas that may once have been densely inhabited. In China, that stage has not yet been reached. True, there are large cities here, but as in India the great majority of the people (nearly 75 percent) still live on—and from—the land. This means that the map of population distribution reflects the livability and productivity of China's basins, lowlands, and plains. Compare Figs. 10-3A and 10-8, and China's continuing dependence on soil, water, and warmth will be evident.

The population map also suggests that in certain areas, limitations of the environment are being overcome. Industrialization in the Northeast, irrigation in the Inner Mongolia Autonomous Region, and oil-well drilling in Xinjiang have enabled millions of Chinese to migrate from China Proper into these frontier zones, where they now outnumber the indigenous minorities.

EASTERN LOWLANDS

Physiographically, it is appropriate to distinguish between the Middle and Lower Basins of the Huang and Chang rivers; however, as Fig. 10-8 shows, in human terms they form one large region. The Eastern Lowlands incorporate the North China Plain, including Beijing and Tianjin, and the productive Middle and Lower Chang basins, with Nanjing, Wuhan, and the great metropolis of Shanghai. In every respect, this is China's **core area**. Here lies the greatest population concentration, the highest percentage of urbanization (about 50 percent), enormous farm production, burgeoning industrial complexes, and the most intensive communications networks.

Chang Basin

Unlike the shallow, muddy, loess-carrying Huang (hence *Yellow* River and *Yellow* Sea), the Chang Jiang is China's most navigable waterway. Oceangoing ships can sail over 600 miles (1,000 km) up the river through Nanjing (3.0 million) to the Wuhan conurbation (Wuhan [4.4 million] is short for Wuchang, Hanyang, and Hankou); boats of up to 1,000 tons can travel twice that distance to Chongqing (3.5 million) in the Sichuan Basin (Fig. 10-9). Several of the Chang's tributaries are also navigable, and a total of more than 18,500 miles (30,000 km) of water transport routes operate in the entire Chang Basin. Therefore, the Chang Jiang is one of China's leading transportation corridors; with its tributaries it attracts the trade of a vast area, including nearly all of middle China and sizeable parts of the adjacent north and south. Funneled down the Chang, most of this enormous volume of trade is transshipped at Shanghai, whose metropolitan population size of over 15 million reflects the productivity of this hinterland.

Early in Chinese history, when the Chang's basin was being opened up and rice and wheat cultivation began, a canal was built to link this granary to the northern core of old China. Over 1,000 miles (1,600 km) in length, this was the longest artificial waterway in the world, but during the nineteenth century it fell into disrepair. Known as the Grand Canal, it

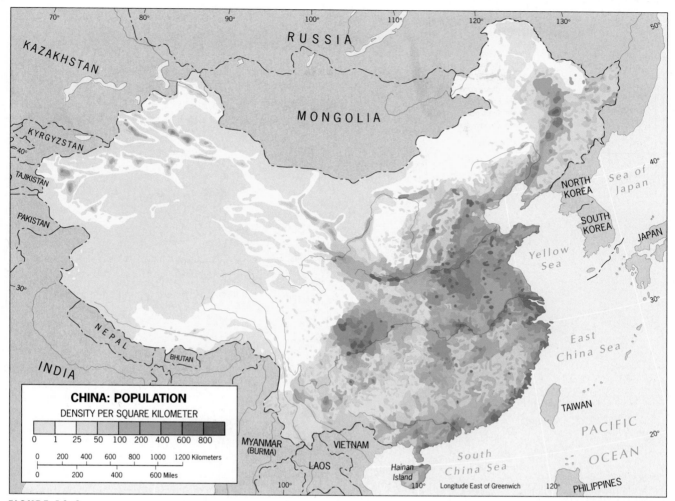

CHINA: POPULATION
DENSITY PER SQUARE KILOMETER
0 1 25 50 100 200 400 600 800

0   200   400   600   800   1000   1200 Kilometers
0   200   400   600 Miles

FIGURE 10-8

was dredged and rebuilt during the period when the Nationalists held control over eastern China. After 1949, the communist regime continued this restoration effort, and much of the canal is now once again open to barge traffic, supplementing the huge fleet of vessels that hauls domestic interregional trade along the east coast (Fig. 10-9).

The bulk of China's internal trade in agricultural as well as industrial products is either derived from or distributed to the Chang region. International trade also goes principally through Shanghai, whose port normally handles half of the country's overseas tonnage; the rest is split among China's other leading ports, including Tianjin on the Bohai Gulf in the north and Guangzhou on the Pearl River in the south.

Shanghai   Shanghai, just a regional town until the mid-nineteenth century, rose to prominence as a result of its selection as a treaty port by the British. Ever since, its un-

equaled locational advantages have sustained its position as China's leading city in almost every respect. This metropolis lies on the Huangpu River at one corner of the Chang delta, an area of about 20,000 square miles (50,000 sq km) containing more than 50 million people (Fig. 10-9, inset map). Some two-thirds of these are farmers who produce food, silk filaments, and cotton for the city's industries.

Thus Shanghai has as its immediate hinterland one of the most densely populated areas on earth, and beyond the delta lies what has to be the most populous region in the world to be served by one major outlet. During the nineteenth century and until the war with Japan, the principal exports to pass through Shanghai were tea and silk; large quantities of cotton textiles and opium were imported. At that time, Shanghai's prominence was undisputed, and it handled two-thirds of all of China's external trade. But its fortunes suffered during and after World War II. First, the Nationalists blockaded

CHINA PROPER

POPULATION
- • 50,000–250,000
- • 250,000–1,000,000
- ● 1,000,000–5,000,000
- ⬤ Over 5,000,000

National capital is underlined

Canal
Railroad

INLAND WATERWAYS
- Ocean-going ships
- Large river steamers
- Small river steamers
- Steamboats

FIGURE 10-9

the port (in 1949) and conducted bombing raids on it; then the new Beijing government decided to disperse its industries up-country, thereby reducing their vulnerability to attack. Meanwhile, safer Tianjin had taken over as the leading port. However, Shanghai's unparalleled situational advan-

tages promised a comeback, which soon occurred. In the 1960s, the port regained its dominant position, and the industrial complex (textiles, food processing, metals, shipyards, rubber, and chemicals) resumed its expansion. Today, Shanghai ranks as the world's fifth-largest urban concentration,

As part of China's urban modernization, the waterfront (Bund) of Shanghai in 1993 was under massive reconstruction. The view is upstream along the Huangpu River, tributary of the Chang Jiang. Here the British built their colonial headquarters, and Victorian-era buildings can still be seen. New skyscrapers are beginning to tower over this downtown scene.

and its population is growing about five times faster than the national rate of increase.

North China Plain

As Fig. 10-10 indicates, both rice and wheat are grown in the Lower Chang Basin. This component of the Eastern Lowlands produces a variety of crops, but northward, temperatures as well as precipitation diminish (shown as a line on the map that marks the northern limit of rice cultivation). Here we enter the North China Plain, where winter wheat and barley are planted to the south of Beijing (spring wheat is raised to the north); in the spring, other crops follow the winter wheat. Millet, sorghum (gaoliang), soybeans, corn, a variety of fruits and vegetables, tobacco, and cotton are also cultivated in this central zone of China Proper. This part of China was subdivided into very small parcels of land, and the communist regime has effected a major reorganization of landholding. At the same time, an enormous effort has been made to control the flood problem that has afflicted the Huang Basin for uncounted centuries and to expand the land area under irrigation.

The North China Plain is one of the world's most heavily populated agricultural areas, with more than 1,400 people per square mile (540 people per sq km) of cultivated land. Here, the ultimate hope of the Beijing government lay less in land redistribution than in raising yields through improved fertilization, expanded irrigation facilities, and the more intensive use of labor. A series of dams on the Huang River now somewhat reduces the flood danger, but outside the irrigated areas the ever-present problem of rainfall variability and drought persists. The North China Plain has not pro-

duced any substantial food surplus even under normal circumstances; thus when the weather turns unfavorable, the situation soon becomes precarious. The specter of famine may have receded, but the food situation is still uncertain in this very critical part of China Proper.

The two major cities in this portion of the Eastern Lowlands are the historic capital, Beijing, and the port city of Tianjin on the Bohai Gulf, both positioned near the northern edge of the North China Plain. In common with many of China's harbor sites, that of the river port of Tianjin is not particularly good; but this city is well situated to serve the northern sector of the Eastern Lowlands, the nearby capital, the Upper Huang Basin, and Inner Mongolia beyond. Like Shanghai, Tianjin had its modern start as a treaty port, but the city's major growth awaited communist rule. For many decades it had remained a center for light industry and a flood-prone harbor; after 1949, a new artificial port was constructed and flood canals were dug. More importantly, Tianjin was chosen as a site for major industrial development, and large investments were made in the chemical industry (in which Tianjin now leads China), in such basic industries as iron and steel production, in heavy machinery manufacturing, and in textiles. Today, with a population approaching 11 million, Tianjin is the country's third-largest metropolis and the center of one of its leading industrial complexes.

Beijing Beijing, in contrast, has chiefly remained the political, educational, and cultural center of China. Although industrial development also occurred here after the communist takeover, its dimensions have not been comparable to Tianjin. The communist administration did, however, greatly expand the municipal area of Beijing (as we noted), which is

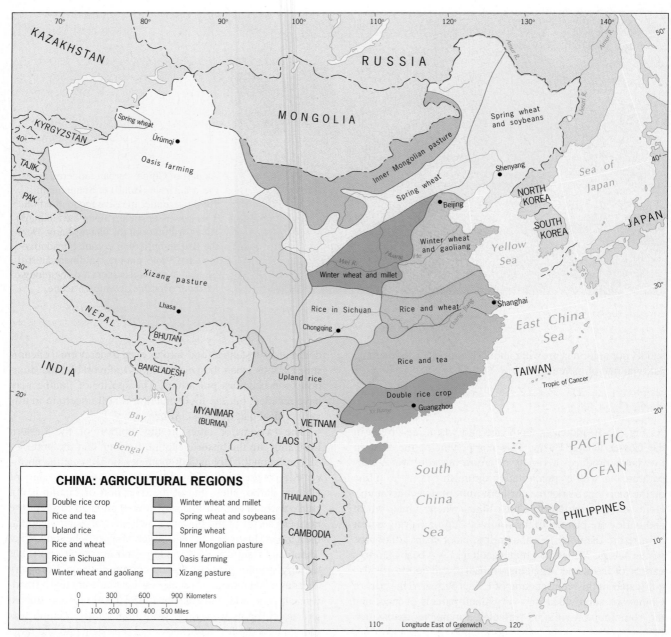

FIGURE 10-10

not controlled by the province of Hebei but is directly under the central government's authority. In one direction, Beijing was enlarged all the way to the Great Wall—30 miles (50 km) to the north—so that the "urban" area includes hundreds of thousands of farmers; not surprisingly, an enormous total population has been circumscribed, enough to rank Beijing, with its 12.4 million inhabitants, as the seventh-largest metropolitan concentration on earth.

Just as Shanghai symbolizes bustling, crowded, industrial, trade-oriented China, so Beijing represents Chinese history, power, and order. Beijing's main thoroughfare, reputedly the longest city street in the world, straddles Tiananmen Square and the (adjacent) so-called Forbidden City, the treasure-laden preserve of the rulers and elite of old. The great Gate of Heavenly Peace overlooks both ancient city and modern square, scene of the birth of communist rule. As the map shows, Beijing has a noteworthy relative location. The city lies close to the northern margin of the North China Plain and not far from the Great Wall and the Ming Tombs (repositories of great treasures from the Ming Dynasty, partially unearthed but holding other, still-buried riches). And nearby was made China's greatest archeological discovery, Peking Man.

Beijing's Gate of Heavenly Peace, scene of Mao's famous 1949 proclamation, on another historic occasion: the raising of the anti-government "Goddess of Liberty" statue on May 31, 1989—five days before the Red Army savagely crushed the student occupation of Tiananmen Square.

SHAANXI AND SICHUAN

Westward of the Eastern Lowlands, the population map reveals two major concentrations. The northern one of these, centered on the city of Xian, is based on the Loess Plateau, the densely inhabited, fertile area where millet, corn, and winter wheat are cultivated. Xian (3.3 million) is one of the country's most historic cities, dating back to Chinese culture's earliest times. Its modern geography is related to its situation as a gateway to the west, as a safe interior city when foreigners invaded, as a haven for the communists when the struggle with the Nationalists broke out, and as a prominent industrial center when the communist planners took over. Almost all of Xian's architectural history has been destroyed, and the city today is an industrial complex with large textile factories, steel plants, coal-based power plants, and other facilities.

As the capital of Shaanxi Province, Xian has political and cultural importance as well. Nearby, archeologists in the 1970s unearthed one of China's most remarkable finds. An emperor's burial ground, more than 2,000 years old, contained an army of 6,000 life-sized horses and men, wagons and weapons, fashioned in terra cotta and lined up in formation,

ready for action at the deceased ruler's command. More modern reminders of Xian's relative location are its surviving pagodas and its large, well-attended mosque.

Southward lies another great population agglomeration, physiographically in the Sichuan Basin and politically in Sichuan Province. Chengdu (3.4 million), capital of the province, is no match for Xian, but it was one of the most active centers of the pro-democracy movement of the late 1980s. Sichuan's fertile, well-watered soils yield an amazing variety of crops: in addition to grains (rice, wheat, corn), there are soybeans, tea, sweet potatoes, sugar cane, and many different fruits and vegetables.

As the maps indicate, after the Chang River leaves the Sichuan Basin and before it reaches its Middle Basin, it flows through a zone of narrow valleys and gorges. The Gezhouba Dam here has already been completed, controlling floods and producing large amounts of electricity. Now Chinese planners are embarked on a gargantuan dam project known as "Three Gorges," the world's largest. The Three Gorges segment of the Chang valley is a series of canyons of majestic dimensions, that has inspired Chinese painters and poets for many centuries. The new dam, more than 600 feet high, would submerge the scenic cliffs and displace more than a

Xian was China's formative center, a great city of cultural strength and political power, the eastern terminus of the Silk Road. Its heritage was severely damaged during the communist era, but the city's historic importance is undiminished. This is Big Goose Pagoda, one of Xian's revered landmarks.

million inhabitants of this part of the valley. A debate over the project has raged in China for years, with environmentalists pitted against developers. In 1993, the venture was being delayed but not because of environmental considerations: China did not have the billions of dollars needed to pay for construction. But as China's economy prospers, the time will come when these funds are available, and then economics is likely to prevail over environmental concerns.

Surface Links and Energy Sources

When we compare the map of population distribution with that of energy resources and communications, it is clear that both Xian, west of the North China Plain, and Chengdu, west of the Lower and Middle Chang basins, are the centers of more than farming areas (Fig. 10-11). China has huge coal reserves and large oilfields, and both Shaanxi and Sichuan provinces are well endowed. In the southeastern part of Sichuan, moreover, lie important reserves of natural gas.

As Fig. 10-11 shows, energy resources are widely dispersed throughout eastern and northern China, and additional discoveries are likely; the shallow seas off the China coast are especially promising. Pipelines connect oilfields in the Northeast to the industrial region of Beijing-Tianjin, and the development of pipeline links is continuing in Xinjiang as well. Unlike nearby Japan, China has its own substantial energy resources.

China's railroad network (also shown in Fig. 10-11) is inadequate. Sun Yat-sen had envisioned a China whose distant regions would be connected by a well-planned rail system, but his design could not be executed because the country's instability and lack of capital held it back. Indeed, much

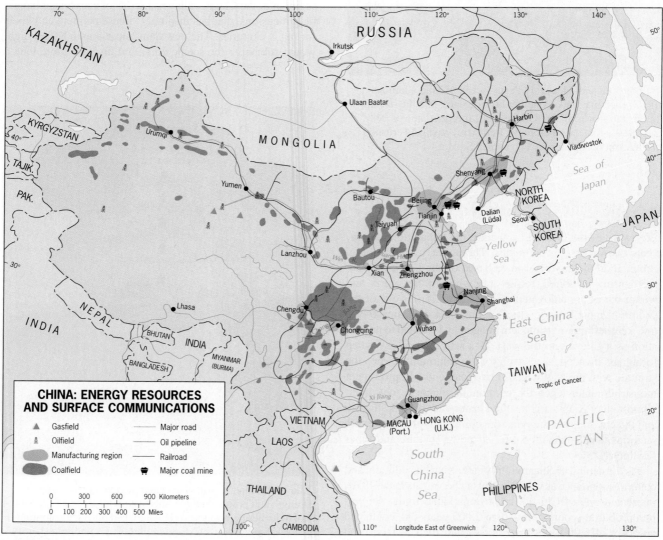

FIGURE 10-11

of China's network, notably in the Northeast, was laid out and built under Japanese and (later) Soviet engineers. A national scheme did not evolve prior to the communist period; the first railroad bridges across the Chang River (at Wuhan and Chongqing) were not built until the 1950s. Double-tracking is proceeding, but the system is constantly strained by the country's growing needs. China does not yet have a web of superhighways with trucks competing against the railroads for cargoes; road travel is slow and inefficient, with the roads clogged with vehicles ranging from bicycles and horse-drawn wagons to cars and buses. Expanding and improving China's rail network remains a national priority in this developing country with its substantial distances, far-flung resources, and huge population.

THE SOUTH

The population map tells us that southern China does not have a population cluster comparable to those of the Eastern Lowlands. China's South is important in other ways, however, for two principal reasons.

First, as we noted in Chapter 4 (pp. 244-245), momentous economic developments are taking place here, notably in Guangdong Province. These are based not on plentiful local mineral or fuel resources, but on the availability of a large, comparatively cheaply employable, skilled, unskilled, and trainable labor force. Beijing's economic reforms have had more impact here than anywhere else in the country. Already, Guangdong and especially its capital, Guangzhou (4.1 million), are beehives of activity, contributing enormously to China's overall economic growth rate; and on the Chinese side of the border with Hong Kong, the name Shenzhen has become synonymous with China's new era of commercialism. And soon, Hong Kong itself will become part not so much of China but of this burgeoning economic complex. If China's communist rulers in Beijing do not impede this progress by precipitating a political crisis over Hong Kong's reintegration, this part of China's South will be a guidepost to the China of the future.

Inland from Guangdong Province, though, lies a very different China—and a second reason why the South is especially significant. Note that Guangdong Province, in this direction, is not bounded by a Chinese province, but by an

The Guangxi Zhuang Autonomous Region lies between economically mushrooming Guangdong Province to the east and Yunnan to the west. This Region is designated for minorities, and here in Yangshuo, on market day, farmers are selling their goods. But many younger residents are moving east, attracted by the opportunities of the Pacific Rim.

Autonomous Region: Guangxi Zhuang. Why this should be so is revealed by Fig. 10-12, the map of China's ethnic minorities. This map constitutes another way to distinguish Han China, or China Proper, from the rest of the country. In Guangxi Zhuang Autonomous Region, China takes on a very different appearance. The people, their villages, their fields—the entire cultural landscape is more Southeast Asian than Han Chinese (see photo p. 535). And the landscape shows this contrast in other ways. Population pressure is great, soil erosion is severe, and the poverty and stagnation of the region stand in strong contrast to rapidly developing Guangdong.

Still farther westward in the corner between Laos and Myanmar (Burma) lies Yunnan Province, where Han Chinese and Southeast Asian ethnic minorities share the land. As

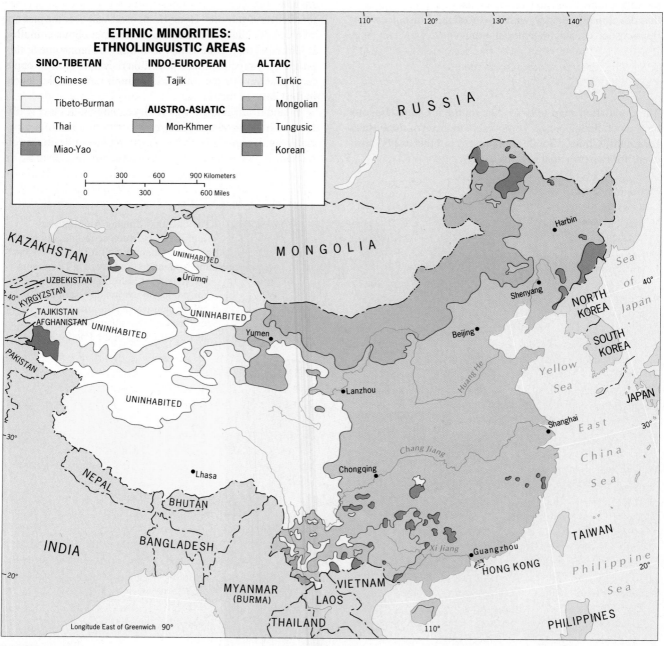

FIGURE 10-12

the map shows, Yunnan is a mosaic of ethnic groups; the capital, Kunming (2.0 million) is the center of Chinese presence here. A road journey through Yunnan's high-relief topography is a trip across a cultural jigsaw: one valley or basin is inhabited by a particular ethnic community, the next by Han Chinese, and the one thereafter by still another group. Interpreting the cultural geography here is quite a challenge.

Earlier we noted that South China ranges environmentally from subtropical to tropical, providing a year-round growing season in many areas and allowing the double-cropping of rice. Vegetables, fruits, and other crops are grown in great variety, but the problem is that the South has relatively little level land. Hillslopes are terraced almost everywhere (see photo p. 517), and farms, per unit area, are very productive. But that is not enough to make this region self-sufficient in food, and grain imports are always required.

THE NORTHEAST

As Fig. 10-7 confirms, China's northeastern zone is large as well as northerly, almost all of it lying poleward of the 40th parallel. There are three provinces here: Liaoning in the south, facing the Yellow Sea; Jilin in the heart of the region; and Heilongjiang, by far the largest, in the north. This is the region colonized by the Japanese and called Manchukuo when they controlled it; it also was known as Manchuria (see box p. 522), the original base of the Manchus who invaded China and, once assimilated, established the last true dynasty over the country. But the Chinese themselves do not like the Manchu label and prefer to call this part of China simply the Northeast (*Dongbei*).

When you take the ferry from Tianjin and sail up the Liaodong Gulf toward the mouth of the Liao River, your first impression is that this lowland is a continuation of the North China Plain. Near the coast, farms look as they do near the mouth of the Huang River; to the east, the Liaodong Peninsula looks very much like the Shandong Peninsula across the water. That impression, however, soon disappears. Travel northward, and the Northeast China Plain (Fig. 10-4) reveals the effects of cold weather and thin soils. Farmlands are patchy; smokestacks rise, it seems, everywhere. This is industrial, not primarily farming, country. But the landscape looks like a Rustbelt: equipment often is old and outdated, transport systems inadequate, buildings in disrepair (see the photo on p. 539). Stream, land, and air pollution is dreadful. The Northeast in some ways resembles industrial zones in former communist Eastern Europe.

The Northeast has seen many ups and downs. Although Japanese colonialism was ruthless and exploitive, Japan did build railroads, roads, bridges, factories, and other components of the regional infrastructure. (At one time, *half* of the entire railroad mileage in China was in the Northeast.) After the Japanese were ousted, the Soviets entered the area and looted it of machinery, equipment, and other goods. During the late 1940s, the Northeast was a ravaged frontier. But then the communists took power, and they made the industrial development of the Northeast a priority. From the 1950s until the 1970s, the Northeast led the nation in manufacturing growth. The population of the region, just a few million in the 1940s, mushroomed to more than 100 million. Towns and cities grew by leaps and bounds.

All this was based on the Northeast's considerable mineral wealth (Fig. 10-13). Iron ore deposits and coalfields lie concentrated in the Liao Basin; aluminum ore, feroalloys, lead, zinc, and other metals are plentiful, too. As the map shows, the region also is well endowed with oil reserves; the largest of these lie in the Daqing Reserve between Harbin (3.3 million) and Qiqihar (1.7 million). The communist planners made Changchun (2.5 million), capital of Jilin Province, the Northeast's automobile manufacturing center. Shenyang (5.3 million), capital of Liaoning Province, became the leading steelmaking complex. Fushun (1.6 million) and Anshan (1.7 million) were assigned other functions, and the communists encouraged Han Chinese to migrate to the Northeast to find employment there. To attract them, the government built apartments, schools, hospitals, recreational facilities, and even old-age homes, all to speed the industrial development of the region.

For a time, it worked. The Northeast during the 1970s produced fully one-quarter of the entire country's industrial output. Stimulated by the needs of the growing cities and towns, farms expanded and crops diversified. The Songhua River basin in the hinterland of Harbin, capital of Heilongjiang Province, became a productive agricultural zone despite the environmental limitations here. The Northeast was a shining example of what communist planning could accomplish.

Then came the economic restructuring of the 1980s. Under the new, market-driven economic order, the very state companies that had led China's industrial resurgence were unable to adapt. Suddenly they were aging, inefficient, obsolete. Official policy encouraged restructuring and privatization, but the huge state factories could not (and much of their labor force would not) comply. Today, the Northeast's contribution to China's industrial output is a mere 12 percent as the production of factories in the Southeast, from Shanghai to Guangdong, rises annually.

Economic geographers, however, will not write off the prospects of this northeastern corner of China; this region, as we noted, has had its downturns before. The Northeast still contains a storehouse of resources, including extensive stands of oak and other hardwood forests on the mountain slopes that encircle the Northeast China Plain. The locational advantages of the major centers are undiminished; Harbin, for example, lies at the convergence of five railroads and is situated at the head of navigation on the Songhua River, link to the northeast corner of the region and, beyond the Amur River, to Russia's Khabarovsk.

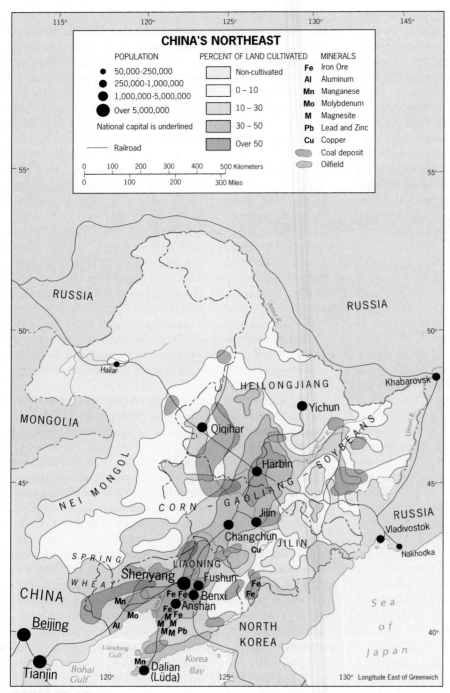

FIGURE 10-13

When the Koreas reunite, producing what may become the western Pacific Rim's second Japan, and when the Russian Far East takes off (Vladivostok and Nakhodka are the natural ports for the upper Northeast), this northern frontier of China may yet reverse its decline and enter still another era of economic success.

RETROSPECT ON THE REGIONS OF CHINA PROPER

In the preceding pages we have noted some of the numerous spatial patterns in China's human geography. Northern China grows wheat for food and cotton for sale; South China grows

The Northeast is China's Rustbelt, its outdated machinery and obsolete equipment left behind in the rush to the Pacific Rim. Shenyang, once the heart of the Northeast's industrial juggernaut, now suffers from decay and unemployment. Many of the old pollution-spewing smokestacks no longer function.

rice and tea. Northern and northwestern China is the land of the ox and even the camel; southern China uses the water buffalo. Northern, interior China is mired in the communist dogma of the past; southern, coastal China is swept up in the changes of the Pacific Rim.

Beijing's enforced order and imposed uniformity stand in glaring contrast to the southeastern city of Guangzhou, where the social atmosphere is lively, open, even boisterous—reminiscent, it is said, of the Canton of old. And so the regional contrasts in Han China come to the fore: entrepreneurs in Shenzhen who complain of bureaucratic hurdles created by Beijing; workers in Guangzhou who complain about being "colonized" by the old men in the capital; investors in Shanghai who feel disadvantaged by government policies that seem to favor other areas; minorities in Yunnan who begin to make their voices heard in provincial and national capitals. Tiananmen Square notwithstanding, power in China is devolving to the provinces. What this will mean for the empire of the future is, as yet, anybody's guess.

We now complete our survey of China's regional human geography by turning from populous, intensively utilized China Proper to the lands of the north and west that are under Han China's rule—a frontier zone of vast expanses, rugged terrain, unsettled political boundaries, and spotty and sometimes tenuous human occupance.

INNER MONGOLIA

The area identified in Fig. 10-7 as the Nei Mongol (Inner Mongolia) Autonomous Region is a case in point. Here, Chinese farmers long competed with nomadic horse- and camel-riding Mongols for control over an area that consists of vast expanses of steppe and desert as well as a few riverine ribbons of settlement (including the Huang River where it crosses the Ordos Desert) and oases of farmland. In effect, this region is a transition zone between China Proper and the independent buffer state of Mongolia (see box pp. 540-541). When Inner Mongolia was made an Autonomous Region, Mongols may have been in the majority here; but today, the A.R.'s population of about 22 million is overwhelmingly Han Chinese. Fewer than 5 million Mongols inhabit the region today, most of them concentrated in the northern border area. (Note, however, that this number is twice as large as the entire population of the neighboring independent state of Mongolia.)

Since the onset of Chinese communist rule, this region has changed in many ways. Its boundaries have repeatedly been modified, assigning territory to neighboring Chinese provinces. The capital, Hohhot (1.1 million), has been connected by rail to China's network, and there is also a line connecting it to Mongolia and its capital, Ulaan Baatar (630,000). West of Hohhot, the city of Baotou (1.4 million) has become the region's principal industrial center, based on iron ores and coal discovered in its hinterland. Coupled with the ribbon of towns and farms along the Huang River's great bend, these developments stand in stark contrast to the tents and herds of the ethnic Mongols, many of whom still move with their animals along time-honored nomadic routes.

XINJIANG

As Fig. 10-7 shows, an elongated Chinese province, Gansu, borders Inner Mongolia to the west and south. A look at surface communications indicates why Gansu is there: it con-

MONGOLIA: BUFFER STATE BETWEEN FRONTIERS

Between China's Inner Mongolia and Xinjiang to the south, and Russia's Eastern Frontier region to the north, lies a vast, landlocked, isolated country called Mongolia (Fig. 10-14). With only 2.4 million inhabitants in an area larger than Alaska, Mongolia is a steppe- and desert-dominated vacuum between two of the world's most power-

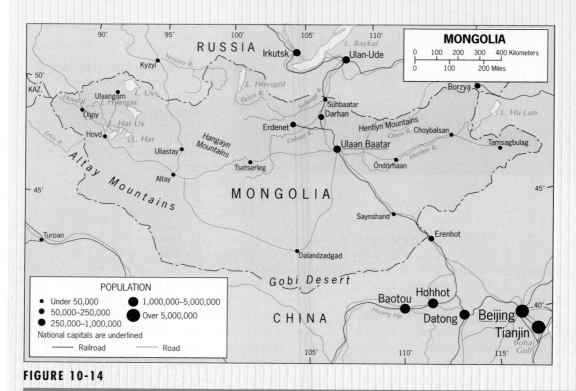

FIGURE 10-14

tains the vital Hexi Corridor, a road and rail route that links Xian (and thus Han China via the provincial capital, Lanzhou [1.8 million]) to Xinjiang, China's potentially vital northwestern corner.

Xinjiang is China's largest single administrative area, an Autonomous Region covering more than one-sixth of the entire country. It borders Russia, Kazakhstan, Kyrgyzstan, Tajikistan, the Wakhan Corridor of Afghanistan, Pakistan, and Jammu and Kashmir (where China has a boundary dispute with India). As the map of ethnic minorities (Fig. 10-12) reveals, this is a region of many and diverse peoples.

Since 1949, the Chinese have made a major effort to develop the limited agricultural potential of the Tarim Basin. Canals and *qanats* were built, oases enlarged, and the acreage of productive farmland has at least quadrupled from what it was 45 years ago. The long-neglected northern rim of the Tarim Basin, in particular, has been brought into the sphere of development, and fields of cotton and wheat now attest to the success of the program.

The Junggar Basin, although it contains only about one-third of the Xinjiang Uygur Autonomous Region's approximately 16 million inhabitants, has a number of assets of importance to China. First, it has long been the site of strategic east-west routes. Second, the main westward rail link toward Kazakhstan runs from Xian in eastern China through Yumen in Gansu and Ürümqi in Junggar. Third, the Junggar Basin is proving to contain sizeable oilfields (as is the Tarim Basin), notably around the town of Karamay near the border with Kazakhstan. Pipelines have been laid all the way east to Yumen and Lanzhou, where refineries have been built. Thus it is not altogether surprising that the Autonomous Region's capital of Ürümqi (1.7 million) lies in the less populous but strategically important northern part of Xinjiang.

Xinjiang's population is now more than one-third Chinese (up from 5 percent in 1949) as a result of Beijing's determination to integrate this distant region into the national framework; however, the ethnic Chinese proportion stopped growing here during the mid-1980s and is now stable. The

ful countries. Mongolia's historical geography is tied to Russia and the former Soviet Union as well as to China. Mongolia was part of the great Mongol Empire, source of those great waves of horsemen who rode into the Slavic heartland seven centuries ago. The Mongols also invaded China, establishing the short-lived Yuan Dynasty (1279–1368). Eventually, however, the Mongol domain fell under Chinese control. From the late 1600s until 1911, Mongolia was part of the Chinese Empire.

The Mongolians seized independence in 1911 while China was in revolutionary chaos. In those days, the area of present-day Mongolia was known as Outer Mongolia; the ill-defined region to its southeast and east was called Inner Mongolia. As the Chinese tried once again to assert their hegemony over Outer Mongolia, the Russian Revolution broke out. This gave the (Outer) Mongolians another chance to hold off the Chinese: in the early 1920s the country became a People's Republic on the Soviet model, with Soviet protection. Inner Mongolia remained a Chinese-dominated frontier and became the Nei Mongol Autonomous Region under Beijing. The border between the two Mongolias is still heavily militarized.

Today, with its vast deserts, grassy plains, forest-clad mountains, and fish-filled lakes, Mongolia lies uneasily between larger, unstable neighbors, neither of which can permit its incorporation into the other. Despite its historic associations, ethnic affinities, and cultural involvement with China, Mongolia's development for some 70 years

was guided by the Soviets. The country was made a member of COMECON (the now-defunct, Soviet-led economic union constituted by its then Eastern European satellite countries); the Soviet Communist Party controlled its political life; Soviet armed forces were stationed within its borders. The capital, Ulaan Baatar, the country's largest city, lies close to the Lake Baykal subregion of Russia's Eastern Frontier. China seems far away from here, across the forbidding Gobi Desert.

But Mongolia's cultural and economic landscapes are East Asian, despite the Soviet/Russian veneer, and the winds of political change have reached here, too. The Mongolian alphabet, abolished under Soviet educational regulations, has recently been reinstated. Public demonstrations in support of political reforms led to democratic elections. Mongolian nationalism is on the rise, but what are its options? Russia's eastern zone is not immune to the instability that now marks its western and southern periphery; and China remains a communist state politically.

Mongolia today is a landlocked, isolated, vulnerable **buffer state**, wedged between larger, more populous, more powerful neighbors. Most of its people still ride their horses and herd their livestock across the Wyoming-like countryside. Below the surface, however, lie valuable minerals. In the north, Mongolian territory leaves only a narrow corridor of Russian soil between it and Lake Baykal. Will Mongolia escape the fate of those other inner-Asian buffers, Tibet and Afghanistan?

majority of the people in Xinjiang are Muslim Uygurs, Kazakhs, and Kyrgyz, with cultural affinities across the border to all five of the republics of Turkestan (see Fig. 7-18). The Uygurs, most numerous among these peoples (about 50 percent of the population), have for centuries been concentrated in the oases of the Tarim Basin, and the mobile Kazakhs and Kyrgyz circulate along nomadic routes that cross international borders.

We also should contemplate the potential strategic significance of Xinjiang. China's space program and nuclear weapons development program long have been based here, and this region of China lies closest to the Russian heartland. The prospect of a confrontation between China and Russia seems remote today, but the two giants have historic, territorial, and ideological differences. Should a power struggle develop, Xinjiang would be one of China's greatest assets. More immediately, however, the challenge may be one of continued control and stability. The transformation of Turkestan (see Chapter 7) already is affecting this remote Chinese fron-

tier, where notions of independence and Islamic revival have long been suppressed.

XIZANG (TIBET)

Earlier we noted the harsh physical environments that dominate this colony of China. It is a vast area designated an Autonomous Region, but in fact Tibet is an occupied society. In terms of human geography, we should take note of two areas.

The first of these is the core area of the Tibetan culture, between the Himalayas to the south and the Transhimalayas lying not far to the north. In this area, some valleys lie below 7,000 feet (2,100 m); the climate is comparatively mild, and some cultivation is possible. Here lies Tibet's main population cluster, including the crossroads capital of Lhasa (155,000). The Chinese government has made investments to develop these valleys, which contain excellent sites for hydroelectric power projects (some have been put to use in a few light industries) and several promising mineral deposits. The second area of interest is the Qaidam Basin in the north. This basin lies thousands of feet below the surrounding Kunlun and Altun mountains and has always contained a concentration of nomadic pastoralists. Recently, however, exploration has revealed the presence of oilfields and coal reserves below the surface of the Qaidam Basin, and the development of these resources is now underway.

Politically, Xizang was pressed into the Chinese fold during the 1950s, first through frontier settlement and economic interference and then in 1958 by force of arms after Tibetan villagers tried to resist the Chinese presence. Tibetan society had been organized around the fortress-like monasteries of Buddhist monks who paid allegiance to their supreme leader, the Dalai Lama. The Chinese wanted to modernize this feudal system, but the Tibetans clung to their traditions. In 1958, they proved no match for the Chinese armed forces; the Dalai Lama was ousted, and the monasteries were emptied. The Chinese destroyed much of Tibet's cultural heritage, looting the region of its religious treasures and works of art. Their harsh rule took a severe toll on Tibetan society, but after Mao's death in 1976 the Chinese relaxed their tight control. Although amends were made (religious treasures were returned to Xizang, monastery reconstruction was permitted, and Buddhist religious life resumed), pro-independence rioting has been frequent since 1987, and the Chinese have again tightened their grip. Since its annexation in 1965, Xizang has been administered as an Autonomous Region; although it is large in territorial size, Tibet's population remains slightly more than 2 million.

Visiting Chinese-occupied Tibet is a depressing experience for the neutral observer. Tibetan domestic architecture has beauty and expressiveness, but the Chinese have built ugly, gray structures to house the occupiers—often directly in front of Tibetan monasteries and shrines. This juxtaposition of Chinese banality and Tibetan civility seems calculated to perpetuate hostility, an anachronism in this supposedly postcolonial world. During the rule of Mao Zedong, the Chinese officially regarded Tibet as a territorial cushion against India (then allied with the Soviet Union). The demise of the Soviet Empire has not softened Beijing's stand.

◆ THE COMMUNIST TRANSFORMATION

When China's communist era began, now nearly a half-century ago, the country was wracked by war. Its economy lay in ruins. Famines threatened. Population growth soared. The Nationalists had fled to Taiwan, but the new communist regime still faced opposition in many corners of the country.

Almost immediately after taking power, the communists launched massive programs of reconstruction and reform. The most urgent needs lay in agriculture; China was hungry and dependent on foreign sources for emergency food supplies. Using the Soviet model as their guide, the Chinese rulers began a program to expropriate all landowners and to redistribute the land among landless villagers. The next stage was the **collectivization** of agriculture, begun in 1955. All the peasants in a village would pool their land and their labor. Compensation was in proportion to the size of each share and to the labor contributed by each family. The farmers still had the right to withdraw their land from the collective if they wished. At first, participation in this new program was slow, but collectivization was enforced by coercion, often brutally. By the end of 1956, compliance was nearly total everywhere in Han China (the ethnic minorities were at first exluded from the national plan). Overall farm output increased, but not as much as the planners in Beijing had hoped.

In 1958, the program of collectivization was carried a step further. In less than one year, more than 120 million peasant households, already organized into collectives, were apportioned into about 26,500 People's Communes designed to contain about 20,000 people each. The private landholdings of the socialist collective system were abolished. Now, **communization** would signify China's *Great Leap Forward* (as government propaganda called it) from socialism to communism. The impact of this scheme on people's lives was unprecedented. Adults in the communes, men and women, were organized into hierarchies of "production teams" with military designations such as companies, batallions, and brigades. Communal quarters were built, men and women segregated, children sent off to boarding schools. Households, according to official propaganda, would be things of the past. Wage systems were abolished in favor of allocations of free food and clothing plus a small stipend.

The commune system extended beyond the farm communes. Workers by the thousands were organized to tackle projects such as flood-control levees, irrigation dams, and roads. Fences and hedges between former collectives were torn down, fields were consolidated. Some villages were leveled, others enlarged as communes. In the rural areas, China's cultural landscape changed dramatically.

After a decade of communist rule and reorganization, China was a changed country, and not only in its rural areas. Beijing's rulers benefited from the nearly unanimous realization that some form of communal organization was needed to coordinate farm production—but this was not enough to save the communists' program. It was all done with too much haste, too little planning, and too little preparation for the effects of the social dislocation caused by the communization of so many millions. Resentment rose as opposition was harshly put down (even a mild complaint overheard by a Communist Party member could lead to a jail sentence). Soon the regime was forced to allow families to reunite and to restore private plots to some peasants.

KONGFUZI (CONFUCIUS)

Confucius (*Kongfuzi* or *Kongzi* in pinyin spelling) was China's most influential philosopher and teacher, whose ideas dominated Chinese life and thought for over 20 centuries. But the leaders who took control in 1949 considered Kongzi ideals incompatible with communist doctrine, and the diminution of Kongzi principles was one of Mao Zedong's primary objectives. Kongfuzi left his followers a wealth of "sayings," many of which were frequently quoted as part of daily life in China; Mao's ubiquitous "Red Book" of quotations was part of the campaign to erase that tradition.

Kongfuzi was born in 551 B.C. and died in 479 B.C. He was one of many philosophers who lived and wrote during China's classical (Zhou dynastic) period. The Kongzi school of thought was one of several to arise during this era; the philosophies of Daoism (Taoism) also emerged at this time. Kongfuzi was appalled at the suffering of the ordinary people in China, the political conflicts, and the harsh rule by feudal lords. In his teaching he urged that the poor assert themselves and demand the reasons for their treatment at the hands of their rulers (thereby undermining the absolutism of government in China); he also tutored the indigent as well as the privileged, giving the poor an education that had hitherto been denied them and ending the aristocracy's exclusive access to the knowledge that constituted power.

Kongfuzi, therefore, was a revolutionary in his time—but he was no prophet. Indeed, he had an aversion to supernatural mysticism and argued that human virtues and abilities should determine a person's position and responsibilities in society. In those days, it was believed that China's aristocratic rulers had divine ancestors and governed in accordance with the wishes of those godly connections. Kongfuzi proposed that the dynastic rulers give the reins of state to ministers chosen for their competence and merit. This was another Kongzi heresy, but in time the idea came to be accepted and practiced.

His earthly philosophies notwithstanding, Kongfuzi took on the mantle of a spiritual leader after his death. His ideas spread to Korea, Japan, and Southeast Asia, and temples were built in his honor all over China. As so often happens, Kongfuzi was a leader whose teachings were far ahead of his time. His thoughts emerged from the mass of philosophical writing of his day later to become guiding principles during the formative Han Dynasty. Kongfuzi had written that the state should not exist for the pleasure and power of the aristocratic elite; it should be a cooperative system, and its principal goal should be the well-being and happiness of the people. As time went on, a mass of Kongzi writings evolved; many of these Kongfuzi never wrote, but they were attributed to him nonetheless.

At the heart of this body of literature lie the *Confucian Classics*, 13 texts that became the focus of education in China for 2,000 years. In the fields of government, law, literature, religion, morality, and in every conceivable way, the *Classics* were the Chinese civilization's guide. The entire national system of education (including the state examinations through which everyone, poor or privileged, could achieve entry into the arena of political power) was based on the *Classics*. Kongfuzi was a champion of the family as the foundation of Chinese culture, and the *Classics* prescribe a respect for parents and the aged that was a hallmark of Chinese society. It has been said that to be Chinese—whether Buddhist, Christian, or even communist—one would have to be a follower of Kongfuzi; hardly any conversation of substance could be held without reference to some of his principles.

When the Western powers penetrated China, Kongzi philosophy came face to face with practical Western education. For the first time, a segment of China's people (initially small) began to call for reform and modernization, especially in teaching. Kongzi principles could guide an isolated China, but they were found wanting in the new age of competition. The Manchus resisted change, and during their brief tenure the Nationalists under Sun Yat-sen tried to combine Kongzi and Western knowledge into a neo-Kongzi philosophy. But it was left to the communists, beginning in 1949, to attempt to substitute an entirely new set of principles to guide Chinese society. Kongzi thought was attacked on all fronts, the *Classics* were abandoned, ideological indoctrination pervaded the new education, and even the family was assaulted during the early years of communization. However, it proved impossible to eradicate two millennia of cultural conditioning in a few decades. Thus the fading spirit of Kongfuzi will haunt physical and mental landscapes in China for generations to come.

China's communist boss during this period, Mao Zedong, had other plans for his country. He wanted to rid China of its Kongzi (Confucian) traditions, for centuries a guiding force in Chinese culture and society (see the box above), and to substitute the principles of communism as a philosophical basis for the new Chinese order. In so doing, he miscalculated the strength of Chinese traditions and ensured the failure of his socioeconomic program.

China's industries, too, were reorganized into state-owned, communal enterprises. Soviet technicians (and large loans), coupled with major mineral and energy-resource discoveries, for a time drove China's industrial machine rapidly forward. Huge investments were made in the establishment of heavy industries and in the dispersal of manufacturing from the East and Northeast toward more western locales. During the 1960s and 1970s, China's Northeast boomed, its industrial output growing prodigiously. Workers in huge numbers moved to the burgeoning cities from Shenyang to Harbin, where the state-run factories readily employed them. On the map, China's communist experiment seemed an unrivaled success.

In Beijing, however, the communist inner circle quarreled over the directions the country should take. Mao Zedong regarded the Soviet Union as weak in its application of communist doctrine, and "revisionist" Russian technicians and advisers were forced to leave China. Mao also suspected many of his subordinates of "revisionist" views, and so he launched the so-called *Great Proletarian Cultural Revolution*, in effect a campaign for a return to strict and orthodox communism. Schools were closed because they ostensibly promoted "bourgeois" ideas. Millions of young people found themselves at loose ends, ready to be recruited into the Red Guards, a paramilitary organization that was organized to combat the revisionists. In the power struggle that ensued, thousands of Communist Party leaders and other officials lost their positions and were replaced by people more loyal to Mao. Soon, the Red Guards were fighting cadres organized to support local leaders, and eventually other pro-Mao "brigades" challenged the Red Guards themselves. China, once again, was in chaos. Perhaps as many as 20 million citizens died in another of communism's dreadful impacts on those under its heel.

PRAGMATISM AND REFORM

The discord outlasted Mao himself. The Cultural Revolution was launched in 1966; Mao died in 1976, leaving China on the verge of civil war, its leaders divided into orthodox Maoist and reformist camps. The civil war was averted, and the power struggle ended in victory for those who called themselves "pragmatic moderates" in the sometimes confusing language of communist dogma. The country's deputy premier, soon to take the reigns of power, was a man whose name will forever be etched on China's historical landscape: Deng Xiaoping.

Deng was a moderate, but he was no unbridled reformer. Communist regimes are fond of exhortations and slogans, and Deng now proclaimed the *Four Modernizations* as China's goal between the early 1980s and 2000. (Although Deng is usually credited with this slogan, it was Zhou Enlai who introduced it as long ago as 1964, only to see it engulfed by the Cultural Revolution.) Deng's version of the Four Modernizations would spur China's development in four areas: (1) agriculture; (2) the armed forces; (3) industry; and (4) science, technology, and medicine. State planning would guide China toward specific goals, but Deng also proposed that Mao's policy of self-reliance be abandoned in favor of foreign trade and interaction, the purchase of foreign technology and machinery, and the use of foreign scientists. Moreover, capitalist incentives would be allowed to spur production, and education would again be based on merit and achievement rather than political attitudes and connections.

But slogans are easily written and less easily fulfilled. In the post-Mao period, China became a more open society as foreigners entered the country in large numbers in many capacities, and thousands of Chinese students were sent to foreign universities to gain the skills China needs. Meanwhile, the contradictions between political orthodoxy and economic reform led to the tragic events of 1989, and the struggle between hardliners and reformers intensified anew. In the 1990s, China's contentious rulers face unaccustomed scrutiny on such issues as the export of prison-labor products, human rights, forced abortions, and international weapons dealing. And even as China benefits economically from its government's new monetary policies and the Four Modernizations proceed, the social costs of freer-enterprise economics are rising—from the cadres of the unemployed in the streets of the once-thriving Northeast to the new underclasses in the bustling cities on the southeastern coast.

◆ EMERGING CHINA

China's communist era has witnessed the metamorphosis of the world's most populous nation. Despite the dislocations of collectivization, communization, the Cultural Revolution, power struggles, policy reversals, and civil conflicts, China as a communist state managed to stave off famine and hunger, control its population growth, improve its overall infrastructure and economy, strengthen its military capacity, and enhance its position in the world. Now, China is poised to absorb one of the **economic tigers** on the western Pacific Rim (Hong Kong), has rapidly growing trade with neighbors near and far, and doubles its national wealth every eight years at present economic growth rates. And its government still espouses communist dogma.

GEOGRAPHY OF CONTRADICTIONS

How to manipulate the coexistence of communist politics and market economics? That was the key question confronting Deng Xiaoping and his comrades when they took power in

FIGURE 10-15

1979. In terms of ideology, this objective would seem unattainable; an economic "open door" policy would surely lead to rising pressures for political democracy.

But Deng thought otherwise. If China's economic experiments could be spatially separated from the bulk of the country, the political impact would be kept at bay. At the outset, the new economic policies would apply mainly to China's bridgehead on the Pacific Rim, leaving most of the vast country comparatively unaffected. Accordingly, the government introduced a complicated system of *Special Economic Zones*, so-called *Open Cities*, and *Open Coastal Areas* that would attract technologies and investments from abroad and transform the economic geography of eastern China (Fig. 10-15).

In these economic zones, investors are offered many incentives. Taxes are low. Import and export regulations are eased. Land leases are simplified. The hiring of labor under contract is allowed. Products made in the economic zones may be sold on foreign markets and, under some restrictions, in China as well. And profits made here may be sent back to the investors' home countries.

ECONOMIC ZONES

As Fig. 10-15 shows, five **Special Economic Zones (SEZs)** were established:

1. *Shenzhen,* adjacent to booming Hong Kong
2. *Zhuhai,* on the estuary of the Pearl River below Guangzhou and just above the former Portuguese city of Macau
3. *Shantou,* directly opposite the southern end of productive Taiwan
4. *Xiamen,* also across from Taiwan
5. *Hainan* Island, China's southernmost province, located across the Gulf of Tonkin from northern Vietnam.

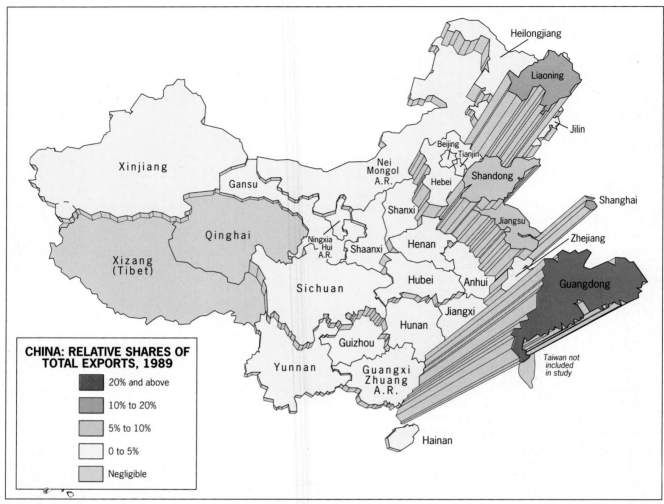

FIGURE 10-16

Three of the five SEZs lie in Guangdong Province, which as Fig. 10-16 underscores, is the crucible of Chinese activity on the emerging Pacific Rim of Austrasia (see also Chapter 4, pp. 244-245). The last to be authorized was Hainan Island, now known not only for its free-enterprise economy but also for its high-volume smuggling operations. But, in fact, four of the SEZs initially were less successful than the Deng regime had anticipated. Only Shenzhen took off in spectacular fashion because of its situation just across the now-porous border from Hong Kong. Shenzhen was simply a fishing village of 20,000 in 1980; in 1994 its population was passing 2.5 million, more than a hundred-fold increase! Shenzhen's skyline rivals Hong Kong's; its economy is booming; more than 1,000 companies, using low-cost local labor, manufacture items ranging from toys to sophisticated electronic equipment. Retailing, tourism, even a stock market now thrive here. As we noted in Chapter 4, Shenzhen is the focus of China's new Pacific Rim era.

The proximity and accessibility of Hong Kong gave Shenzhen an advantage none of the other SEZs has. Shen-

zhen thus serves as the model for SEZ development, but it is unrealistic to expect the others to grow comparably. Obviously, Xiamen and Shantou would benefit from the normalization of relations with Taiwan across the Taiwan Strait. But Shenzhen, too, will face a moment of truth when Hong Kong is reabsorbed into China in 1997. Will the boom continue after this British colony is no longer?

Fourteen **Open Cities**, all on China's Pacific coast, were designated under the new economic program. These are, from north to south (Fig. 10-15):

1.	Dalian (Lüda)	Liaoning Province
2.	Qinhuangdao	Hebei Province
3.	Tianjin	Tianjin Shi
4.	Yantai	Shandong Province
5.	Qingdao	Shandong Province
6.	Lianyungang	Jiangsu Province
7.	Nantong	Jiangsu Province
8.	Shanghai	Shanghai Shi
9.	Ningbo	Zhejiang Province

10. Wenzhou	Zhejiang Province
11. Fuzhou	Fujian Province
12. Guangzhou	Guangdong Province
13. Zhanjiang	Guangdong Province
14. Beihai	Guangxi Zhuang A.R.

But designating a city as "open" does not automatically stimulate its economic development. When most of these places experienced minimal "open-door" growth, the government decided to focus its attention and assistance on four of them: Dalian (near the tip of the Liaodong Peninsula), Tianjin (near Beijing), Shanghai (China's largest city), and Guangzhou (capital of fast-growing Guangdong Province).

Six **Open Coastal Areas** form yet another geographic layer of the new economic system. Economic incentives here are less far-reaching than in the SEZs or the Open Cities, but foreign investment is nonetheless encouraged. These Open Coastal Areas comprise parts of China's many Pacific-coast deltas and peninsulas: the Liaodong Peninsula of the lower Northeast, the Shandong Peninsula off the North China Plain, the delta area of the Chang River around Shanghai, the coastal zone around the SEZ of Xiamen, the margins of the Pearl River estuary, and the delta and peninsula around Zhanjiang (Fig. 10-15).

It was and remains the intent of the Chinese regime to open virtually the entire Pacific littoral to market-driven economic activities. Although none of the other economic zones has become a second Shenzhen, China's economic-growth figures attest to the success of the policy. And when the inevitable clash between political reformers, energized by the new order, and the Communist conservatives occurred in 1989, the Communist apparatus managed to keep control. In the mid-1990s, China was still following a course the dying Soviet Union could not: "perestroika" *without* "glasnost."

◆ CHINA'S POPULATION CHALLENGE

As we have observed previously, high rates of population growth inhibit the national economic development of emerging countries. China, the most populous country in the world, therefore faces a special problem. Even a modest rate of population growth adds millions to its citizenry every year.

During the early years of the communist era, Mao's regime refused to institute measures to slow China's population growth. Any such policies, Mao stated, played into the hands of the world's imperialists. Numbers meant strength. In 1974, when the world's countries convened a conference in Bucharest, Romania to discuss world population problems, the representative of the People's Republic of China urged developing countries not to follow advice from the developed world to adopt policies designed to reduce growth rates. Such advice, he said, constituted a plot to deprive emerging countries of the one advantage they have: fast-growing populations.

But there were leaders in China who realized that their country's progress would be stymied by rapid population growth, and who tried to stem the tide through local propaganda and education. Some of these leaders lost their positions as "revisionists," but their point was not lost on China's leaders of the future. When Deng rose to power, China embarked on a vigorous population-control program. In the early 1970s, the annual rate of natural increase was about 3 percent; by the mid-1980s, it had been lowered to 1.2 percent. Families were instructed to have one child only and were penalized for having more—by losing tax advantages, educational opportunities, even housing privileges.

Billboards and posters are often part of the Chinese public landscape, particularly in the South. Here, in the city of Chengdu in central Sichuan Province, the message concerning the official population policy is clear enough—in English (!) as well as Chinese (on the portion of the billboard not shown).

Although it may be said that the one-child policy gained general acceptance, the program also had some serious negative consequences. The number of abortions soared, sometimes late during pregnancies and at times coerced by party enforcers. Female infanticide also rose as families wanted their one child to be a boy. Reports of such dreadful byproducts of the population-control campaign leaked to the outside world, and during the late 1980s, the policy was relaxed somewhat. Certain farming and fishing families were allowed to have a second child to fulfill future labor needs, and rural families could request permission to have a second child if their first baby was a girl. (Ethnic minority groups were mostly exempt.)

As a result, China's population growth rate has inched upward again, and is now about 1.3 percent (adding 15 million people to the nation every year). It had been the objective of China's economic planners to reach 1.2 billion people by the year 2000, but as our extrapolation in Appendix A shows, the turn-of-the-century figure is more likely to be close to 1.3 billion. That means an additional 100 million people to be housed, educated, treated, and, eventually, employed. Against that background, China's current economic surge looks rather less promising.

Nevertheless, China's economic achievements are ensuring its capacity to sustain progress despite its huge population challenge. According to official statistics, China's farmers increased their output by an amazing 60 percent during the 1980s, partly as a result of capitalist incentives. Market economics achieved what the Great Leap Forward could not. The question is whether China's autocratic regime can continue to coexist with its economic machine.

URBAN CHINA

The great majority of Chinese—nearly three-quarters of the population—live in rural areas, in tens of thousands of villages surrounded by farmlands. There, in the countryside, the evidence of three turbulent decades is still visible: traditional dwellings in disrepair, ugly brick commune buildings rising above village roofs, fading slogans on compound walls. But China's urban population is rising. In 1993, an estimated 300 million Chinese resided in cities and towns—a mere 26 percent of the total population but numbering more than the combined population of the United States and Canada.

China's cities are crowded, congested, polluted, massive, often bleak agglomerations of urban humanity. In some ways they resemble the cities of England during the early years of the Industrial Revolution. Soot-belching smokestacks rise like a forest from city blocks; factories and workshops large and small vie for space with apartment buildings, schools, and architectural treasures of the past. Soviet-style public buildings bear witness to the early communist period. No zoning regulations seem to exist, nor any protection for the urban environment. Turn a streetcorner, and the deafening roar of machines drowns out all else; walk along the street and as one roar fades, another ear-splitting racket replaces it. Across the street is a school, and one wonders how students can learn under such conditions day after day. Gaze upward and the sky is shrouded by an ever-present cloud of pollution so thick that the air is actually dangerous to breathe. Cross a bridge and look down: here is another vivid portrait of China's industrial march. The water below is little more than an industrial sewer, filled with waste and colored gray by tons of toxic chemicals.

By the standards of suburbanized Western cities, China's metropolises are bleak and severe places. In the streets, waves of bicycle riders vie with buses and trucks for space; humans carry and pull loads that seem to be beyond their capacity to do so. There are large public squares, but few leafy parks. The great majority of the people are dressed in workers' clothing. Luxuries are few—except in the bustling new centers on the Pacific coast.

And yet China's great cities are crucibles of the new China. Beijing's skyline reflects the post-Soviet age as modern buildings rise above the old. Shanghai's old colonial frontage is flanked by new highrises, and the city's entire waterfront is under reconstruction. Guangzhou does not yet have suburbs on the American model, but its outskirts mark the impact of a new economic order. Cities farther west—Xian, Chengdu, Kunming—are less affected, but they, too, are beginning to change. Inevitably, the government's new economic policies will draw larger numbers of people from the rural areas to the cities. Efforts will be made to slow this flow, but China will urbanize just as Europe did two centuries ago.

As Fig. 10-9 shows, China's largest cities (like those in India) remain single centers, widely dispersed, interacting with and interconnected to large hinterlands. For so populous a country, Shanghai, Beijing, and Guangzhou are not enormous in size, and distances between these and other major cities are substantial. Only along the Chang Jiang is there evidence of megalopolitan development (see Chapter 3). Wuhan is the major interior center here, along with Nanjing to the east and Chongqing to the west. Urbanization is a mirror of development, and China's cities reveal the distance this gigantic, emerging country still has to travel.

CHINESE OVERSEAS

China for long periods of its history was a closed society, cut off from the outside world and locked behind a *Bamboo Curtain*, as foreigners called it. But Chinese have been venturing overseas for centuries, often risking their lives to escape to Hong Kong or, in earlier times, points beyond. As free and voluntary migrants and as desperate exiles, they settled in distant lands. There are "Chinatowns" from Washington, D.C. to Melbourne, Australia. But the largest number of overseas Chinese, as many as 30 million, live in Southeast Asia.

One of Shanghai's more prosperous residential districts is Hongkou, once the Japanese "concession" in the city. The tiled roofs, tree-lined streets, and good repair of most buildings indicate the neighborhood's relative well-being.

We will discuss these overseas Chinese when we study the Southeast Asian geographic realm in Chapter 11, but in the present context it is important to note their regional origins. Among Chinese who migrated to Hong Kong, the city and environs of Shanghai were early sources: textile-makers from there fled from communism during the late 1940s, and their knowledge, experience, and resources helped create Hong Kong's first economic boom.

But the two provinces that have always been China's most outward-looking and that have produced the largest number of emigrants are Guangdong (with its back door to Hong Kong) and neighboring Fujian, across from Taiwan. Boats have sailed from these coasts for centuries, carrying migrants southward to Vietnam, Thailand, Indonesia, Malaysia, and elsewhere in the neighboring realm. In Singapore, Chinese became the majority and now control the state.

Even during the days of the Bamboo Curtain, the overseas Chinese managed to maintain contact with the home folk. The village or clan of origin was not soon (if ever) forgotten, and news as well as assistance always flowed between the ethnic core areas in Southeast Asia and mainland China, often via Hong Kong or Taiwan. Today, those links

are freer and more easily sustained, and they have become an important ingredient in the rise of Guangdong and Fujian on China's Pacific Rim. China's overseas diaspora are wealthy far beyond their numbers. They have made Singapore one of the Pacific Rim's economic tigers. They have taken control of large parts of the commercial and manufacturing sectors of such countries as Thailand, Indonesia, Malaysia, and even the Philippines. They save their money carefully. Now, with China's new opportunities beckoning, they have the capacity to invest in their ancestral homeland, benefiting from its new economic policies. The growing interconnections between China's overseas entrepreneurs and the old sources of emigration are helping forge a new China on the Pacific Rim.

◆ CHINA IN THE WORLD TODAY— AND TOMORROW

Two centuries ago, Napoleon remarked that China was a giant asleep; whoever would awaken the Chinese giant would be sorry. Today, China is awake. European colonialism, Japanese aggression, and communist ideology combined to stir China into action. China's course has been erratic, but as the twenty-first century looms, China at last seems poised to take the world's center stage.

A quarter-century ago, China was engulfed by the Cultural Revolution, its borders sealed, its Soviet communist advisers ousted, its economy stagnant. Now, its government has proved that it can overcome internal dissent and overpower political reform movements; its borders are open to investors and tourists; its advisors are capitalists; and its economy defied the recent global recession by growing at a prodigious rate. In 1991, China and the (then) Soviet Union signed a trade agreement that included the shipment of grain, meat, and peanuts from China to Russia—an amazing reversal of food-supply directions. China, a food *exporter*!

China and Russia (now heir to the Soviet agreement) are neighbors, and cross-border trade from the Northeast to Russia's Far East is growing, boosting especially the markets of Khabarovsk and Vladivostok, both connected by water to Chinese cities. But neighboring Russia is not China's leading trade partner. China today trades with Japan, the United States, Germany, Canada, and other countries the world over. Japan (not counting the trade that flows through Hong Kong) has become China's leading trade partner, and the United States ranks second. Trade with the United States was boosted by China's status as a "most-favored nation" under U.S. laws, but the events of 1989, as well as concern about human-rights abuses in China, have generated opposition in Congress to this standing. Should China's status be revoked, the new and growing exchange relationship between the two Pacific powers would suffer.

What does China trade? In the early 1990s, textiles, oil, chemicals, light industrial products of many kinds, and weapons were the leading exports; imports were mainly various types of machinery, transport equipment, and iron and steel. But the range of exports as well as imports is changing. Some high-tech products are appearing among exports, and 300 million urban Chinese are buying air conditioners, electronic goods, automobiles, and other products for which Japan is the best-situated supplier. China's global trade network is expanding in ways totally unforeseen by Mao Zedong and his isolationist comrades.

THE GEOGRAPHIC REALM

As we noted in Chapter 4, a new geographic region—*Austrasia*—is forming along the western margin of the Pacific Ocean, and the geographic realm centered on China is undergoing change. China's present and future relations with territories on its periphery will determine how the new framework develops. In 1993, problems were affecting the scheduled transfer of Hong Kong to China because of belated British efforts to democratize the colony. After ruling Hong Kong for nearly 150 years (most of the last century under a lease agreement), the British in 1992 decided to liberalize its government—just five years before the transfer to China. The Beijing regime regarded this as a provocation, and what had appeared to be a smooth transition now promised trouble.

Relations with Taiwan also remain difficult. Taiwan is one of the western Pacific Rim's most successful economies, and the country has the talents, skills, and investment capital mainland China needs. Hong Kong has been a "back door" for Taiwanese investment in China, but in 1997 that avenue will close. Both China and Taiwan put obstacles in the way of mutually profitable normalization: China by insisting on Taiwan's nonrecognition as an independent country and Taiwan by obstructing linkages to the mainland. Perhaps we should look at the map and remind ourselves of two small groups of islands, Quemoy and Matsu, on the very doorstep of the SEZ of Xiamen and the Open City of Fuzhou, respectively. The potential for political strife still inhibits the realization of economic opportunities.

And there remains the unresolved challenge of divided Korea. In Chapter 4 we noted the two Koreas' economic-geographic complementarity. For decades, however, South Korea has been in the forefront of Pacific Rim development (aided by the United States), whereas North Korea has remained an orthodox communist dictatorship and a client of China and the former Soviet Union. In the early 1990s, China and South Korea resumed diplomatic relations, a move that underscored Beijing's economic pragmatism. North Korea's communist regime found itself increasingly isolated, its capital, Pyongyang, a Havana-like relic of communist bombast. More seriously, concern has risen that North Korea might be reinforcing its isolation through the development of nuclear-weapons capacity. The unification of the two Koreas is an elusive goal, and North Korea endures as an anachronism on the Pacific Rim—and on China's doorstep.

THE FUTURE

For all its recent progress, China remains a developing country plagued by poverty; its per capita GNP is only slightly above that of the South Asian geographic realm (U.S. $370 in 1990). Today, China is on the move, and its motion is creating regional disparities not seen since the communists took power in 1949. The China of Shenzhen is another world from that of, say, Chengdu or Xian. To the centrifugal forces arising from the contradiction between communist governance and capitalist economics will be added those resulting from regional inequalities. Further challenges will lie in the minority areas, in the far west (Turkestan), and in disputed boundaries and border territories. China still claims parts of India; India claims areas of Tibet (Xizang) adjacent to Kashmir; and China has unresolved territorial issues with Kazakhstan and Russia. And there also remains the question of Mongolia, on which China in the past has staked claims.

When an empire or nation-state experiences a time of internal division, its government may use territorial issues to engender nationalism. Should China's communist regime confront domestic discord caused by its economic reforms, the possibility of such action exists. Under present circumstances, such a campaign is not necessary, and in 1992 China actually settled with Russia several border disputes it had long had with the former Soviet Union. But the potential for territorial challenges remains—from the Russian Far East to Indian Assam.

What will China's position be in the world of the twenty-first century? Its regional geography provides some clues. China contains a wide range of natural resources, from fuels to metals and from fertile soils to hardwood forests. China's human resources are enormous, its peoples' skills and capacities demonstrated within China as well as outside its borders. China's neighbors range from productive Japan to potential-laden Vietnam and from Russia's developing Far East to Turkestan's frontiers; on Asia's Pacific Rim, China occupies a central location. Barring a Soviet-type organizational collapse, China will become the giant of East Asia, and not only in a demographic sense.

Thus China is likely to become more than an economic force of world proportions; China also appears on course to achieve global superpower stature. So long as it retains its autocratic form of government (which made possible the imposition of draconian population policies and comprehensive economic experiments without the inconvenience of electoral consultation), China will be able to practice the kind of state capitalism that—in another guise—made South Korea an economic power. And unlike Japan or South Korea, China has no constraints on its military power. China's armies served as security forces during the pro-democracy turbulence of 1989; the armed forces are enormous, and they are modernizing. China is a nuclear power adjoined by other nuclear powers, and its self-interest will compel continued preparedness.

Already, China has confirmed that self-interest, not the wishes of presently more powerful competitors, guides its strategic decision making. Despite U.S. objections, for instance, China sold its Silkworm missile systems to Southwest Asian buyers. (The United States shortly thereafter sold fighter jets to Taiwan.) When we look forward to a modernizing, mobilizing China in the coming century, it is not difficult to hear the echo of Napoleon's famous words.

THE REALM IN TRANSITION

1. Regional disparities are growing rapidly in developing China and will test the fabric of the national state.
2. Relations among China, the last colonial Hong Kong government, and the United Kingdom are worsening rather than improving as the fateful year 1997 approaches. Will something unforeseen happen *prior* to 1997?
3. People, goods, and ideas are diffusing into China's western regions from across Turkestan's borders. Here, too, are many of China's most sensitive strategic weapons operations. Are these processes compatible?
4. China's latent claims to Russian and Indian territory and to a vast zone of the South China Sea indicate that Beijing's expansionism has not ended.
5. China's economic engine is driven substantially by investments from Taiwan and from thousands of overseas Chinese, mostly in Southeast Asia. Will such investments eventually translate into political influence?

◆ PRONUNCIATION GUIDE

Akihito (ah-kee-HEE-toh)
Alashan (ahl-ah-SHAHN)
Altay (AL-tye)
Altun (ahl-TUN)*
Amur (uh-MOOR)
Annam (uh-NAHM)
Anshan (ahn-SHAHN)
Assam (uh-SAHM)
Austrasia (aw-STRAY-zhuh)
Baotou (bao-TOH)
Beihai (bay-HYE)
Beijing (bay-ZHING)
Bohai (bwoh-HYE)
Cadre (KAH-dray)
Canton (kan-TONN)
Chang Jiang (chung jee-AHNG)
Changchun (CHAHNG-CHOON)*
Chengdu (chung-DOO)
Chiang Kai-shek (jee-AHNG
  kye-SHECK)
Chongqing (chong-CHING)
Confucius (kun-FEW-shuss)
Da Hinggan (dah-hing-GAHN)
Dalai Lama (dah-lye LAHMA)
Dalian (dah-lee-ENN)
Daoism (DAU-ism)
Daqing (dah-CHING)
Deng Xiaoping (DUNG shau-PING)
Diaspora (dye-ASP-poar-ruh)
Dongbei (dung-BAY)
Dzungaria (joong-GAH-ree-uh)
Endoreic (en-doh-RAY-ick)
Fujian (foo-jee-ENN)
Fushun (foo-SHUN)*
Fuzhou (foo-ZHOH)
Gansu (gahn-SOO)
Gaoliang (gow-lee-AHNG)
Gezhouba (guh-JOH-bah)
Glasnost (GLUZZ-nost)
Gobi (GOH-bee)
Graben (GRAH-ben)
Guangdong (gwahng-DUNG)
Guangxi Zhuang (gwahng-shee
  JWAHNG)
Guangzhou (gwahng-JOH)
Guilin (gway-LIN)
Guizhou (gway-JOH)
Haikou (HYE-KOH)

*"U" or final "u" pronounced as in "put"

Hainan (HYE-NAHN)
Han (HAHN)
Hangzhou (hahng-JOH)
Hankou (hahn-KOH)
Hanoi (han-NOY)
Hanyang (hahn-YAHNG)
Hebei (huh-BAY)
Hegemony (heh-JEH-muh-nee)
Heilongjiang (hay-long-jee-AHNG)
Hexi (huh-SHEE)
Hohhot (huh-HOO-tuh)
Hongkou (hong-KOH)
Huang [He] (HWAHNG [huh])
Huangpu (hwahng-POO)
Hunan (hoo-NAHN)
Jiangsu (jee-ahng-SOO)
Jiangxi (jee-ahng-SHEE)
Jilin (jee-LIN)
Junggar (JOONG-gahr)
Kaifeng (kye-FUNG)
Karamay (kah-RAH-may)
Kazakh (KUZZ-uck)
Kazakhstan (KUZZ-uck-stahn)
Khabarovsk (kuh-BAHR-uffsk)
Kirghiz (keer-GEEZE)
Kongfuzi (kung-FOODZEE)
Kongzi (KUNG-dzee)
Korea (kuh-REE-uh)
Kublai Khan (koob-lye
  KAHN)
Kunlun (KOON-LOON)
Kunming (koon-MING)
Kyrgyz (keer-GEEZE)
Kyrgyzstan (KEER-geeze-stahn)
Lanzhou (lahn-JOH)
Lhasa (LAH-suh)
Lianyungang (lee-en-
  yoong-GAHNG)
Liao (lee-AU)
Liaodong (lee-au-DUNG)
Liaoning (lee-au-NING)
Littoral (LITT-uh-rull)
Loess (LERSS)
Lüda (LOO-dah)
Macau [Macao] (muh-KAU)
Mackinder, Halford (muh-KIN-der,
  HAL-ferd)
Malaysia (muh-LAY-zhuh)
Manchu (man-CHOO)
Manchukuo (mahn-JOH-kwoh)
Manchuria (man-CHOORY-uh)

Matsu (mah-TSOO)
Mao Zedong (MAU zee-DUNG)
Molybdenum (muh-LIB-dun-um)
Mongol (MUNG-goal)
Mongolia (mung-GOH-lee-uh)
Myanmar (mee-ahn-MAH)
Nakhodka (nuh-KAUGHT-kuh)
Nanjing (nahn-ZHING)
Nantong (nahn-TOONG)
Nei Mongol (nay-MUNG-goal)
Ningbo (ning-BWOH)
Ningxia Hui (NING-shee-AH
  HWAY)
Ordos (ORD-uss)
Pamirs (pah-MEERZ)
Perestroika (perra-STROY-kuh)
Peking (pea-KING)
Philippines (FILL-uh-peenz)
Pinyin (pin-YIN)
Playa (PLY-uh)
Pyongyang (pea-AWHNG-yahng)
Qaidam (CHYE-DAHM)
Qanats (KAH-nahts)
Quemoy (keh-MOY)
Qin (CHIN)
Qing (CHING)
Qingdao (ching-DAU)
Qinghai (ching-HYE)
Qinhuangdao (chin-hwahng-DAU)
Qinling (chin-LING)
Qiqihar (chee-CHEE-har)
Ruijin (rway-JEEN)
Ryukyu (ree-YOO-kyoo)
Shaanxi (shahn-SHEE)
Sha Mian (shah mee-AHN)
Shan (SHAHN)
Shandong (shahn-DUNG)
Shanghai (shang-HYE)
Shantou (SHAHN-TOH)
Shanxi (shahn-SHEE)
Shenyang (shun-YAHNG)
Shenzhen (shun-ZHEN)
Shi (SHEE)
Sichuan (zeh-CHWAHN)
Sinicization (sine-ih-sye-ZAY-shun)
Song (SUNG)
Songhua (SUNG-hwah)
Spykman (SPIKE-mun)
Sui (SWAY)
Sun Yat-sen (SOON yaht-SENN)
Taiwan (tye-WAHN)

Tajikistan (tah-JEEK-ih-stahn)
Taklimakan (tahk-luh-muh-KAHN)
Tarim (TAH-REEM)
Thailand (TYE-land)
Tiananmen (TYAHN-un-men)
Tianjin (tyahn-JEEN)
Tian Shan (TYAHN SHAHN)
Tibet (tuh-BETT)
Tonkin (TAHN-KIN)
Ulaan Baatar (oo-lahn BAH-tor)
Ürümqi (oo-ROOM-chee)
Uygur (WEE-ghoor)
Versailles (vair-SYE)

Vietnam (vee-et-NAHM)
Vladivostok (vlad-uh-vuh-STAHK)
Wakhan (wah-KAHN)
Wei (WAY)
Wenzhou (whunn-JOH)
Wuchang (woo-CHAHNG)
Wuhan (woo-HAHN)
Wuzhou (woo-JOH)
Xiamen (shah-MEN)
Xian (shee-AHN)
Xi Jiang (SHEE jee-AHNG)
Xinjiang (shin-jee-AHNG)
Xizang (sheedz-AHNG)

Yanan (yen-AHN)
Yangshuo (YAHNG-SHWOH)
Yangtze (YANG-dzee)
Yantai (yahn-TYE)
Yuan (YOO-ahn)
Yumen (YOO-mun)
Yunnan (yoon-NAHN)
Zhanjiang (JAHN-jee-AHNG)
Zhejiang (JEJ-ee-AHNG)
Zhou (JOH)
Zhou Enlai (JOH en-lye)
Zhuhai (joo-HYE)

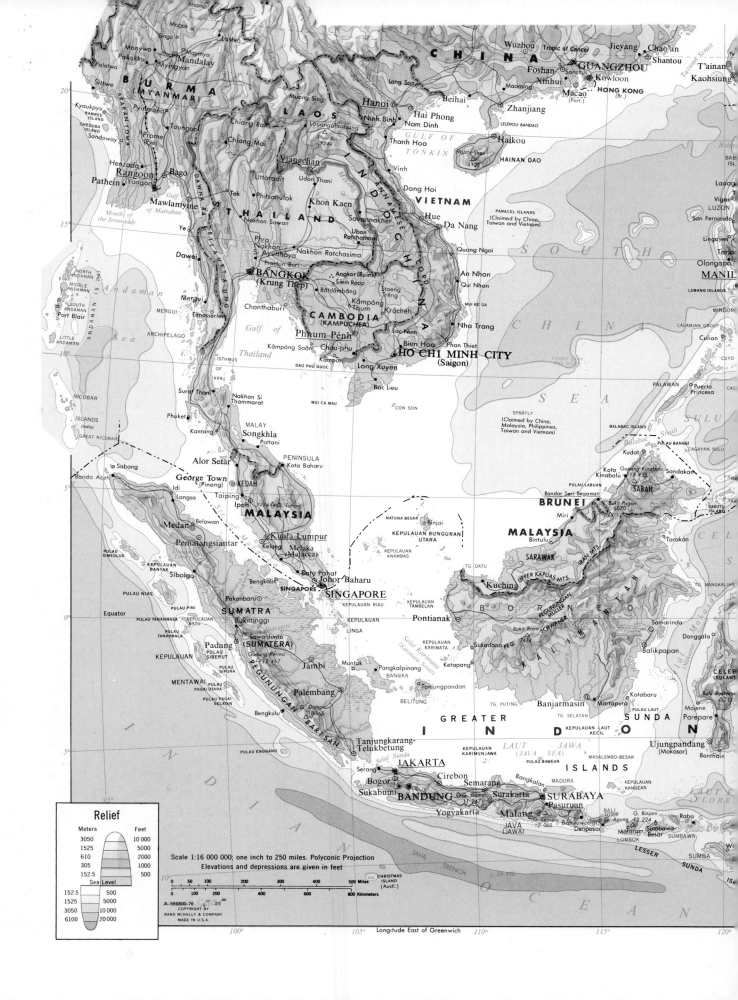

CHAPTER 11

SOUTHEAST ASIA: BETWEEN THE GIANTS

IDEAS & CONCEPTS

Shatter belt (3)
Buffer zone
Political geography
Genetic boundary classification
Territorial morphology
   Compact
   Elongated
   Fragmented

Prorupt
Perforated
Domino theory
Insurgent state model
Southeast Asia city model
Territorial sea
Exclusive economic zone (EEZ)
Maritime boundaries

REGIONS

Mainland
   Indochina
Peninsular/Insular
   Malaysia
   Indonesia

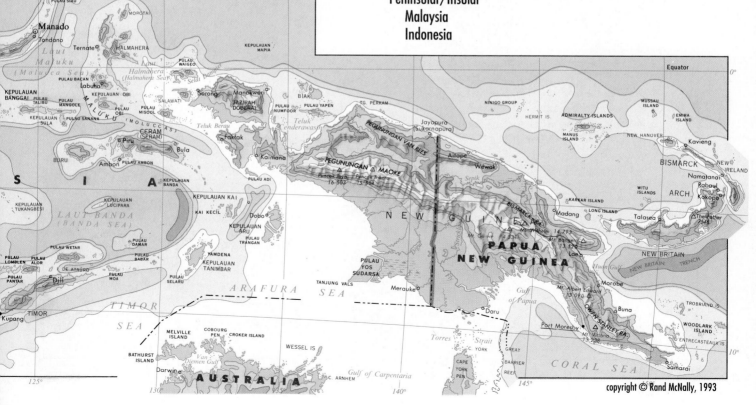

Southeast Asia . . . the very name roils American emotions. Here the United States owned its only major colony. Here American foreign policy suffered its most disastrous failure. Here the United States fought the only war it ever lost. Here Washington's worst Cold War fears failed to materialize. And here, today, the United States has an opportunity to share in the economic expansion now taking place—an opportunity long denied it because of the aftermath of the Indochina War (1964–1975).

Southeast Asia is a realm of peninsulas and islands, a corner of Asia bounded by India on the northwest and China on the northeast. Its western coasts are washed by the Indian Ocean, and to the east stretches the vast Pacific. From all these directions, Southeast Asia has been penetrated by outside forces. From India came traders; from China, settlers. From across the Indian Ocean came the Arabs to engage in commerce and the Europeans in pursuit of empires. And from across the Pacific came the Americans. Southeast Asia has been the scene of countless contests for power and primacy—the competitors have come from near and far. Like Eastern Europe, Southeast Asia is a geographic realm of great cultural diversity, and it is also a shatter belt.

The concepts of **shatter belt** and **buffer zone** go hand in hand, and the map of Southeast Asia will remind you somewhat of Eastern Europe: a mosaic of smaller countries on the periphery of one of the world's largest states. Just as

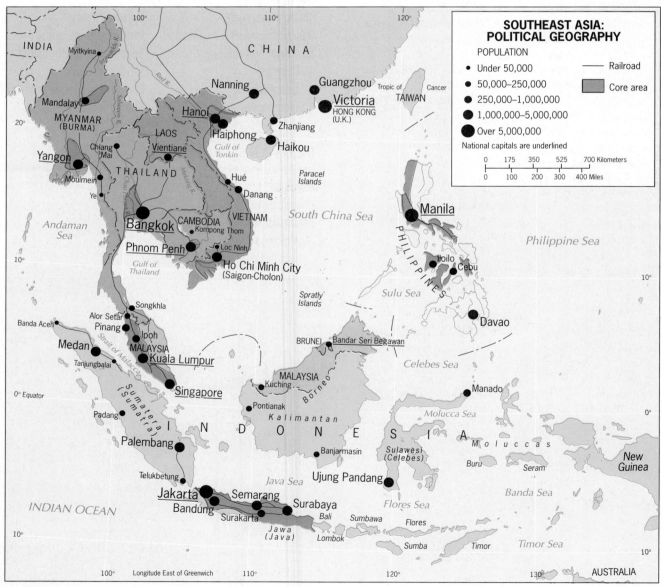

FIGURE 11-1

pressures and stresses from without and within fractured the political geography of Eastern Europe, Southeast Asia's limited space also shows the effects of divisive forces. As in Eastern Europe, the map has changed even in recent times. In 1965, Singapore broke away from Malaysia and became the realm's smallest political entity territorially, a contemporary city-state. Even more recently, a boundary disappeared in 1976 when North and South Vietnam were united. Again, as in Eastern Europe, boundary disputes continue, and so do ethnic tensions.

Because the politico-geographical map (Fig. 11-1) is so complicated, it should be studied attentively. One good way to strengthen your mental map of this realm is to follow the mainland coastline from west to east. The westernmost state in the realm is Myanmar (called Burma before 1989), the only country in Southeast Asia that borders both India and China. Myanmar shares the "neck" of the Malay Peninsula with Thailand, heart of the mainland region. The south of the peninsula is part of Malaysia—except for Singapore, at the very tip of it. Facing the Gulf of Thailand is Cambodia. Still moving generally eastward, we reach Vietnam, a strip of land that extends all the way to the Chinese border. And surrounded by its neighbors is landlocked Laos, remote and isolated. This leaves the islands that constitute insular Southeast Asia: the Philippines in the north and Indonesia in the south, and between them the offshore portion of Malaysia, situated on the largely Indonesian island of Borneo. Completing the map is Brunei, smallest country in the realm in terms of population but, as we will see, important in the regional picture.

These are the countries of a realm in which there is no dominant state—no Brazil, no China, no India—although Indonesia is its population giant. Southeast Asia is a realm of mountain barriers, unproductive uplands, rugged coastlines, and far-flung islands—as well as fertile valleys and deltas, rich volcanic soils, and productive plains (Fig. 11-3). No single dominant core of indigenous development emerged here. Cultural diversity prevails: the realm is a mosaic of ethnic and linguistic groups, of various and different religions, of contrasting ways of life. Its historical geography, still etched on the map, is one of foreign intervention and competition. Just a half-century ago, at the end of World War II, only Thailand was an independent state, itself a buffer between French colonialism to the east and British power to the west. The colonial era has ended, but its aftermath still hangs heavily over fragmented Southeast Asia.

◆ POPULATION PATTERNS

Compared to the huge population numbers and densities in the habitable regions of South Asia and China, demographic totals for the countries of Southeast Asia seem minor. Landlocked Laos, territorially quite a large country,

TEN MAJOR GEOGRAPHIC QUALITIES OF SOUTHEAST ASIA

1. The Southeast Asian realm is fragmented into numerous peninsulas and islands.
2. Southeast Asia, like Eastern Europe, exhibits the characteristics of a shatter belt. Pressures on this realm from external sources have always been strong.
3. Southeast Asia exhibits intense cultural fragmentation, reflected by complex ethnic, linguistic, and religious geographies.
4. The legacies of powerful foreign influences (Asian as well as non-Asian) continue to mark the cultural landscapes of Southeast Asia.
5. Southeast Asia's politico-geographical traditions involve frequent balkanization, instability, and conflict.
6. Population in Southeast Asia tends to be strongly clustered, even in rural areas.
7. Compared to neighboring regions, mainland Southeast Asia's physiologic population densities remain relatively low.
8. Rapid population growth has prevailed in the island regions of Southeast Asia, notably in the Philippines, during much of the twentieth century.
9. Intraregional communications in Southeast Asia remain inferior. External connections are often more effective than internal linkages.
10. The boundaries of the Southeast Asian realm are problematic. Transitions occur into the adjoining South Asian, Chinese, and Pacific realms.

has a population of just 4.7 million. Even the more densely inhabited coastal areas of Southeast Asia have fewer people and smaller agglomerations than elsewhere in southern and eastern Asia. There is nothing in this realm to compare to the enormous dimensions of human clusters in India's Ganges lowland or the North China Plain. The entire pattern is different: Southeast Asia's fewer dense population clusters are relatively small and lie separated from one another by areas of much sparser human settlement. Nonetheless, when everything is added up, this realm does exhibit some substantial numbers: the 1994 Southeast Asian population total of 469 million is about 20 million higher than all of North Africa/Southwest Asia, 89 percent of Subsaharan Africa's total, and 81 percent of Europe's.

Why do Southeast Asia's population patterns differ from those of its giant neighbors to the north and west? When

◆   F O C U S   O N   A   S Y S T E M A T I C   F I E L D

POLITICAL GEOGRAPHY

Southeast Asia is a laboratory for the study of **political geography**. This systematic field, one of the oldest in geography, focuses on the spatial expressions of political behavior. Boundaries on land and on the oceans, the roles of capital cities, power relationships among states, administrative systems, voter behavior, conflicts over resources, and even matters involving outer space have politico-geographical dimensions. We have already been introduced to aspects of political geography when we studied the possible unification of Western Europe (*supranationalism*), the historic shattering of Eastern Europe (*balkanization*), the support given by Pakistan to Muslims in Kashmir (*irredentism*), and the competition for hegemony over parts of Eurasia (*geopolitics*).

The field of political geography grew from geographers' interest in the spatial nature of the national state. What are the ingredients of a nation-state? Why do some states survive over many centuries, while others (such as the Ottoman Empire, the Austro-Hungarian Empire, and now the Soviet Empire) collapse? About a century ago, Friedrich Ratzel (1844–1904) proposed a theory that likened the nation-state to a biological organism. Just as an organism is born, grows, matures, and eventually dies, Ratzel argued, states go through stages of birth (around a culture hearth or core area), expansion (perhaps by colonization), maturity (stability), and eventual collapse. Only the sporadic absorption of new land and people, he suggested, could stave off the state's decline. This was a blueprint for imperialism!

Later, political geographers realized that every state has ties that bind it (*centripetal forces*) and stresses that tend to break it apart (*centrifugal forces*).

When the centripetal or unifying forces are much stronger, the state succeeds; when the centrifugal or divisive forces prevail, the state fails (*devolution*). In Japan today the centripetal forces are very strong; but in South Africa centrifugal forces dominate. Identifying and measuring these forces, obvious as they may be, is one of the challenges of political geography.

We sometimes call countries *nations*, but many countries contain more than one nation, which is why it is better to call them *states*. It might be appropriate to call Finland a nation, but Nigeria (as we noted in Chapter 8) contains three nations and many minority peoples. States that are also pluralistic societies, such as Russia, exhibit powerful centrifugal forces that must be overcome. Even Canada confronts cultural division, spatially expressed, that could lead to its fragmentation; thus Canada is not yet a nation.

The most basic device in political geography is the world political map (Fig. I-12). This map reveals the enormous range in the sizes of states; some are microstates (or ministates), while others are giants. The map also suggests why the United Nations officially recognizes a group of Geographically Disadvantaged States (GDS): more than two dozen states have no *maritime boundaries* and are landlocked. As we noted in the case of Bolivia, this can have a disastrous effect on the fortunes of a country. Another aspect of the world map is the *territorial morphology* or physical shape of states. In this chapter we examine in some detail the shapes of Southeast Asian states, and discuss what effect their morphology may have had on their development.

States have capitals, core areas, administrative divisions, and boundaries. Boundaries are sensitive parts of the anatomy of a state: just as people are territorial about their individual properties, so nations and states are sensitive about their territories and limits. Boundaries, in effect, are contracts between neighboring states. The *definition* of a boundary is likely to be found in an elaborate treaty that verbally describes its precise location. Cartographers then perform the *delimitation* (official mapping) of what the treaty stipulates. Certain boundaries are actually placed on the ground as fences, walls, or other artificial barriers; this represents the *demarcation* of the boundary. The world map shows that some boundaries have a sinuous form, while others are straight lines. Boundaries can therefore be classified as *geometric* (straight-line or curved), *physiographic* (coinciding with rivers or mountain crests), or *anthropogeographic* (marking breaks or transitions in the cultural landscape).

Another way to view boundaries has to do with their evolution or genesis. This **genetic boundary classification** was established by Richard Hartshorne (1899–1992), a leading American political geographer. Hartshorne reasoned that certain boundaries were defined and delimited before the present-day human landscape developed. In Fig. 11-2 (upper-left map), the boundary between Malaysia and Indonesia on the island of Borneo is an example of this *antecedent* type. Most of this border passes through sparsely inhabited tropical rainforest, and the break in settlement can even be detected on the world population map (Fig. I-9). A second category of boundaries evolved as the cultural landscape of an area took shape. These *subsequent* boundaries are exemplified by the map in the upper right of Fig.

GENETIC POLITICAL BOUNDARY TYPES

ANTECEDENT

SUBSEQUENT

SUPERIMPOSED

RELICT

FIGURE 11-2

11-2, which shows the border between China and Vietnam, the result of a long-term process of adjustment and modification.

Some boundaries are forcibly drawn across a unified cultural landscape. Such a *superimposed* boundary exists in the center of the island of New Guinea and separates Indonesia's West Irian from the country of Papua New Guinea (Fig. 11-2, lower-left map). West Irian, which is mostly peopled by ethnic Papuans, was a part of the Netherlands East Indies that did not receive independence as Indonesia in 1949. After many tension-filled years, the Indonesians finally invaded this territory in 1962 to drive out the remaining Dutch; following U.N. mediation and an eventual plebiscite, West Irian was formally attached to Indonesia in 1969—thereby perpetuating the boundary that the colonial administrators had originally superimposed on New Guinea in the early nineteenth century. The fourth genetic boundary type is the *relict* boundary— a border that has ceased to function, but whose imprints are still evident on the cultural landscape. The boundary between former North Vietnam and South Vietnam (Fig. 11-2, lower-right map) is a classic example: once demarcated militarily, it has had relict status since 1976 following the reunification of Vietnam in the aftermath of the Indochina War (1964–1975).

As this chapter emphasizes, Southeast Asia is fertile ground for the study of political geography.

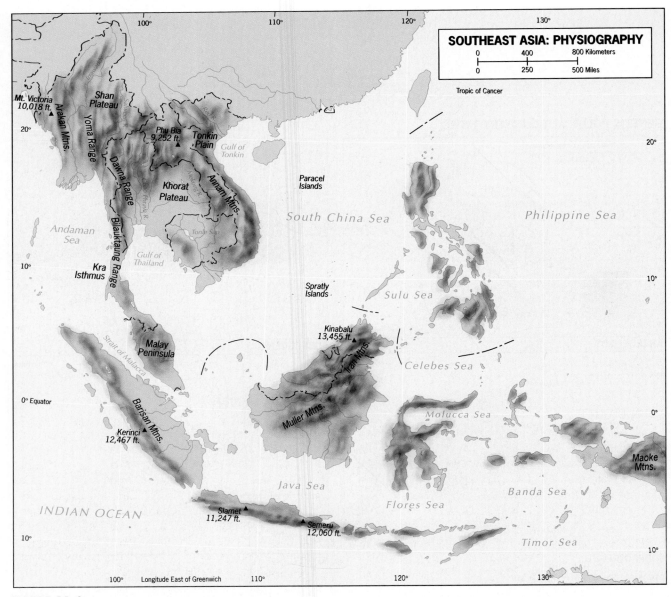

FIGURE 11-3

there is such population pressure and such land shortage in adjacent realms, why has Southeast Asia not been flooded by waves of immigrants? Several factors have combined to inhibit large-scale invasions. In the first place, physical obstacles hinder travel along the overland routes into Southeast Asia. In discussing the Indian subcontinent, we noted the barrier effect of the densely forested hills and mountains that lie along the border between northeastern India and northwestern Myanmar (Burma). North of Myanmar lies forbidding Xizang (Tibet), and northeast of Myanmar and north of Laos is the high Yunnan Plateau. Transit is easier into northern Vietnam via southeastern China, and along this avenue considerable contact and migration have indeed occurred. Moreover, spatial interaction within Southeast Asia

itself is not enhanced by the rugged, somewhat parallel ridges that hinder east-west communications between the fertile valleys of the realm (Fig. 11-3).

Second—and the population map reflects this—there are limits to the agricultural opportunities in Southeast Asia. Much of the realm is covered by dense tropical rainforests, part of an ecological complex whose effect on human settlement we have previously observed in other low-latitude zones of the world. Except in certain locales, the soils of mainland Southeast Asia are excessively *leached* (diluted of chemical nutrients) by generally heavy rains. In the areas of monsoonal rainfall regimes and savanna climate (the latter prevailing across much of the mainland interior), there is a dry season. But the limitations imposed by savanna

The Irrawaddy River is to Myanmar (Burma) what the Ganges-Brahmaputra is to Bangladesh and the Mekong to Vietnam—a lifeline of water for vast farmlands. Note that the road (right) lies atop an artificial levee above the floodplain basin.

conditions on agriculture are all too familiar: high evapotranspiration rates, long droughts, hard-baked soils, meager fertility, high runoff, and erosional problems. These add up to anything but a peasant farmer's paradise in these parts of Southeast Asia. Food production, therefore, comes mainly from environmentally favored areas. When this realm is at peace, these areas yield enough food not only to feed the population but also to export large quantities to hungry neighbors.

THE CLUSTERS

Population in Southeast Asia has become concentrated in three kinds of favorable natural environments. First, there are the valleys and deltas of Southeast Asia's major rivers, where alluvial soils have been formed. Four major rivers stand out. In Myanmar, the Irrawaddy rises near the border with China's Xizang (Tibet) and creates a delta (see photo above) as it empties into the Andaman Sea. In Thailand, the Chao Phraya traverses the length of the country and flows into the Gulf of Thailand. In southern Vietnam, there is the extensive delta of the great Mekong River, which rises in the high mountains on the Qinghai-Xizang border in interior China and crosses the entire Indochinese peninsula. And in northern Vietnam lies the Red River, whose lowland (the Tonkin Plain) is probably the most densely settled area in mainland Southeast Asia. Each of these four

river basins contains one of the realm's major mainland population clusters (see Fig. I-9).

Second, volcanic mountains abound in Southeast Asia—at least in the island archipelagoes. In certain parts of these island chains, the conditions are right for the formation of deep, dark, and rich volcanic soils, especially across much of Jawa (Java).* The population map indicates the significance of this fertility: the island of Jawa is one of the world's most densely populated and intensively cultivated areas. On Jawa's productive land live about 110 million people, approximately 60 percent of the inhabitants of all the islands of Indonesia and almost a quarter of the population of the entire realm.

Another look at Fig. I-9 indicates that one additional area remains to be accounted for: the belt of comparatively dense population that extends along the western coast of the lower Malay Peninsula, apparently unrelated to either alluvial or volcanic soils. This represents the third basis for population agglomeration in Southeast Asia—the plantation economy. Actually, plantations were introduced by the European colonizers throughout most of insular (as well as parts of coastal mainland) Southeast Asia, but nowhere did they so thoroughly transform the economic geography as in Malaya.

*As in Africa, names and spellings have changed with independence. In this chapter, the contemporary spellings will be used, except when reference is made to the colonial period. Thus Indonesia's four major islands are Jawa, Sumatera, Kalimantan (the Indonesian part of Borneo), and Sulawesi. The Dutch called them Java, Sumatra, Borneo, and Celebes.

Rubber trees were planted on tens of thousands of acres, and colonial Malaya became the world's leading exporter of this product. Undoubtedly, modern Malaysia would not have developed so populous a core area without its plantation economy (which is also marked by recent increases in palm oil production).

Where there are no alluvial soils, no volcanic soils, and no productive plantations, as in the rainforested areas and the steep-sloped uplands where there is little level land, far fewer people manage to make a living. Here the practice of shifting subsistence cultivation prevails (see Fig. 6-2), sometimes augmented by hunting, fishing, and the gathering of wild nuts, berries, and the like. This practice resembles that in tropical-interior South America and Equatorial Africa; because land that was once cleared for cultivation must be left alone for years in order to regenerate, only sparse populations can be sustained. Here in the forests and uplands of Southeast Asia, nonsedentary cultivators still live in considerable isolation from the peoples of the core areas. The forests are dense and difficult to penetrate, distances are great, and steep slopes only add to the obstacles that constrain efficient surface communications.

Over a major part of their combined length, the political boundaries of continental Southeast Asia traverse these rather thinly peopled inland areas. But here are the roots of some of the region's political troubles: these unstable interior zones have never been effectively integrated into the states of which they are part. The people who live in the forested hills may not be well disposed toward those who occupy the dominant core areas. Therefore these frontier-like inner reaches of Southeast Asia—with their protective isolation and distance from the seats of political power— are fertile ground for revolutionary activities if not for agriculture (which can turn illicit, too, as in the case of the huge opium poppy harvests of the notorious "Golden Triangle" where the borders of Myanmar, Laos, and Thailand converge). Thus the integration of highland and lowland economies, and the improvement of connections between national cores and peripheries, are very important tasks for Southeast Asian governments.

◆ INDOCHINA

The French colonialists called their Southeast Asian possessions Indochina, and the name is appropriate for the bulk of the mainland region because it suggests the two leading Asian influences that have affected the realm for the past 2,000 years. Overland immigration was mainly southward from southern China, resulting from the expansion of the Chinese Empire. The Indians came from the west by way of the seas, as their trading ships plied the coasts and settlers from India founded colonies on Southeast Asian shores

in the Malay Peninsula, on the lower Mekong plain, and on Jawa and Borneo.

With the migrants from the Indian subcontinent came their faiths: first Hinduism and Buddhism, later Islam. The Muslim religion was also promoted by the growing number of Arab traders who appeared on the scene, and Islam became the dominant religion in Indonesia (where nearly 90 percent of the population adheres to it today). But in Myanmar, Thailand, and Cambodia, Buddhism remained supreme, and in all three countries, the overwhelming majority of the people are now adherents. In culturally diverse Malaysia, the Malays are Muslims (to be a Malay is to be a Muslim), and almost all Chinese are Buddhists; but most Malaysians of Indian ancestry continue to adhere to the Hindu way of life. Although Southeast Asia has generated its own local cultural expressions, most of what remains in tangible form has resulted from the infusion of foreign elements. For instance, the main temple at Angkor Wat (see photo at right), constructed in Cambodia during the twelfth century, remains a monument to Indian architecture of that time.

The *Indo* part of Indochina, then, refers to the cultural imprints from South Asia: the Hindu presence, the importance of the Buddhist faith (which came to Southeast Asia via Sri Lanka [Ceylon] and its seafaring merchants), the influences of Indian architecture and art (especially sculpture), writing and literature, and social structures and patterns.

The Chinese role in Southeast Asia has been substantial as well. Chinese emperors coveted Southeast Asian lands, and China's power at times reached deeply into the realm. Social and political upheavals in China sent millions of Sinicized people southward. Chinese traders, pilgrims, sailors, fishermen, and others sailed from southeastern China to the coasts of Southeast Asia and established settlements there. Over time, those settlements attracted additional Chinese emigrants, and Chinese influence in the realm grew. Not surprisingly, relations between the Chinese settlers and the earlier inhabitants of Southeast Asia have at times been strained, even violent. The Chinese presence in Southeast Asia is long-term, but the invasion has continued into modern times. The economic power of Chinese minorities and their role in the political life of the area have led to conflicts.

The Chinese initially profited from the arrival of the Europeans, who stimulated the growth of agriculture, trade, and industries; here they found opportunities they did not have at home. The Chinese established rubber holdings, found jobs on the docks and in the mines, cleared the bush, and transported goods in their sampans. They brought with them skills that proved to be very useful, and as tailors, shoemakers, blacksmiths, and fishermen, they prospered. The Chinese also proved to be astute in business; soon, they not only dominated the region's retail trade but also held prominent positions in banking, industry, and shipping. Thus

The temple complex of Angkor Wat in Cambodia is a reminder of six centuries of Khmer Empire—the beginning of a Cambodian state. Angkor Wat is a Hindu structure, one of Hinduism's greatest architectural and artistic expressions.

their importance has always been far out of proportion to their modest numbers in Southeast Asia. The Europeans used them for their own designs but found the Chinese to be stubborn competitors at times—so much so that eventually they tried to impose restrictions on Chinese immigration. The United States, when it took control of the Philippines, also sought to stop the influx of Chinese into those islands.

When the European colonial powers withdrew and Southeast Asia's independent states emerged, Chinese population sectors ranged from nearly 50 percent of the total in Malaysia (in 1963) to barely over 1 percent in Myanmar (Fig. 11-4). In Singapore, Chinese today constitute 78 percent of the population of 2.8 million; when Singapore seceded from Malaysia in 1965, the Chinese component in the latter was reduced to about 35 percent. In Indonesia, the percentage of Chinese in the total population is not high (no more than 3 percent), but the Indonesian population is so large that even this small percentage indicates a Chinese sector of more than 5 million. In Thailand, on the other hand, many Chinese have married Thais, and the Chinese minority of about 12 percent has become a cornerstone of Thai society, dominant in trade and commerce.

In general, Southeast Asia's Chinese communities remained quite aloof and formed their own separate societies in the cities and towns. They kept their culture and language alive by maintaining social clubs, schools, and even residential suburbs that, in practice if not by law, were Chinese in character. There was a time when they were in the middle between the Europeans and Southeast Asians and when the hostility of the local people was directed toward white people as well as toward the Chinese. Since the withdrawal of the Europeans, however, the Chinese have become the main target of this antagonism, which remained strong because of the Chinese involvement in money lending, banking, and trade monopolies. Moreover, there is the specter of an imagined or real Chinese political imperialism along Southeast Asia's northern flanks.

The *china* in Indochina, therefore, represents a diversity of penetrations. The source of most of the old invasions was in southern China (Fig. 11-4), and Chinese territorial consolidation provided the impetus for successive immigrations. Mongoloid racial features carried southward from East Asia mixed with the preexisting Malay stock to produce a transition from Chinese-like people in the northern

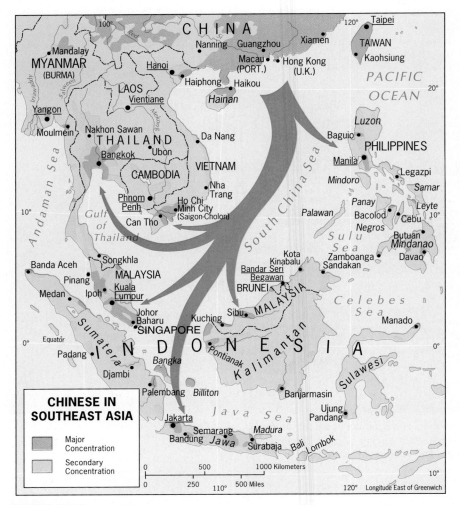

FIGURE 11-4

mainland to darker-skinned Malay types in the distant Indonesian east. Although Indian cultural influences remained strong, Chinese modes of dress, plastic arts, types of houses and boats, and other cultural attributes were adopted throughout Southeast Asia. During the past century, and especially during the last half-century, renewed Chinese immigration brought skills and energies that propelled these minorities to positions of comparative wealth and power in this realm.

◆ THE ETHNIC MOSAIC

It is revealing that much of the mainland region of Southeast Asia is often referred to as Indochina, a name that reflects external influences but fails to reveal indigenous cultural-geographical roots. Perhaps this is so because alien forces have been so powerful in the realm. Yet despite the cultural influences of India and China and notwithstanding the European colonial era, the great majority of the realm's

peoples have regional identities. When you study Fig. 11-5, view it as you saw Fig. 1-6, the language map of Europe. In Southeast Asia, as in Europe, there is a very rough correspondence between language, culture, and the political framework. But just as there are German speakers in France, Hungarians in Romania, and Austrians in Italy, so there are Siamese (the people of Thailand) in Myanmar, Khmers (of Cambodia) in Thailand, and Malays in Vietnam. Like the Europeans, all these peoples belong to a single geographical race, the Asian (or Asiatic). But here, too, different nations exist.

The largest of these is the Indonesian people, the inhabitants of the great archipelago that extends from Sumatera west of the Malay Peninsula to the Moluccas in the east, and from the lesser Sunda Islands in the south to the Philippines in the north. Collectively, all these peoples—the Filipinos, Malays, and Indonesians—are known as Indonesians, but they have been divided by history and politics. Within Indonesia itself, there are the Jawanese (the people of Jawa [Java]), the Bataks of Sumatera, the Balinese, and

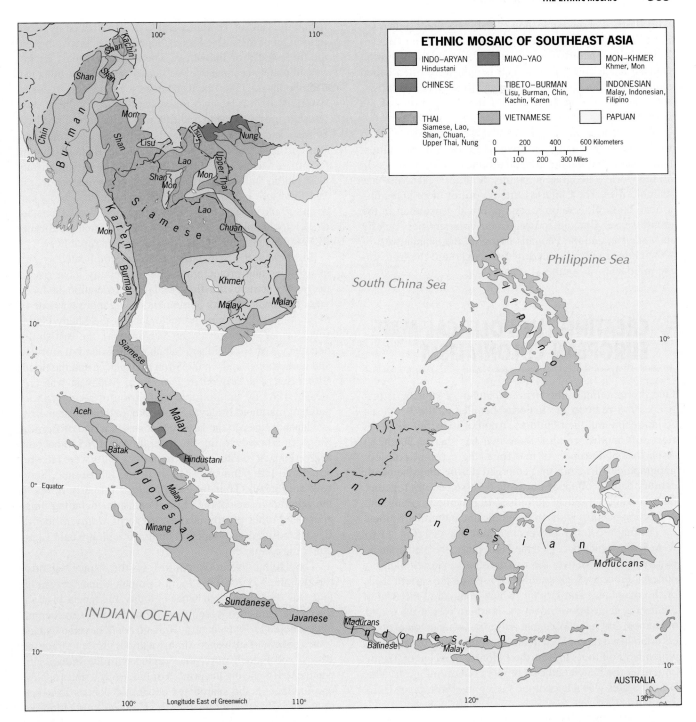

FIGURE 11-5

the Moluccans of the east, among many other discrete groups (Fig. 11-5). In the Philippines, too, island insularity and contrasting ways of life are reflected in the cultural mosaic. Near Muslim Indonesia lie Filipino islands where the people do not adhere to the faith of the majority (Roman Catholicism) but rather to Islam.

On the mainland north of the Malay-occupied peninsula, the map shows four regionally dominant peoples: the Vietnamese of Vietnam, the Khmer of Cambodia, the Siamese of Thailand, and the Burmans of Myanmar (Burma). In a very general way this distribution is mirrored by the political framework, although this does not apply to Laos. As with all maps at this scale, however, details are lost. In Vietnam, as elsewhere, there are people (non-Vietnamese) living in comparative isolation, not part of the majority. Under French colonial rule such people were often called *montagnards*, people of the uplands, living a life of subsistence quite different from that of the rice-growers of the river plains.

We will look more closely at the ethnic mosaic when we focus on the realm's individual states, but it is important to keep in mind the general outlines of Fig. 11-5. As in all the world's geographic realms, Southeast Asia also has its minorities. The Chinese form minorities more in the historical, cultural, economic, and political sense than in the ethnic sense. But the Indian-Hindu communities, chiefly in Malaysia, and the Papuans living on the islands west of New Guinea (islands ruled by Indonesia) are the true ethnic minorities of Southeast Asia.

◆ CREATING THE POLITICAL MAP: EUROPEAN COLONIALISM

One shortcoming of the term *Indochina*, obviously, is that it does not refer to the majority peoples of the Southeast Asian realm and their cultures. Another is its failure to reflect still another cultural force that has shaped Southeast Asia: that of Europe. As in Africa, the colonial politico-geographical structure that emerged during the nineteenth century had the effect of throwing diverse peoples together under a single political administration while dividing and separating peoples with strong ethnic and cultural affinities.

The major colonial powers in Southeast Asia (Fig. 11-6) were Spain, which occupied the Philippines; the Netherlands, which took control over the vast archipelago now called Indonesia; France, which established itself on the eastern flank of the mainland; and Britain, which conquered the Malay Peninsula, gained power over the northern part of the island of Borneo, and ruled Myanmar (Burma). Other colonial powers established footholds, but not for long—except Portugal, which held on to its half of the Indonesian island of Timor until after the Dutch had been thrown out of Indonesia. The United States was a latecomer, taking the Philippines from Spain in 1898.

Naturally, the colonial powers divided their possessions into administrative units as they did in Africa and elsewhere. In some instances, these political entities became independent states when the colonial power withdrew. France, one of the mainland's leading colonial powers, divided its Southeast Asian empire into five units. Three of these lay along the east coast: Tonkin in the north next to China, centered on the basin of the Red River; Cochin China in the south, with the Mekong delta as its focus; and between these two, Annam. The other two French territories were Cambodia,

facing the Gulf of Thailand, and Laos, landlocked in the interior. Out of these five French dependencies there emerged the three states of Indochina. The three east-coast territories ultimately became one such state, Vietnam; the other two (Cambodia and Laos) each achieved separate independence.

The British ruled two major entities in Southeast Asia (Burma and Malaya) in addition to a large part of northern Borneo and many small islands in the South China Sea. Burma (now Myanmar) was attached to Britain's Indian empire; it was governed from 1886 until 1937 from distant New Delhi. But when British India became independent in 1947 and split into several countries, Myanmar was not part of the grand design that created East and West Pakistan, Ceylon (Sri Lanka), and India. Instead, Myanmar (as Burma) in 1948 was given the status of a sovereign republic.

In Malaya, the British developed a complicated system of colonies and protectorates that eventually gave rise to the equally complex, far-flung Malaysian federation. Included were the former Straits Settlements (Singapore was one of these colonies), the nine protectorates on the Malay Peninsula (former sultanates of the Muslim era), the British dependencies of Sarawak and Sabah on the island of Borneo, and numerous islands in the Strait of Malacca and the South China Sea. The original federation of Malaysia was created in 1963 by the political unification of recently independent mainland Malaya, Singapore, and the former British dependencies on the largely Indonesian island of Borneo. Singapore, however, left the federation in 1965 to become a sovereign city-state, and the remaining units were later restructured into peninsular Malaysia and, on Borneo, Sarawak and Sabah. Thus the term *Malaya* properly refers to the geographic area of the Malay Peninsula, including Singapore and other nearby islands; the term *Malaysia* identifies the politico-geographical entity of which Kuala Lumpur is the capital city.

The Hollanders took control of the "spice islands" through their Dutch East India Company, and the wealth that was extracted from what is today Indonesia brought the Netherlands its Golden Age. From the mid-seventeenth to the late-eighteenth century, Holland could develop its East Indies sphere of influence almost without challenge, for the British and French were preoccupied with the Indian subcontinent. By playing the princes of Indonesia's states against one another in the search for economic concessions and political influence, by placing the Chinese in positions of responsibility, and by imposing systems of forced labor in areas directly under its control, the company had a ruinous effect on the Indonesian societies it subjugated. Java (Jawa), the most populous and productive island, became the focus of Dutch administration; from its capital at Batavia (now Jakarta), the company extended its sphere of influence into Sumatra (Sumatera), Dutch Borneo (Kalimantan), Celebes (Sulawesi), and the smaller islands of the East Indies. This was not accomplished overnight, and the struggle for territorial control was carried on long after the Dutch East

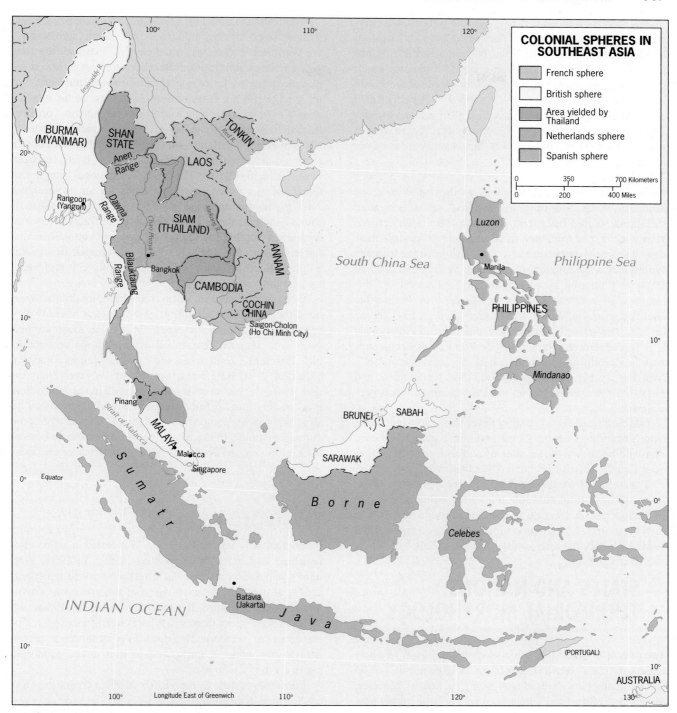

FIGURE 11-6

India Company had yielded its administration to the Netherlands government. Dutch colonialism thus threw a girdle around Indonesia's more than 13,000 islands, paving the way for the creation of the realm's largest nation-state in terms of population (191 million today) and territory.

In the colonial tutelage of Southeast Asia, the Philippines, long under Spanish domination, had a unique experi-

ence. From as early as 1571, the islands north of Indonesia were under Spain's control (they were named for its King Philip II). Spanish rule began at a time when Islam was reaching the southern Philippines via northern Borneo. The Spaniards spread their Roman Catholic faith with great zeal, and between them the soldiers and the priests consolidated Hispanic dominance over the mostly Malay population.

Manila, a city founded in 1571, became a profitable way-station on the route between southern China and western Mexico (Acapulco usually was the trans-Pacific destination for the galleons leaving Manila's port). There was much profit to be made, but the indigenous people shared little in it. Great landholdings were awarded to loyal Spanish civil servants and to men of the church. Oppression eventually yielded revolution, and Spain was confronted with a major uprising when the Spanish-American War broke out elsewhere in 1898.

As part of the settlement of that war, the United States replaced Spain in Manila. That was not the end of the revolution, however. The Filipinos now took up arms against their new foreign ruler, and not until 1905, after terrible losses of life, did American forces manage to "pacify" their new dominion. Subsequently, U.S. administration in the Philippines was more progressive than Spain's had been. In 1934, the Philippine Independence Law was passed, providing for a 10-year transition to sovereignty. But before this could be arranged, World War II intervened. In 1941, Japan invaded the islands, temporarily ousting the Americans; U.S. forces returned three years later and, with strong support from Filipino forces, defeated the Japanese in 1945. Now the agenda for independence could be resumed, and in 1946 the Sovereign Republic of the Philippines was proclaimed.

One Southeast Asian country never was colonized: Thailand. Situated between the French and British colonial spheres, Thailand became a buffer state of convenience, an indigenous kingdom left unconquered by consensus.

Today, all of Southeast Asia's states are independent. But centuries of colonial rule left strong cultural imprints. In their urban landscapes, in their educational systems, in their civil service, and in countless other ways, this realm's states still carry the marks of their colonial past.

◆ STATES AND NATIONS: TERRITORIAL MORPHOLOGY

The map of Southeast Asia is a geopolitical jigsaw of many varied parts. Indonesia is broken into a large number of islands. Vietnam is a sliver of land facing the South China Sea. Myanmar (Burma) and Thailand consist of large, bulky northern areas that extend southward into narrow ribbons on the Malay Peninsula.

A state's physical shape, its **territorial morphology**, is among the factors that affect its cohesion and political viability. A state that consists of several separate parts located far from each other obviously faces problems that do not afflict a state with a single, contiguous territory. When we study a state, therefore, it is useful to keep in mind its morphology.

The technical terms for states' territorial shapes are easily understood. Certain countries—for example, Cambodia—are **compact**. Others, as in the case of Vietnam, are long and narrow or **elongated**. Still others are broken into separate pieces, as are Indonesia, the Philippines, and Malaysia; these countries are **fragmented**. And a few of the world's countries have large main territories with narrow corridors leading from these; Southeast Asia has two such countries—Myanmar and Thailand—and they are called **prorupt** (a term you may not find in the dictionary, but which has been in use since the 1950s). There is one additional category of territorial morphology, but it does not occur in Southeast Asia: the **perforated** state. When one state completely surrounds another, the encircling state is perforated. Look at the map of South Africa again (p. 454), and you will see that the kingdom of Lesotho is totally surrounded by the Republic of South Africa; South Africa, therefore, is perforated by Lesotho.

One point of caution: states' morphologies do not *determine* their cohesion, viability, unity, or lack thereof; they can, however, *influence* these conditions. For example, Belgium is a compact state, but it is strongly divided culturally. Belgium's compactness has not ensured any national cohesion; but if Belgium's two principal regions, Flanders and Wallonia, had been separate islands in the North Sea, it is unlikely that a unified Belgium would have existed. In Belgium's case, compactness of shape helped overcome cultural disunity. In the pages that follow, we take note of the morphology of Southeast Asia's countries as we assess their geography.

ELONGATED VIETNAM

Consider this: when the Indochina War ended in 1975, Vietnam had *half* the population it has today. In 1994, Vietnam's population reached 72.3 million, with fully 60 percent of it under 21 years of age. For the great majority of Vietnamese, therefore, the terrible war of the 1960s and 1970s is history, not memory. What concerns most Vietnamese now is to overcome two decades of isolation, to reconnect to the world at large, and to join in the economic boom on the Pacific Rim.

Former President George Bush said that Operation Desert Storm in Kuwait and Iraq (1991) had laid to rest, once and for all, the ghost of the Indochina War. He may have been mistaken. Today, Americans, some of them of Vietnamese ancestry, are again able to enter Vietnam legally, and some are doing so. This opening, made possible by Bush himself, will begin the process of normalization.

Travel up the crowded road from the northern port of Haiphong to the capital, Hanoi, or sail up the river to the southern metropolis officially known as Ho Chi Minh City—but called Saigon by almost everyone there—and you are quickly reminded of the cultural effects of Vietnam's elongation. Vietnam was delimited by the French colonizers as a 1,200-mile (2,000-km) strip of land extending from the Chinese border in the north to the tip of the Mekong delta in the south. Substantially smaller than California, this coastal belt was the domain of the Vietnamese (Fig. 11-5). The French recognized that Vietnam, whose average width is under 150 miles (240 km), was no homogenous colony, so they divided it into three units: (1) Tonkin, land of the Red River delta and centered on Hanoi in the north; (2) Cochin China, region of the Mekong delta and centered on Saigon in the south; and (3) Annam, focused on the ancient city of Hué, in the middle.

The Vietnamese (or *Annamese*, also *Annamites*, after their cultural heartland) speak the same language, although the northerners can quickly be distinguished from southerners by their accent. As elsewhere in their colonial empire, the French introduced their language as the *lingua franca* of Indochina, but their tenure was cut short by the Japanese, who invaded Vietnam in 1940. During the Japanese occupation, Vietnamese nationalism became a powerful force, and after the Japanese defeat in 1945, the French could not regain control. In 1954, the French suffered a disastrous final trouncing on the battlefield at Dien Bien Phu, and were ousted.

But Vietnam did not become a unified state even after its forces routed the colonizers. Separate regimes took control: a communist one in Hanoi and a noncommunist version in Saigon. Vietnam's pronounced elongation had made things difficult for the French; now it played its role during the postcolonial period. On the regional map, note that Vietnam is widest in the north and in the south, with a very narrow "waist" in its middle zone. North and South were worlds apart.

Many Americans still remember the way in which the United States became involved in the inevitable conflict between communists and noncommunists in Vietnam during the 1960s and 1970s. At first, American military advisors were sent to Saigon to help the shaky regime there cope with communist insurgents. When the tide turned against the South, military forces and equipment were committed to halt the further spread of communism (see the box entitled "Domino Theory"); at one time, more than half a million U.S. soldiers were in the country.

The conduct of the war, its mounting casualties, and its apparent futility created severe social tensions in the United States. If you had been a student during that period, your college experience would have been radically different from that of today. Protest rallies, "teach-ins," anti-draft demonstrations, "sit-ins," marches, strikes, and even hostage-taking disrupted campus life. Several student protesters were killed by the National Guard on a Midwest campus in 1970. The Indochina War threatened the stability of American society. It drove an American President (Lyndon Johnson) from office in 1968 and destroyed the electoral chances of his

DOMINO THEORY

During the Indochina War (1964–1975), it was United States policy to contain communist expansion by supporting the efforts of the government of South Vietnam to defeat communist insurgents. Soon, the war engulfed North Vietnam as U.S. bombers attacked targets north of the border between North and South. And in the later phases of the war, conflict spilled over into Laos and Cambodia. In addition, U.S. warplanes took off from bases in Thailand. Like dominoes, one country after another fell to the ravages of the war or was threatened.

Some scholars warned that this domino effect could eventually affect not only Thailand but also Malaysia, Indonesia, and Burma (today Myanmar): the whole Southeast Asian realm, they predicted, could be destabilized. But, as we know, that did not happen. The war remained confined to Indochina. And the domino "theory" seemed invalid.

But is the theory totally without merit? Unfortunately, some political geographers to this day make the mistake of defining this idea in terms of communist activity. Communist insurgency, though, is only one way a country may be destabilized (as is happening today in Peru). But right-wing rebellion (Nicaragua's Contras), ethnic conflict (Bosnia-Herzegovina), religious extremism (Algeria), and even economic and environmental causes can create havoc in a country. Properly defined, the **domino theory** holds that destabilization from any cause in one country can result in the collapse of order in a neighboring country, starting a chain of events that can affect a series of contiguous states in turn.

In fact, any visitor to Laos and especially to Cambodia will see the disastrous long-term impact of the "Vietnam" war on these countries and societies; these dominoes certainly fell. Today, Indochina is relatively stable (although Cambodia suffers from sporadic strife). Now, the dominoes are falling in another shatter belt, Eastern Europe. Look at the map again: the struggle in former Yugoslavia has moved from Slovenia to Croatia, on to Bosnia-Herzegovina and Serbia-Montenegro, and threatens to engulf Kosovo, Macedonia, Albania, and perhaps even Greece and Turkey. In 1993–1994, it was U.S. and U.N. policy to attempt to contain ex-Yugoslavia's conflicts and prevent them from spreading to Kosovo and beyond. There may be something to the domino theory after all.

Vice President (Hubert Humphrey), who would not disavow it.

In 1975, the Saigon government fell and the United States was ousted—just two decades after the French defeat at Dien Bien Phu. When the last helicopter left from the roof of the U.S. embassy in Saigon amid scenes of desperation and desertion, it marked the end of a sequence of events that closely conformed to the **insurgent state model** outlined in Chapter 6 (p. 325).

Vietnam Today

After the war's end, the Soviets, not the neighboring Chinese, became Vietnam's patrons. As Vietnam was finally unified in 1976 under a dogmatic communist regime, renewed conflict broke out, this time involving China (boundary issues in the north were the cause). There was more to this, however. The Soviets exploited the historic distrust between Chinese and Vietnamese because Moscow was itself at odds with Beijing, and so life for Vietnam's Chinese minorities became much more difficult. Many Chinese (the number will never be known) joined in the tragic and disastrous exodus of *boat people* who sailed from Vietnam's coasts in small, often unseaworthy boats. Of the estimated 2 million of these refugees, more than half perished from storms, exposure, pirates, starvation, and sinkings. The great majority of those who survived were brought to the United States.

While accepting hundreds of thousands of Vietnamese refugees, the United States simultaneously placed a strong embargo on Vietnam, a policy that isolated the communist country and stifled its economy. When the Soviet Union collapsed in 1991, Vietnam was in a desperate position—with one saving grace. With two major river deltas and plenty of fertile farmland, Vietnam can produce large harvests of rice, enough to feed its own population *and* export

One of Vietnam's major problems in the near future is its decayed infrastructure. Crowded chaotic streets, dilapidated buildings, inadequate (and, in Saigon, intermittent) electricity supply, poor telephone service, and other obstacles will hamper development. This typical street scene reflects "Ho Chi Minh City" as it is today: people sleep ten or more to a room in the French-era buildings in which no repairs have been done for decades. But the people are well-educated and capable, awaiting their opportunity on the Pacific Rim.

Hanoi, the capital of Vietnam, lags behind Ho Chi Minh City (Saigon-Cholon) in almost every way, except political dogma: the old communist regime still rules from here, far from the more open, progressive south.

to foreign markets. At least, therefore, Vietnam could survive its boycott.

Now there are signs that the embargo is ending. In 1993, French President Mitterrand visited Hanoi and Saigon, and European and East Asian businesspeople were establishing offices in the country. The prohibition against American visitors to Vietnam has been lifted. Consumer goods from Japan, Taiwan, Malaysia and elsewhere are flowing into the cities. The urban agglomeration of Saigon-Cholon (Cholon is the city's Chinatown) now contains as many as 7 million people, its streets choked with bicycles, mopeds, handcarts, buses, and a few cars (see photo at left). Hanoi is lagging far behind with 1.2 million inhabitants and far less of the bustle that characterizes Saigon (see photo above). In both North and South, however, infrastructure is poor. Saigon is overcrowded, dilapidated, inefficient, and burdened by corrupt and stagnant communist management. Teachers, policemen, politicians, and bureaucrats from the North are identified by their accent, and frustrated Southerners often say that after the French and the Americans and the Russians, they now have still another group of outsiders to contend with—the Northerners.

And indeed, Saigon and the South generally are well ahead of distant Hanoi and the North. The people are comparatively well educated, energetic, and productive when they have the chance. They are willing to work for wages even lower than those of Hong Kong. The first new buildings are going up, the first new factories are being built. Soon, the western Pacific Rim will have a new economic tiger on the block.

COMPACT CAMBODIA

Former French Indochina contained two additional entities—Cambodia and Laos. In Cambodia, the French possessed one of the greatest treasures of Hinduism: the city of Angkor, capital of the ancient Khmer Empire, and the temple complex known as Angkor Wat. The Khmer Empire prevailed here from the ninth to the fifteenth centuries, and Angkor Wat was built by King Suryavarman II to symbolize the universe in accordance with the precepts of Hindu cosmology. When the French took control of this area during the 1860s, Angkor lay in ruins, sacked by invading Vietnamese and Thai armies. The French created a protectorate, began the restoration of the shrines, established Cambodia's permanent boundaries, and restored the monarchy under their supervision.

Geographically, Cambodia enjoys several advantages, notably its compact shape. Compact states enclose a maximum of territory within a minimum of boundary and are without peninsulas, islands, or other remote extensions of the national spatial framework. Cambodia had the further advantage of strong ethnic and cultural homogeneity: 90 percent of its people are Khmers, with the rest equally divided between Vietnamese and Chinese.

As Fig. 11-1 shows, Southeast Asia's greatest river, the Mekong, enters Cambodia from Laos and crosses it from north to south, creating a great bend before flowing into southern Vietnam. Phnom Penh (1.2 million), the country's present capital, lies on the river's western bank where it makes that bend. The ancient capital of Angkor lies in the northwest, not far from Tonle Sap, a major interior lake that drains into the Mekong.

Cambodia fell victim to the Indochina War in the most dreadful way, and neither its compactness nor its isolation could save it. The war first spilled over from Vietnam into its eastern border areas. Then, in 1970, the last king was deposed by military rulers; in 1975, that government was itself overthrown by communist revolutionaries, the so-called Khmer Rouge. These new rulers embarked on a course of terror and destruction in order to reconstruct *Kampuchea* (as they called Cambodia) as a rural society. They drove townspeople into the countryside where they had no place to live or work, emptied hospitals and sent the sick and dying into the streets, outlawed religion and family life, and killed as many as 2 million Cambodians (out of a total of 7.5 million) in the process. In the late 1970s, Vietnam, victorious in its own war, invaded Cambodia to drive the Khmer Rouge away. But this led to new terror, and a stream of refugees crossed the border into Thailand.

Once a self-sufficient country that could feed others, Cambodia now must import food. The economy is dominantly agricultural, and most people grow rice and beans for subsistence. Its population, severely reduced by war, again surpassed 7 million around 1990 (it stands at 9.5 million today). But the country's future remains uncertain, with powerful adversaries on its borders and armed factions within them.

Laos

North of Cambodia lies Southeast Asia's only landlocked country, Laos. Interior and isolated, Laos changed little during 60 years of French colonial administration (1893–1953). Then, along with other French-ruled areas, Laos became an independent state. Soon its well-entrenched, traditional kingdom fell victim to rivalries between traditionalists and communists, and the old order collapsed.

Laos has no fewer than five neighbors, one of which is the East Asian giant, China. A long stretch of its western boundary is formed by the Mekong River, and the important sensitive border with Vietnam to the east lies in moun-

tainous terrain. With 4.7 million people (about half of them ethnic Lao, related to the Thai of Thailand), Laos lies surrounded by comparatively powerful states. The country has no railroads, just a few miles of paved roads, and very little industry; it is only 16 percent urbanized (the capital, Vientiane, contains only about 500,000 people). Laos remains the region's poorest and most vulnerable entity.

PRORUPT NEIGHBORS

Quite a different spatial form is represented by the configurations of Myanmar and Thailand. The main territories of these two states, which contain their respective core areas, are essentially compact—but to the south they share sections of the slender Malay Peninsula. These peninsular portions are long and even narrower, and (as noted earlier) states with such extensions leading away from the main body of territory are referred to as *prorupt* states. Obviously, Thailand is the best example: its proruption extends nearly 600 miles (1,000 km) southward from just west of the capital, Bangkok (Fig. 11-7). Where the Thailand-Myanmar boundary (terminating at the Kra Isthmus) runs along the peninsula, the Thai proruption is in places less than 20 miles wide. Naturally, such proruptions can be troublesome, especially when they are as lengthy as this. In the entire state of Thailand, no area lies farther from the core or from Bangkok than the southern extremity of its very tenuous proruption. But at least Thailand's railroad network extends all the way to the Malaysian border; in the case of Myanmar, not only does the railway terminate more than 300 miles (500 km) short of the end of the proruption, but there is not even a permanent road over its southernmost 150 miles (Fig. 11-7).

The territorial shape of Thailand and Myanmar may be similar, but their internal geography is quite different. To a large degree, this is the result of differences in relative location. Thailand, as the map shows, occupies the heart of Southeast Asia's mainland region. Historically, it formed a buffer state between French colonial holdings to the east and the British sphere to the west. Indochina to the east became a complex of French dependencies; Myanmar to the west fell under British sway (Fig. 11-6).

Thailand

Thailand's morphology should be viewed in relation to its latitudinal extent. There are humid, nearly equatorial conditions in its southern proruption and more marginally tropical, savanna-like environments over much of the mainland to the north (Fig. I-7). The heart of Thailand is a great low-lying plain, with extensive alluvial flatlands watered by the many tributaries of the Chao Phraya River. Irrigation systems guide the waters to paddies (ricefields) not reached by those streams. In the east lies the Khorat Plateau, where

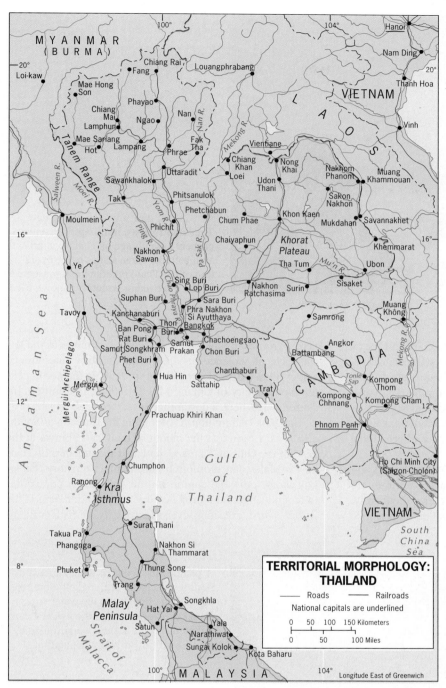

FIGURE 11-7

the soils are poorer, moisture is less available, and population is much less dense. Ethnically, Thailand's 58 million people are about 80 percent Thai (or Siamese—the country used to be known as Siam). About 12 percent are Chinese, and the remainder are various minorities living in the northern mountains, in the eastern borderlands, and in the southern peninsula. The Thais are concentrated in the country's core area (Fig. 11-1). The Chinese are heavily clustered in the urban areas, especially in and around Bangkok, where they dominate trade and commerce. Thailand remains a bastion of Buddhism: no less than 95 percent of the population adheres to this faith.

Thailand's economy has been growing rapidly, but so is its population (government policy strongly encourages family planning). Thailand is the world's leading rice exporter (followed by the United States and Vietnam), and

The Grand Palace in Bangkok is a magnificent assemblage of Thai royal art and architecture. Three stages in the evolution of the Thai kingdom are represented by the three major structures shown here.

another farm product, tapioca roots, exported to Europe as animal feed, also has earned Thailand substantial income. Corn is exported from the drier interior savannalands. Teak from northern forests, rubber from trees on the peninsula, and tin from mines in the west are additional sources of foreign revenue. The discovery of oil and natural gas in the Gulf of Thailand has helped the country's international balance of payments.

Political stability has been a problem. Thailand is a kingdom that traces its roots over many centuries, but military intervention in government affairs has occurred frequently. In 1993, the burned-out hulks of government buildings stood across from Bangkok's Grand Palace, testimony to the latest wave of violence resulting from military interference and civilian (particularly student) protest. In the north, centered on the city of Chiang Mai (242,000), intermittent insurgency has been a destabilizing factor.

Bangkok (8.5 million) is becoming the Calcutta of Southeast Asia. Located at the mouth of the Chao Phraya, at the head of the Gulf of Thailand, and near the juncture of Thailand's Malaysian proruption and its main territory, Bangkok is not only the capital but also the main port, chief industrial center, and primate city. Air pollution here is the world's worst, exceeding even Mexico City's. Much of Bang-

A picturesque scene, it would appear at first, in the canals of Bangkok—but note the garbage floating in the water (right). Bangkok is often called the world's most polluted city, its air and water fouled by a huge and fast-growing population.

kok lies at or below sea level (land subsidence is a severe problem), and the city is riddled by a network of canals with busy boat traffic—and with islands of floating garbage. Road traffic is so choked that circulation frequently fails. Yet Bangkok is graced by numerous magnificent palaces and shrines, and tourism is a major industry here.

By the standards of the developing world, Thailand is a relatively free country, although rules of law are strict, especially where drugs are concerned, and penalties are severe. On the other hand, the monarchy never permitted the showing of the film *The King and I*, a fanciful tale of the royal past, and possession of a videotape of it remains forbidden. And the Thais want to be called Thais, not Siamese. The word *Siam* means golden-yellow in Chinese, who used this term because of the gold-swathed temples and shrines in this Buddhist country. The Thais prefer to be known by the name they have used for centuries, not the one given and then adopted by foreigners.

Myanmar (Burma)

Myanmar stands in contrast to Thailand in many ways. Not only is it Thailand's western neighbor, but Myanmar also is the realm's northernmost country and possesses the longest of all borders with China. Here, too, the territorial morphology of the state is further complicated by a shift in the Burmese core area that took place during colonial times. Prior to the colonial period, the focus of the embryonic state was in the so-called *dry zone* between the Arakan Mountains and the Shan Plateau (Fig. 11-3). Mandalay in Upper Myanmar, today containing 740,000 people, was the main urban node. Then the British developed the rice potential of the Irrawaddy delta, and Rangoon (now called Yangon, with 3.9 million residents), a less important city occupied earlier by the British, became the new capital. The old and new core areas are connected by the Irrawaddy in its function as a water route, but the center of gravity in modern Myanmar now lies in the south.

The lowland core area is the domain of the majority Burman population, but the surrounding areas are occupied by 11 other peoples (Fig. 11-5), all of whom have come under the domination of the majority since the British unification of colonial Burma. The Karens live to the southeast of the Irrawaddy delta, in Myanmar's proruption, and in adjacent Thailand. The Shans mostly inhabit the eastern plateau that borders northwestern Thailand, the Kachins live near the Chinese boundary in the far north, and the Chins reside in the highlands along the Indian border in the extreme west.

Other groups also form part of the country's complex population of 44.1 million, and this heterogeneity was further intensified during the period of Britain's India-connected administration (1886–1937) when more than 1 million Indians entered the country. These Indians came as shopkeepers, money lenders, and commercial agents, and

their presence heightened Burmese resentment against British policies. During World War II, the national division of Myanmar was brought sharply into focus when the lowland Burmans welcomed the Japanese intrusion, whereas the peoples of the surrounding hill country, who had seen less of the British maladministration and who had little sympathy for their Burman compatriots anyway, generally remained pro-England.

The political geography of Myanmar constitutes a particularly good example of the role and effect of territorial shape and internal state structure. This is a prorupt state, with its southern extension into the upper Malay Peninsula; in addition, Myanmar's core area is surrounded on the west, north, and east by a horseshoe of great mountains—mountains where many of the country's 11 minority peoples had their homelands before British occupation and subsequent Burman control. In 1976, nine of these indigenous peoples, opposed to the Burman government, formed a union representing about 8 million people in the country. What these peoples have demanded is the right to self-determination in their own homelands. The Karens, for instance, a nation of 3 million (about 7 percent of Myanmar's population), have proclaimed that they wish to create an autonomous territory within a federal Myanmar. The Shans of the far north, a mountain people totaling about 4 million, demand similar rights. Even though rebellious minorities have suffered a string of setbacks since 1985, these centrifugal forces bedevil the central government, which constantly seeks to establish tighter authority over its outlying areas.

The economic geography of Myanmar is potentially more encouraging. The country can still feed itself from its Irrawaddy paddies (although isolationist trade policies have ended its days as a leading rice exporter); the Salween River, in a parallel valley east of the Irrawaddy, has fertile and productive ricefields as well. With its warm monsoonal climate, Myanmar's fine alluvial soils also yield harvests of sugar cane, beans and other vegetables, and, in drier areas, cotton and peanuts. Known mineral deposits have only begun to be exploited; tin has been mined for many years, and there is petroleum in the lower Irrawaddy valley not far from Yangon.

More than anything else, however, the repressive national policies of a brutal military government have all but halted Myanmar's development in recent years. The state needs internal consensus and liberalization—and far better political leadership than that of the entrenched present regime, which has brought impoverished and exhausted Myanmar into the ranks of the world's poorest countries.

CHALLENGES OF FRAGMENTATION

On the peninsulas and islands of Southeast Asia's southern and eastern periphery lie five of the realm's 10 states (Fig. 11-1). Few regions in the world contain so diverse a set of countries. Malaysia, the former British colony, consists of two major areas separated by hundreds of miles of South China Sea. The realm's southernmost state, Indonesia, sprawls across thousands of islands from Sumatera in the west to New Guinea in the east. North of the Indonesian archipelago lies the Philippines, a nation that once was a U.S. colony. These are three of the most severely fragmented states on earth, and each has faced the challenges that such politico-spatial division brings. This insular region of Southeast Asia also contains two small but important sovereign entities: a city-state and a sultanate. The city-state is Singapore, once a part of Malaysia (and one instance in which internal centrifugal forces were too great to be overcome). The sultanate is Brunei, an oil-rich Muslim territory on the island of Borneo that seems transplanted from the Persian Gulf. Few parts of the world are more varied or more interesting geographically.

Divided Malaysia

Malaysia's ethnic and cultural divisions are etched into its landscapes. The Malays are traditionally a rural people. They originated in this region and displaced earlier aboriginal peoples, now no longer significant numerically. The Malays, who constitute over half of the Malaysian population of 19.7 million, possess a strong cultural unity expressed in a common language, adherence to the Muslim faith, and a sense of territoriality that arises from their Malayan origins and their collective view of Chinese, South Asian, and other foreign intruders. Although they have held control over the government, the Malays often express a fear of the more aggressive, commercially oriented, and urbanized Chinese minority (who constitute 30 percent of the population).

Malay-Chinese differences worsened during World War II, when the Japanese (who occupied the area) elevated the Malays into positions of authority but ruthlessly persecuted the Chinese, driving many of them into the forested interior where they founded a communist-inspired resistance movement that long continued to destabilize the region. The British returned, then yielded after a system of interracial cooperation had been achieved. But social tensions continued and produced racial clashes in Kuala Lumpur. Today the landscape of this capital city (of 2.1 million) reflects Malaysia's cultural mosaic (see photo below); but trade and commerce are mainly in the hands of the Chinese, around whose shops and businesses the city's life revolves. Rising now above the traditional and colonial townscapes are the skyscrapers that symbolize Kuala Lumpur's new economic power (see the box entitled "The Southeast Asian City" for an overview of the urban experience in this realm). Malaysia's economy has grown rapidly since 1980, and major investments are being made here.

The map is essential to any appraisal of Malaysia as a politico-geographical entity. The Malay Peninsula was Britain's most important colonial possession in Southeast Asia; by comparison, Britain's holdings on the Indonesian archipelago (Fig. 11-6) were quite neglected. The British focused their attention on Malaya and created a substantial economy there. The Strait of Malacca, between the Malay Peninsula and Sumatera, became one of the world's busiest and most strategic waterways, and Singapore, at the southern end of it, a prized possession. The map confirms Singapore's locational advantage at the entrance to the strait and near the southern end of the South China Sea.*

*As one of Austrasia's leading economic tigers, Singapore is profiled in Chapter 4 on pp. 245-247.

Kuala Lumpur: the Islamic, the colonial, and the modern. The Kuala Lumpur-Kelang area is not yet designated as an economic tiger, but the rate of economic growth is high here and Malaysia's role in the regional system is growing.

THE SOUTHEAST ASIAN CITY

In their survey of urbanization trends in this realm, Thomas Leinbach and Richard Ulack have offered some noteworthy generalizations about the larger cities of postcolonial Southeast Asia. First, these cities are all experiencing rapid growth; between 1950 and 1990, the region's urban population almost doubled in relative size (from 15 to 28 percent) and quintupled in absolute numbers (26 to 130 million). Second, despite a waning overall presence, foreigners still play a decisive role in the commercial lives of cities, with the Japanese influence particularly prominent today. Third, the most recent episode of urban agglomeration has been heavily concentrated in the large coastal cities, reinforcing their old colonial-era dominance through renewal of their functions as collection-distribution nodes for interior hinterlands as well as leading ports for external trade and shipping. And fourth, they exhibit similar internal land-use patterns, a spatial structuring worth examining in some detail.

The general intraurban pattern of residential and non-residential activities is summarized in the **Southeast Asia city model** developed by Terence McGee in his book, *The Southeast Asian City* (Fig. 11-8). The old colonial port zone, its functions renewed in the postcolonial period, is the city's focus, together with the largely commercial district that surrounds it. Although no formal central business district (CBD) is evident, its elements are present as separate clusters within the land-use belt beyond the port: the government zone, the Western commercial zone (a colonialist remnant that is practically a CBD by itself), the alien commercial zone—usually dominated by Chinese merchants whose residences are attached to their places of business—and the mixed land-use zone that contains miscellaneous economic activities, including light industry. The other non-residential areas are the market-gardening zone at the urban periphery and, still farther from the city, an industrial estate (park) of recent vintage. The residential zones in McGee's model are quite reminiscent of the Griffin-Ford model of the "Latin" American city (Fig. 6-8, p. 316) which, as we saw, could be extended to the Third World city in general.

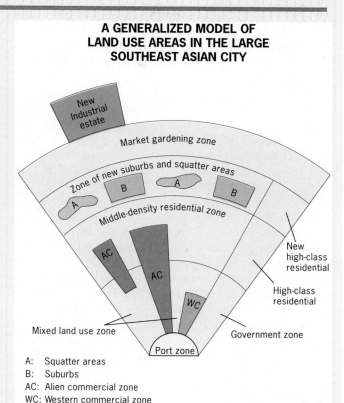

A GENERALIZED MODEL OF LAND USE AREAS IN THE LARGE SOUTHEAST ASIAN CITY

A: Squatter areas
B: Suburbs
AC: Alien commercial zone
WC: Western commercial zone

FIGURE 11-8

Among the similarities between the two are the hybrid sector/ring framework; an elite residential sector that includes new suburbanization; an inner-city zone of comfortable middle-income housing (with new suburban offshoots in the McGee schema); and peripheral concentrations of low-income squatter settlement. The differences are relatively minor and can partly be accounted for by local cultural and historical variations. If the Griffin-Ford model can be viewed as a generalization of the spatial organization of the Third World city, then the McGee model illustrates the departures that occur in coastal cities that were laid out as major colonial ports but continue their development within a now-independent country.

The core area of Malaysia lies on the western side of the peninsula (Fig. 11-1). Here are the capital, the best surface communications, the plantations, the industries, and most of the mines. The Malays of this area, moreover, have political decision-making power by virtue of their majority over the Chinese. But there is no strong alliance between the Malays of the mainland peninsula and the majority population in the territories of Sarawak and Sabah, which

PINANG: A FUTURE SINGAPORE?

Near the northern end of Malaysia on the Malay Peninsula's western coast lies the island of Pinang (Fig. 11-1), connected to the mainland by a long bridge. Like Singapore, located at the opposite end of that Malaysian coast, Pinang is home to a dominantly Chinese population. It was a thriving port in the colonial era and has become a major center for the high-technology manufacturing of components for the international computer industry. Ultramodern industrial parks dot Pinang's landscape, and among the names they emblazon are Intel, Advanced Micro Devices, SONY, Phillips, Motorola, and Hitachi.

Pinang's development is no fluke; it is part of a careful strategy to carve a niche for this island province in the emerging western Pacific Rim of Australasia (see Chapter 4). The emphasis in Pinang today is on building a highly skilled labor force that is locally nurtured in technical schools as well as training facilities financed by the multinational corporations themselves. This effort has become so successful that Pinang is now a leading partner in the formation of a regional-scale *growth triangle*—together with the nearby Indonesian island of Sumatera and southern Thailand to the north. In this context, Pinang supplies sophisticated computer manufacturing know-how while its less advanced neighbors provide a comparatively cheap, lower-skilled work force. This venture will nudge Pinang ever closer to economic tiger status, and some in this part of Southeast Asia have already taken to calling the island province a new Singapore.

constitute the offshore component of Malaysia on the island of Borneo. Thus the balance of power among Malaysia's culturally and geographically separated sectors is fragile.

Nonetheless, Malaysia has prospered. Effective economic leadership has yielded returns in the form of steady development, even in the face of risks: the decline of natural rubber and variations in the price of tin on world markets were offset in the 1980s by the expansion of palm oil production on plantations (which produce an inexpensive type of shortening that is demanded by the manufacturers of processed foods in developed countries). Sabah and Sarawak have been yielding increasing amounts of petroleum, and economic diversification has brought greater security.

Foreign manufacturers, attracted by the skills and wages of Malaysia's workforce, have located hundreds of factories there, notably in the electronics industry (see the box entitled "Pinang: A Future Singapore?"). Divided Malaysia has to date surmounted its geographic obstacles.

Archipelagic Indonesia

The very term *archipelago* denotes fragmentation, and Indonesia is the world's most expansive archipelagic state. Spread across some 13,000 islands, Indonesia's 190.9 million inhabitants live separated and clustered—separated by water and clustered on islands large and small. The map of Indonesia (Fig. 11-1) requires some attention. There are five large islands, of which one, easternmost New Guinea, is shared with the independent state of Papua New Guinea (to be discussed in the vignette on the Pacific realm that follows this chapter). The other four major islands are collectively known as the Greater Sunda Islands: Jawa (Java), the most populous and important; Sumatera (Sumatra) in the west; Kalimantan (the Indonesian portion of Borneo) in the center; and wishbone-shaped Sulawesi (Celebes) to the east. Extending eastward from Jawa are the Lesser Sunda Islands, including Bali and, near the eastern end, Timor. Another important island chain within Indonesia is constituted by the Moluccan Islands, which lie between Sulawesi and New Guinea. The central water body of Indonesia is the Java Sea.

Dutch Colonialism When the Dutch arrived to colonize this archipelago, it was by no means a unified country. There had already been a Hindu invasion, and Buddhism had also made inroads among the local peoples. But in the sixteenth century, Islam arrived and became entrenched as the dominant religion, spreading rapidly before the seventeenth-century arrival of the Dutch colonizers (Indonesia today remains the world's largest Muslim country). Not only was there religious division, but there also was no political unity. Thus the Dutch could take advantage of rivalries among local rulers—and did so with the help of Chinese, whom they brought to the "Netherlands East Indies" to assist in the colonial administration.

The Dutch exploited their East Indies possessions ruthlessly, thereby sowing the seeds of Indonesian nationalism. Throughout the first half of this century, the Indonesian drive for self-determination intensified; following World War II, independence was proclaimed in 1945. The Dutch then fought a losing battle to regain their East Indies, finally yielding in 1949. The easternmost territory of Irian Jaya (West Irian) on New Guinea was awarded to Indonesia in 1969; and formerly Portuguese Timor became Indonesia's twenty-seventh province in 1976, only after Indonesian armed forces had invaded the area.

BRUNEI

Brunei is an anomaly in Southeast Asia—an oil-exporting Islamic sultanate far from the Persian Gulf. Located on the north coast of Borneo, sandwiched between Malaysian Sarawak and Sabah (see Fig. 11-1), the Brunei sultanate is a former British-protected remnant of a much larger Islamic kingdom that once controlled all of Borneo and areas beyond. Brunei achieved full independence in 1984. With a mere 2,225 square miles (5,700 sq km)—slightly larger than Delaware—and only 310,000 people, Brunei is dwarfed by the other political entities of Southeast Asia. But the discovery of oil in 1929 (and natural gas in 1965) heralded a new age for this remote territory.

Today, Brunei is one of the largest oil producers in the British Commonwealth, and recent offshore discoveries suggest that production will increase. As a result, the population is growing rapidly by immigration (70 percent of Brunei's residents are Malay, 19 percent Chinese), and the sultanate enjoys one of the highest standards of living in Southeast Asia (gross national product per capita in 1989—the latest year for which data are available—was U.S. $13,290). Most of the people live near the oilfields in the western corner of the country and in the capital in the east—Bandar Seri Begawan. The evidence of a development boom can be seen in modern apartment houses, shopping centers, and hotels—a sharp contrast to many other towns on Borneo. There are some marked internal contrasts as well: Brunei's interior still remains an area of subsistence agriculture and rural isolation, virtually untouched by the modernization of the burgeoning coastal zone.

The Dutch had chosen Jawa as their colonial headquarters and established their capital city (Batavia, now renamed Jakarta [11.2 million]) there. Today, Jawa is still the undisputed core area of Indonesia; with about 110 million inhabitants, Jawa is also one of the world's most densely peopled places (at 2,150 people per square mile, it has a higher population density than Bangladesh's) and one of the most agriculturally productive. In no other way—population, urbanization, communications, productivity—can any of the other islands compare. Sumatera, much larger in size, has just over 40 million inhabitants; this island also ranks second in economic terms, with large rubber plantations in its eastern coastal lowland. Kalimantan (the Indonesian name for their portion of the island of Borneo) is huge but sparsely populated by only about 10 million residents; Indonesia shares Borneo not only with Malaysia but also with the Islamic Sultanate of Brunei (see box above). And Sulawesi, east of Borneo, has an estimated 14

Jawa, smallest of Indonesia's four main islands, contains more people than all the rest of Indonesia combined. Land is fertile but at a premium, so every slope that can be cultivated is terraced. Volcanic soils here are rich and fertile, and rice production (double cropping is common) is very large.

million inhabitants, barely more than one-tenth of Jawa. Thus Jawa dominates—a situation that in a fragmented domain can produce problems. Distance and water inhibit circulation and contact, and Indonesia has coped with secessionist uprisings on several of its islands. Old differences and disagreements tend to reemerge after the common enemy has been defeated.

Unity in Diversity Viewed from this perspective, the persistence of Indonesia as a unified state is another politico-geographical wonder on a par with post-colonial India. Wide waters and high mountains have helped perpetuate cultural distinctions and differences; centrifugal political forces have been powerful and, on more than one occasion, have nearly pulled the country apart. But Indonesia's national integration appears to have strengthened, as the country's motto—*Unity in Diversity*—underscores. With more than 300 discrete ethnic clusters, over 250 individual languages, and just about every religion practiced on earth, Indonesian nationalism has faced enormous odds that are overcome, to some extent, by development based on the archipelago's considerable resource base. This includes sizeable petroleum reserves, large rubber and palm oil plantations (Sumatera shares the nearby Malay Peninsula's environments), extensive lumber resources, major tin deposits offshore from eastern Sumatera, and soils that produce tea, coffee, and other cash crops. However, Indonesia's large population continues to grow at a yearly rate of almost 2 percent (which means a doubling time of only 40 years), a longer-term threat to the country's future; significantly, rice and wheat have already become prominent among annual imports.

What Indonesia has achieved is etched against the country's continuing cultural complexity. There are dozens of distinct aboriginal cultures; virtually every coastal community has its own roots and traditions. And the majority, the rice-growing Indonesians, include not only the numerous Jawanese—who are Muslims largely in name only and have their own cultural identity—but also the Sudanese (who constitute 15 percent of Indonesia's population), the Madurese (5 percent), and others. Perhaps the best impression of the cultural mosaic comes from the string of islands that extends eastward from Jawa to Timor (Fig. 11-1). The rice-growers of Bali are mainly Hindus; the population of Lombok is mainly Muslim, with some Balinese Hindu immigrants. Sumbawa is a Muslim community, but the next island, Flores, is mostly Roman Catholic. On Timor, Protestant groups predominate, and this island remains marked by its long-time division into a Dutch-controlled and a Portuguese-owned sector (Fig. 11-6).

An independence movement on former Portuguese Timor was subdued by invading Indonesian forces, a campaign that was followed by severe dislocation and famine. Indonesia's relationship with its easternmost island sector of New Guinea is similarly unstable. The people of New Guinea are Papuans, not Indonesians (Fig. 11-5). They traded one foreign master (the Dutch) for another, and there has been resistance to Indonesia's administration. With the Papuan state of Papua New Guinea across the border on the eastern half of the island of New Guinea (see map p. 586), Indonesia—now itself branded as a colonial power—faces yet another challenge.

Philippine Fragmentation

After Indonesia and Vietnam, the Philippines, with 66.8 million people, is Southeast Asia's next most populous state (Fig. 11-9). However, few of the generalizations that can be made about the realm would apply without qualification to this island-chain country, and the Philippines' location relative to the mainstreams of change in this part of the world has had much to do with this. The islands, inhabited by peoples of Malay ancestry with Indonesian strains, shared with much of the rest of Southeast Asia an early period of Hindu cultural influence, strongest in the south and southwest and diminishing northward. Next came a Chinese invasion, felt more strongly on the largest island of Luzon in the northern part of the Philippine archipelago. Islam's arrival was delayed somewhat by the position of the Philippines well to the east of the mainland and to the north of the Indonesian islands. The southern Muslim beachheads, however, were soon overwhelmed by the Spanish invasion of the sixteenth century; today the Philippines, adjacent to the world's largest Muslim state (Indonesia), is 84 percent Roman Catholic, 10 percent Protestant, and only 5 percent Muslim.

Out of the Philippine melting pot, where Mongoloid-Malay, Arab, Chinese, Japanese, Spanish, and American elements have met and mixed, has emerged the distinctive culture of the Filipino. It is not a homogeneous or a unified culture, but in Southeast Asia it is in many ways unique. One example of its absorptive qualities is demonstrated by the way the Chinese infusion has been accommodated: although the "pure" Chinese minority numbers less than 2 percent of the population (far lower than in most Southeast Asian countries), a much larger portion of the Philippine population carries a decidedly Chinese ethnic imprint. What has happened is that the Chinese have intermarried, producing a sort of Chinese-mestizo element that constitutes more than 10 percent of the total population. In another cultural sphere, the country's ethnic mixture and variety are paralleled by its great linguistic diversity. Nearly 90 Malay languages, major and minor, are spoken by the 67 million people of the Philippines; only about 1 percent still use Spanish. Visayan is the language most commonly spoken, and more than 50 percent of the population is able to use English. At independence in 1946, the largest of the Malay languages, Tagalog or Pilipino, was adopted as the country's official language, and its general use is strongly promoted

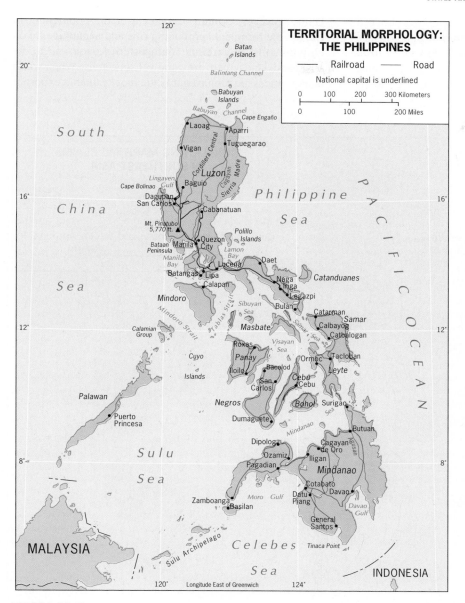

TERRITORIAL MORPHOLOGY: THE PHILIPPINES

—— Railroad   —— Road

National capital is underlined

0   100   200   300 Kilometers

0   100   200 Miles

FIGURE 11-9

through the educational system. English is learned as a subsidiary language and remains the chief *lingua franca*; an English-Tagalog hybrid ("Taglish") is increasingly heard today, remarkably cutting across all levels of society.

The widespread use of English in the Philippines, of course, results from a half-century of American rule and influence, beginning in 1898 when the islands were ceded to the United States by Spain under the terms of the treaty that followed the Spanish-American War. The United States took over a country in open revolt against its former colonial master and proceeded to destroy the Filipino independence struggle, now directed against the new foreign rulers. It is a measure of the subsequent success of U.S. adminis-

tration in the Philippines that this was the only dependency in Southeast Asia that during World War II sided against the Japanese in favor of the colonial power. United States rule had its good and bad features, but the Americans did initiate reforms that were long overdue, and they were already in the process of negotiating a future independence for the Philippines when the war intervened in 1941.

The Philippines' population, concentrated where the good farmlands lie in the plains, is densest in three general areas (Fig. I-9): (1) the northwestern and south-central part of Luzon (metropolitan Manila [10.7 million], with nearly one-sixth of the national population, lies at the southern end of this zone); (2) the southeastern proruption of Luzon;

and (3) the islands surrounding the Visayan Sea between Luzon and Mindanao. The Philippine archipelago consists of over 7,000 mostly mountainous islands, of which Luzon and Mindanao are the two largest (accounting for almost two-thirds of the total area). About a dozen of these is-

lands, however, contain 95 percent of the population. In Luzon, the farmlands producing rice and sugarcane lie on alluvial soils; in extreme southeastern Luzon and in the Visayan Islands, there are good volcanic soils. When world market prices are high, sugar is the most valuable export of

FIGURE 11-10

the agriculture-dominated Philippines; timber, copra, and coconut oil are also major exports, but most Filipino farmers are busy raising the subsistence crops, rice and corn. As in the other Southeast Asian countries, there is a considerable range of supporting food crops.

Unlike the other countries in this realm, the Philippine state now faces a severe overpopulation problem. Its annual rate of natural increase currently stands at 2.4 percent, which yields a doubling time of only 28 years. Undoubtedly, a contributing factor in recent population growth is the influence of the Roman Catholic church, one of the developing world's most conservative on the issue of birth control and family planning. A disturbing parallel trend today is the proliferation of poverty and malnutrition in many parts of the Philippines, which portends disaster if current demographic forecasts become reality.

◆ SOUTHEAST ASIA'S SEAS

Southeast Asia is a realm of coastlines, peninsulas, and islands. For countless centuries, traders, migrants, and travelers have plied these waters.

The twentieth century has witnessed a process that political geographers call the "scramble for the oceans." For many hundreds of years, coastal states claimed as their own a narrow stretch of coastal water, that they called their **territorial sea**. Then, early in this century, those coastal claims began to widen—from 3 nautical miles to 6 and even 12. After World War II, states demanded rights over even more expansive zones and not only claimed the water surface but also what lay beneath. Repeated United Nations Law of the Sea (UNCLOS) conferences were convened to stem the tide, but with only limited success. In the 1980s, states were given the right to claim a 12-mile territorial sea and, beyond this, an **exclusive economic zone (EEZ)** extending up to 200 nautical miles from the coast. The resources in and under this EEZ (such as fish, oil, and manganese nodules) belong to the coastal state.

Consider the impact of this in Southeast Asia, where many seas and other waters are much less than 400 nautical miles (the EEZ's of two adjacent states) wide, and some are less than 24 miles (two territorial seas) in width. As in the case of Europe (see map, p. 71), **maritime boundaries** had to be drawn in order to allocate the realm's waters (Fig. 11-10).

But boundaries can create conflicts, and they certainly have done so in Southeast Asia. When seas are narrower than 400 nautical miles, the *median line* concept is used to divide them. But defining and delimiting that line can be problematic. In Southeast Asia, the South China Sea is becoming an arena of conflict. Note the Spratly Islands in

Fig. 11-10, which are claimed (and in some cases occupied) by China, Malaysia, the Philippines, Vietnam, and Taiwan. Farther north, the Paracel Islands now under control of China are claimed on historic grounds by Vietnam.

Because the ownership of these islands is not settled, EEZ boundaries in the South China Sea cannot be defined or delimited. This, however, has not prevented China from claiming virtually all of the South China Sea—nor from "granting" an oil exploration concession to a foreign company in waters that may well belong, at least in part, to Vietnam. As in the case of Europe's North Sea, the greatest potential for oil and natural gas production lies right in the area where maritime claims meet.

So the maritime map of Southeast Asia is almost as ambiguous as the territorial one; the Paracels and Spratlys are only two of many actual and potential sites of maritime disputes in this embattled realm. The oceans are the last frontier of international competition and expansion, and Southeast Asia's maritime map will be redrawn many times before consensus is reached.

THE REALM IN TRANSITION

1. Indonesia, the realm's giant, functions as a colonial power on the huge island of New Guinea. In so doing, it extends its influence from one geographic realm into another. This will pose a challenge in the future.

2. Pacific Rim developments are occurring in most Southeast Asian countries. Certain places are seen to have great potential: Pinang, Malaysia is sometimes referred to as a "Singapore of the Future."

3. Environmental damage in Southeast Asia is increasing at an alarming rate. Forest loss is accelerating. Water supplies are polluted and in some areas already inadequate. Rapid economic growth is worsening the situation.

4. The end of Vietnam's isolation is opening an additional sector of the growing economic complex on the western Pacific Rim.

5. Potential energy resources in the realm's maritime areas are incompletely known. Critical marine boundary issues are in urgent need of negotiation and settlement as the prospects for conflict are rising.

◆ PRONUNCIATION GUIDE

Acapulco   (ah-kah-PULL-koh)

Andaman   (ANN-duh-mun)

Angkor [Wat]   (ANG-kor [WOT])

Annam   (uh-NAHM)

Annam[ese]/[ite]   (anna-[MEEZE]/[MITE])

Arakan   (ah-ruh-KAHN)

Archipelago   (ark-uh-PELL-uh-goh)

Bali[nese]   (BAH-lee [NEEZE])

Banda   (BAHN-duh)

Bandar Seri Begawan   (BUN-dahr SERRY buh-GAH-wun)

Bataks   (buh-TAHKS)

Batavia   (buh-TAY-vee-uh)

Borneo   (BOAR-nee-oh)

Bosnia-Herzegovina   (BOZ-nee-uh-hert-suh-go-VEE-nuh)

Brunei   (broo-NYE)

Burmans   (BURR-munz)

Cambodia   (kam-BOH-dee-uh)

Celebes   (SELL-uh-beeze)

Chao Phraya   (CHOW pruh-yah)

Chiang Mai   (chee-AHNG mye)

Cholon   (choh-LONN)

Cochin   (KOH-chin)

Copra   (KOH-pruh)

Croatia   (kroh-AY-shuh)

Dien Bien Phu   (d'yen b'yen FOOH)

Filipinos   (filla-PEA-noze)

Flores   (FLAW-rihss)

Haiphong   (hye-FONG)

Hanoi   (han-NOY)

Hartshorne   (HARTSS-horn)

Hegemony   (heh-JEH-muh-nee)

Ho Chi Minh   [see Minh, Ho Chi]

Hué   (HWAY)

Indonesia   (indo-NEE-zhuh)

Irian Jaya   (IH-ree-ahn JYE-uh)

Irrawaddy   (ih-ruh-WODDY)

Jakarta   (juh-KAHR-tuh)

Java   (JAH-vuh)

Jawa   (JAH-vuh)

Kachins   (kuh-CHINZ)

Kalimantan   (kalla-MAN-tan)

Kampuchea   (kahm-pooh-CHEE-uh)

Karens   (kuh-RENZ)

Kashmir   (KASH-meer)

Kelang   (kuh-LAHNG)

Khmer [Rouge]   (kuh-MAIR [ROOZH])

Khorat   (koh-RAHT)

Kompong Thom   (kahm-pong TOM)

Kosovo   (KAW-suh-voh)

Kra Isthmus   (KRAH ISS-muss)

Kuala Lumpur   (KWAHL-uh LOOM-poor)*

Kuwait   (koo-WAIT)

Lao   (LAU)

Laos   (LAUSS)

Laotian   (lay-OH-shun)

Lesotho   (leh-SOO-too)

Lingua franca   (LEEN-gwuh FRUNK-uh)

Lombok   (LAHM-bahk)

Luzon   (loo-ZAHN)

Macedonia   (massa-DOH-nee-uh)

Madurese   (muh-dooh-REECE)

Malacca   (muh-LAH-kuh)

Malay   (muh-LAY)

Malaya   (muh-LAY-uh)

Malaysia   (muh-LAY-zhuh)

Mandalay   (man-duh-LAY)

Mekong   (MAY-kong)

Mestizo   (meh-STEE-zoh)

Mindanao   (min-duh-NOW)

Minh, Ho Chi   (MINN, hoh chee)

Moluccans   (muh-LUCK-unz)

Montagnards   (MON-tun-yardz)

Myanmar   (mee-ahn-MAH)

New Guinea   (noo-GHINNY)

Nicaragua   (nick-uh-RAH-gwuh)

Papua[ns]   (pahp-OO-uh[-unz])

Paracel   (para-SELL)

Philippines   (FILL-uh-peenz)

Phnom Penh   (puh-NOM PEN)

Pilipino   (pill-uh-PEA-noh)

Pinang   (puh-NANG)

Prorupt   (pro-RUPPT)

Qinghai   (ching-HYE)

Ratzel, Friedrich   (RAHT-sull, FREED-rish)

Sabah   (SAHB-ah)

Saigon   (sye-GAHN)

Salween   (SAL-ween)

Sarawak   (suh-RAH-wahk)

Shan   (SHAHN)

Siam   (sye-AMM)

Singapore   (SING-uh-poar)

Sinicized   (SYE-nuh-sized)

Spratlys   (SPRAT-leeze)

Sri Lanka   (sree-LAHNG-kuh)

Sulawesi   (soo-luh-WAY-see)

Sultanate   (SULL-tuh-nut)

Sumatera   (suh-MAH-tuh-ruh)

Sumatra   (suh-MAH-truh)

Sumbawa   (soom-BAH-wuh)

Sunda[nese]   (SOON-duh [NEEZE])

Suryavarman   (soory-AHVA-mun)

Tagalog   (tuh-GAH-log)

Thailand   (TYE-land)

Timor   (TEE-more)

Tonkin   (TAHN-kin)

Tonle Sap   (tahn-lay SAP)

Vientiane   (vyen-TYAHN)

Vietnam   (vee-et-NAHM)

Visayan   (vuh-SYE-un)

Wallonia   (wah-LOANY-uh)

Xizang   (sheedz-AHNG)

Yangon   (yahn-KOH)

Yunnan   (yoon-NAHN)

*Each "u" in Lumpur pronounced as in "look."

THE PACIFIC REALM

Between the Americas to the east and Austrasia to the west lies the vast Pacific Ocean, larger than all the world's land areas combined. In this great ocean lie tens of thousands of islands, some large (New Guinea is by far the biggest), most small (many are uninhabited). Despite the preponderance of water, this fragmented, culturally complex realm does possess regional identities. It includes the Hawaiian Islands, Tahiti, Fiji, Tonga, Samoa—fabled names in a world apart.

Indonesia and the Philippines are not part of the Pacific realm; neither are Australia and New Zealand. Before the European invasion, Australia and New Zealand would have been included—Australia as a discrete Pacific region on the basis of its indigenous black population, and New Zealand because its Maori population has Polynesian affinities. But black Australians and Maori New Zealanders have been engulfed by the Europeanization of their countries, and the regional geography of Australia and New Zealand today is decidedly not Pacific. Only on the island of New Guinea do Pacific peoples remain the dominant cultural element. Although the realm's contents lie frag-

IDEAS & CONCEPTS

High-island culture
Low-island culture
Maritime environment

mented, scattered, and remote, the Pacific World does possess three distinct regions: *Melanesia*, *Micronesia*, and *Polynesia*.

◆ MELANESIA

New Guinea lies at the western end of a Pacific region that extends eastward to Fiji and includes the Solomon Islands, Vanuatu, and New Caledonia (Fig. P-1). These islands are inhabited by Melanesian peoples who have very dark skins

New Guinea, by far the largest island in the Pacific realm, is divided politically between Indonesia, which controls the western half, and the independent state of Papua New Guinea to the east. This village on the Sepik River well represents the local cultural landscape.

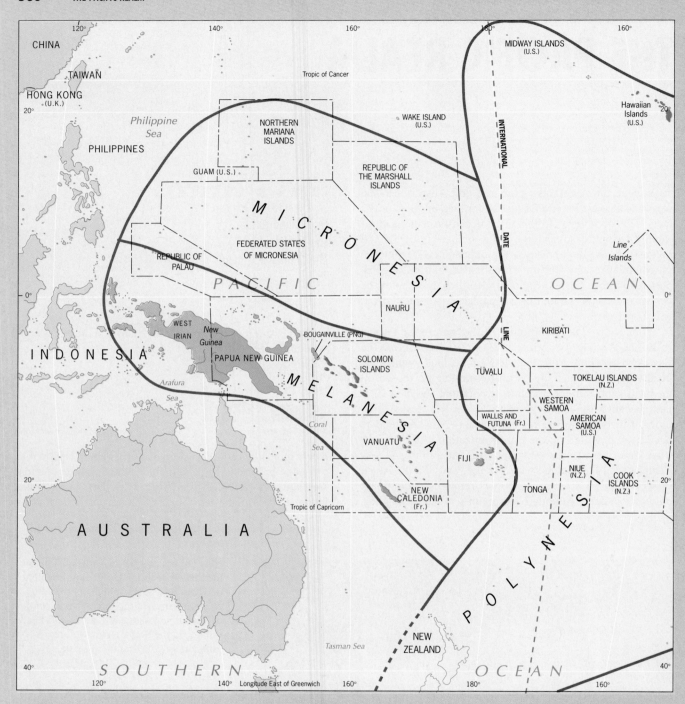

FIGURE P-1

and dark hair (*melas* means black); the region as a whole is called Melanesia. Some cultural geographers include the Papuan peoples of New Guinea in the Melanesian race, but others suggest that the Papuans are more closely related to the aboriginal (indigenous) Australians. In any case, Melanesia is by far the most populous Pacific region (roughly 5.5 million in 1994). New Guinea alone has a population of almost 6 million (although statistics are unreliable), but this island is divided into two halves by a superimposed geometric boundary (see lower-left map p. 559) that separates non-Melanesian West Irian, now an Indonesian province, from independent Papua New Guinea (P.N.G.).

PACIFIC REGIONS

| 0 | 400 | 800 | 1200 Kilometers |
| 0 | | 400 | 800 Miles |

With 4.0 million inhabitants today, P.N.G. became a sovereign state in 1975 after nearly a century of British and Australian administration. It is one of the world's poorest and least developed countries, with much of the mountainous interior—where the Papuan population is clustered in tiny villages—hardly touched by the changes that trans- formed neighboring Australia. The largest town and capital, Port Moresby, has more than 250,000 residents; only about one-eighth of the people of P.N.G. live in urban areas, but the current rate of urbanization is phenomenal. Although English is used by the educated minority, about 57 percent of the population remains illiterate, and over 700 languages

TEN MAJOR GEOGRAPHIC QUALITIES OF THE PACIFIC REALM

1. The Pacific realm's total area is the largest of all geographic realms. Its land area, however, is among the smallest.

2. The bulk of the land area of the Pacific realm lies on the island of New Guinea.

3. Papua New Guinea, with a population of 4 million, alone contains over three-fifths of the Pacific realm's population.

4. The Pacific realm consists of three regions: Melanesia (including New Guinea), Micronesia, and Polynesia.

5. The Pacific realm is one of the most fragmented of all world realms.

6. The Pacific realm's islands and cultures may be divided into volcanic *high-island* cultures and coral *low-island* cultures.

7. The Hawaiian Islands, the fiftieth State of the United States, lie in the northern sector of Polynesia. As in New Zealand, Hawaiian indigenous culture has been submerged by Westernization.

8. In Polynesia, local culture is nearly everywhere severely strained by external influences.

9. Indigenous Polynesian culture exhibits a remarkable consistency and uniformity throughout the Polynesian region, its enormous dimensions and dispersal notwithstanding.

10. The Pacific realm is in politico-geographical transition as islands attain independence or redirect their political associations.

are spoken by the Papuan and Melanesian communities. The Melanesians are concentrated in the northern and eastern coastal areas of the country, and here, as in the other islands of this region, they grow root crops and bananas for subsistence. However, recent discoveries of major mineral deposits (copper, gold, and oil) point to the country's bright development potential; there is also enormous scope for increasing the output of such profitable export crops as palm oil, coffee, and cocoa.

As it attempts to move forward, however, Papua New Guinea is beset by centrifugal forces that have now given rise to an active insurgency on Bougainville, the country's easternmost island (Fig. P-1). In 1992, attacking P.N.G. forces pursued Bougainville-based rebels onto a nearby island belonging to the neighboring Solomon Islands, heightening tensions between the two countries. Although P.N.G. and the Solomons subsequently signed an agreement to halt further violence, the latter's open support of the independence-seeking rebels in their complaints to the U.N. Commission on Human Rights has infuriated the Port Moresby government.

The enduring Bougainville problem is a legacy of the colonial era. The island's last master was Australia, which, rather than redraw the old colonial boundary, insisted that Bougainville's unenthusiastic inhabitants join the newly created state of Papua New Guinea (rather than the Solomons) in 1975; the Australian decision may well have been influenced by the fact that Bougainville contained one of the world's largest copper mines. In the mid-1990s, insurgents and P.N.G. soldiers continue fighting on Bougainville—while many island-nations in this part of the South Pacific regard the conflict as the gravest threat to regional stability since World War II ended half a century ago.

◆ MICRONESIA

North of Melanesia and east of the Philippines lie the islands that constitute the region known as Micronesia (Fig. P-1). In this case, the name (*micro* means small) refers to the size of the islands, not the physical appearance of the population. The 2,000-plus islands of Micronesia are not only tiny (many of them no larger than 1 square mile), but they are also much lower-lying, on an average, than those of Melanesia. There are some volcanic islands (*high islands*, as the people call them), but they are outnumbered by islands composed of coral, the *low islands* that barely lie above sea level. Guam, with 210 square miles (550 sq km), is Micronesia's largest island, and no island elevation anywhere in Micronesia reaches 3,300 feet (1,000 m).

Until 1986, the region was largely a U.S. Trust Territory (the last of the post-World War II trusteeships supervised by the United Nations), but that status has now changed. The former U.S.-administered territory is today divided into four island groups: the Northern Mariana Islands, the Republic of Palau, the Federated States of Micronesia, and the Republic of the Marshall Islands (Fig. P-1). The last two political units have achieved limited sovereignty under the continued guidance and financial support of the United States. The Northern Marianas chose to become a U.S. commonwealth, along the lines of Puerto Rico. But Palau continues as the last of the trust territories, scheduled to become independent in "free association" with the United States when 75 percent of its citizens agree to the terms of a U.S. sovereignty compact; in referenda held thus far, Palauan voters have not yet given their approval.

The Micronesians are not nearly as numerous as the Melanesians (totaling only about 450,000), but they nevertheless constitute a distinct racial group in the Pacific realm.

◆ POLYNESIA

To the east of Micronesia and Melanesia lies the heart of the Pacific, enclosed by a great triangle stretching from the Hawaiian Islands to Chile's Easter Island to New Zealand. This is Polynesia (Fig. P-1), a region of numerous islands (*poly* means many), ranging from volcanic mountains rising above the Pacific's waters (Mauna Kea on Hawaii reaches nearly 13,800 feet [over 4,200 m]), clothed by luxuriant tropical forests and drenched by well over 100 inches of rainfall each year, to low coral atolls where a few palm trees form the only vegetation and where drought is a persistent problem. The Polynesians have somewhat lighter-colored skin and wavier hair than do the other peoples of the Pacific realm; they are often also described as having an excellent physique. Anthropologists differentiate between these original Polynesians and a second group, the Neo-Hawaiians, who are a blend of Polynesian, European, and Asian ancestries. In the U.S. State of Hawaii—actually an archipelago of more than 130 islands—Polynesian culture has not only been Europeanized but also Orientalized.

Its vastness and the diversity of its natural environments notwithstanding, Polynesia clearly constitutes a geographic region within the Pacific realm (population: 1.9 million). Polynesian culture, though spatially fragmented, exhibits a remarkable consistency and uniformity from one island to the next, from one end of this widely dispersed region to the other; this consistency is particularly expressed in vocabularies, technologies, housing, and art forms. The Polynesians are uniquely adapted to their **maritime environment**, and long before European sailing ships began to arrive in their waters, Polynesian seafarers had learned to navigate their wide expanses of ocean in huge double canoes as long as 150 feet (45 m). They traveled hundreds of miles to favorite fishing zones and engaged in inter-island barter trade, using maps constructed from bamboo sticks and cowrie shells and navigating by the stars. However, modern descriptions of a Pacific Polynesian paradise of emerald seas, lush landscapes, and gentle people distort harsh realities. Polynesian society was forced to accommodate much loss of life at sea when storms claimed their boats, families were ripped apart by accident as well as migration, hunger and starvation afflicted the inhabitants of smaller islands, and the island communities were often embroiled in violent conflicts and cruel retributions.

The political geography of Polynesia is complex. The Hawaiian Islands, in 1959, became the fiftieth State to join the United States. The State's population is now 1.2 million, with over 80 percent living on the island of Oahu. There, the superimposition of cultures is exquisitely illustrated by the panorama of Honolulu's skyscrapers against the famous extinct volcano at nearby Diamond Head (see photo p. 590). The Kingdom of Tonga became an independent country in 1970 after seven decades as a British protec-

Pacific islands are "low" or "high," the latter consisting of volcanic rocks protruding above the ocean surface. This high island, Bora Bora, is one of Polynesia's most scenic. The Polynesians have always been the great seafarers of the Pacific.

Culturally, it is useful to distinguish their communities as **high-island cultures** based on the better-watered volcanic islands where agriculture is the mainstay, or **low-island cultures** based on the sometimes drought-plagued coral islands where fishing is the chief mode of subsistence. These numerous Micronesian communities have developed a large number of locally spoken languages, many of them mutually unintelligible.

But these islands do not exist in isolation. There is a certain complementarity between the islands of farmers and the islands of fishing people, and Micronesians—especially the low-islanders—are skilled seafarers. The trade for food and basic needs encourages circulation; sometimes the threat of a devastating typhoon (hurricane) compels the low-islanders to seek the safety of higher ground elsewhere. Thus movement in the Pacific realm has always been by water. Even today many islanders are expert boaters, often choosing a water route when a road is also available.

Oahu, once the domain of King Kamehameha's subjects, became an American possession in 1893, then a part of a U.S. State less than 40 years ago. The Hawaiian Islands today are an American outpost, Polynesian culture and language submerged under the modernization reflected by this photo of Honolulu's Waikiki waterfront.

torate; the British-administered Ellice Islands were renamed Tuvalu, and along with the Gilbert Islands to the north (now renamed Kiribati), they received independence from Britain in 1978. Other islands continued under French control (including the Marquesas Islands and Tahiti), under New Zealand's administration (Rarotonga), and under British, U.S., and Chilean flags.

In the process of politico-geographical fragmentation, Polynesian culture has been dealt some severe blows. Land developers, hotel builders, and tourist dollars have set Tahiti on a course along which Hawaii has already traveled far. The Americanization of eastern Samoa has created a new society quite different from the old. Polynesia has lost much of its ancient cultural consistency; today, the region is a patchwork of new and old—the new often bleak and barren with the old under intensifying pressure.

In the mid-1990s, the diverse island states and dependencies of the Pacific realm are cooperating in the search for solutions to their many shared problems. A key development is the recent emergence of the South Pacific Forum,

whose 13 members include Australia, the Cook Islands, Fiji, Kiribati, Nauru, New Zealand, Niue, Papua New Guinea, the Solomon Islands, Tonga, Tuvalu, Vanuatu, and Western Samoa. The Forum's agencies have helped foster regional cooperation with respect to fisheries, tourism, and air and sea transportation. Today, this organization is moving actively toward discussion and confrontation of the realm's major political issues: increasing Western aid to accelerate economic development, expediting the termination of colonialism in French New Caledonia and the American trust territories (almost completed in Micronesia, as we saw), and the banning of nuclear weapons testing, particularly by France. The South Pacific Forum is headquartered in the modern, centrally located Fijian capital of Suva, and is proving to be an effective mechanism for bringing together Melanesians, Micronesians, Polynesians, and their transplanted European counterparts from Australia and New Zealand to define their mutual problems, discuss their common causes, and debate the future course of their unique oceanic realm.

◆ PRONUNCIATION GUIDE

Atoll   (AY-toal)
Bougainville   (BOO-gun-vil)
Ellice   (ELL-uss)
Fiji   (FEE-jee)
Fijian   (fuh-JEE-un)
Guam   (GWAHM)
Hawaii   (huh-WAH-ee)
Irian   (IH-ree-ahn)
Kiribati   (KIH-ruh-bahss)
Maori   (MAU-ree/MAH-aw-ree)
Mariana   (marry-ANNA)
Marquesas   (mahr-KAY-suzz)

Mauna Kea   (mau-nuh KAY-uh)
Melanesia   (mella-NEE-zhuh)
Micronesia   (mye-kroh-NEE-zhuh)
Nauru   (nah-OO-roo)
New Caledonia   (noo kalla-DOAN-yuh)
New Guinea   (noo GHINNY)
Niue   (nee-OOH-ay)
Oahu   (uh-WAH-hoo)
Palau   (puh-LAU)
Papua New Guinea   (pahp-OO-uh noo-GHINNY)

Papuan   (pahp-OO-un)
Polynesia   (polla-NEE-zhuh)
Port Moresby   (port MORZ-bee)
Rarotonga   (rarra-TAHNG-guh)
Samoa   (suh-MOH-uh)
Suva   (SOO-vuh)
Tahiti   (tuh-HEET-ee)
Tonga   (TAHNG-guh)
Tuvalu   (too-VAHL-oo)
Vanuatu   (vahn-uh-WAH-too)
Waikiki   (wye-kuh-KEE)

APPENDIX A

AREA AND DEMOGRAPHIC DATA FOR THE WORLD'S STATES

(Smallest Microstates Omitted)

	Area (1000 sq mi)	Population (millions) 1994	2000	2010	Annual Rate of Natural Incr. (Percent)	Doubling Time (years)	Life Expectancy at Birth (years)	Percent Urban Population	1994 Population Density (per sq mi)	1990 Gross Nat'l Product Per Capita (U.S. dollars)
World	**51,510.8**	**5,602.8**	**6,205.7**	**7,111.5**	**1.7**	**41**	**65**	**43**	**109**	**3,790**
Developed Realms	19,579.3	1,163.8	1,190.4	1,236.5	0.5	148	74	73	59	17,900
Developing Realms	31,931.5	4,439.0	5,015.4	5,875.1	2.0	34	63	34	139	810
Europe	**2,261.0**	**579.7**	**586.9**	**591.7**	**0.2**	**338**	**75**	**73**	**256**	**12,990**
Albania	11.1	3.4	3.8	3.9	1.9	36	73	36	307	—
Austria	32.4	7.9	8.0	8.2	0.1	495	76	55	244	19,240
Belarus	80.2	10.3	10.5	11.1	0.3	217	72	67	129	3,110
Belgium	11.8	10.1	10.2	9.7	0.2	347	76	95	854	15,440
Bosnia-Herzegovina	19.7	4.3	4.5	4.4	0.8	90	72	36	218	—
Bulgaria	42.8	8.9	8.8	8.8	0.0		72	68	207	2,210
Croatia	21.8	4.6	4.6	4.8	0.1	1386	72	51	209	—
Czech Republic	30.6	10.4	10.6	10.8	0.1	1386	72	79	340	—
Denmark	16.6	5.2	5.2	5.1	0.1	753	75	85	312	22,090
Estonia	17.4	1.6	1.6	1.7	0.2	365	70	71	91	3,830
Finland	130.1	5.1	5.2	5.0	0.3	224	75	62	39	26,070
France	211.2	57.3	58.8	58.8	0.4	169	77	73	272	19,480
Germany	137.8	80.5	80.2	78.2	-0.1		75	90	584	20,750
Greece	50.9	10.3	10.4	10.4	0.1	990	75	58	203	6,000
Hungary	35.9	10.3	10.2	10.5	-0.2		70	63	287	2,780
Iceland	39.8	0.3	0.3	0.3	1.2	58	78	90	7	21,150
Ireland	27.1	3.6	3.7	3.4	0.6	122	74	56	132	9,550
Italy	116.3	58.1	58.3	56.4	0.1	1386	76	72	499	16,850
Latvia	24.9	2.7	2.7	2.9	0.1	630	70	71	109	3,410
Liechtenstein	0.06	0.1	0.1	0.1	0.6	110	70	—	506	—
Lithuania	25.2	3.8	3.9	4.1	0.4	2.0	72	69	150	2,710
Luxembourg	1.0	0.4	0.4	0.4	0.3	239	74	78	390	28,770
Macedonia	9.9	2.0	2.1	2.2	1.0	70	72	54	201	—
Malta	0.3	0.4	0.4	0.4	0.8	92	76	85	1,221	6,630
Moldova	13.0	4.4	4.7	5.2	0.8	88	69	48	342	2,170
Netherlands	15.9	15.3	15.8	16.6	0.5	147	77	89	965	17,330
Norway	125.2	4.3	4.4	4.5	0.4	193	77	71	34	23,120
Poland	120.7	38.7	39.5	41.3	0.4	187	71	61	320	1,700
Portugal	34.3	10.5	10.6	10.8	0.1	533	74	30	306	4,890
Romania	91.7	23.2	23.4	24.0	0.1	578	70	54	253	1,640

	Area (1000 sq mi)	Population (millions)			Annual Rate of Natural Incr. (Percent)	Doubling Time (years)	Life Expectancy at Birth (years)	Percent Urban Population	1994 Population Density (per sq mi)	1990 Gross Nat'l Product Per Capita (U.S. dollars)
		1994	2000	2010						
Serbia-										
Montenegro	26.9	10.1	10.4	10.8	0.5	141	72	47	375	—
Slovakia	18.8	5.4	5.6	5.9	0.5	154	71	67	287	—
Slovenia	7.8	1.9	2.0	2.1	0.3	267	73	49	248	—
Spain	194.9	38.7	39.1	40.1	0.2	433	77	91	198	10,920
Sweden	173.7	8.7	8.9	8.9	0.3	210	78	83	50	23,860
Switzerland	15.9	6.9	7.0	6.9	0.3	231	77	60	435	32,790
Ukraine	233.1	52.2	52.4	53.3	0.1	1155	71	68	224	2,340
United Kingdom	94.2	58.1	59.0	59.9	0.3	257	76	90	617	16,070
Russia	**6,592.8**	**150.0**	**152.1**	**162.3**	**0.2**	**301**	**69**	**74**	**23**	**1,820**
Armenia	11.5	3.6	4.0	4.5	1.8	40	72	68	315	2,150
Azerbaijan	33.4	7.4	8.3	9.5	2.0	36	70	53	222	1,670
Georgia	26.9	5.6	5.9	6.1	0.9	80	72	56	207	1,640
North America	**7,509.9**	**287.4**	**301.1**	**327.6**	**0.8**	**89**	**75**	**75**	**38**	**21,580**
Canada	3,831.0	27.8	29.1	32.1	0.8	89	77	78	7	20,450
United States	3,678.9	259.6	272.0	295.5	0.8	89	75	75	71	21,700
Japan	**145.7**	**125.2**	**127.6**	**129.4**	**0.3**	**217**	**79**	**77**	**859**	**25,430**
Australia- New Zealand	**3,069.9**	**21.6**	**22.7**	**25.3**	**0.9**	**81**	**76**	**85**	**7**	**16,366**
Australia	2,966.2	18.1	19.0	21.5	0.8	83	76	85	6	17,080
New Zealand	103.7	3.5	3.7	3.8	1.0	71	75	84	34	12,680
Middle America	**1,055.0**	**159.7**	**183.4**	**208.6**	**2.3**	**32**	**68**	**63**	**151**	**2,143**
Bahamas	5.4	0.3	0.3	0.3	1.5	47	73	75	50	11,510
Barbados	0.2	0.3	0.3	0.3	0.7	102	73	32	1,308	6,540
Belize	8.9	0.2	0.3	0.3	3.1	22	70	52	27	1,970
Costa Rica	19.6	3.3	3.8	4.5	2.4	29	77	45	170	1,910
Cuba	44.2	11.1	11.9	12.3	1.1	62	76	73	251	—
Dominican Republic	18.8	7.8	9.0	9.9	2.0	30	68	58	416	820
El Salvador	8.7	5.9	7.0	7.8	2.9	24	65	48	678	1,100
Guadeloupe	0.7	0.4	0.4	0.4	1.4	50	75	48	571	—
Guatemala	42.0	10.3	12.4	15.8	3.1	22	63	39	246	900
Haiti	10.7	6.8	8.1	9.4	2.9	24	55	29	636	370
Honduras	43.3	5.8	7.0	8.7	3.2	22	64	44	134	590
Jamaica	4.2	2.6	2.9	3.1	2.0	35	73	51	621	1,510
Martinique	0.4	0.4	0.4	0.4	1.2	59	78	82	947	—
Mexico	761.6	91.8	105.4	119.5	2.3	30	69	71	121	2,490
Netherlands Antilles	0.4	0.2	0.2	0.2	1.3	55	74	53	492	—
Nicaragua	50.2	4.4	5.2	6.4	3.1	23	62	57	87	340
Panama	29.8	2.5	2.8	3.2	1.9	37	73	53	85	1,830
Puerto Rico	3.4	3.8	4.1	3.9	1.2	59	74	72	1,120	6,470
Saint Lucia	0.2	0.2	0.2	0.2	1.7	40	72	46	807	1,900
St. Vincent and the Grenadines	0.2	0.1	0.1	0.1	1.6	43	72	21	594	1,610
Trinidad and Tobago	2.0	1.3	1.4	1.5	1.4	50	70	64	649	3,470
Virgin Is.	0.1	0.1	0.1	0.1	1.7	41	76	39	1,024	—

	Area (1000 sq mi)	Population (millions)			Annual Rate of Natural Incr. (Percent)	Doubling Time (years)	Life Expectancy at Birth (years)	Percent Urban Population	1994 Population Density (per sq mi)	1990 Gross Nat'l Product Per Capita (U.S. dollars)
		1994	2000	2010						
South America	**6,875.0**	**311.5**	**349.2**	**399.4**	**1.9**	**36**	**67**	**74**	**45**	**2,180**
Argentina	1,068.3	33.9	36.5	40.2	1.2	56	70	86	32	2,370
Bolivia	424.2	8.2	9.6	11.3	2.7	26	61	51	19	620
Brazil	3,286.5	156.5	174.9	200.2	1.9	37	65	74	48	2,680
Chile	292.3	14.1	15.6	17.2	1.8	39	74	85	48	1,940
Colombia	439.7	35.6	40.1	45.6	2.0	35	71	68	81	1,240
Ecuador	109.5	10.5	12.1	14.5	2.4	29	67	55	96	960
French Guiana	35.1	0.1	0.1	0.2	2.2	31	74	81	3	—
Guyana	83.0	0.8	0.9	1.0	1.8	39	64	35	10	370
Paraguay	157.1	4.8	5.6	6.9	2.7	25	67	43	30	1,110
Peru	496.2	23.5	26.7	31.0	2.2	32	61	70	47	1,160
Suriname	63.0	0.5	0.5	0.6	2.0	34	70	48	7	3,050
Uruguay	68.0	3.2	3.3	3.5	0.8	83	72	89	47	2,860
Venezuela	352.1	19.8	23.1	27.3	2.5	27	70	84	56	2,610
North Africa/Southwest Asia	**7,729.2**	**448.1**	**528.1**	**672.1**	**2.7**	**27**	**63**	**49**	**58**	**—**
Afghanistan	250.0	17.7	20.7	34.5	2.6	27	42	18	71	—
Algeria	919.6	27.5	32.5	37.9	2.8	25	66	50	30	2,060
Bahrain	0.3	0.6	0.6	0.8	2.4	29	72	81	1,854	6,910
Cyprus	3.6	0.7	0.8	0.8	1.1	66	76	62	203	8,040
Djibouti	8.9	0.5	0.5	0.7	2.9	24	48	79	52	—
Egypt	386.7	58.4	67.5	81.3	2.4	28	60	45	151	600
Eritrea*	45.4	2.8	3.3	4.3	2.9	24	48	16	62	—
Iran	636.3	63.7	77.6	105.0	3.3	21	64	54	100	2,450
Iraq	167.9	19.6	24.3	34.1	3.7	19	67	73	117	—
Israel	8.0	5.4	5.9	6.9	1.5	45	76	91	674	10,970
Gaza	0.1	0.8	1.0	1.3	4.6	15	—	—	7,587	—
West Bank	2.3	1.7	2.1	2.4	3.6	19	—	—	746	—
Jordan	35.5	3.8	4.6	6.4	3.4	20	71	70	107	1,240
Kazakhstan	1,049.2	17.4	18.9	21.9	1.4	50	69	58	17	2,470
Kuwait	6.9	1.5	1.7	3.2	3.0	23	74	—	212	—
Kyrgyzstan	76.6	4.7	5.4	6.6	2.2	31	68	38	61	1,550
Lebanon	4.0	3.6	4.1	4.9	2.1	33	68	84	896	—
Libya	679.4	4.8	5.7	7.1	3.0	23	68	83	7	5,410
Morocco	275.1	27.7	32.0	36.0	2.4	29	64	46	101	950
Oman	82.0	1.7	2.1	3.0	3.5	20	66	11	21	5,650
Qatar	4.3	0.5	0.6	0.7	2.5	28	71	90	118	15,860
Saudi Arabia	830.0	17.2	21.1	31.1	3.5	20	65	77	21	7,070
Somalia	246.2	8.8	10.5	13.9	2.9	24	46	24	36	150
Sudan	976.5	28.2	33.9	42.2	3.1	22	53	20	29	400
Syria	71.5	14.8	18.5	25.6	3.8	18	65	50	207	990
Tajikistan	55.3	5.9	7.1	9.1	3.2	22	69	31	106	1,050
Tunisia	63.2	8.8	9.9	11.3	2.1	33	66	53	139	1,420
Turkey	301.4	61.9	70.5	81.2	2.2	32	66	59	205	1,630
Turkmenistan	188.5	4.1	4.8	5.5	2.7	26	65	45	22	1,700
United Arab Emirates	32.3	2.7	3.1	4.9	2.8	25	71	78	82	19,860
Uzbekistan	172.7	22.5	26.4	32.8	2.7	25	69	40	130	1,350
Yemen	203.9	11.1	13.6	19.0	3.5	20	49	25	55	540
Subsaharan Africa	**8,158.0**	**528.8**	**634.2**	**854.2**	**3.1**	**23**	**52**	**26**	**65**	**529**
Angola	481.4	9.4	11.1	14.9	2.8	25	44	26	20	620
Benin	43.5	5.3	6.4	8.9	3.1	23	47	39	122	360
Botswana	231.8	1.4	1.7	2.4	3.1	23	59	24	6	2,040
Burkina Faso	105.9	10.2	12.4	17.0	3.3	21	52	18	96	330
Burundi	10.8	6.2	7.5	10.1	3.2	21	52	5	574	210

*Estimated data (country created May, 1993)

	Area (1000 sq mi)	Population (millions)			Annual Rate of Natural Incr. (Percent)	Doubling Time (years)	Life Expectancy at Birth (years)	Percent Urban Population	1994 Population Density (per sq mi)	1990 Gross Nat'l Product Per Capita (U.S. dollars)
		1994	2000	2010						
Cameroon	183.6	13.5	16.3	23.1	3.2	22	57	42	73	940
Cape Verde Is.	1.6	0.4	0.5	0.7	3.3	21	61	33	269	890
Central African Republic	240.5	3.3	3.9	4.9	2.6	27	47	43	14	390
Chad	495.8	5.5	6.4	7.7	2.5	28	46	30	11	190
Comoros Is.	0.7	0.5	0.7	0.9	3.5	20	56	26	756	480
Congo	132.1	2.5	3.0	3.9	2.9	24	54	41	19	1,010
Equatorial Guinea	10.8	0.4	0.5	0.6	2.6	26	50	28	36	330
Ethiopia	426.4	54.5	67.6	94.0	2.8	25	47	12	128	120
Gabon	103.4	1.2	1.3	1.4	2.5	28	53	43	11	3,220
Gambia	4.4	1.0	1.1	1.6	2.6	27	44	22	217	260
Ghana	92.1	17.0	20.5	26.9	3.2	22	54	32	185	390
Guinea	94.9	8.2	9.5	11.6	2.5	28	42	22	86	480
Guinea-Bissau	14.0	1.0	1.2	1.5	2.0	35	42	27	75	180
Ivory Coast	124.5	13.9	17.2	25.5	3.6	19	54	43	112	730
Kenya	225.0	28.1	34.9	44.8	3.7	19	61	22	125	370
Lesotho	11.7	2.0	2.4	3.1	2.9	24	58	19	170	470
Liberia	43.0	3.0	3.6	5.5	3.2	22	55	44	69	—
Madagascar	226.7	12.7	15.4	21.3	3.2	22	55	23	56	230
Malawi	45.8	9.3	11.4	14.9	3.5	20	49	15	204	200
Mali	478.8	9.1	10.8	14.2	3.0	23	45	22	19	270
Mauritania	398.0	2.2	2.6	3.5	2.8	25	48	41	6	500
Mauritius	0.8	1.1	1.2	1.3	1.5	48	69	41	1,407	2,250
Moçambique	302.3	17.5	20.5	26.6	2.7	26	48	23	58	80
Namibia	318.3	1.5	1.9	2.9	3.1	22	60	27	5	—
Niger	489.2	8.9	10.7	15.1	3.2	22	45	15	18	310
Nigeria	356.7	95.6	113.9	152.2	3.0	23	49	16	268	370
Reunion	1.0	0.6	0.7	0.8	1.8	38	71	62	641	—
Rwanda	10.2	8.3	10.1	14.4	3.4	20	50	7	809	310
São Tome and Principe	0.4	0.1	0.2	0.2	2.5	28	66	38	333	380
Senegal	75.8	8.4	9.9	13.1	2.8	25	48	37	111	710
Sierra Leone	27.7	4.7	5.4	7.3	2.6	27	43	30	168	240
South Africa	471.4	43.9	51.4	66.0	2.6	26	64	56	93	2,520
Swaziland	6.7	0.9	1.1	1.5	3.2	22	55	23	131	820
Tanzania	364.9	29.4	36.1	50.2	3.5	20	52	21	81	120
Togo	21.9	4.1	5.1	7.1	3.7	19	55	24	187	410
Uganda	91.1	18.8	23.3	32.5	3.7	19	51	10	206	220
Zaïre	905.6	40.3	48.6	65.6	3.1	22	52	40	45	230
Zambia	290.6	9.0	11.3	15.5	3.8	18	53	49	31	420
Zimbabwe	150.8	11.0	13.2	17.0	3.1	22	60	26	73	640
South Asia	**1,701.1**	**1,204.6**	**1,370.0**	**1,585.5**	**2.2**	**33**	**58**	**24**	**832**	**337**
Bangladesh	55.6	116.8	134.3	165.1	2.4	29	53	14	2,100	200
Bhutan	18.2	0.8	0.9	1.1	2.0	35	47	13	44	190
India	1,237.1	918.6	1,035.7	1,172.1	2.0	34	59	26	743	350
Maldives	0.1	0.2	0.3	0.4	3.4	20	61	28	2,374	440
Nepal	54.4	20.9	24.2	30.2	2.5	28	50	8	383	170
Pakistan	310.4	129.2	154.7	195.1	3.1	23	56	28	416	380
Sri Lanka	25.3	18.2	19.9	21.4	1.5	46	71	22	718	470
Chinese Realm	**4,394.6**	**1,295.5**	**1,398.6**	**1,534.8**	**1.3**	**54**	**70**	**30**	**459**	**—**
China	3,691.5	1,196.2	1,292.5	1,420.3	1.3	53	70	26	324	370
Hong Kong	0.4	5.8	6.1	6.3	0.7	99	78	100	14,572	11,540
Korea, North	46.5	23.1	25.7	28.5	1.9	37	69	64	496	—
Korea, South	38.0	45.2	48.2	51.7	1.1	65	71	74	1,190	5,400
Macau	0.01	0.5	0.5	0.6	1.3	52	79	97	48,669	—
Mongolia	604.3	2.4	2.8	3.5	2.8	25	65	42	4	112
Taiwan	13.9	21.3	22.8	24.0	1.1	62	74	71	1,532	8,815

	Area (1000 sq mi)	Population (millions)			Annual Rate of Natural Incr. (Percent)	Doubling Time (years)	Life Expectancy at Birth (years)	Percent Urban Population	1994 Population Density (per sq mi)	1990 Gross Nat'l Product Per Capita (U.S. dollars)
		1994	2000	2010						
Southeast Asia	**1,734.4**	**469.2**	**526.7**	**591.9**	**1.9**	**37**	**62**	**29**	**423**	**933**
Brunei	2.2	0.3	0.3	0.4	2.5	28	71	59	131	—
Cambodia	69.9	9.5	10.8	10.5	2.2	32	49	13	135	200
Indonesia	741.1	190.9	211.6	238.8	1.7	40	61	31	258	560
Laos	91.4	4.7	5.6	7.2	2.9	24	50	16	51	200
Malaysia	127.3	19.7	22.9	27.1	2.5	27	71	35	155	2,340
Myanmar (Burma)	261.2	44.1	49.5	57.7	1.9	36	58	24	169	—
Philippines	115.8	66.8	77.2	85.5	2.4	28	65	43	577	730
Singapore	0.2	2.8	3.1	3.2	1.4	51	75	100	14,206	12,310
Thailand	198.1	58.0	63.1	69.2	1.4	48	67	18	293	1,420
Vietnam	127.2	72.3	82.6	92.4	2.2	31	64	20	569	—
Pacific Realm	**212.4**	**6.1**	**7.0**	**8.5**	**2.3**	**30**	**56**	**19**	**44**	**969**
Fiji	7.1	0.8	0.9	0.9	2.0	35	61	39	110	1,770
Fed. States of Micronesia	0.3	0.1	0.1	0.1	2.3	31	—	—	—	—
French Polynesia	1.5	0.2	0.2	0.3	2.3	31	69	58	144	—
New Caledonia	7.4	0.2	0.2	0.2	1.8	39	73	59	25	—
Papua-New Guinea	178.3	4.0	4.6	5.7	2.3	31	54	13	23	860
Solomon Islands	11.0	0.4	0.5	0.6	3.6	20	61	9	35	580
Vanuatu	5.7	0.2	0.2	0.3	3.1	22	70	18	33	1,060
Western Samoa	1.1	0.2	0.2	0.3	2.8	25	67	21	187	730

MAP READING AND INTERPRETATION

As can be seen throughout this book, maps are tools that are very useful in gaining an understanding of patterns in geographic space. In fact, they constitute an important visual or *graphic communication* medium whereby encoded spatial messages are transmitted from the cartographer (mapmaker) to the map reader. Of course, this shorthand is necessary because the real world is so complex that a great deal of geographic information must be compressed into the small confines of maps that can fit onto the pages of this book. At the same time, cartographers must carefully choose which information to include; these decisions force them to omit many things in order to prevent cluttering a map with less relevant information. For instance, Fig. B shows several city blocks in central London but avoids mapping individual

buildings because they would interfere with the main information being presented—the spatial distribution of cholera deaths.

◆ MAP READING

Deciphering the coded messages contained in the maps of this book—map reading—is not difficult, and it becomes quite easy with a little experience in using this "language" of geography. The need to miniaturize portions of the world on small maps is discussed in the section on map scale (pp. 7-8), and two additional contrasting examples are provided in Figs. A and B. Orientation, or direction, on maps can usually be discerned by reference to the geographic grid of

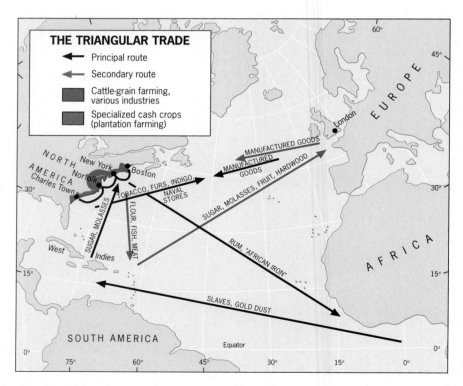

FIGURE A

latitude and longitude. *Latitude* is measured from 0° to 90° north and south of the equator (parallels of latitude are always drawn in an east-west direction), with the *equator* being 0° and the North and South Poles being 90°N and 90°S, respectively. Meridians of *longitude* (always drawn north-south) are measured 180° east and west of the *prime meridian* (0°), which passes through the Greenwich Observatory in London; the 180th meridian, for the most part, serves as the *international date line* that lies in the middle of the Pacific Ocean. Inspection of Fig. A shows that north is not automatically at the top of a map; instead, the direction of north curves along every meridian, with all such lines of longitude converging at the North Pole. The many minor directional distortions in this map are unavoidable: it is geometrically impossible to transfer the grid of a three-dimensional sphere (globe) onto a two-dimensional flat map. Therefore, compromises in the form of *map projections* must be devised in which, for example, properties such as areal size and distance are preserved but directional constancy is sacrificed.

◆ MAP SYMBOLS

Once the background mechanics of scale and orientation are understood, the main task of decoding the map's content can proceed. The content of most maps in this book is organized within the framework of point, line, and area symbols, which are made especially clear through the use of color. These symbols are usually identified in the map's *legend,* as in Fig. A. Occasionally, the map designer omits the legend but must tell the reader verbally in a caption or within the text what the map is about. Figure B, for instance, is a map of cholera deaths in the London neighborhood of Soho during the outbreak of 1854, with each Ⓟ symbol representing a municipal water pump and each red dot the location of a cholera fatality. *Point symbols* are shown as dots on the map and can tell us two things: the location of each phenomenon and, sometimes, its quantity. The cities of New York and London in Fig. A and the dot pattern of cholera fatalities in Fig. B (with each red dot symbolizing one death) are examples.

Line symbols connect places between which some sort of movement or flow is occurring. The "triangular trade" among Britain and its seventeenth-century Atlantic colonies (Fig. A) is a good example: each leg of these trading routes is clearly mapped, the goods moving along are identified, and the more heavily traveled principal routes are differentiated from the secondary ones. *Area symbols* are used to classify two-dimensional spaces and thus provide the cartographic basis for regionalization schemes (as can be seen in Fig. I-7 on p. 16-17). Such classifications can be

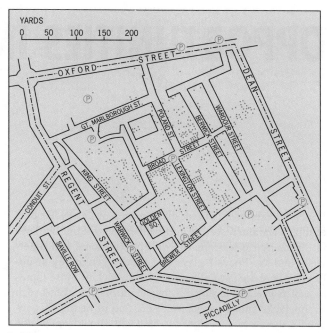

FIGURE B

developed at many levels of generalization. In Fig. A, blue and beige areas broadly offset land from ocean surfaces; more specifically, red and green area symbols along the eastern North American coast delimit a pair of regions that specialized in different types of commercial agriculture. Area symbols may also be used to communicate quantitative information: for example, the tan-colored zones in Fig. I-6 (p. 14-15) delimit semiarid areas within which annual precipitation averages from 12 to 20 inches (30 to 50 cm).

◆ MAP INTERPRETATION

The explanation of cartographic patterns is one of the geographer's most important tasks. Although that task is performed for you throughout this book, readers should be aware that today's practitioners use many sophisticated techniques and machines to analyze vast quantities of areal data. (For some insights into computer mapping, see p. 345.) These modern methods notwithstanding, geographic inquiry still focuses on the search for meaningful *spatial relationships.* This longstanding concern of the discipline is classically demonstrated in Fig. B. By showing on his map that cholera fatalities clustered around municipal water pumps, Dr. John Snow was able to persuade city authorities to shut them off; almost immediately, the number of new disease victims dwindled to zero, thereby confirming Snow's theory that contaminated drinking water was crucial in the spread of cholera.

OPPORTUNITIES IN GEOGRAPHY

The chapters of this book give an idea of the wide range of topics and interests that geographers pursue, particularly in the many discussions of concepts (regional and otherwise) and the Focus Essays that survey geography's subfields. There are specializations within each of those topics as well as in regional studies—but an introductory book such as this lacks enough space to discuss all of them. This appendix, therefore, is designed to help you, should you decide to major or minor in geography and/or to consider it as a career option.

◆ AREAS OF SPECIALIZATION

As in all disciplines, areas of concentration or specialization change over time. In North American geography early in the twentieth century, there was a period when most geographers were physical geographers, and the physical landscape was the main object of geographic analysis. Then the pendulum swung toward human (cultural) geography, and students everywhere focused on the imprints of human activity on the surface of the earth. Still later the analysis of spatial organization became a major concern. In the meantime, geography's attraction for some students lay in technical areas: in cartography, in remote sensing, in computer-assisted spatial data analysis, and, most recently, in geographic information systems.

All this meant that geography posed (and continues to pose) a challenge to its professionals. New developments require that we keep up to date, but we must also continue to build on established foundations.

REGIONAL GEOGRAPHY

One of these established foundations, of course, is regional geography, which encompasses a large group of specializations. Some geographers specialize in the theory of regions: how they should be defined, how they are structured, and how their internal components work. This leads in the direction of *regional science*, and some geographers have preferred to call themselves regional scientists. But make no mistake: regional science is regional geography.

Another, older approach to regional geography involves specialization in an area of the world ranging in size from a geographic realm to a single region or even a state or part of a state. There was a time when regional geographers, because of the interdisciplinary nature of their knowledge, were sought after by government agencies. Courses in regional geography abounded in universities' geography departments; regional geographers played key roles in international studies and research programs. But then the drive to make geography a more rigorous science and to search for universal (rather than regional) truths contributed to a decline in regional geography. The results were not long in coming, and lately you have probably seen the issue of "geographic illiteracy" discussed in newspapers and magazines. Now the pendulum is swinging back again, and regional geography and regional specialization are reviving. This is a propitious time to consider regional geography as a professional field.

YOUR PERSONAL INTERESTS

Geography, as we have pointed out, is united by several bonds, of which regional geography is but one. Regional geography exemplifies the spatial view that all geographers hold; the spatial approach to study and research binds physical and human geographers, regionalists, and topical specialists. Another unifying theme is an abiding interest in the relationships between human societies and natural environments. We have referred to that topic frequently in this book; as an area of specialization it has gone through difficult times. Perhaps more than anything, geography remains a field of *synthesis*, of understanding interrelationships.

Geography also is a field science, using "field" in another context. In the past, almost all major geography departments required a student's participation in a "field camp" as part of a master's degree program; thus many undergraduate programs included field experience. It was one of those bonding practices in which students and faculty with diverse interests met, worked together, and learned from one another. Today, few field camps of this sort are offered, but that does not change what geography is all about. If you see an opportunity for field experience with professional geographers—even just a one-day reconnaissance—take it. But realize this: a few days in the field with geographic instruction may hook you for life.

Some geographers, in fact, are far better field-data gatherers than analysts or writers. In this respect they are not alone: this also happens in archeology, geology, and biology (among other field disciplines). This does not mean that these field workers do not contribute significantly to knowledge. Often on a research team some of the members are better in the field and others excel in subsequent analysis. From all points of view, however, fieldwork is important.

Geography, then, is practiced in the field and in the office, in physical and human contexts, in generality and detail. Small wonder that so many areas of specialization have developed! If you check the undergraduate catalogue of your college or university, you will see some of these specializations listed as semester-length courses. But no geography department, no matter how large, could offer them all.

How does an area of specialization develop, and how can one become a part of it? The way in which geographic specializations have developed tells us much about the entire discipline. Some major areas, now old and established, began as research and theory building by one scholar and his or her students. These graduate students dispersed to the faculties of other universities and began teaching what they had learned. Thus, for example, did Carl Sauer's cultural geography spread from the University of California at Berkeley.

It is one of the joys of geography that the basics and methods, once learned, are applicable to so many features of the human and physical world. Geographers have specialized in areas as disparate as shopping-center location and glacier movement, tourism and coastal erosion, real estate and wildlife, retirement communities and sports. Many of these specializations began with the interests and energies of one scholar. Some thrived and grew into major geographic pursuits; others remained one-person shows, but with potential. When you discuss your own interests with a faculty advisor, you may refer to a university where you would like to do graduate work. "Oh yes," the answer may be, "Professor X is in their geography department, working on just that." Or perhaps your advisor will suggest another university where a member of the faculty is known to be working on the topic in which you are interested. That is the time to write a letter of inquiry. What is the professor working on now? Are graduate students involved? Are research funds available? What are the career prospects after graduation?

The Association of American Geographers or **AAG** (1710 16th Street, N.W., Washington, D.C. 20009-3198 [(202) 234-1450]) recognizes more than 40 so-called Specialty Groups. In academic year 1993–1994, the Specialty Group roster included the following branches of geography:

Africa
Aging and the Aged
American Ethnic Geography
American Indians
Applied
Asia
Biblical
Biogeography
Canadian Studies
Cartography
China
Climate
Coastal and Marine
Contemporary Agriculture and Rural Land Use
Cultural
Cultural Ecology
Energy and Environment
Environmental Perception and Behavioral Geography
Europe
Geographic Information Systems
Geographic Perspectives on Women
Geography Education
Geomorphology
Hazards
Historical
Human Rights
Industrial
Latin America
Mathematical Models and Quantitative Methods
Medical
Microcomputers
Political
Population
Recreation, Tourism, and Sport
Regional Development and Planning
Religions and Belief Systems
Remote Sensing
Rural Development
Russia, Central Eurasia, and East Europe
Socialist Geography
Transportation
Urban
Water Resources

If you write or call the AAG at the address or telephone number given above, they will be glad to send you the newest listing of these Specialty Groups, which also includes the address of the current chairperson of each group. We encourage you to contact the person who chairs the group(s) you are interested in. These elected leaders are ready, enthusiastic, and willing to provide you with the information you are looking for.

All this may seem far in the future. Still, the time to start planning for graduate school is now. Applications for admission and financial assistance must be made shortly after the *beginning* of your senior year! That makes your junior year a year of decision.

◆ AN UNDERGRADUATE PROGRAM

The most important concern for any geography major or minor is basic education and training in the field. An undergraduate curriculum contains all or several of the following courses (titles may vary):

1. Introduction to Physical Geography I (climatology, elementary oceanography)
2. Introduction to Physical Geography II (natural landscapes, landforms, soils, elementary biogeography)
3. Introduction to Human Geography (principles and topics of cultural and economic geography)
4. World Regional Geography (major world realms)

These beginning courses are followed by more specialized courses, including both substantive and methodological ones:

5. Introduction to Quantitative Methods of Analysis
6. Introductory Cartography
7. Analysis of Remotely Sensed Data
8. Cultural Geography
9. Political Geography
10. Urban Geography
11. Economic Geography
12. Historical Geography
13. Geomorphology
14. Geography of United States–Canada, Europe, and/or other major world realms

You can see how Physical Geography II would be followed by Geomorphology, and how Human Geography now divides into such areas as intermediate and/or advanced cultural and economic geography. As you progress, the focus becomes even more specialized. Thus Economic Geography may be followed by:

15. Industrial Geography
16. Transportation Geography
17. Agricultural Geography

At the same time, regional concentrations may come into sharper focus:

18. Geography of Western Europe (or other regions)

In these more advanced courses, you will use the technical know-how from courses numbered **5**, **6**, and **7** (and perhaps others). Now you can avail yourself of the opportunity to develop these skills further. Many departments offer such courses as these:

19. Advanced Quantitative Methods (involving numerous computer applications)
20. Advanced Cartography (involving geographic information systems)
21. Advanced Satellite Imagery Interpretation

From this list (which represents only part of a comprehensive curriculum), it is evident that you cannot, even in four years of undergraduate study, register for all courses. The geography major in many universities requires a minimum of only 30 (semester-hour) credits—just 10 courses, fewer than half those listed here. This is another reason to begin thinking about specialization at an early stage.

Because of the number and variety of possible geography courses, most departments require that their majors complete a core program that includes courses in substantive areas as well as theory and methods. That core program is important, and you should not be tempted to put off these courses until your last semesters. What you learn in the core program will make what follows (or should follow) much more meaningful.

You should also be aware of the flexibility of many undergraduate programs, something that can be especially important to geographers. Imagine that you are majoring in geography and develop an interest in Southeast Asia. But the regional specialization of the geographers in your department may be focused somewhere else—say, Africa. However, courses on Southeast Asia are indeed offered by other departments, such as anthropology, history, or political science. If you are going to be a regional specialist, those courses will be very useful and should be part of your curriculum—but you are not able to receive geography credit for them. After successfully completing those courses, however, you may be able to register for an independent study or reading course in the geography of Southeast Asia if a faculty member is willing to guide you. Always discuss such matters with the undergraduate advisor or chairperson of your geography department.

◆ LOOKING AHEAD

By now, as a geography major, you will be thinking of the future—either in terms of graduate school or a salaried job.

In this connection, if there is one important lesson to keep in mind, it is to *plan ahead* (a redundancy for emphasis!). The choice of graduate school is one of the most important you will make in your life. The professional preparation you acquire as a graduate student will affect your competitiveness in the job market for years to come.

CHOOSING A GRADUATE SCHOOL

Your choice of a graduate school hinges on several factors, and the geography program it offers is one of them. Possibly, you are constrained by residency factors and your choice involves the schools of only one state. Your undergraduate record and grade point average affect the options. As a geographer, you may have strong feelings in favor of or against particular parts of the country. And although some schools may offer you financial support, others may not.

Certainly the programs and specializations of the prospective graduate department are extremely important. If you have settled on your own area of interest, it is wise to find a department that offers opportunities in that direction. If you have yet to decide, it is best to select a large department with several options. Some students are so impressed by the work and writings of a particular geographer that they go to his or her university solely to learn from and work with that scholar.

In every case, information and preparation are crucial. Many a prospective graduate student has arrived on campus eager to begin work with a favorite professor, only to find that the professor is away on a sabbatical leave!

Fortunately, information can be acquired with little difficulty. One of the most useful publications of the AAG is the *Guide to Programs of Geography in the United States and Canada*, published each fall. A copy of this annual directory should be available in the office of your geography department, but if you plan to enter graduate school, a personal copy would be an asset (the AAG charged $28 for nonmembers in 1993). Not only does the *Guide* describe the programs, requirements, financial aid, and other aspects of geography departments in North America, but it also lists all faculty members and their current research and teaching specializations. Moreover, it also contains a complete listing of the Association's approximately 7,000 members and their specializations. You will find the *Guide* indispensable in your decision-making.

You may discover one particular department that stands out as the most interesting, and most appropriate for you. But do not limit yourself to one school; after careful investigation, it is best to rank a half-dozen schools (or more), write to all of them for admission application forms, and apply to several. Multiple applications are costly, but the investment is worth it.

ASSISTANTSHIPS AND SCHOLARSHIPS

One reason to apply to several universities has to do with financial (and other) support for which you may be eligible. If you have a reasonably well-rounded undergraduate program behind you and a good record of achievement, you are eligible to become a teaching assistant (TA) in a graduate department. Such assistantships usually offer full or partial tuition plus a monthly stipend (during the nine-month academic year). Conditions vary, but this position may make it possible for you to attend a university that would otherwise be out of reach. A tuition waiver alone can be worth more than $10,000 annually. Application for an assistantship is made directly to the department; you should write to the department contact listed in the latest AAG *Guide*, who will either respond directly to you or forward your inquiry to the departmental committee that evaluates applications.

What does a TA do? The responsibilities vary, but often teaching assistants are expected to lead discussion sections (of a larger class taught by a professor), laboratories, or other classes. They prepare and grade examinations and help undergraduates tackle problems arising from their courses. This is an excellent way to determine your own ability and interest in a teaching career.

In some instances, especially in larger geography departments, research assistantships are available. When a member of the faculty is awarded a large grant (e.g., by the National Science Foundation) for a research project, that grant may make possible the appointment of one or more research assistants (RAs). These individuals perform tasks generated by the project and are rewarded by a modest salary (usually comparable to that of a TA). Normally, RAs do not receive tuition waivers, but sometimes the department and the graduate school can arrange a waiver to make the research assistantship more attractive to better students. Usually, RAs are chosen from among graduate students already on campus who have proven their interest and ability. Sometimes, though, an incoming student is appointed. Always ask about opportunities.

Geography students are also among those eligible for many scholarships and fellowships offered by universities and off-campus organizations. When you write your introductory letter, be sure to inquire about other forms of financial aid.

◆ JOBS FOR GEOGRAPHERS

Upon completing your bachelor's degree, you may decide to take a job rather than going on to graduate school. Again, this is a decision best made early in your junior year, for two main reasons: (1) so that you can tailor your curriculum for a vocational objective, and (2) so that you can start searching for a job well before graduation.

INTERNSHIPS

A very good way to enter the job market—and to become familiar with the working environment—is by taking an internship in an agency, office, or firm. Many organizations find it useful to have interns. Internships help organizations train beginning professionals and give companies an opportunity to observe the performance of trainees. Many an intern has ultimately been employed by his or her organization. Some employers have even suggested what courses the intern should take in the next academic year to improve future performance. For example, an urban or regional planning agency that employs an intern might suggest that the intern add a relevant advanced cartography course or urban planning course to his or her program of study.

Some organizations will appoint interns on a continuing basis, say, two afternoons a week around the year; others make full-time, summer-only appointments available. Occasionally, an internship can be linked to a departmental curriculum, yielding academic credit as well as vocational experience. Your department undergraduate advisor or the chairperson will be the best source of assistance.

One of the most interesting internship programs is offered by the National Geographic Society in Washington, D.C. Every year, the Society invites three groups of about eight interns each to work with the permanent staff of its various departments. Application forms are available in every geography department in the United States, and competition is strong. The application itself is a useful exercise, as it tells you what the Society (and other organizations) look for in your qualifications.

PROFESSIONAL OPPORTUNITIES

A term you will sometimes see in connection with jobs is *applied geography.* This, one supposes, distinguishes geography teaching from practical geography. In fact, however, all professional geography in education, business, government, and elsewhere is "applied." In the past a large majority of geography graduates became teachers—in elementary and high schools, colleges, and universities. More recently, geographers have entered other arenas in increasing numbers. In part, this is related to the decline in geographic education in schools, but it also reflects the growing recognition of geographic skills by employers in business and government.

Nevertheless, what you as a geographer can contribute to a business is not yet as clear to the managers of many companies as it should be. The anybody-can-do-geography attitude is a form of ignorance you will undoubtedly con-

front. (This is so even in precollegiate education, where geography was submerged in social studies—and often taught by teachers who had never taken a course in the discipline!)

So, once employed, you may have to prove not only yourself but also the usefulness of the skills and capacities you bring to your job. There is a positive side to this. Many employers, once dubious about hiring a geographer, learn how geography can contribute—and become enthusiastic users of geographic talent. Where one geographer is employed, whether in a travel firm, publishing house, or planning office, you will soon find more.

Geographers are employed in business, government, and education. Planning is a profession that employs many geographers. Employment in business has grown in recent years (see below). Governments at national, state, and local levels have always been major employers of geographers. And in education, where geography was long in decline, the demand for geography teachers will grow again.

For more detailed information about employment, you should write to the AAG (at the address given earlier) for the inexpensive booklet entitled *Careers in Geography*; check with your geography department, whose office is very likely to have a copy.

BUSINESS AND GEOGRAPHY

With their education in global and international affairs, their knowledge of specialized areas of interest to business, and their training in cartography, methods of quantitative analysis, and writing, geographers with a bachelor's degree are attractive business-employment prospects. Some undergraduate students have already chosen the type of business they will enter; for them, there are departments that offer concentrations in their areas of interest. Several geography departments, for instance, offer a curriculum that concentrates on tourism and travel—a business in which geographic skills are especially useful.

These days, many companies want graduates with a strong knowledge of international affairs and fluency in at least one foreign language in addition to their skills in other areas. The business world is quite different from academia, and the transition is not always easy. Your employer will want to use your abilities to enhance profits. You may at first be placed in a job where your geographic skills are not immediately applicable, and it will be up to you to look for opportunities to do so.

One of our students was in such a situation some years ago. She was one of more than a dozen new employees doing what was essentially clerical work. (Some companies will use this type of work to determine a new employee's punctuality, work habits, adaptability, and productivity.) One day she heard that the company was considering establishing a branch to sell its product in East Africa. The student had a regional interest in East Africa as an undergraduate and had even taken a year of Swahili-language training. On her own time she wrote a carefully documented memorandum to the company's president and vice presidents, describing factors that should be taken into consideration in the projected expansion. That report evinced her regional skills, locational insights, and knowledge of the local market and transport problems, along with the probable cultural reaction to the company's product, the country's political circumstances, and (last but significantly) this employee's ability to present such issues effectively. She supported her report with good maps and several illustrations. Soon she received a special assignment to participate in the planning process, and her rise in the company's ranks had begun. She had seized the opportunity and demonstrated the utility of her geographic skills.

Geography graduates have established themselves in businesses of all kinds: banking, international trade, manufacturing, retailing, and many more. Should you join a large firm, you may be pleasantly surprised by the number of other geographers who hold jobs there—not under the title of geographer but under countless other titles ranging from analyst and cartographer to market researcher and program manager. These are positions for which the appointees have competed with other graduates, including business graduates. As we noted earlier, once an employer sees the assets a geographer can bring, the role of geographers in the company is assured.

GOVERNMENT AND GEOGRAPHY

Government has long been a major employer of geographers at the national and state as well as local level. *Careers in Geography* (the AAG booklet) estimates that at least 2,500 geographers are working for governments, about half of them for the federal government. In the U.S. State Department, for instance, there is an Office of the Geographer staffed by professional geographers. Other agencies where geographers are employed include the Defense Mapping Agency, the Bureau of the Census, the U.S. Geological Survey, the Central Intelligence Agency, and the Army Corps of Engineers. Still other employers are the Library of Congress, the National Science Foundation, and the Smithsonian Institution. Many other branches of government also have positions for which geographers are eligible.

Opportunities also exist at state and local levels. Several states now have their own Office of the State Geographer; all states have agencies engaged in planning, resource analysis, environmental protection, and transportation policymaking. All these agencies need geographers who have skills in cartography, remote sensing, database analysis, and the operation of geographic information systems.

Securing a position in government requires early action. If you want a job with the federal government, start at the beginning of your senior year. Every state capital and many other large cities have a Federal Job Information Center (FJIC). (The Washington office is at 1900 E Street, N.W., Washington, D.C. 20415.) You may request information about a particular agency and its job opportunities; it is appropriate to write directly to the personnel office of the agency or agencies in which you are interested. You may also write to the Office of Personnel Management (OPM), Washington, D.C. 20415, which has offices in most large cities.

PLANNING AND GEOGRAPHY

Planning has become one of geography's allied professions. The planning process is a complex one comprising people trained in many fields. Geographers, with their cartographic, locational, regional, and analytical skills, are sought after by planning agencies. Many an undergraduate student gets that first professional opportunity as an intern in a planning office.

Planning is done by many agencies and offices at levels ranging from the federal to the municipal. Cities have planning offices, as do regional authorities. Working in a planning office can be a very rewarding experience because it involves the solving of social and economic problems, the conservation and protection of the environment, the weighing of diverse and often conflicting arguments and viewpoints, and much interaction with workers trained in other fields. Planning is a superb learning experience.

A career in planning can be much enhanced by a background in geography, but you will have to adjust your undergraduate curriculum to include courses in such areas as public administration, public finance, and other related fields. Thus a career in planning itself requires early planning on your part. At many universities, the geography department is closely associated with the planning department, and your faculty advisor can inform you about course requirements. But if you have your eye on a particular office or agency, you should also request information from its director about desired and required skills.

Planning is by no means a monopoly of government. Government-related organizations such as the Agency for International Development (AID), the World Bank, and the International Monetary Fund (IMF) have planning offices, as do nongovernmental organizations such as banks, airline companies, industrial firms, and multinational corporations. In the private sector the opportunities for planners have been expanding, and you may wish to explore them. An important organization is the planners' equivalent of the geographers' AAG: the American Planning Association, 1776 Massachusetts Avenue, N.W., Washington, D.C. 20036. The AAG's *Careers in Geography* provides additional information on this expanding field and where you might pursue its study.

TEACHING

If you are presently a freshman or sophomore, your graduation may coincide with the end of a long decline in the geography-teaching profession. Just 30 years ago, teaching geography in elementary or high school was the goal of thousands of undergraduate geography majors. But then began the merging of geography into the hybrid field called "social studies," and prospective teachers no longer needed to have any training in geography in many states. In Florida, for example, teachers were formerly required to take courses in regional geography and conservation (as taught in geography departments), but in the 1970s those requirements were dropped. Education planners fell victim to the myth that the teaching of geography does not require any training. What was left of geography often was taught by teachers whose own fields were history, civics, or even basketball coaching!

If you read the daily newspapers, you have seen reports of the predictable results. "Geographic illiteracy" has become a common complaint (often made by the same education planners who pushed geography into the social-studies program and eliminated teacher-education requirements). Now states are returning to the education requirement so that teachers will learn some geography. And geography is returning to elementary and high school curricula—assisted in particular by the National Geographic Society, which supports state-level Geographic Alliances of educators and academic geographers. The need for geography teachers will soon be on the upswing again.

So this may be a good time to consider teaching geography as a career. You should do some research, however, because states vary in their progressiveness in this area. Opportunities will become more widespread as the 1990s proceed. You should visit the School of Education in your college or university and ask questions about this. Also, write not only the AAG but also the National Council for Geographic Education (NCGE) at Indiana University of Pennsylvania, Indiana, PA 15705. Your state geographic society or Geographic Alliance also may be helpful; ask your department advisor or chairperson for details.

The opportunities in geography are many, but often they are not as obvious as those in other fields. You will find that good and timely preparation produces results and, frequently, unexpected rewards.

The discussion in this appendix has provided a comprehensive answer to the oft-asked question, "What can one do with geography?" If you wish to explore this question further, along with the AAG's *Careers in Geography* we recommend *On Becoming a Professional Geographer*, a book edited by Martin S. Kenzer (Merrill/ Macmillan, 1989).

We wish you every success in all your future endeavors. And, speaking for the entire community of professional geographers, we would be delighted to have you join our ranks, should you choose a geography-related career.

REFERENCES AND FURTHER READINGS

(Bullets [•] denote basic introductory works)

INTRODUCTION

Abler, Ronald F., Marcus, Melvin G., & Olson, Judy M., eds. *Geography's Inner Worlds: Pervasive Themes in Contemporary American Geography* (New Brunswick, N.J.: Rutgers University Press, 1992).

Berry, Brian J.L. "Urbanization," in Billie Lee Turner II, ed., *The Earth as Transformed by Human Action: Global and Regional Changes in the Biosphere Over the Past 300 Years* (New York: Cambridge University Press, 1990), pp. 103–119.

Boyd, Andrew. *An Atlas of World Affairs* (London & New York: Routledge, 9 rev. ed., 1992).

Brunn, Stanley D. & Williams, Jack F., eds. *Cities of the World: World Regional Urban Development* (New York: Harper Collins, 2 rev. ed., 1993).

Chisholm, Michael. *Modern World Development* (Totowa, N.J.: Barnes & Noble, 1982).

• de Blij, H.J. *Human Geography: Culture, Society, and Space* (New York: John Wiley & Sons, 4 rev. ed., 1993).

• de Blij, H.J. & Muller, Peter O. *Physical Geography of the Global Environment* (New York: John Wiley & Sons, 1993).

Dickinson, Robert E. *The Regional Concept* (London: Routledge & Kegan Paul, 1976).

Fenneman, Nevin M. "The Circumference of Geography," *Annals of the Association of American Geographers*, (1919): 3–11.

Freeman, Michael. *Atlas of the World Economy* (New York: Simon & Schuster, 1991).

• Glassner, Martin I. & de Blij, H.J. *Systematic Political Geography* (New York: John Wiley & Sons, 4 rev. ed., 1989).

• *Goode's World Atlas* (Chicago: Rand McNally, 19 rev. ed., 1994).

Haggett, Peter. *Locational Analysis in Human Geography* (London: Edward Arnold, 1965).

Haggett, Peter. *The Geographer's Art* (New York: Blackwell, 1990).

• Harris, Chauncy D., chief ed. *A Geographical Bibliography for American Libraries* (Washington, D.C.: Association of American Geographers/National Geographic Society, 1985).

Hartshorn, Truman A. *Interpreting the City: An Urban Geography* (New York: John Wiley & Sons, 2 rev. ed., 1992). Quotation from pp. 1–3.

Hunter, Brian, ed. *The Statesman's Year-Book* (New York: St. Martin's Press, annual).

Jackson, John Brinckerhoff. *Discovering the Vernacular Landscape* (New Haven: Yale University Press, 1984).

James, Preston E. & Martin, Geoffrey J. *All Possible Worlds: A History of Geographical Ideas* (New York: John Wiley & Sons, 3 rev. ed., 1993).

Johnston, R.J., et al., eds. *The Dictionary of Human Geography* (Cambridge, Mass.: Blackwell, 3 rev. ed., 1993).

Kroeber, Alfred L. & Kluckhohn, Clyde. "Culture: A Critical Review of Concepts and Definitions," *Papers of the Peabody Museum of American Archaeology and Ethnology*, 47 (1952).

Lanegran, David A. & Palm, Risa, eds. *An Invitation to Geography* (New York: McGraw-Hill, 2 rev. ed., 1978).

Livingstone, David. *The Geographical Tradition* (Cambridge, Mass.: Blackwell, 1992).

Morrill, Richard L. "The Nature, Unity and Value of Geography," *The Professional Geographer*, 35 (1983): 1–9.

Muehrcke, Phillip C. & Muehrcke, Juliana C. *Map Use: Reading-Analysis-Interpretation* (Madison, Wis.: JP Publications, 3 rev. ed., 1992).

Paddison, Ronan & Morris, Arthur S. *Regionalism and the Regional Question* (New York: Blackwell, 1988).

• Pattison, William D. "The Four Traditions of Geography," *Journal of Geography*, 63 (1964): 211–216.

Sauer, Carl O. "Cultural Geography," *Encyclopedia of the Social Sciences*, Vol. 6 (New York: Macmillan, 1931), pp. 621–623.

Shortridge, James R. *The Middle West: Its Meaning in American Culture* (Lawrence, Kan.: University Press of Kansas, 1989).

Small, John & Witherick, Michael, eds. *A Modern Dictionary of Geography* (London & New York: Routledge, 2 rev. ed., 1989).

Turner, Billie Lee II, ed. *The Earth as Transformed by Human Action: Global and Regional Changes in the Biosphere Over the Past 300 Years* (New York: Cambridge University Press, 1990).

Wegener, Alfred. *The Origin of Continents and Oceans* (New York: Dover, reprint of the 1915 original, trans. John Biram, 1966).

• Wheeler, James O., Muller, Peter O., & Thrall, Grant I. *Economic Geography* (New York: John Wiley & Sons, 3 rev. ed., 1995).

Whittlesey, Derwent S., et al. "The Regional Concept and the Regional Method," in James, Preston E. and Jones, Clarence F., eds., *American Geography: Inventory and Prospect* (Syracuse, N.Y.: Syracuse University Press, 1954), pp. 19–68.

World Urbanization Prospects 1950-2025 (The 1992 Revision) (New York: United Nations, Population Division, Department of Economic and Social Development, 1993).

CHAPTER 1

Beckinsale, Monica & Beckinsale, Robert P. *Southern Europe: A Systematic Geographical Study* (New York: Holmes & Meier, 1975).

• Bloom, Arthur L. *Geomorphology: A Systematic Analysis of Late Cenozoic Landforms* (Englewood Cliffs, N.J.: Prentice-Hall, 1978).

Burtenshaw, David, et al. *The European City: A Western Perspective* (New York: John Wiley & Sons, 1991).

• Butzer, Karl W. *Geomorphology From the Earth* (New York: Harper & Row, 1976).

Chisholm, Michael. *Rural Settlement and Land Use: An Essay in Location* (London: Hutchinson University Library, 3 rev. ed., 1979).

• Chorley, Richard J., et al. *Geomorphology* (London & New York: Methuen, 1985).

Clayton, Keith M. & Kormoss, I.B.F., eds. *Oxford Regional Economic Atlas of Western Europe* (London: Oxford University Press, 1971).

Clout, Hugh D., ed. *Regional Development in Western Europe* (New York: John Wiley & Sons, 3 rev. ed., 1987).

Clout, Hugh D., et al. *Western Europe: Geographical Perspectives* (London & New York: Longman, 2 rev. ed., 1989).

Cohen, Leonard J. *Broken Bonds: The Rise and Fall of Yugoslavia* (Boulder, Col.: Westview Press, 1993).

Cole, John P. & Cole, Francis J. *The Geography of the European Community* (London & New York: Routledge, 1993).

Cole, John P. & Cole, Francis J. "The New Germany," *Focus*, 41 (Fall 1991): 1–6.

Dawson, Andrew H. *The Geography of European Integration: A Common European Home?* (New York: Wiley/Belhaven, 1993).

de Blij, H.J. & Muller, Peter O. *Physical Geography of the Global Environment* (New York: John Wiley & Sons, 1993).

Demko, George J., ed. *Regional Development Problems and Policies in Eastern and Western Europe* (New York: St. Martin's Press, 1984).

• Diem, Aubrey. *Western Europe: A Geographical Analysis* (New York: John Wiley & Sons, 1979).

Diuk, Nadia and Karatnycky, Adrian. *New Nations Rising: The Fall of the Soviets and the Challenge of Independence.* (New York: John Wiley and Sons, 1993.)

Embleton, Clifford, ed. *Geomorphology of Europe* (New York: Wiley-Interscience, 1984).

Glebe, Günther & O'Loughlin, John, eds. *Foreign Minorities in Continental European Cities* (Wiesbaden, West Germany: Franz Steiner Verlag, 1987).

• Gottmann, Jean. *A Geography of Europe* (New York: Holt, Rinehart & Winston, 4 rev. ed., 1969).

Haggett, Peter. *Locational Analysis in Human Geography* (London: Edward Arnold, 1965). Definition from p. 19.

Hall, Ray & Ogden, Philip. *Europe's Population in the 1970s and 1980s* (Cambridge, U.K.: Cambridge University Press, 1985).

Hancock, M. Donald & Welsh, Helga A., eds. *German Unification: Process and Outcomes* (Boulder, Col.: Westview Press, 1993).

Harris, Chauncy D. "Unification of Germany in 1990," *Geographical Review*, 81 (1991): 170–182.

• Hoffman, George W., ed. *Europe in the 1990's: A Geographic Analysis* (New York: John Wiley & Sons [6 rev. ed. of *A Geography of Europe: Problems and Prospects*], 1989).

Houston, James M. *A Social Geography of Europe* (London: Gerald Duckworth, 1963).

Ilbery, Brian W. *Western Europe: A Systematic Human Geography* (New York: Oxford University Press, 2 rev. ed., 1986).

Jefferson, Mark. "The Law of the Primate City," *Geographical Review*, 29 (1939). Quotation from p. 226.

Johnston, Ronald J. & Gardiner, Vince, eds. *The Changing Geography of the United Kingdom* (London & New York: Routledge, 2 rev. ed., 1991).

• Jordan, Terry G. *The European Culture Area: A Systematic Geography* (New York: Harper & Row, 2 rev. ed., 1989).

Knox, Paul L. *The Geography of Western Europe: A Socio-Economic Survey* (Totowa, N.J.: Barnes & Noble, 1984).

Krause, Axel. *Inside the New Europe* (New York: HarperCollins, 1991).

Lawday, David. "Europe's [Mediterranean] Sun Belt Also Rises," *Newsweek*, July 18, 1988, pp. 27–29.

Lewis, Flora. *Europe: Road to Unity* (New York: Touchstone/Simon & Schuster, 1991).

Magas, Branka. *The Destruction of Yugoslavia* (London & New York: Verso, 1993).

• McDonald, James R. *The European Scene: A Geographic Perspective* (Englewood Cliffs, N.J.: Prentice-Hall, 1992).

• Mellor, Roy E.H. & Smith, E. Alistair. *Europe: A Geographical Survey of the Continent* (New York: St. Martin's Press, 1979).

Murphy, Alexander B. "The Emerging Europe of the 1990s," *Geographical Review*, 81 (1991): 1–17.

Nagorski, Andrew. *The Birth of Freedom: Shaping Lives and Societies in the New Eastern Europe* (New York: Simon & Schuster, 1993).

O'Loughlin, John V. & van der Wusten, Herman, eds. *The New Political Geography of Eastern Europe* (London & New York: Belhaven/Wiley, 1993).

Park, Chris. *Acid Rain: Rhetoric and Reality* (London & New York: Routledge, 1989).

Pinder, David, ed. *Western Europe: Challenge and Change* (New York: Guilford, 1991).

Pounds, Norman J.G. *An Historical Geography of Europe, 1800–1914* (New York: Cambridge University Press, 1985).

Rodwin, Lloyd & Sazanami, Hidehiko. *Industrial Change and Regional Economic Transformation: The Experience of Western Europe* (New York: HarperCollins Academic, 1991).

Rugg, Dean S. *Eastern Europe* (London & New York: Longman, 1986).

Sommers, Lawrence M. "Cities of Western Europe," in Brunn, Stanley D. & Williams, Jack F., eds., *Cities of the World: World Regional Urban Development* (New York: Harper & Row, 1983), pp. 84–121. Quotation from p. 97.

Turnock, David. *Eastern Europe: An Economic and Political Geography* (London & New York: Routledge, 1989).

Turnock, David. *The Human Geography of Eastern Europe* (London & New York: Routledge, 1989).

Wheeler, James O. & Muller, Peter O. *Economic Geography* (New York: John Wiley & Sons, 2 rev. ed., 1986), Chapter 13.

White, Paul. *The West European City: A Social Geography* (London & New York: Longman, 1984).

Williams, Allan M. *The European Community: The Contradictions of Integration* (Cambridge, Mass.: Blackwell, 1991).

Williams, Allan M. *The West European Economy: A Geography of Post-War Development* (Savage, Md.: Rowman & Littlefield, 1988).

BIBLIOGRAPHICAL FOOTNOTE
Many geographies take as their subject individual countries (or groups of countries such as the British Isles and the Alpine states), both for Europe and all the other world realms. A good beginning is a continuing series of sketches in *Focus*, published four times a year by the American Geographical Society in New York City. The Van Nostrand *Searchlight Se-*

ries of paperbacks (many dating back to the 1960s) includes discussions of several of the world's nations and areas; unfortunately, these are out of print in the 1990s (and therefore not cited in this book), but they can still be found in many libraries. Also refer to Praeger's *Country Profies Series* (many now published by its branch, Westview Press of Boulder, Colorado) and Routledge's (formerly Methuen's) *Advanced Geographies Series,* which is supplemented by that publisher's many other regional titles. Additional information on accessing the geographical literature is found for each chapter in the *Study Guide* that accompanies this book.

CHAPTER 2

Allworth, Edward, ed. *Ethnic Russia in the U.S.S.R.: The Dilemma of Dominance* (Elmsford, N.Y.: Pergamon Press, 1980).

• Bater, James H. *The Soviet Scene: A Geographical Perspective* (London & New York: Routledge, 1989).

Bater, James H. & French, Richard A., eds. *Studies in Russian Historical Geography* (New York: Academic Press, 2 vols., 1983).

Bobrick, Benson. *East of the Sun: The Epic Conquest and Tragic History of Siberia* (New York: Poseidon Press, 1992).

Bradshaw, Michael J., ed. *The Soviet Union: A New Regional Geography?* (New York: Wiley/Longman, 1991).

• Brown, Archie, et al., eds. *Cambridge Encyclopedia of Russia and the Soviet Union* (New York: Cambridge University Press, 1982).

Chew, Allen F. *Atlas of Russian History: Eleven Centuries of Changing Borders* (New Haven: Yale University Press, 1967).

Clem, Ralph S. "Russians and Others: Ethnic Tensions in the Soviet Union," *Focus,* September-October, 1980.

• Cole, John P. *Geography of the Soviet Union* (Stoneham, Mass.: Butterworth, 1984).

• Critchfield, Howard J. *General Climatology* (Englewood Cliffs, N.J.: Prentice-Hall, 4 rev. ed., 1983).

• de Blij, H.J. & Muller, Peter O. *Physical Geography of the Global Environment* (New York: John Wiley & Sons, 1993).

Dellenbrant, Jan A. *The Soviet Regional Dilemma: People, Planning, and Natural Resources* (Armonk, N.Y.: M.E. Sharpe, 1986).

Demko, George J. & Fuchs, Roland J., eds. *Geographical Studies on the Soviet Union: Essays in Honor of Chauncy D. Harris* (Chicago: University of Chicago, Department of Geography, Research Paper No. 211, 1984).

Dienes, Leslie. *Soviet Asia: Economic Development and National Policy Choices* (Boulder, Col.: Westview Press, 1987).

Diuk, Nadia and Karatnycky, Adrian. *New Nations Rising: The Fall of the Soviets and the Challenge of Independence.* (New York: John Wiley and Sons, 1993.)

Feshbach, Murray & Friendly, Alfred, Jr. *Ecocide in the USSR: Health and Nature Under Siege* (New York: Basic Books, 1992).

Fisher, Lois. *Survival in Russia: Chaos and Hope in Everyday Life* (Boulder, Col.: Westview Press, 1993).

French, Richard A. & Hamilton, F. E. Ian, eds. *The Socialist City: Spatial Structure and Urban Policy* (New York: John Wiley & Sons, 1979).

Hajda, Lubomyr & Beissinger, Mark, eds. *The Nationalities Factor in Soviet Politics and Society* (Boulder, Col.: Westview Press, 1990).

Horensma, Pier. *The Soviet Arctic* (London & New York: Routledge, 1991).

Hosking, Geoffrey. *The Awakening of the Soviet Union* (Cambridge, Mass.: Harvard University Press, 1989).

• Howe, G. Melvyn. *The Soviet Union: A Geographical Study* (London & New York: Longman, 2 rev. ed., 1986).

Jensen, Robert G., et al., eds. *Soviet Natural Resources in the World Economy* (Chicago: University of Chicago Press, 1983).

Karklins, Rasma. *Ethnic Relations in the U.S.S.R.: The Perspective From Below* (Winchester, Mass.: Allen & Unwin, 1986).

Lewis, Robert A., ed. *Geographic Perspectives on Soviet Central Asia* (London & New York: Routledge, 1992).

Lydolph, Paul E. *Climates of the Soviet Union* (Amsterdam: Elsevier Scientific Publishing Co., 1977).

• Lydolph, Paul E. *Geography of the U.S.S.R.* (Elkhart Lake, Wis.: Misty Valley Publishing, 1990).

Mackinder, Halford J. *Democratic Ideals and Reality* (New York: Holt, 1919).

• Mather, John R. *Climatology: Fundamentals and Applications* (New York: McGraw-Hill, 1974).

Mellor, Roy E.H. *The Soviet Union and Its Geographical Problems* (London: Macmillan, 1982).

Nahaylo, Bohdan & Swoboda, Victor. *Soviet Disunion: A History of the Nationalities Problem in the U.S.S.R.* (New York: Free Press, 1990).

Nijman, Jan. *The Geopolitics of Power and Conflict: Superpowers in the International System, 1945–1992* (London & New York: Belhaven Press, 1993).

Parker, William H. *The Soviet Union* (London & New York: Longman, 2 rev. ed., 1983).

Peterson, D.J. *Troubled Lands: The Legacy of Soviet Environmental Destruction* (Boulder, Col.: Westview Press, 1993).

Pryde, Philip R. *Environmental Management in the Soviet Union* (New York: Cambridge University Press, 1991).

Remnick, David. *Lenin's Tomb: The Last Days of the Soviet Empire* (New York: Random House, 1993).

Rodgers, Allan, ed. *The Soviet Far East: Geographical Perspectives on Development* (London & New York: Routledge, 1990).

"Russia Reborn: A Survey of Russia," *The Economist,* December 5, 1992, Special Insert, pp. 3–26.

Sagers, Matthew J. & Green, Milford B. The *Transportation of Soviet Energy Resources* (Totowa, N.J.: Rowman & Littlefield, 1986).

Sagers, Matthew J. & Shabad, Theodore. *The Chemical Industry in the U.S.S.R.: An Economic Geography* (Boulder, Col.: Westview Press, 1990).

Smith, Hedrick. *The New Russians* (New York: Avon Trade Paperbacks, updated version, 1991).

Stewart, John Massey, ed. *The Soviet Environment: Problems, Policies and Politics* (New York: Cambridge University Press, 1992).

• Symons, Leslie, ed. *The Soviet Union: A Systematic Geography* (London & New York: Routledge, 2 rev. ed., 1990).

"The New Russia: Special Report," *Time,* December 7, 1992, pp. 32–69.

Treadgold, Donald W. *Twentieth-Century Russia* (Boulder, Col.: Westview Press, 7 rev. ed., 1990).

U.S.S.R. Energy Atlas (Washington: U.S. Central Intelligence Agency, 1985).

Wixman, Ronald. *The Peoples of the U.S.S.R.: An Ethnographic Handbook* (Armonk, N.Y.: M.E. Sharpe, 1984).

Wood, Alan, ed. *Siberia: Problems and Prospects for Regional Development* (London & New York: Methuen, 1987).

Wood, Alan & French, Richard A., eds. *The Development of Siberia: People and Resources* (Basingstoke, UK: Macmillan, 1989).

Zum Brunnen, Craig & Osleeb, Jeffrey P. *The Soviet Iron and Steel Industry* (Totowa, N.J.: Rowman & Littlefield, 1986).

TRANSCAUCASIA

Academy of Sciences of the Georgian S.S.R. *Soviet Georgia: Its Geography, History, and Economy* (Moscow: Progress Publishers, 1967).

Akiner, Shirin. *Islamic Peoples of the Soviet Union* (London: Kegan Paul International, 2 rev. ed., 1983).

"Armenia and Azerbaijan: Still Fighting," *The Economist*, January 9, 1993, 47–48.

Bater, James H. & French, Richard A., eds. *Studies in Russian Historical Geography* (New York: Academic Press, 2 vols., 1983).

Bohlen, Celestine. "Blockade and Winter Deepen Misery in Armenia," *New York Times*, February 7, 1993, 4.

Bonner, Raymond. "War in Caucasus Shows Ethnic Hate's Front Line," *New York Times*, August 2, 1993, A4.

Bradshaw, Michael J., ed. *The Soviet Union: A New Regional Geography?* (New York: Wiley/Longman, 1991).

Chew, Allen F. *Atlas of Russian History: Eleven Centuries of Changing Borders* (New Haven: Yale University Press, 1967).

Clem, Ralph S. "Russians and Others: Ethnic Tensions in the Soviet Union," *Focus,* September-October, 1980.

• Cole, John P. *Geography of the Soviet Union* (Stoneham, Mass.: Butterworth, 1984).

"Georgia: Divided They Fight," *The Economist*, March 2, 1991, 52.

Geiger, Bernhard. *Peoples and Languages of the Caucasus* (The Hague, Neth.; Mouton, 1959).

Hajda, Lubomyr & Beissinger, Mark, eds. *The Nationalities Factor in Soviet Politics and Society* (Boulder, Colo.: Westview Press, 1990).

Hockstader, Lee. "The Armenians and Azerbaijanis Go Over the Edge: An Ethnic War Threatens to Draw in the Region," *Washington Post Weekly*, Sept. 20–26, 1993, 17.

• Howe, G. Melvyn. *The Soviet Union: A Geographical Study* (London & New York: Longman, 2 rev. ed., 1986).

Karklins, Rasma. *Ethnic Relations in the U.S.S.R.: The Perspective from Below* (Winchester, Mass.: Allen & Unwin, 1986).

Kozlov, V. *The Peoples of the Soviet Union* (London: Hutchinson University Library, 1988).

Lang, David Marshall. *The Georgians* (New York: Praeger, 1966).

Lewis, Robert A., ed. *Geographic Perspectives on Soviet Central Asia* (London & New York: Routledge, 1992).

• Lydolph, Paul E. *Geography of the U.S.S.R.* (Elkhart Lake, Wisc.: Misty Valley Publishing, 1990).

Nahaylo, Bohdan & Swoboda, Victor. *Soviet Disunion: A History of the Nationalities Problem in the U.S.S.R.* (New York: Free Press, 1990).

Ratnieks, H. "Power Reserve Beyond the Caucasus," *Geographical Magazine*, Vol. 48, No. 7, 1976, 397–398.

Schmemann, Serge. "War and Politics Blocking Prosperity for Azerbaijan: New Oil Rich Nation Is Struggling for Stability," *New York Times,* July 9, 1993, A1, A6.

Shaginyan, Marietta. *Journey Through Soviet Armenia* (Moscow: Foreign Languages Publishing House, 1954).

Shapiro, Margaret. " 'Welcome to Hell' in Armenia: War and Blockades Have Left the Country Freezing to Death in the Dark," *Washington Post Weekly*, February 8–14, 1993, 12.

"Soviet Transcaucasus: A Mess on a Map," *The Economist*, September 28, 1991, 56.

Symons, Leslie, ed. *The Soviet Union: A Systematic Geography* (London & New York: Routledge, 2 rev. ed., 1990).

"The Transcaucasus: In the Mountains Between Russia and Iraq Seven Wars are Going On, Almost Unnoticed by the Outside World," *The Economist*, July 24, 1993, 55–56.

Wixman, Ronald. *Language Aspects of the Ethnic Patterns and Processes in the North Caucasus* (Chicago: University of Chicago, Department of Geography, Research paper No. 191, 1980).

Wixman, Ronald. *The Peoples of the U.S.S.R.: An Ethnographic Handbook* (Armonk, N.Y.: M.E. Sharpe, 1984).

CHAPTER 3

Adams, John S. "Residential Structure of Midwestern Cities," *Annals of the Association of American Geographers*, 60 (1970): 37–62. Model diagram adapted from p. 56.

Allen, James P. & Turner, Eugene J. *We the People: An Atlas of America's Ethnic Diversity* (New York: Macmillan, 1987).

Atlas of North America: Space Age Portrait of a Continent (Washington: National Geographic Society, 1985).

• Atwood, Wallace W. *The Physiographic Provinces of North America* (New York: Ginn, 1940).

Bell, Daniel. *The Coming of Postindustrial Society* (New York: Basic Books, 1973). Quotation taken from p. xvi of the 1976 [paperback] Foreword.

Berry, Brian J.L. "The Decline of the Aging Metropolis: Cultural Bases and Social Process," in Sternlieb, George & Hughes, James W., eds., *Post-Industrial America: Metropolitan Decline and Inter-Regional Job Shifts* (New Brunswick, N.J.: Center for Urban Policy Research, Rutgers University, 1975), pp. 175–185.

• Birdsall, Stephen S. & Florin, John W. *Regional Landscapes of the United States and Canada* (New York: John Wiley & Sons, 4 rev. ed., 1992).

Bone, Robert M. *The Geography of the Canadian North: Issues and Challenges* (Toronto: Oxford University Press, 1992).

Borchert, John R. "American Metropolitan Evolution," *Geographical Review*, 57 (1967): 301–332.

Borchert, John R. "Futures of American Cities," in Hart, John Fraser, ed., *Our Changing Cities* (Baltimore: Johns Hopkins University Press, 1991), pp. 218–250.

"California: The Endangered Dream," *Time*, Special Issue, November 18, 1991.

"Canada: Back on Track," *Time*, December 21, 1992, pp. 48–49.

• Christian, Charles M. & Harper, Robert A., eds. *Modern Metropolitan Systems* (Columbus: Charles E. Merrill, 1982).

• Clark, David. *Post-Industrial America: A Geographical Perspective* (London & New York: Methuen, 1985).

Conzen, Michael P., ed. *The Making of the American Landscape* (New York: HarperCollins Academic, 1990).

"For Want of Glue: A Survey of Canada," *The Economist*, June 29, 1991, 20-page supplement.

Garreau, Joel. *Edge City: Life on the New Frontier* (New York: Doubleday, 1991).

Garreau, Joel. *The Nine Nations of North America* (Boston: Houghton Mifflin, 1981). Quotation from pp. 1–2.

Gastil, Raymond D. *Cultural Regions of the United States* (Seattle: University of Washington Press, 1975).

Gottmann, Jean. *Megalopolis: The Urbanized Northeastern Seaboard of the United States* (New York: Twentieth Century Fund, 1961).

Granatstein, J.L. & Hillmer, Norman. *For Better or For Worse: Canada and the United States to the 1990s* (Toronto: Copp Clark Pittman Ltd., 1991).

Harris, Chauncy D. & Ullman, Edward L. "The Nature of Cities," *Annals of the American Academy of Political and Social Science*, 242 (1945): 7–17.

Harris, R. Cole. "Regionalism and the Canadian Archipelago," in McCann, Lawrence D., ed., *Heartland and Hinterland: A Geography of Canada* (Scarborough, Ont.: Prentice-Hall Canada, 1982), pp. 458–484.

Hart, John Fraser. *The Land That Feeds Us* (New York: W.W. Norton, 1991).

Hart, John Fraser, ed. *Our Changing Cities* (Baltimore: Johns Hopkins University Press, 1991).

• Hartshorn, Truman A. *Interpreting the City: An Urban Geography* (New York: John Wiley & Sons, 2 rev. ed., 1992).

Herzog, Lawrence A. *Where North Meets South: Cities, Space, and Politics on the U.S.-Mexico Border* (Austin, Tex.: University of Texas Press, 1990).

Historical Atlas of the United States (Washington: National Geographic Society, centennial ed., 1988).

• Hunt, Charles B. *Natural Regions of the United States and Canada* (San Francisco: W.H. Freeman, 2 rev. ed., 1974).

Janelle, Donald G., ed. *Geographical Snapshots of North America* (New York: Guilford Press, 1992).

Knox, Paul L. *Urbanization: An Introduction to Urban Geography* (Englewood Cliffs, N.J.: Prentice-Hall, 1994).

Knox, Paul L., et al. *The United States: A Contemporary Human Geography* (London & New York: Longman, 1988).

Malcolm, Andrew H. *The Canadians* (New York: Times Books, 1985).

Mayer, Harold M. "Geography in City and Regional Planning," in Frazier, John W., ed., *Applied Geography: Selected Perspectives* (Englewood Cliffs, N.J.: Prentice-Hall, 1982). Quotation from p. 27.

• McCann, Lawrence D., ed. *Heartland and Hinterland: A Geography of Canada* (Scarborough, Ont.: Prentice-Hall Canada, 1982).

• McKnight, Tom L. *Regional Geography of the United States and Canada* (Englewood Cliffs, N.J.: Prentice-Hall, 1992).

Meinig, Donald W. *The Shaping of America: A Geographical Perspective on 500 Years of History; Volume 1, Atlantic America, 1492–1800* (New Haven: Yale University Press, 1986).

Meinig, Donald W. *The Shaping of America: A Geographical Perspective on 500 Years of History; Volume 2, Continental America, 1800–1867* (New Haven: Yale University Press, 1993).

• Mitchell, Robert D. & Groves, Paul A., eds. *North America: The Historical Geography of a Changing Continent* (Totowa, N.J.: Rowman & Littlefield, 1987).

Muller, Peter O. *Contemporary Suburban America* (Englewood Cliffs, N.J.: Prentice-Hall, 1981).

Noble, Allen G., ed. *To Build in a New Land: Ethnic Landscapes in North America* (Baltimore: Johns Hopkins University Press, 1992).

• Paterson, John H. *North America: A Geography of Canada and the United States* (New York: Oxford University Press, 8 rev. ed., 1989).

• Putnam, Donald F. & Putnam, Robert G. *Canada: A Regional Analysis* (Toronto: Dent, 2 rev. ed., 1979).

Robinson, J. Lewis. *Concepts and Themes in the Regional Geography of Canada* (Vancouver: Talonbooks, 1983).

Rooney, John F. Jr., et al., eds. *This Remarkable Continent: An Atlas of United States and Canadian Society and Cultures* (College Station, Tex.: Texas A&M University Press, 1982).

Vance, James E., Jr. *The Continuing City: Urban Morphology in Western Civilization* (Baltimore: Johns Hopkins University Press, 1990).

Vance, James E., Jr. *This Scene of Man: The Role and Structure of the City in the Geography of Western Civilization* (New York: Harper's College Press, 1977). Urban realms model discussed on pp. 411–416.

Wheeler, James O. & Muller, Peter O. *Economic Geography* (New York: John Wiley & Sons, 2 rev. ed., 1986). Map adapted from p. 330.

Wilson, Alexander. *The Culture of Nature: North American Landscape from Disney to the Exxon Valdez* (Cambridge, Mass.: Blackwell, 1992).

Woodward, C. Vann. "The South Tomorrow," *Time*, "The South Today"—Special Issue, September 27, 1976. Quotation from p. 99.

Yeates, Maurice H. *Main Street: Windsor to Quebec City* (Toronto: Macmillan of Canada, 1975).

• Yeates, Maurice H. *The North American City* (New York: Harper & Row, 4 rev. ed., 1990). Canadian urban evolution model on pp. 60–67.

• Zelinsky, Wilbur. *The Cultural Geography of the United States: A Revised Edition* (Englewood Cliffs, N.J.: Prentice-Hall, 2 rev. ed., 1992).

CHAPTER 4

Amsden, Alice. *Asia's Next Giant: Late Industrialization in South Korea* (New York: Oxford University Press, 1989).

• Barrett, Rees D. & Ford, Roslyn A. *Patterns in the Human Geography of Australia* (South Melbourne: Macmillan of Australia, 1987).

Borthwick, Mark. *Pacific Century: The Emergence of Modern Pacific Asia* (Boulder, Col.: Westview Press, 1992).

Britton, Steve, et al., eds. *Changing Places in New Zealand: A Geography of Restructuring* (Dunedin, N. Z.: Allied Press, 1992).

• Burks, Ardath W. *Japan: A Postindustrial Power* (Boulder, Col.: Westview Press, 3 rev. ed., 1991).

Chapman, Graham P. & Baker, Kathleen M., eds. *The Changing Geography of Asia* (London & New York: Routledge, 1992).

• Chisholm, Michael. *Modern World Development* (Totowa, N. J.: Barnes & Noble, 1982).

Chiu, T. N. & So, C. L., eds. *A Geography of Hong Kong* (New York: Oxford University Press, 2 rev. ed., 1987).

Chowdhury, Anis & Islam, Iyanatul. *The Newly Industrializing Economies of East Asia* (London & New York: Routledge, 1993).

• Conkling, Edgar & McConnell, James. "The World's New Economic Powerhouse," *Focus*, January 1985, pp. 2–7.

Corey, Kenneth E. "Singapore: The Planned New City of the Pacific Rim," in Blakely, Edward J. & Stimson, Robert J., eds., *New Cities of the Pacific Rim* (Berkeley: University of California, Institute of Urban and Regional Development, Monograph 43, 1992).

Cybriwsky, Roman A. *Tokyo: The Changing Profile of an Urban Giant* (Boston: G. K. Hall, 1991).

de Souza, Anthony R. & Porter, Philip W. *The Underdevelopment and Modernization of the Third World* (Washington, D.C.: Association of American Geographers, Commission on College Geography, Resource Paper No. 28, 1974).

Dickenson, John P., et al. *A Geography of the Third World* (London & New York: Methuen, 1983).

Dirlik, Arif, ed. *What Is in a Rim? Critical Perspectives on the Pacific Region Idea* (Boulder, Col.: Westview Press, 1993).

Dixon, Chris. *South East Asia in the World Economy* (New York: Cambridge University Press, 1991).

• Drakakis-Smith, David. *Pacific Asia* (London & New York: Routledge, 1992).

Drakakis-Smith, David & Dixon, Chris, eds. *Economic and Social Development in Pacific Asia* (London & New York: Routledge, 1993).

"Environment and Development in Australia," *Australian Geographer*, 19 (May 1988): 3–220.

Eyre, John D. *Nagoya: The Changing Geography of a Japanese Regional Metropolis* (Chapel Hill: University of North Carolina, Studies in Geography No. 17, 1982).

Goldstein, Steven M., ed. *Minidragons: Fragile Economic Miracles in the Pacific* (Boulder, Col.: Westview Press, 1991).

Harris, Chauncy D. "The Urban and Industrial Transformation of Japan," *Geographical Review*, 72 (January 1982): 50–89.

• Heathcote, Ronald L., ed. *The Australian Experience: Essays in Australian Land Settlement and Resource Management* (Melbourne: Longman Cheshire, 1988).

Ho, Samuel P. S. *Economic Development of Taiwan* (New Haven, Conn.: Yale University Press, 1978).

Hoare, James & Pares, Susan. *Korea: An Introduction* (London & New York: Routledge, 1988).

• Hodder, Rupert. *The West Pacific Rim: An Introduction* (New York: John Wiley & Sons, 1992).

Inglis, C., et al., eds. *Asians in Australia: The Dynamics of Migration and Settlement* (Sydney: Allen & Unwin, 1992).

Jao, Y.C. & Leung, Chi-Keung, eds. *China's Special Economic Zones: Policies, Problems and Prospects* (New York: Oxford University Press, 1986).

Japan: An Illustrated Encyclopedia, 2 vols. (New York: Kodansha International, 1993).

• Jeans, Dennis N., ed. *Australia: A Geography. Volume 1: The Natural Environment* (Sydney: Sydney University Press, 1986); *Volume 2: Space and Society* (Sydney: Sydney University Press, 1987).

Kelly, Ian. *Hong Kong: A Political-Geographic Analysis* (Honolulu: University Press of Hawaii, 1987).

• Kornhauser, David H. *Japan: Geographical Background to Urban-Industrial Development* (London & New York: Longman, 2 rev. ed., 1982).

Lewis, George J. *Human Migration: A Geographical Perspective* (New York: St. Martin's Press, 1982).

Li, Victor Hao. "The New Orient Express," *World Monitor Magazine*, November 1988, pp. 24–35.

Linder, Staffan B. *The Pacific Century: Economic and Political Consequences of Asian-Pacific Dynamism* (Stanford, Calif.: Stanford University Press, 1986).

Lines, William J. *Taming the Great South Land: A History of the Conquest of Nature in Australia* (Berkeley: University of California Press, 1992).

Lo, Chor-Pang. *Hong Kong* (New York: Wiley/Belhaven, 1992).

MacDonald, Donald. *A Geography of Modern Japan* (Ashford, U.K.: Paul Norbury, 1985).

Morton, Harry & Johnston, Carol M. *The Farthest Corner: New Zealand—A Twice Discovered Land* (Honolulu: University of Hawaii Press, 1989).

Murata, Kiyogi & Ota, Isamu, eds. *An Industrial Geography of Japan* (New York: St. Martin's Press, 1980).

• Noh, Toshio & Kimura, John C., eds. *Japan: A Regional Geography of An Island Nation* (Tokyo: Teikoku-Shoin, 1985).

Ogawa, Naohiro, et al., eds. *Human Resources in Development Along the Asia-Pacific Rim* (Singapore: Oxford University Press, 1993).

Powell, Joseph M. *An Historical Geography of Modern Australia: The Restive Fringe* (New York: Cambridge University Press, 1988).

Reischauer, Edwin O. *The Japanese Today: Change and Continuity* (Cambridge, Mass.: Belknap/Harvard University Press, 1988).

Rich, David C. *The Industrial Geography of Australia* (Sydney: Methuen, 1987).

Rigg, Jonathan. *South-East Asia: A Region in Transition* (New York: HarperCollins Academic, 1990).

Rostow, Walt W. *The Stages of Economic Growth* (New York: Cambridge University Press, 2 rev. ed., 1971).

Sandhu, Kernial S. & Wheatley, Paul, eds. *Management of Success: The Moulding of Modern Singapore* (Singapore: Institute of Southeast Asian Studies, 1990).

Shenon, Philip. "Missing Out on [Southeast Asia's] Glittering Market," *New York Times*, September 12, 1993, 1F, 6F.

Shibusawa, Masahide, et al. *Pacific Asia in the 1990s* (London & New York: Routledge, 1991).

Smith, Michael, et al. *Asia's New Industrial World* (London & New York: Methuen, 1985).

Terrill, Ross. *The Australians* (New York: Simon & Schuster, 1987).

Theroux, Paul. "Going to See the Dragon: A Journey Through South China, Land of Capitalist Miracles, Where Yesterday's Rice Paddy Becomes Tomorrow's Metropolis, and a Thousand Factories Bloom," *Harper's Magazine*, October 1993, 33–56.

• Trewartha, Glenn T. *Japan: A Geography* (Madison: University of Wisconsin Press, 2 rev. ed., 1965).

Vogel, Ezra F. *The Four Little Dragons: The Spread of Industrialization in East Asia* (Cambridge, Mass.: Harvard University Press, 1991).

Yeung, Yue-Man & Xu-Wei Hu, eds. *China's Coastal Cities: Catalysts for Modernization* (Honolulu: University of Hawaii Press, 1992).

CHAPTER 5

• Augelli, John P. "The Rimland-Mainland Concept of Culture Areas in Middle America," *Annals of the Association of American Geographers*, 52 (1962): pp. 119–129.

• Blakemore, Harold & Smith, Clifford T., eds. *Latin America: Geographical Perspectives* (London & New York: Methuen, 2 rev. ed., 1983).

• Blouet, Brian W. & Blouet, Olwyn M., eds. *Latin America and the Caribbean: A Systematic and Regional Survey* (New York: John Wiley & Sons, 2 rev. ed., 1993).

Boehm, Richard G. & Visser, Sent, eds. *Latin America: Case Studies* (Dubuque, Iowa: Kendall/Hunt, 1984).

Booth, John A. & Walker, Thomas W. *Understanding Central America* (Boulder, Col.: Westview Press, 2 rev. ed., 1993).

• Butlin, Robin. *Historical Geography: Through the Gates of Space and Time* (London & New York: Routledge, 1993).

Butzer, Karl W., ed. *The Americas Before and After 1492: Current Geographical Research* (Cambridge, Mass.: Blackwell, 1992).

Caufield, Catherine. *In the Rainforest: Report from a Strange, Beautiful, Imperiled World* (Chicago: University of Chicago Press, rev. ed., 1991).

• Clark, Andrew H. "Historical Geography," in James, Preston E. & Jones, Clarence F., eds., *American Geography: Inventory and Prospect* (Syracuse, N.Y.: Syracuse University Press, 1954), pp. 70–105.

Collier, Simon, et al., eds. *The Cambridge Encyclopedia of Latin America and the Caribbean* (New York: Cambridge University Press, 2 rev. ed., 1992).

Crow, John A. *The Epic of Latin America* (Berkeley: University of California Press, 4 rev. ed., 1992).

Crowley, William & Griffin, Ernst C. "Political Upheaval in Central America," *Focus*, September-October 1983.

Davidson, William V. & Parsons, James, eds. *Historical Geography of Latin America* (Baton Rouge: Louisiana State University Press, 1980).

Dunkerley, James. *Power in the Isthmus: A Political History of Modern Central America* (New York: Verso/Routledge, 1989).

Farah, Douglas. "The Last Decade: Central America is Staggering Under Its '80s Legacy," *Washington Post Weekly*, June 14–20, 1993, pp. 6–7.

• Gilbert, Alan. *Latin America* (London & New York: Routledge, 1990).

Golden, Tim. "In Free Trade, Mexico Sees an Economy in U.S. Image," *New York Times*, July 23, 1992, pp. A1, C17.

Gourou, Pierre. *The Tropical World: Its Social and Economic Conditions and Its Future Status* (London & New York: Longman, 5 rev. ed., trans. Stanley H. Beaver, 1980).

Griffin, Ernst C. & Ford, Larry R. "Cities of Latin America," in Brunn, Stanley D. & Williams, Jack F., eds., *Cities of the World: World Regional Urban Development* (New York: HarperCollins, 2 rev. ed., 1993), pp. 224–265.

Helms, Mary W. *Middle America: A Culture History of Heartland and Frontiers* (Englewood Cliffs, N.J.: Prentice-Hall, 1975).

Herzog, Lawrence A. *Where North Meets South: Cities, Space, and Politics on the U.S.-Mexico Border* (Austin, Tex.: University of Texas Press, 1990).

Huntington, Ellsworth. *Mainsprings of Civilization* (New York: John Wiley & Sons, 1945).

• James, Preston E. & Minkel, Clarence W. *Latin America* (New York: John Wiley & Sons, 5 rev. ed., 1986), Chaps. 2–16.

Knight, Franklin W. & Palmer, Colin A., eds. *The Modern Caribbean* (Chapel Hill, N.C.: University of North Carolina Press, 1989).

Krauss, Clifford. *Inside Central America: Its People, Politics, and History* (New York: Touchstone/Simon & Schuster, 1991).

Kurlansky, Mark. *A Continent of Islands: Searching for the Caribbean Destiny* (Reading, Mass.: Addison-Wesley, 1992).

Levy, Daniel C. & Szekely, Gabriel. *Mexico: Paradoxes of Stability and Change* (Boulder, Col.: Westview Press, 2 rev. ed., 1987).

Lockhart, Douglas G., et al., eds. *The Development Process in Small Island States* (London & New York: Routledge, 1993).

• Lowenthal, David. *West Indian Societies* (New York: Oxford University Press, 1972).

"[Maquiladora] Manufacturing in Mexico: On Uncle Sam's Coat-Tails," *The Economist*, September 16, 1989, p. 82.

MacPherson, John. *Caribbean Lands* (London & New York: Longman, 4 rev. ed., 1980).

McCullough, David G. *The Path Between the Seas: The Creation of the Panama Canal, 1870–1914* (New York: Simon & Schuster, 1977).

Meinig, Donald W. "Symbolic Landscapes: Some Idealizations of American Communities," in Meinig, Donald W., ed., *The Interpretation of Ordinary Landscapes: Geographical Essays* (New York: Oxford University Press, 1979), pp. 164–192.

Miller, E. Willard & Miller, Ruby M. *The Third World—Caribbean Region* (Monticello, Ill.: Vance Bibliographies, No. 2918, 1990).

Miller, E. Willard & Miller, Ruby M. *The Third World—Central America* (Monticello, Ill.: Vance Bibliographies, No. 2841, 1990).

Miller, E. Willard & Miller, Ruby M. *The Third World—Latin America: Economic Aspects* (Monticello, Ill.: Vance Bibliographies, No. 2839, 1990).

Miller, E. Willard & Miller, Ruby M. *The Third World—Mexico* (Monticello, Ill.: Vance Bibliographies, No. 2844, 1990).

Myers, Norman. *The Primary Source: Tropical Forests and Our Future* (New York: W.W. Norton, 1984).

Pearce, Douglas. *Tourism Today: A Geographical Analysis* (London & New York: Longman, 1986).

Pick, James B. & Butler, Edgar W. *The Mexico Handbook: Economic and Demographic Maps and Statistics* (Boulder, Col.: Westview Press, 1993).

Potter, Robert B., ed. *Urbanization, Planning and Development in the Caribbean* (London & New York: Mansell, 1989).

Richardson, Bonham C. *The Caribbean in the Wider World, 1492–1992: A Regional Geography* (New York: Cambridge University Press, 1992).

Riding, Alan. *Distant Neighbors: A Portrait of the Mexicans* (New York: Alfred A. Knopf, 1985).

Sargent, Charles S., Jr. "The Latin American City," in Blouet, Brian W. & Blouet, Olwyn M., eds., *Latin America and the Caribbean: A Systematic and Regional Survey* (New York: John Wiley & Sons, 2 rev. ed., 1993), pp. 172–216. Quotation from p. 191; diagram adapted from p. 192.

Sauer, Carl O. "Foreword to Historical Geography," *Annals of the Association of American Geographers*, 31 (1941): 1–24.

Scott, Ian. *Urban and Spatial Development in Mexico* (Baltimore: Johns Hopkins University Press, 1982).

Sealey, Neil. *Caribbean World: A Complete Geography* (New York: Cambridge University Press, 1992).

Sklair, Leslie. *Assembling for Development: Maquila Industry in Mexico and the United States* (Boulder, Col.: Westview Press, 1989).

Stanislawski, Dan. "Early Spanish Town Planning in the New World," *Geographical Review*, 37 (1947): 94–105.

Turner, Billie Lee II. *Once Beneath the Forest* (Boulder, Col.: Westview Press, 1983).

Uchitelle, Louis. "America's Newest Industrial Belt: Northern Mexico," *New York Times*, March 21, 1993, pp. F1, F4.

Ward, Peter M. *Mexico City* (Boston: G.K. Hall, 1990).

Watts, David. *The West Indies: Patterns of Development, Culture and Environmental Change Since 1492* (New York: Cambridge University Press, 1987).

Weaver, Muriel P. *The Aztecs, Maya, and Their Predecessors: Archaeology of Middle America* (New York: Academic Press, 2 rev. ed., 1981).

• West, Robert C., Augelli, John P., et al. *Middle America: Its Lands and Peoples* (Englewood Cliffs, N.J.: Prentice-Hall, 3 rev. ed., 1989).

Wilken, Gene C. *Good Farmers: Traditional Agricultural Resource Management in Mexico and Central America* (Berkeley & Los Angeles: University of California Press, 1987).

CHAPTER 6

"Antarctica: Is Any Place Safe From Mankind?" *Time*, Cover Story, January 15, 1990, pp. 56–62.

Augelli, John P. "The Controversial Image of Latin America: A Geographer's View," *Journal of Geography*, 62 (1963): pp. 103–112. Quotation from p. 111.

Berry, Brian J.L., et al. *The Global Economy: Resource Use, Locational Choice, and International Trade* (Englewood Cliffs, N.J.: Prentice-Hall, 1993).

• Blakemore, Harold & Smith, Clifford T., eds. *Latin America: Geographical Perspectives* (London & New York: Methuen, 2 rev. ed., 1983).

• Blouet, Brian W. & Blouet, Olwyn M., eds. *Latin America and the Caribbean: A Systematic and Regional Survey* (New York: John Wiley & Sons, 2 rev. ed., 1993).

Boehm, Richard G. & Visser, Sent, eds. *Latin America: Case Studies* (Dubuque, Iowa: Kendall/Hunt, 1984).

Brawer, Moshe. *Atlas of South America* (New York: Simon & Schuster, 1991).

• Bromley, Rosemary D.F. & Bromley, Ray. *South American Development: A Geographical Introduction* (New York: Cambridge University Press, 2 rev. ed., 1988).

Brunn, Stanley D. & Williams, Jack F., eds. *Cities of the World: World Regional Urban Development* (New York: Harper-Collins, 2 rev. ed., 1993).

Butzer, Karl W., ed. *The Americas Before and After 1492: Current Geographical Research* (Cambridge, Mass.: Blackwell, 1992).

Caviedes, César N. *The Southern Cone: Realities of the Authoritarian State* (Totowa, N.J.: Rowman & Allanheld, 1984).

Child, Jack. *Geopolitics and Conflict in South America: Quarrels Among Neighbors* (Westport, Conn.: Praeger, 1985).

"Cocaine Wars: South America's Bloody Business," *Time*, February 25, 1985, pp. 26–35.

Collier, Simon, et al., eds. *The Cambridge Encyclopedia of Latin America and the Caribbean* (New York: Cambridge University Press, 2 rev. ed., 1992).

Crow, John A. *The Epic of Latin America* (Berkeley: University of California Press, 4 rev. ed., 1992).

de Blij, H.J. "A Regional Geography of Antarctica and the Southern Ocean," *University of Miami Law Review*, 33 (1978): 299–314.

Denevan, William M., ed. *Hispanic Lands and Peoples: Selected Writings of James J. Parsons* (Boulder, Col.: Westview Press, 1988).

Drakakis-Smith, David. *The Third World City* (Boston: Routledge & Kegan Paul, 1987).

Enders, Thomas O. & Mattione, Richard P. *Latin America: The Crisis of Debt and Growth* (Washington, D.C.: Brookings Institution, 1984).

Freeman, Michael. *Atlas of World Economy* (New York: Simon & Schuster, 1991).

Gilbert, Alan. *Latin America* (London & New York: Routledge, 1990).

Gilbert, Alan & Gugler, Josef. *Cities, Poverty and Development: Urbanization in the Third World* (New York: Oxford University Press, 2 rev. ed., 1992).

Griffin, Ernst C. & Ford, Larry R. "Cities of Latin America," in Brunn, Stanley D. & Williams, Jack F., eds. *Cities of the World: World Regional Urban Development* (New York: HarperCollins, 2 rev. ed., 1993), pp. 224–265.

• Griffin, Ernst C. & Ford, Larry R. "A Model of Latin American City Structure," *Geographical Review*, 70 (1980): 397–422. Model diagram adapted from p. 406.

Grigg, David B. *The Agricultural Systems of the World: An Evolutionary Approach* (London: Cambridge University Press, 1974).

Gwynne, Robert N. *Industrialization and Urbanization in Latin America* (Baltimore: Johns Hopkins University Press, 1986).

Hartshorn, Truman A. (& Alexander, John W.). *Economic Geography* (Englewood Cliffs, N.J.: Prentice-Hall, 3 rev. ed., 1988).

Hudson, Tim. "South American High: A Geography of Cocaine," *Focus*, January 1985, pp. 22–29.

• James, Preston E. & Minkel, Clarence W. *Latin America* (New York: John Wiley & Sons, 5 rev. ed., 1986), Chaps. 17–28, 37.

Knox, Paul L. & Agnew, John A. *The Geography of the World-Economy* (London & New York: Routledge, 1989).

Lowder, Stella. *The Geography of Third World Cities* (Totowa, N.J.: Barnes & Noble, 1986).

McColl, Robert W. "The Insurgent State: Territorial Bases of Revolution," *Annals of the Association of American Geographers*, 59 (1969): 613–631.

Maos, Jacob O. *The Spatial Organization of New Land Settlement in Latin America* (Boulder, Col.: Westview Press, 1984).

Miller, E. Willard & Miller, Ruby M. *The Third World—Colombia, Venezuela, Guyana, Suriname, French Guiana* (No. 2912); *Argentina and Uruguay* (No. 2913); *Brazil and Paraguay* (No. 2914); *Chile, Bolivia, Peru, Ecuador* (No. 2915) (Monticello, Ill.: Vance Bibliographies, 1990).

Miller, E. Willard & Miller, Ruby M. *The Third World—Latin America: Economic Aspects* (Monticello, Ill.: Vance Bibliographies, No. 2839, 1990).

Mörner, Magnus. *The Andean Past: Land, Societies and Conflict* (New York: Columbia University Press, 1985).

Morris, Arthur S. *Latin America: Economic Development and Regional Differentiation* (Totowa, N.J.: Barnes & Noble, 1981).

• Morris, Arthur S. *South America* (Totowa, N.J.: Barnes & Noble, 3 rev. ed., 1987).

Nash, Nathaniel C. "Chile: Japan's Backdoor to the West," *New York Times*, April 15, 1993, pp. C1, C6.

Nash, Nathaniel C. "Squalid Slums Grow as People Flood Latin America's Cities," *New York Times*, October 11, 1992, pp. 1, 10.

Odell, Peter R. & Preston, David A. *Economies and Societies in Latin America* (Chichester, U.K.: John Wiley & Sons, 2 rev. ed., 1978).

Potter, Robert B. *Third World Urbanization: Contemporary Issues in Geography* (New York: Oxford University Press, 1991).

• Preston, David A., ed. *Latin American Development: Geographical Perspectives* (London & New York: Longman, 1987).

Sanchez-Albornoz, Nicolas. *The Population of Latin America: A History* (Berkeley: University of California Press, 1974).

Steward, Julian H. & Faron, Louis C., eds. *Native Peoples of South America* (New York: McGraw-Hill, 1959).

Sugden, David E. *Arctic and Antarctic: A Modern Geographical Synthesis* (Totowa, N.J.: Barnes & Noble, 1982).

Theroux, Paul. *The Old Patagonian Express: By Train Through the Americas* (Boston: Houghton Mifflin, 1979).

Wagley, Charles. *The Latin American Tradition: Essays on the Unity and Diversity of Latin American Culture* (New York: Columbia University Press, 1968).

• Wheeler, James O. & Muller, Peter O. *Economic Geography* (New York: John Wiley & Sons, 2 rev. ed., 1986).

Wilkie, Richard W. *Latin American Population and Urbanization Analysis: Maps and Statistics, 1950–1982* (Westwood, Calif.: UCLA Latin American Center, 1984).

BRAZIL

• Becker, Bertha K. & Egler, Claudio A.G. *Brazil: A New Regional Power in the World-Economy—A Regional Geography* (New York: Cambridge University Press, 1992).

• Bromley, Rosemary D.F. & Bromley, Ray. *South American Development: A Geographical Introduction* (New York: Cambridge University Press, 2 rev. ed., 1988).

Burns, E. Bradford. *A History of Brazil* (New York: Columbia University Press, 2 rev. ed., 1980).

de Castro, Josué. *Death in the Northeast* (New York: Vintage Books, 1969).

• Dickenson, John P. *Brazil* (London & New York: Longman, 1983).

Dickinson, Robert E., ed. *The Geophysiology of Amazonia: Vegetation and Climate Interactions* (New York: John Wiley & Sons, 1987).

Fearnside, Philip M. *Human Carrying Capacity of the Brazilian Rainforest* (New York: Columbia University Press, 1986).

Font, Mauricio A. *Coffee, Contention, and Change in the Making of Modern Brazil* (Cambridge, Mass.: Blackwell, 1990).

Foresta, Ronald A. "Amazonia and the Politics of Geopolitics," *Geographical Review*, 82 (1992): 128–142.

Godfrey, Brian J. "Modernizing the Brazilian City," *Geographical Review*, 81 (1991): 18–34.

Hall, Anthony L. *Developing Amazonia: Deforestation and Social Conflict in Brazil's Carajás Programme* (Manchester, U.K.: Manchester University Press, 1989).

Haller, Archibald O. "A Socioeconomic Regionalization of Brazil," *Geographical Review*, 72 (1982): 450–464.

Hemming, John, ed. *Change in the Amazon Basin: Volume I, Man's Impact on Forests and Rivers; Volume II, The Frontier After a Decade of Colonization* (Manchester, U.K.: Manchester University Press, 1985).

• Henshall, Janet D. & Momsen, Richard P. *A Geography of Brazilian Development* (Boulder, Col.: Westview Press, 1974).

• James, Preston E. & Minkel, Clarence W. *Latin America* (New York: John Wiley & Sons, 5 rev. ed., 1986), Chaps. 29–36.

Kandell, Jonathan. *Passage Through El Dorado* (New York: William Morrow, 1984).

Katzman, Martin T. *Cities and Frontiers in Brazil: Regional Dimensions of Economic Development* (Cambridge, Mass.: Harvard University Press, 1977).

Lisansky, Judith. *Migrants to Amazonia: Spontaneous Colonization in the Brazilian Frontier* (Boulder, Col.: Westview Press, 1989).

Lobb, C. Gary. "Brazil," in Blouet, Brian W. & Blouet, Olwyn M., eds. *Latin America and the Caribbean: A Systematic and Regional Survey* (New York: John Wiley & Sons, 2 rev. ed., 1993), pp. 357–389.

• Margolis, Mac. *The Last New World: The Conquest of the Amazon Frontier* (New York: W.W. Norton, 1992).

Merrick, Thomas W. & Graham, Douglas H. *Population and Economic Development in Brazil: 1800 to the Present* (Baltimore: Johns Hopkins University Press, 1979).

Monmonier, Mark S. *Computer-Assisted Cartography: Principles and Prospects* (Englewood Cliffs, N.J.: Prentice-Hall, 1982).

Moran, Emilio F., ed. *The Dilemma of Amazonian Development* (Boulder, Col.: Westview Press, 1983).

Roberts, J. Timmons. "Squatters and Urban Growth in Amazonia," *Geographical Review*, 82 (1992): 441–457.

Schneider, Ronald M. *Order and Progress: A Political History of Brazil* (Boulder, Col.: Westview Press, 1991).

Schumann, Debra A. & Partridge, William L., eds. *The Human Ecology of Tropical Settlement* (Boulder, Col.: Westview Press, 1987).

Smith, Nigel J.H. *Rainforest Corridors: The Transamazonian Colonization Scheme* (Berkeley: University of California Press, 1982).

Stepan, Alfred, ed. *Democratizing Brazil: Problems of Transition and Consolidation* (New York: Oxford University Press, 1989).

"Survey of Brazil: The Blessed and the Cursed," *The Economist*, December 7, 1991, Special Insert, pp. 3–22.

"Torching the Amazon: Can the Rain Forest Be Saved?," *Time*, September 18, 1989, Cover Story, pp. 76–85.

Uys, Errol L. *Brazil* (New York: Simon & Schuster, 1986).

Wilkie, Richard W. *Latin American Population and Urbanization Analysis: Maps and Statistics, 1950-1982* (Westwood, Calif.: UCLA Latin American Center, 1984).

Wood, Charles H. & Magno de Carvalho, José A. *The Demography of Inequalities in Brazil* (New York: Cambridge University Press, 1988).

CHAPTER 7

Agnew, Clive & Anderson, Ewan. *Water Resources in the Arid Realm* (London & New York: Routledge, 1992).

Ahmed, Akbar S. *Discovering Islam: Making Sense of Muslim History and Society* (Boston: Routledge & Kegan Paul, 1987).

Ajami, Fouad. *The Arab Predicament: Arab Political Thought and Practice Since 1967* (New York: Cambridge University Press, updated ed., 1992).

Akiner, Shirin, ed. *Cultural Change and Continuity in Central Asia* (London & New York: Kegan Paul International, 1992).

Barakat, Halim. *The Arab World: Society, Culture, and State* (Berkeley: University of California Press, 1993).

• Beaumont, Peter et al. *The Middle East: A Geographical Study* (New York: Wiley/Halsted, 2 rev. ed., 1988).

• Blake, Gerald H., et al. *The Cambridge Atlas of the Middle East and North Africa* (Cambridge, UK: Cambridge University Press, 1988).

"Central Asia: The Silk Road Catches Fire," *The Economist*, December 26, 1992, pp. 44–46.

• Chapman, Graham P. & Baker, Kathleen M., eds. *The Changing Geography of Africa and the Middle East* (London & New York: Routledge, 1992).

• Chapman, Graham P. & Baker, Kathleen M., eds. *The Changing Geography of Asia* (London & New York: Routledge, 1992).

Cloudsley-Thompson, J.L., ed. *Sahara Desert* (Oxford, U.K.: Pergamon Press, 1984).

Cohen, Saul B. "Middle East Geopolitical Transformation: The Disappearance of a Shatter Belt," *Journal of Geography*, 91 (January/February, 1992): 2–10.

• Cook, Earl F. *Man, Energy, Society* (San Francisco: W.H. Freeman, 1976).

• Cressey, George B. *Crossroads: Land and Life in Southwest Asia* (Philadelphia: J.B. Lippincott, 1960).

Diuk, Nadia and Karatnycky, Adrian. *New Nations Rising: The Fall of the Soviets and the Challenge of Independence.* (New York: John Wiley and Sons, 1993.)

Dohrs, Fred E. & Sommers, Lawrence M., eds. *Cultural Geography: Selected Readings* (New York: Thomas Y. Crowell, 1967).

• Drysdale, Alasdair & Blake, Gerald H. *The Middle East and North Africa: A Political Geography* (New York: Oxford University Press, 1985).

Esposito, John L. *The Islamic Threat: Myth or Reality?* (New York: Oxford University Press, 1992).

• Fisher, William B. *The Middle East: A Physical, Social and Regional Geography* (London & New York: Methuen, 7 rev. ed., 1978).

Freeman-Grenville, G.S.P. *The Historical Atlas of the Middle East* (New York: Simon & Schuster, 1993).

Fromkin, David. *A Peace to End All Peace: Creating the Modern Middle East, 1914–1922* (New York: Henry Holt, 1989).

Fromkin, David. "How the Modern Middle East Map Came to be Drawn," *Smithsonian Magazine*, May 1991, pp. 132–148.

Fuller, Graham, et al. *Turkey's New Geopolitics: From the Balkans to Western China* (Boulder, Col.: Westview Press, 1993).

• Gould, Peter R. *Spatial Diffusion* (Washington: Association of American Geographers, Commission on College Geography, Resource Paper No. 4, 1969).

Gould, Peter R. *The Slow Plague: A Geography of the AIDS Pandemic* (Cambridge, Mass.: Blackwell, 1993).

Heathcote, Ronald L. *The Arid Lands: Their Use and Abuse* (London & New York: Longman, 1983).

• Held, Colbert C. *Middle East Patterns: Places, Peoples, and Politics* (Boulder, Col.: Westview Press, 2 rev. ed., 1993).

Hourani, Albert H. *A History of the Arab Peoples* (Cambridge, Mass.: Belknap/Harvard University Press, 1991).

"Islam's Path East," *Aramco World Magazine*, November-December, 1991, pp. 1–67.

Joffé, George, ed. *North Africa: Nation, State and Region* (London & New York: Routledge, 1993).

Khoury, Philip S. & Kostiner, Joseph, eds. *Tribes and State Formation in the Middle East* (Berkeley: University of California Press, 1991).

Kotlyakov, V.M. "The Aral Sea Basin: A Critical Environmental Zone," *Environment*, January-February, 1991, pp. 4–9, 36–38.

Kreyenbroek, Philip G. & Sperl, Stefan, eds. *The Kurds: A Contemporary Overview* (London & New York: Routledge, 1992).

Lamb, David. *The Arabs: Journeys Beyond the Mirage* (New York: Random House, 1987).

Lemarchand, Philippe, ed. *The Arab World, The Gulf, and the Middle East: An Atlas* (Boulder, Col.: Westview Press, 1992).

Lewis, Robert A., ed. *Geographic Perspectives on Soviet Central Asia* (London & New York: Routledge, 1992).

• Longrigg, Stephen H. *The Middle East: A Social Geography* (Chicago: Aldine, 2 rev. ed., 1970).

Mikesell, Marvin W. "Tradition and Innovation in Cultural Geography," *Annals of the Association of American Geographers*, 68 (1978): 1–16.

Miller, E. Willard & Miller, Ruby M. *The Third World—Islam, Muslims, Arab States* (Monticello, Ill.: Vance Bibliographies, No. 2735, 1989).

Miller, E. Willard & Miller, Ruby M. *The Third World—Middle East* (Monticello, Ill.: Vance Bibliographies, No. 2730, 1989).

Miller, Judith. "The Islamic Wave," *New York Times Magazine*, May 31, 1992, pp. 23–26, 38, 40, 42.

• Mostyn, Trevor, ed. *The Cambridge Encyclopedia of the Middle East and North Africa* (New York: Cambridge University Press, 1988).

"Muslims in the U.S.S.R.," *Aramco World Magazine*, January-February, 1990, 1–43.

Newman, David. *Population, Settlement, and Conflict: Israel and the West Bank* (New York: Cambridge University Press, 1991).

Omran, Abdel R. & Roudi, Farzaneh. "The Middle East Population Puzzle," *Population Bulletin*, 48 (July 1993): 1–40.

Peters, Joan. *From Time Immemorial: The Origins of the Arab-Israeli Conflict Over Palestine* (New York: Harper & Row, 1984).

Prescott, J.R.V. *Political Frontiers and Boundaries* (Winchester, Mass.: Allen & Unwin, 1987).

Rahman, Mushtaqur, ed. *Muslim World: Geography and Development* (Lanham, Md.: University Press of America, 1987).

Robinson, Francis. *Atlas of the Islamic World Since 1500* (New York: Facts on File, 1982).

Rogers, Everett M. *Diffusion of Innovations* (New York: Free Press, 3 rev. ed., 1983).

Rumer, Boris Z. *Soviet Central Asia: "A Tragic Experiment"* (Winchester, Mass.: Unwin Hyman, 1989).

Sayigh, Yusif A. *Elusive Development: From Dependence to Self-Reliance in the Arab Region* (London & New York: Routledge, 1991).

• Spencer, Joseph E. "The Growth of Cultural Geography," *American Behavioral Scientist*, 22 (1978): 79–92.

Starr, Joyce R. & Stoll, Daniel C., eds. *The Politics of Scarcity: Water in the Middle East* (Boulder, Col.: Westview Press, 1988).

Wagner, Philip L. & Mikesell, Marvin W., eds. *Readings in Cultural Geography* (Chicago: University of Chicago Press, 1962).

Wagstaff, J. Malcolm, ed. *Landscape and Culture: Geographical and Archaeological Perspectives* (New York: Basil Blackwell, 1987).

Wixman, Ronald. *The Peoples of the U.S.S.R.: An Ethnographic Handbook* (Armonk, N.Y.: M.E. Sharpe, 1984).

Wright, Robin B. *Sacred Rage: The Crusade of Modern Islam* (New York: Linden Press/Simon & Schuster, 1985).

CHAPTER 8

"Africa: The Scramble for Existence," *Time*, Cover Story, September 7, 1992, pp. 40–46.

Barnett, Tony & Blaikie, Piers. *AIDS in Africa: Its Present and Future Impact* (New York: Guilford Press, 1992).

• Bell, Morag. *Contemporary Africa: Development, Culture and the State* (White Plains, N.Y.: Longman, 1986).

• Best, Alan C.G. & de Blij, H.J. *African Survey* (New York: John Wiley & Sons, 1977).

Binns, Tony. *Tropical Africa* (London & New York: Routledge, 1994).

Black, Richard & Robinson, Vaughan, eds. *Geography and Refugees: Patterns and Processes of Change* (New York: Wiley/Belhaven, 1993).

Bohannan, Paul & Curtin, Philip D. *Africa and Africans* (Prospect Heights, Ill.: Waveland Press, 3 rev. ed., 1991).

Chapman, Graham P. & Baker, Kathleen M., eds. *The Changing Geography of Africa and the Middle East* (London & New York: Routledge, 1992).

Christopher, Anthony J. *Colonial Africa: An Historical Geography* (Totowa, N.J.: Barnes & Noble, 1984).

Cliff, Andrew D. & Haggett, Peter. *Atlas of Disease Distributions* (Cambridge, Mass.: Blackwell, 1989).

Coquery-Vidrovitch, Catherine. *Africa: Endurance and Change South of the Sahara* (Berkeley: University of California Press, trans. David Maisel, 1989).

Curtin, Philip D. *The Atlantic Slave Trade* (Madison: University of Wisconsin Press, 1969).

Davidson, Basil. *The Black Man's Burden: Africa and the Curse of the Nation-State* (New York: Times Books/Random House, 1992).

• de Blij, H.J. "Africa's Geomosaic Under Stress," *Journal of Geography*, 90 (January/February, 1991): 2–9.

Drakakis-Smith, David. *Urban and Regional Change in Southern Africa* (London & New York: Routledge, 1992).

Foster, Harold D. *Health, Disease and the Environment* (New York: John Wiley & Sons, 1992).

Freeman-Grenville, G.S.P. *The New Atlas of African History* (New York: Simon & Schuster, 1991).

Gesler, Wilbert M. *The Cultural Geography of Health Care* (Pittsburgh: University of Pittsburgh Press, 1991).

Gleave, Michael B., ed. *Tropical African Development: Geographical Perspectives* (New York: Wiley/Longman, 1992).

Goliber, Thomas J. "Africa's Expanding Population: Old Problems, New Policies," *Population Bulletin*, Vol. 44, November 1989, pp. 1–49.

• Gould, Peter R. *The Slow Plague: A Geography of the AIDS Pandemic* (Cambridge, Mass.: Blackwell, 1993).

Gourou, Pierre. *The Tropical World: Its Social and Economic Conditions and Its Future Status* (London & New York: Longman, 5 rev. ed., trans. Stanley H. Beaver, 1980).

Griffiths, Ieuan L.L., ed. *An Atlas of African Affairs* (London & New York: Methuen, 2 rev. ed., 1984).

• Grove, Alfred T. *The Changing Geography of Africa* (New York: Oxford University Press, 1989).

Harrison Church, Ronald J. *West Africa: A Study of the Environment and Man's Use of It* (London: Longman, 8 rev. ed., 1980).

Hodd, Michael. *Economies of Africa: Geography, Population, History, Stability, Structure, Performance, Forecasts* (Boston: G.K. Hall, 1991).

Howe, G. Melvyn. *A World Geography of Human Diseases* (New York: Academic Press, 1977).

Howe, G. Melvin, ed. *Global Geocancerology: World Geography of Human Cancers* (Edinburgh, U.K.: Churchill Livingstone, 1986).

Huke, Robert E. "The Green Revolution," *Journal of Geography*, 84 (1985): 248–254.

• Knight, C. Gregory & Newman, James L., eds. *Contemporary Africa: Geography and Change* (Englewood Cliffs, N.J.: Prentice-Hall, 1976).

Lamb, David. *The Africans* (New York: Random House, 1983).

Learmonth, Andrew T.A. *Disease Ecology: An Introduction* (New York: Blackwell, 1988).

Lewis, Laurence A. & Berry, Leonard. *African Environments and Resources* (Winchester, Mass.: Unwin Hyman, 1988).

Martin, Esmond B. & de Blij, H.J., eds. *African Perspectives: An Exchange of Essays on the Economic Geography of Nine African States* (London & New York: Methuen, 1981).

• Meade, Melinda S., et al. *Medical Geography* (New York: Guilford Press, 1987).

Mehretu, Assefa. *Regional Disparity in Sub-Saharan Africa: Structural Readjustment of Uneven Development* (Boulder, Col.: Westview Press, 1989).

Middleton, John, ed. *Encyclopedia of Sub-Saharan Africa* (New York: Simon & Schuster, 4 vols., 1994).

Miller, E. Willard & Miller, Ruby M. *The Third World—Africa: Economic Aspects* (Monticello, Ill.: Vance Bibliographies, No. 2968, 1990).

Miller, E. Willard & Miller, Ruby M. *The Third World—Africa: West Africa and Nigeria* (Nos. 2974 & 2975); *East Africa* (No.

2976); *Central Africa* (No. 2977); *Southern Africa* (No. 2978) (Monticello, Ill.: Vance Bibliographies, 1990).

• Moon, Graham & Jones, Kelvyn. *Health, Disease and Society: An Introduction to Medical Geography* (New York: Routledge & Kegan Paul, 1988).

Mortimore, Michael. *Adapting to Drought: Farmers, Famines and Desertification in West Africa* (New York: Cambridge University Press, 1989).

• Mountjoy, Alan & Hilling, David. *Africa: Geography and Development* (Totowa, N.J.: Barnes & Noble, 1987).

Murdock, George P. *Africa: Its Peoples and Their Culture History* (New York: McGraw-Hill, 1959).

O'Connor, Anthony M. *Poverty in Africa: A Geographical Approach* (New York: Columbia University Press, 1991).

O'Connor, Anthony M. *The Geography of Tropical African Development: A Study of Spatial Patterns of Economic Change Since Independence* (Elmsford, N.Y.: Pergamon, 2 rev. ed., 1978).

Oliver, Roland. *The African Experience* (London: Weidenfeld & Nicolson, 1991).

Oliver, Roland & Crowder, Michael, eds. *The Cambridge Encyclopedia of Africa* (Cambridge, U.K.: Cambridge University Press, 1981).

Pakenham, Thomas. *The Scramble for Africa: The White Man's Conquest of the Dark Continent From 1876 to 1912* (New York: Random House, 1991).

Phillips, David R. *Health and Health Care in the Third World* (New York: John Wiley & Sons, 1990).

Pritchard, J.M. *Landform and Landscape in Africa* (London: Edward Arnold, 1979).

• Senior, Michael & Okunrotifa, P. *A Regional Geography of Africa* (London & New York: Longman, 1983).

Shannon, Gary W., Pyle, Gerald F., & Bashshur, Rashid L. *The Geography of AIDS: Origins and Course of an Epidemic* (New York: Guilford Press, 1991).

Siddle, David & Swindell, Ken. *Rural Change in Tropical Africa: From Colonies to Nation-States* (Cambridge, Mass.: Blackwell, 1990).

Stock, Robert F. *Cholera in Africa: Diffusion of the Disease 1970–1975, with Particular Emphasis on West Africa* (London: International African Institute, 1976).

Stren, Richard & White, Rodney, eds. *African Cities in Crisis: Managing Rapid Urban Growth* (Boulder, Col.: Westview Press, 1989).

Udo, Reuben K. *The Human Geography of Tropical Africa* (Exeter, N.H.: Heinemann Educational Books, 1982).

Whitaker, Jennifer S. *How Can Africa Survive?* (New York: Harper & Row, 1988).

World Bank. *Sub-Saharan Africa: From Crisis to Sustainable Growth* (Washington, D.C.: World Bank, 1989).

SOUTH AFRICA

"After Apartheid: A Survey of South Africa," *The Economist*, November 3, 1990, 26 pp.

Berger, Peter L. & Godsell, Bobby, eds. *A Future South Africa: Visions, Strategies, and Realities* (Boulder, Col.: Westview Press, 1989).

• Best, Alan C.G. & de Blij, H.J. *African Survey* (New York: John Wiley & Sons, 1977), Chap. 20.

Blumenfeld, Jesmond, ed. *South Africa in Crisis* (Beckenham, U.K.: Croom Helm, 1987).

Board, Christopher, et al. "The Structure of the South African Space-Economy: An Integrated Approach," *Regional Studies*, 4 (1970): 357–392.

• Christopher, Anthony J. *South Africa* (London & New York: Longman, 1982).

Christopher, Anthony J. *South Africa: The Impact of Past Geographies* (Cape Town: Juta & Co., 1984).

• Cole, Monica M. *South Africa* (London: Methuen, 2 rev. ed., 1966).

Crush, J. "The Southern African Regional Formation: A Geographical Prspective," *Tijdschrift Voor Economische en Sociale Geografie*, 73 (1982): 200–212.

de Blij, H.J. "Africa's Geomosaic Under Stress," *Journal of Geography*, 90 (January/February, 1991): 2–9.

• Drakakis-Smith, David. *Urban and Regional Change in Southern Africa* (London & New York: Routledge, 1992).

Drechsel, Willem & Waslander, Christine, eds. *South Africa in Focus* (London & New York: Kegan Paul International, 1993).

Fair, Thomas J.D. *South Africa: Spatial Frameworks for Development* (Cape Town: Juta & Co., 1982).

Leach, Graham. *The Afrikaners* (London: Macmillan, 1989).

Lelyveld, Joseph. *Move Your Shadow: South Africa, Black and White* (New York: Times Books, 1985).

Lemon, Anthony. "Toward the 'New South Africa,'" *Journal of Geography*, 90 (1991): 254–263.

Mandy, Nigel. *A City Divided: Johannesburg and Soweto* (New York: St. Martin's Press, 1985).

MacLeod, Scott. "Birth of a Nation," *Time*, June 14, 1993, 34–38.

McCarthy, J. & Smit, D. *South African City: Theory in Analysis and Planning* (Cape Town: Juta & Co., 1984).

Meredith, Martin. *In the Name of Apartheid: South Africa in the Postwar Period* (New York: Harper & Row, 1988).

Miller, E. Willard & Miller, Ruby M. *The Third World—Africa: Southern Africa* (Monticello, Ill.: Vance Bibliographies, No. 2978, 1990).

Murdock, George P. *Africa: Its Peoples and Their Culture History* (New York: McGraw-Hill, 1959).

Pirie, Gordon. "The Decivilizing Rails: Railways and Underdevelopment in Southern Africa," *Tijdschrift Voor Economische en Sociale Geografie*, 73 (1982): 221–228.

Rogerson, C.M., guest ed. "Urbanization in South Africa: Special Issue," *Urban Geography*, 9 (September-October 1989): 549–653.

Saul, John S. *South Africa: Apartheid and After* (Boulder, Colo.: Westview Press, 1990).

• Smith, David M. *Apartheid in South Africa* (Cambridge, U.K.: Cambridge University Press, 3 rev. ed., 1990).

• Smith, David, M. ed. *The Apartheid City and Beyond: Urbanization and Social Change in South Africa* (London & New York: Routledge, 1992).

"The Final Lap: A Survey of South Africa," *The Economist*, March 20, 1993, 26 pp.

• Thompson, Leonard. *A History of South Africa* (New Haven: Yale University Press, 1990).

Tomlinson, Richard. *Urbanization in Post-Apartheid South Africa* (New York: HarperCollins Academic, 1990).

Western, John C. *Outcast Cape Town* (Minneapolis: University of Minnesota Press, 1981).

CHAPTER 9

Bhardwaj, Surinder M. *Hindu Places of Pilgrimage in India: A Study in Cultural Geography* (Berkeley & Los Angeles: University of California Press, 1973).

Brush, John E. "Spatial Patterns of Population in Indian Cities," in Dwyer, Denis J., ed., *The City in the Third World* (New York: Barnes & Noble, 1974), pp. 105–132.

Burki, Shahid J. *Pakistan: A Nation in the Making* (Boulder, Col.: Westview Press, 1986).

Chapman, Graham P. & Baker, Kathleen M., eds. *The Changing Geography of Asia* (London & New York: Routledge, 1992).

Costa, Frank J., et al., eds. *Asian Urbanization: Problems and Processes* (Forestburgh, N.Y.: Lubrecht and Cramer, 1988).

Costa, Frank J., et al., eds. *Urbanization in Asia: Spatial Dimensions and Policy Issues* (Honolulu: University of Hawaii Press, 1989).

Crossette, Barbara. *India: Facing the Twenty-First Century* (Bloomington, IN: Indiana University Press, 1993).

Dumont, Louis. *Homo Hierarchus: The Caste System and Its Implications* (Chicago: University of Chicago Press, 1970).

• Dutt, Ashok K. "Cities of South Asia," in Brunn, Stanley D. & Williams, Jack F., eds., *Cities of the World: World Regional Urban Development* (New York: HarperCollins, 2 rev. ed., 1993), pp. 351-387.

• Dutt, Ashok K. & Geib, Margaret. *An Atlas of South Asia* (Boulder, Col.: Westview Press, 1987).

Dutt, Ashok K., et al. *India: Resources, Potentialities, and Planning* (Dubuque, Iowa: Kendall/Hunt, 1972).

Er-Rashid, Haroun. *Geography of Bangladesh* (Boulder, Col.: Westview Press, 1977).

• Farmer, Bertram H. *An Introduction to South Asia* (London & New York: Routledge, 2 rev. ed, 1993).

Frater, Alexander. *Chasing the Monsoon: A Modern Pilgrimage Through India* (New York: Alfred A. Knopf, 1991).

Gargan, Edward A. "Rising Tide of Hindu Hostility is Worrying India's Muslims," *New York Times*, September 17, 1993, A1, A5.

Gupte, Pranay. *Vengeance: India After the Assassination of Indira Gandhi* (New York: W.W. Norton, 1985).

Hennayake, Shantha K. & Duncan, James S. "A Disputed Homeland: Sri Lanka's Civil War," *Focus*, Spring 1987, pp. 20–27.

"Islam's Path East," *Aramco World Magazine*, November-December, 1991, pp. 1–67.

Ives, Jack D. & Messerli, Bruno. *Himalayan Dilemma: Reconciling Development and Conservation* (London & New York: Routledge, 1989).

Johnson, Basil L.C. *Bangladesh* (Totowa, N.J.: Barnes & Noble, 2 rev. ed., 1982).

• Johnson, Basil L.C. *Development in South Asia* (New York: Viking Penguin, 1983).

Johnson, Basil L.C. *India: Resources and Development* (Totowa, N.J.: Barnes & Noble, 1979).

• Jones, Huw R. *A Population Geography* (New York: Guilford Press, 2 rev. ed., 1991).

Karan, Pradyumna P. *Bhutan* (Lexington: University Press of Kentucky, 1967).

Karan, Pradyumna P. *Nepal: A Cultural and Physical Geography* (Lexington: University Press of Kentucky, 1960).

Kosinski, Leszek A. & Elahi, K. Maudood, eds. *Population Redistribution and Development in South Asia* (Hingham, Mass.: D. Reidel, 1985).

Lall, Arthur S. *The Emergence of Modern India* (New York: Columbia University Press, 1981).

Lapierre, Dominique. *The City of Joy* [*Calcutta*] (Garden City, N.Y.: Doubleday, trans. Kathryn Spink, 1985).

Lipner, Julius J. *Hinduism* (London & New York: Routledge, 1994).

Lukacs, John R., ed. *The People of South Asia: The Biological Anthropology of India, Pakistan, and Nepal* (New York: Plenum Press, 1984).

McGowan, William. *Only Man Is Vile: The Tragedy of Sri Lanka* (New York: Farrar, Straus & Giroux, 1992).

Manogaran, Chelvadurai. *Ethnic Conflict and Reconciliation in Sri Lanka* (Honolulu: University of Hawaii Press, 1987).

Miller, E. Willard & Miller, Ruby M. *The Third World—India, Sri Lanka, and Nepal* (Monticello, Ill.: Vance Bibliographies, No. 2654, 1989).

Miller, E. Willard & Miller, Ruby M. *The Third World—Southern Asia and Southeastern Asia* (Monticello, Ill.: Vance Bibliographies, No. 2649, 1989).

• Muthiah, S., ed. *An Atlas of India* (New York: Oxford University Press, 1992).

Nadkarni, M.V., et al, eds. *India: The Emerging Challenges* (Newbury Park, Calif.: Sage Publications, 1991).

Naipaul, V.S. *India: A Million Mutinies Now* (New York: Viking Press, 1990).

• Newman, James L. & Matzke, Gordon E. *Population: Patterns, Dynamics, and Prospects* (Englewood Cliffs, N.J.: Prentice-Hall, 1984).

• Noble, Allen G. & Dutt, Ashok K., eds. *India: Cultural Patterns and Processes* (Boulder, Col.: Westview Press, 1982).

Novak, James J. *Bangladesh: Reflections on the Water* (Bloomington, IN: Indiana University Press, 1993).

Parnwell, Michael. *Population Movements and the Third World* (London & New York: Routledge, 1993).

Ramachandram, R. *Urbanization and Urban Systems in India* (New Delhi: Oxford University Press, 1989).

Robinson, Francis, ed. *The Cambridge Encyclopedia of India, Pakistan, Bangladesh, Sri Lanka, Nepal, Bhutan, and the Maldives* (New York: Cambridge University Press, 1989).

Schwartzberg, Joseph E. *A Historical Atlas of South Asia: 2nd Impression, With Additional Material* (New York: Oxford University Press, 1992).

Siddiqi, A. *Pakistan: Its Resources and Development* (Hong Kong: Asian Research Service, 1985).

Sopher, David E., ed. *An Exploration of India: Geographical Perspectives on Society and Culture* (Ithaca, N.Y.: Cornell University Press, 1980).

• Spate, Oskar H.K. & Learmonth, Andrew T.A. *India and Pakistan: A General and Regional Geography* (London: Methuen, 2 vols., 1971).

• Spencer, Joseph E. & Thomas, William L. Jr. *Asia, East by South: A Cultural Geography* (New York: John Wiley & Sons, 2 rev. ed., 1971).

Sukhwal, B.L. *India: Economic Resource Base and Contemporary Political Patterns* (New York: Envoy Press, 1987).

Tambiah, Stanley J. *Sri Lanka: Ethnic Fratricide and the Dismantling of Democracy* (Chicago: University of Chicago Press, 1986).

• Weeks, John R. *Population: An Introduction to Concepts and Issues* (Belmont, Calif.: Wadsworth, 5 rev. ed., 1992).

Weisman, Steven R. "India: Always Inventing Itself," *New York Times Magazine*, December 11, 1988, pp. 50–51, 62, 64, 110–111.

Wolpert, Stanley. *India* (Berkeley: University of California Press, 1991).

CHAPTER 10

Barnett, A. Doak. *China's Far West: Four Decades of Change* (Boulder, Col.: Westview Press, 1993).

Buchanan, Keith, et al. *China: The Land and People* (New York: Crown, 1981).

• Cannon, Terry & Jenkins, Alan, eds. *The Geography of Contemporary China: The Impact of Deng Xiaoping's Decade* (London & New York: Routledge, 1990).

Chapman, Graham P. & Baker, Kathleen M., eds. *The Changing Geography of Asia* (London & New York: Routledge, 1992).

Cheng, Chu-Yuan. *Behind the Tiananmen Square Massacre: Social, Political, and Economic Ferment in China* (Boulder, Col.: Westview Press, 1990).

• Cressey, George B. *Asia's Lands and Peoples: A Geography of One-Third of the Earth and Two-Thirds of Its People* (New York: McGraw-Hill, 3 rev. ed., 1963).

Fairbank, John King. *China: A New History* (Cambridge, Mass.: Belknap/Harvard University Press, 1992).

Geelan, P.J.M. & Twitchett, D.C., eds. *The Times Atlas of China* (New York: Van Nostrand, 2 rev. ed., 1984).

• Ginsburg, Norton S., ed. *The Pattern of Asia* (Englewood Cliffs, N.J.: Prentice-Hall, 1958).

Ginsburg, Norton S. & Lalor, Bernard A., eds. *China: The 80s Era* (Boulder, Col.: Westview Press, 1984).

Goodman, David, ed. *China's Regional Development* (London & New York: Routledge, 1989).

Guldin, Gregory E., ed. *Urbanizing China* (Westport, Conn.: Greenwood Press, 1992).

Hillel, Daniel. "Lash of the Dragon: China's Yellow River Remains Untamed," *Natural History*, August 1991, pp. 28–37.

Hook, Brian, ed. *The Cambridge Encyclopedia of China* (New York: Cambridge University Press, 1982).

Hsieh, Chiao-min & Hsieh, Jean Kan. *China: A Provincial Atlas* (New York: Macmillan, 1992).

Information China (Elmsford, N.Y.: Pergamon Press, 3 vols., 1988).

Jao, Y.C. & Leung, Chi-Keung, eds. *China's Special Economic Zones: Policies, Problems and Prospects* (New York: Oxford University Press, 1986).

• Jingzhi Sun, ed. *The Economic Geography of China* (New York: Oxford University Press, 1988).

Kirkby, Richard J.R. *Urbanization in China: Town and Country in a Developing Economy, 1949–2000 A.D.* (New York: Columbia University Press, 1985).

Knapp, Ronald G., ed. *Chinese Landscapes: The Village as Place* (Honolulu: University of Hawaii Press, 1992).

Lattimore, Owen. *Inner Asian Frontiers of China* (New York: American Geographical Society, 1940).

• Leeming, Frank. *The Changing Geography of China* (Cambridge, Mass.: Blackwell, 1993).

Ma, Laurence J.C. & Hanten, Edward W., eds. *Urban Development in Modern China* (Boulder, Col.: Westview Press, 1981).

• Mackerras, Colin & Yorke, Amanda. *The Cambridge Handbook of Contemporary China* (New York: Cambridge University Press, 1991).

Miller, E. Willard & Miller, Ruby M. *The Third World—Far East* (Monticello, Ill.: Vance Bibliographies, No. 2499, 1988).

Murphey, Rhoads. *The Fading of the Maoist Vision: City and Country in China's Development* (London & New York: Methuen, 1980).

Murphey, Rhoads, et al., eds. *The Chinese: Adapting the Past, Building the Future* (Ann Arbor, Mich.: University of Michigan, Center for Chinese Studies, 1986).

Pannell, Clifton W. "Regional Shifts in China's Industrial Output," *The Professional Geographer*, 40 (February 1988): 19–31.

Pannell, Clifton W., ed. *East Asia: Geographical and Historical Approaches to Foreign Area Studies* (Dubuque, Iowa: Kendall/Hunt, 1983).

• Pannell, Clifton W. & Ma, Laurence J.C. *China: The Geography of Development and Modernization* (New York: Halsted Press/V.H. Winston, 1983).

Pannell, Clifton W. & Torguson, Jeffrey S. "Interpreting Spatial Patterns From the 1990 China Census," *Geographical Review*, 81 (1991): 304–317.

Salisbury, Harrison E. *The Long March: The Untold Story* (New York: Harper & Row, 1985).

Schinz, Alfred. *Cities in China* (Berlin: Gebrüder Borntraeger, 1989).

• Selya, Roger Mark, ed. *The Geography of China, 1975–1991: An Annotated Bibliography* (East Lansing, Mich.: Asian Studies Center, Michigan State University, 1993).

Shenggen Fan. *Regional Productivity Growth in China's Agriculture* (Boulder, Col.: Westview Press, 1990).

Sit, Victor F.S., ed. *Chinese Cities: The Growth of the Metropolis Since 1949* (New York: Oxford University Press, 1985).

Sivin, Nathan, ed. *The Contemporary Atlas of China* (Boston: Houghton Mifflin, 1988).

Smil, Vaclav. *The Bad Earth: Environmental Degradation in China* (Armonk, N.Y.: M.E. Sharpe, 1984).

Smil, Vaclav. "China's Food," *Scientific American*, December 1985, pp. 116–124.

• Smith, Christopher J. *China: People and Places in the Land of One Billion* (Boulder, Col.: Westview Press, 1991).

"Special China Issue," *Focus*, Spring 1992, 37 pp.

Spence, Jonathan D. *The Search for Modern China* (New York: W.W. Norton, 1990).

• Spencer, Joseph E. & Thomas, William L. Jr. *Asia, East by South: A Cultural Geography* (New York: John Wiley & Sons, 2 rev. ed., 1971).

Terrill, Ross. *China in Our Time: The Epic Saga of the People's Republic From the Communist Victory to Tiananmen Square and Beyond* (New York: Simon & Schuster, 1992).

Theroux, Paul. "Going to See the Dragon: A Journey Through South China, Land of Capitalist Miracles, Where Yesterday's Rice Paddy Becomes Tomorrow's Metropolis, and a Thousand Factories Bloom," *Harper's Magazine*, October 1993, 33–56.

Theroux, Paul. *Riding the Iron Rooster: By Train Through China* (New York: G.P. Putnam's Sons, 1988).

• Tregear, Thomas R. *China: A Geographical Survey* (New York: John Wiley & Sons/Halsted Press, 1980).

Veeck, Gregory, ed. *The Uneven Landscape: Geographic Studies in Post-Reform China* (Baton Rouge, La.: Geoscience Publications, 1991).

Wittwer, Sylvan, et al. *Feeding a Billion: Frontiers of Chinese Agriculture* (East Lansing, Mich.: Michigan State University Press, 1987).

Wu Dunn, Cheryl. "As China Leaps Ahead, The Poor Slip Behind," *New York Times*, May 23, 1993, p. E-3.

Yeung, Yue-Man. *Changing Cities of Pacific Asia: A Scholarly Interpretation* (Hong Kong: The Chinese University Press, 1990).

• Yeung, Yue-Man & Xu-Wei Hu, eds. *China's Coastal Cities: Catalysts for Modernization* (Honolulu: University of Hawaii Press, 1992).

Zhao Songqiao. *Physical Geography of China* (New York & Beijing: John Wiley & Sons/Science Press, 1986).

CHAPTER 11

Black, Richard & Robinson, Vaughan, eds. *Geography and Refugees: Patterns and Processes of Change* (New York: Wiley/Belhaven, 1993).

Borthwick, Mark. *Pacific Century: The Emergence of Modern Pacific Asia* (Boulder, Col.: Westview Press, 1992).

Broek, Jan O.M. "Diversity and Unity in Southeast Asia," *Geographical Review*, 34 (1944): 175–195.

Burling, Robbins. *Hill Farms and Padi Fields: Life in Mainland Southeast Asia* (Englewood Cliffs, N.J.: Prentice-Hall, 1965).

Chapman, Graham P. & Baker, Kathleen M., eds. *The Changing Geography of Asia* (London & New York: Routledge, 1992).

Chisholm, Michael & Smith, David M., eds. *Shared Space, Divided Space: Essays on Conflict and Territorial Organization* (New York: HarperCollins Academic, 1990).

Cho, George. *The Malaysian Economy: Spatial Perspectives* (London & New York: Routledge, 1990).

Costa, Frank J., et al., eds. *Urbanization in Asia: Spatial Dimensions and Policy Issues* (Honolulu: University of Hawaii Press, 1989).

• Dixon, Chris. *South East Asia in the World Economy* (New York: Cambridge University Press, 1991).

Drakakis-Smith, David. *Pacific Asia* (London & New York: Routledge, 1992).

• Dutt, Ashok K., ed. *Southeast Asia: Realm of Contrasts* (Boulder, Col.: Westview Press, 3 rev. ed., 1985).

• Dwyer, Denis J., ed. *South East Asian Development* (New York: Wiley/Longman, 1990).

• Fisher, Charles A. *Southeast Asia: A Social, Economic and Political Geography* (New York: E.P. Dutton, 2 rev. ed., 1966).

• Fryer, Donald W. *Emerging South-East Asia: A Study in Growth and Stagnation* (New York: John Wiley & Sons, 2 rev. ed., 1979).

Ginsburg, Norton S., et al., eds. *The Extended Metropolis: Settlement Transition in Asia* (Honolulu: University of Hawaii Press, 1991).

• Glassner, Martin I. & de Blij, H.J. *Systematic Political Geography* (New York: John Wiley & Sons, 4 rev. ed., 1989).

"Good Morning, Vietnam," *Time*, February 15, 1993, pp. 42–44.

Hardjono, Joan, ed. *Indonesia: Resources, Ecology, and Environment* (Singapore: Oxford University Press, 1991).

Hart, Gillian, et al., eds. *Agrarian Transformations: Local Processes and the State in Southeast Asia* (Berkeley: University of California Press, 1989).

Hartshorne, Richard. "Suggestions on the Terminology of Political Boundaries," *Annals of the Association of American Geographers*, 26 (1936): 56–57.

• Hill, Ronald D., ed. *South-East Asia: A Systematic Geography* (New York: Oxford University Press, 1979).

Hussey, Antonia. "Rapid Industrialization in Thailand, 1986–1991," *Geographical Review*, 83 (1993): 14–28.

"Islam's Path East," *Aramco World Magazine*, November-December, 1991, pp. 1–67.

Karnow, Stanley. *In Our Image: America's Empire in the Philippines* (New York: Random House, 1989).

Leinbach, Thomas R. & Sien, Chia Lin, eds. *South-East Asian Transport* (New York: Oxford University Press, 1989).

Leinbach, Thomas R. & Ulack, Richard. "Cities of Southeast Asia," in Brunn, Stanley D. & Williams, Jack F., eds., *Cities of the World: World Regional Urban Development* (New York: HarperCollins, 2 rev. ed., 1993), pp. 389–429.

Leitner, Helga & Sheppard, Eric S. "Indonesia: Internal Conditions, the Global Economy, and Regional Development," *Journal of Geography*, 86 (November/December, 1987): 282–291.

McCloud, Donald G. *System and Process in Southeast Asia: The Evolution of a Region* (Boulder, Col.: Westview Press, 1986).

• McGee, Terence G. *The Southeast Asian City: A Social Geography* (New York: Praeger, 1967). Diagram adapted from p. 128.

Mellor, Roy E.H. *Nation, State, and Territory: A Political Geography* (London & New York: Routledge, 1989).

Miller, E. Willard & Miller, Ruby M. *The Third World—Southern Asia and Southeastern Asia* (Monticello, Ill.: Vance Bibliographies, No. 2649, 1989).

Parnwell, Michael. *Population Movements and the Third World* (London & New York: Routledge, 1993).

"Powerhouse Penang [Pinang]," *The Economist*, May 22, 1993, p. 39.

• Prescott, J.R.V. *Political Frontiers and Boundaries* (Winchester, Mass.: Allen & Unwin, 1987).

• Rigg, Jonathan. *Southeast Asia: A Region in Transition—A Thematic Human Geography of the ASEAN Region* (New York: HarperCollins Academic, 1991).

Rumley, Dennis & Minghi, Julian V., eds. *The Geography of Border Landscapes* (London & New York: Routledge, 1991).

SarDesai, D.R. *Southeast Asia: Past and Present* (Boulder, Col.: Westview Press, 2 rev. ed., 1989).

Shibusawa, Masahide, et al. *Pacific Asia in the 1990s* (London & New York: Routledge, 1991).

"South-East Asian Economies: Dreams of Gold," *The Economist*, March 20, 1993, pp. 21–24.

• Spencer, Joseph E. & Thomas, William L. Jr. *Asia, East by South: A Cultural Geography* (New York: John Wiley & Sons, 2 rev. ed., 1971).

Taylor, Peter J. & House, John W., eds. *Political Geography: Recent Advances and Future Directions* (Totowa, N.J.: Barnes & Noble, 1984).

• Ulack, Richard & Pauer, Gyula. *Atlas of Southeast Asia* (New York: Macmillan, 1988).

White, Gilbert F. "The Mekong River Plan," *Scientific American*, April 1963, pp. 49–60.

Yeung, Yue-Man. *Changing Cities of Pacific Asia: A Scholarly Interpretation* (Hong Kong: The Chinese University Press, 1990).

PACIFIC REALM

Bayliss-Smith, Timothy P., et al. *Islands, Islanders and the World: The Colonial and Post-Colonial Experience of Eastern Fiji* (New York: Cambridge University Press, 1988).

• Brookfield, Harold C., ed. *The Pacific in Transition: Geographical Perspectives on Adaptation and Change* (New York: St. Martin's Press, 1973).

Brookfield, Harold C. & Hart, Doreen. *Melanesia: A Geographical Interpretation of an Island World* (New York: Barnes & Noble, 1971).

Bunge, Frederica M. & Cooke, Melinda W., eds. *Oceania: A Regional Study* (Washington, D.C.: U.S. Government Printing Office, 1984).

Campbell, Ian C. *A History of the Pacific Islands* (Berkeley: University of California Press, 1990).

Carter, John, ed. *Pacific Islands Yearbook* (Sydney, Australia: Pacific Publications, annual).

Couper, Alastair D., ed. *Development and Social Change in the Pacific Islands* (London & New York: Routledge, 1989).

• Freeman, Otis W., ed. *Geography of the Pacific* (New York: John Wiley & Sons, 1951).

Friis, Herman R., ed. *The Pacific Basin: A History of Its Geographical Exploration* (New York: American Geographical Society, Special Publication No. 38, 1967).

Gourevitch, Peter A., guest ed. "The Pacific Region: Challenges to Policy and Theory," *Annals of the American Academy of Political and Social Science*, 505 (September 1989).

Grossman, Lawrence S. *Peasants, Subsistence Ecology, and Development in the Highlands of Papua New Guinea* (Princeton, N.J.: Princeton University Press, 1984).

Howard, A., ed. *Polynesia: Readings on a Culture Area* (Scranton, Penn.: Chandler, 1971).

• Howlett, Diana. *Papua New Guinea: Geography and Change* (Melbourne, Australia: Thomas Nelson, 1973).

Howse, Derek, ed. *Background to Discovery: Pacific Exploration from Dampier to Cook* (Berkeley: University of California Press, 1990).

Kissling, Christopher C., ed. *Transport and Communications for Pacific Microstates: Issues in Organization and Management* (Suva, Fiji: University of the South Pacific, Institute of Pacific Studies, 1984).

Kluge, P.F. *The Edge of Paradise: America in Micronesia* (New York: Random House, 1991).

Levison, Michael, et al. *The Settlement of Polynesia: A Computer Simulation* (Minneapolis: University of Minnesota Press, 1973).

Lockhart, Douglas G., et al., eds. *The Development Process in Small Island States* (London & New York: Routledge, 1993).

Mitchell, Andrew. *The Fragile South Pacific: An Ecological Odyssey* (Austin: University of Texas Press, 1991).

• "Mobility and Identity in the Island Pacific," Special Issue, *Pacific Viewpoint*, 26, No. 1 (1985).

Morgan, Joseph R., ed. *Hawaii* (Boulder, Col.: Westview Press, 1983).

Nunn, Patrick. *Oceanic Islands* (Cambridge, Mass.: Blackwell, 1993).

• Oliver, Douglas L. *The Pacific Islands* (Honolulu: University of Hawaii Press, 3 rev. ed., 1989).

• Peake, Martin. *Pacific People and Society* (New York: Cambridge University Press, 1992).

Quanchi, Max. *Pacific People and Change* (New York: Cambridge University Press, 1992).

Robillard, Albert B., ed. *Social Change in the Pacific Islands* (London & New York: Kegan Paul International, 1991).

Sager, Robert J. "The Pacific Islands: A New Geography," *Focus*, Summer 1988, pp. 10–14.

Segal, Gerald. *Rethinking the Pacific* (New York: Oxford University Press, 1990).

Spate, Oskar H.K. *The Spanish Lake: A History of the Pacific Since Magellan, Vol. I* (Beckenham, U.K.: Croom Helm, 1979).

Spate, Oskar H.K. *The Pacific Since Magellan, Vol. II: Monopolists and Freebooters* (Minneapolis: University of Minnesota Press, 1983).

Spate, Oskar H.K. *The Pacific Since Magellan, Vol. III: Paradise Found and Lost* (Minneapolis: University of Minnesota Press, 1989).

Theroux, Paul. *The Happy Isles of Oceania: Paddling the Pacific* (New York: G.P. Putnam's Sons, 1992).

Vayda, Andrew P., ed. *Peoples and Cultures of the Pacific* (New York: Natural History Press, 1968).

Ward, R. Gerard, ed. *Man in the Pacific Islands: Essays on Geographical Change in the Pacific* (New York: Oxford University Press, 1972).

Wurm, Stephen A. & Hattori, Shiro, eds. *Language Atlas of the Pacific Area* (Canberra, Australia: Australian Academy of the Humanities & Japan Academy, 1983).

GLOSSARY

Absolute location The position or place of a certain item on the surface of the earth as expressed in degrees, minutes, and seconds of **latitude**,* 0° to 90° north or south of the equator, and **longitude,** 0° to 180° east or west of the *prime meridian* passing through Greenwich, England (a suburb of London).

Accessibility The degree of ease with which it is possible to reach a certain location from other locations. Accessibility varies from place to place and can be measured.

Acculturation Cultural modification resulting from intercultural borrowing. In cultural geography, the term is used to designate the change that occurs in the culture of indigenous peoples when contact is made with a society that is technologically more advanced.

Acid rain A growing environmental peril whereby acidified rainwater severely damages plant and animal life. Caused by the oxides of sulfur and nitrogen that are released into the atmosphere when coal, oil, and natural gas are burned, especially in major manufacturing zones.

Agglomerated (nucleated) settlement A compact, closely packed settlement (usually a hamlet or larger village) sharply demarcated from adjoining farmlands.

Agglomeration Process involving the clustering or concentrating of people or activities. Often refers to manufacturing plants and businesses that benefit from close proximity because they share skilled-labor pools and technological and financial amenities.

Agrarian Relating to the use of land in rural communities, or to agricultural societies in general.

Agriculture The purposeful tending of crops and livestock in order to produce food and fiber.

Alluvial Refers to the mud, silt, and sand (*alluvium*) deposited by rivers and streams. *Alluvial plains* adjoin many larger rivers; they consist of such renewable deposits that are laid down during floods, creating fertile and productive soils. Alluvial **deltas** mark the mouths of rivers such as the Mississippi and the Nile.

Altiplano High-elevation plateau, basin, or valley between even higher mountain ranges. In the Andes Mountains of South America, altiplanos lie at 10,000 feet (3000 m) and even higher.

*Words in boldface type are defined elsewhere in this Glossary.

Altitudinal zonation Vertical regions defined by physical-environmental zones at various elevations, particularly in the highlands of South and Middle America. See *puna, tierra caliente, tierra fría, tierra helada,* and *tierra templada.*

Antecedent boundary A political boundary that existed before the **cultural landscape** emerged and stayed in place while people moved in to occupy the surrounding area. An example is the 49th parallel boundary, dividing the United States and Canada between the Pacific Ocean and Lake of the Woods in northernmost Minnesota.

Anthracite coal Highest carbon content coal (therefore of the highest quality), that was formed under conditions of high pressure and temperature that eliminated most impurities. Anthracite burns almost without smoke and produces high heat.

Apartheid Literally, "apartness." The Afrikaans term given to South Africa's policies of racial separation, and the highly segregated socio-geographical patterns they have produced—a system now being dismantled.

Aquaculture The use of a river segment or an artificial body of water such as a pond for the raising and harvesting of food products, including fish, shellfish, and even seaweed. Japan is among the world's leaders in aquaculture.

Aquifer An underground reservoir of water contained within a porous, water-bearing rock layer.

Arable Literally, cultivable. Land fit for cultivation by one farming method or another.

Archipelago A set of islands grouped closely together, usually elongated into a chain.

Area A term that refers to a part of the earth's surface with less specificity than **region**. For example, *urban area* alludes very generally to a place where urban development has taken place, whereas *urban region* requires certain specific criteria upon which a delimitation is based (e.g., the spatial extent of commuting or the built townscape).

Areal interdependence A term related to **functional specialization**. When one area produces certain goods or has certain raw materials or resources and another area has a different set of resources and produces different goods, their needs may be *complementary*; by exchanging raw materials and products, they can satisfy each other's requirements. The concepts of areal interdependence and **complementarity** are related: both have to do with exchange opportunities between regions.

Arithmetic density   A country's population, expressed as an average per unit area (square mile or square kilometer), without regard for its distribution or the limits of **arable** land—see also **physiologic density**.

Aryan   From the Sanskrit Arya ("noble"), a name applied to an ancient people who spoke an Indo-European language and who moved into northern India from the northwest. Although properly a language-related term, Aryan has assumed additional meanings, especially racial ones.

Atmosphere   The earth's envelope of gases that rests on the oceans and land surface and penetrates the open spaces within soils. This layer of nitrogen (78 percent), oxygen (21 percent), and traces of other gases is densest at the earth's surface and thins with altitude. It is held against the planet by the force of gravity.

Austrasia   New name for the western Pacific Rim where a significant regional realignment is now taking place. Includes rapidly-developing countries and parts of countries lining the Pacific from Japan's Hokkaido in the north to New Zealand in the south.

Autocratic   An autocratic government holds absolute power; rule is often by one person or a small group of persons who control the country by despotic means.

Balkanization   The fragmentation of a region into smaller, often hostile political units.

Barrio   Term meaning "neighborhood" in Spanish. Usually refers to an urban community in a Middle or South American city; also applied to low-income, inner-city concentrations of Hispanics in such southwestern U.S. cities as Los Angeles.

Bauxite   Aluminum ore; an earthy, reddish-colored material that usually contains some iron as well. Soil-forming processes such as leaching and redeposition of aluminum and iron compounds contribute to bauxite formation, and many deposits exist at shallow depths in the wet tropics.

Birth rate   The *crude birth rate* is expressed as the annual number of births per 1000 individuals within a given population.

Bituminous coal   Softer coal of lesser quality than **anthracite** (more impurities remain) but of higher grade than **lignite**. Usually found in relatively undisturbed, extensive horizontal layers, often close enough to the surface to permit strip-mining. When heated and converted to coking coal or *coke*, it is used to make iron and steel.

Break-of-bulk point   A location along a transport route where goods must be transferred from one carrier to another. In a port, the cargoes of oceangoing ships are unloaded and put on trains, trucks, or perhaps smaller river boats for inland distribution.

Buffer zone   A set of countries separating ideological or political adversaries. In southern Asia, Afghanistan, Nepal, and Bhutan were parts of a buffer zone between British and Russian-Chinese imperial spheres. Thailand was a *buffer state* between British and French colonial domains in mainland Southeast Asia.

Caliente   See *tierra caliente*.

Cartel   An international syndicate formed to promote common interests in some economic sphere through the formulation of joint pricing policies and the limitation of market options for consumers. The Organization of Petroleum Exporting Countries (OPEC) is a classic example.

Cartogram   A specially transformed map not based on traditional representations of scale or area; Fig. I-10 is a classic example, mapping the world in *population-space*.

Cartography   The art and science of making maps, including data compilation, layout, and design. Also concerned with the interpretation of mapped patterns.

Caste system   The strict social segregation of people—specifically in India's Hindu society—on the basis of ancestry and occupation.

Cay   Pronounced *kee*. A low-lying small island usually composed of coral and sand. Often part of an island chain such as the Florida Keys or the Bahamas archipelago.

Central business district (CBD)   The downtown heart of a central city, the CBD is marked by high land values, a concentration of business and commerce, and the clustering of the tallest buildings.

Centrality   The strength of an urban center in its capacity to attract producers and consumers to its facilities; a city's "reach" into the surrounding region.

Centrifugal forces   A term employed to designate forces that tend to divide a country—such as internal religious, linguistic, ethnic, or ideological differences.

Centripetal forces   Forces that unite and bind a country together—such as a strong national culture, shared ideological objectives, and a common faith.

Charismatic   Personal qualities of certain leaders that enable them to capture and hold the popular imagination, to secure the allegiance and even the devotion of the masses. Gandhi, Mao Zedong, and Franklin D. Roosevelt are good examples in this century.

China Proper   The eastern portion of China that contains most of the country's huge population, stretching from the Amur River in the far Northeast to the southern border with Vietnam. On Fig. 10-4, regions ① through ④.

City-state   An independent political entity consisting of a single city with (and sometimes without) an immediate **hinterland**. The ancient city-states of Greece have their modern equivalent in Singapore.

Climate   The long-term conditions (over at least 30 years) of aggregate weather over a region, summarized by averages and measures of variability; a synthesis of the succession of weather events we have learned to expect at any given location.

Coal   See **anthracite** and **bituminous** coal.

Collectivization   The reorganization of a country's agriculture that involves the expropriation of private holdings and their incorporation into relatively large-scale units, which are farmed and administered cooperatively by those who live there. This system transformed agriculture in the former Soviet Union, and went beyond the Soviet model in China's program of communization.

Colonialism See **imperialism**.

Common Market Name given to a group of 12 European countries that belong to a **supranational** association to promote their economic interests (see Fig. 1-8). Official name is *European Community* (*EC*).

Compact state A politico-geographical term to describe a state that possesses a roughly circular, oval, or rectangular territory in which the distance from the geometric center to any point on the boundary exhibits little variance. Cambodia, Uruguay, and Poland are examples of this shape category.

Complementarity Regional complementarity exists when two regions, through an exchange of raw materials and/or finished products, can specifically satisfy each other's demands.

Concentric zone model A structural model of the American central city that suggests the existence of five concentric land-use rings arranged around a common center (see Fig. 3-1A).

Condominium In political geography, this denotes the shared administration of a territory by two governments.

Coniferous forest A forest of cone-bearing, needleleaf evergreen trees with straight trunks and short branches, including spruce, fir, and pine.

Contagious diffusion The distance-controlled spreading of an idea, innovation, or some other item through a local population by contact from person to person—analogous to the communication of a contagious illness.

Contiguous A word of some importance to geographers that means, literally, to be in contact with, adjoining, or adjacent. Sometimes we hear the continental (*conterminous*) United States minus Alaska referred to as contiguous. Alaska is not contiguous to these "lower 48" states because Canada lies in between; neither is Hawaii, separated by over 2000 miles of ocean.

Continental drift The slow movement of continents controlled by the processes associated with **plate tectonics**.

Continental shelf Beyond the coastlines of the continents the surface beneath the water, in many offshore areas, declines very gently until the depth of about 660 feet (200 m). Beyond the 660-foot line the sea bottom usually drops off sharply, along the *continental slope*, toward the much deeper mid-oceanic basin. The submerged continental margin is called the continental shelf, and it extends from the shoreline to the upper edge of the continental slope.

Continentality The variation of the continental effect on air temperatures in the interior portions of the world's landmasses. The greater the distance from the moderating influence of an ocean, the greater the extreme in summer and winter temperatures. Continental interiors, of course, tend to be dry as well when the distance from oceanic moisture sources becomes considerable.

Conurbation General term used to identify large multi-metropolitan complexes formed by the coalescence of two or more major urban areas. The Boston-Washington **Megalopolis** along the U.S northeastern seaboard is an outstanding example.

Copra Meat of the coconut; fruit of the coconut palm.

Cordillera Mountain chain consisting of sets of parallel ranges, especially the Andes in northwestern South America.

Core area In geography, a term with several connotations. Core refers to the center, heart, or focus. The core area of a **nation-state** is constituted by the national heartland—the largest population cluster, the most productive region, the area with greatest **centrality** and **accessibility**, probably containing the capital city as well.

Core-periphery relationships The contrasting spatial characteristics of, and linkages between, the *have* (core) and *have-not* (periphery) components of a national or regional system.

Corridor In general, refers to a spatial entity in which human activity is organized in a linear manner, as along a major transport route or in a valley confined by highlands. Specific meaning in politico-geographical context is a land extension that connects an otherwise **landlocked** state to the sea. History has seen several such corridors come and go. Poland once had a corridor (it now has a lengthy coastline); Bolivia lost a corridor to the Pacific Ocean between Peru and Chile.

Cultural diffusion The process of spreading and adoption of a cultural element, from its place of origin across a wider area.

Cultural ecology The multiple interactions and relationships between a culture and its natural environment.

Cultural landscape The forms and artifacts sequentially placed on the physical landscape by the activities of various human occupants. By this progressive imprinting of the human presence, the physical landscape is modified into the cultural landscape, forming an interacting unity between the two.

Cultural pluralism A society in which two or more population groups, each practicing its own **culture**, live adjacent to one another without mixing inside a single **state**.

Culture The sum total of the knowledge, attitudes, and habitual behavior patterns shared and transmitted by the members of a society. This is anthropolgist Ralph Linton's definition; hundreds of others exist.

Culture area A distinct, culturally discrete spatial unit; a region within which certain cultural norms prevail.

Culture hearth Heartland, source area, innovation center; place of origin of a major culture.

Cyclical movement Movement—for example, **nomadic migration**—that has a closed route repeated annually or seasonally.

Death rate The *crude death rate* is expressed as the annual number of deaths per 1000 individuals within a given population.

Deciduous A deciduous tree loses its leaves at the beginning of winter or the start of the dry season.

Definition In political geography, the written legal description (in a treaty-like document) of a boundary between two countries or territories—see **delimitation**.

Deglomeration Deconcentration.

Delimitation In political geography, the translation of the written terms of a boundary treaty (the **definition**) into an official cartographic representation.

Delta **Alluvial** lowland at the mouth of a river, formed when the river deposits its alluvial load on reaching the sea. Often triangular in shape—hence the use of the Greek letter whose symbol is Δ.

Demarcation In political geography, the actual placing of a political boundary on the landscape by means of barriers, fences, walls, or other markers.

Demographic transition model Multi-stage model, based on Western Europe's experience, of changes in population growth exhibited by countries undergoing industrialization. High **birth rates** and **death rates** are followed by plunging death rates, producing a huge net population gain; this is followed by the convergence of birth and death rates at a low overall level. See Fig. 9-2.

Demography The interdisciplinary study of population— especially **birth rates** and **death rates**, growth patterns, longevity, **migration**, and related characteristics.

Density of population The number of people per unit area. Also see **arithmetic density** and **physiologic density** measures.

Desert An arid area supporting sparse vegetation, receiving less than 10 inches (25 cm) of precipitation per year. Usually exhibits extremes of heat and cold because the moderating influence of moisture is absent.

Desertification The process of **desert** expansion into neighboring **steppelands** as a result of human degradation of fragile semiarid environments.

Determinism See **environmental determinism**.

Development The economic, social, and institutional growth of national **states**.

Devolution The process whereby regions within a **state** demand and gain political stength and growing autonomy at the expense of the central government.

Dhows Wooden boats with characteristic triangular sail, plying the seas between Arabian and East African coasts.

Diffusion The spatial spreading or dissemination of a culture element (such as a technological innovation) or some other phenomenon (e.g., a disease outbreak). See also **contagious**, **expansion**, **hierarchical**, and **relocation diffusion**.

Dispersed settlement In contrast to **agglomerated** or **nucleated** settlement, dispersed settlement is characterized by a much lower **density of population** and the wide spacing of individual homesteads (especially in rural North America).

Distance decay The various degenerative effects of distance on human spatial structures and interactions.

Diurnal Daily.

Divided capital In political geography, a country whose administrative functions are carried on in more than one city is said to have divided capitals.

Domestication The transformation of a wild animal or wild plant into a domesticated animal or a cultivated crop to gain control over food production. A necessary evolutionary step in the development of humankind—the invention of **agriculture**.

Double cropping The planting, cultivation, and harvesting of two crops successively within a single year on the same plot of farmland.

Doubling time The time required for a population to double in size.

Ecology Strictly speaking, this refers to the study of the many interrelationships between all forms of life and the natural environments in which they have evolved and continue to develop. The study of *ecosystems* focuses on the interactions between specific organisms and their environments. See also **cultural ecology**.

Economic tiger One of the burgeoning beehive countries of the Pacific Rim of **Australasia**. Using postwar Japan as a model, these countries have experienced significant modernization, industrialization, and Western-style economic growth since 1980. The four leading economic tigers are South Korea, Taiwan, Hong Kong, and Singapore.

Economies of scale The savings that accrue from large-scale production whereby the unit cost of manufacturing decreases as the level of operation enlarges. Supermarkets operate on this principle and are able to charge lower prices than small grocery stores.

Ecumene The habitable portions of the earth's surface where permanent human settlements have arisen.

Elite A small but influential upper-echelon social class whose power and privilege give it control over a country's political, economic, and cultural life.

Elongated state A **state** whose territory is decidedly long and narrow in that its length is at least six times greater than its average width. Chile and Vietnam are two classic examples on the world political map.

Emigrant A person migrating away from a country or area; an out-migrant.

Empirical Relating to the real world, as opposed to theoretical abstraction.

Enclave A piece of territory that is surrounded by another political unit of which it is not a part.

Entrepôt A place, usually a port city, where goods are imported, stored, and transshipped; a **break-of-bulk point**.

Environmental determinism The view that the natural environment has a controlling influence over various aspects of human life, including cultural development. Also referred to as *environmentalism*.

Epidemic A local or regional outbreak of a disease.

Erosion A combination of gradational forces that shape the earth's surface landforms. Running water, wind action, and the force of moving ice combine to wear away soil and rock. Human activities often speed erosional processes, such as through the destruction of natural vegetation, careless farming practices, and overgrazing by livestock.

Escarpment A cliff or steep slope; frequently marks the edge of a plateau.

Estuary The widening mouth of a river as it reaches the sea. An estuary forms when the margin of the land has subsided somewhat (or as ocean levels rise following glaciation periods) and seawater has invaded the river's lowest portion.

European state model A **state** consisting of a legally defined territory inhabited by a population governed from a capital city by a representative government.

Evapotranspiration The loss of moisture to the **atmosphere** through the combined processes of evaporation from the soil and transpiration by plants.

Exclave A bounded (non-island) piece of territory that is part of a particular **state** but lies separated from it by the territory of another state.

Exclusive Economic Zone (EEZ) An oceanic zone extending up to 200 nautical miles from a shoreline, within which the coastal state can control fishing, mineral exploration, and additional activities by all other countries.

Expansion diffusion The spreading of an innovation or an idea through a fixed population in such a way that the number of those adopting grows continuously larger, resulting in an expanding area of dissemination.

Extraterritoriality Politico-geographical concept suggesting that the property of one **state** lying within the boundaries of another actually forms an extension of the first state.

Favela Shantytown on the outskirts or even well within an urban area in Brazil.

Fazenda Coffee plantation in Brazil.

Federal state A political framework wherein a central government represents the various entities within a **nation-state** where they have common interests—defense, foreign affairs, and the like—yet allows these various entities to retain their own identities and to have their own laws, policies, and customs in certain spheres.

Federation See **federal state.**

Ferroalloy A metallic mineral smelted with iron to produce steel of a particular quality. Manganese, for example, provides steel with tensile strength and the ability to withstand abrasives. Other ferroalloys are nickel, chromium, cobalt, molybdenum, and tungsten.

Fertile Crescent Crescent-shaped zone of productive lands extending from near the southeastern Mediterranean coast through Lebanon and Syria to the **alluvial** lowlands of Mesopotamia (in Iraq). Once more fertile than today, this is one of the world's great source areas of agricultural and other innovations.

Feudalism Prevailing politico-geographical system in Europe during the Middle Ages when land was owned by the nobility and was worked by **peasants** and serfs. Feudalism also existed in other parts of the world, and the system persisted into this century in Ethiopia and Iran, among other places.

Fjord Narrow, steep-sided, elongated, and inundated coastal valley deepened by glacier ice that has since melted away, leaving the sea to penetrate.

Floodplain Low-lying area adjacent to a mature river, often covered by **alluvial** deposits and subject to the river's floods.

Forced migration Human **migration** flows in which the movers have no choice but to relocate.

Formal region A type of region marked by a certain degree of homogeneity in one or more phenomena; also called *uniform* region or *homogeneous* region.

Forward capital Capital city positioned in actually or potentially contested territory, usually near an international border; it confirms the **state's** determination to maintain its presence in the region in contention.

Four Motors of Europe Rhône-Alpes (France), Baden-Württemberg (Germany), Catalonia (Spain), and Lombardy (Italy). Each is a high-technology-driven region marked by exceptional industrial vitality and economic success—not only within Europe but on the global scene as well.

Fragmented state A **state** whose territory consists of several separated parts, not a **contiguous** whole. The individual parts may be isolated from each other by the land area of other states or by international waters.

Francophone Describes a country or region where other languages are also spoken, but where French is the *lingua franca* or the language of the **elite**. Quebec is Francophone Canada.

Fría See *tierra fría*.

Frontier Zone of advance penetration, usually of contention; an area not yet fully integrated into a national **state**.

Functional region A region marked less by its sameness than its dynamic internal structure; because it usually focuses on a central node, also called *nodal* or *focal* region.

Functional specialization The production of particular goods or services as a dominant activity in a particular location.

Geographic realm The basic **spatial** unit in our world regionalization scheme. Each realm is defined in terms of a synthesis of its total human geography—a composite of its leading cultural, economic, historical, political, and appropriate environmental features.

Geometric boundaries Political boundaries **defined** and **delimited** (and occasionally **demarcated**) as straight lines or arcs.

Geomorphology The geographic study of the configuration of the earth's solid surface—the world's landscapes and their constituent landforms.

Ghetto An intraurban region marked by a particular ethnic character. Often an inner-city poverty zone, such as the black ghetto in the large central city of the United States. Ghetto residents are involuntarily segregated from other income and racial groups.

Glaciation See **Pleistocene Epoch.**

Green Revolution The successful recent development of higher yield, fast-growing varieties of rice and other cereals in certain developing countries. This led to increased production per unit area and a temporary narrowing of the gap between population growth and food needs; today, unfortunately, that gap is widening again.

Gross national product (GNP) The total value of all goods and services produced in a country during a given year.

Growing season The number of days between the last frost in the spring and the first frost of the fall.

Growth pole An urban center with certain attributes that, if augmented by a measure of investment support, will stimulate regional economic development in its **hinterland**.

Hacienda Literally, a large estate in a Spanish-speaking country. Sometimes equated with **plantation**, but there are important differences between these two types of agricultural enterprise (see pp. 278-279).

Heartland theory The hypothesis, proposed by British geographer Halford Mackinder during the first two decades of this century, that any political power based in the heart of Eurasia (see Fig. 2-15) could gain sufficient strength to eventually dominate the world. Further, since Eastern Europe controlled access to the Eurasian interior, its ruler would command the vast "heartland" to the east.

Hegemony The political dominance of a country (or even a region) by another country. The former Soviet Union's postwar grip on Eastern Europe, which lasted from 1945 to 1990, was a classic example.

Helada See *tierra helada.*

Hierarchical diffusion A form of **diffusion** in which an idea or innovation spreads by "trickling down" from larger to smaller adoption units. An urban **hierarchy** is usually involved, encouraging the leapfrogging of innovations over wide areas, with geographic distance a less important influence.

Hierarchy An order or gradation of phenomena, with each level or rank subordinate to the one above it and superior to the one below. The levels in a national urban hierarchy are constituted by hamlets, villages, towns, cities, and (frequently) the **primate city**.

High seas Areas of the oceans away from land, beyond national jurisdiction, open and free for all to use.

Highveld A term used in Southern Africa to identify the high, grass-covered plateau that dominates much of the region. *Veld* means "grassland" in Dutch and Afrikaans. The lowest-lying areas in South Africa are called *lowveld*; areas that lie at intermediate elevations are the *middleveld*.

Hinterland Literally, "country behind," a term that applies to a surrounding area served by an urban center. That center is the focus of goods and services produced for its hinterland and is its dominant urban influence as well. In the case of a port city, the hinterland also includes the inland area whose trade flows through that port.

Humus Dark-colored upper layer of a soil that consists of decomposed and decaying organic matter such as leaves and branches, nutrient-rich and giving the soil a high fertility.

Hydrologic cycle The system of exchange involving water in its various forms as it continually circulates among the **atmosphere**, the oceans, and above and below the land surface.

Iconography The identity of a region as expressed through its cherished symbols; its particular **cultural landscape** and personality.

Immigrant A person migrating into a particular country or area; an in-migrant.

Imperialism The drive toward the creation and expansion of a colonial empire and, once established, its perpetuation.

Indentured workers Contract laborers who sell their services for a stipulated period of time.

Infrastructure The foundations of a society: urban centers, transport networks, communications, energy distribution systems, farms, factories, mines, and such facilities as schools, hospitals, postal services, and police and armed forces.

Insular Having the qualities and properties of an island. Real islands are not alone in possessing such properties of **isolation**: an **oasis** in the middle of a **desert** also has qualities of insularity.

Insurgent state Territorial embodiment of a successful guerrilla movement. The establishment by anti-government insurgents of a territorial base in which they exercise full control; thus, a state within a **state**.

Intermontane Literally, "between mountains." The location can bestow certain qualities of natural protection or **isolation** to a community.

Internal migration **Migration** flow within a **nation-state**, such as ongoing westward and southward movements in the United States.

International migration **Migration** flow involving movement across international boundaries.

Intervening opportunity The presence of a nearer opportunity that greatly diminishes the attractiveness of sites farther away.

Irredentism A policy of cultural extension and potential political expansion aimed at a national group living in a neighboring country.

Irrigation The artificial watering of croplands. In Egypt's Nile valley, *basin irrigation* is an ancient method that involved the use of floodwaters that were trapped in basins on the floodplain and released in stages to augment rainfall. Today's *perennial irrigation* requires the construction of dams and irrigation canals for year-round water supply.

Isobar A line connecting points of equal atmospheric pressure.

Isohyet A line connecting points of equal rainfall total.

Isolation The condition of being geographically cut off or far removed from mainstreams of thought and action. It also denotes a lack of receptivity to outside influences, caused at least partially by inaccessibility.

Isoline A line connecting points of equal value.

Isotherm A line connecting points of equal temperature.

Isthmus A **land bridge**; a comparatively narrow link between larger bodies of land. Central America forms such a link between North and South America.

Juxtaposition Contrasting places in close proximity to one another.

Land alienation One society or culture group taking land from another. In Subsaharan Africa, for example, European colonialists took land from indigenous Africans and put it to new uses, fencing it off and restricting settlement.

Land bridge A narrow **isthmian** link between two large land-masses. They are temporary features—at least in terms of geologic time—subject to appearance and disappearance as the land or sea-level rises and falls.

Landlocked An interior country or **state** that is surrounded by land. Without coasts, a landlocked state is at a disadvantage in a number of ways—in terms of **accessibility** to international trade routes, and in the scramble for possession of areas of the **continental shelf** and control of the **exclusive economic zone** beyond.

Land reform The spatial reorganization of **agriculture** through the allocation of farmland (often expropriated from landlords) to **peasants** and tenants who never owned land. Also, the consolidation of excessively fragmented farmland into more productive, perhaps cooperatively-run farm units.

Late Cenozoic Ice Age The latest in a series of ice ages that mark the earth's environmental history, lasting from ca. 3.5 million to 10,000 years ago. As many as 24 separate advances and retreats of continental icesheets took place, the height of activity occurring during the **Pleistocene Epoch** (ca. 2 million to 10,000 years ago).

Latitude Lines of latitude are **parallels** that are aligned east-west across the globe, from 0° latitude at the equator to 90° north and south latitude at the poles. Areas of low latitude, therefore, lie near the equator in the tropics; high latitudes are those in the north polar (Arctic) and south polar (Antarctic) regions.

Leeward The protected or downwind side of a topographic barrier with respect to the winds that flow across it.

Lignite Also called brown coal, a low-grade variety of coal somewhat higher in fuel content than peat but not nearly as good as the next higher grade, **bituminous coal**. Lignite cannot be used for most industrial processes, but it is important as a residential fuel in certain parts of the world.

Lingua franca The term derives from "Frankish language," and applied to a tongue spoken in ancient Mediterranean ports that consisted of a mixture of Italian, French, Greek, Spanish, and even some Arabic. Today it refers to a "common language," a second language that can be spoken and understood by many peoples, although they speak other languages at home.

Littoral Coastal; along the shore.

Llanos Name given to the **savanna**-like grasslands of the Orinoco River's wide basin in the interior parts of Colombia and Venezuela.

Location theory A logical attempt to explain the locational pattern of an economic activity and the manner in which its producing areas are interrelated. The agricultural location theory contained in the **Von Thünen model** is a leading example.

Loess Deposit of very fine silt or dust that is laid down after having been windborne for a considerable distance. Loess is notable for its fertility under **irrigation** and its ability to stand in steep vertical walls when **eroded** by a river or (as in China's Loess Plateau) excavated for cave-type human dwellings.

Longitude Angular distance (0° to 180°) east or west as measured from the prime **meridian** (0°) that passes through the Greenwich Observatory in suburban London, England. For much of its length in the mid-Pacific Ocean, the 180th meridian functions as the *international date line*.

Maghreb The region lying in the northwestern corner of Africa, consisting of the countries of Morocco, Algeria, and Tunisia.

Main Street Canada's dominant **conurbation** that is home to more than 60 percent of the country's inhabitants; stretches southwestward from Quebec City in the middle St. Lawrence valley to Windsor on the Detroit River.

Maquiladora The term given to modern industrial plants in Mexico's northern (U.S.) border zone. These foreign-owned factories assemble imported components and/or raw materials, and then export finished manufactures, mainly to the United States. Most import duties are minimized, bringing jobs to Mexico and the advantages of low wage rates to the foreign entrepreneurs.

Marchland A **frontier** or area of uncertain boundaries that is subject to various national claims and an unstable political history. The term refers to the movement of various national armies across such zones.

Megalopolis Term used to designate large coalescing supercities that are forming in diverse parts of the world; used specifically to refer to the Boston-Washington multi-metropolitan corridor on the northeastern seaboard of the United States, but the term is now used generically with a lower-case *m* as a synonym for **conurbation**.

Mercantilism Protectionist policy of European **states** during the sixteenth to the eighteenth centuries that promoted a state's economic position in the contest with other countries. The acquisition of gold and silver and the maintenance of a favorable trade balance (more exports than imports) were central to the policy.

Meridian Line of **longitude**, aligned north-south across the globe, that together with **parallels** of **latitude** forms the global grid system. All meridians converge at both poles and are at their maximum distances from each other at the equator.

Mestizo The root of this word is the Latin for *mixed*; it means a person of mixed white and Amerindian ancestry.

Metropolitan area See **urban (metropolitan) area**.

Mexica Empire The name the ancient Aztecs gave to the domain over which they held **hegemony** on the north-central mainland of Middle America.

Migration A change in residence intended to be permanent. See also **forced, internal, international,** and **voluntary migration**.

Migratory movement Human relocation movement from a source to a destination without a return journey, as opposed to **cyclical movement**.

Model   An idealized representation of reality built to demonstrate certain of its properties. A **spatial** model focuses on a geographical dimension of the real world.

Monotheism   The belief in, and worship of, a single god.

Monsoon   Refers to the seasonal reversal of wind (and moisture) flows in certain parts of the subtropics and lower-middle latitudes. The *dry monsoon* occurs during the cool season when dry offshore winds prevail. The *wet monsoon* occurs in the hot summer months, which produce onshore winds that bring large amounts of rainfall. The air-pressure differential over land and sea is the triggering mechanism. Monsoons make their greatest regional impact in the coastal and near-coastal zones of South Asia, Southeast Asia, and China.

Mulatto   A person of mixed African (black) and European (white) ancestry.

Multinationals   Internationally active corporations that can strongly influence the economic and political affairs of many countries they operate in.

Multiple nuclei model   The Harris-Ullman model that showed the mid-twentieth-century American central city to consist of several land-use zones arranged around nuclear growth points (see Fig. 3-1C).

Nation   Legally a term encompassing all the citizens of a **state**, it also has other connotations. Most definitions now tend to refer to a group of tightly-knit people possessing bonds of language, ethnicity, religion, and other shared cultural attributes. Such homogeneity actually prevails within very few states.

Nation-state   A country whose population possesses a substantial degree of cultural homogeneity and unity. The ideal form to which most **nations** and **states** aspire—a political unit wherein the territorial state coincides with the area settled by a certain national group or people.

Natural increase rate   Population growth measured as the excess of live births over deaths per 1000 individuals per year. Natural increase of a population does not reflect either **emigrant** or **immigrant** movements.

Natural resource   Any valued element of (or means to an end using) the environment; includes minerals, water, vegetation, and soil.

Nautical mile   By international agreement, the nautical mile—the standard measure at sea—is 6076.12 feet in length, equivalent to approximately 1.15 statute miles (1.85 km).

Network (transport)   The entire regional system of transportation connections and nodes through which movement can occur.

Nomadism   **Cyclical movement** among a definite set of places. Nomadic peoples mostly are **pastoralists**.

Norden   A regional appellation for the Northern European countries of Denmark, Norway, Sweden, Finland, Estonia, and Iceland.

Nucleated settlement   See **agglomerated settlement**.

Nucleation   Cluster; agglomeration.

Oasis   An area, small or large, where the supply of water permits the transformation of the surrounding **desert** into a green crop-land; the most important focus of human activity for miles around. It may also encompass a densely populated **corridor** along a major river where **irrigation** projects stabilize the water supply—as along the Nile in Egypt, which can be viewed as an elongated chain of oases.

Occidental   Western. See **Oriental**.

Oligarchy   Political system involving rule by a small minority, an often corrupt **elite**.

Oriental   The root of the word oriental is from the Latin for *rise*. Thus it has to do with the direction in which one sees the sun "rise"—the east; oriental therefore means Eastern. Occidental originates from the Latin for *fall*, or the "setting" of the sun in the west; occidental therefore means Western.

Orographic precipitation   Mountain-induced precipitation, especially where air masses are forced over topographic barriers. Downwind areas beyond such a mountain range experience the relative dryness known as the **rain shadow effect**.

Pacific Rim   A far-flung group of countries and parts of countries (extending clockwise on the map from New Zealand to Chile) sharing the following criteria: they face the Pacific Ocean; they evince relatively high levels of economic development, industrialization, and urbanization; their imports and exports mainly move across Pacific waters.

Pacific Ring of Fire   Zone of crustal instability along tectonic plate boundaries, marked by earthquakes and volcanic activity, that rings the Pacific Ocean basin.

Paddies (paddyfields)   Ricefields.

Pandemic   An outbreak of a disease that spreads worldwide.

Pangaea   A vast, singular landmass consisting of most of the areas of the present continents (including the Americas, Eurasia, Africa, Australia, and Antarctica), which existed until near the end of the Mesozoic era when **plate** divergence and **continental drift** broke it apart. The "northern" segment of this Pangaean supercontinent is called Laurasia, the "southern" part Gondwana (see Fig. 8-3).

Parallel   An east-west line of **latitude** that is intersected at right angles by **meridians** of **longitude**.

Páramos   See *puna*.

Pastoralism   A form of agricultural activity that involves the raising of livestock. Many peoples described as herders actually pursue mixed **agriculture**, in that they may also fish, hunt, or even grow a few crops. But pastoral peoples' lives do revolve around their animals.

Peasants   In a **stratified** society, peasants are the lowest class of people who depend on **agriculture** for a living. But they often own no land at all and must survive as tenants or day workers.

Peninsula   A comparatively narrow, finger-like stretch of land extending from the main landmass into the sea. Florida and Korea are examples.

Peon (*peone*)   Term used in Middle and South America to identify people who often live in serfdom to a wealthy landowner; landless **peasants** in continuous indebtedness.

Per capita Capita means *individual*. Income, production, or some other measure is often given per individual.

Perforated state A **state** whose territory completely surrounds that of another state. South Africa, which encloses Lesotho and is perforated by it, is a classic example (see Fig. 8-13).

Periodic market Village market that opens every third or fourth day or at some other regular interval. Part of a regional network of similar markets in a preindustrial, rural setting where goods are brought to market on foot (or perhaps by bicycle) and barter remains a major mode of exchange.

Permafrost Permanently frozen water in the soil and bedrock (of cold environments), as much as 1000 feet (300 m) in depth, producing the effect of completely frozen ground. Can thaw near surface during brief warm season.

Physiographic political boundaries Political boundaries that coincide with prominent physical features in the natural landscape—such as rivers or the crest ridges of mountain ranges.

Physiographic region (province) A region within which there prevails substantial natural-landscape homogeneity, expressed by a certain degree of uniformity in surface **relief**, **climate**, vegetation, and soils.

Physiologic density The number of people per unit area of **arable** land.

Pidgin A language that consists of words borrowed and adapted from other languages. Originally developed from commerce among peoples speaking different languages.

Pilgrimage A journey to a place of great religious significance by an individual or by a group of people (such as a pilgrimage to Mecca for Muslims).

Plantation A large estate owned by an individual, family, or corporation and organized to produce a cash crop. Almost all plantations were established within the tropics; in recent decades, many have been divided into smaller holdings or reorganized as cooperatives.

Plate tectonics Plates are bonded portions of the earth's mantle and crust, averaging 60 miles (100 km) in thickness. More than a dozen such plates exist (see Fig. I-3), most of continental proportions, and they are in motion. Where they meet one slides under the other, crumpling the surface crust and producing significant volcanic and earthquake activity. A major mountain-building force.

Pleistocene Epoch Recent period of geologic time that spans the rise of humankind, beginning about 2 million years ago. Marked by glaciations (repeated advances of continental icesheets) and milder interglaciations (icesheet retreats). Although the last 10,000 years are known as the *Recent Epoch*, Pleistocene-like conditions seem to be continuing and the present is probably another Pleistocene interglaciation; the glaciers likely will return. See also **Late Cenozoic Ice Age**.

Plural society See **cultural pluralism**.

Polder Land reclaimed from the sea adjacent to shore by constructing dikes and pumping out the water. Technique used widely along the coast of the Netherlands since the Middle Ages to add badly needed additional space for settlements and farms.

Pollution The release of a substance, through human activity, which chemically, physically, or biologically alters the air or water it is discharged into. Such a discharge negatively impacts the environment, with possible harmful effects on living organisms—including humans.

Population (age-sex) structure Graphic representation (profile) of a population according to age and sex, as in Fig. 9-3.

Population explosion The rapid growth of the world's human population during the past century, attended by ever-shorter **doubling times** and accelerating *rates* of increase.

Postindustrial economy Emerging economy, in the United States and a handful of other highly advanced countries, as traditional industry is overshadowed by a higher-technology productive complex dominated by services, information-related, and managerial activities.

Primary economic activity Activities engaged in the direct extraction of **natural resources** from the environment—such as mining, fishing, lumbering, and especially a**griculture**.

Primate city A country's largest city—ranking atop the urban **hierarchy**—most expressive of the national culture and usually (but not always) the capital city as well.

Process Causal force that shapes a **spatial** pattern as it unfolds over time.

Proletariat Lower-income working class in a community or society. People who own no capital or means of production and who live by selling their labor.

Prorupt A type of **state** territorial shape that exhibits a narrow, elongated land extension leading away from the main body of territory. Thailand is a leading example (see Fig. 11-7).

Protectorate In Britain's system of colonial administration, the protectorate was a designation that involved the guarantee of certain rights (such as the restriction of European settlement and **land alienation**) to peoples who had been placed under the control of the Crown.

Puna In Andean South America, the highest-lying habitable **altitudinal zone**—ca. 12,000 to 15,000 feet (3,600 to 4,500 m)—between the tree line (upper limit of the *tierra fría*) and the snow line (lower limit of the *tierra helada*). Too cold and barren to support anything but the grazing of sheep and other hardy livestock. Also known as the *páramos*.

Push-pull concept The idea that **migration** flows are simultaneously stimulated by conditions in the source area, which tend to drive people away, and by the perceived attractiveness of the destination.

Qanat In **desert** zones, particularly in Iran and western China, an underground tunnel built to carry **irrigation** water by gravity flow from nearby mountains (where **orographic precipitation** occurs) to the arid flatlands below.

Quaternary economic activity Activities engaged in the collection, processing, and manipulation of *information*.

Rain shadow effect   The relative dryness in areas downwind of mountain ranges caused by **orographic precipitation**, wherein moist air masses are forced to deposit most of their water content in the highlands.

Realm   See **geographic realm**.

Region   A commonly used term and a geographic concept of central importance. An **area** on the earth's surface marked by certain properties.

Regional complementarity   See **complementarity**.

Relative location   The regional position or **situation** of a place relative to the position of other places. Distance, **accessibility**, and connectivity affect relative location.

Relict boundary   A political boundary that has ceased to function, but the imprint of which can still be detected on the **cultural landscape**.

Relief   Vertical difference between the highest and lowest elevations within a particular area.

Relocation diffusion   Sequential **diffusion** process in which the items being diffused are transmitted by their carrier agents as they evacuate the old areas and relocate to new ones. The most common form of relocation diffusion involves the spreading of innovations by a **migrating** population.

Rural density   A measure that indicates the number of persons per unit area living in the rural areas of a country, outside of the urban concentrations.

Sahel   Semiarid zone extending across most of Africa between the southern margins of the arid Sahara and the moister tropical savanna and forest zone to the south. Chronic drought, **desertification**, and overgrazing have contributed to severe famines in this area since 1970.

Savanna   Tropical grassland containing widely spaced trees; also the name given to the tropical wet-and-dry climate (*Aw*).

Scale   Representation of a real-world phenomenon at a certain level of reduction or generalization. In **cartography**, the ratio of map distance to ground distance; indicated on a map as bar graph, representative fraction, and/or verbal statement.

Scale economies   See **economies of scale**.

Secondary economic activity   Activities that process raw materials and transform them into finished industrial products. The *manufacturing* sector.

Sector model   A structural model of the American central city that suggests land-use areas conform to a wedge-shaped pattern focused on the downtown core (see Fig. 3-1B).

Sedentary   Permanently attached to a particular area; a population fixed in its location. The opposite of **nomadic**.

Selva   Tropical rainforest.

Separate development   The spatial expression of South Africa's "grand" **apartheid** scheme, wherein nonwhite groups were required to settle in segregated "homelands." During the 1980s, the implementation of this plan collapsed, and its dismantling—together with the system of apartheid—is now well underway.

Sequent occupance   The notion that successive societies leave their cultural imprints on a place, each contributing to the cumulative **cultural landscape**.

Shantytown   Unplanned slum development on the margins of cities in the developing realms, dominated by crude dwellings and shelters mostly made of scrap wood, iron, and even pieces of cardboard.

Sharecropping   Relationship between a large landowner and farmers on the land wherein the farmers pay rent for the land they farm by giving the landlord a share of the annual harvest.

Shatter belt   Region caught between stronger, colliding external cultural-political forces, under persistent stress and often fragmented by aggressive rivals. Eastern Europe and Southeast Asia are classic examples.

Shield   A large, stable, relatively flat expanse of very old rocks that forms the geologic core of a continental landmass; North America's Canadian Shield is an example.

Shifting agriculture   Cultivation of crops in recently cut and burned tropical forest clearings, soon to be abandoned in favor of newly cleared nearby forest land. Also known as *slash-and-burn agriculture*.

Sinicization   Giving a Chinese cultural imprint; Chinese **acculturation**.

Site   The internal locational attributes of an urban center, including its local spatial organization and physical setting.

Situation   The external locational attributes of an urban center; its **relative location** or regional position with reference to other nonlocal places.

Slash-and-burn agriculture   See **shifting agriculture**.

Southern Cone   The southern, mid-latitude portion of South America constituted by the countries of Chile, Argentina, and Uruguay; often included as well is the southernmost part of Brazil, south of the Tropic of Capricorn (23½° S).

Spatial   Pertaining to space on the earth's surface. Synonym for *geographic(al)*.

Spatial diffusion   See **diffusion**.

Spatial interaction   See **complementarity**, **transferability**, and **intervening opportunity**.

Spatial model   See **model**.

Spatial morphology   See **territorial morphology**.

Spatial process   See **process**.

Squatter settlement   See **shantytown**.

State   A politically organized territory that is administered by a sovereign government and is recognized by a significant portion of the international community. A state must also contain a permanent resident population, an organized economy, and a functioning internal circulation system.

State capitalism   Government-controlled corporations competing under free market conditions, usually in a tightly regimented society.

Steppe   Semiarid grassland; short-grass prairie. Also, the name given to the semiarid climate type (*BS*).

Stratification (social)   In a layered or stratified society, the population is divided into a **hierarchy** of social classes. In an industrialized society, the **proletariat** is at the lower end; **elites** that possess capital and control the means of production are at the upper level. In the traditional **caste system** of Hindu India, the "untouchables" form the lowest class or caste, whereas the still-wealthy remnants of the princely class are at the top.

Subsequent boundary   A political boundary that developed contemporaneously with the evolution of the major elements of the **cultural landscape** through which it passes.

Subsistence   Existing on the minimum necessities to sustain life; spending most of one's time in pursuit of survival.

Suburban downtown   Significant concentration of diversified economic activities around a highly **accessible** suburban location, including retailing, light industry, and a variety of major corporate and commercial operations. Late-twentieth-century coequal to the American central city's **central business district (CBD)**.

Superimposed boundary   A political boundary emplaced by powerful outsiders on a developed human landscape. Usually ignores pre-existing cultural-spatial patterns, such as the border that now divides North and South Korea.

Supranational   A venture involving three or more national **states**—political, economic, and/or cultural cooperation to promote shared objectives. Europe's Economic Community or **Common Market** is one such organization.

System   Any group of objects or institutions and their mutual interactions. Geography treats systems that are expressed **spatially**, such as **regions**.

Systematic geography   Topical geography: cultural, political, economic geography, and the like.

Takeoff   Economic concept to identify a stage in a country's **development** when conditions are set for a domestic Industrial Revolution, which occurred in Britain in the late eighteenth century and in Japan in the late nineteenth following the Meiji Restoration.

Tectonics   See **plate tectonics**.

Tell   The lower slopes and narrow coastal plains along the Atlas Mountains in northwesternmost Africa, where the majority of the **Maghreb** region's population is clustered.

Templada   See *tierra templada*.

Terracing   The transformation of a hillside or mountain slope into a step-like sequence of horizontal fields for intensive cultivation (see photo p. 517).

Territoriality   A country's or more local community's sense of property and attachment toward its territory, as expressed by its determination to keep it inviolable and strongly defended.

Territorial morphology   A **state's** geographical shape, which can have a decisive impact on its spatial cohesion and political viability. A **compact** shape is most desirable; among the less effi-

cient shapes are those exhibited by **elongated**, **fragmented**, **perforated**, and **prorupt** states.

Territorial sea   Zone of seawater adjacent to a country's coast, held to be part of the national territory and treated as a segment of the sovereign **state**.

Tertiary economic activity   Activities that engage in *services*—such as transportation, banking, retailing, education, and routine office-based jobs.

Theocracy   A **state** whose government is under the control of a ruler who is deemed to be divinely guided or under the control of a group of religious leaders, as in post-Khomeini Iran. The opposite of the theocratic state is the **secular** state.

Tierra caliente   The lowest of the **altitudinal zones** into which the human settlement of Middle and South America is classified according to elevation. The *caliente* is the hot humid coastal plain and adjacent slopes up to 2,500 feet (750 m) above sea level. The natural vegetation is the dense and luxuriant tropical rainforest; the crops are sugar, bananas, cacao, and rice in the lower areas, and coffee, tobacco, and corn along the higher slopes.

Tierra fría   The cold, high-lying **altitudinal zone** of settlement in Andean South America, extending from about 6,000 feet (1850 m) in elevation up to nearly 12,000 feet (3600 m). **Coniferous** trees stand here; upward they change into scrub and grassland. There are also important pastures within the *fría*, and wheat can be cultivated. Several major population clusters in western South America lie at these altitudes.

Tierra helada   The highest and coldest **altitudinal zone** in Andean South America (lying above 15,000 feet [4,500 m]), an uninhabitable environment of permanent snow and ice that extends upward to the Andes' highest peaks of more than 20,000 feet (6,000 m).

Tierra templada   The intermediate **altitudinal zone** of settlement in Middle and South America, lying between 2,500 feet (750 m) and 6,000 feet (1850 m) in elevation. This is the "temperate" zone, with moderate temperatures compared to the *tierra caliente* below. Crops include coffee, tobacco, corn, and some wheat.

Time-space convergence   The increasing nearness of places that occurs as modern transportation breakthroughs progressively reduce the time-distance between them. The trip by boat from New York to San Francisco before the Civil War took months. After 1870, the transcontinental railroad cut the travel time to less than two weeks; by 1930, trains made the journey in three days. After 1945, propeller planes made the trip in about 12 hours; and by 1960, non-stop jet planes achieved today's travel time of five hours.

Totalitarian   A government whose leaders rule by absolute control, tolerating no differences of political opinion.

Transculturation   Cultural borrowing that occurs when different cultures of approximately equal complexity and technological level come into close contact. In **acculturation**, by contrast, an indigenous society's culture is modified by contact with a technologically superior society.

Transferability   The capacity to move a good from one place to another at a bearable cost; the ease with which a commodity may be transported.

Transition zone   An area of **spatial** change where the peripheries of two adjacent realms or regions join; marked by a gradual shift (rather than a sharp break) in the characteristics that distinguish these neighboring geographic entities from one another.

Tropical savanna   See **savanna**.

Tsunami   A seismic (earthquake-generated) sea wave that can attain gigantic proportions and cause coastal devastation.

Underdeveloped countries (UDCs)   Countries that, by various measures, suffer seriously from negative economic and social conditions, including low **per capita** incomes, poor nutrition, inadequate health, and related disadvantaged circumstances.

Unitary state   A **nation-state** that has a centralized government and administration that exercises power equally over all parts of the state.

Urbanization   A term with several connotations. The proportion of a country's population living in urban places is its level of urbanization. The **process** of urbanization involves the movement to, and the clustering of, people in towns and cities—a major force in every geographic realm today. Another kind of urbanization occurs when an expanding city absorbs rural countryside and transforms it into suburbs; in the case of cities in the developing world, this also generates peripheral **shantytowns**.

Urban (metropolitan) area   The entire built-up, non-rural area and its population, including the most recently constructed suburban appendages. Provides a better picture of the dimensions and population of such an area than the delimited municipality (central city) that forms its heart.

Urban realms model   A spatial generalization of the large, late-twentieth-century city in the United States. It is shown to be a widely dispersed, multi-centered metropolis consisting of increasingly independent zones or *realms*, each focused on its own **suburban downtown**; the only exception is the shrunken central realm, which is focused on the **central business district** (see Fig. 3-11).

Veld   Open grassland on the South African plateau, becoming mixed with scrub at lower elevations where it is called *bushveld*. As in Middle and South America, there is an **altitudinal zonation** into **highveld**, *middleveld*, and *lowveld*.

Voluntary migration   Population movement in which people relocate in response to perceived opportunity, not because they are forced to move.

Von Thünen model   Explains the location of agricultural activities in a commercial, profit-making economy. A **process** of spatial competition allocates various farming activities into concentric rings around a central market city, with profit-earning capability the determining force in how far a crop locates from the market. The original (1826) *Isolated State model* now applies to the continental scale (see Fig. 1-4).

Water table   When precipitation falls on the soil, some of the water is drawn downward through the pores in the soil and rock under the force of gravity. Below the surface it reaches a level where it can go no further; there it joins water that already saturates the rock completely. This water that "stands" underground is *groundwater*, and the upper level of the zone of saturation is the *water table*.

Windward   The exposed, upwind side of a topographic barrier that faces the winds that flow across it.

World geographic realm   See **geographic realm**.

LIST OF MAPS AND FIGURES

LIST OF PHOTO CREDITS

Introduction
Page 3: Randy Taylor/Sygma. **Page 6:** Craig Aurness/West Light. **Page 8:** Steve McCurry/Magnum Photos, Inc. **Page 19:** Robert Caputo/Stock, Boston. **Page 20:** M & E Bernheim/Woodfin Camp & Associates. **Page 26 (top):** Earth Satellite Corporation. **Page 26 (bottom):** © CIRES, 1992. **Page 28:** Barrie Rokeach/The Image Bank. **Page 29:** Peter Loud/Leo de Wys, Inc. **Page 38 (top):** David Ball/The Picture Cube. **Page 38 (bottom):** Jeffrey Alford/Asia Access.

Chapter 1
Page 49: Dennis Hallinan/FPG International. **Page 52, top:** Steve Vidler/Leo de Wys, Inc. **Page 52, bottom:** Connie Coleman/AllStock, Inc. **Page 54:** /Leo de Wys, Inc. **Page 62:** DeRichemond/The Image Works. **Page 63:** Nagele/FPG International. **Page 64:** /Tom Craig. **Page 73:** Laurie Sparham/Matrix. **Page 77:** Thyssen Hutte/Superstock. **Page 79:** A. Tannenbaum/Sygma. **Page 82:** /Q.A. Photos. **Page 87:** Farrell Graham/Photo Researchers. **Page 94:** Federico Pagni/Pictor. **Page 101:** Jan Hausbrandt/Black Star. **Page 106:** Joachim Messerschmidt/Bruce Coleman, Inc. **Page 109:** Sovfoto/Eastfoto. **Page 112:** Ed Hille/Matrix. **Page 114:** Novosti/Sovfoto/Eastfoto.

Chapter 2
Page 127: Bob Krist/Leo de Wys, Inc. **Page 129:** David Turnley/Black Star. **Page 153 (top):** James Nachtwey/Magnum Photos, Inc. **Page 153 (bottom):** Harm J. de Blij. **Page 155:** Lada Cais/Sovfoto/Eastfoto. **Page 156:** Peter Turnley/Black Star. **Page 158:** Novosti/Gamma Liaison. **Page 159:** /Sovfoto/Eastfoto. **Page 161:** Sygma.

Vignette: Transcaucasia Page 167: Jeremy Nicholl/Woodfin Camp & Associates. **Page 168:** Chip Hires/Gamma Liaison. **Page 169:** Robert Nickelsberg/Gamma Liaison.

Chapter 3
Page 175: Ted Speigel/Black Star. **Page 178:** Manfred Gottschalk/West Light. **Page 181, top:** /Gamma Liaison. **Page 181, bottom:** /Gamma Liaison. **Page 190, top:** James D. Wilson/Woodfin Camp & Associates. **Page 190, bottom:** Mitch Kezar/Black Star. **Page 191:** Tom Carroll/FPG International. **Page 193:** Christopher Morris/Black Star. **Page 198:** Arthur C. Smith III/Grant Heilman Photography. **Page 200, top:** Richard Kalvarf/Magnum Photos, Inc. **Page 200, bottom:** Courtesy The North Star Steel Company. **Page 202:** Brian Stablyk/AllStock, Inc. **Page 207:** Rhoda Sidney/The Image Works. **Page 211:** ©James Blank. **Page 213:** Paul Fusco/Magnum Photos, Inc. **Page 215:** J. Blank/H. Armstrong Roberts.

Chapter 4
Page 225: Scott Rutherford/Black Star. **Page 227:** Steve Vidler/Leo de Wys, Inc. **Page 231:** Shinichi Kanno/FPG International. **Page 234:** Earth Satellite Corporation. **Page 235:** Robert Perron/f/STOP Pictures. **Page 236:** Ken Straiton/The Stock Market. **Page 238:** Jon Burbank/The Image Works. **Page 240:** Alain Avrard/Photo Researchers. **Page 242:** Jeff Heger/FPG International. **Page 244:** Will & Deni McIntyre/Photo Researchers. **Page 245:** Kees/Sygma. **Page 247:** Yoichiro Miyazaki/FPG International. **Page 251:** Shinichi Kanno/FPG International. **Page 254:** Fridmar Damm/Leo de Wys, Inc. **Page 260:** Harm J. de Blij.

Chapter 5
Page 275: Steve Vidler/Leo de Wys, Inc. **Page 281:** James Blair/National Geographic Society. **Page 282:** © Jim Forbes. **Page 283:** Harm J. de Blij. **Page 286:** Juergen Schmitt/The Image Bank. **Page 288:** © Jay Lurie/Bruce Coleman, Inc. **Page 290:** Byron Augustin/Tom Stack & Associates. **Page 294:** © Peter Poulides. **Page 297:** Chuck Mason/The Image Bank.

Chapter 6
Page 308: K. Scholz/H. Armstrong Roberts. **Page 310:** Harm J. de Blij. **Page 317:** Gerhard Gscheilde/The Image Bank. **Page 319:** H. Sutton/

H. Armstrong Roberts. **Page 326:** Steve Vidler/Leo de Wys, Inc. **Page 330 (top):** Harm J. de Blij. **Page 330 (bottom):** Courtesy Entidad Binacional Yacyreta, Argentina. **Page 332:** M. Thomas/Gamma Liaison.

Vignette: Emerging Brazil Page 336: Claus Meyer/Black Star. **Page 339:** Harm J. de Blij. **Page 341:** Stephanie Maze/ Woodfin Camp & Associates. **Page 342:** S. Jorge/The Image Bank. **Page 344:** Bulcao JL/Gamma Liaison.

Chapter 7
Page 362: Kurgan-Lisnet/Gamma Liaison. **Page 364:** Manfred Gottschalk/Tom Stack & Associates. **Page 365:** Steve Vidler/Leo de Wys, Inc. **Page 374:** Dana Downie/AllStock, Inc. **Page 376:** DMSP Archive, NSIDC/CIRES, University of Colorado, Boulder. **Page 378:** Earth Satellite Corporation. **Page 380:** M. Timothy O'Keefe/Bruce Coleman, Inc. **Page 384:** Steve McCurry/Magnum Photos, Inc. **Page 391:** Richard Nowitz. **Page 395:** Harm J. deBlij. **Page 398 (top):** Dallas & John Heaton/West Light. **Page 398 (bottom):** © Ed Kashi. **Page 401:** Tom McHugh/Photo Researchers. **Page 403:** Patrick Robert/Sygma.

Chapter 8
Page 417: Harm J. de Blij. **Page 420:** Owen Franken/Stock, Boston. **Page 423:** Ray Wilkinson/Woodfin Camp & Associates. **Page 424:** COMSTOCK, Inc. **Page 426:** Harm J. de Blij. **Page 429:** Andreas Pistolesi/The Image Bank. **Page 438:** M & E Bernheim/Woodfin Camp & Associates. **Page 439:** P. Roberts/Sygma. **Page 440:** © Jason Lauré. **Page 443:** Guido Alberto Rossi/The Image Bank. **Page 445:** Shaw McCutcheon/Bruce Coleman, Inc. **Page 448:** Marc & Evelyn Bernheim/Woodfin Camp & Associates.

Vignette: South Africa Page 454: Patricia Lanza/Bruce Coleman, Inc. **Page 457:** Nicholas Devore III/Photographers Aspen. **Page 459:** Denis Farrell/AP/Wide World Photos.

Chapter 9
Page 464: J. Hiebeler/Leo de Wys, Inc. **Page 472:** R & S Michaud/Woodfin Camp & Associates. **Page 473:** Robert Weinreb/Bruce Coleman, Inc. **Page 475:** Bettmann Archive. **Page 478:** Franke Keating/Photo Researchers. **Page 482:** Porterfield/Chickering/Photo Researchers. **Page 484:** Baldev/Sygma. **Page 491 (top):** Jacques Jangoux/Tony Stone Images. **Page 491 (bottom):** Porterfield/Chickering/Photo Researchers. **Page 492 (rt):** Steve McCurry/Magnum Photos, Inc. **Page 492 (lt.):** Steve McCurry/Magnum Photos, Inc. **Page 495:** P. Menzel/Stock, Boston. **Page 498:** James P. Blair/National Geographical Society. **Page 503:** Bill Wassman/The Stock Market.

Chapter 10
Page 513: Cheng Zhishan/Sovfoto/eastfoto. **Page 517:** Brian Brake/Photo Researchers. **Page 518:** M. Jacob/The Image Works. **Page 521:** Charles Bowman/Leo de Wys, Inc. **Page 524:** Harm J. de Blij. **Page 531:** Forrest Anderson/Gamma Liaison. **Page 533 (top):** Peter Charlesworth/JB Pictures. **Page 533 (bottom):** A. Michaud/FPG International. **Page 535:** Noboru Komine/Photo Researchers. **Page 539:** Larry Mulvehill/Photo Researchers. **Page 547:** Harm J. de Blij. **Page 549:** Noboru Komine/Photo Researchers.

Chapter 11
Page 561: Kal Muller/Woodfin Camp & Associates. **Page 563:** Paul John Miller/Black Star. **Page 570:** Harm J. de Blij. **Page 571:** Robert Nickelberg/Gamma Liaison. **Page 574 top:** Tom Owen Edmunds/The Image Bank. **Page 574 bottom:** C. B. & D. W. Frith/Bruce Coleman, Inc. **Page 576:** Porterfield/Chickering/Photo Researchers. **Page 579:** Gary Milburn/Tom Stack & Associates.

Vignette: The Pacific Realm Page 585: Christopher Arnesen/AllStock, Inc. **Page 589:** John Bryson/The Image Bank. **Page 590:** Douglas Peebles/West Light.

GEOGRAPHICAL INDEX (GAZETTEER)

All entries refer to map contents.

INDEX